Mass

$1 \text{ g} = 10^{-3} \text{ kg}$
$1 \text{ kg} = 10^3 \text{ g}$
$1 \text{ u} = 1.66 \times 10^{-24} \text{ g} = 1.66 \times 10^{-27} \text{ kg}$
$1 \text{ metric ton} = 1000 \text{ kg}$

Length

$1 \text{ nm} = 10^{-9} \text{ m}$
$1 \text{ cm} = 10^{-2} \text{ m} = 0.394 \text{ in.}$
$1 \text{ m} = 10^{-3} \text{ km} = 3.28 \text{ ft} = 39.4 \text{ in.}$
$1 \text{ km} = 10^3 \text{ m} = 0.621 \text{ mi}$
$1 \text{ in.} = 2.54 \text{ cm} = 2.54 \times 10^{-2} \text{ m}$
$1 \text{ ft} = 0.305 \text{ m} = 30.5 \text{ cm}$
$1 \text{ mi} = 5280 \text{ ft} = 1609 \text{ m} = 1.609 \text{ km}$

Area

$1 \text{ cm}^2 = 10^{-4} \text{ m}^2 = 0.155\,0 \text{ in}^2$
$\quad = 1.08 \times 10^{-3} \text{ ft}^2$
$1 \text{ m}^2 = 10^4 \text{ cm}^2 = 10.76 \text{ ft}^2 = 1550 \text{ in}^2$
$1 \text{ in}^2 = 6.94 \times 10^{-3} \text{ ft}^2 = 6.45 \text{ cm}^2$
$\quad = 6.45 \times 10^{-4} \text{ m}^2$
$1 \text{ ft}^2 = 144 \text{ in}^2 = 9.29 \times 10^{-2} \text{ m}^2 = 929 \text{ cm}^2$

Volume

$1 \text{ cm}^3 = 10^{-6} \text{ m}^3 = 3.53 \times 10^{-5} \text{ ft}^3$
$\quad = 6.10 \times 10^{-2} \text{ in}^3$
$1 \text{ m}^3 = 10^6 \text{ cm}^3 = 10^3 \text{ L} = 35.3 \text{ ft}^3$
$\quad = 6.10 \times 10^4 \text{ in}^3 = 264 \text{ gal}$
$1 \text{ liter} = 10^3 \text{ cm}^3 = 10^{-3} \text{ m}^3 = 1.056 \text{ qt}$
$\quad = 0.264 \text{ gal} = 0.035\,3 \text{ ft}^3$
$1 \text{ in}^3 = 5.79 \times 10^{-4} \text{ ft}^3 = 16.4 \text{ cm}^3$
$\quad = 1.64 \times 10^{-5} \text{ m}^3$
$1 \text{ ft}^3 = 1728 \text{ in}^3 = 7.48 \text{ gal} = 0.028\,3 \text{ m}^3$
$\quad = 28.3 \text{ L}$
$1 \text{ qt} = 2 \text{ pt} = 946 \text{ cm}^3 = 0.946 \text{ L}$
$1 \text{ gal} = 4 \text{ qt} = 231 \text{ in}^3 = 0.134 \text{ ft}^3 = 3.785 \text{ L}$

Time

$1 \text{ h} = 60 \text{ min} = 3600 \text{ s}$
$1 \text{ day} = 24 \text{ h} = 1440 \text{ min} = 8.64 \times 10^4 \text{ s}$
$1 \text{ y} = 365 \text{ days} = 8.76 \times 10^3 \text{ h}$
$\quad = 5.26 \times 10^5 \text{ min} = 3.16 \times 10^7 \text{ s}$

Angle

$1 \text{ rad} = 57.3°$

$1° = 0.0175 \text{ rad}$	$60° = \pi/3 \text{ rad}$
$15° = \pi/12 \text{ rad}$	$90° = \pi/2 \text{ rad}$
$30° = \pi/6 \text{ rad}$	$180° = \pi \text{ rad}$
$45° = \pi/4 \text{ rad}$	$360° = 2\pi \text{ rad}$

$1 \text{ rev/min} = (\pi/30) \text{ rad/s} = 0.104\,7 \text{ rad/s}$

Speed

$1 \text{ m/s} = 3.60 \text{ km/h} = 3.28 \text{ ft/s}$
$\quad = 2.24 \text{ mi/h}$
$1 \text{ km/h} = 0.278 \text{ m/s} = 0.621 \text{ mi/h}$
$\quad = 0.911 \text{ ft/s}$
$1 \text{ ft/s} = 0.682 \text{ mi/h} = 0.305 \text{ m/s}$
$\quad = 1.10 \text{ km/h}$
$1 \text{ mi/h} = 1.467 \text{ ft/s} = 1.609 \text{ km/h}$
$\quad = 0.447 \text{ m/s}$
$60 \text{ mi/h} = 88 \text{ ft/s}$

Force

$1 \text{ N} = 0.225 \text{ lb}$
$1 \text{ lb} = 4.45 \text{ N}$
Equivalent weight of a mass of 1 kg
$\quad$ on Earth's surface $= 2.2 \text{ lb} = 9.8 \text{ N}$

Pressure

$1 \text{ Pa (N/m}^2) = 1.45 \times 10^{-4} \text{ lb/in}^2$
$\quad = 7.5 \times 10^{-3} \text{ torr (mm Hg)}$
$1 \text{ torr (mm Hg)} = 133 \text{ Pa (N/m}^2)$
$\quad = 0.02 \text{ lb/in}^2$
$1 \text{ atm} = 14.7 \text{ lb/in}^2 = 1.013 \times 10^5 \text{ N/m}^2$
$\quad = 30 \text{ in. Hg} = 76 \text{ cm Hg}$
$1 \text{ lb/in}^2 = 6.90 \times 10^3 \text{ Pa (N/m}^2)$
$1 \text{ bar} = 10^5 \text{ Pa}$
$1 \text{ millibar} = 10^2 \text{ Pa}$

Energy

$1 \text{ J} = 0.738 \text{ ft} \cdot \text{lb} = 0.239 \text{ cal}$
$\quad = 9.48 \times 10^{-4} \text{ Btu} = 6.24 \times 10^{18} \text{ eV}$
$1 \text{ kcal} = 4186 \text{ J} = 3.968 \text{ Btu}$
$1 \text{ Btu} = 1055 \text{ J} = 778 \text{ ft} \cdot \text{lb} = 0.252 \text{ kcal}$
$1 \text{ cal} = 4.186 \text{ J} = 3.97 \times 10^{-3} \text{ Btu}$
$\quad = 3.09 \text{ ft} \cdot \text{lb}$
$1 \text{ ft} \cdot \text{lb} = 1.36 \text{ J} = 1.29 \times 10^{-3} \text{ Btu}$
$1 \text{ eV} = 1.60 \times 10^{-19} \text{ J}$
$1 \text{ kWh} = 3.6 \times 10^6 \text{ J}$

Power

$1 \text{ W} = 0.738 \text{ ft} \cdot \text{lb/s} = 1.34 \times 10^{-3} \text{ hp}$
$\quad = 3.41 \text{ Btu/h}$
$1 \text{ ft} \cdot \text{lb/s} = 1.36 \text{ W} = 1.82 \times 10^{-3} \text{ hp}$
$1 \text{ hp} = 550 \text{ ft} \cdot \text{lb/s} = 745.7 \text{ W}$
$\quad = 2545 \text{ Btu/h}$

Mass–Energy Equivalents

$1 \text{ u} = 1.66 \times 10^{-27} \text{ kg} \leftrightarrow 931.5 \text{ MeV}$
$1 \text{ electron mass} = 9.11 \times 10^{-31} \text{ kg}$
$\quad = 5.49 \times 10^{-4} \text{ u} \leftrightarrow 0.511 \text{ MeV}$
$1 \text{ proton mass} = 1.672\,62 \times 10^{-27} \text{ kg}$
$\quad = 1.007\,276 \text{ u} \leftrightarrow 938.27 \text{ MeV}$
$1 \text{ neutron mass} = 1.674\,93 \times 10^{-27} \text{ kg}$
$\quad = 1.008\,665 \text{ u} \leftrightarrow 939.57 \text{ MeV}$

Temperature

$T_F = \frac{9}{5} T_C + 32$
$T_C = \frac{5}{9}(T_F - 32)$
$T = T_C + 273$

cgs Force

$1 \text{ dyne} = 10^{-5} \text{ N} = 2.25 \times 10^{-6} \text{ lb}$

cgs Energy

$1 \text{ erg} = 10^{-7} \text{ J} = 7.38 \times 10^{-6} \text{ ft} \cdot \text{lb}$

College Physics Essentials

Mechanics
Thermodynamics
Waves

Eighth Edition
(Volume One)

College Physics Essentials

Mechanics
Thermodynamics
Waves

Eighth Edition
(Volume One)

Jerry D. Wilson
Anthony J. Buffa
Bo Lou

CRC Press
Taylor & Francis Group
Boca Raton   London   New York

CRC Press is an imprint of the
Taylor & Francis Group, an **informa** business

CRC Press
Taylor & Francis Group
6000 Broken Sound Parkway NW, Suite 300
Boca Raton, FL 33487-2742

First issued in paperback 2022

© 2020 by Taylor & Francis Group, LLC
CRC Press is an imprint of Taylor & Francis Group, an Informa business

No claim to original U.S. Government works

ISBN-13: 978-1-138-47632-5 (hbk)
ISBN-13: 978-1-03-233728-9 (pbk)
DOI: 10.1201/9780429323362

Publisher's Note
The publisher has gone to great lengths to ensure the quality of this reprint but points out that some imperfections in the original copies may be apparent.

Visit the Taylor & Francis Web site at
http://www.taylorandfrancis.com

and the CRC Press Web site at
http://www.crcpress.com

Contents

Authors

Jerry D. Wilson is a professor emeritus of physics and former chair of the Division of Biological and Physical Sciences at Lander University in Greenwood, South Carolina. He received his BS degree from Ohio University, MS degree from Union College, and, in 1970, a PhD from Ohio University. He earned his MS degree while employed as a materials behavior physicist by the General Electric Corporation. As a doctoral graduate student, Professor Wilson held the faculty rank of instructor and began teaching physical science courses. During this time, he coauthored a physical science text. In conjunction with his teaching career, Professor Wilson has continued his writing and has authored or coauthored six titles.

Anthony J. Buffa is an emeritus professor of physics at California Polytechnic State University, San Luis Obispo, where he was also a research associate at the Radioanalytical Facility. During his career, he taught courses ranging from introductory physical science to quantum mechanics, while he developed and revised many laboratory experiments. He received his BS in physics from Rensselaer Polytechnic Institute and both his MS and PhD in physics from the University of Illinois, Urbana-Champaign. In retirement, Professor Buffa continues to be involved in teaching and writing. Combining physics with his interests in art and architecture, Dr. Buffa develops his own artwork and sketches, which he uses to increase his effectiveness in teaching. In his spare time, he enjoys his children and grandsons, gardening, walking the dog, and traveling with his wife.

Bo Lou is professor in the Department of Physical Sciences at Ferris State University, and earned his PhD in physics from Emory University. He teaches a variety of undergraduate physics courses and is the coauthor of several physics textbooks. He emphasizes the importance of conceptual understanding of the basic laws and principles of physics and their practical applications to the real world.

1

Measurement and Problem Solving*

Is it a first down? Let's measure!

---

* The mathematics needed in this chapter involves scientific (powers-of-10) notation and trigonometry relationships. You may want to review these topics in Appendix I.

Is it first and ten in the chapter-opening photo? A measurement is needed, as with many other things in our lives. Length measurements tell us how far it is between cities, how tall you are, and as in the photo, if it's first down and ten (yards to go). Time measurements tell you how long it is until the class ends, when the semester or quarter begins, and how old you are. Drugs taken because of illnesses are given in measured doses. Lives depend on various measurements made by doctors, medical technologists, and pharmacists in the diagnosis and treatment of disease.

Measurements enable us to compute quantities and solve problems. Units of measurement are also important in measurements and problem solving. For example, in finding the volume of a rectangular box, if you measure its dimensions in inches, the volume would have units of in^3 (cubic inches); if measured in centimeters, then the units would be cm^3 (cubic centimeters). Measurement and problem solving are part of our lives. They play a particularly central role in our attempts to describe and understand the physical world, as will be seen in this chapter.

1.1 Why and How We Measure

Imagine that someone is giving you directions to their house. Would you find it helpful to be told, "Drive along Main Street for a little while, and turn right at one of the lights. Then keep going for quite a long way?" Or would you want to deal with a bank that sent you a statement at the end of the month saying, "You still have some money left in your account. Not a great deal, though."

Measurement is important to all of us. It is one of the concrete ways in which we deal with our world. This concept is particularly true in physics. *Physics is concerned with the description and understanding of nature*, and measurement is one of its most important tools.

There are ways of describing the physical world that do not involve measurement. For instance, we might talk about the color of a flower or a dress. But the perception of color is subjective; it may vary from one person to another. Indeed, many people are color-blind and cannot tell certain colors apart. Light received by our eyes can be described in terms of wavelengths and frequencies. Different wavelengths are associated with different colors because of the physiological response of our eyes to light. But unlike the sensations or perceptions of color, wavelengths can be measured. They are the same for everyone. In other words, measurements are objective. *Physics attempts to describe and understand nature in an objective way through measurement.*

1.1.1 Standard Units

Measurements are expressed in terms of **units**. As you are probably aware, a large variety of units are used to express measured values. Some of the earliest units of measurement, such as the foot, were originally referenced to parts of the human body. Even today, the hand is still used as a unit to measure the height of horses. One hand is equal to 4 inches (in.). If a

unit becomes officially accepted, it is called a **standard unit**. Traditionally, a government or international body establishes standard units.

A group of standard units and their combinations is called a **system of units**. Two major systems of units are in use today – the metric system and the British system. The latter is still widely used in the United States but has virtually disappeared in the rest of the world, having been replaced by the metric system.

Different units in the same system or units of different systems can be used to describe the same thing. For example, your height can be expressed in inches, feet, centimeters, meters – or even miles, for that matter (although this unit would not be very convenient). It is always possible to convert from one unit to another, and such conversions are sometimes necessary. However, it is best, and certainly most practical, to work consistently within the same system of units, as will be seen.

1.2 SI Units of Length, Mass, and Time

Length, mass, and time are fundamental or base physical quantities that are used to describe a great many quantities and phenomena. In fact, the topics of mechanics (the study of motion and force) covered in the first part of this book require *only* these physical quantities. The system of units used by scientists to represent these and other quantities is based on the metric system.

Historically, the metric system was the outgrowth of proposals for a more uniform system of weights and measures in France during the seventeenth and eighteenth centuries. The modern version of the metric system is called the **International System of Units**, officially abbreviated as **SI** (from the French *Système International des Unités*).

The SI includes *base quantities* and *derived quantities*, which are described by base units and derived units, respectively. **Base units**, such as the meter (m), the kilogram (kg), and the second (s) are defined by standards. Other quantities that are expressed in terms of combinations of base units are called **derived units**. [Think of the Olympic 100-m dash: the distance is measured in meters (m) and the amount of time in seconds (s). To express how fast, or the rate the athlete runs, the derived unit of meters per second (m/s) is used, which represents distance traveled per unit of time, or length per time.]

1.2.1 Length

Length is the base quantity used to measure distances or dimensions in space. We commonly say that length is the distance between two points. But the distance between any two points depends on how the space between them is traversed, which may be in a straight or a curved path.

The SI unit of length is the **meter (m)**. The meter was originally defined as 1/10 000 000 of the distance from the North Pole to the equator along a meridian running through Paris

(▼ **Figure 1.1a**).* A portion of this meridian between Dunkirk, France, and Barcelona, Spain, was surveyed to establish the standard length, which was assigned the name *metre*, from the Greek word *metron*, meaning "a measure." (The American spelling is *meter*.) A meter is 39.37 in. – slightly longer than a yard (3.37 in. longer).

The length of the meter was initially preserved in the form of a material standard: the distance between two marks on a metal bar (made of a platinum-iridium alloy) that was stored under controlled conditions in France and called the Meter of the Archives. However, it is not desirable to have a reference standard that changes with external conditions, such as temperature. In 1983, the meter was redefined in terms of a more accurate standard, an unvarying property of light: the length of the path traveled by light in a vacuum during an interval of 1/299 792 458 of a second (Figure 1.1b). Light travels 299 792 458 m in one second, and the speed of light in a vacuum is $c = 299\,792\,458$ m/s. (c is the common symbol for the speed of light.) Thus, light travels 1 m in 1/299 792 458 s. Note that the length standard is referenced to time, which can be measured with great accuracy.

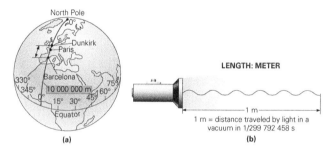

▲ FIGURE 1.1   **The SI length standard: the meter (a)** The meter was originally defined as 1/10 000 000 of the distance from the North Pole to the equator along a meridian running through Paris. A metal bar (called the Meter of the Archives) was constructed as a standard. **(b)** The meter is currently defined in terms of the speed of light.

1.2.2 Mass

Mass is the base quantity used to describe amounts of matter. The more massive an object, the more matter it contains. The SI unit of mass is the **kilogram (kg)**. The kilogram was originally defined in terms of a specific volume of water; that is, a cube 0.10 m (10 cm) on a side (thereby associating the mass standard with the length standard). However, the kilogram is now referenced to a specific material standard: the mass of a prototype platinum-iridium cylinder kept at the International Bureau of Weights and Measures in Sevres, France (▶ **Figure 1.2**). The United States has a duplicate of the prototype cylinder. The duplicate serves as a

reference for secondary standards that are used in everyday life and commerce. It is hoped that the kilogram may eventually be referenced to something other than a material standard.†

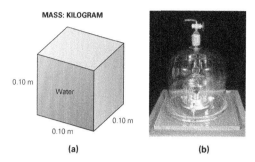

▲ FIGURE 1.2   **The SI mass standard: the kilogram (a)** The kilogram was originally defined in terms of a specific volume of water, that of a cube 0.10 m (10 cm) on a side, thereby associating the mass standard with the length standard. **(b)** This is a replica showing the size and shape of the kilogram standard. The replica is in a glass dome. The standard for the kilogram unit of mass is a cylinder of platinum-iridium alloy kept by the International Bureau of Weights and Measures near Paris. A duplicate kept at the National Institute of Standards and Technology (NIST) serves as the mass standard for the United States.

You may have noticed that the phrase *weights and measures* is generally used instead of *masses and measures*. In the SI, mass is a base quantity, but in the British system, weight is used to describe amounts of mass – for example, weight in pounds instead of mass in kilograms. The weight of an object is the gravitational attraction that the Earth exerts on the object. For example, when you weigh yourself on a scale, your weight is a measure of the downward gravitational force exerted on your mass by the Earth. Weight is a measure of mass in this way near the Earth's surface, because weight and mass are directly proportional to each other.

But treating weight as a base quantity creates some problems. A base quantity should have the same value everywhere. This is the case with mass – an object has the same mass, or amount of matter, regardless of its location. *But this is not true of weight.* For example, the weight of an object on the Moon is less than its weight on the Earth (one-sixth as much). This is because the gravitational attraction exerted on an object by the Moon (the object's weight on the Moon) is less than that exerted by the Earth. That is, an object with a given amount of mass has a particular weight on the Earth, but on the Moon, the same amount of mass will weigh only about one-sixth as much. Therefore, the weight is not a base quantity in SI.

For now, keep in mind that in a given location, such as on the Earth's surface, *weight is related to mass, but they are not the same*. Since the weight of an object of a certain mass can vary with location, it is much more practical to take mass as the base quantity, as the SI does. Base quantities should remain the same

---

* Note that this book and most physicists have adopted the practice of writing large numbers with a thin space for three-digit groups – for example, 10 000 000 (not 10,000,000). This is done to avoid confusion with the European practice of using a comma as a decimal point. For instance, 3.141 in the United States would be written as 3,141 in Europe. Long decimal numbers, such as 0.537 842 365, may also be separated, for consistency. Thin spaces are generally used for numbers with more than four digits on either side of the decimal point.

† While in the final production stages of this book (2019), the International Committee for Weights and Measures has adopted a new definition of the kilogram based on a fundamental constant, Planck's constant, $6.626\,070\,15 \times 10^{-34}$ when expressed in units of kg·m²/s. Planck's constant is discussed later in Chapter 27.

regardless of where they are measured, under normal or standard conditions. The distinction between mass and weight will be more fully explained in a later chapter. Our discussion until then will be chiefly concerned with mass.

1.2.3 Time

Time is a difficult concept to define. A common definition is that time is the continuous, forward flow of events. This statement is not so much a definition as an observation that time has never been known to run backward, as it might appear to do when you view a film run backward in a projector. Time is sometimes said to be a fourth dimension, accompanying the three dimensions of space (*x*, *y*, *z*, *t*). That is, if something exists in space, it also exists in time. In any case, events can be used to mark time measurements. The events are analogous to the marks on a meterstick used for measurements of length.

The SI unit of time is the **second (s)**. The solar "clock" was originally used to define the second. A solar day is the interval of time that elapses between two successive crossings of the same longitude line (meridian) by the Sun. A second was fixed as 1/86 400 of this apparent solar day (1 day = 24 h = 1440 min = 86 400 s). However, the elliptical path of the Earth's motion around the Sun causes apparent solar days to vary in length.

As a more precise standard, an average, or mean, solar day was computed from the lengths of the apparent solar days during a solar year. In 1956, the second was referenced to this mean solar day. But the mean solar day is not exactly the same for each yearly period because of minor variations in the Earth's motions and a very small but steady slowing of its rate of rotation due to tidal friction. So scientists kept looking for something better.

In 1967, an atomic standard was adopted as a better reference. The second was defined by the radiation frequency of the cesium-133 atom. This "atomic clock" used a beam of cesium atoms to maintain our time standard, with a variation of about 1 s in 300 years. In 1999, another cesium-133 atomic clock was adopted, the atomic fountain clock, which, as the name implies, is based on the radiation frequency of a fountain of cesium atoms rather than a beam (▶ **Figure 1.3**). The variation of this "timepiece" is less than 1 second per 20 million years!*

1.2.4 SI Base Units

The SI has seven *base units* for seven base quantities, which are assumed to be mutually independent. In addition to the meter, kilogram, and second for (1) length, (2) mass, and (3) time, SI units include (4) electric current (charge/second) in amperes (A), (5) temperature in kelvins (K), (6) amount of substance in moles (mol), and (7) luminous intensity in candelas (cd). See ▶ **Table 1.1**.

The foregoing quantities are thought to compose the smallest number of base quantities needed for a complete description of everything observed or measured in nature.

---

* An even more precise clock, the 3D quantum gas clock, is expected to neither gain nor lose a second in about 100 trillion years!

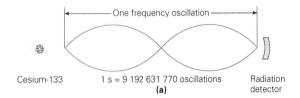

▲ **FIGURE 1.3   The SI time standard: the second** The second was once defined in terms of the average solar day. **(a)** It is now defined by the frequency of the radiation associated with an atomic transition. **(b)** The atomic fountain "clock" shown here, at NIST, is the time standard for the United States. The variation of this "timepiece" is less than 1 s per 20 million years.

TABLE 1.1   The Seven Base Units of the SI

Name of Unit (abbreviation)	Property Measured
meter (m)	length
kilogram (kg)	mass
second (s)	time
ampere (A)	electric current
kelvin (K)	temperature
mole (mol)	amount of substance
candela (cd)	luminous intensity

1.3 More about the Metric System

The metric system involving the standard units of length, mass, and time, now incorporated into the SI, was once called the **mks system** (for *meter-kilogram-second*). Another metric system that has been used in dealing with relatively small quantities is the **cgs system** (for *centimeter-gram-second*). In the United States, the system still generally in use is the British (or English) engineering system, in which the standard units of length, mass, and time are foot, slug, and second, respectively. You may not have heard of the slug, because as mentioned earlier, gravitational force (weight) is commonly used instead of mass – pounds instead of slugs – to describe quantities of matter. As a result, the British system is sometimes called the **fps system** (for *foot-pound-second*).

The metric system is predominant throughout the world and is coming into increasing use in the United States. Because it is simpler mathematically, the SI is the preferred system of units for science and technology. SI units are used throughout most of this book. All quantities can be expressed in SI units. However, some units from other systems are accepted for limited use as a matter of practicality – for example, the time unit of hour and the temperature unit of **degree Celsius** (°C). British units will sometimes be used in the early chapters for comparison purposes, since these units are still employed in everyday activities and many practical applications.

The increasing worldwide use of the metric system means that you should be familiar with it. One of the greatest advantages of the metric system is that it is a decimal, or base-10, system. This means that larger or smaller units may be obtained by multiplying or dividing by powers of 10. A list of some multiples and corresponding prefixes for metric units is given in ▼ **Table 1.2**.

TABLE 1.2 Some Multiples and Prefixes for Metric Units[a]

Multiple[b]	Prefix (and abbreviation)	Pronunciation
10^{12}	tera- (T)	ter'a (as in *terrace*)
10^{9}	giga- (G)	jig'a (*jig* as in *jiggle*, *a* as in *about*)
10^{6}	**mega- (M)**	meg'a (as in *megaphone*)
10^{3}	**kilo- (k)**	kil'o (as in *kilowatt*)
10^{2}	hecto- (h)	hek'to (*heck-toe*)
10	deka- (da)	dek'a (*deck* plus *a* as in *about*)
10^{-1}	deci- (d)	des'i (as in *decimal*)
10^{-2}	**centi- (c)**	sen'ti (as in *sentimental*)
10^{-3}	**milli- (m)**	mil'li (as in *military*)
10^{-6}	**micro- (μ)**	mi'kro (as in *microphone*)
10^{-9}	nano- (n)	nan'o (*an* as in *annual*)
10^{-12}	pico- (p)	pe'ko (*peek-oh*)
10^{-15}	femto- (f)	fem'to (*fem* as in *feminine*)
10^{-18}	atto- (a)	at'toe (as in *anatomy*)

[a] For example, 1 gram (g) multiplied by 1000, or 10^{3}, is 1 kilogram (kg); 1 gram multiplied by 1/1000, or 10^{-3}, is 1 milligram (mg).

[b] The most commonly used prefixes are printed in boldface. Note that the abbreviations for the multiples 10^{6} and greater are capitalized, whereas the abbreviations for the smaller multiples are lowercased.

For metric measurements, the prefixes *micro-*, *milli-*, *centi-*, *kilo-*, and *mega-* are the ones most commonly used – for example, microsecond (μs), millimeter (mm), centimeter (cm), kilometer (km), and megabyte (MB) as for computer disk or digital memory sizes. The decimal characteristics of the metric system make it convenient to change measurements from one size of metric unit to another. In the British system, different conversion factors must be used, such as 16 for converting pounds to ounces and 12 for converting feet to inches, whereas in the metric system, the conversion factors are multiples of 10. For example, 100 (10^{2}) to convert meters to centimeters (1 m = 100 cm) and 1000 (10^{3}) to convert meters to millimeters (1 m = 1000 mm).

You are already familiar with one base-10 system – US currency. Just as a meter can be divided into 10 decimeters, 100 centimeters, or 1000 millimeters, the "base unit" of the dollar can be broken down into 10 "decidollars" (dimes), 100 "centidollars" (cents), or 1000 "millidollars" (tenths of a cent, or mills, used in figuring property taxes and bond levies). Since all the metric prefixes are powers of 10, there are no metric analogues for quarters or nickels.

The official metric prefixes help eliminate confusion. For example, in the United States, a billion is a thousand million (10^{9}); in Great Britain, a billion is a million million (10^{12}). The use of metric prefixes eliminates any confusion, since *giga-* indicates 10^{9} and *tera-* stands for 10^{12}. You will probably be hearing more about *nano-*, the prefix that indicates 10^{-9}, with respect to nanotechnology (*nanotech* for short). In general, nanotechnology is any technology done on the nanometer scale. A nanometer (nm) is one billionth (10^{-9}) of a meter, about the width of three to four atoms. Basically, nanotechnology involves the manufacture or building of things one atom or molecule at a time, so the nanometer is the appropriate scale. One atom or molecule at a time? That may sound a bit farfetched, but it's not (see ▼ **Figure 1.4**).

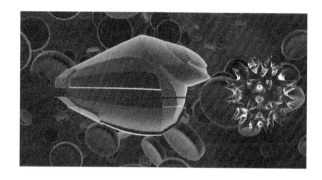

▲ FIGURE 1.4 **Nano robot-killing virus** A nanobot measures only a few atoms across and is measured in nanometers, or a millionth of a millimeter. The small size gives them the ability to interact at the bacteria and virus level. Operating as a swarm, these tiny robots have the promise to do some really incredible things.

It is difficult for us to grasp or visualize the new concept of nanotechnology. Even so, keep in mind that a nanometer is one billionth of a meter. The diameter of an average human hair is about 40 000 nm – huge compared with the new nano-applications. The future should be an exciting nano-time.

1.3.1 Volume

In the SI, the standard unit of volume is the cubic meter (m^3) – the three-dimensional derived unit of the meter base unit. Because this unit is rather large, it is often more convenient to use the nonstandard unit of volume (or capacity) of a cube 10 cm on a side. This volume was given the name *litre*, which is spelled **liter (L)** in the United States. The volume of a liter is 1000 cm^3 ($10 \times 10 \times 10$ cm).

Since 1 L = 1000 mL (milliliters, mL) it follows that 1 mL = 1 cm³. See ▼ **Figure 1.5a.** (The cubic centimeter is sometimes abbreviated as cc, particularly in chemistry and biology. Also, the milliliter is sometimes abbreviated as ml, but the capital L is preferred [mL] so as not to be confused with the numeral one, 1.)

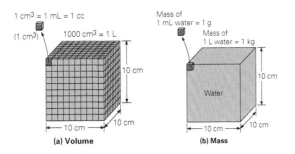

▲ **FIGURE 1.5** **The liter and the kilogram** Other metric units are derived from the meter. **(a)** A unit of volume (capacity) was taken to be the volume of a cube 10 cm, or 10 m, on a side and was given the name *liter* (L). **(b)** The mass of a liter of water is 1 kg. Note that the decimeter cube contains 1000 cm³, or 1000 mL. Thus, 1 cm³, or 1 mL, of water has a mass of 1 g.

Recall from Figure 1.2 that the standard unit of mass, the kilogram, was originally defined to be the mass of a cubic volume of water 10 cm, or 0.10 m, on a side, or the mass of one liter (1 L) of water at 4 °C. That is, 1 L *of water has a mass of* 1 kg (Figure 1.5b). Also, since 1 kg = 1000 g and 1 L = 1000 cm³ (= 1000 mL), then 1 cm³ (or 1 mL) *of water has a mass* of 1 g.

EXAMPLE 1.1: THE METRIC TON (OR TONNE) – ANOTHER UNIT OF MASS

As discussed, the metric unit of mass was originally related to length, with a liter (1000 cm³) of water having a mass of 1 kg. The standard metric unit of volume is the cubic meter (m³) and this volume of water was used to define a larger unit of mass called the metric ton (or tonne, as it is sometimes spelled). A metric ton is equivalent to how many kilograms?

THINKING IT THROUGH. A cubic meter is a relatively large volume and holds a large amount of water (more than a cubic yard; why?). The key is to find how many cubic volumes measuring 10 cm on a side (liters) are in a cubic meter. A large number would be expected.

SOLUTION

Each liter of water has a mass of 1 kg, so we need to find out how many liters are in 1 m³. Since there are 100 cm in a meter, a cubic meter is simply a cube with sides 100 cm in length. Therefore, a cubic meter (1 m³) has a volume of 10^2 cm × 10^2 cm × 10^2 cm = 10^6 cm³. Since 1 L has a volume of 10^3 cm³, there must be $(10^6$ cm³)/(10^3 cm³/L) = 1000 L in 1 m³. Thus, 1 metric ton is equivalent to 1000 kg.

Note that this line of reasoning can be expressed very concisely in a single ratio:

$$\frac{1\,m^3}{1\,L} = \frac{100\,cm \times 100\,cm \times 100\,cm}{10\,cm \times 10\,cm \times 10\,cm} = 1000 \quad or \quad 1\,m^3 = 1000\,L$$

FOLLOW-UP EXERCISE. What would be the length of the sides of a cube that contained a metric kiloton of water? (*Answers to all Follow-Up Exercises are given in Appendix V at the back of the book.*)

Because the metric system is coming into increasing use in the United States, you may find it helpful to have an idea of how metric and British units compare. The relative sizes of some units are illustrated in ▼ **Figure 1.6.** The mathematical conversion from one unit to another will be discussed shortly.

1.4 Unit Analysis

The fundamental or base quantities used in physical descriptions are called *dimensions*. For example, length, mass, and time are dimensions. You could measure the distance between two points and express it in units of meters, centimeters, or feet, but the quantity would still have the dimension of length.

Dimensions provide a procedure by which the consistency of equations may be checked. In practice, it is convenient to use specific units, such as m, kg, and s (see ▶ **Table 1.3**). Such units

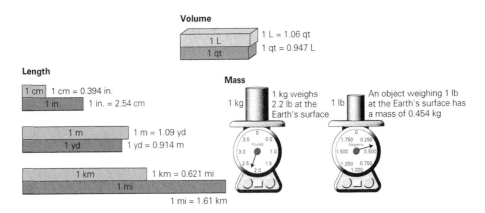

▲ **FIGURE 1.6** **Comparison of some SI and British units** The bars illustrate the relative magnitudes of each pair of units. (*Note:* The comparison scales are different in each case.)

can be treated as algebraic quantities and be canceled. Using units to check equations is called unit analysis, which shows the consistency of units and whether an equation is dimensionally correct.

TABLE 1.3 Some Units of Common Quantities

Quantity	Unit
Mass (m)	kg
Time (t)	s
Length (L)	m
Area (A)	m^2
Volume (V)	m^3
Velocity (v)	m/s
Acceleration (a)	m/s^2

You have used equations and know that an equation is a mathematical equality. Since physical quantities used in equations have units, *the two sides of an equation must be equal not only in numerical value (magnitude), but also in units (dimensions).* For example, suppose you had the length quantities $L = 3.0$ m and $W = 4.0$ m. Inserting these values into the equation $A = L \times W$ gives 3.0 m $\times 4.0$ m $= 12$ m^2. Both sides of the equation are numerically equal ($3 \times 4 = 12$), and both sides have the same units, m $\times$ m $= m^2 = $ (length)^2. If an equation is correct by unit analysis, it must be dimensionally correct. Example 1.2 demonstrates the further use of unit analysis.

EXAMPLE 1.2: CHECKING DIMENSIONS – UNIT ANALYSIS

A professor puts two equations on the board: (a) $v = v_0 + at$ and (b) $x = v/(2a)$, where x is distance in meters (m); v and v_0 are velocities in meters/second (m/s); a is acceleration in (meters/second)/second, or meters/second^2 (m/s^2); and t is time in seconds (s). Are the equations dimensionally correct? Use unit analysis to find out.

THINKING IT THROUGH. Simply insert the units for the quantities in each equation, cancel, and check the units on both sides.

SOLUTION

(a) The equation is

$$v = v_0 + at$$

Inserting units for the physical quantities gives (**Table 1.3**)

$$\frac{m}{s} = \frac{m}{s} + \left(\frac{m}{s^2} \times s\right) \quad \text{or} \quad \frac{m}{s} = \frac{m}{s} + \left(\frac{m}{s \times s} \times s\right)$$

Notice that units cancel like numbers in a fraction. Then, simplifying,

$$\frac{m}{s} = \frac{m}{s} + \frac{m}{s} \quad \text{(dimensionally correct)}$$

The equation is dimensionally correct, since the units on each side are meters per second. (The equation is also a correct relationship, as will be seen in Chapter 2.)

(b) Using unit analysis, the equation

$$x = \frac{v}{2a}$$

is

$$m = \frac{\left(\dfrac{m}{s}\right)}{\left(\dfrac{m}{s^2}\right)} = \frac{m}{s} \times \frac{s^2}{m} \quad \text{or} \quad m = s$$

(not dimensionally correct)

The meter (m) is not the same unit as the second (s), so in this case, the equation is not dimensionally correct (length $\neq$ time), and therefore is also not physically correct.

FOLLOW-UP EXERCISE. Is the equation $ax = v^2$ dimensionally correct? *(Answers to all Follow-Up Exercises are given in Appendix V at the back of the book.)*

Unit analysis will tell if an equation is dimensionally correct, but a dimensionally correct equation may not correctly express the physical relationship of quantities. For example, in terms of units, the equation $x = at^2$ is

$$m = (m/s^2)(s^2) = m$$

This equation is dimensionally correct (length $=$ length). But, as will be learned in Chapter 2, it is not *physically* correct. The correct form of the equation – both dimensionally and physically – is $x = (\frac{1}{2})at^2$. [The fraction ($\frac{1}{2}$) has no dimensions; it is an exact dimensionless number.] Unit analysis *cannot* tell you if an equation is physically correct, only whether or not it is dimensionally consistent.

1.4.1 Determining the Units of Quantities

Another aspect of unit analysis that is very important in physics is the determination of the units of quantities from defining equations. For example, the **density** (ρ) of an object (represented by the Greek letter rho, ρ) is defined by the equation

$$\rho = \frac{m}{V} \quad \left(\text{units of } \frac{kg}{m^3}\right) \tag{1.1}$$

where m is the object's mass and V its volume. (Density is the mass per unit volume and is a measure of the compactness of the mass of an object or substance.) In SI units, mass is measured in kilograms and volume in cubic meters, which gives the derived SI unit for density as kilograms per cubic meter (kg/m^3).

How about π? Does it have units? The relationship between the circumference (c) and the diameter (d) of a circle is given by the equation $c = \pi d$, so $\pi = c/d$. If the lengths are measured in meters, then unitwise,

$$\pi = \frac{c}{d}\left(\frac{\cancel{m}}{\cancel{m}}\right)$$

Thus, π has no units. It is unitless, or a dimensionless constant.

1.5 Unit Conversions

Because units in different systems, or even different units in the same system, can be used to express the same quantity, it is sometimes necessary to convert the units of a quantity from one unit to another. For example, we may need to convert feet to yards or convert inches to centimeters. You already know how to do many unit conversions. If a fence is 24 ft long, what is its length in yards? Your immediate answer is 8 yards.

How did you do this conversion? Well, you must have known a relationship between the units of foot and yard. That is, you know that 1 yd = 3 ft. This is what is called an *equivalence statement*. As was seen in Section 1.4, the numerical values and units on both sides of an equation must be the same. In equivalence statements, we commonly use an equal sign to indicate that 1 yd and 3 ft stand for the *same*, or *equivalent*, *length*. The numbers are different because they stand for different *units* of length.

Conversion factors are used to mathematically change units. The equivalence statements can be expressed in the form of ratios – for example, 1 yd/3 ft or 3 ft/1 yd. (The "1" is often omitted in the denominators of such ratios for convenience – for example, 3 ft/yd.) To understand why such ratios are useful, note the expression 1 yd = 3 ft in ratio form:

$$\frac{1\,\text{yd}}{3\,\text{ft}} = \frac{3\,\text{ft}}{3\,\text{ft}} = 1 \quad \text{or} \quad \frac{3\,\text{ft}}{1\,\text{yd}} = \frac{1\,\text{yd}}{1\,\text{yd}} = 1$$

As can be seen, a conversion factor has an actual value of unity or one – and you can multiply any quantity by one without changing its value or size. Thus, *a conversion factor simply lets you express a quantity in terms of other units without changing its physical value or size.*

The manner in which 24 ft is converted to yards may be expressed mathematically as follows:

$$24\,\cancel{\text{ft}} \times \frac{1\,\text{yd}}{3\,\cancel{\text{ft}}} = 8\,\text{yd} \quad (\text{"ft" units cancel})$$

Using the appropriate conversion factor form, the "ft" units cancel, as shown by the slash marks, giving the correct unit.

Suppose you are asked to convert 5.0 in. to centimeters. You may not know the conversion factor in this case, but it can be obtained from a table (such as the one that appears inside the front cover of this book). The needed relationships are 1 in. = 2.54 cm or 1 cm = 0.394 in. It makes no difference which of these equivalence statements you use. The question, once you have expressed the equivalence statement as a ratio conversion factor, is whether to multiply or divide by that factor to make the conversion. *In doing unit conversions, take advantage of unit analysis – that is, let the units determine the appropriate form of conversion factor.*

Note that the equivalence statements can give rise to two forms of the conversion factors: 1 in./2.54 cm and 2.54 cm/in. When changing inches to centimeters, the appropriate form for multiplying is 2.54 cm/in. When changing centimeters to inches, use the form 0.394 in./cm. For example,

$$5.0\,\cancel{\text{in.}} \times \frac{2.54\,\text{cm}}{\cancel{\text{in.}}} = 12.7\,\text{cm}$$

$$12.7\,\cancel{\text{cm}} \times \frac{0.394\,\text{in.}}{\cancel{\text{cm}}} = 5.0\,\text{in.}$$

The multiplication of conversion factors in canceling units is usually more convenient than division.

A few commonly used equivalence statements are not dimensionally or physically correct; for example, consider 1 kg = 2.2 lb, which is used for quickly determining the weight of an object near the Earth's surface given its mass. The kilogram is a unit of mass, and the pound is a unit of weight (force). This means that 1 kg is *equivalent* to 2.2 lb; that is, a 1-kg *mass* has a *weight* of 2.2 lb. Since mass and weight are directly proportional, the dimensionally incorrect conversion factor 1 kg/2.2 lb may be used (but *only* near the Earth's surface).

EXAMPLE 1.3: CONVERTING UNITS –
USE OF CONVERSION FACTORS

The world record of men's pole vault is 6.16 m while the female record is 5.06 m. What is the difference in these heights in feet?

THINKING IT THROUGH. After using the correct conversion factor, the rest is arithmetic.

SOLUTION

From the conversion table, 1 m = 3.28 ft, so converting the heights to feet:

$$6.16\,\cancel{\text{m}} \times \frac{3.28\,\text{ft}}{\cancel{\text{m}}} = 20.2\,\text{ft}$$

$$5.06\,\cancel{\text{m}} \times \frac{3.28\,\text{ft}}{\cancel{\text{m}}} = 16.6\,\text{ft}$$

And the difference in height is $\Delta h = 20.2\,\text{ft} - 16.6\,\text{ft} = 3.6\,\text{ft}$.

Another approach would be to subtract the heights in meters and have only a single conversion:

$$6.16\,\text{m} - 5.06\,\text{m} = 1.10\,\text{m} \times \frac{3.28\,\text{ft}}{\text{m}} = 3.6\,\text{ft}$$

Another foot-meter conversion is shown in ▼ **Figure 1.7a**. Is it correct?

(a)   (b)

▲ FIGURE 1.7   **Unit conversion** Signs sometimes list both the British and metric units, as shown here for elevation **(a)** and speed **(b)**.

FOLLOW-UP EXERCISE. Rather than use a single conversion factor from the table, use commonly known factors to convert a 30-day month to seconds. *(Answers to all Follow-Up Exercises are given in Appendix V at the back of the book.)*

**EXAMPLE 1.4: MORE CONVERSIONS –
A REALLY LONG CAPILLARY SYSTEM**

Capillaries, the smallest blood vessels of the body, connect the arterial system with the venous system and supply our tissues with oxygen and nutrients (▼ **Figure 1.8**). It is estimated that if all of the capillaries of an average adult were unwound and spread out end to end, they would extend to a length of about 64 000 km. (a) How many miles is this length? (b) Compare this length with the circumference of the Earth.

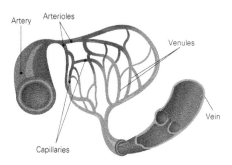

▲ FIGURE 1.8   **Capillary system** Capillaries connect the arterial and venous systems in our bodies. They are the smallest blood vessels, but their total length is impressive. See Example 1.4.

THINKING IT THROUGH. (a) This conversion is straightforward – just use the appropriate conversion factor. (b) How is the circumference of a circle or sphere calculated? There is an equation to do so, but the radius or diameter of the Earth must be known. (If you do not remember one of these values, see the solar system data table inside the back cover of this book.)

SOLUTION

(a) From the conversion table, 1 km = 0.621 mi, so

$$\frac{64\,000\,\cancel{km} \times 0.621\,\text{mi}}{1\,\cancel{km}} = 40\,000\,\text{mi} \quad (\text{rounded off})$$

(b) A length of 40 000 mi is substantial. To see how this length compares with the circumference (*c*) of the Earth, recall that the radius of the Earth is approximately 4000 mi, so the diameter (*d*) is 8000 mi. The circumference of a circle is given by $c = \pi d$ (Appendix IC), and

$$c = \pi d \approx 3 \times 8000\,\text{mi} \approx 24\,000\,\text{mi} \quad (\text{rounded off})$$

[To make a general comparison, $\pi\,(= 3.14...)$ is rounded off to 3. The ≈ symbol means "approximately equal to."]
So,

$$\frac{\text{Capillary length}}{\text{Earth's circumference}} \approx \frac{40\,000\,\text{mi}}{24\,000\,\text{mi}} = 1.7$$

The capillaries of your body have a total length that would extend about 1.7 times around the world. Wow!

FOLLOW-UP EXERCISE. Taking the average distance between the East Coast and West Coast of the continental United States to be 4800 km, how many times would the total length of your body's capillaries cross the country?

Quantities may also be divided by conversion factors. For example,

$$12\,\text{in.} \Big/ \left(\frac{1\,\text{in.}}{2.54\,\text{cm}}\right) = 12\,\text{in.} \times (2.54\,\text{cm/in.}) = 30\,\text{cm}$$

Using the multiplication form saves the step of inverting the ratio.

**EXAMPLE 1.5: CONVERTING UNITS OF AREA –
CHOOSING THE CORRECT CONVERSION FACTOR**

A hall bulletin board has an area of 2.5 m². What is this area in (a) square centimeters (cm²) and (b) square inches (in²)?

THINKING IT THROUGH. This problem is a conversion of area units, and we know that 1 m = 100 cm. So, some squaring must be done to get square meters related to square centimeters.

SOLUTION

A common error in such conversions is the use of incorrect conversion factors. Because 1 m = 100 cm, it is sometimes assumed that 1 m² = 100 cm², which is *wrong*. The correct area conversion factor may be obtained directly from the correct linear conversion factor, 100 cm/1 m, or 10^2 cm/1 m, by *squaring* the linear conversion factor:

$$\left(\frac{10^2\,\text{cm}}{1\,\text{m}}\right)^2 = \frac{10^4\,\text{cm}^2}{1\,\text{m}^2}$$

Hence, $1\,\text{m}^2 = 10^4\,\text{cm}^2\ (= 10\,000\,\text{cm}^2)$.

(a) Then using the conversion factor explicitly squared:

$$2.5\,\text{m}^2 \times \left(\frac{10^2\,\text{cm}}{1\,\text{m}}\right)^2 = 2.5\,\text{m}^2 \times \frac{10^4\,\text{cm}^2}{1\,\text{m}^2} = 2.5 \times 10^4\,\text{cm}^2$$

(b) Using the cm^2 result found in (a), by a similar procedure,

$$2.54 \times 10^4\,\text{cm}^2 \left(\frac{0.394\,\text{in.}}{\text{cm}}\right)^2$$
$$= 2.54 \times 10^4\,\text{cm}^2 \times \left(\frac{0.155\,\text{in}^2}{\text{cm}^2}\right) = 3.94 \times 10^3\,\text{in}^2$$

FOLLOW-UP EXERCISE. How many cubic centimeters are in $1\,\text{m}^3$?

EXAMPLE 1.6: THE BETTER DEAL

A grocery store has a sale on sodas. A 2-liter bottle sells for \$1.35, and the price of a half-gallon bottle is \$1.32. Which is the better buy?

THINKING IT THROUGH. The answer is obtained by knowing the price per common volume. This means that liters must be converted to quarts or vice versa. (The cancellation slashes will now be omitted as being understood.)

SOLUTION

To get a common volume, let's convert liters to quarts using the conversion factor given inside the front cover of the book (1 L = 1.056 qt):

$$2.0\,\text{L}\left(\frac{1.056\,\text{qt}}{\text{L}}\right) = 2.1\,\text{qt}$$

Then in terms of quarts, the base price of the liquid in the 2.0-L bottle is

$$\frac{\$1.35}{2.1\,\text{qt}} = \frac{\$0.64}{\text{qt}}$$

Similarly for the half-gallon (2 qt) volume:

$$\frac{\$1.32}{2.0\,\text{qt}} = \frac{\$0.66}{\text{qt}}$$

So, the better buy is the 2-liter bottle, even though its price is higher. (Keep in mind a liter is larger than a quart.)

FOLLOW-UP EXERCISE. Work the Example the other way – changing quarts to liters – to see if the result is the same.

1.6 Significant Figures

Most of the time, you will be given numerical data when asked to solve a problem. In general, such data are either exact numbers or measured numbers (quantities). **Exact numbers** are numbers without any uncertainty. This category includes numbers such as the 100 used to calculate a percentage and the 2 in the equation $r = d/2$ relating the radius and diameter of a circle. **Measured numbers** are numbers obtained from measurement processes and thus generally have some degree of uncertainty.

When calculations are done with measured numbers, the uncertainty of measurement is *propagated*, or carried along, by the mathematical operations. The question of how to report a result arises. For example, suppose that you are asked to find time (t) from the equation $x = vt$ and are given that $x = 5.3$ m and $v = 1.67$ m/s. That is,

$$t = \frac{x}{v} = \frac{5.3\,\text{m}}{1.67\,\text{m/s}} = ?$$

Doing the division operation on a calculator yields a result such as 3.173 652 695 s (▼ **Figure 1.9**). How many figures, or digits, should you report in the answer?

▲ **FIGURE 1.9** **Significant figures and insignificant figures** For the division operation 5.3/1.67, a calculator with a floating decimal point gives many digits. A calculated quantity can be no more accurate than the least accurate quantity involved in the calculation, so this result should be rounded off to two significant figures – that is, 3.2.

A widely used procedure for estimating uncertainty involves the use of **significant figures (sf)**, sometimes called *significant digits*. The degree of accuracy of a measured quantity depends on how finely divided the measuring scale of the instrument is. For example, you might measure the length of an object as 2.5 cm with one instrument and 2.54 cm with another. The second instrument with a finer scale provides more significant figures and thus a greater degree of accuracy.

Basically, *the significant figures in any measurement are the digits that are known with certainty, plus one digit that is uncertain.* This set of digits is usually defined as all of the digits that can be read directly from the instrument used to make the measurement, plus one uncertain digit that is obtained by estimating the fraction of the smallest division of the instrument's scale.

The quantities 2.5 and 2.54 cm have two and three significant figures, respectively. This is rather evident. However, some confusion may arise when a quantity contains one or more zeros. For example, how many significant figures does the quantity 0.0254 m have? What about 104.6 m, or 2705.0 m? In such cases, the following rules will be used to determine significant figures:

1. Zeros at the beginning of a number are not significant. They merely locate the decimal point. For example,

    0.0254 m has three significant figures (2, 5, 4)

2. Zeros within a number are significant. For example,

    104.6 m has four significant figures (1, 0, 4, 6)

3. Zeros at the end of a number after the decimal point are significant. For example,

    2705.0 m has five significant figures (2, 7, 0, 5, 0)

4. In whole numbers without a decimal point that end in one or more zeros (trailing zeros) – for example, 500 kg – the zeros may or may not be significant. In such cases, it is not clear which zeros serve only to locate the decimal point and which are actually part of the measurement. For example, if the first zero after the 5 in 500 kg is the estimated digit in the measurement, then there are only two significant figures. Similarly, if the last zero is the estimated digit (500 kg), then there are three significant figures. This ambiguity may be removed by using scientific (powers-of-10) notation:

    5.0×10^2 kg has two significant figures
    5.00×10^2 kg has three significant figures

This notation is helpful in expressing the results of calculations with the proper numbers of significant figures, as will be seen shortly. (Appendix I includes a review of scientific notations.)

(*Note:* To avoid confusion regarding numbers having trailing zeros used as given quantities in text examples and exercises, the trailing zeros will be considered significant. For example, assume that a time of 20 s has two significant figures, even if it is not written out as 2.0×10^1 s.)

It is important to report the results of mathematical operations with the proper number of significant figures. This is accomplished by using rules for (1) multiplication and division and (2) addition and subtraction. To obtain the proper number of significant

figures, the results are rounded off. Here are some general rules that will be used for mathematical operations and rounding.

1.6.1 Significant Figures in Calculations

1. When multiplying and dividing quantities, leave as many significant figures in the answer as there are in the quantity with the least number of significant figures.
2. When adding or subtracting quantities, leave the same number of decimal places (rounded) in the answer as there are in the quantity with the least number of decimal places.

1.6.2 Rules for Rounding*

1. If the first digit to be dropped is less than 5, leave the preceding digit as is.
2. If the first digit to be dropped is 5 or greater, increase the preceding digit by one.

The rules for significant figures mean that the result of a calculation can be no more accurate than the least accurate quantity used. That is, accuracy cannot be gained performing mathematical operations. Thus, the result that should be reported for the division operation discussed at the beginning of this section is

$$\underset{(3\,\text{sf})}{\overset{(2\,\text{sf})}{\frac{5.3\,\text{m}}{1.67\,\text{m/s}}}} = 3.2\,\text{s} \quad (2\,\text{sf})$$

The result is rounded off to two significant figures (see Figure 1.9). Applications of these rules are shown in the following Examples.

EXAMPLE 1.7: USING SIGNIFICANT FIGURES IN MULTIPLICATION AND DIVISION – ROUNDING APPLICATIONS

The following operations are performed and the results rounded off to the proper number of significant figures:*

Multiplication:

$$\underset{(2\,\text{sf})\quad(3\,\text{sf})}{2.4\,\text{m} \times 3.65\,\text{m}} = 8.76\,\text{m}^2 = 8.8\,\text{m}^2 \quad \text{(rounded to two sf)}$$

Division:

$$\underset{(3\,\text{sf})}{\overset{(4\,\text{sf})}{\frac{725.0\,\text{m}}{0.125\,\text{s}}}} = 5800\,\text{m/s} = 5.80 \times 10^3\,\text{m/s}$$

(represented with three sf; why?)

---

* It should be noted that these rounding rules give an approximation of accuracy, as opposed to the results provided by more advanced statistical methods.

FOLLOW-UP EXERCISE. Perform the following operations, and express the answers in the standard powers-of-10 notation (one digit to the left of the decimal point) with the proper number of significant figures: (a) $(2.0 \times 10^5 \text{ kg})(0.035 \times 10^2 \text{ kg})$ and (b) $(148 \times 10^{-6} \text{ m})/(0.4906 \times 10^{-6} \text{ m})$.

EXAMPLE 1.8: USING SIGNIFICANT FIGURES IN ADDITION AND SUBTRACTION – APPLICATION OF RULES

The following operations are performed by finding the number that has the least number of decimal places. (Units have been omitted for convenience.)

Addition:

In the numbers to be added, note that 23.1 has the least number of decimal places (one):

$$
\begin{array}{r}
23.1 \\
0.546 \\
+1.45 \\
\hline
25.096
\end{array}
\quad \xrightarrow{\text{rounding off}} \quad 25.1
$$

Subtraction:

The same rounding procedure is used. Here, 157 has the least number of decimal places (none).

$$
\begin{array}{r}
157 \\
-5.5 \\
\hline
151.5
\end{array}
\quad \xrightarrow{\text{rounding off}} \quad 152
$$

FOLLOW-UP EXERCISE. Given the numbers 23.15, 0.546, and 1.058: (a) add the first two numbers and (b) subtract the last number from the first.

Suppose that you have to deal with mixed operations – multiplication and/or division *and* addition and/or subtraction. What do you do in this case? Just follow the regular rules for order of algebraic operations and observe significant figures as you go.*

The number of digits reported in a result depends on the number of digits in the given data. The rules for rounding will generally be observed in this book. However, there will be exceptions that may make a difference, as explained in the following Problem-Solving Hint.

1.6.3 Problem-Solving Hint: The "Correct" Answer

When working problems, you naturally strive to get the correct answer and will probably want to check your answers against those listed in the "Answers to Odd-Numbered Questions

---

* Order of operations: (1) calculations done from left to right, (2) calculations inside parentheses, (3) multiplication and division, (4) addition and subtraction.

---

and Exercises" section in Appendix VI the back of the book. However, on occasion, your answer may differ slightly from that given, even though you have solved the problem correctly. There are several reasons why this could occur.

It is best to round off only the final result of a multipart calculation, but this practice is not always convenient in elaborate calculations. Sometimes, the results of intermediate steps are important in themselves and need to be rounded off to the appropriate number of digits as if each were a final answer. Similarly, Examples in this book are often worked in steps to show the stages in the *reasoning* of the solution. The results obtained when the results of intermediate steps are rounded off may differ slightly from those obtained when only the final answer is rounded.

Rounding differences may also occur when using conversion factors. For example, in changing 5.0 mi to kilometers using the conversion factor listed inside the front cover of this book in different forms,

$$(5.0 \text{ mi})\left(\frac{1.609 \text{ km}}{1 \text{ mi}}\right) = (8.045 \text{ km}) = 8.0 \text{ km} \quad \text{(two significant figures)}$$

and

$$(5.0 \text{ mi})\left(\frac{1 \text{ km}}{0.621 \text{ mi}}\right) = (8.051 \text{ km}) = 8.1 \text{ km} \quad \text{(two significant figures)}$$

The difference arises because of rounding of the conversion factors. Actually, 1 km = 0.6214 mi, so 1 mi = (1/0.6214) km = 1.609 269 km ≈ 1.609 km. (Try repeating these conversions with the unrounded factors and see what you get.) To avoid rounding differences in conversions, the multiplication form of a conversion factor will generally be used, as in the first of the foregoing equations, unless there is a convenient exact factor, such as 1 min/60 s.

Slight differences in answers may also occur when different methods are used to solve a problem. Keep in mind that when solving a problem, *if your answer differs from that in the text in only the last digit, the disparity is most likely the result of a rounding difference for an alternative method of solution being used.*

1.7 Problem Solving

An important aspect of physics is problem solving. In general, this involves the application of physical principles and equations to data from a particular situation in order to find some unknown or wanted quantity. There is no universal method for approaching problem solving that will automatically produce a solution. However, although there is no magic formula for problem solving, there are some sound practices that can be very useful. The steps in the following procedure are intended to provide you with a framework that can be applied to solving most of the problems you will encounter during your course of study. (Modifications may be made to suit your own style.)

These steps will generally be used in dealing with the Example problems throughout the text. Additional problem-solving hints will be given where appropriate.

1.7.1 General Problem-Solving Steps

1. *Read the problem carefully and analyze it.* What is given, and what is wanted?
2. *Where appropriate, draw a diagram as an aid in visualizing and analyzing the physical situation of the problem.* This step may not be necessary in every case, but it is often useful.
3. *Write down the given data and what is to be found. Make sure the data are expressed in the same system of units (usually SI).* If necessary, use the unit conversion procedure learned earlier in the chapter. Some data may not be given explicitly. For example, if a car "starts from rest," its initial speed is zero ($v_0 = 0$); in some instances, you may be expected to know certain quantities, such as the acceleration due to gravity, g, or can look them up in tables.
4. *Determine which principle(s) and equation(s) are applicable to the situation, and how they can be used to get from the information given to what is to be found.* You may have to devise a strategy that involves several steps. Also, try to simplify equations as much as possible through algebraic manipulation. The fewer calculations you do, the less likely you are to make a mistake – so don't put in numbers until you have to.
5. *Substitute the given quantities (data) into the equation(s) and perform calculations.* Report the result with the proper units and proper number of significant figures.
6. *Consider whether the results are reasonable.* Does the answer have an appropriate magnitude? (This means, is it in the right ballpark?) For example, if a person's calculated mass turns out to be 4.60×10^2 kg, the result should be questioned, since a mass of 460 kg has a weight of 1010 lb. (Also, in motion problems, direction may be important.) ▶ **Figure 1.10** summarizes the main steps in the form of a flowchart.

In general, there are three types of examples in this book, as listed in ▶ **Table 1.4**. The preceding steps are applicable to the two types, because they include calculations. Conceptual Examples, in general, do not follow these steps, being primarily conceptual in nature.

In reading the Worked Examples and Integrated Examples, you should be able to recognize the general application or flow of the preceding steps. This format will be used throughout the text. Let's take an Example and an Integrated Example as illustrations. Comments will be made in these Examples to point out the problem-solving approach and steps that

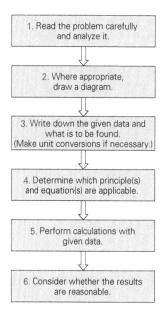

▲ **FIGURE 1.10 A flowchart for the suggested problem-solving procedure.**

TABLE 1.4 Types of Examples

(Worked) **Example** – primarily mathematical in nature
  Sections: **Thinking It Through**
      Solution

Integrated Example – (a) conceptual multiple choice, (b) mathematical follow-up
  Sections: **(a) Conceptual Reasoning**
      (b) Quantitative Reasoning and Solution

Conceptual Example – in general, needs only reasoning to obtain the answer, although some simple math may be required at times to justify the reasoning
  Sections: **Reasoning and Answer**

will not be made in the text Examples but should be understood. Since no physical principles have really been covered, math and trig problems will be used, which should serve as a good review.

EXAMPLE 1.9: FINDING THE OUTSIDE SURFACE AREA OF A CYLINDRICAL CONTAINER

A closed cylindrical container used to store material from a manufacturing process has an outside radius of 50.0 cm and a height of 1.30 m. What is the total outside surface area of the container?

THINKING IT THROUGH. (In this type of Example, the Thinking It Through section generally combines problem-solving steps 1 and 2 given previously.)

It should be noted immediately that the length units are given in mixed units, so a unit conversion will be in order. To visualize and analyze the cylinder, drawing a diagram is helpful (▶ **Figure 1.11**). With this information in mind, proceed to finding the solution, using the expression for the area of a cylinder (the combined areas of the circular ends and the cylinder's side).

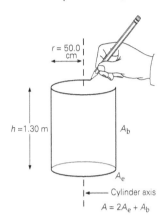

▲ FIGURE 1.11   **A helpful step in problem solving** Drawing a diagram helps you visualize and better understand the situation. See Example 1.9.

SOLUTION

Writing what is given and what is to be found (step 3 in our procedure):

Given:

$$r = 50.0 \text{ cm}$$
$$h = 1.30 \text{ m}$$

Find: A (the total outside surface area of the cylinder)

First, let's tend to the mixed units. You should be able in this case to immediately write $r = 50.0 \text{ cm} = 0.500 \text{ m}$. But often conversions are not obvious, so going through the unit conversion for illustration:

$$r = (50.0 \text{ cm}) \left(\frac{1 \text{ m}}{100 \text{ cm}} \right) = 0.500 \text{ m}$$

The general equations for areas (and volumes) of commonly shaped objects can be easily looked up (given in Appendix IC). Looking at Figure 1.11, note that the outside surface area of a cylinder consists of that of two circular ends and that of a rectangle (the body of the cylinder laid out flat). Equations for the areas of these common shapes are generally remembered. So the area of the two ends would be

$$2A_e = 2 \times \pi r^2$$

(2 times the area of the circular end; area of a circle $= \pi r^2$)

and the area of the body of the cylinder is

$$A_b = 2\pi r \times h \quad \text{(circumference of circular end times height)}$$

Then the total area A is

$$A = 2A_e + A_b = 2\pi r^2 + 2\pi rh$$

The data could be put into the equation, but sometimes an equation may be simplified to save some calculation steps.

$$A = 2\pi r(r + h) = 2\pi(0.500 \text{ m})(0.500 \text{ m} + 1.30 \text{ m})$$
$$= \pi(1.80 \text{ m}^2) = 5.65 \text{ m}^2$$

The result appears reasonable considering the cylinder's dimensions.

FOLLOW-UP EXERCISE. If the wall thickness of the cylinder's side and ends is 1.00 cm, what is the inside volume of the cylinder?

INTEGRATED EXAMPLE 1.10: SIDES AND ANGLES*

(a) A gardener has a rectangular plot measuring 3.0 m × 4.0 m. She wishes to use half of this area to make a triangular flower bed. Of the two types of triangles shown in ▼ **Figure 1.12**, which should she use to do this: (1) the right triangle, (2) the isosceles triangle – two sides equal, or (3) either one? (b) In laying out the flower bed, the gardener decides to use a right triangle. Wishing to line the sides with rows of stone, she wants to know the total length (L) of the triangle sides. She would also like to know the values of the acute angles of the triangle. Can you help her so she doesn't have to do physical measurements? (Appendix I includes a review of trigonometric relationships.)

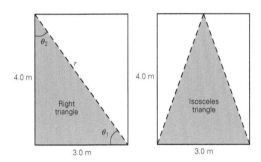

▲ FIGURE 1.12   **A flower bed project** Two types of triangles for a new flower bed. See Example 1.10.

(a) **CONCEPTUAL REASONING.** The rectangular plot has a total area of 3.0 m × 4.0 m $= 12 \text{ m}^2$. It is obvious that the right triangle (on the left) in Figure 1.12 divides the plot in half. This is not as obvious for the isosceles triangle, but with a little study you should see that the outside areas could be arranged such that their combined area would be the same as that of the shaded isosceles triangle. So the isosceles triangle also divides the plot in half, and the answer is (3). [This could be proven mathematically by computing the areas of the triangles. Area $= (\frac{1}{2})$ (altitude × base).]

(b) **QUANTITATIVE REASONING AND SOLUTION.** To find the total length of the sides, the length of the hypotenuse of the triangle is needed. This can be done using the Pythagorean theorem, $x^2 + y^2 = r^2$, and

$$r = \sqrt{x^2 + y^2} = \sqrt{(3.0 \text{ m})^2 + (4.0 \text{ m})^2} = \sqrt{25 \text{ m}^2} = 5.0 \text{ m}$$

---

* Here and throughout the text, angles will be considered exact; that is, they do not determine the number of significant figures.

(Or directly, you may have noticed that this is a 3-4-5 right triangle.) Then,

$$L = 3.0\,\text{m} + 4.0\,\text{m} + 5.0\,\text{m} = 12.0\,\text{m}$$

The acute angles of the triangle can be found by using trigonometry. Referring to the angles in Figure 1.12,

$$\tan\theta_1 = \frac{\text{Side opposite}}{\text{Side adjacent}} = \frac{4.0\,\text{m}}{3.0\,\text{m}}$$

and

$$\theta_1 = \tan^{-1}\left(\frac{4.0\,\text{m}}{3.0\,\text{m}}\right) = 53°$$

Similarly,

$$\theta_2 = \tan^{-1}\left(\frac{3.0\,\text{m}}{4.0\,\text{m}}\right) = 37°$$

The two angles add to 90° as would be expected with the right-angle triangle ($53° + 37° = 90°$).

FOLLOW-UP EXERCISE. What is the total length of the sides and the interior angles for the isosceles triangle in Figure 1.12?

These Examples illustrate how the problem-solving steps are woven into finding the solution of a problem. You will see this pattern throughout the solved Examples in the text, although not as explicitly explained. Try to develop your problem-solving skills in a similar manner.

1.7.2 Approximation and Order-of-Magnitude Calculations

At times when solving a problem, you may not be interested in an exact answer, but want only an estimate or a "ballpark" figure. Approximations can be made by rounding off quantities so as to make the calculations easier, and perhaps obtainable without the use of a calculator. For example, suppose you want to get an idea of the area of a circle with radius $r = 9.5$ cm. Then, rounding $9.5\,\text{cm} \approx 10\,\text{cm}$, and $\pi \approx 3$ instead of 3.14,

$$A = \pi r^2 \approx 3(10\,\text{cm})^2 = 300\,\text{cm}^2$$

(Note that significant figures are not a concern in calculations involving approximations.) The answer is not exact, but it is a good approximation. Compute the exact answer and see.

Powers-of-10, or scientific, notation is particularly convenient in making estimates or approximations in what are called **order-of-magnitude calculations**. *Order of magnitude* means that a quantity is expressed to the power of 10 closest to the actual value. For example, in the foregoing calculation,

approximating $9.5\,\text{cm} \approx 10\,\text{cm}$ is expressing 9.5 as 10^1, and we say that the radius is *on the order of* 10 cm. Expressing a distance of $75\,\text{km} \approx 10^2\,\text{km}$ indicates that the distance is on the order of 10^2 km. The radius of the Earth is $6.4 \times 10^3\,\text{km} \approx 10^4\,\text{km}$, or on the order of 10^4 km. A nanostructure with a width of $8.2 \times 10^{-9}\,\text{m}$ is on the order of 10^{-8} m, or 10 nm. (Why an exponent of -8?)

An order-of-magnitude calculation gives only an estimate, of course. But this estimate may be enough to provide you with a better grasp or understanding of a physical situation. Usually, the result of an order-of-magnitude calculation is precise within a power of 10, or *within an order of magnitude*. That is, the number (prefix) multiplied by the power of 10 is somewhere between 1 and 10. For example, if a length result of 10^5 km were obtained, it would be expected that the exact answer was somewhere between 1×10^5 km and 10×10^5 km.

EXAMPLE 1.11: ORDER-OF-MAGNITUDE CALCULATION – DRAWING BLOOD

A medical technologist draws 15 cc (cubic centimeter) of blood from a patient's vein. Back in the lab, it is determined that this volume of blood has a mass of 16 g. Estimate the density of the blood in SI units.

THINKING IT THROUGH. The data are given in cgs (centimeter-gram-second) units, which are often used for practicality when dealing with small, whole-number quantities in some situations. The cc abbreviation is commonly used in the medical and chemistry fields for cm^3. Density (ρ) is mass per unit volume, where $\rho = m/V$ (Section 1.4).

SOLUTION

First, changing to SI standard units:

Given:

$$m = (16\,\text{g})\left(\frac{1\,\text{kg}}{1000\,\text{g}}\right) = 1.6 \times 10^{-2}\,\text{kg} \approx 10^{-2}\,\text{kg}$$

$$V = (15\,\text{cm}^3)\left(\frac{1\,\text{m}}{10^2\,\text{cm}}\right)^3 = 1.5 \times 10^{-5}\,\text{m}^3 \approx 10^{-5}\,\text{m}^3$$

Find: estimate of ρ (density)

So, we have

$$\rho = \frac{m}{V} \approx \frac{10^{-2}\,\text{kg}}{10^{-5}\,\text{m}^3} = 10^3\,\text{kg/m}^3$$

This result is quite close to the average density of whole blood, $1.05 \times 10^3\,\text{kg/m}^3$.

FOLLOW-UP EXERCISE. A patient receives 750 cc of whole blood in a transfusion. Estimate the mass of the blood, in standard units.

EXAMPLE 1.12: HOW MANY RED CELLS ARE IN YOUR BLOOD?

The blood volume in the human body varies with a person's age, body size, and sex. On average, this volume is about 5.0 L. A typical value of red blood cells (erythrocytes) per volume is 5 000 000 (5.0×10^6) cells per cubic millimeter. Estimate the number of red blood cells in an average human body.

THINKING IT THROUGH. The red blood cell count in cells per cubic millimeter is sort of a red blood cell "number density." Multiplying this figure by the total volume of blood [(cells/volume) × total volume] will give the total number of cells. But note that the volumes must have the same units. First let's start by converting 5.0 L to cubic meters (m^3): $1 L = 10^{-3} m^3$. (See inside front cover.)

SOLUTION

Given:

$$V = 5.0\,L$$

$$= (5.0\,L)\left(10^{-3}\,\frac{m^3}{L}\right)$$

$$= 5.0 \times 10^{-3}\,m^3 \simeq 10^{-2}\,m^3 \text{ (for an estimation)}$$

$$\text{cells/volume} = 5.0 \times 10^6\,\frac{\text{cells}}{mm^3} \approx 10^7\,\frac{\text{cells}}{mm^3}$$

Find: the approximate number of red blood cells in the body

Then, changing to cubic meters,

$$\frac{\text{cells}}{\text{volume}} \simeq 10^7\,\frac{\text{cells}}{mm^3}\left(\frac{10^3\,mm}{1\,m}\right)^3 \approx 10^{16}\,\frac{\text{cells}}{m^3}$$

(*Note:* The conversion factor for liters to cubic meters was obtained directly from the conversion tables, but there is no conversion factor given for converting cubic millimeters to cubic meters, so a known conversion factor is cubed.) Then,

$$\left(\frac{\text{cells}}{\text{volume}}\right)(\text{total volume}) \approx \left(10^{16}\,\frac{\text{cells}}{m^3}\right)(10^{-2}\,m^3)$$

$$= 10^{14}\text{ red blood cells}$$

That's a bunch of cells. Red blood cells (erythrocytes) are one of the most abundant cells in the human body.

FOLLOW-UP EXERCISE. The average number of white blood cells (leukocytes) in human blood is normally 5000 to 10 000 cells per cubic millimeter. Estimate the number of white blood cells you have in your body.

Chapter 1 Review

- **SI units of length, mass, and time**. The meter (m), the kilogram (kg), and the second (s), respectively.
- **Liter (L).** A volume of 10 cm × 10 cm × 10 cm = 1000 cm^3 or 1000 mL. A liter of water has a mass of 1 kg or 1000 g. Therefore, 1 cm^3 or 1 mL has a mass of 1 g.

- **Unit analysis.** Unit analysis can be used to determine if an equation is dimensionally correct. Unit analysis can also be used to find the unit of a quantity.
- **Significant figures (digits).** The digits that are known with certainty, plus one digit that is uncertain, in a measured value.
- **Problem solving.** Problems should be worked using a consistent procedure. Order-of-magnitude calculations may be done when an estimated value is desired.
- **Density (ρ).** The mass per unit volume of an object or substance, which is a measure of the compactness of the material it contains:

$$\rho = \frac{m}{V} \quad \left(\text{density} = \frac{\text{mass}}{\text{volume}}\right) \qquad (1.1)$$

End of Chapter Questions and Exercises

Multiple Choice Questions

1.2 SI Units of Length, Mass, and Time

1. How many base units are there in the SI: (a) 3, (b) 5, (c) 7, or (d) 9?
2. The only SI standard represented by material standard or artifact is the (a) meter, (b) kilogram, (c) second, (d) electric charge.
3. Which of the following is the SI base unit for mass: (a) pound, (b) gram, (c) kilogram, or (d) metric ton?

1.3 More about the Metric System

4. The prefix *giga-* means (a) 10^{-9}, (b) 10^9, (c) 10^{-6}, (d) 10^6.
5. The prefix *micro-* means (a) 10^6, (b) 10^{-6}, (c) 10^3, (d) 10^{-3}.
6. A new technology is concerned with objects the size of what metric prefix: (a) *nano-*, (b) *micro-*, (c) *mega-*, or (d) *giga-*?
7. Which of the following has the greatest volume: (a) 1 L, (b) 1 qt, (c) 2000 μL, or (d) 2000 mL?
8. Which of the following metric prefixes is the smallest: (a) *micro-*, (b) *centi-*, (c) *nano-*, or (d) *milli-*?

1.4 Unit Analysis

9. Both sides of an equation are equal in (a) numerical value, (b) units, (c) dimensions, (d) all of the preceding.
10. Unit analysis of an equation cannot tell you if (a) the equation is dimensionally correct, (b) the equation is physically correct, (c) the numerical value is correct, (d) both (b) and (c).

1.5 Unit Conversions

11. A good way to ensure proper unit conversion is (a) to use another measurement instrument, (b) always work in the same system of units, (c) use unit analysis, (d) have someone check your math.
12. You often see 1 kg = 2.2 lb. This expression means that (a) 1 kg is equivalent to 2.2 lb, (b) this is a true equation, (c) 1 lb = 2.2 kg, (d) none of the preceding.
13. You have a quantity of water and wish to express this in volume units that give the largest value. Which of the

following units should be used: (a) in³, (b) mL, (c) μL, or (d) cm³?

1.6 Significant Figures

14. Which of the following has the greatest number of significant figures: (a) 103.07, (b) 124.5, (c) 0.09916, or (d) 5.408×10^5?

15. Which of the following numbers has four significant figures: (a) 140.05, (b) 276.02, (c) 0.004 006, or (d) 0.073 004?

16. In a multiplication and/or division operation involving the numbers 15 437, 201.08, and 408.0×10^5, the result should have how many significant figures: (a) 3, (b) 4, (c) 5, or (d) any number?

1.7 Problem Solving

17. An important step in problem solving before mathematically solving an equation is (a) checking units, (b) checking significant figures, (c) checking with a friend, (d) checking to see if the result will be reasonable.

18. An important final step in problem solving before reporting an answer is (a) saving your calculations, (b) reading the problem again, (c) seeing if the answer is reasonable, (d) checking your results with another student.

19. In order-of-magnitude calculations, you should (a) pay close attention to significant figures, (b) work primarily in the British system, (c) get results within a factor of 100, (d) express a quantity to the power of 10 closest to the actual value.

Conceptual Questions

1.2 SI Units of Length, Mass, and Time

1. Why is weight not a base quantity?

2. What replaced the original definition of the second and why? Is the replacement still used?

3. Give a couple of major differences between the SI and the British system.

1.3 More about the Metric System

4. If a fellow student tells you he saw a 3-cm-long ladybug, would you believe him? How about another student saying she caught a 10-kg salmon?

5. Explain why 1 mL is equivalent to 1 cm³.

6. Explain why a metric ton is equivalent to 1000 kg.

1.4 Unit Analysis

7. Can unit analysis tell you whether you have used the correct equation in solving a problem? Explain.

8. The equation for the area of a circle from two sources is given as $A = \pi r^2$ and $A = \pi d^2/2$. Can unit analysis tell you which is correct? Explain.

9. How might unit analysis help determine the units of a quantity?

1.5 Unit Conversions

10. Are an equation and an equivalence statement the same? Explain.

11. Does it make any difference whether you multiply or divide by a conversion factor? Explain.

12. A popular saying is "Give him an inch and he'll take a mile." What would be the equivalent numerical values and units in the metric system?

1.6 Significant Figures

13. What is the purpose of significant figures?

14. Are all the significant figures reported for a measured value accurately known? Explain.

15. How are the number of significant figures determined for the results of calculations involving (a) multiplication, (b) division, (c) addition, and (d) subtraction?

1.7 Problem Solving

16. What are the main steps in the problem-solving procedure suggested in this chapter?

17. When you do order-of-magnitude calculations, should you be concerned about significant figures? Explain.

18. When doing an order-of-magnitude calculation, how accurate can you expect the answer to be? Explain.

19. Is the following statement reasonable? It took 300 L of gasoline to fill the car's tank. (Justify your answer.)

20. Is the following statement reasonable? A car traveling 30 km/h through a school speed zone exceeds the speed limit of 25 mi/h. (Justify your answer.)

Exercises*

Integrated Exercises (IEs) are two-part exercises. The first part typically requires a conceptual answer choice based on basic principles and reasoning. The following part is quantitative calculations associated with the conceptual choice made in the first part of the exercise.

1.3 More about the Metric System

1. • The metric system is a decimal (base-10) system, and the British system is, in part, a duodecimal (base-12) system. Discuss the ramifications if our monetary system had a duodecimal base. What would be the possible values of our coins if this were the case?

2. • (a) In the British system, 16 oz = 1 pt and 16 oz = 1 lb. Is something wrong here? Explain. (b) Here's an old one: A pound of feathers weighs more than a pound of gold. How can that be? [*Hint:* Look up *ounce* in the dictionary.]

3. • Convert the following: (a) 40 000 000 bytes to MB, (b) 0.5722 mL to L, (c) 2.684 m to cm, and (d) 5500 bucks to kilobucks.

4. •• A water filled rectangular container measures 25 cm × 35 cm × 55 cm is filled with water. What is the mass of this volume of water?

5. •• (a) What volume in liters is a cube 20 cm on a side? (b) If the cube is filled with water, what is the mass of the water?

---

* The bullets denote the degree of difficulty of the exercises: •, simple; ••, medium; and •••, more difficult.

1.4 Unit Analysis

6. • Show that the equation $x = x_0 + vt$, where v is velocity, x and x_0 are lengths, and t is time, is dimensionally correct.

7. • If x refers to distance, v_0 and v to velocities, a to acceleration, and t to time, which of the following equations is dimensionally correct: (a) $x = v_0 t + at^3$, (b) $v^2 = v_0^2 + 2at$, (c) $x = at + vt^2$, or (d) $v^2 = v_0^2 + 2ax$?

8. • Use SI unit analysis to show that the equation for the surface area of a sphere $A = 4\pi r^2$, where A is the area and r is the radius of a sphere, is dimensionally correct.

9. •• The general equation for a parabola is $y = ax^2 + bx + c$, where a, b, and c are constants. What are the units of each constant if y and x are in meters?

10. •• You are told that the volume of a sphere is given by $V = \pi d^3/6$, where V is the volume and d is the diameter of the sphere. Is this equation dimensionally correct?

11. •• The units for **pressure (*p*)** in terms of SI base units are known to be $kg/(m\cdot s^2)$. For a physics class assignment, a student derives an expression for the pressure exerted by the wind on a wall in terms of the air density (ρ) and wind speed (v) and her result is $p = \rho v^2$. Use SI unit analysis to show that her result is dimensionally consistent. Does this prove that this relationship is physically correct?

12. •• Is the equation for the area of a trapezoid, $A = 1/2[a(b_1 + b_2)]$, where a is the height and b_1 and b_2 are the bases, dimensionally correct (▼**Figure 1.13**)?

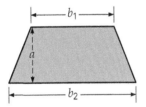

▲ FIGURE 1.13 **The area of a trapezoid** See Exercise 12.

13. ••• Newton's second law of motion (Section 4.3) is expressed by the equation $F = ma$, where F represents force, m is mass, and a is acceleration. (a) The SI unit of force is, appropriately, called the newton (N). What are the units of the newton in terms of base quantities? (b) An equation for force associated with uniform circular motion (Section 7.3) is $F = mv^2/r$, where v is speed and r is the radius of the circular path. Does this equation give the same units for the newton?

14. ••• Einstein's famous mass-energy equivalence is expressed by the equation $E = mc^2$, where E is energy, m is mass, and c is the speed of light. (a) What are the SI units of energy? (b) Another equation for energy is $E = mgh$, where m is mass, g is the acceleration due to gravity, and h is height. Does this equation give the same units as in part (a)?

1.5 Unit Conversion

15. • Figure 1.7a (top) shows the elevation of a location in both feet and meters. Is the conversion correct?

16. IE • (a) If you wanted to express your height with the largest number, which units would you use: (1) meters, (2) feet, (3) inches, or (4) centimeters? Why? (b) If you are 6.00 ft tall, what is your height in centimeters?

17. • If the capillaries of an average adult were unwound and spread out end to end, they would extend to a length over 40 000 mi (Figure 1.8). If you are 1.75 m tall, how many times your height would the capillary length equal?

18. IE • (a) Compared with a 2-liter soda bottle, a half-gallon soda bottle holds (1) more, (2) the same amount of, or (3) less soda. (b) Verify your answer for part (a).

19. • (a) A football field is 300 ft long and 160 ft wide. What are the field's dimensions in meters? (b) A football is 11.0 to $11(\frac{1}{4})$ in. long. What is its length in centimeters?

20. • How many (a) quarts and (b) gallons are there in 10.0 L?

21. • A submarine is submerged 175 fathoms below the surface. What is its depth in meters? (A *fathom* is an old nautical measurement equal to 2 yd.)

22. •• The sailfish is the fastest fish in the world – able to swim at a speed of 68 miles per hour (▼ **Figure 1.14**). (a) What is this speed expressed in m/s? (b) How long would it take the fish to travel the length of a 300-ft football field at this speed? [*Hint:* $v = d/t$.]

▲ FIGURE 1.14 **Fastest fish** See Exercise 22.

23. IE •• (a) Which of the following represents the greatest speed: (1) 1 m/s, (2) 1 km/h, (3) 1 ft/s, or (4) 1 mi/h? (b) Express the speed 15.0 m/s in mi/h.

24. •• An automobile speedometer is shown in ▶ **Figure 1.15**. (a) What would be the equivalent scale readings (for each empty box) in kilometers per hour? (b) What would be the 70-mi/h speed limit in kilometers per hour?

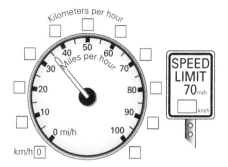

▲ FIGURE 1.15   **Speedometer readings** See Exercise 24.

25. •• A person weighs 170 lb. (a) What is his mass in kilograms? (b) Assuming the density of the average human body is about that of water (which is true), 1,000 kg/m³ estimate his body's volume in both cubic meters and liters. Explain why the smaller unit of the liter is more appropriate (convenient) for describing a volume of this size.

26. •• If the components of the human circulatory system (arteries, veins, and capillaries) were completely extended and placed end to end, the length would be on the order of 100 000 km. Would the length of the circulatory system reach around the circumference of the Moon? If so, how many times?

27. •• The human heartbeat, as determined by the pulse rate, is normally about 60 beats/min. If the heart pumps 75 mL of blood per beat, what volume of blood is pumped in one day in liters?

28. •• Some common product labels are shown in ▼ **Figure 1.16**. From the units on the labels, find (a) the number of milliliters in 2 fl. oz and (b) the number of ounces in 100 g.

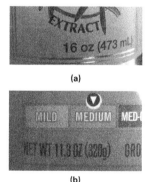

(a)

(b)

▲ FIGURE 1.16   **Conversion factors** See Exercise 28. [Note there are fluid and mass ounces (oz.)]

29. •• ▶ **Figure 1.17** is a picture of red blood cells seen under a scanning electron microscope. Normally, women possess about 4.5 million of these cells in each cubic millimeter of blood. If the blood flow to the heart is 250 mL/min, how many red blood cells does a woman's heart receive each second?

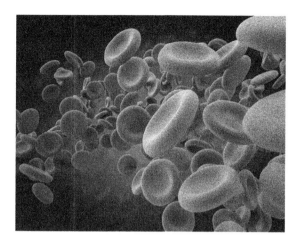

▲ FIGURE 1.17   **Red blood cells** See Exercise 29.

30. •• A student was 18 in. long when she was born. She is now 5 ft 6 in. tall and 20 years old. How many centimeters a year did she grow on average?

31. ••• How many minutes of arc does the Earth rotate in 1 min of time?

32. ••• The density of metal mercury is 13.6 g/cm³. (a) What is this density as expressed in kilograms per cubic meter? (b) How many kilograms of mercury would be required to fill a 0.250-L container?

1.6 Significant Figures

33. • Express the length 50 500 μm (micrometers) in centimeters, decimeters, and meters, to three significant figures.

34. • Using a meterstick, a student measures a length and reports it to be 0.8755 m. What is the smallest division on the meterstick scale?

35. • Determine the number of significant figures in the following measured numbers: (a) 1.007 m, (b) 8.03 cm, (c) 16.272 kg, (d) 0.015 μs (microseconds).

36. • Round the following numbers to two significant figures: (a) 95.61, (b) 0.00208, (c) 9438, (d) 0.000344.

37. •• The cover of your physics book measures 0.274 m long and 0.222 m wide. What is its area in square meters?

38. •• The interior storage compartment of a restaurant refrigerator measures 1.3 m high, 1.05 m wide, and 67 cm deep. Determine its volume in cubic feet.

39. IE •• The top of a rectangular table measures 1.245 m by 0.760 m. (a) The smallest division on the scale of the measurement instrument is (1) m, (2) cm, (3) mm. Why? (b) What is the area of the tabletop?

40. IE •• The outside dimensions of a cylindrical soda can are reported as 12.559 cm for the diameter and 5.62 cm for the height. (a) How many significant figures will the total outside area have: (1) two, (2) three, (3) four, or (4) five? Why? (b) What is the total outside surface area of the can in square centimeters?

41. •• Express the following calculations using the proper number of significant figures: (a) $12.634 + 2.1$, (b) $13.5 - 2.134$, (c) $(0.25 \text{ m})^2$, (d) $\sqrt{2.37/3.5}$.

42. IE ••• In doing a problem, a student adds 46.9 m and 5.72 m and then subtracts 38 m from the result. (a) How many decimal places will the final answer have: (1) zero, (2) one, or (3) two? Why? (b) What is the final answer?

43. ••• Work this exercise by the two given procedures as directed, commenting on and explaining any difference in the answers. Use your calculator for the calculations. Compute $p = mv$, where $v = x/t$, given $x = 8.5$ m, $t = 2.7$ s, and $m = 0.66$ kg. (a) First compute v and then p. (b) Compute $p = mx/t$ without an intermediate step. (c) Are the results the same? If not, why?

1.7 Problem Solving

44. • A corner construction lot has the shape of a right triangle. If the two sides perpendicular to each other are 37 m long and 42.3 m long, what is the length of the hypotenuse?

45. •• The lightest solid material is silica aerogel, which has a typical density of only about 0.10 g/cm³. The molecular structure of silica aerogel is typically 95% empty space. What is the mass of 1 m³ of silica aerogel?

46. •• A *cord* of wood is a volume of cut wood equal to a stack 8.0 ft long, 4.0 ft wide, and 4.0 ft high. How many cords are there in 3.0 m³?

47. •• Nutrition Facts labels now appear on most foods. An abbreviated label concerned with fat is shown in ▼ **Figure 1.18**. When burned in the body, each gram of fat supplies 9 Calories. (A food Calorie is really a kilocalorie, as will be learned in Chapter 11.) (a) What percentage of the Calories in one serving is supplied by fat? (b) You may notice that our answer doesn't agree with the listed Total Fat percentage in Figure 1.18. This is because the given Percent Daily Values are the percentages of the maximum recommended amounts of nutrients (in grams) contained in a 2000-Calorie diet. What are the maximum recommended amounts of total fat for a 2000-Calorie diet?

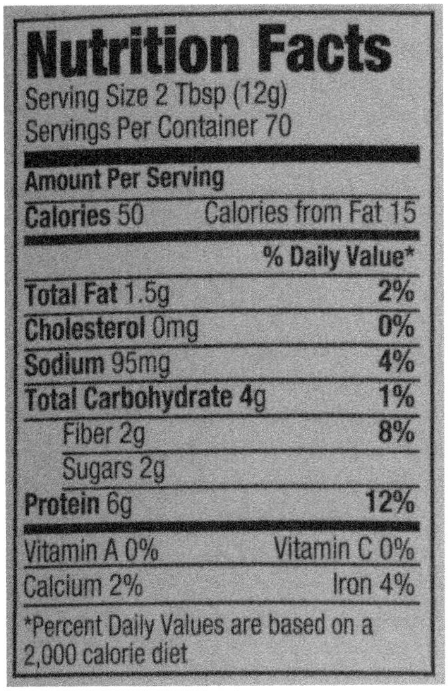

48. •• The thickness of the numbered pages of a textbook is measured to be 3.75 cm. (a) If the last page of the book is numbered 860, what is the average thickness of a page? (b) Repeat the calculation by using order-of-magnitude calculations.

49. •• The mass of the Earth is 5.98×10^{24} kg. What is the average density of the Earth in standard units?

50. IE •• To go to a football stadium from your house, you first drive 1000 m north, then 500 m west, and finally 1500 m south. (a) Relative to your home, the football stadium is (1) north of west, (2) south of east, (3) north of east, (4) south of west. (b) What is the straight-line distance from your house to the stadium?

51. •• Two chains of length 1.0 m are used to support a lamp, as shown in ▼ **Figure 1.19**. The distance between the two chains along the ceiling is 1.0 m. What is the vertical distance from the lamp to the ceiling?

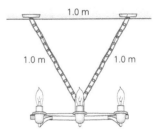

▲ FIGURE 1.19 **Support the lamp** See Exercise 51.

52. •• Tony's Pizza Palace sells a medium 9.0-in. (diameter) pizza for $7.95, and a large 12-in. pizza for $13.50. Which pizza is the better buy?

53. •• The Channel Tunnel, or "Chunnel," which runs under the English Channel between Great Britain and France, is 31 mi long. (There are actually three separate tunnels.) A shuttle train that carries passengers through the tunnel travels with an average speed of 75 mi/h. On average, how long, in minutes, does the shuttle take to make a one-way trip through the Chunnel?

54. •• Human adult blood contains, on average, 7000/mm³ white blood cells (leukocytes) and 250 000/mm³ platelets (thrombocytes). If a person has a blood volume of 5.0 L, estimate the total number of white cells and platelets in the blood.

Nutrition Facts

Serving Size 2 Tbsp (12g)
Servings Per Container 70

Amount Per Serving

Calories 50   Calories from Fat 15

	% Daily Value*
Total Fat 1.5g	**2%**
Cholesterol 0mg	**0%**
Sodium 95mg	**4%**
Total Carbohydrate 4g	**1%**
Fiber 2g	**8%**
Sugars 2g	
Protein 6g	**12%**

Vitamin A 0%	Vitamin C 0%
Calcium 2%	Iron 4%

*Percent Daily Values are based on a 2,000 calorie diet

▲ FIGURE 1.18 **Nutrition facts** See Exercise 47.

55. •• The average number of hairs on the normal human scalp is 125 000. A healthy person loses about 65 hairs per day. (New hair from the hair follicle pushes the old hair out.) (a) How many hairs are lost in one month? (b) Pattern baldness (top-of-the-head hair loss) affects about 35 million men in the United States. If an average of 15% of the scalp is bald, how many hairs are lost per year by one of these "bald is beautiful" people?

56. IE •• A car is driven 13 mi east and then a certain distance due north, ending up at a position 25° north of east of its initial position. (a) The distance traveled by the car due north is (1) less than, (2) equal to, or (3) greater than 13 mi. Why? (b) What distance due north does the car travel?

57. IE ••• At the Indianapolis 500 time trials, each car makes four consecutive laps, with its overall or average speed determining that car's place on race day. Each lap covers 2.5 mi (exact). During a practice run, cautiously and gradually taking his car faster and faster, a driver records the following average speeds for each successive lap: 160 mi/h, 180 mi/h, 200 mi/h, and 220 mi/h. (a) Will his average speed be (1) exactly the average of these speeds (190 mi/h), (2) greater than 190 mi/h, or (3) less than 190 mi/h? Explain. (b) To corroborate your conceptual reasoning, calculate the car's average speed.

58. ••• Approximately 118 mi wide, 307 mi long, and averaging 279 ft in depth, Lake Michigan is the second-largest Great Lake by volume. Estimate its volume of water in cubic meters.

59. IE ••• In the Tour de France, a bicyclist races up two successive (straight) hills of different slope and length. The first is 2.00 km long at an angle of 5° above the horizontal. This is immediately followed by one 3.00 km long at 7°. (a) What will be the overall (net) angle from start to finish: (1) smaller than 5°, (2) between 5° and 7°, or (3) greater than 7°? (b) Calculate the actual overall (net) angle of rise experienced by this racer from start to finish, to corroborate your reasoning in part (a).

60. ••• A student wants to determine the distance from the lakeshore to a small island (▼ **Figure 1.20**). He first draws a 50-m line parallel to the shore. Then, he goes to the ends of the line and measures the angles of the lines of sight from the island relative to the line he has drawn. The angles are 30° and 40°. How far is the island from the shore?

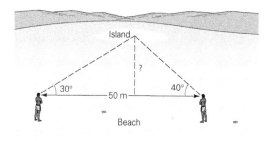

▲ FIGURE 1.20 **Measuring with lines of sight** See Exercise 60.

2

Kinematics: Description of Motion*

The length of cheetah's stride is partially determined by its speed and the acceleration due to gravity.

---

* The mathematics needed in this chapter involves general algebraic equation manipulation. You may want to review this in Appendix I.

The cheetah is running at full stride (completely airborne) in the chapter-opening photo. This fastest of all land animals is capable of attaining speeds up to 113 km/h, or 70 mi/h. The sense of motion in the photograph is so strong that you can almost feel the air rushing by. And yet this sense of motion is an illusion. Motion takes place in time, but the photo can "freeze" only a single instant. You'll find that without the dimension of time, motion cannot be described at all.

The description of motion involves the representation of a restless world. Nothing is ever perfectly still. You may sit, apparently at rest, but your blood flows, and air moves into and out of your lungs. The air is composed of gas molecules moving at different speeds and in different directions. And while experiencing stillness, you, your chair, the building you are in, and the air you breathe are all rotating and revolving through space with the Earth, part of a solar system in a spiraling galaxy in an expanding universe.

The branch of physics concerned with the study of motion and what produces and affects motion is called **mechanics**. The roots of mechanics and of human interest in motion go back to early civilizations. The study of the motions of heavenly bodies, or *celestial mechanics*, grew out of measuring time and location. Several early Greek scientists, notably Aristotle, put forth theories of motion that were useful descriptions, but were later proved to be incomplete or incorrect. Our currently accepted concepts of motion were formulated in large part by Galileo (1564–1642) and Isaac Newton (1642–1727).

Mechanics is usually divided into two parts: (1) kinematics and (2) dynamics. **Kinematics** deals with the *description* of the motion of objects, without consideration of what causes the motion. **Dynamics** analyzes the *causes* of motion. This chapter covers kinematics and reduces the description of motion to its simplest terms by considering linear motion; that is, motion in a straight line. You'll learn to analyze changes in motion – speeding up, slowing down, and stopping. Along the way, a particularly interesting case of accelerated motion will be presented: free fall (motion under the influence of gravity only).

2.1 Distance and Speed: Scalar Quantities

2.1.1 Distance

Motion is observed all around us. But what is motion? This question seems simple; however, you might have some difficulty giving an immediate answer. (And, it's not fair to use forms of the verb *to move* to describe motion.) After a little thought, you should be able to conclude that **motion** (or moving) *involves changing position*. Motion can be described in part by specifying *how far* something travels in changing position – that is, the distance it travels. **Distance** is simply the *total path length* traversed in moving from one location to another. For example, you may drive to school from your hometown and express the distance traveled in miles or kilometers. In general, the distance between two points depends on the path traveled (▶ **Figure 2.1**).

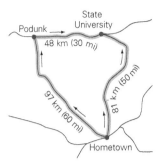

▲ FIGURE 2.1 **Distance – total path length** In driving to State University from Hometown, one student may take the shortest route and travel a distance of 81 km (50 mi). Another student takes a longer route in order to visit a friend in Podunk before returning to school. The longer trip is in two segments, but the distance traveled is the total length, 97 km + 48 km = 145 km (90 mi).

Along with many other quantities in physics, distance is a scalar quantity. A **scalar quantity** is one with only magnitude or size. That is, a *scalar* has only a numerical value and units, such as 160 km or 100 mi. Distance tells you the magnitude only – how far, but not the direction. Other examples of scalars are quantities such as 10 s (time), 3.0 kg (mass), and 20 °C (temperature). Some scalars may have negative values; for example, –10 °F.

2.1.2 Speed

When something is in motion, its position changes with time. That is, it moves a certain distance in a given amount of time. Both *length* and *time* are therefore important quantities in describing motion. For example, imagine a car and a pedestrian moving down a street and traveling a distance of one block. You would expect the car to travel faster and thus to cover the same distance in a shorter time than the person. A length-time relationship can be expressed by using the *rate* at which distance is traveled, or speed.

Average speed ($\overline{s}$) is the distance d traveled – that is, the actual path length – divided by the total time Δt elapsed in traveling that distance:

$$\text{average speed} = \frac{\text{distance traveled}}{\text{total time to travel that distance}} \quad (2.1)$$

$$\overline{s} = \frac{d}{\Delta t} = \frac{d}{t_2 - t_1}$$

SI unit of speed: meters per second (m/s).

A symbol with a bar over it is commonly used to denote an average. The Greek letter delta (Δ) is used to represent a change or difference in a quantity; in this case, the time difference between the beginning (t_1) and end (t_2) of a trip, or the elapsed total time.

The SI standard unit of speed is meters per second (m/s, length/time), although kilometers per hour (km/h) is used in many everyday applications. The British standard unit is feet per second (ft/s), but a commonly used unit is miles per hour (mi/h). Often, the initial time could be taken to be zero, $t_1 = 0$, as in resetting a stopwatch, and thus Equation 2.1 may be written $\bar{s} = d/t$, where it is understood that t is the total time.

Since distance is a scalar (as is time), speed is also a scalar. The distance does not have to be in a straight line (see Figure 2.1). For example, you probably have computed the average speed of an automobile trip by using the distance obtained from the starting and ending odometer readings. Suppose these readings were 17 455 and 17 775 km, respectively, for a 4.0-h trip. Subtracting the readings gives a total traveled distance d of 320 km, so the average speed of the trip is $d/t = 320\ \text{km}/4.0\ \text{h} = 80\ \text{km/h}$ (or about 50 mi/h).

Average speed gives a general description of motion over a time interval Δt. In the case of the auto trip with an average speed of 80 km/h, the car's speed wasn't always 80 km/h. With various stops and starts on the trip, the car must have been moving more slowly than the average speed at various times. It therefore had to be moving more rapidly than the average speed another part of the time. With an average speed, you don't know how fast the car was moving at any particular instant of time during the trip. By analogy, the average test score of a class doesn't tell you the score of any particular student.

On the other hand, **instantaneous speed** tells how fast something is moving *at a particular instant of time*. That is, when $\Delta t \rightarrow 0$ (the time interval approaches zero), which represents an instant of time. The speedometer of a car gives an approximate instantaneous speed. For example, the speedometer shown in ▼ **Figure 2.2** indicates a speed of about 50 mi/h, or 80 km/h. If the car travels with constant speed (so the speedometer reading does not change), then the average and instantaneous speeds will be equal. (Do you agree? Think of the previous average test score analogy. What if all of the students in the class got the same score?)

▲ FIGURE 2.2 **Instantaneous speed** The speedometer of a car gives the speed over a very short interval of time, so its reading approaches the instantaneous speed. Note the speeds are given in mi/h and km/h.

EXAMPLE 2.1: SLOW MOTION – ROVER MOVES ALONG

In January 2004, a Mars Exploration Rover touched down on the surface of Mars and rolled out for exploration (▼ **Figure 2.3**). The average speed of the Rover on flat, hard ground was 5.0 cm/s. (a) Assuming the Rover traveled continuously over this terrain at its average speed, how much time would it take to travel 2.0 m nonstop in a straight line? (b) However, in order to ensure a safe drive, the Rover was equipped with hazard avoidance software that caused it to stop and assess its location every few seconds. It was programmed to drive at its average speed for 10 s, then stop and observe the terrain for 20 s before moving onward for another 10 s and repeating the cycle. Taking its programming into account, what would be the Rover's average speed in traveling the 2.0 m? (There were actually two Rovers on this mission, named Spirit and Opportunity. Both rovers functioned for over four years on the Red Planet.)

▲ FIGURE 2.3 **A Mars exploration rover** Twin Rovers landed on opposite sides of the Martian planet. See Example 2.1.

THINKING IT THROUGH. (a) Because the average speed and the distance are known, the time can be computed from the equation for average speed (Equation 2.1). (b) Here, to calculate the average speed, the *total* time, which includes stops, must be used.

SOLUTION

Listing the data in symbol form (cm/s is converted directly to m/s):

Given:

(a) $\bar{s} = 5.0\ \text{cm/s} = 0.050\ \text{m/s}$
$d = 2.0\ \text{m}$

(b) cycles of 10-s travel, 20-s stops

Find:

(a) Δt (time to travel distance)
(b) $\bar{s}$ (average speed)

(a) Rearranging Equation 2.1, $\bar{s} = (d/\Delta t)$ to solve for time,

$$\Delta t = \frac{d}{\bar{s}} = \frac{2.0\ \text{m}}{0.050\ \text{m/s}} = 40\ \text{s}$$

So it takes the Rover 40 s to travel a path length of 2.0 m.

(b) Here, the total time for the 2.0-m distance is needed. In each 10-s interval, a distance of $(0.050 \text{ m/s}) \times (10 \text{ s}) = 0.50 \text{ m}$ would be traveled. So, the total time would be four 10-s intervals for actual travel and three 20-s intervals of stopping, giving $\Delta t = 4 \times (10 \text{ s}) + 3 \times (20 \text{ s}) = 100 \text{ s}$. Then

$$\bar{s} = \frac{d}{\Delta t} = \frac{d}{t_2 - t_1} = \frac{2.0 \text{ m}}{100 \text{ s}} = 0.020 \text{ m/s} = 2.0 \text{ cm/s}$$

FOLLOW-UP EXERCISE. Suppose the Rover's programming was for 5.0 s of travel and for 10-s stops. How long would it take to travel the 2.0 m in this case? *(Answers to all Follow-Up Exercises are given in Appendix V at the back of the book.)*

2.2 One-Dimensional Displacement and Velocity: Vector Quantities

2.2.1 Displacement

For straight-line, or linear, motion, it is convenient to specify position by using the familiar two-dimensional Cartesian coordinate system with x- and y-axes at right angles, as shown in ▼ **Figure 2.4**. A straight-line path can be in any direction relative to the axes, but for convenience, the coordinate axes are usually oriented so that the motion is along one of them.

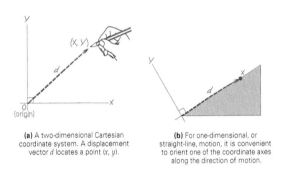

(a) A two-dimensional Cartesian coordinate system. A displacement vector *d* locates a point (x, y).

(b) For one-dimensional, or straight-line, motion, it is convenient to orient one of the coordinate axes along the direction of motion.

▲ FIGURE 2.4 **Cartesian coordinates and one-dimensional displacement** **(a)** A two-dimensional Cartesian coordinate system. A displacement vector *d* locates a point (x, y). **(b)** For one-dimensional, or straight-line, motion, it is convenient to orient one of the coordinate axes along the direction of motion.

As discussed in the previous section, distance is a scalar quantity with only magnitude (and units). However, to more completely describe motion, more information can be given by adding a *direction*. This information is particularly easy to convey for a change of position in a straight line. **Displacement** is defined as the straight-line distance between two points, along with the direction directly from the starting point to the final position. Unlike distance (a scalar), displacement can have either positive or negative values, with the signs indicating the directions along a coordinate axis.

As such, displacement is a **vector quantity**. In other words, a *vector* has both magnitude and direction. For example, when

describing the displacement of an airplane as 25 km north, this is a *vector* description (magnitude and direction). Other vector quantities include acceleration and force, as will be learned later.

There is an algebra that applies to vectors, which involves how to specify and deal with both the magnitude and direction of vectors. This is done relatively easily in one dimension by using positive (+) and negative (−) signs to indicate directions. To illustrate this with displacements, consider the situation shown in ▼ **Figure 2.5**, where x_1 and x_2 indicate the initial and final positions, respectively, on the x-axis as a student moves in a straight line from his locker to the physics lab. As can be seen in Figure 2.5a, the scalar distance traveled is 8.0 m. To specify displacement (a vector) between x_1 and x_2, we use the expression

$$\Delta x = x_2 - x_1 \qquad (2.2)$$

where Δ is again used to represent a change in a quantity. Then, as in Figure 2.5b,

$$\Delta x = x_2 - x_1 = +9.0 \text{ m} - (+1.0 \text{ m}) = +8.0 \text{ m}$$

where the + signs indicate the positions on the positive x-axis. Hence, the student's displacement (magnitude and direction) is 8.0 m in the +x-direction, as indicated by the positive (+) result in Figure 2.5b. (As in "regular" mathematics, the positive sign is often omitted, as being understood, so this displacement can be written as $\Delta x = 8.0 \text{ m}$ instead of $\Delta x = +8.0 \text{ m}$.)

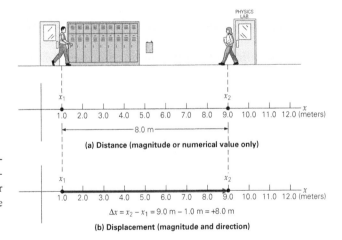

(a) Distance (magnitude or numerical value only)

$\Delta x = x_2 - x_1 = 9.0 \text{ m} - 1.0 \text{ m} = +8.0 \text{ m}$

(b) Displacement (magnitude and direction)

▲ FIGURE 2.5 **Distance (scalar) and displacement (vector) (a)** The distance (straight-line path) between the student on the left and the physics lab is 8.0 m and is a scalar quantity. **(b)** To indicate displacement, x_1 and x_2 specify the initial and final positions, respectively. The displacement is then $\Delta x = x_2 - x_1 = +9.0 \text{ m} - (+1.0 \text{ m}) = +8.0 \text{ m}$, that is, 8.0 m in the +x-direction.

Vector quantities in this book are usually indicated by boldface type with an over arrow; for example, a displacement vector is indicated by $\vec{d}$ or $\vec{x}$, and a velocity vector is indicated by $\vec{v}$. However, when working in one dimension, this notation is not necessarily needed. Instead, positive (+) and negative (−) signs can be used to indicate the only two possible directions. The *x*-axis is commonly used for horizontal motions, and a positive (+) sign is taken to indicate the direction to the right, or in the "+*x*-direction," and a negative (−) sign indicates the direction to the left, or in the "−*x*-direction."

Keep in mind that these signs only "point" in *particular directions*. An object moving along the negative *x*-axis toward the origin would be moving in the positive *x*-direction. How about an object moving along the positive *x*-axis toward the origin? If you said in the negative *x*-direction, you are correct.

Suppose the other student in Figure 2.5 walks from the physics lab (her initial position is different, $x_1 = +9.0$ m) to the end of the lockers (the final position is now $x_2 = +1.0$ m). Her displacement would be

$$\Delta x = x_2 - x_1 = +1.0 \text{ m} - (+9.0 \text{ m}) = -8.0 \text{ m}$$

The negative sign indicates that the direction of her displacement was in the −*x*-direction or to the left in the figure. In this case, we say that the two students' displacements are equal (in magnitude) and opposite (in direction).

2.2.2 Velocity

As has been learned, speed, like the distance it incorporates, is a scalar quantity – it has magnitude only. Another more descriptive quantity used to describe motion is *velocity*. Speed and velocity are often used synonymously in everyday conversation, but the terms have different meanings in physics. Speed is a scalar, and velocity is a vector – velocity has both magnitude and direction. Unlike speed, one-dimensional velocities can have both positive and negative values, indicating the only two possible directions (as with displacement).

Velocity tells how fast something is moving *and* in which direction it is moving. And just as there are average and instantaneous speeds, there are average and instantaneous velocities involving vector displacements. The **average velocity** is the displacement divided by the total travel time. In one dimension, this involves just motion along one axis, which is taken to be the *x*-axis. In this case,

$$\text{average velocity} = \frac{\text{displacement}}{\text{total travel time}} \quad (2.3)$$

$$\bar{v} = \frac{\Delta x}{\Delta t} = \frac{x_2 - x_1}{t_2 - t_1}$$

SI unit of velocity meters per second (m/s)

Another common form of this equation is

$$\bar{v} = \frac{\Delta x}{\Delta t} = \frac{(x_2 - x_1)}{(t_2 - t_1)} = \frac{(x - x_o)}{(t - t_o)} = \frac{(x - x_o)}{t}$$

or, after rearranging,

$$x = x_o + \bar{v}t \quad (2.3)$$

where x_o is the initial position, x is the final position, and $\Delta t = t$ with $t_o = 0$. See Section 2.3 for more on this notation.

In the case of more than one displacement (such as for successive displacements), the average velocity is equal to the *total or net* displacement divided by the total time. The total displacement is found by adding the displacements algebraically by including the directional signs.

You might be wondering whether there is a relationship between average speed and average velocity. A quick look at Figure 2.5 will show that if all the motion is in one direction – that is, there is no reversal of direction – the distance is equal to the magnitude of the displacement. Then the average speed is equal to the magnitude of the average velocity. *However, be careful.* This set of relationships is not true if there is a reversal of direction, as Example 2.2 shows.

EXAMPLE 2.2: THERE AND BACK – AVERAGE VELOCITIES

A jogger jogs from one end to the other of a straight 300-m track in 2.50 min and then jogs back to the starting point in 3.30 min. What was the jogger's average velocity (a) in jogging to the far end of the track, (b) coming back to the starting point, and (c) for the total jog?

THINKING IT THROUGH. The average velocities are computed from the defining equation. Note that the times given are the Δt's associated with the particular displacements.

SOLUTION

Given:

> $\Delta x_1 = +300$ m (taking the initial direction as positive)
> $\Delta x_2 = -300$ m (then the direction of the return trip must be negative)
> $\Delta t_1 = (2.50 \text{ min}) \times (60 \text{ s/min}) = 150$ s
> $\Delta t_2 = (3.30 \text{ min}) \times (60 \text{ s/min}) = 198$ s

Find:

Average velocities for
(a) the first leg of the jog
(b) the return jog
(c) the total jog

(a) The jogger's average velocity for the trip down the track is found using Equation 2.3:

$$\bar{v}_1 = \frac{\Delta x_1}{\Delta t_1} = \frac{+300 \text{ m}}{150 \text{ s}} = +2.00 \text{ m/s}$$

(b) Similarly, for the return trip,

$$\bar{v}_2 = \frac{\Delta x_2}{\Delta t_2} = \frac{-300 \text{ m}}{198 \text{ s}} = -1.52 \text{ m/s}$$

(c) For the total trip, there are two displacements to consider, down and back, so these are added together to get the total displacement, and then divided by the total time:

$$\bar{v}_3 = \frac{\Delta x_1 + \Delta x_2}{\Delta t_1 + \Delta t_2} = \frac{300 \text{ m} + (-300 \text{ m})}{150 \text{ s} + 198 \text{ s}} = 0 \text{ m/s}$$

The average velocity for the total trip is zero! Do you see why? Recall from the definition of displacement that the magnitude of displacement is the straight-line distance between two points. The displacement from one point back to the same point is zero; hence the average velocity is zero. (See ▼ **Figure 2.6**.)

▲ **FIGURE 2.6 Back home again!** Despite having covered nearly 110 m on the base paths, at the moment the runner slides through the batter's box (his original position) into home plate, his displacement is zero – at least, if he is a right-handed batter. No matter how fast he ran the bases, his average velocity for the round trip is zero.

The total or net displacement for this case could have been found by simply taking $\Delta x = x_{\text{final}} - x_{\text{initial}} = 0 - 0 = 0$, where the initial and final positions are taken to be the origin, but it was done in parts here for illustration purposes.

FOLLOW-UP EXERCISE. Find the jogger's average speed for each of the cases in this Example and compare these with the magnitudes of the respective average velocities. [Will the average speed for part (c) be zero?] *(Answers to all Follow-Up Exercises are given in Appendix V at the back of the book.)*

As Example 2.2 shows, average velocity provides only an overall description of motion. One way to take a closer look at motion is to take smaller time intervals; that is, to let the observation time interval (Δt) become smaller and smaller. As with speed, when Δt approaches zero, an instantaneous velocity is obtained, which describes how fast something is moving and in which direction *at a particular instant of time.*

Instantaneous velocity is defined mathematically as

$$v = \lim_{\Delta t \to 0} \frac{\Delta x}{\Delta t} \tag{2.4}$$

This expression is read as "the instantaneous velocity is equal to the limit of $\Delta x/\Delta t$ as Δt goes to zero." The time interval does

not ever equal zero (why?), but *approaches* zero. Instantaneous velocity is technically still an average velocity, but over such a very small Δt that it is essentially an average "at an instant in time," which is why it is called the *instantaneous* velocity.

Uniform motion means motion with a constant or uniform velocity (constant magnitude *and* constant direction). As a one-dimensional example of this, the car in ▼ **Figure 2.7** has a uniform velocity. It travels the same distance in the same direction and experiences the same displacement in equal time intervals (50 km each hour) and the direction of its motion does not change. Hence, the magnitudes of the average velocity and instantaneous velocity are equal in this case. The average of a constant is equal to that constant.

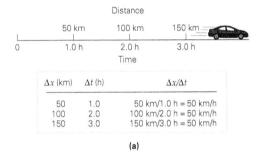

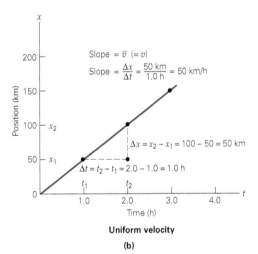

▲ **FIGURE 2.7 Uniform linear motion – constant velocity** In uniform linear motion, an object travels at a constant velocity, covering the same distance in the same direction during equal time intervals. **(a)** Here, a car travels 50 km in the same direction each hour. **(b)** An x-versus-t plot is a straight line, since equal distances are covered in equal times. The numerical value of the slope of the line is equal to the magnitude of the velocity, and the sign of the slope gives its direction. (The average velocity equals the instantaneous velocity in this case. Why?)

2.2.3 Graphical Analysis

Graphical analysis is often helpful in understanding motion and its related quantities. For example, the motion of the car in Figure 2.7a may be represented on a plot of position versus time, or x versus t. As can be seen from Figure 2.7b, a straight line is obtained for a uniform, or constant, velocity on such a graph.

Recall from Cartesian graphs of y versus x that the slope of a straight line is given by $\Delta y/\Delta x$. Here, with a plot of x versus t, the slope of the line, $\Delta x/\Delta t$, is therefore equal to the average velocity $\bar{v} = \Delta x/\Delta t$. For uniform motion, this value is equal to the instantaneous velocity. That is, $\bar{v} = v$. (Why?) The numerical value of the slope is the magnitude of the velocity, and the sign of the slope gives the direction. A positive slope indicates that x increases with time, so the motion is in the positive x-direction. (The positive sign is often omitted as being understood, which will be done in general from here on.)

Suppose that a plot of position versus time for a car's motion is a straight line with a negative slope, as in ▼ **Figure 2.8**. What does this indicate? As the figure shows, the position (x) values get smaller with time at a constant rate, indicating that the car is traveling in uniform motion, but now in the negative x-direction which correlates with the negative value of the slope.

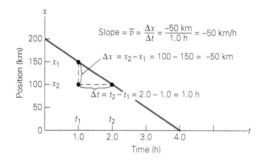

▲ FIGURE 2.8 **Position-versus-time graph for an object in uniform motion in the negative x-direction** A straight line on an x-versus-t plot with a negative slope indicates uniform motion in the negative x-direction. Note that the object's location changes at a constant rate. At $t = 4.0$ h, the object is at $x = 0$. How would the graph look if the motion continues for $t > 4.0$ h?

In most instances, the motion of an object is *nonuniform*, meaning that different distances are covered in equal intervals of time. An x-versus-t plot for such motion in one dimension is a curved line, as illustrated in ▶ **Figure 2.9**. The average velocity of the object during any interval of time is the slope of a straight line between the two points on the curve that correspond to the starting and ending times of the interval. In the figure, since $\bar{v} = \Delta x/\Delta t$, the average velocity of the total trip is the slope of the straight line joining the beginning and ending points of the curve.

The instantaneous velocity is equal to the slope of the tangent line to the curve at the time of interest. Five typical tangent lines are shown in Figure 2.9. At (1), the slope is positive, and the motion is therefore in the $+x$-direction. At (2), the slope of a horizontal tangent line is zero, so there is no motion for an instant. That is, the object has instantaneously stopped ($v = 0$) at that time. At (3), the slope is negative, so the object is moving in the $-x$-direction. Thus, the object stopped in the process of changing direction at point (2). What is happening at points (4) and (5)?

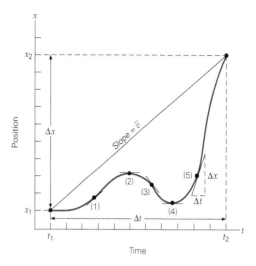

▲ FIGURE 2.9 **Position-versus-time graph for an object in nonuniform linear motion** For a nonuniform velocity, an x-versus-t plot is a curved line. The slope of the line between two points is the average velocity between these positions, and the instantaneous velocity at any time t is the slope of a line tangent to the curve at that point. Five tangent lines are shown, with the intervals for $\Delta x/\Delta t$ shown in the fifth tangent line. Can you describe the object's motion in words?

By drawing various tangent lines along the curve, it can be seen that their slopes vary in magnitude and direction (sign), indicating that the instantaneous velocity is changing with time. An object in nonuniform motion can speed up, slow down, or change direction. How nonuniform motion is described is the topic of Section 2.3.

2.3 Acceleration

The basic description of motion involving the time rate of change of position (and direction) is called *velocity*. Going one step further, we can consider how this *rate of change* itself changes. Suppose an object is moving at a constant velocity and then the velocity changes. Such a change in velocity is called an *acceleration*. The gas pedal on an automobile is commonly called the *accelerator*. When you press down on the accelerator, the car speeds up; when you let up on the accelerator, the car slows down. In either case, there is a change in velocity with time. **Acceleration** is defined as the time rate of change of velocity.

Analogous to average velocity, the **average acceleration** is defined as the change in velocity divided by the time taken to make the change:

$$\text{average acceleration} = \frac{\text{change in velocity}}{\text{time to make the change}} \quad (2.5)$$

$$\bar{a} = \frac{\Delta v}{\Delta t} = \frac{v_2 - v_1}{t_2 - t_2} = \frac{v - v_\text{o}}{t - t_\text{o}}$$

SI unit of acceleration: meters per second squared (m/s^2).

Note that the initial and final variables have been changed to a more commonly used notation. That is, v_o and t_o are the initial or original velocity and time, respectively, and v and t are the general velocity and time at some point in the future, such as when you want to know the velocity v at a particular time t. (This may or may not be the final velocity of a particular situation. There may be an acceleration after this time.)

From $\Delta v/\Delta t$, the SI units of acceleration can be seen to be meters per second (Δv) per second (Δt); that is, (m/s)/s or m/(s·s), which is commonly expressed as meters per second squared (m/s^2). In the British system, the units are feet per second squared (ft/s^2).

Because velocity is a vector quantity, so is acceleration, as acceleration represents a change in velocity. Since velocity has both magnitude and direction, a change in velocity may involve changes in either or both of these factors. Thus an acceleration may result from a change in *speed* (magnitude), a change in *direction*, or a change in *both*, as illustrated in ▼ **Figure 2.10**.

For straight-line, linear motion, positive and negative signs will be used to indicate the directions of velocity and acceleration, as was done for linear displacements. Equation 2.5 is commonly simplified and written as

$$\bar{a} = \frac{v - v_0}{t} \tag{2.6}$$

where t_0 is taken to be zero. (v_0 may not be zero, so it cannot generally be omitted.)

Analogous to instantaneous velocity, **instantaneous acceleration** is the acceleration at a particular instant of time. This quantity is expressed mathematically as

$$a = \lim_{\Delta t \to 0} \frac{\Delta v}{\Delta t} \tag{2.7}$$

The conditions of the time interval approaching zero are the same here as described for instantaneous velocity.

Since both velocity and acceleration are vector quantities and the positive and negative signs are only indicating the direction of the vector, a positive acceleration does not necessarily mean an increase in speed, or vice versa. The direction of the velocity is also important. ▶ **Figure 2.11** illustrates the four scenarios.

**EXAMPLE 2.3: SLOWING IT DOWN –
AVERAGE ACCELERATION**

A couple in a sport-utility vehicle (SUV) are traveling at 90 km/h on a straight highway. The driver sees an accident in the distance and slows down to 40 km/h in 5.0 s. What is the average acceleration of the SUV?

THINKING IT THROUGH. To find the average acceleration, the variables as defined in Equation 2.6 must be given, and they are.

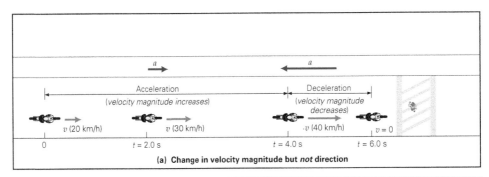

(a) **Change in velocity magnitude but *not* direction**

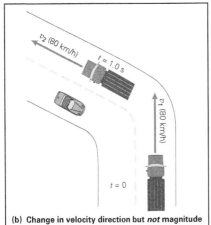

(b) **Change in velocity direction but *not* magnitude**

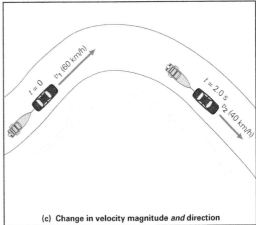

(c) **Change in velocity magnitude *and* direction**

▲ FIGURE 2.10 **Acceleration – the time rate of change of velocity** Since velocity is a vector quantity, with magnitude and direction, an acceleration can occur when there is **(a)** a change in magnitude, but not direction; **(b)** a change in direction, but not magnitude; or **(c)** a change in both magnitude and direction.

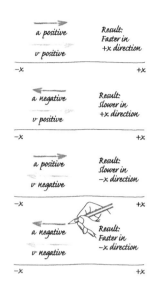

▲ FIGURE 2.11 **Signs of velocity and acceleration** Since both velocity and acceleration are vector quantities and the positive and negative signs are only indicating the direction of the vector, a positive acceleration does not necessarily mean an increase in speed, or vice versa. The direction of the velocity is also important.

SOLUTION

Listing the data and converting units,

Given:

$$v_o = (90 \text{ km/h})\left(\frac{0.278 \text{ m/s}}{1 \text{ km/h}}\right) = 25 \text{ m/s}$$

$$v = (40 \text{ km/h})\left(\frac{0.278 \text{ m/s}}{1 \text{ km/h}}\right) = 11 \text{ m/s}$$

$$t = 5.0 \text{ s}$$

Find:

$\bar{a}$ (average acceleration)

[Here, the instantaneous velocities are assumed to be in the positive direction, and conversions to standard units (meters per second) are made right away, since it is noted that the speed is given in km/h. In general, standard units should be used.]

Given the initial and final velocities and the time interval, the average acceleration can be found by using Equation 2.6:

$$\bar{a} = \frac{v - v_o}{t} = \frac{11 \text{ m/s} - (25 \text{ m/s})}{5.0 \text{ s}} = -2.8 \text{ m/s}^2$$

The negative sign indicates the direction of the (vector) acceleration. In this case, the acceleration is opposite to the direction of velocity and the car slows. Such an acceleration is sometimes called a *deceleration*, since the car is slowing. (*Note:* This is why v_o cannot arbitrarily be set to zero, because as shown here there may be motion, and $v_o \neq 0$ at $t_o = 0$.)

FOLLOW-UP EXERCISE. Does a negative acceleration necessarily mean that a moving object is slowing down (decelerating) or that its speed is decreasing? [*Hint:* See Figure 2.11.]

2.3.1 Constant Acceleration

Although acceleration can vary with time, our study of motion will generally be restricted to constant accelerations for simplicity. (An important constant acceleration is the acceleration due to gravity near the Earth's surface, which will be considered in the next section.) Since for a constant acceleration, the average acceleration is equal to the constant value ($\bar{a} = a$), the bar over the acceleration in Equation 2.6 may be omitted. Thus, for a constant acceleration, the equation relating velocity, acceleration, and time is commonly written (rearranging Equation 2.6) as follows:

$$v = v_o + at \quad \text{(constant acceleration only)}$$

(Note that the term "*at*" represents the *change* in velocity, since $at = v - v_o = \Delta v$.)

**EXAMPLE 2.4: FAST START, SLOW STOP –
MOTION WITH CONSTANT ACCELERATION**

A drag racer starting from rest accelerates in a straight line at a constant rate of 5.5 m/s² for 6.0 s. (a) What is the racer's velocity at the end of this time? (b) If a parachute deployed at this time causes the racer to slow down uniformly at a rate of 2.4 m/s², how long will the racer take to come to a stop?

THINKING IT THROUGH. The racer first speeds up and then slows down, so close attention must be given to the directional signs of the vector quantities. Choose a coordinate system with the positive direction in the direction of the initial velocity. (Draw a sketch of the situation for yourself.) The answers can then be found by using the appropriate equations. Note that there are two different parts to the motion, with two different accelerations. Let's distinguish these phases with subscripts of 1 and 2.

SOLUTION

Taking the initial motion to be in the positive direction, we have the following data:

Given:

(a) $v_o = 0$ (at rest)
$a_1 = 5.5 \text{ m/s}^2$
$t_1 = 6.0 \text{ s}$

(b) $v_o = v_1$ [from part (a)]
$v_2 = 0$ (comes to stop)
$a_2 = -2.4 \text{ m/s}^2$ (opposite direction of v_o)

Find:

(a) v_1 (final velocity for first part)
(b) t_2 (time for second part)

The data have been listed in two parts. This practice helps avoid confusion with symbols. Note that the final velocity v_1 that is to be found in part (a) becomes the initial velocity v_o for part (b).

(a) To find the final velocity v_1, Equation 2.8 may be used directly:

$$v_1 = v_o + a_1 t_1 = 0 + (5.5 \text{ m/s}^2)(6.0 \text{ s}) = 33 \text{ m/s}$$

(b) Here we want to find time, so solving Equation 2.6 for t_2 and using $v_o = v_1 = 33$ m/s from part (a),

$$t_2 = \frac{v_2 - v_o}{a_2} = \frac{0 - (33 \text{ m/s})}{-2.4 \text{ m/s}^2} = 14 \text{ s}$$

Note that the time comes out positive, as it should. Why?

FOLLOW-UP EXERCISE. What is the racer's instantaneous velocity 10 s after the parachute is deployed?

Motions with constant accelerations are easy to represent graphically by plotting instantaneous velocity versus time. In this case, the v-versus-t plot is a straight line, the slope of which is equal to the acceleration, as illustrated in ▼ **Figure 2.12**. Note that Equation 2.8 can be written as $v = at + v_o$, which, as you may recognize, has the form of an equation of a straight line, $y = mx + b$ (slope m and intercept b). In Figure 2.12a, the motion is in the positive direction, and the acceleration term adds to the velocity after $t = 0$, as illustrated by the vertical arrows at the right of the graph. Here, the slope is positive ($a > 0$).

In Figure 2.12b, the negative slope ($a < 0$) indicates a negative acceleration that produces a slowing down, or deceleration. However, Figure 2.12c illustrates how a negative acceleration can speed things up (for motion in the negative direction). The situation in Figure 2.12d is slightly more complex. Can you explain what is happening there?

When an object moves at a constant acceleration, its velocity changes by the same amount in each equal time interval. For example, if the acceleration is 10 m/s² in the same direction as that of the initial velocity, the object's velocity increases by 10 m/s each second. As an example of this, suppose that the object has an initial velocity v_o of 20 m/s in a particular direction at $t_o = 0$. Then, for $t = 0, 1.0, 2.0, 3.0,$ and 4.0 s, the magnitudes of the velocities are 20, 30, 40, 50, and 60 m/s, respectively.

The average velocity may be computed in the regular manner (Equation 2.3), or you may recognize that the uniformly increasing series of numbers 20, 30, 40, 50, and 60 has an average value of 40 (the midway value of the series) and $\bar{v} = 40$ m/s. Note that the average of the initial and final values also gives the average of the series – that is, $(20 + 60)/2 = 40$. Only when the velocity changes at a uniform rate because of a constant acceleration is $\bar{v}$ then the average of the initial and final velocities:

$$\bar{v} = \frac{v + v_o}{2} \quad \text{(constant acceleration only)} \tag{2.9}$$

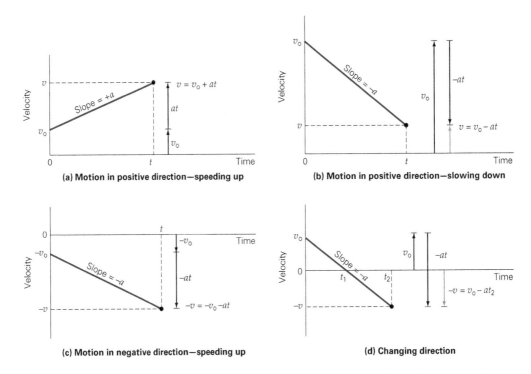

▲ **FIGURE 2.12 Velocity-versus-time graphs for motions with constant accelerations** The slope of a v-versus-t plot is the acceleration. **(a)** A positive slope indicates an increase in the velocity in the positive direction. The vertical arrows to the right indicate how the acceleration adds velocity to the initial velocity, v_o. **(b)** A negative slope indicates a decrease in the initial velocity, v_o, or a deceleration. **(c)** Here a negative slope indicates a negative acceleration, but the initial velocity is in the negative direction, $v_o < 0$, so the speed of the object increases in that direction. **(d)** The situation here is initially similar to that of part (b) but ends up resembling that in part (c). Can you explain what happened at time t_1?

EXAMPLE 2.5: ON THE WATER – USING MULTIPLE EQUATIONS

A motorboat starting from rest on a lake accelerates in a straight line at a constant rate of 3.0 m/s² for 8.0 s. How far does the boat travel during this time?

THINKING IT THROUGH. We have only one equation for distance (Equation 2.3, $x = x_o + \bar{v}t$), but this equation cannot be used directly. The average velocity must first be found, so multiple equations and steps are involved.

SOLUTION

Reading the example, summarizing the given data, and identifying what is to be found (assuming the boat to accelerate in the +x-direction) gives the following:

Given:

$x_o = 0$
$v_o = 0$
$a = 3.0$ m/s²
$t = 8.0$ s

Find:

x (distance)

(Note that all of the units are standard.)

In analyzing the problem, one might reason: To find x, Equation 2.3 needs to be used in the form $x = x_o + \bar{v}t$. (The average velocity $\bar{v}$ must be used because the velocity is changing and thus not constant.) With time given, the solution to the problem then involves finding $\bar{v}$. Then by Equation 2.9, $\bar{v} = (v + v_o)/2$, and with $v_o = 0$, only the final velocity v is needed to solve the problem. Equation 2.8, $v = v_o + at$, can be used to calculate v from the given data. So it follows that:

The velocity of the boat at the end of 8.0 s is

$$v = v_o + at = 0 + (3.0 \text{ m/s}^2)(8.0 \text{ s}) = 24 \text{ m/s}$$

The average velocity over that time interval is

$$\bar{v} = \frac{v + v_o}{2} = \frac{24 \text{ m/s} + 0}{2} = 12 \text{ m/s}$$

Finally, the magnitude of the displacement, which in this case is the same as the distance traveled, is given by Equation 2.3 (choosing the boat's initial location at the origin, so $x_o = 0$):

$$x = \bar{v}t = (12 \text{ m/s})(8.0 \text{ s}) = 96 \text{ m}$$

FOLLOW-UP EXERCISE. (Sneak preview.) In Section 2.4, the following equation will be derived: $x = v_o t + \frac{1}{2}at^2$. Use the data in this Example to see if this equation gives the distance traveled.

2.4 Kinematic Equations (Constant Acceleration)

The description of motion in one dimension with constant acceleration requires only three basic equations. From previous sections, these equations are

$$x = x_o + \bar{v}t \tag{2.3}$$

$$\bar{v} = \frac{v + v_o}{2} \quad \text{(constant acceleration only)} \tag{2.9}$$

$$v = v_o + at \quad \text{(constant acceleration only} \tag{2.8}$$

(Keep in mind that the first equation, Equation 2.3, is general and is not limited to situations in which there is constant acceleration, as are the latter two equations.)

However, as Example 2.5 showed, the description of motion in some instances requires multiple applications of these equations, which may not be obvious at first. It would be helpful if there were a way to reduce the number of operations in solving kinematic problems, and there is – by combining equations algebraically.

For instance, suppose an expression that gives location x in terms of time and acceleration is wanted, rather than one in terms of time and average velocity (as in Equation 2.3). Eliminating $\bar{v}$ from Equation 2.3 by substituting for $\bar{v}$ from Equation 2.9 into Equation 2.3,

$$x = x_o + \bar{v}t \tag{2.3}$$

and

$$\bar{v} = \frac{v + v_o}{2} \tag{2.9}$$

and on substituting,

$$x = x_o + \tfrac{1}{2}(v + v_o)t \quad \text{(constant acceleration only)} \tag{2.10}$$

Then, substituting for v from Equation 2.8 ($v = v_o + at$) gives

$$x = x_o + \tfrac{1}{2}(v_o + at + v_o)t$$

Simplifying,

$$x = x_o + v_o t + \tfrac{1}{2}at^2 \quad \text{(constant acceleration only)} \tag{2.11}$$

Essentially, this series of steps was done in Example 2.5. The combined equation allows the displacement of the motorboat in that Example to be computed directly:

$$x - x_o = \Delta x = v_o t + \tfrac{1}{2}at^2 = 0 + \tfrac{1}{2}(3.0 \text{ m/s}^2)(8.0 \text{ s})^2 = 96 \text{ m}$$

Much easier, isn't it?

We may want an expression that gives velocity as a function of position x rather than time (as in Equation 2.8). In this case,

t can be eliminated from Equation 2.8 by using Equation 2.10 in the form

$$v + v_0 = 2\frac{(x - x_0)}{t}$$

Then, multiplying this equation by Equation 2.8 in the form $(v - v_0) = at$ gives

$$(v + v_0)(v - v_0) = 2a(x - x_0)$$

and using the relationship $v^2 - v_0^2 = (v + v_0)(v - v_0)$,

$$v^2 = v_0^2 + 2a(x - x_0) \quad \text{(constant acceleration only)} \quad (2.12)$$

2.4.1 Problem-Solving Hint

Students in introductory physics courses are sometimes overwhelmed by the various kinematic equations. Keep in mind that equations and mathematics are the tools of physics. As any mechanic or carpenter will tell you, tools make your work easier as long as you are familiar with them and know how to use them. The same is true for physics tools. Summarizing the equations for linear motion with *constant* acceleration:

$$v = v_0 + at \quad (2.8)$$

$$x = x_0 + \tfrac{1}{2}(v + v_0)t \quad (2.10)$$

$$x = x_0 + v_0 t + \tfrac{1}{2}at^2 \quad (2.11)$$

$$v^2 = v_0^2 + 2a(x - x_0) \quad (2.12)$$

This set of equations is used to solve the majority of kinematic problems. (Occasionally, there may be interest in average speed or velocity, and for that Equation 2.3 can be used.)

Note that each of the equations in the list has four or five variables. All but one of the variables in an equation must be known in order to be able to solve for what you are trying to find. Generally, an equation with the unknown or wanted quantity is chosen. But, as pointed out, the other variables in the equation must be known. If they are not, then the wrong equation was chosen or another equation must be used to find the variables. (Another possibility is that not enough data are given to solve the problem, but that is not the case in this textbook.)

Always try to understand and visualize a problem. Listing the data as described in the suggested problem-solving procedure in Section 1.7 may help you decide which equation to use, by determining the known and unknown variables. Remember this approach as you work through the remaining Examples in the chapter. Also, don't overlook any *implied data or restrictive conditions*, as illustrated in the following examples.

EXAMPLE 2.6: MOVING APART – WHERE ARE THEY NOW?

Two riders on dune buggies sit 10 m apart on a long, straight track, facing in opposite directions. Starting at the same time, both riders accelerate at a constant rate of 2.0 m/s². How far apart will the dune buggies be at the end of 3.0 s?

THINKING IT THROUGH. The dune buggies are initially 10 m apart, so they can be positioned anywhere on the x-axis. It is convenient to place one at the origin so that one initial position (x_0) is zero. A sketch of the situation is shown in ▼ **Figure 2.13**.

SOLUTION

Listing the data

Given:

$$x_{0_A} = 0$$
$$a_A = -2.0 \text{ m/s}^2$$
$$t = 3.0 \text{ s}$$
$$x_{0_B} = 10 \text{ m}$$
$$a_B = 2.0 \text{ m/s}^2$$

Find:

separation distance at $t = 3.0$ s

The displacement of each vehicle is given by Equation 2.11 [the only displacement (Δx) equation with acceleration (a)]: $x = x_0 + v_0 t + \tfrac{1}{2}at^2$. But there is no v_0 in the Given list. Some implied data must have been missed. It should be quickly noted that $v_0 = 0$ for both vehicles, so

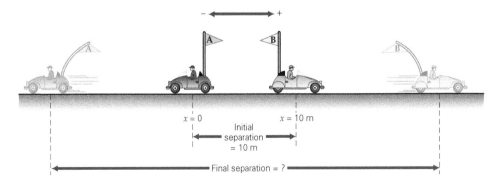

▲ FIGURE 2.13   **Away they go!** Two dune buggies accelerate away from each other. How far apart are they at a later time? See Example 2.6.

$$x_A = x_{o_A} + v_{o_A}t + \tfrac{1}{2}a_A t^2 = 0 + 0 + \tfrac{1}{2}(-2.0 \text{ m/s}^2)(3.0 \text{ s})^2 = -9.0 \text{ m}$$

And for buggy B with a nonzero x_o,

$$x_B = x_{o_B} + v_{o_B}t + \tfrac{1}{2}a_B t^2 = 10 \text{ m} + 0 + \tfrac{1}{2}(2.0 \text{ m/s}^2)(3.0 \text{ s})^2 = 19 \text{ m}$$

Hence vehicle A is 9.0 m to the left of the origin on the $-x$-axis, whereas vehicle B is at a position of 19 m to the right of the origin on the $+x$-axis. And so, the separation distance between the two dune buggies is $19 \text{ m} + 9 \text{ m} = 28 \text{ m}$.

FOLLOW-UP EXERCISE. Would it make any difference in the separation distance if vehicle B had been initially put at the origin instead of vehicle A? Try it and find out.

EXAMPLE 2.7: PUTTING ON THE BRAKES – VEHICLE STOPPING DISTANCE

The stopping distance of a vehicle is an important safety factor. This distance depends on the initial speed (v_o) and the braking capacity, which produces the deceleration, a, assumed to be constant. Express the stopping distance x in terms of these quantities.

THINKING IT THROUGH. The signs of the velocity and acceleration are taken to be positive and negative respectively, indicating they are in opposite directions, so the car comes to a stop. Again, a kinematic equation is required, and the appropriate one may be better determined by listing what is given and what is to be found. Notice that the distance x is wanted, and time is not involved.

SOLUTION

Here the quantities are variables and represented in symbol form:

Given:

$v_o\ (> 0)$
$-a$ (it is a deceleration)
$v = 0$ (car comes to stop)
$x_o = 0$ (car taken to be initially at the origin)

Find:

stopping distance x (in terms of the given variables)

Again, it is helpful to make a sketch of the situation, particularly when vector quantities are involved (▼ **Figure 2.14**). Since

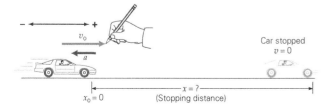

▲ FIGURE 2.14 **Vehicle stopping distance** A sketch to help visualize the situation for Example 2.7.

Equation 2.12 has the variables we want, it should allow us to find the stopping distance x. Expressing the negative acceleration explicitly and assuming $x_o = 0$ gives

$$v^2 = v_o^2 - 2ax$$

Since the vehicle comes to a stop ($v = 0$), solving for x:

$$x = \frac{v_o^2}{2a}$$

This equation gives x expressed in terms of the vehicle's initial speed and stopping acceleration. Notice that the stopping distance x is proportional to the *square* of the initial speed. Doubling the initial speed therefore increases the stopping distance by a factor of 4 (for the same deceleration). That is, if the stopping distance is x_1 for an initial speed of v_1, then for a twofold increase in the initial speed ($v_2 = 2v_1$), the stopping distance would increase fourfold:

$$x_1 = \frac{v_1^2}{2a}$$
$$x_2 = \frac{v_2^2}{2a} = \frac{(2v_1)^2}{2a} = 4\left(\frac{v_1^2}{2a}\right) = 4x_1$$

The same result can be obtained by using ratios:

$$\frac{x_2}{x_1} = \frac{v_2^2}{v_1^2} = \left(\frac{v_2}{v_1}\right)^2 = 2^2 = 4$$

Do you think this consideration is important in setting speed limits, for example, in school zones? (The driver's reaction time should also be considered. A method for approximating a person's reaction time is given in Section 2.5.)

FOLLOW-UP EXERCISE. Tests have shown that the Chevy Blazer has an average braking deceleration of 7.5 m/s², while that of a Toyota Celica is 9.2 m/s². Suppose these two vehicles are being driven down a straight, level road at 97 km/h (60 mi/h), with the Celica in front of the Blazer. A cat runs across the road ahead of them, and both drivers apply their brakes at the same time and come to safe stops (not hitting the cat). Assuming constant deceleration and the same reaction times for both drivers, what is the minimum safe tailgating distance for the Blazer so that there won't be a rear-end collision with the Celica when the two vehicles come to a stop?

2.4.2 Graphical Analysis of Kinematic Equations

As was shown in Figure 2.12, plots of v versus t give straight-line graphs where the slopes are values of the constant accelerations. There is another interesting aspect of v-versus-t graphs. Consider the one shown in ▶ **Figure 2.15a**, particularly the shaded area under the curve. Suppose we calculate the area of the shaded triangle, where, in general, $A = \tfrac{1}{2}ab$ [Area $= \tfrac{1}{2}$(altitude)(base)].

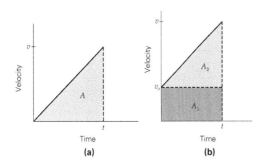

▲ **FIGURE 2.15** **Velocity-versus-time graphs, one more time (a)** In the straight-line plot for a constant acceleration, the area under the curve is equal to Δx, the displacement covered. **(b)** If v_0 is not zero, the displacement is still given by the area under the curve Δx, but here divided into two parts, areas A_1 and A_2.

For the graph in Figure 2.15a, the altitude is v and the base is t, so $A = \frac{1}{2}vt$. But from the equation $v = v_0 + at$, we have $v = at$, where $v_0 = 0$ (zero intercept on graph). Therefore,

$$A = \tfrac{1}{2}vt = \tfrac{1}{2}(at)t = \tfrac{1}{2}at^2 = \Delta x$$

Hence, Δx is equal to the area under a v-versus-t curve.

Now look at Figure 2.15b. Here, there is a nonzero value of v_0 at $t = 0$, so the object is initially moving. Consider the two shaded areas. We know that the area of the triangle is $A_2 = \frac{1}{2}at^2$, and the area of the rectangle can be seen (with $x_0 = 0$) to be $A_1 = v_0 t$. Adding these areas to get the total area yields

$$A_1 + A_2 = v_0 t + \tfrac{1}{2}at^2 = \Delta x$$

This is just Equation 2.11, which is equal to the area under the v-versus-t curve.

2.5 Free Fall

One of the more common cases of constant acceleration is the acceleration due to gravity near the Earth's surface. When an object is dropped, its initial velocity (at the instant it is released) is zero. At a later time while falling, it has a nonzero velocity. There has been a change in velocity and thus, by definition, an acceleration. **This acceleration due to gravity (g)** near the Earth's surface has an approximate magnitude of

$$g = 9.80\,\text{m/s}^2 \text{ (acceleration due to gravity)}$$

(or 980 cm/s²), and is directed downward (toward the center of the Earth). In British units, the value of g is about 32.2 ft/s². The values given here for g are only approximate because the acceleration due to gravity varies slightly at different locations as a result of differences in elevation and regional average mass densities of the Earth. These small variations will be ignored in this book unless otherwise noted. (Gravitation is studied in more detail in Section 7.5.) Air resistance is another factor that affects (reduces) the acceleration of a falling object, but it too

will be ignored here for simplicity. (The frictional effect of air resistance will be considered in Section 4.6.)

Objects in motion solely under the influence of gravity are said to be in **free fall**. The words *free fall* may bring to mind dropped objects. However, the term applies to any motion under the sole influence of gravity. Objects released from rest, thrown upward or downward, are all in free fall once they are released. That is, after $t = 0$ (the time of release), only gravity is influencing the motion. The set of equations for motion in one dimension with constant acceleration given in the last section can be used to describe free fall.

The acceleration due to gravity, g, has the same value for all free-falling objects, regardless of their mass or weight. It was once thought that heavier bodies accelerate faster than lighter bodies. This concept was part of Aristotle's theory of motion. You can easily observe that a coin accelerates faster than a piece of paper when dropped simultaneously from the same height. But in this case, air resistance plays a noticeable role. If the paper is crumpled into a compact ball, it gives the coin a much better race. Similarly, a feather "floats" down much more slowly than a coin falls. However, in a near-vacuum, where there is negligible air resistance, the feather and the coin have the same acceleration – the acceleration due to gravity (▼ **Figure 2.16**).

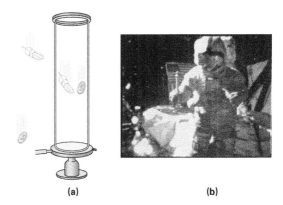

▲ **FIGURE 2.16** **Free fall and air resistance (a)** When dropped simultaneously from the same height, a feather falls more slowly than a coin, because of air resistance. But when both objects are dropped in an evacuated container with a good partial vacuum, where air resistance is negligible, the feather and the coin both have the same constant acceleration. **(b)** Astronaut David Scott performed an experiment on the Moon in 1971 by simultaneously dropping a feather and a hammer from the same height. He did not need a vacuum pump. The Moon has no atmosphere and therefore no air resistance. The hammer and the feather reached the lunar surface at the same time.

At the end of the last Apollo 15 moon walk in 1971, Commander David Scott performed a live demonstration for the television cameras. He held out a 1.32-kg aluminum geological hammer and a 0.03-kg falcon feather and dropped them simultaneously from approximately the same height (approximately 1.6 m). Because they were essentially in a vacuum, there was no air resistance and the feather fell at the same rate as the hammer. Both objects, however, had a smaller acceleration and fell at a slower rate than on

Earth. The acceleration due to gravity near the Moon's surface is only about one-sixth of that near the Earth's surface ($g_M \approx g/6$).

Currently accepted ideas about the motion of falling bodies are due in large part to Galileo. He challenged Aristotle's theory and experimentally investigated the motion of such objects. Legend has it that Galileo studied the accelerations of falling bodies by dropping objects of different weights from the top of the Leaning Tower of Pisa.

It is customary to use y to represent the vertical direction and to take upward as positive (as with the vertical y-axis of Cartesian coordinates). Because the acceleration due to gravity is always downward, it is in the negative y-direction. This negative acceleration, $a = -g = -9.80$ m/s^2, should be substituted into the equations of motion. (Remember g is the *magnitude* of the gravitational acceleration near the surface of the Earth so it has a positive value; $g = 9.80$ m/s^2.) However, the relationship $a = -g$ may be expressed explicitly in the equations for linear motion directional convenience:

$$v = v_o - gt \tag{2.8'}$$

$$y = y_o + v_o t - \tfrac{1}{2} g t^2 \tag{2.11'}$$
(free-fall equations with $a_y = -g$)

$$v^2 = v_o^2 - 2g(y - y_o) \tag{2.12'}$$

Equation 2.10 applies to free fall as well, but it does not contain g:

$$y = y_o + \tfrac{1}{2}(v + v_o)t \tag{2.10'}$$

The origin ($y = 0$) of the frame of reference is usually taken to be at the initial position of the object. You must be explicit about the directions of vector quantities. The location y and the velocities v and v_o may be positive (up) or negative (down), but the acceleration due to gravity is always downward (negative). Writing $-g$ explicitly in the equations is a reminder of its direction. Then the value of g is simply inserted as 9.80 m/s^2.

The use of these equations and the sign convention (with $-g$ explicitly expressed in the equations) are illustrated in the following Examples. (This convention will be used throughout the text.)

EXAMPLE 2.8: A STONE THROWN DOWNWARD – THE KINEMATIC EQUATIONS REVISITED

A boy on a bridge throws a stone vertically downward with an initial speed of 14.7 m/s toward the river below. If the stone hits the water 2.00 s later, what is the height of the bridge above the water?

THINKING IT THROUGH. This is a free-fall problem, but note that the initial velocity is downward, which is taken as the negative direction. It is important to express this factor explicitly. Draw a sketch to help you analyze the situation if needed.

SOLUTION

As usual, first writing what is given and what is to be found:

Given:

$$v_o = -14.7 \text{m/s}$$
$$t = 2.00 \text{ s}$$
$$g\ (= 9.80 \text{ m/s}^2)$$

Find:

y (bridge height above water)

Notice that g is listed as a positive number, since by our convention the directional negative sign has already been put into the previous kinematic equations.

Which equation(s) will provide the solution using the given data? It should be evident that the distance the stone travels in an amount of time t is given directly by Equation 2.11'. Taking $y_o = 0$:

$$\begin{aligned}
y &= v_o t - \tfrac{1}{2} g t^2 \\
&= (-14.7 \text{ m/s})(2.00 \text{ s}) - \tfrac{1}{2}(9.80 \text{ m/s}^2)(2.00 \text{ s})^2 \\
&= -29.4 \text{ m} - 19.6 \text{ m} = -49.0 \text{ m}
\end{aligned}$$

The negative sign indicates that the rock's displacement is downward, as it should be. Thus the height bridge is 49.0 m.

FOLLOW-UP EXERCISE. How much longer would it take for the stone to reach the river if the boy in this Example had dropped the ball rather than thrown it?

Reaction time is the time it takes a person to notice, think, and act in response to a situation – for example, the time between first observing and then responding to an obstruction on the road ahead by applying the brakes. Reaction time varies with the complexity of the situation (and with the individual). In general, the largest part of a person's reaction time is spent thinking, but practice in dealing with a given situation can reduce this time. The following Example gives a simple method for measuring reaction time.

EXAMPLE 2.9: MEASURING REACTION TIME – FREE FALL

A person's reaction time can be measured by having another person drop a ruler (without warning) through the first person's thumb and forefinger, as shown in ▶ **Figure 2.17**. After observing the unexpected release, the first person grasps the falling ruler as quickly as possible, and the length of the ruler below the top of the finger is noted. Suppose the ruler descends 18.0 cm before it is caught. What is the person's reaction time?

THINKING IT THROUGH. Both distance and time are involved. This observation indicates which kinematic equation should be used.

▲ FIGURE 2.17 **Reaction time** A person's reaction time can be measured by having the person grasp a dropped ruler. See Example 2.9.

SOLUTION

Notice that only the distance of fall is given. However, a couple of other things are known, such as v_o and g. So, taking $y_o = 0$:

Given:

$y = -18.0 \text{ cm} = -0.180 \text{ m}$
$v_o = 0$
$g \, (= 9.80 \text{ m/s}^2)$

Find:

t (reaction time)

(Note that the distance y has been converted to meters. Why?) It can be seen that Equation 2.11′ applies here (with $v_o = 0$), giving

$$y = -\tfrac{1}{2} g t^2$$

Then solving for t,

$$t = \sqrt{\frac{2y}{-g}} = \sqrt{\frac{2(-0.180 \text{ m})}{-9.80 \text{ m/s}^2}} = 0.192 \text{ s}$$

Try this experiment with a fellow student and measure your reaction time. Why do you think another person besides you should drop the ruler?

FOLLOW-UP EXERCISE. A popular trick is to substitute a crisp dollar bill lengthwise for the ruler in Figure 2.17, telling the person that he or she can have the dollar if able to catch it. Is this proposal a good deal? (The length of a dollar is 15.7 cm.)

Here are some interesting facts about free-fall motion of an object thrown upward in the absence of air resistance. First, if the object returns to its launch elevation, the times of flight upward and downward are the same. Similarly, note that at the very top of the trajectory, the object's velocity is zero for an instant, but the acceleration (even at the top) remains a constant 9.8 m/s² downward. It is a common misconception that at the

top of the trajectory the acceleration is zero. If this were the case, the object would remain there, as if gravity had been turned off!

Finally, the object returns to the starting point with the same speed as that at which it was launched. (The velocities have the same magnitude, but are opposite in direction.)

EXAMPLE 2.10: FREE FALL UP AND DOWN – USING IMPLICIT DATA

A worker on a scaffold in front of a billboard throws a ball straight up. The ball has an initial speed of 11.2 m/s when it leaves the worker's hand at the top of the billboard (▼ **Figure 2.18**). (a) What is the maximum height the ball reaches relative to the top of the billboard? (b) How long does it take the ball to reach this height? (c) What is the position of the ball at $t = 2.00$ s?

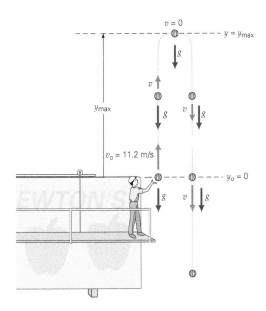

▲ FIGURE 2.18 **Free fall up and down** Note the lengths of the velocity and acceleration vectors at different times. (The upward and downward paths of the ball are horizontally displaced for illustration purposes.) See Example 2.10.

THINKING IT THROUGH. In part (a), only the upward part of the motion has to be considered. Note that the ball stops (zero instantaneous velocity) at the maximum height, which allows this height to be determined. In part (b), knowing the maximum height allows the determination of the upward time of flight. In part (c), the distance-time equation (Equation 2.11′) applies for any time and therefore allows calculation of the position (y) of the ball relative to the launch point at $t = 2.00$ s.

SOLUTION

It might appear that all that is given is the initial velocity v_o at time t_o. However, a couple of other pieces of information are implied that should be recognized. One, of course, is the acceleration g, and the other is the velocity at the maximum height where the ball stops. Here, in changing direction, the velocity of the ball is momentarily zero, so (again taking $y_o = 0$):

Given:

$v_o = 11.2$ m/s
$g \, (= 9.80 \text{ m/s}^2)$
$v = 0$ (at y_{max})
$t = 2.00$ s [for part (c)]

Find:

(a) y_{max} (maximum height)
(b) t_u (time upward)
(c) y (at $t = 2.00$ s)

(a) Notice that the height ($y_o = 0$) is referenced to the top of the billboard. For this part of the problem, we need be concerned with only the upward motion – the ball is thrown upward and stops at its maximum height y_{max}. With $v = 0$ at this height, y_{max} may be found directly from Equation 2.12′,

$$v^2 = 0 = v_o^2 - 2gy_{max}$$

So,

$$y_{max} = \frac{v_o^2}{2g} = \frac{(11.2 \text{ m/s})^2}{2(9.80 \text{ m/s}^2)} = 6.40 \text{ m}$$

relative to the top of the billboard ($y_o = 0$; see Figure 2.18).

(b) The time the ball travels upward to its maximum height is designated t_u. This is the time it takes for the ball to reach y_{max}, where $v = 0$. Since v_o and v are known, the time t_u can be found directly from Equation 2.8′,

$$v = 0 = v_o - gt_u$$

So,

$$t_u = \frac{v_o}{g} = \frac{11.2 \text{ m/s}}{9.80 \text{ m/s}^2} = 1.14 \text{ s}$$

(c) The height of the ball at $t = 2.00$ s is given directly by Equation 2.11′:

$$y = v_o t - \tfrac{1}{2} gt^2$$
$$= (11.2 \text{ m/s})(2.00 \text{ s}) - \tfrac{1}{2}(9.80 \text{ m/s}^2)(2.00 \text{ s})^2$$
$$= 22.4 \text{ m} - 19.6 \text{ m} = 2.8 \text{ m}$$

Note that this height is 2.8 m above, or measured upward from, the reference point ($y_o = 0$). The ball has reached its maximum height in 1.14 s and is on the way back down. Considered from another reference point, the situation in part (c) can be analyzed by imagining dropping a ball from a height of y_{max} above the top of the billboard with $v_o = 0$ and asking how far it falls in a time $t = 2.00$ s $- t_u = 2.00$ s $- 1.14$ s $= 0.86$ s. The answer is (this time with $y_o = 0$ at the maximum height)

$$y = v_o t - \tfrac{1}{2} gt^2 = 0 - \tfrac{1}{2}(9.80 \text{ m/s}^2)(0.86 \text{ s})^2 = -3.6 \text{ m}$$

This height is the same as the position found previously, but is measured with respect to the maximum height as the reference point; that is,

$$y_{max} - 3.6 \text{ m} = 6.4 \text{ m} - 3.6 \text{ m} = 2.8 \text{ m}$$

above the starting point.

FOLLOW-UP EXERCISE. At what height does the ball in this Example have a speed of 5.00 m/s? [*Hint:* The ball attains this height twice – once on the way up, and once on the way down.]

2.5.1 Problem-Solving Hint

When working vertical projectile problems involving motions up and down, it is often convenient to divide the problem into two parts and consider each part separately. As seen in Example 2.10, for the upward part of the motion, the velocity is zero at the maximum height. A quantity of zero usually simplifies the calculations. Similarly, the downward part of the motion is analogous to that of an object dropped from a height where the initial velocity can be taken as zero.

However, as Example 2.10 shows, the appropriate equations may be used directly for any position or time of the motion. For instance, note in part (c) that the height was found directly for a time *after* the ball had reached the maximum height. The velocity of the ball at that time could also have been found directly from Equation 2.8′, $v = v_o - gt$.

Also, note that the initial position was consistently taken as $y_o = 0$. This assumption is taken for convenience when the situation involves only one object (then $y_o = 0$ at $t_o = 0$). Using this convention can save a lot of time in writing and solving equations.

The same is true with only one object in horizontal motion: You can usually take $x_o = 0$ at $t_o = 0$. There are a couple of exceptions to this case, however. The first is if the problem specifies the object to be initially located at a position other than $x_o = 0$, and the second is if the problem involves two objects, as in Example 2.6. In the latter case, if one object is taken to be initially at the origin, the other's initial position is not zero.

EXAMPLE 2.11: LUNAR LANDING

A Lunar Lander makes a descent toward a level plain on the Moon. It descends slowly by using retro (braking) rockets. At a height of 6.0 m above the surface, the rockets are shut down with the Lander having a downward speed of 1.5 m/s. What is the speed of the Lander just before touching down?

THINKING IT THROUGH. This appears to be analogous to a simple free-fall problem of throwing an object downward – and it is, but the situation takes place on the Moon. It was noted previously that the acceleration due to gravity on the Moon, g_M, is one-sixth of that on the Earth, g_E. (No problem with air resistance on the Moon – it has no atmosphere.)

SOLUTION

Given:

$y = -6.0$ m ($y_0 = 0$ where rockets are shut down)
$v_0 = -1.5$ m/s
$g_M = g_E / 6$
$\quad = (9.8 \text{ m/s}^2)/6 = 1.6 \text{ m/s}^2$

Find:

v (just before touching down)

Then Equation 2.12′ can be used:

$$v^2 = v_0^2 - 2g_M y = (-1.5 \text{ m/s})^2 - 2(1.6 \text{ m/s}^2)(-6.0 \text{ m}) = 21 \text{ m}^2/\text{s}^2$$

So,

$$v^2 = 21 \text{ m}^2/\text{s}^2 \text{ and } v = \sqrt{21 \text{ m}^2/\text{s}^2} = \pm 4.6 \text{ m/s}$$

This is the velocity, which we know is downward, so the negative root is selected, and $v = -4.6$ m/s, and the speed is 4.6 m/s.

FOLLOW-UP EXERCISE. From the 6.0-m height, how long did the Lander's descent take?

Chapter 2 Review

- **Motion** involves a change of position; it can be described in terms of the distance moved (a scalar) or the displacement (a vector).
- A **scalar** quantity has magnitude (value and units) only; a **vector** quantity has magnitude *and* direction.
- **Average speed** ($\bar{s}$) (a scalar) is the distance traveled divided by the total time:

$$\text{average speed} = \frac{\text{distance traveled}}{\text{total time to travel that distance}}$$
$$\bar{s} = \frac{d}{\Delta t} = \frac{d}{t_2 - t_1} \tag{2.1}$$

- **Average velocity** (a vector) is the displacement divided by the total travel time:

$$\text{average velocity} = \frac{\text{displacement}}{\text{total travel time}}$$
$$\bar{v} = \frac{\Delta x}{\Delta t} = \frac{x_2 - x_1}{t_2 - t_1} \quad \text{or} \quad x = x_0 + \bar{v}t \tag{2.3}$$

- **Instantaneous velocity** (a vector) describes how fast something is moving and in what direction at a particular instant of time.
- **Acceleration** is the time rate of change of velocity and hence is a vector quantity:

$$\text{average acceleration} = \frac{\text{change in velocity}}{\text{time to make the change}}$$

$$\bar{a} = \frac{\Delta v}{\Delta v} = \frac{v_2 - v_1}{t_2 - t_1} \tag{2.5}$$

- The kinematic equations for *constant* acceleration:

$$\bar{v} = \frac{v + v_0}{2} \tag{2.9}$$

$$v = v_0 + at \tag{2.8}$$

$$x = x_0 + \tfrac{1}{2}(v + v_0)t \tag{2.10}$$

$$x = x_0 + v_0 t + \tfrac{1}{2}at^2 \tag{2.11}$$

$$v^2 = v_0^2 + 2a(x - x_0) \tag{2.12}$$

- An object in **free fall** has a constant acceleration of magnitude $g = 9.80$ m/s^2 (acceleration due to gravity) near the surface of the Earth.

End of Chapter Questions and Exercises

Multiple Choice Questions

2.1 Distance and Speed: Scalar Quantities and

2.2 One-Dimensional Displacement and Velocity: Vector Quantities

1. A scalar quantity has (a) only magnitude, (b) only direction, (c) both magnitude and direction.
2. Which of the following is always true about the magnitude of the displacement: (a) it is greater than the distance traveled; (b) it is equal to the distance traveled; (c) it is less than the distance traveled; or (d) it is less than or equal to the distance traveled?
3. A vector quantity has (a) only magnitude, (b) only direction, (c) both direction and magnitude.
4. What can be said about average speed relative to the magnitude of the average velocity? (a) greater than, (b) equal to, (c) less than, (d) both (a) and (b).
5. Distance is to displacement as (a) a scalar is to a scalar, (b) a vector is to a scalar, (c) a scalar is to a vector, (d) a vector is to a vector.

2.3 Acceleration

6. On a position-versus-time plot for an object that has a constant acceleration, the graph is (a) a horizontal line, (b) a non-horizontal and non-vertical straight line, (c) a vertical line, (d) a curve.
7. An acceleration may result from (a) an increase in speed, (b) a decrease in speed, (c) a change of direction, (d) all of the preceding.
8. A negative acceleration can cause (a) an increase in speed, (b) a decrease in speed, (c) either (a) or (b).

9. The gas pedal of an automobile is commonly referred to as the *accelerator*. Which of the following might also be called an accelerator: (a) the brakes, (b) the steering wheel, (c) the gear shift, or (d) all of the preceding?

10. For a constant acceleration, what changes uniformly? (a) acceleration, (b) velocity, (c) displacement, (d) distance.

11. Which one of the following is true for a deceleration? (a) The velocity remains constant. (b) The acceleration is negative. (c) The acceleration is in the direction opposite to the velocity. (d) The acceleration is zero.

2.4 Kinematic Equations (Constant Acceleration)

12. For a constant linear acceleration, the velocity-versus-time graph is (a) a horizontal line, (b) a vertical line, (c) a non-horizontal and non-vertical straight line, (d) a curved line.

13. For a constant linear acceleration, the position-versus-time graph would be (a) a horizontal line, (b) a vertical line, (c) a non-horizontal and non-vertical straight line, (d) a curve.

2.5 Free Fall

14. An object is thrown vertically upward. Which of the following statements is true: (a) its velocity changes nonuniformly; (b) its maximum height is independent of the initial velocity; (c) its travel time upward is slightly greater than its travel time downward; or (d) its speed on returning to its starting point is the same as its initial speed?

15. The free-fall motion described in this section applies to (a) an object dropped from rest, (b) an object thrown vertically downward, (c) an object thrown vertically upward, (d) all of the preceding.

16. A dropped object in free fall (a) falls 9.8 m each second, (b) falls 9.8 m during the first second, (c) has an increase in speed of 9.8 m/s each second, (d) has an increase in acceleration of 9.8 m/s^2 each second.

17. An object is thrown straight upward. At its maximum height, (a) its velocity is zero, (b) its acceleration is zero, (c) both (a) and (b), (d) neither (a) and (b).

18. When an object is thrown vertically upward, it is accelerating on (a) the way up, (b) the way down, (c) both (a) and (b), (d) neither (a) and (b).

Conceptual Questions

2.1 Distance and Speed: Scalar Quantities
and
2.2 One-Dimensional Displacement and Velocity: Vector Quantities

1. Can the displacement of a person's trip be zero, yet the distance involved in the trip be nonzero? How about the reverse situation? Explain.

2. You are told that a person has walked 750 m. What can you safely say about the person's final position relative to the starting point?

3. If the displacement of an object is 300 m north, what can you say about the distance traveled by the object?

4. Speed is the magnitude of velocity. Is average speed the magnitude of average velocity? Explain.

5. The average velocity of a jogger on a straight track is computed to be +5 km/h. Is it possible for the jogger's instantaneous velocity to be negative at any time during the jog? Explain.

2.3 Acceleration

6. A car is traveling at a constant speed of 60 mi/h on a circular track. Is the car accelerating? Explain.

7. Does a fast-moving object always have greater acceleration than a slower object? Give a few examples, and explain.

8. A classmate states that a negative acceleration always means that a moving object is decelerating. Is this statement true? Explain.

9. Describe the motions of the two objects that have the velocity-versus-time plots shown in ▼ **Figure 2.19**.

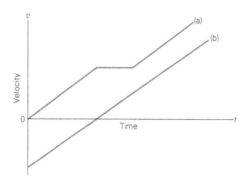

▲ FIGURE 2.19 **Description of motion** See Conceptual Question 9.

10. An object traveling at a constant velocity v_0 experiences a constant acceleration in the same direction for a period of time t. Then an acceleration of equal magnitude is experienced in the opposite direction of v_0 for the same period of time t. What is the object's final velocity?

2.4 Kinematic Equations (Constant Acceleration)

11. If an object's velocity-versus-time graph is a horizontal line, what can you say about the object's acceleration?

12. How many variables must be known to solve a kinematic equation?

13. Consider Equation 2.12, $v^2 = v_0^2 + 2a(x - x_0)$. An object starts from rest ($v_0 = 0$) and accelerates. Since v is squared and therefore always positive, can the acceleration be negative? Explain.

2.5 Free Fall

14. When a ball is thrown upward, what are its velocity and acceleration at its highest point?

15. If the instantaneous velocity of an object is zero, is the acceleration necessarily zero?

16. Imagine you are in space far away from any planet, and you throw a ball as you would on the Earth. Describe the ball's motion.

17. A person drops a stone from the window of a building. One second later, she drops another stone. How does the distance between the stones vary with time?

Exercises*

Integrated Exercises (IEs) are two-part exercises. The first part typically requires a conceptual answer choice based on physical thinking and basic principles. The following part is quantitative calculations associated with the conceptual choice made in the first part of the exercise.

2.1 Distance and Speed: Scalar Quantities
and
2.2 One-Dimensional Displacement and Velocity: Vector Quantities

1. • What is the magnitude of the displacement of a car that travels half a lap along a circle that has a radius of 150 m? How about when the car travels a full lap?

2. • A motorist travels 80 km at 100 km/h, and 50 km at 75 km/h. What is the average speed for the trip?

3. • An Olympic sprinter can run 100 yards in 9.0 s. At the same rate, how long would it take the sprinter to run 100 m?

4. •• A hospital nurse walks 25 m to a patient's room at the end of the hall in 0.50 min. She talks with the patient for 4.0 min, and then walks back to the nursing station at the same rate she came. What was the nurse's average speed?

5. •• A train goes one way and back on a straight, level track. The first half of the trip is 300 km and is traveled at a speed of 75 km/h. After a 0.50 h layover, the train returns the 300 km at a speed of 85 km/h. What is the train's (a) average speed and (b) average velocity?

6. IE •• A car travels three-quarters of a lap on a circular track of radius R. (a) The magnitude of the displacement is (1) less than R, (2) greater than R, but less than $2R$, (3) greater than $2R$. (b) If $R = 50$ m, what is the magnitude of the displacement?

7. •• The interstate distance between two cities is 150 km. (a) If you drive the distance at the legal speed limit of 65 mi/h, how long would the trip take? (b) Suppose on the return trip you pushed it up to 80 mi/h (and didn't get caught). How much time would you save?

8. IE •• A race car travels a complete lap on a circular track of radius 500 m in 50 s. (a) The average velocity of the race car is (1) zero, (2) 100 m/s, (3) 200 m/s, (4) none of the preceding. Why? (b) What is the average speed of the race car?

9. IE •• A student runs 30 m east, 40 m north, and 50 m west. (a) The magnitude of the student's net displacement is (1) between 0 and 20 m, (2) between 20 m and 40 m, (3) between 40 m and 60 m. (b) What is his net displacement?

10. •• A student throws a ball vertically upward such that it travels 7.1 m to its maximum height. If the ball is caught at the initial height 2.4 s after being thrown, (a) what is the ball's average speed, and (b) what is its average velocity?

11. •• An insect crawls along the edge of a rectangular swimming pool of length 27 m and width 21 m (▼ **Figure 2.20**). If it crawls from corner A to corner B in 30 min, (a) what is its average speed, and (b) what is the magnitude of its average velocity?

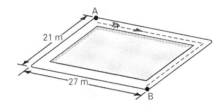

▲ FIGURE 2.20 **Speed versus velocity** See Exercise 11. (Not drawn to scale; insect is displaced for clarity.)

12. •• A plot of position versus time is shown in ▼ **Figure 2.21** for an object in linear motion. (a) What are the average velocities for the segments AB, BC, CD, DE, EF, FG, and BG? (b) State whether the motion is uniform or nonuniform in each case. (c) What is the instantaneous velocity at point D?

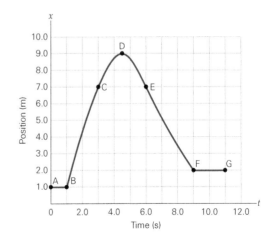

▲ FIGURE 2.21 **Position versus time** See Exercise 12.

13. •• In demonstrating a dance step, a person moves in one dimension, as shown in ▶ **Figure 2.22**. What are (a) the average speed and (b) the average velocity for each phase

of the motion? (c) What are the instantaneous velocities at $t = 1.0$ s, 2.5 s, 4.5 s, and 6.0 s? (d) What is the average velocity for the interval between $t = 4.5$ s and $t = 9.0$ s? [*Hint:* Recall that the overall displacement is the displacement between the starting point and the ending point.]

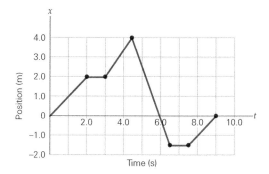

▲ **FIGURE 2.22** **Position versus time** See Exercise 13.

14. •• A high school kicker makes a 30.0-yd field goal attempt (in American football) and hits the crossbar at a height of 10.0 ft. (a) What is the net displacement of the football from the time it leaves the ground until it hits the crossbar? (b) Assuming the football took 2.50 s to hit the crossbar, what was its average velocity? (c) Explain why you *cannot* determine its average speed from these data.

15. •• The location of a moving particle at a particular time is given by $x = at - bt^2$, where $a = 10$ m/s and $b = 0.50$ m/s². (a) Where is the particle at $t = 0$? (b) What is the particle's displacement for the time interval $t_1 = 2.0$ s and $t_2 = 4.0$ s?

16. •• The displacement of an object is given as a function of time by $x = 3t^2$ m. What is the magnitude of the average velocity for (a) $\Delta t = 2.0$ s − 0, and (b) $\Delta t = 4.0$ s − 2.0 s?

17. ••• A student driving home for the holidays starts at 8:00 AM to make the 675-km trip, practically all of which is on nonurban interstate highways. If she wants to arrive home no later than 3:00 PM, what must be her minimum average speed? Will she have to exceed the 65-mi/h speed limit?

18. ••• A regional airline flight consists of two legs with an intermediate stop. The airplane flies 400 km due north from airport A to airport B. From there, it flies 300 km due east to its final destination at airport C. (a) What is the plane's displacement from its starting point? (b) If the first leg takes 45 min and the second leg 30 min, what is the average velocity for the trip? (c) What is the average speed for the trip? (d) Why is the average speed not the same as the magnitude for the average velocity?

2.3 Acceleration

19. • An automobile traveling at 15.0 km/h along a straight, level road accelerates to 65.0 km/h in 6.00 s. What is the magnitude of the auto's average acceleration?

20. • A sports car can accelerate from 0 to 60 mi/h in 3.9 s. What is the magnitude of the average acceleration of the car in meters per second squared?

21. • If the sports car in Exercise 20 can accelerate at a rate of 7.2 m/s², how long does the car take to accelerate from 0 to 60 mi/h?

22. IE •• A couple is traveling by car down a straight highway at 40 km/h. They see an accident in the distance, so the driver applies the brakes, and in 5.0 s the car uniformly slows down to rest. (a) The direction of the acceleration vector is (1) in the same direction as, (2) opposite to, (3) at 90° relative to the velocity vector. Why? (b) By how much must the velocity change each second from the start of braking to the car's complete stop?

23. •• A paramedic drives an ambulance at a constant speed of 75 km/h on a straight street for ten city blocks. Because of heavy traffic, the driver slows to 30 km/h in 6.0 s and travels two more blocks. What was the average acceleration of the vehicle?

24. •• After landing, a jetliner on a straight runway taxis to a stop at an average velocity of −35.0 km/h. If the plane takes 7.00 s to come to rest, what are the plane's initial velocity and acceleration?

25. •• What is the acceleration for each graph segment in ▼ **Figure 2.23**? Describe the motion of the object over the total time interval.

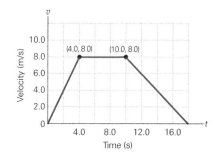

▲ **FIGURE 2.23** **Velocity versus time** See Exercises 25 and 43.

26. •• ▶ **Figure 2.24** shows a plot of velocity versus time for an object in linear motion. (a) Compute the acceleration for each phase of motion. (b) Describe how the object moves during the last time segment.

27. •• A car initially traveling to the right at a steady speed of 25 m/s for 5.0 s applies its brakes and slows at a constant rate of 5.0 m/s² for 3.0 s. It then continues traveling to the right at a steady but slower speed with no additional braking for another 6.0 s. (a) To help with

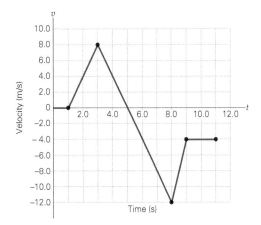

▲ **FIGURE 2.24 Velocity versus time** See Exercises 26 and 45.

the calculations, make a sketch of the car's velocity versus time, being sure to show all three time intervals. (b) What is its velocity after the 3.0 s of braking? (c) What was its displacement during the total 14.0 s of its motion? (d) What was its average speed for the 14.0 s?

28. ••• A train normally travels at a uniform speed of 72 km/h on a long stretch of straight, level track. On a particular day, the train must make a 2.0-min stop at a station along this track. If the train decelerates at a uniform rate of 1.0 m/s² and, after the stop, accelerates at a rate of 0.50 m/s², how much time is lost because of stopping at the station?

2.4 Kinematic Equations (Constant Acceleration)

29. • At a sports car rally, a car starting from rest accelerates uniformly at a rate of 9.0 m/s² over a straight-line distance of 100 m. The time to beat in this event is 4.5 s. Does the driver beat this time? If not, what must the minimum acceleration be to do so?

30. • A car accelerates from rest at a constant rate of 2.0 m/s² for 5.0 s. (a) What is the speed of the car at the end of that time? (b) How far does the car travel in this time?

31. • A car traveling at 25 mi/h is to stop on a 35-m-long shoulder of the road. (a) What is the required magnitude of the minimum acceleration? (b) How much time will elapse during this minimum deceleration until the car stops?

32. • A motorboat traveling on a straight course slows uniformly from 60 km/h to 40 km/h in a distance of 50 m. What is the boat's acceleration?

33. •• The driver of a pickup truck going 100 km/h applies the brakes, giving the truck a uniform deceleration of 6.50 m/s² while it travels 20.0 m. (a) What is the speed of the truck in kilometers per hour at the end of this distance? (b) How much time has elapsed?

34. •• A roller coaster car traveling at a constant speed of 20.0 m/s on a level track comes to a straight incline with a constant slope. While going up the incline, the car has a constant acceleration of 0.750 m/s² in

magnitude. (a) What is the speed of the car at 10.0 s on the incline? (b) How far has the car traveled up the incline at this time?

35. •• A rocket car is traveling at a constant speed of 250 km/h on a salt flat. The driver gives the car a reverse thrust, and the car experiences a continuous and constant deceleration of 8.25 m/s². How much time elapses until the car is 175 m from the point where the reverse thrust is applied? Describe the situation for your answer.

36. •• Two identical cars capable of accelerating at 3.00 m/s² are racing on a straight track with running starts. Car A has an initial speed of 2.50 m/s; car B starts with speed of 5.0 m/s. (a) What is the separation of the two cars after 10 s? (b) Which car is moving faster after 10 s?

37. •• According to Newton's laws of motion (which will be studied in Chapter 4), a frictionless 30° incline should provide an acceleration of 4.90 m/s² down the incline. A student with a stopwatch finds that an object, starting from rest, slides down a 15.00-m very smooth incline in exactly 3.00 s. Is the incline frictionless?

38. IE •• An object moves in the +x-direction at a speed of 40 m/s. As it passes through the origin, it starts to experience a constant acceleration of 3.5 m/s² in the −x-direction. (a) What will happen next? (1) The object will reverse its direction of travel at the origin; (2) the object will keep traveling in the +x-direction; (3) the object will travel in the +x-direction and then reverse its direction. Why? (b) How much time elapses before the object returns to the origin? (c) What is the velocity of the object when it returns to the origin?

39. •• A rifle bullet with a muzzle speed of 330 m/s is fired directly into a special dense material that stops the bullet in 25.0 cm. Assuming the bullet's deceleration to be constant, what is its magnitude?

40. •• The speed limit in a school zone is 40 km/h (about 25 mi/h). A driver traveling at this speed sees a child run onto the road 13 m ahead of his car. He applies the brakes, and the car decelerates at a uniform rate of 8.0 m/s². If the driver's reaction time is 0.25 s, will the car stop before hitting the child?

41. •• Assuming a reaction time of 0.50 s for the driver in Exercise 40, will the car stop before hitting the child?

42. •• A bullet traveling horizontally at a speed of 350 m/s hits a board perpendicular to its surface, passes through and emerges on the other side at a speed of 210 m/s. If the board is 4.00 cm thick, how long does the bullet take to pass through it?

43. •• (a) Show that the area under the curve of a velocity-versus-time plot for a constant acceleration is equal to the displacement. [*Hint:* The area of a triangle is *ab*/2, or one-half the altitude times the base.] (b) Compute the displacement traveled for the motion represented by Figure 2.23.

44. •• An object initially at rest experiences an acceleration of 1.5 m/s² for 6.0 s and then travels at that constant velocity for another 8.0 s. What is the object's average velocity over the 14-s interval?

45. ••• Figure 2.24 shows a plot of velocity versus time for an object in linear motion. (a) What are the instantaneous velocities at $t = 8.0$ s and $t = 11.0$ s? (b) Compute the final displacement of the object. (c) Compute the total distance the object travels.

46. IE ••• (a) A car traveling at a speed of v can brake to an emergency stop in a distance x. Assuming all other driving conditions are similar, if the traveling speed of the car doubles, the stopping distance will be (1) $\sqrt{2}x$, (2) $2x$, (3) $4x$. (b) A driver traveling at 40.0 km/h in a school zone can brake to an emergency stop in 3.00 m. What would be the braking distance if the car were traveling at 60.0 km/h?

2.5 Free Fall

47. • A student drops a ball from the top of a tall building; the ball takes 2.8 s to reach the ground. (a) What was the ball's speed just before hitting the ground? (b) What is the height of the building?

48. IE • The time it takes for an object dropped from the top of cliff A to hit the water in the lake below is twice the time it takes for another object dropped from the top of cliff B to reach the lake. (a) The height of cliff A is (1) one-half, (2) two times, (3) four times that of cliff B. (b) If it takes 1.80 s for the object to fall from cliff A to the water, what are the heights of cliffs A and B?

49. • For the motion of a dropped object in free fall, sketch the general forms of the graphs of (a) v versus t and (b) y versus t.

50. • You can perform a popular trick by dropping a dollar bill (lengthwise) through the thumb and forefinger of a fellow student. Tell your fellow student to grab the dollar bill as fast as possible, and he or she can have the dollar if able to catch it. (The length of a dollar is 15.7 cm, and the average human reaction time is about 0.20 s. See Figure 2.17.) Is this proposal a good deal? Justify your answer.

51. • A juggler tosses a ball vertically a certain distance. How much higher must the ball be tossed so as to spend twice as much time in the air?

52. • A boy throws a stone straight upward with an initial speed of 15.0 m/s. What maximum height will the stone reach before falling back down?

53. • In Exercise 52, what would be the maximum height of the stone if the boy and the stone were on the surface of the Moon, where the acceleration due to gravity is only one-sixth of that of the Earth's?

54. •• The Petronas Twin Towers in Malaysia and the Chicago Sears Tower have heights of about 452 m and 443 m, respectively. If objects were dropped from the top of each, what would be the difference in the time it takes the objects to reach the ground?

55. •• In an air bag test, a car traveling at 100 km/h is remotely driven into a brick wall. Suppose an identical car is dropped onto a hard surface. From what height would the car have to be dropped to have the same impact as that with the brick wall?

56. •• You throw a stone vertically upward with an initial speed of 6.0 m/s from a third-story office window. If the window is 12 m above the ground, find (a) the time the stone is in flight and (b) the speed of the stone just before it hits the ground.

57. IE •• A Super Ball is dropped from a height of 4.00 m. Assuming the ball rebounds with 95% of its impact speed, (a) the ball would bounce to (1) less than 95%, (2) equal to 95%, or (3) more than 95% of the initial height? (b) How high will the ball go?

58. •• In ▼ **Figure 2.25**, a student at a window on the second floor of a dorm sees his math professor walking on the sidewalk beside the building. He drops a water balloon from 18.0 m above the ground when the professor is 1.0 m from the point directly beneath the window. If the professor is 1.70 m tall and walks at a rate of 0.450 m/s, does the balloon hit her? If not, how close does it come?

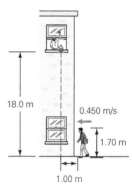

▲ FIGURE 2.25 **Hit the professor** See Exercise 58. (This figure is not drawn to scale.)

59. •• A photographer in a helicopter ascending vertically at a constant rate of 12.5 m/s accidentally drops a camera out the window when the helicopter is 60.0 m above the ground. (a) How long will the camera take to reach the ground? (b) What will its speed be when it hits?

60. IE •• The acceleration due to gravity on the Moon is about one-sixth of that on the Earth. (a) If an object were dropped from the same height on the Moon and on the Earth, the time it would take to reach the surface on the Moon is (1) $\sqrt{6}$, (2) 6, or (3) 36 times the time it would take on the Earth. (b) For a projectile with an initial velocity of 18.0 m/s upward, what would be the maximum height and the total time of flight on the Moon and on the Earth?

61. ••• It takes 0.210 s for a dropped object to pass a window that is 1.35 m tall. From what height above the top of the window was the object released? (See ▼ **Figure 2.26**.)

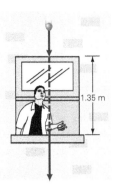

▲ **FIGURE 2.26**   **From where did it come?** See Exercise 61.

62. ••• A tennis ball is dropped from a height of 10.0 m. It rebounds off the floor and comes up to a height of only 4.00 m on its first rebound. (Ignore the small amount of time the ball is in contact with the floor.) (a) Determine the ball's speed just before it hits the floor on the way down. (b) Determine the ball's speed as it leaves the floor on its way up to its first rebound height. (c) How long is the ball in the air from the time it is dropped until the time it reaches its maximum height on the first rebound?

63. ••• A car and a motorcycle start from rest at the same time on a straight track, but the motorcycle is 25.0 m behind the car (▼ **Figure 2.27**). The car accelerates at a uniform rate of 3.70 m/s^2 and the motorcycle at a uniform rate of 4.40 m/s^2. (a) How much time elapses before the motorcycle overtakes the car? (b) How far will each have traveled during that time? (c) How far ahead of the car will the motorcycle be 2.00 s later? (Both vehicles are still accelerating.)

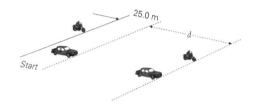

▲ **FIGURE 2.27**   **A tie race** See Exercise 63. (This figure is not drawn to scale.)

3

Motion in Two Dimensions*

Sailboats can sail into the wind using a zigzag pattern.

Can you sail against the wind? The answer is yes, as shown in the chapter-opening photo. It is called *tacking* by the wise use of vector components. Clearly you cannot sail directly into the wind. However, by using a zigzag path, a sailboat can achieve a net direct path into the wind.

Motion doesn't have to be in a straight line. As will be learned shortly, a generalized version of vectors introduced in Section 2.2 can be used to describe motion in curved paths as well. Such analysis of *curvilinear* motion will eventually allow you to analyze the behavior of batted balls, planets circling the Sun, and even the motions of electrons in atoms.

Two-dimensional curvilinear motion can be analyzed by using rectangular components of motion. Essentially, the curved motion is broken down or resolved into rectangular (x and y) components so the motion can be considered linearly in both dimensions. The kinematic equations introduced in Chapter 2 can be applied to these components. For an object moving in a curved path, for example, the x- and y-coordinates of its motion will give the object's position at any time.

---

* The mathematics needed in this chapter involves trigonometric functions. You may want to review these in Appendix I.

3.1 Components of Motion

In Section 2.1, an object moving in a straight line was considered to be moving along one of the Cartesian axes (*x* or *y*). But what if the motion is not along an axis? For example, consider the situation illustrated in ▼ **Figure 3.1**. Here, three balls are moving uniformly across a tabletop. The ball rolling in a straight line along the side of the table, designated as the *x*-direction, is moving in one dimension. That is, its motion can be described with a single coordinate, *x*. Similarly, the motion of the ball rolling along the end of the table in the *y*-direction can be described by a single *y*-coordinate. However, for this coordinate choice, both *x*- and *y*-coordinates are needed to describe the motion of the ball rolling diagonally across the table; that is, the motion is described *in two dimensions*.

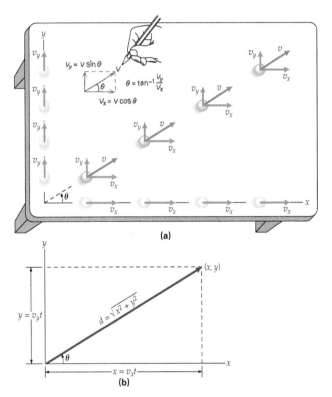

(a)

(b)

▲ **FIGURE 3.1 Components of motion (a)** The velocity (and displacement) for uniform, straight-line motion – that of the diagonally moving ball – may have *x*- and *y*-components (v_x and v_y as shown in the pencil drawing), because of the chosen orientation of the coordinate axes. Note that the velocity and displacement of the ball in the *x*-direction are exactly the same as those that a ball rolling along the *x*-axis with a uniform velocity of v_x would have. A comparable relationship holds true for the ball's motion in the *y*-direction. Since the motion is uniform, the ratio v_y/v_x (and therefore *θ*) is constant. **(b)** The coordinates (*x*, *y*) of the ball's position and the distance *d* the ball has traveled from the origin can be found at any time *t*.

You might observe that if the diagonally moving ball were the only object to consider, the *x*-axis could be chosen to be in the direction of that ball's motion, and the motion would thereby

be reduced to one dimension. This observation is true, but once the coordinate axes are fixed, motions not along the axes must be described with two coordinates (*x*, *y*), or in two dimensions. Also, keep in mind that not all motions in a plane (two dimensions) are in straight lines. Think about the path of a ball you toss to another person. The path is curved for such projectile motion. (This motion will be considered in Section 3.3.) In general, both coordinates are needed.

In considering the motion of the ball moving diagonally across the table in Figure 3.1a, it can be thought of as moving in the *x*- and *y*-directions simultaneously. That is, it has a velocity in the *x*-direction (v_x) and a velocity in the *y*-direction (v_y) at the same time. The combined velocity components describe the actual motion of the ball. If the ball has a constant velocity *v* in a direction at an angle *θ* relative to the *x*-axis, then the velocities in the *x*- and *y*-directions are obtained by resolving, or breaking down, the velocity vector into components of motion in these directions (see the pencil drawing in Figure 3.1a). As this drawing shows, the v_x and v_y components have magnitudes of

$$v_x = v\cos\theta \tag{3.1a}$$

and

$$v_y = v\sin\theta \tag{3.1b}$$

(velocity components with *θ* relative to +*x*-axis)

(Notice that the speed $v = \sqrt{v_x^2 + v_y^2}$, so it is a combination of the velocities in the *x*- and *y*-directions.)

You are familiar with the use of two-dimensional length components in finding the *x*- and *y*-coordinates in a Cartesian system. For the ball rolling on the table, its position (*x*, *y*), or the distance traveled from the origin in each of the component directions at time *t*, is given by (Equation 2.11 with *a* = 0)

$$x = x_\mathrm{o} + v_x t \tag{3.2a}$$

$$y = y_\mathrm{o} + v_y t \tag{3.2b}$$

(under constant velocity thus zero acceleration)

respectively. (Here, the x_o and y_o are the ball's coordinates at *t* = 0, which may be other than zero.) The ball's straight-line distance from the origin at any given time is then $d = \sqrt{x^2 + y^2}$ (Figure 3.1b).

Note that $\tan\theta = v_y/v_x$ (see Figure 3.1a). So the direction of the motion relative to the *x*-axis is given by $\theta = \tan^{-1}(v_y/v_x)$. Also, $\theta = \tan^{-1}(y/x)$.

In this introduction to components of motion, the velocity vector has been taken to be in the first quadrant (0 < *θ* < 90°), where both the *x*- and *y*-components are positive. But, as will be shown in more detail in the next section, vectors may be in any quadrant, and one or both of their components can be negative. Can you tell in which quadrants the v_x- and v_y-components would both be negative?

EXAMPLE 3.1: ON A ROLL – USING COMPONENTS OF MOTION

If the diagonally moving ball in Figure 3.1a has a constant velocity of 0.50 m/s at an angle of 37° relative to the *x*-axis, find how far it travels in 3.0 s by using *x*- and *y*-components of its motion.

THINKING IT THROUGH. Given the magnitude and direction (angle) of the velocity of the ball, the *x*- and *y*-components of the velocity can be found. Then the distance in each direction can be computed. Since the *x*- and *y*-axes are at right angles to each other, the Pythagorean theorem gives the distance of the straight-line path of the ball, as shown in Figure 3.1b. (Note the procedure: Separate the motion into components, calculate what is needed in each direction, and recombine if necessary.)

SOLUTION

Listing the data,

Given:

$v = 0.50$ m/s
$\theta = 37°$
$t = 3.0$ s

Find:

d (distance traveled)

The distance traveled by the ball in terms of its *x*- and *y*-components is given by $d = \sqrt{x^2 + y^2}$. To find *x* and *y* as given by Equation 3.2, we first need to compute the velocity components v_x and v_y (Equation 3.1):

$$v_x = v\cos 37° = (0.50\,\text{m/s})(0.80) = 0.40\,\text{m/s}$$
$$v_y = v\sin 37° = (0.50\,\text{m/s})(0.60) = 0.30\,\text{m/s}$$

Then, taking $x_o = 0$ and $y_o = 0$, the component distances are

$$x = v_x t = (0.40\,\text{m/s})(3.0\,\text{s}) = 1.2\,\text{m}$$

and

$$y = v_y t = (0.30\,\text{m/s})(3.0\,\text{s}) = 0.90\,\text{m}$$

and the distance of the path is

$$d = \sqrt{x^2 + y^2} = \sqrt{(1.2\,\text{m})^2 + (0.90\,\text{m})^2} = 1.5\,\text{m}$$

FOLLOW-UP EXERCISE. Suppose that a ball is rolling diagonally across a table with the same speed as in this Example, but from the lower right corner, which is taken as the origin of the coordinate system, toward the upper left corner at an angle of 37° relative to the $-x$-axis. What would be the velocity components in this case? (Would the distance change?) *(Answers to all Follow-Up Exercises are given in Appendix V at the back of the book.)*

3.1.1 Problem-Solving Hint

For this simple case, the distance can also be obtained directly from $d = vt = (0.50\text{ m/s})(3.0\text{ s}) = 1.5$ m. However, this Example was solved in a more general way to illustrate the use of components of motion. The direct solution would have been evident if the equations had been combined algebraically before calculation, that is, as

$$x = v_x t = (v\cos\theta)t$$

and

$$y = v_y t = (v\sin\theta)t$$

from which it follows that

$$d = \sqrt{x^2 + y^2} = \sqrt{(v\cos\theta)^2 t^2 + (v\sin\theta)^2 t^2}$$
$$= \sqrt{v^2 t^2 (\cos^2\theta + \sin^2\theta)} = vt$$

Before embarking on the first solution strategy that occurs to you, pause for a moment to see whether there might be an easier or more direct way of approaching the problem.

3.1.2 Kinematic Equations for Components of Motion

Example 3.1 involved two-dimensional motion in a plane. With a constant velocity (constant components v_x and v_y), the motion is in a straight line. The motion may also be accelerated. For motion in a plane with a *constant acceleration* that has components a_x and a_y, the displacement and velocity components are given by the kinematic equations of Section 2.4 written separately for the *x*- and *y*-directions:

$$x = x_o + v_{x_o}t + \frac{1}{2}a_x t^2 \tag{3.3a}$$

$$y = y_o + v_{y_o}t + \frac{1}{2}a_y t^2 \tag{3.3b}$$

$$v_x = v_{x_o} + a_x t \tag{3.3c}$$

$$v_y = v_{y_o} + a_y t \tag{3.3d}$$

(for constant acceleration only)

If an object is initially moving with a constant velocity and suddenly experiences an acceleration in the direction of the velocity or opposite to it, it will continue in a straight-line path, either speeding up or slowing down, respectively.

If, however, the acceleration is at some angle other than 0° or 180° to the velocity vector, the motion will be along a curved path. For the motion of an object to be *curvilinear* – that is, to vary from a straight-line path – an acceleration not parallel to the velocity is required. For such a curved path, the ratio of the velocity components varies with time. That is, the direction of the motion, $\theta = \tan^{-1}(v_y/v_x)$, varies with time, because one or both of the velocity components do.

Consider a ball initially moving along the *x*-axis, as illustrated in ▶ **Figure 3.2**. Assume that, starting at a time $t_o = 0$, the ball

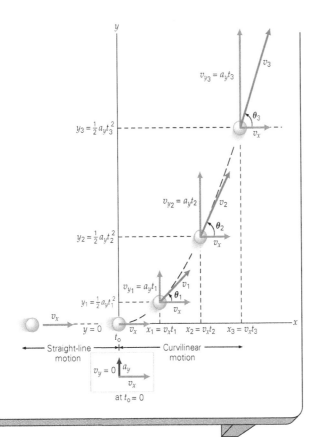

▲ **FIGURE 3.2 Curvilinear motion** An acceleration not parallel to the instantaneous velocity produces a curved path. Here, an acceleration a_y is applied at $t_o = 0$ to a ball initially moving with a constant velocity v_x. The result is a curved path with the velocity components as shown. Notice how v_y increases with time, while v_x remains constant.

receives a constant acceleration a_y in the y-direction. The magnitude of the x-component of the ball's displacement is given by $x = v_x t$, where the $\frac{1}{2} a_x t^2$ term of Equation 3.3a drops out because there is no acceleration in the x-direction ($a_x = 0$). Prior to t_o, the motion was in a straight line along the x-axis. But at any time after t_o, the y-coordinate is not zero, but is given by $y = \frac{1}{2} a_y t^2$ (Equation 3.3b with $y_o = 0$ and $v_{y_o} = 0$). The result is a *curved* path for the ball.

Note that the length (magnitude) of the velocity component v_y changes with time, while that of the v_x component remains constant. The total velocity vector *at any time* is tangent to the curved path of the ball. It is at an angle θ relative to the $+x$-axis, given by $\theta = \tan^{-1}(v_y/v_x)$, which now changes with time, as can be seen in Figure 3.2 and in Example 3.2.

EXAMPLE 3.2: A CURVING PATH – VECTOR COMPONENTS

Suppose that the ball in Figure 3.2 has an initial velocity of 1.50 m/s along the x-axis. Starting at $t_o = 0$, the ball receives an acceleration of 2.80 m/s² in the y-direction. (a) What is the position of the ball 3.00 s after t_o? (b) What is the velocity of the ball at that time?

THINKING IT THROUGH. Keep in mind that the motions in the x- and y-directions can be analyzed independently. For part (a), simply compute the x- and y-positions at the given time, taking into account the acceleration in the y-direction. For part (b), find the component velocities, and vectorially combine them to get the total velocity.

SOLUTION

Referring to Figure 3.2,

Given:

$$v_{x_o} = v_x = 1.50 \text{ m/s}$$
$$v_{y_o} = 0$$
$$a_x = 0$$
$$a_y = 2.80 \text{ m/s}^2$$
$$t = 3.00 \text{ s}$$

Find:

(a) (x, y) (position coordinates)

(b) v (velocity, magnitude and direction)

(a) At 3.00 s after $t_o = 0$, Equations 3.3a and 3.3b tell us that the ball has traveled the following distances from the origin ($x_o = y_o = 0$) in the x- and y-directions:

$$x = v_{x_o} t + \frac{1}{2} a_x t^2 = (1.50 \text{ m/s})(3.00 \text{ s}) + 0 = 4.50 \text{ m}$$

$$y = v_{y_o} t + \frac{1}{2} a_y t^2 = 0 + \frac{1}{2}(2.80 \text{ m/s}^2)(3.00 \text{ s})^2 = 12.6 \text{ m}$$

Thus, the position of the ball is $(x, y) = (4.50 \text{ m}, 12.6 \text{ m})$. If you had computed the distance $d = \sqrt{x^2 + y^2}$, what would have been obtained? [This quantity is the magnitude of the *displacement,* or straight-line distance, from the origin to the $(x, y) = (4.50 \text{ m}, 12.6 \text{ m})$ position.]

(b) The x-component of the velocity is given by Equation 3.3c:

$$v_x = v_{x_o} + a_x t = 1.50 \text{ m/s} + 0 = 1.50 \text{ m/s}$$

(This component is constant, since there is no acceleration in the x-direction.) Similarly, the y-component of the velocity is given by Equation 3.3d:

$$v_y = v_{y_o} + a_y t = 0 + (2.80 \text{ m/s}^2)(3.00 \text{ s}) = 8.40 \text{ m/s}$$

The velocity therefore has a magnitude (speed) of

$$v = \sqrt{v_x^2 + v_y^2} = \sqrt{(1.50 \text{ m/s})^2 + (8.40 \text{ m/s})^2} = 8.53 \text{ m/s}$$

and its direction relative to the $+x$-axis is

$$\theta = \tan^{-1}\left(\frac{v_y}{v_x}\right) = \tan^{-1}\left(\frac{8.40 \text{ m/s}}{1.50 \text{ m/s}}\right) = 79.9°$$

FOLLOW-UP EXERCISE. Suppose that the ball in this Example also received an acceleration of 1.00 m/s² in the $+x$-direction starting at t_o. What would be the position of the ball 3.00 s after t_o in this case?

3.1.3 Problem-Solving Hint

When using the kinematic equations, it is important to note that motion in the x- and y-directions can be analyzed independently – the factor connecting them being time t. That is, you can find (x, y) and/or (v_x, v_y) at a given time t. Also, keep in mind that the initial positions are often set $x_o = 0$ and $y_o = 0$, which means that the object is located at the origin at $t_o = 0$. If the object is actually elsewhere at $t_o = 0$, then the values of x_o and/or y_o would have to be used in the appropriate equations. (See Equations 3.3a and 3.3b.)

3.2 Vector Addition and Subtraction

Many physical quantities, including those describing motion, have a direction associated with them – that is, they are vectors. You have already worked with a few such quantities related to motion (displacement, velocity, and acceleration) and will encounter more during the course of study. A very important technique in the analysis of many physical situations is the addition (and subtraction) of vectors. By adding or combining such quantities (vector addition), the resultant, or net, vector is obtained. This *resultant* vector is the *vector sum.*

You have already been adding vectors. In Section 2.2, displacements in one dimension were added to get the net displacement. In this chapter, vector components of motion in two dimensions will be added to get net effects. Notice that in Example 3.2, the velocity components v_x and v_y were combined to get the resultant velocity.

In this section, vector addition and subtraction in general, along with common vector notation, will be considered. As will be learned, these operations are not the same as scalar or numerical addition and subtraction, with which you are already familiar. Vectors have magnitudes *and* directions, so different rules apply.

In general, there are geometrical (graphical) methods and analytical (computational) methods of vector addition. The geometrical methods are useful in helping you visualize the concepts of vector addition, particularly with a quick sketch. Analytical methods are more commonly used; however, because they are faster and more precise.

Section 3.1 was chiefly about vector components. The notation for the magnitudes of components was, for example, v_x and v_y. To represent vectors, the notation $\vec{A}$ – a boldface symbol with an over arrow – will be used. The magnitude of vector $\vec{A}$ is simply A.

3.2.1 Vector Addition: Geometric Methods

3.2.1.1 Triangle Method

To add two vectors – say, to add $\vec{B}$ to $\vec{A}$ (i.e., to find $\vec{R} = \vec{A} + \vec{B}$) by the triangle method – you first draw $\vec{A}$ on a sheet of graph paper to some scale (▼**Figure 3.3a**). For example, if $\vec{A}$ represents a displacement in meters, a convenient scale is 1 cm : 1 m, or 1 cm of vector length on the graph corresponds to 1 m of displacement. As shown in Figure 3.3b, the direction of the $\vec{A}$ vector is specified as being at an angle θ_A relative to a coordinate axis, usually the x-axis.

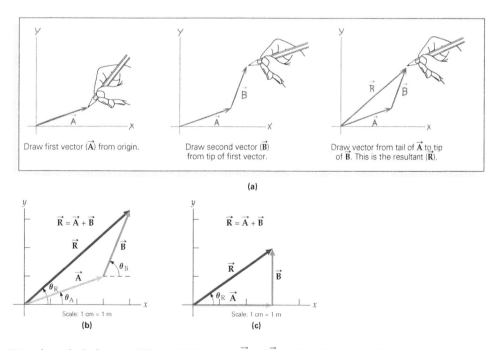

▲ FIGURE 3.3 **Triangle method of vector addition (a)** The vectors $\vec{A}$ and $\vec{B}$ are placed tip to tail. The vector that extends from the tail of $\vec{A}$ to the tip of $\vec{B}$, forming the third side of the triangle, is the resultant or sum $\vec{R} = \vec{A} + \vec{B}$. **(b)** When the vectors are drawn to scale, the magnitude of $\vec{R}$ can be found by measuring the length of $\vec{R}$ and using the scale conversion, and the direction angle θ_R can be measured with a protractor. Analytical methods can also be used. For a non-right triangle, as in part **(b)**, the laws of sines and cosines can be used to determine the magnitude of $\vec{R}$ and θ_R (Appendix I). **(c)** If the vector triangle is a right triangle, $\vec{R}$ is easily obtained via the Pythagorean theorem, and the direction angle is given by an inverse trigonometric function.

Next, draw $\vec{B}$ with its tail starting at the tip of $\vec{A}$. (Thus, this method is also called the *tip-to-tail method*.) The vector from the tail of $\vec{A}$ to the tip of $\vec{B}$ is then the vector sum $\vec{R}$, or the resultant of the two vectors: $\vec{R} = \vec{A} + \vec{B}$.

If the vectors are drawn to scale, the magnitude of $\vec{R}$ can be found by measuring its length and using the scale conversion. In such a graphical approach, the direction angle θ_R is measured with a protractor. If the magnitudes and directions (angles θ) of $\vec{A}$ and $\vec{B}$ are known, the magnitude and direction of $\vec{R}$ can be found analytically by using trigonometric methods. For the non-right triangle in Figure 3.3b, the laws of sines and cosines can be used. (See Appendix I.)

This tip-to-tail method can be extended to any number of vectors. The vector from the tail of the first vector to the tip of the last vector is the resultant or vector sum. For more than two vectors, it is called the *polygon method*.

If the two vectors are perpendicular to each other, the resultant vector is the hypotenuse of the right triangle in Figure 3.3c. It would be much easier to find using the Pythagorean theorem for the magnitude and an inverse trigonometric function to find the direction angle. Notice that $\vec{R}$ is made up of x- and y-vectors $\vec{A}$ and $\vec{B}$. Such x- and y-components are the basis of the convenient analytical component method, which will be discussed shortly.

3.2.1.2 Vector Subtraction

Vector subtraction is a special case of vector addition:

$$\vec{A} - \vec{B} = \vec{A} + (-\vec{B})$$

That is, to subtract $\vec{B}$ from $\vec{A}$, a *negative* $\vec{B}$ is added to $\vec{A}$. In Section 2.2, you learned that a negative sign simply means that the direction of a vector is opposite that of one with a positive sign. The same is true with vectors represented by boldface notation. The vector $-\vec{B}$ has the same magnitude as the vector $\vec{B}$, but is in the opposite direction (▼ **Figure 3.4**). The vector diagram in Figure 3.4 provides a graphical representation of $\vec{A} - \vec{B}$.

3.2.2 Vector Components and the Analytical Component Method

Probably the most widely used analytical method for adding multiple vectors is the **component method**. It will be used again and again throughout the course of our study, so a basic understanding of the method is *essential*. Learn this section well.

3.2.2.1 Adding Rectangular Vector Components

Rectangular components means that vector components are at right 90° angles to each other, usually taken in the rectangular coordinate x- and y-directions. You have already had an introduction to the addition of such components in the discussion of the velocity components of motion in Section 3.1. For a special case, suppose that $\vec{A}$ and $\vec{B}$, two vectors at right angles, are added, as illustrated in ▼ **Figure 3.5a**. The magnitude of $\vec{C}$ is given by the Pythagorean theorem:

$$C = \sqrt{A^2 + B^2} \tag{3.4a}$$

The orientation of $\vec{C}$ relative to the x-axis is given by the angle

$$\theta = \tan^{-1}\left(\frac{B}{A}\right) \tag{3.4b}$$

This notation is how a resultant is expressed in **magnitude-angle form**.

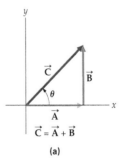

(a)

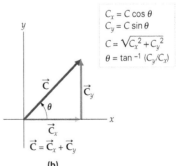

(b)

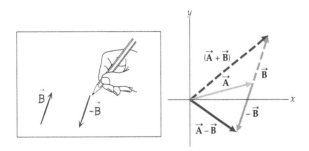

▲ **FIGURE 3.4 Vector subtraction** Vector subtraction is a special case of vector addition; that is, $\vec{A} - \vec{B} = \vec{A} + (-\vec{B})$, where $-\vec{B}$ has the same magnitude as $\vec{B}$, but is in the opposite direction. (See the sketch.) Thus, $\vec{A} + \vec{B}$ is not the same as $\vec{A} - \vec{B}$, in either length or direction. Can you show that $\vec{B} - \vec{A} = -(\vec{A} - \vec{B})$ geometrically?

▲ **FIGURE 3.5 Vector components (a)** The vectors $\vec{A}$ and $\vec{B}$ along the x- and y-axes, respectively, add to give $\vec{C}$. **(b)** A vector $\vec{C}$ may be resolved into rectangular components $\vec{C}_x$ and $\vec{C}_y$.

3.2.2.2 Resolving a Vector into Rectangular Components: Unit Vectors

Resolving a vector into rectangular components is essentially the reverse of adding the rectangular components of the vector. Given a vector $\vec{C}$, Figure 3.5b illustrates how it may be resolved into x- and y-vector components $\vec{C}_x$ and $\vec{C}_y$. Simply complete the vector triangle with x- and y-components. As the diagram shows, the magnitudes, or vector lengths, of these components are given by

$$C_x = C\cos\theta \qquad (3.5a)$$

$$C_y = C\sin\theta \qquad (3.5b)$$

respectively.* (This is similar to $v_x = v\cos\theta$ and $v_y = v\sin\theta$ in Example 3.1.) The angle of direction of $\vec{C}$ can also be expressed in terms of the components, since $\tan\theta = C_y/C_x$, or

$$\theta = \tan^{-1}\left(\frac{C_y}{C_x}\right) \qquad (3.6)$$

(direction of vector from magnitudes of components)

Another way of expressing the magnitude and direction of a vector involves the use of unit vectors. For example, as illustrated in ▼ **Figure 3.6**, a vector $\vec{A}$ can be written as $\vec{A} = A\hat{a}$. The numerical magnitude is represented by A, and $\hat{a}$ is a unit vector, which indicates direction. That is, $\hat{a}$ has a magnitude of unity, or one, with no units, and simply indicates a vector's direction. For example, a velocity along the x-axis can be written $\vec{v} = (4.0\,\text{m/s})\hat{x}$ (i.e., 4.0 m/s magnitude in the $+x$-direction).

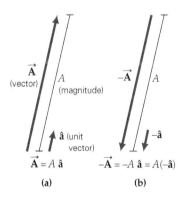

▲ **FIGURE 3.6** **Unit vectors** (a) A unit vector $\hat{a}$ has a magnitude of unity, or one, and thereby simply indicates a vector's direction. Written with the magnitude A, it represents the vector $\vec{A}$, and $\vec{A} = A\hat{a}$. (b) For the vector $-\vec{A}$, the unit vector is $-\hat{a}$, and $-\vec{A} = -A\hat{a} = A(-\hat{a})$.

Note in Figure 3.6 how $-\vec{A}$ would be represented in this notation. Although the negative sign is sometimes put in front

of the numerical magnitude, this quantity is an absolute number or value. The negative actually goes with the unit vector: $-\vec{A} = -A\hat{a} = A(-\hat{a})$. That is, the unit vector is in the $-\hat{a}$ direction (opposite $\hat{a}$). A velocity of $\vec{v} = (-4.0\,\text{m/s})\hat{x}$ has a magnitude of 4.0 m/s in the $-x$-direction, that is, $\vec{v} = (4.0\,\text{m/s})(-\hat{x})$.

This notation can be used to express explicitly the rectangular components of a vector. For example, the ball's displacement from the origin in Example 3.2 could be written $\vec{d} = (4.50\,\text{m})\hat{x} + (12.6\,\text{m})\hat{y}$, where $\hat{x}$ and $\hat{y}$ are unit vectors in the x- and y-directions. In some instances, it may be more convenient to express a general vector in this unit vector **component form**:

$$\vec{C} = C_x\hat{x} + C_y\hat{y} \qquad (3.7)$$

3.2.3 Vector Addition Using Components

The analytical component method of vector addition involves resolving the vectors into rectangular vector components and adding the components for each axis independently. This method is illustrated graphically in ▶ **Figure 3.7** for two vectors $\vec{F}_1$ and $\vec{F}_2$.[†] *The sums of the x- and y-component vectors being added are equal to the corresponding vector components of the resultant vector.*

The same principle applies if you are given three (or more) vectors to add. You could find the resultant by applying the graphical tip-to-tail method. However, this technique involves drawing the vectors to scale and using a protractor to measure angles, which can be time consuming. But in using the component method, you do not have to draw the vectors tip to tail. In fact, it is usually more convenient to put all of the tails together at the origin, as shown in ▶ **Figure 3.8a**. Also, the vectors do not have to be drawn to scale, since the approximate sketch is just a visual aid in applying the analytical method.

Basically, in the component method, the vectors to be added are resolved into their x- and y-components, and the respective components added and then recombined to find the resultant. The resultant of the three vectors in Figure 3.8a is shown in Figure 3.8b. By looking at the x-components, it can be seen that the vector sum of these components is in the $-x$-direction. Similarly, the sum of the y-components is in the $+y$-direction. (Note that $\vec{v}_2$ is in the y-direction and has a zero x-component, just as a vector in the x-direction would have a zero y-component.)

The x- and y-components of the resultant are $\vec{v}_x = \vec{v}_{x_1} + \vec{v}_{x_2} + \vec{v}_{x_3}$ and $\vec{v}_y = \vec{v}_{y_1} + \vec{v}_{y_2} + \vec{v}_{y_3}$. When the numerical values (with positive and negative signs to indicate directions) of the vector components are computed and put into

---

* Figure 3.5b illustrates only a vector in the first quadrant, but the equations hold for all quadrants when vectors are referenced to either the positive or negative x-axis. The directions of the components are indicated by $+$ and $-$ signs, as will be shown shortly.

[†] The symbol $\vec{F}$ is commonly used to denote force, a very important vector quantity that will be studied in Chapter 4. Here, $\vec{F}$ is employed as a general vector, but its use provides familiarity with the notation used in the next chapter, where knowledge of the addition of forces is essential.

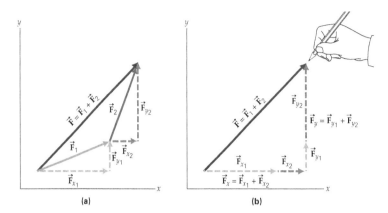

▲ **FIGURE 3.7** **Component addition (a)** In adding vectors by the component method, each vector is first resolved into its *x*- and *y*-component vectors. **(b)** The sums of the *x*- and *y*-components of vectors $\vec{\mathbf{F}}_1$ and $\vec{\mathbf{F}}_2$ are $\vec{\mathbf{F}}_x = \vec{\mathbf{F}}_{x_1} + \vec{\mathbf{F}}_{x_2}$ and $\vec{\mathbf{F}}_y = \vec{\mathbf{F}}_{y_1} + \vec{\mathbf{F}}_{y_2}$, respectively.

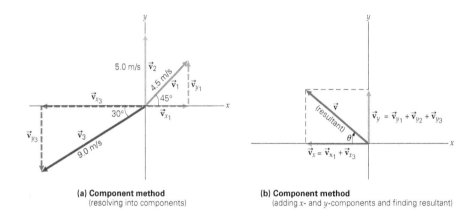

(a) Component method
(resolving into components)

(b) Component method
(adding *x*- and *y*-components and finding resultant)

▲ **FIGURE 3.8** **Component method of vector addition (a)** In the analytical component method, all the vectors to be added ($\vec{\mathbf{v}}_1$, $\vec{\mathbf{v}}_2$ and $\vec{\mathbf{v}}_3$) are first placed with their tails at the origin so that they may be easily resolved into rectangular components. **(b)** The respective summations of all the *x*-components and all the *y*-components are then added to give the components of the resultant $\vec{\mathbf{v}}$.

these equations, you will have values for $v_x < 0$ (negative) and $v_y > 0$ (positive) as shown in Figure 3.8b.

Notice also in Figure 3.8b that the directional angle θ of the resultant is referenced to the *x*-axis, as are the individual vectors in Figure 3.8a. *In adding vectors by the component method, all vectors will be referenced to the nearest x-axis – that is, the +x-axis or −x-axis.* This convention eliminates angles greater than 90° (as occurs when customarily measuring angles counterclockwise from the +*x*-axis) and the use of double-angle formulas, such as cos ($\theta + 90°$). This greatly simplifies calculations. The recommended procedures for adding vectors analytically by the component method can be summarized as follows.

3.2.4 Procedures for Adding Vectors by the Component Method

1. Resolve the vectors to be added into their *x*- and *y*-components. Use the acute angles (angles less than 90°) between the vectors and the *x*-axis to find the magnitudes, and indicate the directions of the components by positive and negative signs (▶ **Figure 3.9**).

2. Add all of the *x*-components together, and all of the *y*-components together algebraically to obtain the *x*- and *y*-components of the resultant, or vector, sum.

3. Express the resultant vector, using:

   (a) The unit vector component form – for example, $\vec{\mathbf{C}} = C_x \hat{\mathbf{x}} + C_y \hat{\mathbf{y}}$ or

   (b) The magnitude-angle form.

For the latter notation, find the magnitude of the resultant by using the summed *x*- and *y*-components and the Pythagorean theorem:

$$C = \sqrt{C_x^2 + C_y^2}$$

Find the angle of direction (relative to the *x*-axis) by taking the inverse tangent ($\tan^{-1}$) of the *absolute value* (i.e., the positive value, ignoring any negative signs) of the ratio of the magnitudes of *y*- and *x*-components:

$$\theta = \tan^{-1}\left|\frac{C_y}{C_x}\right|$$

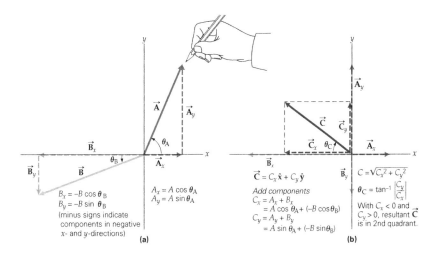

▲ FIGURE 3.9 **Vector addition by the analytical component method** (a) Resolve the vectors into their x- and y-components. **(b)** Add all the x-components and all the y-components together respectively to obtain the x- and y-components of the resultant. Express the resultant in either component form or magnitude-angle form. All angles are referenced to the $+x$- or $-x$-axis to keep them less than 90°.

Designate the quadrant in which the resultant lies. This information is obtained from the signs of the summed components (C_x and C_y) or from a sketch of their addition via the triangle method (see Figure 3.9). The angle θ is the angle between the resultant and the x-axis in that quadrant.

EXAMPLE 3.3: APPLYING THE ANALYTICAL COMPONENT METHOD – SEPARATING AND COMBINING X- AND Y-COMPONENTS

Let's apply the procedural steps of the component method to the addition of the vectors in Figure 3.8a. The vectors with units of meters per second represent velocities.

THINKING IT THROUGH. Follow and learn the steps of the procedure. Basically, the vectors are resolved into components and the respective components are added to get the components of the resultant, which then may be expressed in (unit vector) component form or magnitude-angle form.

SOLUTION

The rectangular components of the vectors are shown in Figure 3.8b. Summing these components and taking the values from Figure 3.8a,

$$\vec{v} = v_x\hat{x} + v_y\hat{y} = \left(v_{x_1} + v_{x_2} + v_{x_3}\right)\hat{x} + \left(v_{y_1} + v_{y_2} + v_{y_3}\right)\hat{y}$$

where

$$v_x = v_{x_1} + v_{x_2} + v_{x_3} = v_1\cos 45° + 0 - v_3\cos 30°$$
$$= (4.5\,\text{m/s})(0.707) - (9.0\,\text{m/s})(0.866) = -4.6\,\text{m/s}$$

and

$$v_y = v_{y_1} + v_{y_2} + v_{y_3} = v_1\sin 45° + v_2 - v_3\sin 30°$$
$$= (4.5\,\text{m/s})(0.707) + (5.0\,\text{m/s}) - (9.0\,\text{m/s})(0.50)$$
$$= 3.7\,\text{m/s}$$

Expressed in tabular form, the components are as follows:

	x-components		y-components
v_{x_1}	$+v_1\cos 45° = +3.2\,\text{m/s}$	v_{y_1}	$+v_1\sin 45° = +3.2\,\text{m/s}$
v_{x_2}	$= 0\,\text{m/s}$	v_{y_2}	$= +5.0\,\text{m/s}$
v_{x_3}	$-v_3\cos 30° = -7.8\,\text{m/s}$	v_{y_3}	$-v_3\sin 30° = -4.5\,\text{m/s}$
Sum :	$v_x = -4.6\,\text{m/s}$		$v_y = +3.7\,\text{m/s}$

The directions of the components are indicated by signs. (The $+$ sign is sometimes omitted as being understood.) Here, $\vec{v}_2$ has no x-component. *Note that in general, for the analytical component method, the x-components are cosine functions and the y-components are sine functions, as long as the angles are referenced to the x-axis.*

In component form, the resultant vector is

$$\vec{v} = (-4.6\,\text{m/s})\hat{x} + (3.7\,\text{m/s})\hat{y}$$

In magnitude-angle form, the resultant velocity has a magnitude of

$$v = \sqrt{v_x^2 + v_y^2} = \sqrt{(-4.6\,\text{m/s})^2 + (3.7\,\text{m/s})^2} = 5.9\,\text{m/s}$$

Since the x-component is negative and the y-component is positive, the resultant lies in the *second quadrant* at an angle of

$$\theta = \tan^{-1}\left|\frac{v_y}{v_x}\right| = \tan^{-1}\left(\frac{3.7\,\text{m/s}}{4.6\,\text{m/s}}\right) = 39°$$

above the $-x$-axis because of the negative x-component (see Figure 3.8b).

FOLLOW-UP EXERCISE. Suppose in this Example that there was an additional velocity vector $\vec{v}_4 = (+4.6\,\text{m/s})\hat{x}$. What would be the resultant of all four vectors in this case?

Although our discussion is limited to motion in two dimensions (in a plane), the component method is easily extended to three dimensions. For a velocity in three dimensions, the vector has x-, y-, and z-components: $\vec{v} = v_x\hat{x} + v_y\hat{y} + v_z\hat{z}$ and magnitude $v = \sqrt{v_x^2 + v_y^2 + v_z^2}$.

EXAMPLE 3.4: FIND THE VECTOR – ADD THEM UP

Given two displacement vectors, $\vec{A}$, with a magnitude of 8.0 m in a direction 45° below the $+x$-axis, and $\vec{B}$, which has an x-component of $+2.0$ m and a y-component of $+4.0$ m. Find a vector $\vec{C}$ so that $\vec{A} + \vec{B} + \vec{C}$ equals a vector $\vec{D}$ that has a magnitude of 6.0 m in the $+y$-direction.

THINKING IT THROUGH. Here again, a sketch helps to clarify the situation and gives a general idea of the attributes of $\vec{C}$. This would be something like ▼ **Figure 3.10**, Make a Sketch and Add Them Up. Note that in part (a) both $\vec{A}$ and $\vec{B}$ have $+x$-components, so $\vec{C}$ would have to have a $-x$-component to cancel these components. (It is given that the resultant $\vec{D}$ points only in the $+y$-direction.) $\vec{B}_y$ and $\vec{D}$ are in the $+y$-direction, but the $\vec{A}_y$-component is larger in the $-y$-direction, so $\vec{C}$ would have to have a $+y$-component. With this information, it can be seen that $\vec{C}$ lies in the second quadrant.

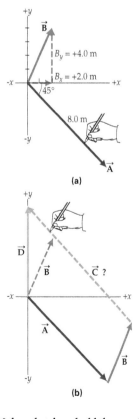

(a)

(b)

▲ FIGURE 3.10 **Make a sketch and add them up (a)** A sketch is made for the vectors $\vec{A}$ and $\vec{B}$. In a vector drawing, the vector lengths are usually set to some scale – for example, 1 cm : 1 m – but in a quick sketch, the vector lengths are estimated. **(b)** By shifting $\vec{B}$ to the tip of $\vec{A}$ and putting in $\vec{D}$, the vector $\vec{C}$ can be found from $\vec{A} + \vec{B} + \vec{C} = \vec{D}$.

A polygon sketch [shown in part (b) of Figure 3.10] confirms this observation.

So $\vec{C}$ has second-quadrant components ($C_x < 0$ and $C_y > 0$) and it has a relatively large magnitude (from the lengths of the vectors in the polygon drawing). This information gives an idea of what we are looking for, making it easier to see if the results from the analytic solution are reasonable.

SOLUTION

Given:

$\vec{A}$: 8.0 m 45° below $+x$ (fourth quadrant)
$\vec{B}_x = (2.0\,\text{m})\hat{x}$ and $\vec{B}_y = (4.0\,\text{m})\hat{y}$

Find:

$\vec{C}$ such that $\vec{A} + \vec{B} + \vec{C} = \vec{D} = (+6.0\,\text{m})\hat{y}$

Setting up the components in tabular form again so they can be easily seen:

x-Components	y-Components
$A_x = A\cos 45° = (8.0\text{ m})$ $(0.707) = +5.7$ m	$A_y = -A\sin 45° = -(8.0\text{ m})$ $(0.707) = -5.7$ m
$B_x = +2.0$ m	$B_y = +4.0$ m
$C_x = ?$	$C_y = ?$
$D_x = 0$	$D_y = +6.0$ m

To find the components of $\vec{C}$, where $\vec{A} + \vec{B} + \vec{C} = \vec{D}$, the x- and y-components are summed separately:

$$x: \vec{A}_x + \vec{B}_x + \vec{C}_x = \vec{D}_x$$

or

$$+5.7\,\text{m} + 2.0\,\text{m} + C_x = 0 \quad \text{and} \quad C_x = -7.7\,\text{m}$$

$$y: \vec{A}_y + \vec{B}_y + \vec{C}_y = \vec{D}_y$$

or

$$-5.7\,\text{m} + 4.0\,\text{m} + C_y = 6.0\,\text{m} \quad \text{and} \quad C_y = +7.7\,\text{m}$$

So,

$$\vec{C} = (-7.7\,\text{m})\hat{x} + (7.7\,\text{m})\hat{y}$$

The result can also be expressed in magnitude-angle form:

$$C = \sqrt{C_x^2 + C_y^2} = \sqrt{(-7.7\,\text{m})^2 + (7.7\,\text{m})^2} = 11\,\text{m}$$

and

$$\theta = \tan^{-1}\left|\frac{C_y}{C_x}\right| = \tan^{-1}\left|\frac{7.7\,\text{m}}{-7.7\,\text{m}}\right| = 45°$$

(above the $-x$-axis; why?)

FOLLOW-UP EXERCISE. Suppose $\vec{D}$ pointed in the opposite direction [that is, $\vec{D} = (-6.0\,\text{m})\hat{y}$]. What would $\vec{C}$ be in this case?

3.3 Projectile Motion

A familiar example of two-dimensional, curvilinear motion is that of an object that is thrown or projected by some means. The motion of a stone thrown across a stream or a golf ball driven off a tee are examples of **projectile motion**. A special case of projectile motion in one dimension occurs when an object is projected vertically upward (or downward or dropped). This case was treated in Section 2.5 in terms of free fall. General two-dimensional projectile motion is in free fall too, because the only acceleration of a projectile is that due to gravity (air resistance neglected). Vector components can be used to analyze projectile motion by simply breaking up the motion into its x- and y-components and treating them separately.

3.3.1 Horizontal Projections

It is instructive to first analyze the motion of an object projected horizontally, or parallel to a level surface. Suppose that you throw an object horizontally with an initial velocity v_{x_0} as in ▼ **Figure 3.11**. Projectile motion is analyzed beginning at the instant of release ($t = 0$). Once the object is released, there is no longer a horizontal acceleration ($a_x = 0$), so throughout the object's path, the horizontal velocity remains constant: $v_x = v_{x_0}$.

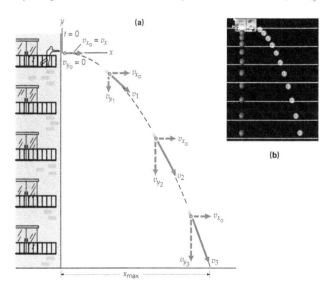

▲ FIGURE 3.11 **Horizontal projection (a)** The velocity components of a projectile launched horizontally show that the projectile travels to the right as it falls downward. Note the increase in v_y. **(b)** The paths of two golf balls: one was projected horizontally at the same time that the other was dropped straight down. The horizontal lines are 15 cm apart, and the interval between flashes was 1/30 s. The vertical motions of the balls are the same. Why? Can you describe the horizontal motion of the pink ball?

According to the equation $x = x_0 + v_x t$ (Equation 3.2a), the projected object would continue to travel in the horizontal direction indefinitely. However, you know that this is not what happens. As soon as the object is projected, it is in free fall in the vertical direction, with $v_{y_0} = 0$ (vertically it behaves as

though it had been dropped) and $a_y = -g$ (upward as the positive direction). In other words, the projected object travels at a uniform velocity in the horizontal direction, while *at the same time* undergoing acceleration in the downward direction under the influence of gravity. The result is a curved path, as illustrated in Figure 3.11. (Compare the motions in Figures 3.11 and 3.2. Do you see any similarities?)

If there were no horizontal motion, the object would simply drop to the ground in a straight line. In fact, the time of flight of the horizontally projected object is *exactly the same as if it were a dropped object falling vertically.*

Note the components of the velocity vector in Figure 3.11a. The length of the horizontal component of the velocity vector remains the same, but the length of the vertical component increases with time. What is the instantaneous velocity at any point along the path? (Think in terms of vector addition, covered in Section 3.2.) Figure 3.11b shows the actual motions of a horizontally projected golf ball and one that is simultaneously dropped from rest. The horizontal reference lines show that the balls fall vertically at the same rate. The only difference is that the horizontally projected ball also travels to the right as it falls.

EXAMPLE 3.5: STARTING AT THE TOP – HORIZONTAL PROJECTION

Suppose that the ball in Figure 3.11a is projected from a height of 25.0 m above the ground and is thrown with an initial horizontal velocity of 8.25 m/s. (a) How long is the ball in flight before striking the ground? (b) How far from the building does the ball strike the ground?

THINKING IT THROUGH. In looking at the components of motion, we see that part (a) involves the time it takes the ball to fall vertically, analogous to a ball dropped from that height. This time is also the time the ball travels in the horizontal direction. The horizontal speed is constant, so the horizontal distance requested in part (b) can be found.

SOLUTION

Writing the data with the origin chosen as the point from which the ball is thrown and downward taken as the negative direction:

Given:

$y = -25.0$ m
$v_{x_0} = 8.25$ m/s
$v_{y_0} = 0$
$a_x = 0$ and $a_y = -g$
($x_0 = 0$ and $y_0 = 0$ because of our choice of axes location.)

Find:

(a) t (time of flight)
(b) x (horizontal distance)

(a) As noted previously, the time of flight is the same as the time it takes for the ball to fall vertically to the ground. To find this time, the equation $y = y_0 + v_{y_0}t - \frac{1}{2}gt^2$ can be used, in which the negative direction of g is expressed explicitly, as was done in Section 2.4. With $v_{y_0} = 0$,

$$y = -\frac{1}{2}gt^2$$

So,

$$t = \sqrt{\frac{2y}{-g}} = \sqrt{\frac{2(-25.0\,\text{m})}{-9.80\,\text{m/s}^2}} = 2.26\,\text{s}$$

(b) The ball travels in the x-direction for the same amount of time it travels in the y-direction (i.e., 2.26 s). Since there is no acceleration in the horizontal direction, the ball travels in this direction with a uniform velocity. Thus, with $x_0 = 0$ and $a_x = 0$,

$$x = v_{x_0}t = (8.25\,\text{m/s})(2.26\,\text{s}) = 18.6\,\text{m}$$

FOLLOW-UP EXERCISE. (a) Choose the axes to be at the base of the building, and show that the resulting equation is the same as in the Example. (b) What is the velocity (in component form) of the ball just before it strikes the ground?

3.3.2 Projections at Arbitrary Angles

The general case of projectile motion involves an object projected at an arbitrary angle θ relative to the horizontal – for example, a golf ball hit by a club (▼ **Figure 3.12**). During projectile motion, the object travels up and down while traveling horizontally with a constant velocity. (Does the ball have acceleration? Yes. At each point of the motion, gravity acts, and $\vec{a} = -g\hat{y}$.)

This motion is also analyzed by using its components. As before, upward is taken as the positive direction and downward as the negative direction. The initial velocity v_0 is first resolved into rectangular components:

$$v_{x_0} = v_0 \cos\theta \tag{3.8a}$$

(initial velocity components)

$$v_{y_0} = v_0 \sin\theta \tag{3.8b}$$

There is no horizontal acceleration and the acceleration due to gravity acts in the $-y$-direction. Thus, the x-component of the

velocity is constant and the y-component varies with time (see Equation 3.3d):

$$v_x = v_{x_0} = v_0 \cos\theta \tag{3.9a}$$

$$v_y = v_{y_0} - gt \tag{3.9b}$$

(projectile motion velocity components)

The components of the instantaneous velocity at various times are illustrated in Figure 3.12. The instantaneous velocity is the vector sum of these components and is tangent to the curved path of the ball at any point. Notice that the ball strikes the ground at the same speed (but with a negative vertical velocity) and at the same angle below the horizontal as it was launched.

Similarly, the displacement components are given by $(x_0 = y_0 = 0)$:

$$x = v_{x_0}t = (v_0 \cos\theta)t \tag{3.10a}$$

$$y = v_{y_0}t - \frac{1}{2}gt^2 = (v_0 \sin\theta)t - \frac{1}{2}gt^2 \tag{3.10b}$$

The curve described by these equations, or the path of motion (trajectory) of the projectile, is called a parabola. The path of projectile motion is commonly observed (▼ **Figure 3.13**).

▲ FIGURE 3.13 **Parabolic path** Water streams from lighted fountains follow parabolic paths.

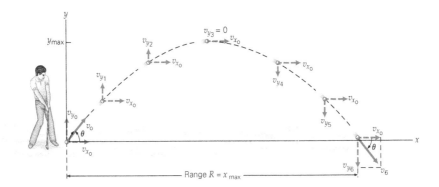

▲ FIGURE 3.12 **Projection at an angle** The velocity components of the ball are shown for various times. Note that $v_y = 0$ at the top of the arc, or at y_{max}. The range R is the maximum horizontal distance, or x_{max}. (Notice that $v_0 = v_6$ in magnitude. Why?)

Note that, as in the case of horizontal projection, *time is the common feature shared by the components of motion.* Aspects of projectile motion that may be of interest in various situations include the time of flight, the maximum height reached, and the range (R), which is the maximum horizontal distance traveled.

EXAMPLE 3.6: TEEING OFF – PROJECTION AT AN ANGLE

Suppose a golf ball is hit off the tee with an initial velocity of 30.0 m/s at an angle of 35° to the horizontal, as in Figure 3.12. (a) What is the maximum height reached by the ball? (b) What is its range?

THINKING IT THROUGH. The maximum height involves the y-component; the procedure for finding this is like that for finding the maximum height of a ball projected vertically upward. The ball travels in the x-direction for the same amount of time it would take for the ball to go up and down.

SOLUTION

Given:

$v_{\text{o}} = 30.0$ m/s
$\theta = 35°$
$a_x = 0$ and $a_y = -g$
$y = 0$ (final position)
($x_{\text{o}} = 0$ and $y_{\text{o}} = 0$ because of our choice of axes location.)

Find:

(a) y_{max}
(b) $R = x_{\text{max}}$

Let us compute $v_{x_{\text{o}}}$ and $v_{y_{\text{o}}}$ explicitly so simplified kinematic equations can be used:

$$v_{x_{\text{o}}} = v_{\text{o}} \cos 35° = (30.0\,\text{m/s})(0.819) = 24.6\,\text{m/s}$$
$$v_{y_{\text{o}}} = v_{\text{o}} \sin 35° = (30.0\,\text{m/s})(0.574) = 17.2\,\text{m/s}$$

(a) Just as for an object thrown vertically upward, $v_y = 0$ at the maximum height (y_{max}). Thus, the time to reach the maximum height (t_{u}) can be found by using Equation 3.3b, with v_y set equal to zero:

$$v_y = 0 = v_{y_{\text{o}}} - gt_{\text{u}}$$

Solving for t_{u},

$$t_{\text{u}} = \frac{v_{y_{\text{o}}}}{g} = \frac{17.2\,\text{m/s}}{9.80\,\text{m/s}^2} = 1.76\,\text{s}$$

(Note that t_{u} represents the amount of time the ball moves upward.)

The maximum height y_{max} is then obtained by substituting t_{u} into Equation 3.10b:

$$y_{\text{max}} = v_{y_{\text{o}}} t_{\text{u}} - \frac{1}{2} g t_{\text{u}}^2$$
$$= (17.2\,\text{m/s})(1.76\,\text{s}) - \frac{1}{2}(9.80\,\text{m/s}^2)(1.76\,\text{s})^2$$
$$= 15.1\,\text{m}$$

The maximum height could also be obtained directly from Equation 2.11′, $v_y^2 = v_{y_{\text{o}}}^2 - 2gy$, with $y = y_{\text{max}}$ and $v_y = 0$. However, the method of solution used here illustrates how the time of flight is obtained.

(b) As in the case of vertical projection, the time in going up is equal to the time in coming down, so the total time of flight is $t = 2t_{\text{u}}$ (to return to the elevation from which the object was projected, $y = y_{\text{o}} = 0$, as can be seen from $y - y_{\text{o}} = v_{y_{\text{o}}} t - \frac{1}{2} g t^2 = 0$, and $t = (2v_{y_{\text{o}}}/g) = 2t_{\text{u}}$).

The range R is equal to the horizontal distance traveled (x_{max}), which is easily found by substituting the total time of flight $t = 2t_{\text{u}} = 2(1.76\,\text{s}) = 3.52\,\text{s}$ into Equation 3.10a:

$$R = x_{\text{max}} = v_x t = v_{x_{\text{o}}}(2t_{\text{u}}) = (24.6\,\text{m/s})(3.52\,\text{s}) = 86.6\,\text{m}$$

FOLLOW-UP EXERCISE. How would the values of maximum height (y_{max}) and the range (x_{max}) compare with those found in this Example if the golf ball had been similarly teed off on the surface of the Moon? [*Hint:* $g_{\text{M}} = g/6$; that is, acceleration due to gravity on the Moon is one-sixth of that on the Earth.] Do not do any numerical calculations. Find the answers by "sight reading" the equations.

The range of a projectile is an important consideration in various applications. This factor is particularly important in sports in which a maximum range is desired, such as golf and javelin throwing.

In general, what is the range of a projectile launched with velocity v_{o} at an angle θ? In order to answer this question, consider the equation used in Example 3.6 to calculate the range, $R = v_x t$. First let's look at the expressions for v_x and t. Since there is no acceleration in the horizontal direction,

$$v_x = v_{x_{\text{o}}} = v_{\text{o}} \cos \theta$$

and the total time t (as shown in Example 3.6) is

$$t = \frac{2v_{y_{\text{o}}}}{g} = \frac{2v_{\text{o}} \sin \theta}{g}$$

and R is given by,

$$R = v_x t = (v_{\text{o}} \cos \theta)\left(\frac{2v_{\text{o}} \sin \theta}{g}\right) = \frac{2v_{\text{o}}^2 \sin \theta \cos \theta}{g}$$

Using the trigonometric identity $\sin 2\theta = 2 \sin \theta \cos \theta$ (see Appendix I),

$$R = \frac{v_{\text{o}}^2 \sin 2\theta}{g} \tag{3.11}$$

(projectile range x_{max} only for $y_{\text{initial}} = y_{\text{final}}$)

Note that the range depends on the magnitude of the initial velocity (or speed), v_{o}, and that the angle of projection, θ, and g are assumed to be constant. Keep in mind that this equation applies only to the special, but common, case of $y_{\text{initial}} = y_{\text{final}}$, that is, when the landing point is at the same height as the launch point.

EXAMPLE 3.7: A THROW FROM THE BRIDGE

A young girl standing on a bridge throws a stone with an initial velocity of 12 m/s at a downward angle of 45° to the horizontal, in an attempt to hit a block of wood floating in the river below (▼ Figure 3.14). If the stone is thrown from a height of 20 m and it just reaches the water when the block is 13 m from the bridge, does the stone hit the block? (Assume that the block does not move appreciably and that it is in the plane of the throw.)

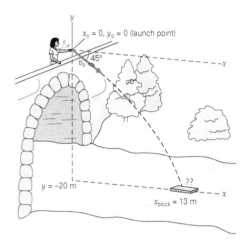

▲ FIGURE 3.14 **A throw from the bridge** Hit or miss? See Example 3.7 for description.

THINKING IT THROUGH. The question is, what is the range of the stone? If this range is the same as the distance between the block and the bridge, then the stone hits the block. To find the range of the stone, we need to find the time of descent (from the y-component of motion) and then use this time to find the distance x_{max}. (Time is the connecting factor.)

SOLUTION

Given:

$$v_o = 12 \text{ m/s}$$
$$\theta = 45° \quad v_{x_o} = v_o \cos 45° = 8.5 \text{m/s}$$
$$y = -20 \text{m} \quad v_{y_o} = -v_o \sin 45° = -8.5 \text{m/s}$$
$$x_{block} = 13 \text{ m} \quad (x_o = y_o = 0)$$

Find:

Range or x_{max} of stone from bridge. (Is it the same as the block's distance from the bridge?)

To find the total time of the travel, $v_y = v_{y_o} - gt$ needs to be used. However, v_y is not known directly (it is not zero when the stone reaches the river). So to use this equation, v_y is needed. This may be found from the kinematic equation Equation 2.11,

$$v_y^2 = v_{y_o}^2 - 2gy$$

as

$$v_y = -\sqrt{(-8.5 \text{m/s})^2 - 2(9.8 \text{m/s}^2)(-20 \text{m})} = -22 \text{m/s}$$

(The negative root is selected because v_y is downward.)

Then solving $v_y = v_{y_o} - gt$ for t,

$$t = \frac{v_{y_o} - v_y}{g} = \frac{-8.5 \text{m/s} - (-22 \text{m/s})}{9.8 \text{m/s}^2} = 1.4 \text{s}$$

The stone's horizontal distance from the bridge at this time is

$$x_{max} = v_{x_o} t = (8.5 \text{m/s})(1.4 \text{s}) = 12 \text{m}$$

So the girl's throw falls short by a meter (the block is at 13 m).

Note that Equation 3.10b, $y = y_o + v_{y_o} t - \frac{1}{2} gt^2$, could have been used to find the time, but this calculation would have involved solving a quadratic equation.

FOLLOW-UP EXERCISE. (a) Why was it assumed that the block was in the plane of the throw? (b) Why wasn't Equation 3.11 used in this Example to find the range? Show that Equation 3.11 works in Example 3.6, but not in Example 3.7, by computing the range in each case and comparing your results with the answers found in the Examples.

CONCEPTUAL EXAMPLE 3.8: WHICH HAS THE GREATER SPEED?

Consider two balls, both thrown with the same initial speed v_o, but one at an angle of 45° above the horizontal and the other at an angle of 45° below the horizontal (▼ Figure 3.15). Determine whether, upon reaching the ground, (a) the ball projected upward will have the greater speed, (b) the ball projected downward will have the greater speed, or (c) both balls will have the same speed. *Clearly establish the reasoning and physical principle(s) used in determining your answer before checking it. That is, why did you select your answer?*

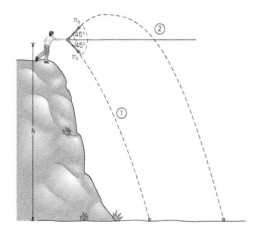

▲ FIGURE 3.15 **Which has the greater speed?** See Example 3.8 for description.

REASONING AND ANSWER. At first, you might think the answer is (b), because this ball is projected downward. But the ball projected upward falls from a greater maximum height, so perhaps the answer is (a). To solve this dilemma, look at the horizontal line in Figure 3.15 between the two velocity vectors that extends beyond the upper trajectory. From this diagram, it can be seen that the trajectories for both balls are the same below this line. Moreover, the downward velocity of the upper ball on reaching this line is v_0 at an angle of 45° below the horizontal (See Figure 3.12). Therefore, relative to the horizontal line and below, the conditions are identical, with the same y-component and the same constant x-component. So the answer is (c).

(This question can be answered easily with the conservation of energy, which you will learn in Chapter 5.)

FOLLOW-UP EXERCISE. Suppose the ball thrown downward was thrown at an angle of −40° and the upward ball at 45°. Which ball would hit the ground with the greater speed in this case?

3.3.3 Problem-Solving Hint

The range of a projectile projected downward, as in Figure 3.15, is found as illustrated in Example 3.7. But what about the range of a projectile projected upward? This case might be thought of as an "extended range" problem. One way to solve it is to divide the trajectory into two parts – (1) the arc above the horizontal line and (2) the downward part below the horizontal line – such that $x_{max} = x_1 + x_2$. You know how to find x_1 (Example 3.6) and x_2 (Example 3.7). Another way to solve the problem is to use $y = y_0 + v_{y_0}t - \frac{1}{2}gt^2$, where y is the final position of the projectile, and solve for t, the total time of flight. You would then use that value in the equation $x = v_0 t$.

Equation 3.11, $R = v_0^2 \sin 2\theta / g$, allows the range to be computed for a particular projection angle and initial velocity on a level surface. However, we are sometimes interested in the maximum range for a given initial velocity – for example, the maximum range of an artillery piece that fires a projectile with a particular muzzle velocity. Is there an optimum angle that gives the maximum range? Under ideal conditions, the answer is yes.

For a particular v_0, the range is a maximum (R_{max}) when $\sin 2\theta = 1$, since this value of θ yields the maximum value of the sine function (which varies from 0 to 1). Thus,

$$R_{max} = \frac{v_0^2}{g} \quad (y_{initial} = y_{final}) \tag{3.12}$$

Because this maximum range is obtained when $\sin 2\theta = 1$ and because $\sin 90° = 1$,

$$2\theta = 90° \quad \text{or} \quad \theta = 45°$$

for the maximum range for a given initial speed when the projectile returns to the elevation from which it was projected. At a greater or smaller angle, for a projectile with the same initial speed, the range will be less, as illustrated in ▶ **Figure 3.16**. Also, the range is the same for angles equally above and below 45°, such as 30° and 60°.

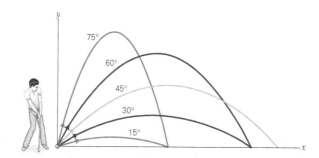

▲ FIGURE 3.16 **Range** For a projectile with a given initial speed, the maximum range is ideally attained with a projection of 45° (no air resistance). For projection angles above and below 45°, the range is shorter, and it is equal for angles equally different from 45° (e.g., 30° and 60°).

Thus, to get the maximum range, a projectile *ideally* should be projected at an angle of 45°. However, up to now, air resistance has been neglected. In actual situations, such as when a baseball is thrown or hit, this factor may have a significant effect. Air resistance reduces the speed of the projectile, thereby reducing the range. As a result, when air resistance is a factor, the angle of projection for maximum range is less than 45°, which gives a greater initial horizontal velocity (▼ **Figure 3.17**). Other factors, such as spin and wind, may also affect the range of a projectile. For example, backspin on a driven golf ball provides lift and the projection angle for the maximum range may be considerably less than 45°.

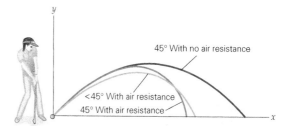

(a)

(b)

▲ FIGURE 3.17 **Air resistance and range (a)** When air resistance is a factor, the angle of projection for maximum range is less than 45°. **(b)** Javelin throw. Because of air resistance, the javelin is thrown at an angle less than 45° in order to achieve maximum range.

Keep in mind that for the maximum range to occur at a projection angle of 45°, the components of initial velocity must be equal – that is, $\tan^{-1}(v_{y_0}/v_{x_0}) = 45°$ and $\tan 45° = 1$, so that $v_{y_0} = v_{x_0}$. However, this condition may not always be physically possible, as Conceptual Example 3.9 shows.

CONCEPTUAL EXAMPLE 3.9: THE LONGEST JUMP – THEORY AND PRACTICE

In a long-jump event, does the jumper normally have a launch angle of (a) less than 45°, (b) exactly 45°, or (c) greater than 45°? Clearly establish the reasoning and physical principle(s) used in determining your answer before checking it. That is, *why* did you select your answer?

REASONING AND ANSWER. Air resistance is not a major factor here (although wind speed is taken into account for record setting in track-and-field events). Therefore, it would seem that in order to achieve maximum range, the jumper would take off at an angle of 45°. But there is another physical consideration. Let's look more closely at the jumper's initial velocity components (▼ Figure 3.18a).

To maximize a long jump, the jumper runs as fast as possible and then pushes upward as strongly as possible to maximize both velocity components. The initial vertical velocity component v_{y_0} depends on the upward push of the jumper's legs, whereas the initial horizontal velocity component v_{x_0} depends mostly on the running speed toward the jump point. In general, a greater velocity can be achieved by running than by jumping, so $v_{x_0} > v_{y_0}$. Then, since $\theta = \tan^{-1}(v_{y_0}/v_{x_0})$, then $\theta < 45°$, where $v_{y_0}/v_{x_0} < 1$ in this case. Hence, the answer is (a). It certainly could not be (c). A typical launch angle for a long jump is 20° to 25°. (If a jumper increased the launch angle to be closer to the ideal 45°, then the running speed would have to decrease, resulting in a decrease in range.)

FOLLOW-UP EXERCISE. When driving in and jumping to score, basketball players seem to be suspended momentarily, or to "hang" in the air (Figure 3.18b). Explain the physics of this effect.

EXAMPLE 3.10: A "SLAP SHOT" – IS IT GOOD?

A hockey player hits a "slap shot" in practice (with no goalie present) when he is 15.0 m directly in front of the net. The net is 1.20 m high, and the puck is initially hit at an angle of 5.00° above the ice with a speed of 35.0 m/s. (a) Determine whether the puck makes it into the net. (b) If it does, determine whether the puck is rising or falling vertically as it crosses the front plane of the net.

THINKING IT THROUGH. First let's make a sketch of the situation using *x–y* coordinates, assuming that the puck is at the origin at the time it is hit and showing the net and its height as in ▼ **Figure 3.19**. Note that the launch angle is exaggerated. An angle of 5.00° is quite small, but then again, the top of the net is not overly high (1.20 m).

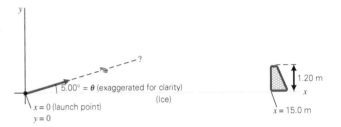

▲ **FIGURE 3.19**   **Slap shot** Is it a goal? See Example 3.10 for description.

To determine whether the shot is of goal quality, we need to know whether the puck's trajectory takes it above the net or into the net. That is, what is the puck's height (*y*) when its horizontal distance is *x* = 15.0 m? Whether the puck is rising or falling at this horizontal distance depends on when the puck reaches its maximum height. The appropriate equation(s) should provide this information, but keep mind that time is the connecting factor between the *x*- and *y*-components.

SOLUTION

Listing the data as usual:

(a)

(b)

▲ **FIGURE 3.18**   **Athletes in action (a)** To maximize a long jump, a jumper runs as fast as possible and then pushes upward as strongly as he can to maximize the velocity components (v_x and v_y). **(b)** When driving in toward the basket and jumping to score, basketball players seem to be suspended momentarily, or "hang" in the air.

Given:

$$x = 15.0 \text{ m}, \ x_\text{o} = 0$$
$$y_\text{net} = 1.20 \text{ m}, \ y_\text{o} = 0$$
$$\theta = 5.00° \text{ and } v_\text{o} = 35.0 \text{ m/s}$$
$$v_{x_\text{o}} = v_\text{o} \cos 5.00° = 34.9 \text{ m/s}$$
$$v_{y_\text{o}} = v_\text{o} \sin 5.00° = 3.05 \text{ m/s}$$

Find:

(a) Whether the puck goes into the net
(b) If so, is it rising or falling?

The vertical location of the puck at any time t is given by $y = v_{y_\text{o}} t - \frac{1}{2} g t^2$, so we need to know how long the puck takes to travel the 15.0 m to the net. The connecting factor of the components is time, so this time can be found from the x motion:

$$x = v_{x_\text{o}} t \quad \text{or} \quad t = \frac{x}{v_{x_\text{o}}} = \frac{15.0 \text{ m}}{34.9 \text{ m/s}} = 0.430 \text{ s}$$

So on reaching the front of the net, the puck is at a height of

$$y = v_{y_\text{o}} t - \frac{1}{2} g t^2 = (3.05 \text{ m/s})(0.430 \text{ s}) - \frac{1}{2}(9.80 \text{ m/s}^2)(0.430 \text{ s})^2$$
$$= 1.31 \text{ m} - 0.906 \text{ m} = 0.40 \text{ m}$$

Goal!

The time (t_u) for the puck to reach its maximum height is given by $v_y = v_{y_\text{o}} - g t_\text{u}$, where $v_y = 0$ and

$$t_\text{u} = \frac{v_{y_\text{o}}}{g} = \frac{3.05 \text{ m/s}}{9.80 \text{ m/s}^2} = 0.311 \text{ s}$$

With the puck reaching the net in 0.430 s, it is descending.

FOLLOW-UP EXERCISE. At what distance from the net did the puck start to descend?

3.4 Relative Velocity (Optional)

Velocity is not absolute but is dependent on the observer. That is, its description is *relative* to the observer's state of motion. If an object is observed moving with a certain velocity, then that velocity must be relative to something else. For example, a bowling ball moves down the alley with a certain velocity, and its velocity is relative to the alley. The motions of objects are often described as being relative to the Earth or ground, which is commonly thought of as a *stationary* frame of reference. In other instances it may be convenient to use a *moving* frame of reference.

Measurements must be made with respect to some reference. This reference is usually taken to be the origin of a coordinate system. The point you designate as the origin of a set of coordinate axes is arbitrary and entirely a matter of choice. For example, you may "attach" the coordinate system to the road or the ground and then measure the displacement or velocity of a car relative to these axes. For a "moving" frame

of reference, the coordinate axes may be attached to a car moving along a highway. In analyzing motion from another reference frame, you do not change the physical situation or what is taking place, only the point of view from which you describe it. Hence, motion is *relative* (to some reference frame), and is referred to as relative velocity. Since velocity is a vector, vector addition and subtraction are helpful in determining relative velocities.

3.4.1 Relative Velocities in One Dimension

When the velocities are linear (along a straight line) in the same or opposite directions and all have the same reference (such as the ground), the relative velocities can be found by using vector subtraction. As an illustration, consider cars moving with constant velocities along a straight, level highway, as in ▶ **Figure 3.20**. The velocities of the cars shown in the figure are *relative to the Earth, or the ground*, as indicated by the reference set of coordinate axes in Figure 3.20a, with motions along the x-axis. They are also relative to the stationary observers standing by the highway and sitting in the parked car A. That is, these observers see the cars as moving with velocities $\vec{v}_\text{B} = +90 \text{ km/h}$ and $\vec{v}_\text{C} = -60 \text{ km/h}$. The relative velocity of two objects is given by the velocity (vector) difference between them. For example, the velocity of car B *relative to car A* is given by

$$\vec{v}_\text{BA} = \vec{v}_\text{B} - \vec{v}_\text{A} = (+90 \text{ km/h})\hat{\mathbf{x}} - 0 = (+90 \text{ km/h})\hat{\mathbf{x}}$$

Thus, a person sitting in car A would see car B move away (in the $+x$-direction) with a speed of 90 km/h. For this linear case, the directions of the velocities are indicated by positive and negative signs (in addition to the subtraction sign in the equation). Similarly, the velocity of car C relative to an observer in car A is

$$\vec{v}_\text{CA} = \vec{v}_\text{C} - \vec{v}_\text{A} = (-60 \text{ km/h})\hat{\mathbf{x}} - 0 = (-60 \text{ km/h})\hat{\mathbf{x}}$$

The person in car A would see car C approaching (in the $-x$-direction) with a speed of 60 km/h.

But suppose that you want to know the velocities of the other cars *relative to car B* (i.e., from the point of view of an observer in car B) or relative to a set of coordinate axes with the origin fixed on car B (Figure 3.20b). Relative to these axes, car B is not moving; it acts as the fixed reference point. The other cars are moving relative to car B. The velocity of car C relative to car B is

$$\vec{v}_\text{CB} = \vec{v}_\text{C} - \vec{v}_\text{B} = (-60 \text{ km/h})\hat{\mathbf{x}} - (+90 \text{ km/h})\hat{\mathbf{x}}$$
$$= (-150 \text{ km/h})\hat{\mathbf{x}}$$

Similarly, car A has a velocity relative to car B of

$$\vec{v}_\text{AB} = \vec{v}_\text{A} - \vec{v}_\text{B} = 0 - (+90 \text{ km/h})\hat{\mathbf{x}} = (-90 \text{ km/h})\hat{\mathbf{x}}$$

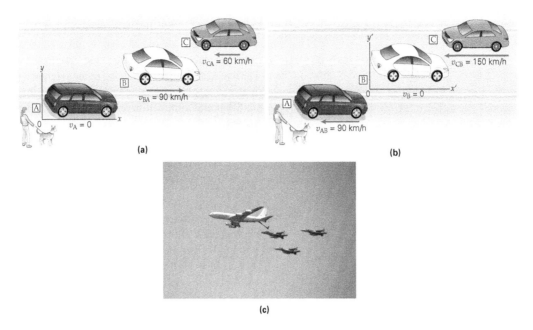

▲ FIGURE 3.20 **Relative velocity** The observed velocity of a car depends on, or is relative to, the frame of reference. The velocities shown in **(a)** are relative to the ground or to the parked car. In **(b)**, the frame of reference is with respect to car B, and he velocities are those that a driver of car B would observe. (See text for description.) **(c)** These aircraft, performing air-to-air refueling, are normally described as traveling at hundreds of kilometers per hour. To what frame of reference do these velocities refer? What is their velocity relative to each other?

Notice that relative to B, the other cars are both moving in the $-x$-direction. That is, car C is approaching car B with a velocity of 150 km/h in the $-x$-direction, and car A appears to be receding from car B with a velocity of 90 km/h in the $-x$-direction. (Imagine yourself in car B, and take that position as stationary. Car C would appear to be coming toward you at a high speed, and car A would be getting farther and farther away, as though it were moving backward relative to you.) Note that in general,

$$\vec{\mathbf{v}}_{AB} = -\vec{\mathbf{v}}_{BA}$$

What about the velocities of cars A and B relative to car C? From the point of view (or reference point) of car C, both cars A and B would appear to be approaching or moving in the $+x$-direction. For the velocity of car B relative to car C,

$$\vec{\mathbf{v}}_{BC} = \vec{\mathbf{v}}_B - \vec{\mathbf{v}}_C = (90\,\text{km/h})\hat{\mathbf{x}} - (-60\,\text{km/h})\hat{\mathbf{x}}$$
$$= (+150\,\text{km/h})\hat{\mathbf{x}}$$

Can you show that $\vec{\mathbf{v}}_{AC} = (+60\,\text{km/h})\hat{\mathbf{x}}$? Also note the situation in Figure 3.20c.

In some instances, velocities do not all have the same reference point. In such cases, relative velocities can be found by means of vector addition. To solve problems of this kind, *it is essential to identify the velocity references with care.*

Let's look first at a one-dimensional (linear) example. Suppose that a straight moving walkway in a major airport moves with a velocity of $\vec{\mathbf{v}}_{wg} = (+1.0\,\text{m/s})\hat{\mathbf{x}}$, where the subscripts indicate the velocity of the walkway (w) relative to the ground (g). A

passenger (p) on the walkway (w) trying to make a flight connection walks with a velocity of $\vec{\mathbf{v}}_{pw} = (+2.0\,\text{m/s})\hat{\mathbf{x}}$ relative to the walkway. What is the passenger's velocity relative to an observer standing next to the walkway (i.e., relative to the ground)?

This velocity, $\vec{\mathbf{v}}_{pg}$, is given by

$$\vec{\mathbf{v}}_{pg} = \vec{\mathbf{v}}_{pw} + \vec{\mathbf{v}}_{wg} = (2.0\,\text{m/s})\hat{\mathbf{x}} + (1.0\,\text{m/s})\hat{\mathbf{x}} = (3.0\,\text{m/s})\hat{\mathbf{x}}$$

Thus, the stationary observer sees the passenger as traveling with speed of 3.0 m/s down the walkway. (Make a sketch and show how the vectors add.) An explanation of the indicator line on the w symbols follows.

3.4.2 Problem-Solving Hint

Notice the pattern of the subscripts in this example. On the right side of the equation, the two inner subscripts out of the four total subscripts are the same (w). Basically, the walkway (w) is used as an intermediate reference frame. The outer subscripts (p and g) are sequentially the same as those for the relative velocity on the left side of the equation. When adding relative velocities, always check to make sure that the subscripts have this relationship – it indicates that you have set up the equation correctly.

What if a passenger got on the walkway going in the opposite direction and walked with the same speed as that of the walkway? Now it is essential to indicate the direction in which the passenger is walking by means of a minus sign:

$\vec{\mathbf{v}}_{pw} = (-1.0\,\text{m/s})\hat{\mathbf{x}}$. In this case, relative to the stationary observer,

$$\vec{\mathbf{v}}_{pg} = \vec{\mathbf{v}}_{pw} + \vec{\mathbf{v}}_{wg} = (-1.0\,\text{m/s})\hat{\mathbf{x}} + (1.0\,\text{m/s})\hat{\mathbf{x}} = 0$$

so the passenger is stationary with respect to the ground, and the walkway acts as a treadmill. (Good physical exercise.)

3.4.3 Relative Velocities in Two Dimensions

Of course, velocities are not always in the same or opposite directions. However, with the knowledge of how to use rectangular components to add or subtract vectors, problems involving relative velocities in two dimensions can be solved, as Examples 3.11 and 3.12 show.

EXAMPLE 3.11: ACROSS AND DOWN THE RIVER – RELATIVE VELOCITY AND COMPONENTS OF MOTION

The current of a 500-m-wide straight river has a flow rate of 2.55 km/h. A motorboat that travels with a constant speed of 8.00 km/h in still water crosses the river (▼ **Figure 3.21**). (a) If the boat's bow points directly across the river toward the opposite shore, what is the velocity of the boat relative to the stationary observer sitting at the corner of the bridge? (b) How far downstream will the boat's landing point be from the point directly opposite its starting point?

THINKING IT THROUGH. Careful designation of the given quantities is very important – the velocity of what, relative to what? Once this is done, part (a) should be straightforward. (See the previous Problem-Solving Hint.) For part (b), kinematics is used, where the time it takes the boat to cross the river is the key.

SOLUTION

As indicated in Figure 3.21, the river's flow velocity ($\vec{\mathbf{v}}_{rs}$, river to shore) is taken to be in the x-direction and the boat's velocity ($\vec{\mathbf{v}}_{br}$, boat to river) to be in the y-direction. Note that the river's flow velocity is *relative to the shore* and that the boat's velocity is *relative to the river*, as indicated by the subscripts. Listing the data,

Given:

$y_{max} = 500\,\text{m}$ (river width)
$\vec{\mathbf{v}}_{rs} = (2.55\,\text{km/h})\hat{\mathbf{x}} = (0.709\,\text{m/s})\hat{\mathbf{x}}$ (velocity of river *relative to shore*)
$\vec{\mathbf{v}}_{br} = (8.00\,\text{km/h})\hat{\mathbf{y}} = (2.22\,\text{m/s})\hat{\mathbf{y}}$ (velocity of boat *relative to river*)

Find:

(a) $\vec{\mathbf{v}}_{bs}$ (velocity of boat *relative to shore*)
(b) x (distance downstream)

Notice that as the boat moves toward the opposite shore, it is also carried downstream by the current. These velocity components would be clearly apparent to the jogger crossing the bridge and to the person sauntering downstream in Figure 3.21. If both observers stay even with the boat, the velocity of each will match one of the components of the boat's velocity. Since the velocity components are constant, the boat travels in a straight line diagonally across the river (much like the ball rolling across the table in Example 3.1).

(a) The velocity of the boat relative to the shore ($\vec{\mathbf{v}}_{bs}$) is given by vector addition. In this case,

$$\vec{\mathbf{v}}_{bs} = \vec{\mathbf{v}}_{br} + \vec{\mathbf{v}}_{rs}$$

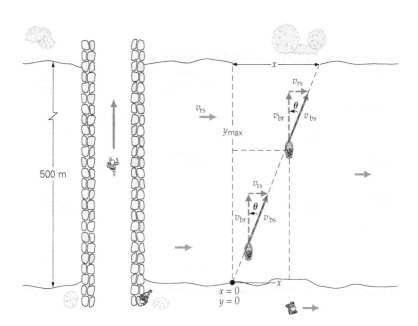

▲ FIGURE 3.21 **Relative velocity and components of motion** As the boat moves across the river, it is carried downstream by the current. (See Example 3.11.)

Since the velocities are not in the same direction and not along one axis, their magnitudes cannot be added directly. Notice in Figure 3.21 that the vectors form a right triangle, so the Pythagorean theorem can be applied to find the magnitude of v_{bs}:

$$v_{bs} = \sqrt{v_{br}^2 + v_{rs}^2} = \sqrt{(2.22\,\text{m/s})^2 + (0.709\,\text{m/s})^2}$$
$$= 2.33\,\text{m/s}$$

The direction of this velocity is defined by

$$\theta = \tan^{-1}\left(\frac{v_{rs}}{v_{br}}\right) = \tan^{-1}\left(\frac{0.709\,\text{m/s}}{2.22\,\text{m/s}}\right) = 17.7°$$

(b) To find the distance x that the current carries the boat downstream, we use components. Note that in the y-direction, $y_{max} = v_{br}t$ and

$$t = \frac{y_{max}}{v_{br}} = \frac{500\,\text{m}}{2.22\,\text{m/s}} = 225\,\text{s}$$

which is the time it takes the boat to cross the river.

During this time, the boat is carried downstream by the current a distance of

$$x = v_{rs}t = (0.709\,\text{m/s})(225\,\text{s}) = 160\,\text{m}$$

FOLLOW-UP EXERCISE. What is the distance traveled by the boat in crossing the river?

EXAMPLE 3.12: FLYING INTO THE WIND – RELATIVE VELOCITY

An airplane with an airspeed of 200 km/h (its speed in still air) flies in a direction such that with a west wind of 50.0 km/h, it travels in a straight line northward. (Wind direction is specified by the direction *from* which the wind blows, so a west wind blows from west to east.) To maintain its course due north, the plane must fly at an angle, as illustrated in ▼ **Figure 3.22**. What is the speed of the plane along its northward path?

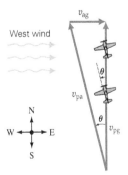

▲ FIGURE 3.22 **Flying into the wind** To fly directly north, the plane's heading (θ-direction) must be west of north.

THINKING IT THROUGH. Here again, the velocity designations are important, but as Figure 3.22 shows, the velocity vectors form a right triangle, and the magnitude of the unknown velocity can be found by using the Pythagorean theorem.

SOLUTION

As always, it is important to identify the reference frame to which the given velocities are relative.

Given:

$\vec{v}_{pa} = 200$ km/h at angle θ (velocity of plane with respect to still air = air speed)

$\vec{v}_{ag} = 50.0$ km/h east (velocity of air with respect to the Earth, or ground = wind speed)

Plane flies due north with velocity $\vec{v}_{pg}$

Find:

$\vec{v}_{pg}$ (ground speed of plane)

The speed of the plane with respect to the Earth, or the ground, v_{pg}, is called the plane's *ground speed*, and v_{pa} is its airspeed. Vectorially, the respective velocities are related by

$$\vec{v}_{pg} = \vec{v}_{pa} + \vec{v}_{ag}$$

If no wind were blowing ($v_{ag} = 0$), the ground speed and airspeed would be equal. However, a headwind (a wind blowing directly toward the plane) would cause a slower ground speed, and a tailwind would cause a faster ground speed. The situation is analogous to that of a boat going upstream versus downstream.

Here, $\vec{v}_{pg}$ is the resultant of the other two vectors, which can be added by the triangle method. Using the Pythagorean theorem to find v_{pg}, noting that v_{pa} is the hypotenuse of the triangle:

$$v_{pg} = \sqrt{v_{pa}^2 - v_{ag}^2} = \sqrt{(200\,\text{km/h})^2 - (50.0\,\text{km/h})^2}$$
$$= 194\,\text{km/h}$$

(Note that it was convenient to use the units of kilometers per hour, since the calculation did not involve any other units.)

FOLLOW-UP EXERCISE. What must be the plane's heading (θ-direction) in this Example for the plane to fly directly north?

Chapter 3 Review

- Motion in two dimensions is analyzed by considering the motion of linear components. The connecting factor between components is time.

Components of Initial Velocity:

$$v_{x_o} = v_o \cos\theta \tag{3.1a}$$

$$v_{y_o} = v_o \sin\theta \tag{3.1b}$$

Components of Displacement (constant acceleration only):

$$x = x_o + v_{x_o}t + \frac{1}{2}a_x t^2 \tag{3.3a}$$

$$y = y_o + v_{y_o}t + \frac{1}{2}a_y t^2 \tag{3.3b}$$

Components of Velocity (constant acceleration only):

$$v_x = v_{x_0} + a_x t \tag{3.3c}$$

$$v_y = v_{y_0} + a_y t \tag{3.3d}$$

- Of the various methods of vector addition, the component method is most useful. A resultant vector can be expressed in **magnitude-angle form** or in **unit vector component form**.

Vector Representation:

$$\left. \begin{array}{l} C = \sqrt{C_x^2 + C_y^2} \\[6pt] \theta = \tan^{-1} \left| \dfrac{C_y}{C_x} \right| \end{array} \right\} \quad \text{(magnitude-angle form)} \tag{3.4a}$$

$$\vec{C} = C_x \hat{\mathbf{x}} + C_y \hat{\mathbf{y}} \quad \text{(component form)} \tag{3.7}$$

- **Projectile motion** is analyzed by considering horizontal and vertical components separately – constant velocity in the horizontal direction and an acceleration due to gravity, g, in the downward vertical direction. (The foregoing equations for constant acceleration then have an acceleration of $a = -g$ instead of a.)
- **Range (R)** is the maximum horizontal distance traveled.

$$R = \frac{v_0^2 \sin 2\theta}{g} \quad \left(\begin{array}{l} \text{projectile range } x_{\max} \\ \text{only for } y_{\text{initial}} = y_{\text{final}} \end{array} \right) \tag{3.11}$$

$$R_{\max} = \frac{v_0^2}{g} \quad (y_{\text{initial}} = y_{\text{final}}) \tag{3.12}$$

- **Relative velocity** is expressed *relative* to a particular reference frame. Vector addition and subtraction are helpful in determining relative velocities to different reference frames.

4. Which one of the following *cannot* be a true statement about an object: (a) It has zero velocity and a nonzero acceleration; (b) it has velocity in the x-direction and acceleration in the y-direction; (c) it has velocity in the y-direction and acceleration in the y-direction; or (d) it has constant velocity and changing acceleration?

3.2 Vector Addition and Subtraction

5. Two linear vectors of magnitudes 3 and 4 are added. The magnitude of the resultant vector is (a) 1, (b) 7, (c) between 1 and 7.
6. The resultant of $\vec{A} - \vec{B}$ is the same as (a) $\vec{B} - \vec{A}$, (b) $-\vec{A} + \vec{B}$, (c) $-(\vec{A} + \vec{B})$, (d) $-(\vec{B} - \vec{A})$.
7. A unit vector has (a) magnitude, (b) direction, (c) neither of these, (d) both of these.

3.3 Projectile Motion

8. If air resistance is neglected, the motion of an object projected at an angle consists of a uniform downward acceleration combined with (a) an equal horizontal acceleration, (b) a uniform horizontal velocity, (c) a constant upward velocity, (d) an acceleration that is always perpendicular to the path of motion.
9. A football is thrown on a long pass. Compared to the ball's initial horizontal velocity component, the velocity at the highest point is (a) greater, (b) less, (c) the same.
10. A football is thrown on a long pass. Compared to the ball's initial vertical velocity, the vertical component of its velocity at the highest point is (a) greater, (b) less, (c) the same.

3.4 Relative Velocity (Optional)

11. You are traveling in a car on a straight, level road going 70 km/h. A car coming toward you appears to be traveling 130 km/h. How fast is the other car going? (a) 130 km/h, (b) 60 km/h, (c) 70 km/h, (d) 80 km/h?
12. Two cars approach each other on a straight, level highway. Car A travels at 60 km/h and car B at 80 km/h. The driver of car B sees car A approaching at a speed of (a) 60 km/h, (b) 80 km/h, (c) 20 km/h, (d) 140 km/h.

End of Chapter Questions and Exercises

Multiple Choice Questions

3.1 Components of Motion

1. On Cartesian axes, the x-component of a vector is generally associated with a (a) cosine, (b) sine, (c) tangent, (d) none of the foregoing.
2. The equation $x = x_0 + v_{x_0} t + \frac{1}{2} a_x t^2$ applies (a) to all kinematic problems, (b) only if v_{y_0} is zero, (c) to constant accelerations, (d) to negative times.
3. For an object in curvilinear motion, (a) the object's velocity components are constant, (b) the y-velocity component is necessarily greater than the x-velocity component, (c) there is an acceleration non-parallel to the object's path, (d) the velocity and acceleration vectors must be at right angles (90°).

Conceptual Questions

3.1 Components of Motion

1. Can the x-component of a vector be greater than the magnitude of the vector? How about the y-component? Explain.
2. Is it possible for an object's velocity to be perpendicular to the object's acceleration? If so, describe the motion.
3. Describe the motion of an object that is initially traveling with a constant velocity and then receives an acceleration of constant magnitude (a) in a direction parallel to the initial velocity, (b) in a direction perpendicular to the initial velocity, and (c) that is always perpendicular to the instantaneous velocity or direction of motion.

3.2 Vector Addition and Subtraction

4. What are the conditions for two vectors to add to zero?

5. Can a vector be less than one of its components? How about equal to one of its components?

6. Can a nonzero vector have a zero x-component? Explain.

7. Is it possible to add a vector quantity to a scalar quantity?

8. Can $\vec{A} + \vec{B}$ equal zero, when $\vec{A}$ and $\vec{B}$ have nonzero magnitudes? Explain.

3.3 Projectile Motion

9. A golf ball is hit on a level fairway. When it lands, its velocity vector has rotated through an angle of 90°. What was the launch angle of the golf ball? [*Hint:* See Figure 3.12.]

10. Figure 3.11b shows a situation of one ball dropping from rest, and at the same time, another ball projected horizontally from the same height. The two balls hit the ground at the same time. Explain.

11. In ▼ **Figure 3.23**, a spring-loaded "cannon" on a wheeled car fires a metal ball vertically. The car is given a push and set in motion horizontally with constant velocity. A pin is pulled with a string to launch the ball, which travels upward and then falls back into the moving cannon every time. Why does the ball always fall back into the cannon? Explain.

▲ FIGURE 3.23 **A ballistics car** See Conceptual Question 11 and Exercise 44.

12. A rifle is sighted-in so that a bullet hits the bull's-eye of a target 1000 m away on the same level. (a) If the same sighting is used to shoot a target uphill, should one aim above, below, or right at the bull's-eye? (b) How about shooting downhill?

3.4 Relative Velocity (Optional)

13. Sitting in a parked bus, you suddenly look up at a bus moving alongside and it appears that you are moving. Why is this? How about with both buses moving in opposite directions?

14. A student walks on a treadmill moving at 4.0 m/s and remains at the same place in the gym. (a) What is the student's velocity relative to the gym floor? (b) What is the student's speed relative to the treadmill?

15. When driving to the basket for a layup, a basketball player usually tosses the ball gently upward relative to herself. Explain why.

16. When you are riding in a fast-moving car, in what direction would you throw an object up so it will return to your hand? Explain.

Exercises*

Integrated Exercises (IEs) are two-part exercises. The first part typically requires a conceptual answer choice based on physical thinking and basic principles. The following part is quantitative calculations associated with the conceptual choice made in the first part of the exercise.

3.1 Components of Motion

1. • An airplane climbs at an angle of 15° with a horizontal component of speed of 200 km/h. (a) What is the plane's actual speed? (b) What is the magnitude of the vertical component of its velocity?

2. IE • A golf ball is hit with an initial speed of 35 m/s at an angle less than 45° above the horizontal. (a) The horizontal velocity component is (1) greater than, (2) equal to, (3) less than the vertical velocity component. Why? (b) If the ball is hit at an angle of 37°, what are the initial horizontal and vertical velocity components?

3. IE • The x- and y-components of an acceleration vector are 3.0 m/s² and 4.0 m/s², respectively. (a) The magnitude of the acceleration vector is (1) less than 3.0 m/s², (2) between 3.0 m/s² and 4.0 m/s², (3) between 4.0 m/s² and 7.0 m/s², (4) equal to 7.0 m/s². (b) What are the magnitude and direction of the acceleration vector?

4. • If the magnitude of a velocity vector is 7.0 m/s and the x-component is 3.0 m/s, what is the y-component?

5. •• The x-component of a velocity vector that has an angle of 37° to the $+x$-axis has a magnitude of 4.8 m/s. (a) What is the magnitude of the velocity? (b) What is the magnitude of the y-component of the velocity?

6. IE •• A student walks 100 m west and 50 m south. (a) To get back to the starting point, the student must walk in a general direction of (1) south of west, (2) north of east, (3) south of east, (4) north of west. (b) What displacement will bring the student back to the starting point?

7. •• A student strolls diagonally across a level rectangular campus plaza, covering the 50-m distance in 1.0 min (▶ **Figure 3.24**). (a) If the diagonal route makes a 37° angle with the long side of the plaza, what would be the distance traveled if the student had walked halfway around the outside of the plaza instead of along the diagonal route? (b) If the student had walked the outside route in 1.0 min at a constant speed, how much time would she have spent on each side?

---

* The bullets denote the degree of difficulty of the exercises: •, simple; ••, medium; and •••, more difficult.

▲ FIGURE 3.24 **Which way?** See Exercise 7.

8. •• A ball rolls at a constant velocity of 1.50 m/s at an angle of 45° below the +x-axis in the fourth quadrant. If we take the ball to be at the origin at $t = 0$ what are its coordinates (x, y) 1.65 s later?

9. •• A ball rolling on a table has a velocity with rectangular components $v_x = 0.60$ m/s and $v_y = 0.80$ m/s. What is the displacement of the ball in an interval of 2.5 s?

10. •• A hot air balloon rises vertically with a speed of 1.5 m/s. At the same time, there is a horizontal 10 km/h wind blowing. In which direction is the balloon moving?

11. **IE** •• During part of its trajectory (which lasts exactly 1 min) a missile travels at a constant speed of 2000 mi/h while maintaining a constant orientation angle of 20° from the vertical. (a) During this phase, what is true about its velocity components: (1) $v_y > v_x$, (2) $v_y = v_x$, or (3) $v_y < v_x$? [*Hint:* Make a sketch and be careful of the angle]. (b) Determine the two velocity components analytically to confirm your choice in part (a) and also calculate how high the missile will rise during this time.

12. •• At the instant a ball rolls off a rooftop it has a horizontal velocity component of +10.0 m/s and a vertical component (downward) of 15.0 m/s. (a) Determine the angle of the roof. (b) What is the ball's speed as it leaves the roof?

13. •• A particle moves at a speed of 3.0 m/s in the +x-direction. Upon reaching the origin, the particle receives a continuous constant acceleration of 0.75 m/s² in the −y-direction. What is the position of the particle 1.0 s later?

3.2 Vector Addition and Subtraction

14. • Using the triangle method, show graphically that (a) $\vec{A} + \vec{B} = \vec{B} + \vec{A}$ and (b) if $\vec{A} - \vec{B} = \vec{C}$, then $\vec{A} = \vec{B} + \vec{C}$.

15. • (a) What is the resultant if $\vec{A} = 3.0\hat{x} + 5.0\hat{y}$ is added to $\vec{B} = 1.0\hat{x} - 3.0\hat{y}$? (b) What are the magnitude and direction of $\vec{A} + \vec{B}$?

16. •• Two boys are pulling a box across a horizontal floor as shown in ▶ **Figure 3.25**. If $F_1 = 50.0$ N and $F_2 = 100$ N, find the resultant (or sum) force by (a) the graphical method and (b) the component method.

17. •• For each of the given vectors, give a vector that, when added to it, yields a *null vector* (a vector with a

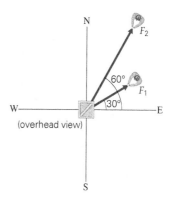

▲ FIGURE 3.25 **Adding force vectors** See Exercises 16 and 35.

magnitude of zero). Express the vector in the form other than that in which it is given (component or magnitude-angle): (a) $\vec{A} = 4.5$ cm, 40° above the +x-axis; (b) $\vec{B} = (2.0$ cm$)\hat{x} - (4.0$ cm$)\hat{y}$; (c) $\vec{C} = 8.0$ cm at an angle of 60° above the −x-axis.

18. **IE** •• (a) If each of the two components (x and y) of a vector are doubled, (1) the vector's magnitude doubles, but the direction remains unchanged; (2) the vector's magnitude remains unchanged, but the direction angle doubles; or (3) both the vector's magnitude and direction angle double. (b) If the x- and y-components of a vector of 10 m at 45° are tripled, what is the new vector?

19. •• Two vectors are given by $\vec{A} = 4.0\hat{x} - 2.0\hat{y}$ and $\vec{B} = 1.0\hat{x} + 5.0\hat{y}$. What is (a) $\vec{A} + \vec{B}$, (b) $\vec{B} - \vec{A}$, and (c) a vector $\vec{C}$ such that $\vec{A} + \vec{B} + \vec{C} = 0$?

20. •• Two brothers are pulling their other brother on a sled (▼ **Figure 3.26**). (a) Find the resultant (or sum) of the vectors $\vec{F}_1$ and $\vec{F}_2$. (b) If $\vec{F}_1$ in the figure were at an angle of 27° instead of 37° with the +x-axis, what would be the resultant (or sum) of $\vec{F}_1$ and $\vec{F}_2$? (N, newton, is the SI unit of force.)

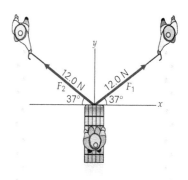

▲ FIGURE 3.26 **Vector addition** See Exercise 20.

21. •• Given two vectors, $\vec{A}$ which has a length of 10.0 and makes an angle of 45° below the $-x$-axis, and $\vec{B}$ which has an x-component of $+2.0$ and a y-component of $+4.0$, (a) sketch the vectors on x-y axes, with all their "tails" starting at the origin, and (b) calculate $\vec{A} + \vec{B}$.

22. •• The velocity of object 1 in component form is $\vec{v}_1 = (+2.0\,\text{m/s})\hat{x} + (-4.0\,\text{m/s})\hat{y}$. Object 2 has twice the speed of object 1 but moves in the opposite direction. (a) Determine the velocity of object 2 in component notation. (b) What is the speed of object 2?

23. •• For the vectors shown in ▼ **Figure 3.27**, determine $\vec{A} + \vec{B} + \vec{C}$.

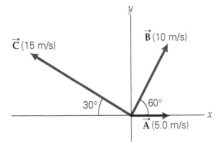

▲ FIGURE 3.27 **Adding vectors** See Exercises 23 and 24.

24. •• For the velocity vectors shown in Figure 3.27, determine $\vec{A} - \vec{B} - \vec{C}$.

25. •• Given two vectors $\vec{A}$ and $\vec{B}$ with magnitudes A and B respectively, you can subtract $\vec{B}$ from $\vec{A}$ to get a third vector $\vec{C} = \vec{A} - \vec{B}$. If the magnitude of $\vec{C}$ is equal to $C = A + B$, what is the relative orientation of vectors $\vec{A}$ and $\vec{B}$?

26. •• In two successive chess moves, a player first moves his queen two squares forward, then moves the queen three steps to the left (from the player's view). Assume each square is 3.0 cm on a side. (a) Using forward (toward the player's opponent) as the $+y$-axis and right as the $+x$-axis, write the queen's net displacement in component form. (b) At what net angle was the queen moved relative to the leftward direction?

27. •• Referring to the parallelogram in ▼ **Figure 3.28**, express $\vec{C}$, $\vec{C} - \vec{B}$ and $(\vec{E} - \vec{D} + \vec{C})$ in terms of $\vec{A}$ and $\vec{B}$.

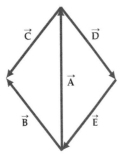

▲ FIGURE 3.28 **Vector combos** See Exercise 27.

28. •• Two force vectors, $\vec{F}_1 = (3.0\,\text{N})\hat{x} - (4.0\,\text{N})\hat{y}$ and $\vec{F}_2 = (-6.0\,\text{N})\hat{x} + (4.5\,\text{N})\hat{y}$, are applied to a particle. What third force $\vec{F}_3$ would make the net, or resultant, force on the particle zero?

29. •• A student works three problems involving the addition of two different vectors, $\vec{F}_1$ and $\vec{F}_2$. He states that the magnitudes of the three resultants are given by (a) $F_1 + F_2$, (b) $F_1 - F_2$, and (c) $\sqrt{F_1^2 + F_2^2}$. Are these results possible? If so, describe the vectors in each case.

30. •• A block weighing 50 N rests on an inclined plane. Its weight is a force directed vertically downward, as illustrated in ▼ **Figure 3.29**. Find the components of the force parallel to the surface of the plane and perpendicular to it.

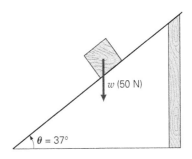

▲ FIGURE 3.29 **Block on an inclined plane** See Exercise 30.

31. •• Two displacements, one with a magnitude of 15.0 m and a second with a magnitude of 20.0 m, can have any angle you want. (a) How would you create the sum of these two vectors so it has the largest magnitude possible? What is that magnitude? (b) How would you orient them so the magnitude of the sum was at its minimum? What value would that be? (c) Generalize the result to any two vectors.

32. ••• A person walks from point A to point B as shown in ▼ **Figure 3.30**. What is the person's displacement relative to point A?

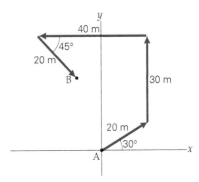

▲ FIGURE 3.30 **Adding displacement vectors** See Exercise 32.

33. IE ••• A meteorologist tracks the movement of a thunderstorm with Doppler radar. At 8:00 PM, the storm was 60 mi northeast of her station. At 10:00 PM, the storm is at 75 mi north. (a) The general direction of the thunderstorm's velocity is (1) south of east, (2) north of west, (3) north of east, (4) south of west. (b) What is the average velocity of the storm?

34. IE ••• A flight controller determines that an airplane is 20.0 mi south of him. Half an hour later, the same plane is 35.0 mi northwest of him. (a) The general direction of the airplane's velocity is (1) east of south, (2) north of west, (3) north of east, (4) west of south. (b) If the plane is flying with constant velocity, what is its velocity during this time?

35. ••• Two students are pulling a box, as shown in Figure 3.25, where $F_1 = 100$ N and $F_2 = 150$ N. What third force would cause the box to be stationary when all three forces are applied?

3.3 Projectile Motion

(*Assume angles to be exact for significant figure purposes.*)

36. • A ball with a horizontal speed of 1.0 m/s rolls off a bench 2.0 m high. (a) How long will the ball take to reach the floor? (b) How far from a point on the floor directly below the edge of the bench will the ball land?

37. • An electron is ejected horizontally at a speed of 1.5×10^6 m/s from the electron gun of an old computer monitor. If the viewing screen is 35 cm from the end of the gun, how far will the electron travel in the vertical direction before hitting the screen? Based on your answer, do you think designers need to worry about this gravitational effect?

38. • A ball rolls horizontally with a speed of 7.6 m/s off the edge of a tall platform. If the ball lands 8.7 m from the point on the ground directly below the edge of the platform, what is the height of the platform?

39. • A ball is projected horizontally with an initial speed of 5.0 m/s. Find its (a) position and (b) velocity at $t = 2.5$ s.

40. • An artillery crew wants to shell a position on level ground 35 km away. If the gun has a muzzle velocity of 770 m/s, to what angle of elevation should the gun be raised?

41. •• A pitcher throws a fastball horizontally at a speed of 140 km/h toward home plate, 18.4 m away. (a) If the batter's combined reaction and swing times total 0.350 s, how long can the batter watch the ball after it has left the pitcher's hand before swinging? (b) In traveling to the plate, how far does the ball drop from its original horizontal line?

42. IE •• Ball A rolls at a constant speed of 0.25 m/s on a table 0.95 m above the floor, and ball B rolls on the floor directly under the first ball with the same speed and direction. (a) When ball A rolls off the table and hits the floor, (1) ball B is ahead of ball A; (2) ball B collides with ball A; (3) ball A is ahead of ball B. Why? (b) When ball A hits the floor, how far from the point directly below the edge of the table will each ball be?

43. •• The pilot of a cargo plane flying 300 km/h at an altitude of 1.5 km wants to drop a load of supplies to campers at a particular location on level ground. Having the designated point in sight, the pilot prepares to drop the supplies. (a) What should the angle be between the horizontal and the pilot's line of sight when the package is released? (b) What is the location of the plane when the supplies hit the ground?

44. •• A wheeled car with a spring-loaded cannon fires a metal ball vertically (Figure 3.24). If the vertical initial speed of the ball is 5.0 m/s as the cannon moves horizontally at a speed of 0.75 m/s, (a) how far from the launch point does the ball fall back into the cannon, and (b) what would happen if the cannon were accelerating?

45. •• A convertible travels down a straight, level road at a slow speed of 13 km/h. A person in the car throws a ball with a speed of 3.6 m/s forward at an angle of 30° above the horizontal, relative to the car. Where is the car when the ball lands?

46. •• A good-guy stuntman is being chased by bad guys on a building's level roof. He comes to the edge and is to jump to the level roof of a lower building 4.0 m below and 5.0 m away. What is the minimum launch speed the stuntman needs to complete the jump? (Landing on the edge is assumed complete.)

47. •• An astronaut on the Moon fires a projectile from a launcher on a level surface so as to get the maximum range. If the launcher gives the projectile a muzzle velocity of 25 m/s, what is the range of the projectile? [*Hint:* The acceleration due to gravity on the Moon is only one-sixth of that on the Earth.]

48. •• In 2004 two Martian probes successfully landed on the Red Planet. The final phase of the landing involved bouncing the probes until they came to rest (they were surrounded by protective inflated "balloons"). During one of the bounces, the telemetry (electronic data sent back to Earth) indicated that the probe took off at 25.0 m/s at an angle of 20° and landed 110 m away (and then bounced again). Assuming the landing region was level, determine the acceleration due to gravity near the Martian surface.

49. •• In laboratory situations, a projectile's range can be used to determine its speed. To see how this is done, suppose a ball rolls off a horizontal table and lands 1.5 m out from the edge of the table. If the tabletop is 90 cm above the floor, determine (a) the time the ball is in the air, and (b) the ball's speed as it left the tabletop.

50. •• A stone thrown off a bridge 20 m above a river has an initial velocity of 12 m/s at an angle of 45° above the horizontal (▶ **Figure 3.31**). (a) What is the range of the stone? (b) At what velocity does the stone strike the water?

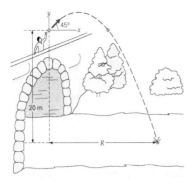

▲ **FIGURE 3.31   A view from the bridge** See Exercise 50.

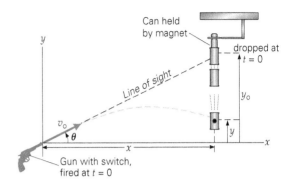

▲ **FIGURE 3.33   A sure shot** See Exercise 54. (Not drawn to scale.)

51. •• If the maximum height reached by a projectile launched on level ground is equal to half the projectile's range, what is the launch angle?

52. •• William Tell is said to have shot an apple off his son's head with an arrow. If the arrow was shot with an initial speed of 55 m/s and the boy was 15 m away, at what launch angle did Bill aim the arrow? (Assume that the arrow and apple are initially at the same height above the ground.)

53. ••• This time, William Tell is shooting at an apple that hangs on a tree (▼ **Figure 3.32**). The apple is a horizontal distance of 20.0 m away and at a height of 4.00 m above the ground. If the arrow is released from a height of 1.00 m above the ground and hits the apple 0.500 s later, what is the arrow's initial velocity?

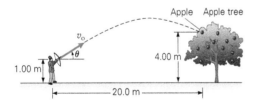

▲ **FIGURE 3.32   Hit the apple** See Exercise 53. (Not drawn to scale.)

54. ••• The apparatus for a popular lecture demonstration is shown in ▶ **Figure 3.33**. A gun is aimed directly at a can, which is released at the same time that the gun is fired. This gun won't miss as long as the initial speed of the bullet is sufficient to reach the falling target before the target hits the floor. Verify this statement, using the figure. [*Hint*: Note that $y_o = x \tan \theta$.]

55. **IE** ••• A shot-putter launches the shot from a vertical distance of 2.0 m off the ground (from just above her ear) at a speed of 12.0 m/s. The initial velocity is at an angle of 20° above the horizontal. Assume the ground is flat. (a) Compared to a projectile launched at the same angle and speed at ground level, would the shot be in the air (1) a longer time, (2) a shorter time, or (3) the same amount of time? (b) Justify your answer explicitly; determine the shot's range and velocity just before impact in unit vector (component) notation.

3.4   Relative Velocity (Optional)

56. • While you are traveling in a car on a straight, level interstate highway at 90 km/h, another car passes you in the same direction; its speedometer reads 120 km/h. (a) What is your velocity relative to the other driver? (b) What is the other car's velocity relative to you?

57. • A shopper is in a hurry to catch a bargain in a department store. She walks up the escalator, rather than letting it carry her, at a speed of 1.0 m/s relative to the escalator. If the escalator is 10 m long and moves at a speed of 0.50 m/s, how long does it take for the shopper to get to the next floor?

58. • A motorboat's speed in still water is 2.0 m/s. The driver wants to go directly across a river with a current speed of 1.5 m/s. At what angle upstream should the boat be steered?

59. • A person riding in the back of a pickup truck traveling at 70 km/h on a straight, level road throws a ball with a speed of 15 km/h relative to the truck in the direction opposite to the truck's motion. What is the velocity of the ball (a) relative to a stationary observer by the side of the road, and (b) relative to the driver of a car moving in the same direction as the truck at a speed of 90 km/h?

60. • In Exercise 59, what are the relative velocities if the ball is thrown in the direction of the truck?

61. •• In a 500-m stretch of a river, the speed of the current is a steady 5.0 m/s. How long does a boat take to finish a round trip (upstream and downstream) if the speed of the boat is 7.5 m/s relative to still water?

62. •• A moving walkway in an airport is 75 m long and moves at a speed of 0.30 m/s. A passenger, after traveling 25 m while standing on the walkway, starts to walk at a speed of 0.50 m/s relative to the surface of the walkway. How long does she take to travel the total distance of the walkway?

63. **IE** •• A swimmer swims north at 0.15 m/s relative to still water across a river that flows at a rate of 0.20 m/s from west to east. (a) The general direction of the swimmer's velocity, relative to the riverbank, is (1) north of

east, (2) south of west, (3) north of west, (4) south of east. (b) Calculate the swimmer's velocity relative to the riverbank.

64. •• A boat that travels at a speed of 6.75 m/s in still water is to go directly across a river and back (▼ **Figure 3.34**). The current flows at 0.50 m/s. (a) At what angle(s) must the boat be steered? (b) How long does it take to make the round trip? (Assume that the boat's speed is constant at all times, and neglect turnaround time.)

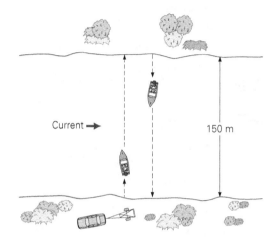

▲ **FIGURE 3.34 Over and back** See Exercise 64. (Not drawn to scale.)

65. •• A pouring rain comes straight down with a raindrop speed of 6.0 m/s. A woman with an umbrella walks eastward at a brisk pace of 1.5 m/s to get home. At what angle should she tilt her umbrella to get the maximum protection from the rain?

66. IE •• It is raining, and there is no wind. When you are sitting in a stationary car, the rain falls straight down relative to the car and the ground. But when you're driving, the rain appears to hit the windshield at an angle. (a) As the velocity of the car increases, this angle (1) also increases, (2) remains the same, (3) decreases. Why? (b) If the raindrops fall straight down at a speed of 10 m/s but appear to make an angle of 25° to the vertical, what is the speed of the car?

67. •• If the flow rate of the current in a straight river is greater than the speed of a boat in the water, the boat cannot make a trip *directly across* the river. Prove this statement.

68. IE •• You are in a fast powerboat that is capable of a sustained steady speed of 20.0 m/s in still water. On a swift, straight section of a river you travel parallel to the bank of the river. You note that you take 15.0 s to go between two trees on the riverbank that are 400 m apart. (a) (1) Are you traveling with the current, (2) are you traveling against the current, or (3) is there no current? (b) If there is a current [reasoned in part (a)], determine its speed.

69. •• An observer by the side of a straight, level, north-south road watches a car (A) moving south at a rate of 75 km/h. A driver in another car (B) going north at 50 km/h also observes car A. (a) What is car A's velocity as observed from car B? (Take north to be positive.) (b) If the roadside observer sees car A brake to a stop in 6.0 s, what constant acceleration would be measured? (c) What constant acceleration would the driver in car B measure for the braking car A?

70. ••• An airplane flies due north with an air speed of 250 km/h. A steady wind at 75 km/h blows eastward. (Air speed is the speed relative to the air.) (a) What is the plane's ground speed (v_{pg})? (b) If the pilot wants to fly due north, what should his heading be?

4

Force and Motion*

The men are applying a force (a push) to get the car moving. It is stuck because of a lack of friction between its tires and the snow.

You don't have to understand any physics to know that what's needed to get the car in the chapter-opening photo moving is a push or a pull. If the frustrated men (or the tow truck that may soon be called) can apply enough *force*, the car will move. But what's keeping the car stuck in the snow? A car relies on *friction*

to move. Here, the problem is most likely that there is not enough friction between the tires and the snowy surface.

Chapters 2 and 3 covered how to analyze motion in terms of kinematics. Now our attention turns to the study of *dynamics* – that is, what *causes* motion and changes in motion. This leads to

---

* The mathematics needed in this chapter involves trigonometric functions. You may want to review them in Appendix I.

the concepts of force and inertia. The study of force and motion occupied many early scientists. The English scientist Isaac Newton (1642–1727; ▼ **Figure 4.1**) summarized the various relationships and principles of those early scientists into three statements, or laws, which not surprisingly are known as *Newton's laws of motion*. These laws sum up the concepts of dynamics. In this chapter, you'll learn what Newton had to say about force and motion.

▲ **FIGURE 4.1   Isaac Newton** (1642–1727), one of the greatest scientific minds of all time, made fundamental contributions to mathematics, astronomy, and several branches of physics, including optics and mechanics. He formulated the laws of motion and universal gravitation (Section 7.5) and was one of the inventors of calculus. He did some of his most profound work when he was in his mid-twenties.

4.1 The Concepts of Force and Net Force

Let's first take a closer look at the meaning of force. It is easy to give examples of forces, but how would you generally define this concept? An operational definition of force is based on observed effects. That is, a force is recognized and described in terms of what it does. From your own experience, you know that *forces can produce changes in motion*. A force can set a stationary object into motion. It can also speed up or slow down a moving object and/or change the direction of its motion. In other words, *a force can produce a change in velocity (speed and/or direction) – that is, an acceleration*. Therefore, an observed *change* in motion, including motion starting from rest, is evidence of a force. This concept leads to a common definition of **force**:

> A force is something that is capable of changing an object's state of motion, that is, changing its velocity or producing an acceleration.

The word *capable* is very significant here. It takes into account the fact that a force may be acting on an object, but its capability to produce a change in motion may be balanced, or canceled, by one or more other forces. The net effect is then zero. Thus, a force may not necessarily produce a change in motion. However, it follows that if a force acts *alone*, the object on which it acts *will* have a change in velocity or an acceleration.

Since a force can produce an acceleration – a vector quantity – force must itself be a vector quantity, with both magnitude and direction. When several forces act on an object, the interest is often in their combined effect – the net force. The net force, $\vec{F}_{net}$, is the vector sum $\Sigma\vec{F}_i$, or resultant, of all the forces acting on an object or system.* Consider the opposite forces illustrated in ▼ **Figure 4.2a**. The net force is zero when forces of equal magnitude act in opposite directions (Figure 4.2b, where signs are used to indicate directions). Such forces are said to be balanced. A nonzero net force is referred to as an unbalanced force (Figure 4.2c). In this case, the situation can be analyzed as though only one force equal to the net force were acting. An unbalanced, or nonzero, net force always produces an acceleration. In some instances, an applied unbalanced force may also deform an object, that is, change its size and/or shape (as will be seen in Section 9.1). A deformation involves a change in motion for some part of an object; hence, there is an acceleration.

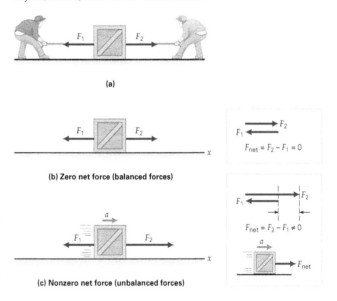

▲ **FIGURE 4.2   Net force (a)** Opposite forces are applied to a crate. **(b)** If the forces are of equal magnitude, the vector resultant, or the net force acting on the crate is zero, the forces acting on the crate are said to be balanced. **(c)** If the forces are unequal in magnitude, the resultant is not zero. A nonzero net force (F_{net}), or unbalanced force, then acts on the crate, producing an acceleration (e.g., setting the crate in motion if it was initially at rest).

---

* In the notation $\Sigma\vec{F}_i$ the Greek letter sigma means the "sum of" the individual forces, as indicated by the *i* subscript: $\Sigma\vec{F}_i = \vec{F}_1 + \vec{F}_2 + \vec{F}_3 + \cdots$, that is, a vector sum. The *i* subscript is sometimes omitted as being understood, $\Sigma\vec{F}$.

Forces are sometimes divided into two types or classes. The more familiar of these classes is *contact forces*. Such forces arise because of physical contact between objects. For example, when you push on a door to open it or throw or kick a ball, you exert a contact force on the door or ball.

The other class of forces is called *action-at-a-distance forces*. Examples of these forces include the gravitational force, the electrical force between two charges, and the magnetic force between two magnets. The Moon is attracted to the Earth and kept in orbit by a gravitational force, but there seems to be nothing physically transmitting that force. (In Chapter 30, the modern view of how such action-at-a-distance forces are thought to be transmitted is given.)

Now, with a better understanding of the concept of force, let's see how force and motion are related through Newton's laws.

4.2 Inertia and Newton's First Law of Motion

The groundwork for Newton's first law of motion was laid by Galileo. In his experimental investigations, Galileo dropped objects to observe motion under the influence of gravity. However, the relatively large acceleration due to gravity causes dropped objects to move quite fast and quite far in a short time. From the kinematic equations in Section 2.4, it can be seen that 3.0 s after being dropped, an object in free fall has a speed of about 29 m/s (64 mi/h) and has fallen a distance of 44 m (about 48 yards, or almost half the length of a football field). Thus, experimental measurements of free-fall distance versus time were particularly difficult to make with the instrumentation available in Galileo's time.

To slow things down so he could study motion, Galileo used balls rolling on inclined planes. He allowed a ball to roll down one inclined plane and then up another with a different degree of incline (▼ **Figure 4.3**). Galileo noted that the ball rolled to approximately the same height in each case, but it rolled farther in the horizontal direction when the angle of incline was smaller. When allowed to roll onto a horizontal surface, the ball traveled a considerable distance and went even farther when the surface was made smoother. Galileo wondered how far the ball would travel if the horizontal surface could be made perfectly smooth (frictionless). Although this situation is impossible to attain experimentally,

Galileo reasoned that in this ideal case with an infinitely long surface, the ball would continue to travel indefinitely with straight-line, uniform motion, since there would be nothing (no net force) to cause its motion to change. (The ball would actually slide, not roll, in this ideal case of the absence of friction.)

According to Aristotle's theory of motion, which had been accepted for about 1900 years prior to Galileo's time, the normal state of a body was to be at rest (with the exception of celestial bodies, which were thought to be naturally in motion). Aristotle no doubt observed that objects moving on a surface tend to slow down and come to rest, so this conclusion would have seemed logical to him. However, from his experiments, Galileo concluded that bodies in motion exhibit the behavior of maintaining that motion, and if an object were initially at rest, it would remain so unless something caused it to move.

Galileo called this tendency of an object to maintain its initial state of motion inertia. That is,

Inertia is the natural tendency of an object to maintain a state of rest or to remain in uniform motion in a straight line (constant velocity).

For example, if you've ever tried to stop a slowly rolling automobile by pushing on it, you felt its resistance to a change in motion. Physicists describe the property of inertia in terms of observed behavior. A comparative example of inertia is illustrated in ▼ **Figure 4.4**. If the two punching bags have the same density (mass per unit volume; see Section 1.4), the larger one has more mass and therefore more inertia, as you would quickly notice when punching each bag.

▲ FIGURE 4.4 **A difference in inertia** The larger punching bag has more mass and hence more inertia, or resistance to a change in motion.

▲ FIGURE 4.3 **Galileo's experiment** A ball rolls farther along the upward incline (but to the same height) as the angle of incline is decreased. On a smooth, horizontal surface, the ball rolls a greater distance before coming to rest. How far would the ball travel on an ideal, perfectly smooth surface? (The ball would slide, rather than roll, in this case because of the absence of friction.)

Newton related the concept of inertia to mass. Originally, he called mass a quantity of matter, but later redefined it as follows:

Mass is a quantitative measure of inertia.

That is, a massive object has more inertia, or more resistance to a change in motion, than does a less massive object. For example, a car has more inertia than a bicycle.

Newton's first law of motion, sometimes called the *law of inertia*, summarizes these observations:

In the absence of an unbalanced applied force ($\vec{F}_{net} = 0$), a body at rest remains at rest, and a body in motion remains in motion with a constant velocity (constant speed and direction).

That is, if the net force acting on an object is zero, then its acceleration is zero. It may be moving with a constant velocity, or be at rest – in both cases, $\Delta\vec{v} = 0$ or $\vec{v}$ = constant.

4.3 Newton's Second Law of Motion

A change in motion, or an acceleration (i.e., a change in velocity – speed and/or direction), is evidence of a net force. All experiments indicate that the acceleration of an object is directly proportional to, and in the direction of, the applied net force; that is, in vector notation,

$$\vec{a} \propto \vec{F}_{net}$$

For example, suppose you separately cued two identical billiard balls. If you hit the second ball twice as hard as the first (i.e., you applied twice as much force), you would expect the acceleration of the second ball to be twice as great as that of the first ball (and still in the direction of the force).

However, as Newton recognized, the inertia or mass of the object also plays a role. For a given net force, the more massive the object, the less its acceleration will be. For example, if you hit two balls of different masses with the same force, the less massive ball would experience a greater acceleration. Specifically, the acceleration is inversely proportional to mass:

$$\vec{a} \propto \frac{\vec{F}_{net}}{m}$$

or in words,

The acceleration of an object is directly proportional to the net force acting on it and inversely proportional to its mass. The direction of the acceleration is in the direction of the applied net force.

▼ **Figure 4.5** presents some illustrations of this principle.

Rewritten as $\vec{F}_{net} \propto m\vec{a}$. **Newton's second law of motion** is commonly expressed in equation form as

$$\vec{F}_{net} = m\vec{a} \quad \text{(Newton's second law)} \tag{4.1}$$

SI unit of force: newton (N) or kilogram-meter per second squared (kg·m/s²)

where $\vec{F}_{net} = \Sigma\vec{F}_i$. Equation 4.1 defines the SI unit of force, which is appropriately called the **newton (N)**.

By unit analysis, Equation 4.1 shows that a newton in base units is defined as $1\,\text{N} = 1\,\text{kg·m/s}^2$. That is, a net force of 1 N gives a mass of 1 kg an acceleration of 1 m/s² (▶ **Figure 4.6**). The British system unit of force is the pound (lb). One pound is equivalent to about 4.5 N (actually, 4.448 N). An average apple weighs about 1 N.

Newton's second law, $\vec{F}_{net} = m\vec{a}$, allows the quantitative analysis of force and motion. It might be thought of as a cause-and-effect relationship, with the force being the cause and acceleration being the motional effect. Notice that if the net force acting on an object is zero, the object's acceleration is zero,

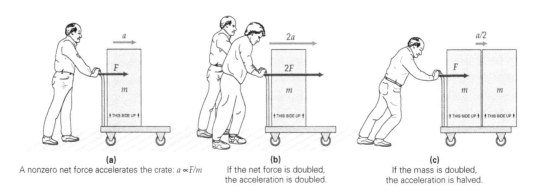

(a)
A nonzero net force accelerates the crate: $a \propto F/m$

(b)
If the net force is doubled, the acceleration is doubled.

(c)
If the mass is doubled, the acceleration is halved.

▲ FIGURE 4.5 **Newton's second law** The relationships among force, acceleration, and mass shown here are expressed by Newton's second law of motion (assuming no friction on the cart wheels, which would slide).

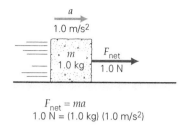

$$F_{net} = ma$$
$$1.0 \text{ N} = (1.0 \text{ kg}) (1.0 \text{ m/s}^2)$$

▲ FIGURE 4.6   **The newton (N)** A net force of 1.0 N acting on a mass of 1.0 kg produces an acceleration of 1.0 m/s².

and it remains at rest or in uniform motion, which is consistent with Newton's first law. For a nonzero net force (an unbalanced force), the resulting acceleration is in the same direction as the net force.

4.3.1 Mass and Weight

Equation 4.1 can be used to relate mass and weight. Recall from Section 1.2 that weight is the gravitational force of attraction that a celestial body exerts on an object. For us, this is mainly the force of the gravitational attraction of the Earth. Its effects are easily demonstrated: when you drop an object, it falls (accelerates) toward the Earth. Since only one force is acting on the object (air resistance neglected, free fall), its weight ($\vec{w}$) is the net force $\vec{F}_{net}$, and the acceleration due to gravity ($\vec{g}$) can be substituted for $\vec{a}$ in Equation 4.1. Therefore in terms of magnitudes,

$$w = mg$$
$$(F_{net} = ma) \tag{4.2}$$

Thus the weight of an object with 1.0 kg of mass is $w = mg = (1.0 \text{ kg})(9.8 \text{ m/s}^2) = 9.8$ N, or 2.2 lb, near the Earth's surface. Although weight and mass are simply related through Equation 4.2, keep in mind that *mass is the fundamental property*. Mass doesn't depend on the value of g, but weight does. As pointed out previously, the acceleration due to gravity on the Moon is about one-sixth that on the Earth. The weight of an object on the Moon would thus be one-sixth of its weight on the Earth, but its mass, which reflects the quantity of matter it contains and its inertia, would be the same in both places.

Newton's second law, along with the fact that $w \propto m$, explains why all objects in free fall have the same acceleration (Section 2.5). Consider, for example, two falling objects, one with twice the mass of the other. The object with twice as much mass would have twice as much weight, or two times as much gravitational force acting on it. But the more massive object also has twice the inertia, so twice as much force is needed to give it the same acceleration. Expressing this relationship mathematically, for the smaller mass (m), the acceleration is $a = F_{net}/m = mg/m = g$, and for the larger mass ($2m$), the acceleration is the same: $a = F_{net}/m = 2mg/(2m) = g$ (▼ Figure 4.7).

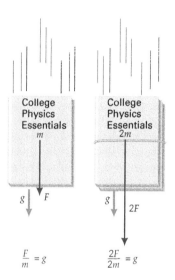

▲ FIGURE 4.7   **Newton's second law and free fall** In free fall, all objects fall with the same constant acceleration g. An object with twice the mass of another has twice as much gravitational force acting on it. But with twice the mass, the object also has twice the inertia, so twice as much force is needed to give it the same acceleration.

EXAMPLE 4.1: NEWTON'S SECOND LAW – FINDING ACCELERATION

A tractor pulls a loaded wagon on a level road with a constant horizontal force of 440 N (▼ Figure 4.8). If the mass of the wagon is 200 kg and that of the load is 75 kg, what is the magnitude of the wagon's acceleration? (Ignore frictional forces.)

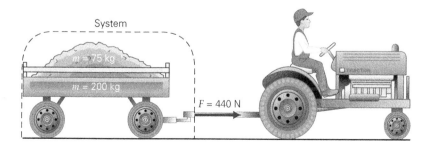

▲ FIGURE 4.8   **Force and acceleration** See Example 4.1 for description.

THINKING IT THROUGH. This problem is a direct application of Newton's second law. The two separate masses (wagon and contents) make up the system.

SOLUTION

Listing the given data and what is to be found:

Given:

$F = 440$ N
$m_1 = 200$ kg (wagon)
$m_2 = 75$ kg (load)

Find:

a (acceleration)

In this case, F is the net force, and the acceleration is given by Equation 4.2, $F_{net} = ma$, where m is the total mass. Solving for the magnitude of a,

$$a = \frac{F_{net}}{m} = \frac{F}{m_1 + m_2} = \frac{440\,\text{N}}{200\,\text{kg} + 75\,\text{kg}} = 1.60 \text{ m/s}^2$$

and the direction of a is in the direction of F_{net} or the direction in which the tractor is pulling. Note that m is the *total* mass of the wagon and its contents. In reality, there would be a total opposing force of friction, $-f$. Suppose there were an effective frictional force of magnitude $f = 140$ N. In this case, the net force would be the vector sum of the force exerted by the tractor and the frictional force. Then the acceleration would be (using directional signs)

$$a = \frac{F_{net}}{m} = \frac{F - f}{m_1 + m_2} = \frac{440\,\text{N} - 140\,\text{N}}{275\,\text{kg}} = 1.09 \text{ m/s}^2$$

Again, the direction of a is in the direction of F_{net}.

With a constant net force, the acceleration is also constant, so the kinematic equations of Section 2.4 can be applied. Suppose the wagon started from rest ($v_0 = 0$). Could you find how far it traveled in 4.00 s? Using the appropriate kinematic equation (Equation 2.11, with $x_0 = 0$) for the case with friction,

$$x = v_0 t + \frac{1}{2} a t^2 = 0 + \frac{1}{2}(1.09\,\text{m/s}^2)(4.00\,\text{s})^2 = 8.72 \text{ m}$$

FOLLOW-UP EXERCISE. Suppose the applied force on the wagon is 550 N. With the same frictional force, what would be the wagon's velocity 4.00 s after starting from rest? (*Answers to all Follow-Up Exercises are given in Appendix V at the back of the book.*)

EXAMPLE 4.2: NEWTON'S SECOND LAW – FINDING MASS

A student weighs 588 N. What is her mass?

THINKING IT THROUGH. Newton's second law allows us to determine an object's mass if we know the object's weight (force), since g is known.

Given:

$w = 588$ N

Find:

m (mass)

SOLUTION

Recall that weight is a (gravitational) force and it is related to the mass of an object by $w = mg$ (Equation 4.2), where g is the acceleration due to gravity (9.80 m/s²). Rearranging the equation,

$$m = \frac{w}{g} = \frac{588\,\text{N}}{9.80\,\text{m/s}^2} = 60.0 \text{ kg}$$

In countries that use the metric system, the kilogram unit of mass is used to express "weight" rather than a force unit. It would be said that this student weighs 60.0 "kilos."

Recall that 1 kg of mass weighs 2.2 lb on the Earth's surface. Then in British units, her weight would be 132 lb.

FOLLOW-UP EXERCISE. (a) A person in Europe is a bit overweight and would like to lose 5.0 "kilos." What would be the equivalent loss in pounds? (b) What is your weight in kilos?

As has been learned, a dynamic system may consist of more than one object. In applications of Newton's second law, it is often advantageous, and sometimes necessary, to isolate a given object within a system. This isolation is possible because *the motion of any part of a system is also described by Newton's second law*, as Example 4.3 shows.

EXAMPLE 4.3: NEWTON'S SECOND LAW – ALL OR PART OF THE SYSTEM?

Two blocks with masses $m_1 = 2.5$ kg and $m_2 = 3.5$ kg rest on a frictionless surface and are connected by a light string (▼ **Figure 4.9**).* A horizontal force (F) of 12.0 N is applied to m_1, as shown in the figure. (a) What is the magnitude of the acceleration of the masses

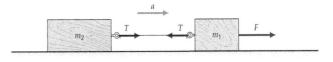

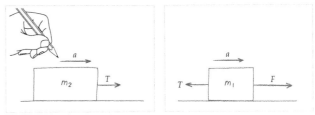

Isolating the masses

▲ **FIGURE 4.9 An accelerated system** See Example 4.3 for description.

---

* When an object is described as being "light," its mass can be ignored in analyzing the situation given in the problem. That is, here the mass of the string is negligible relative to the other masses.

(i.e., of the total system)? (b) What is the magnitude of the force (T) in the string? (When a rope or string is stretched taut, it is said to be under tension.)

THINKING IT THROUGH. It is important to remember that Newton's second law may be applied to a total system or any part of it (a subsystem, so to speak). This capability allows for the analysis of a particular component of a system, if desired. Identification of all of the acting forces is critical, as this Example shows. Then $F_{net} = ma$ is applied to each subsystem or component.

SOLUTION

Carefully listing the data and what is to be found:

Given:

$m_1 = 2.5$ kg
$m_2 = 3.5$ kg
$F = 12.0$ N

Find:

(a) a (acceleration)
(b) T (tension, a force)

Given an applied force, the acceleration of the masses can be found from Newton's second law. It is important to keep in mind that Newton's second law applies to the total system *or to any part of it* – that is, to the total mass ($m_1 + m_2$) or individually to m_1 or m_2. However, *you must be sure to correctly identify the appropriate force or forces in each case.* The net force acting on the combined masses, for example, is not the same as the magnitude of the net force acting on m_2 considered separately, as will be seen.

(a) First, taking the system as a whole (i.e., considering both m_1 and m_2), the net force acting on this system is F. Note that in considering the total system, we are concerned only about the net external force acting on it. The *internal* equal and opposite T forces are not a consideration in this case, since they cancel each other. Since the two objects are connected, they have the save velocity and acceleration. Then, using Newton's second law:

$$a = \frac{F_{net}}{m} = \frac{F}{m_1 + m_2} = \frac{12.0\,\text{N}}{2.5\,\text{kg} + 3.5\,\text{kg}} = 2.0\ \text{m/s}^2$$

The acceleration of both blocks is in the direction of the applied force, as indicated in the figure.

Now, consider each object individually (i.e., consider either m_1 or m_2).

Under tension, a force is exerted on an object by a string is directed along the string. Note in the figure that it is assumed the tension is transmitted *undiminished* through the string. That is, the tension is the same everywhere in the string. Thus, the magnitude of T acting on m_2 is the same as that acting on m_1. This is actually true only if the string has zero mass. Only such idealized *light* (i.e., of negligible mass) strings or ropes will be considered in this book.

Isolating the masses, and referring to Figure 4.9, we can see that there are two forces acting on m_1 (F to the right and T to the left) and only one force acting on m_2 (T to the right).

Taking the direction to the right as positive and applying Newton's second law to m_1:

$$F_{net_1} = F - T = m_1 a \tag{1}$$

to m_2:

$$F_{net_2} = T = m_2 a \tag{2}$$

Substituting (2) into (1), $F - (m_2 a) = m_1 a$ or $F = m_1 a + m_2 a = (m_1 + m_2)a$.
Thus,

$$a = \frac{F}{m_1 + m_2} = \frac{12.0\,\text{N}}{2.5\,\text{kg} + 3.5\,\text{kg}} = 2.0\,\text{m/s}^2$$

which is the same result obtained earlier when both objects are considered as a system.

(b) To find the tension in the string, we can solve either Equation (1) or (2).
From Equation (1): $F - T = m_1 a$ so
$T = F - m_1 a = 12.0$ N $- (2.5$ kg$)(2.0$ m/s$^2) = 7.0$ N
From Equation (2): $T = m_2 a = (3.5$ kg$)(2.0$ m/s$^2) = 7.0$ N.

FOLLOW-UP EXERCISE. Suppose that an additional horizontal force to the left of 3.0 N is applied to m_2 in Figure 4.9. What would be the tension in the connecting string in this case?

4.3.2 The Second Law in Component Form

Not only does Newton's second law hold for any part of a system, but it also applies to each component of the acceleration. For example, a force may be expressed in component notation in two dimensions as follows:

$$\Sigma \vec{F}_i = m\vec{a}$$

and

$$\Sigma (F_x \hat{\mathbf{x}} + F_y \hat{\mathbf{y}}) = m(a_x \hat{\mathbf{x}} + a_y \hat{\mathbf{y}}) = ma_x \hat{\mathbf{x}} + ma_y \hat{\mathbf{y}} \tag{4.3a}$$

Hence, to satisfy both x and y directions independently,

$$\Sigma F_x = ma_x \quad \text{and} \quad \Sigma F_y = ma_y \tag{4.3b}$$

and Newton's second law applies separately to each component of motion. (Also, $\Sigma F_z = ma_z$ in three dimensions.) The components in the equations will be either positive or negative numbers depending on whether along the positive or negative x- or y-axis. Example 4.4 illustrates how the second law is applied using components.

EXAMPLE 4.4: NEWTON'S SECOND LAW – COMPONENTS OF FORCE

A block of mass 0.50 kg travels with a speed of 2.0 m/s in the $+x$-direction on a flat, frictionless surface. On passing through the

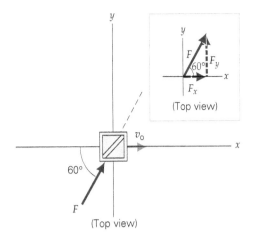

▲ **FIGURE 4.10 Off the straight and narrow** A force is applied to a moving block when it reaches the origin, and the block then begins to deviate from its straight-line path. The vector components are shown in the box.

origin, the block experiences a constant force of 3.0 N at an angle of 60° relative to the x-axis (in the x – y plane) for 1.5 s (▲ **Figure 4.10**, which is a view of the surface from above). What is the velocity of the block at the end of this time?

THINKING IT THROUGH. With the force at an angle to the initial motion, it would appear that the solution is complicated. But note in the insert in Figure 4.10 that the force can be resolved into components. The motion can then be analyzed in each component direction.

SOLUTION

Listing the given data and what is to be found:

Given:

$m = 0.50$ kg
$v_{x_o} = 2.0$ m/s, $v_{y_o} = 0$
$F = 3.0$ N, $\theta = 60°$
$t = 1.5$ s

Find:

$\vec{v}$ (velocity at the end of 1.5 s)

First let's find the magnitudes of the force components:

$$F_x = F\cos 60° = (3.0\,\text{N})(0.500) = 1.5\,\text{N}$$
$$F_y = F\sin 60° = (3.0\,\text{N})(0.866) = 2.6\,\text{N}$$

Then, applying Newton's second law to each direction to find the components of acceleration,

$$a_x = \frac{F_x}{m} = \frac{1.5\,\text{N}}{0.50\,\text{kg}} = 3.0\,\text{m/s}^2$$
$$a_y = \frac{F_y}{m} = \frac{2.6\,\text{N}}{0.50\,\text{kg}} = 5.2\,\text{m/s}^2$$

Next, from the kinematic equation relating velocity and acceleration (Equation 2.8), the magnitudes of the velocity components of the block are given by

$$v_x = v_{x_o} + a_x t = 2.0\,\text{m/s} + (3.0\,\text{m/s}^2)(1.5\,\text{s}) = 6.5\,\text{m/s}$$
$$v_y = v_{y_o} + a_y t = 0 + (5.2\,\text{m/s}^2)(1.5\,\text{s}) = 7.8\,\text{m/s}$$

And, at the end of the 1.5 s, the velocity of the block is

$$\vec{v} = v_x\hat{\mathbf{x}} + v_y\hat{\mathbf{y}} = (6.5\,\text{m/s})\hat{\mathbf{x}} + (7.8\,\text{m/s})\hat{\mathbf{y}}$$

FOLLOW-UP EXERCISE. (a) What is the direction of the velocity at the end of the 1.5 s? (b) If the force were applied at an angle of 30° (rather than 60°) relative to the x-axis, how would the results of this Example be different?

4.4 Newton's Third Law of Motion

Newton formulated a third law that is as far-reaching in its physical significance as the first two laws. For a simple introduction to the third law, consider the forces involved in seatbelt safety. When the brakes are suddenly applied when you are riding in a moving car, because of your inertia you continue to move forward as the car slows. (The frictional force on the seat of your pants is not enough to stop you.) In doing so, you exert forward forces on the seatbelt and shoulder strap. The belt and strap exert corresponding backward reaction forces on you, causing you to slow down with the car. If you hadn't buckled up, you would keep going (Newton's first law) until another backward force, such as that applied by the dashboard or windshield, slowed you down. (In an abrupt collision stop, hopefully the air bags would come into effect.)

Newton recognized that in any force application, there is always a mutual interaction; therefore, forces occur in pairs. An example given by Newton was the following: If you press on a stone with a finger, then the finger is also pressed by, or receives a force from, the stone.

Newton termed the paired forces *action* and *reaction*, and **Newton's third law of motion** is as follows:

> For every force (action), there is an equal and opposite force (reaction).

In symbol notation, Newton's third law may be expressed:

$$\vec{\mathbf{F}}_{12} = -\vec{\mathbf{F}}_{21}$$

That is, $\vec{\mathbf{F}}_{12}$ is the force exerted *on* object 1 *by* object 2, and $-\vec{\mathbf{F}}_{21}$ is the equal and opposite force exerted *on* object 2 *by* object 1. (The negative sign indicates the opposite direction.) *Which force is considered the action or the reaction is arbitrary;* $\vec{F}_{21}$ may be the reaction to $\vec{F}_{12}$ or vice-versa.

At a glance, Newton's third law may seem to contradict Newton's second law: If there are always equal and opposite forces, how can there be a nonzero net force? An important thing to remember about the force pair of the third law is that *the action-reaction forces do not act on the same object.* The second law is concerned with a force (or forces) acting on a particular object (or system). The opposing forces of the third law act on *different* objects. Hence, these forces cannot cancel each other or have a vector sum of zero when the second law is applied to the individual objects.

To illustrate this distinction, consider the situations shown in ▼ Figure 4.11. We often tend to forget reaction forces. For example, in the left portion of Figure 4.11, the obvious force that acts on a block sitting on a table is the Earth's gravitational attraction, which is expressed by the weight *mg.* But *there has to be another force* acting on the block. For the block not to accelerate, the table must exert an upward force N of which the magnitude is equal to the block's weight. Thus,

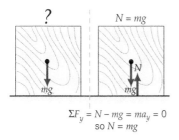

$$\Sigma F_y = N - mg = ma_y = 0$$
$$\text{so } N = mg$$

▲ FIGURE 4.11 **Distinctions between Newton's second and third laws** Newton's second law deals with the forces acting on a particular object. Newton's third law deals with the force pair that acts on different objects. See text for description.

$$\Sigma F_y = +N - mg = ma_y = 0 \quad \text{so} \quad N = mg$$

This upward force N is *not* the reaction force to the weight of the block! It is a reaction force to the block pushing down on the table.

The force that a surface exerts on an object is called a *normal* force and the symbol N is used to denote the force. *Normal* means *perpendicular.* The normal force that a surface exerts on an object is always perpendicular to the surface.

CONCEPTUAL EXAMPLE 4.5: WHERE ARE THE NEWTON'S THIRD LAW FORCE PAIRS?

A woman waiting to cross the street holds a briefcase in her hand as shown in ▶ Figure 4.12a. Identify all of the third law force pairs involving the briefcase in this situation.

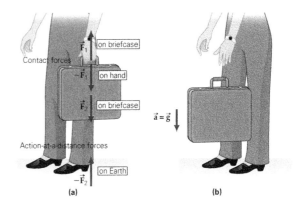

▲ FIGURE 4.12 **Force pairs of Newton's third law (a)** When a person holds a briefcase, there are two force pairs: a contact pair ($\vec{F}_1$ and $-\vec{F}_1$) and an action-at-a-distance (gravity) pair ($\vec{F}_2$ and $-\vec{F}_2$). The net force acting on the briefcase is zero: the upward contact force ($\vec{F}_1$ on the briefcase) balances the downward weight force. Note, however, that the upward contact force and downward weight force are *not* a third law pair. **(b)** Any third law force pairs? See the Follow-Up Exercise.

REASONING AND ANSWER. The briefcase is being held motionless, so its acceleration is zero, and $\Sigma F_y = 0$. Focusing only on the case, two equal and opposite forces acting on it can be identified – the downward weight of the case and the upward applied force by the hand. However, these two forces *cannot* be a third law force pair because they act on the *same* object.

On an overall inspection, you should realize that the reaction force to the upward force of the hand on the briefcase is a downward force on the hand. Then how about the reaction force to the weight of the case? Since weight is the attractive gravitational force on the case by the Earth, the corresponding force on the Earth by the case makes up the third law force pair.

FOLLOW-UP EXERCISE. The woman inadvertently drops her briefcase as illustrated in Figure 4.12b. Are there any third law force pairs in this situation? Explain.

Jet propulsion is yet another example of Newton's third law in action. In the case of a rocket, the rocket and exhaust gases exert equal and opposite forces on each other. As a result, the exhaust gases are accelerated away from the rocket, and the rocket is accelerated in the opposite direction. When a big rocket "blasts off," as in a reusable rocket launch to the International Space Station, it produces a fiery release of exhaust.

A common misconception is that the exhaust gases "push against" the launch pad to accelerate the rocket. If this interpretation were true, there would be no space travel, since there is nothing to "push against" in space. The exhaust gas pushes against the launch pad and thus does not exert a force on the rocket and therefore cannot accelerate it, but it can move or damage the pad. What does accelerate the rocket is the force exerted on the rocket by the expanding exhaust gas. This force is the reaction to the force exerted on the gas by the rocket.

4.5 More on Newton's Laws: Free-Body Diagrams and Translational Equilibrium

Now that you have been introduced to Newton's laws and some applications in analyzing motion, the importance of these laws should be evident. They are so simply stated, yet so far-reaching. The second law is probably the most often applied, because of its mathematical relationship. However, the first and third laws are often used in qualitative analysis, as our continuing study of the different areas of physics will reveal.

In general, we will be concerned with applications that involve constant forces. Constant forces result in constant accelerations and allow the use of the kinematic equations from Section 2.4 in analyzing the motion. When there is a variable force, Newton's second law holds for the *instantaneous* force and acceleration, but the acceleration will vary with time, requiring advanced mathematics to analyze. So in general, our study will be limited to constant accelerations and forces. Several examples of applications of Newton's second law are presented in this section so that you can become familiar with its use. This small but powerful equation will be used again and again throughout the book.

There is still one more item in the problem-solving arsenal that is a great help with force applications – free-body diagrams. These are explained in the following Problem-Solving Strategy.

4.5.1 Problem-Solving Strategy: Free-Body Diagrams

In illustrations of physical situations, sometimes called *space diagrams*, force vectors are drawn at different locations to indicate their points of application. However, presently being concerned with only linear motions, vectors in *free-body diagrams* (FBDs) may be shown as emanating from a common point, which is usually chosen as the origin of the *x–y* axes. One of the axes is generally chosen along the direction of the net force acting on an object, since that is the direction in which the object will accelerate. Also, it is often important to resolve force vectors into components, and properly chosen *x–y* axes simplify this task.

In a free-body diagram, the vector arrows do not have to be drawn exactly to scale. However, the diagram should clearly show whether there is a net force and whether forces balance each other in a particular direction. When the forces aren't balanced, by Newton's second law, there must be an acceleration.

In summary, the general steps in constructing and using free-body diagrams are shown in ▶ **Figure 4.13**, Forces on an Object on an Inclined Plane and Free-Body Diagrams.

1. Make a sketch, or space diagram, of the situation (if one is not already available) and identify the forces acting on each body of the system. A space diagram is an illustration of the physical situation that identifies the force vectors.

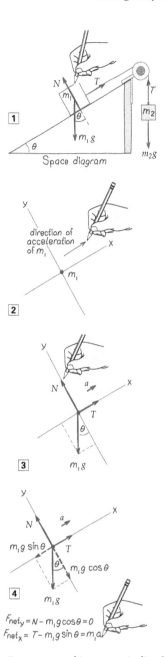

▲ FIGURE 4.13 **Forces on an object on an inclined plane and free-body diagrams** See text for description.

2. Isolate the body for which the free-body diagram is to be constructed. Draw a set of Cartesian axes, with the origin at a point through which the forces act and with one of the axes along the direction of the body's acceleration. (The acceleration will be in the direction of the net force, if there is one.)

3. Draw properly oriented force vectors (including angles) on the diagram, emanating from the origin of the axes. If there is an unbalanced force, assume a direction of acceleration and indicate it with an acceleration vector. Be sure to include only those forces that act on the isolated body of interest.

4. Resolve any forces that are not directed along the *x*- or *y*-axis into *x*- or *y*-components (use plus and minus signs to indicate direction). Use the free-body diagram and force components to analyze the situation in terms of Newton's second law of motion. (*Note*: If you assume that the acceleration is in one direction, and in the solution it comes out with the opposite sign, then the acceleration is actually in the opposite direction from that assumed. For example, if you assume acceleration is in the $+x$-direction, but you get a negative answer, then the acceleration is in the $-x$-direction.)

Free-body diagrams are a particularly useful way of following one of the suggested problem-solving procedures in Section 1.7: Draw a diagram as an aid in visualizing and analyzing the physical situation of the problem. *Make it a practice to draw free-body diagrams for force problems, as done in the following Examples.*

EXAMPLE 4.6: UP OR DOWN? MOTION ON A FRICTIONLESS INCLINED PLANE

Two masses are connected by a light string running over a light pulley of negligible friction, as illustrated in the space diagram (1) of Figure 4.13. One mass ($m_1 = 5.0$ kg) is on a frictionless 20° inclined plane, and the other ($m_2 = 1.5$ kg) is freely suspended. What is the acceleration of the masses?

THINKING IT THROUGH. Apply the preceding Problem-Solving Strategy.

SOLUTION

Following the usual procedure of listing the data and what is to be found:

Given:

> $m_1 = 5.0$ kg
> $m_2 = 1.5$ kg
> $\theta = 20°$

Find:

> *a* (acceleration)

(To help visualize the forces involved, isolate m_1 and m_2 and draw free-body diagrams for each mass.) For mass m_1, there are three concurrent forces (forces acting through a common point). These forces are *T*, its weight $m_1 g$, and *N*, where *T* is the tension force of the string on m_1 and *N* is the normal force of the plane on the block. (See 3 in Figure 4.13.) The forces are shown as emanating from their common point of action. (Recall that a vector can be moved as long as its direction and magnitude are not changed.)

Start by assuming that m_1 accelerates up the plane, which is taken to be in the $+x$-direction. (It makes no difference whether it is assumed that m_1 accelerates up or down the plane, as will be seen shortly.) Notice that $m_1 g$ (the weight) is broken down into components. The *x*-component is opposite to the assumed direction of acceleration, and the *y*-component acts perpendicularly to the plane and is balanced by the normal force *N*. (There is no acceleration in the *y*-direction, so there is no net force in this direction.)

Then, applying Newton's second law in component form (Equation 4.3b) to m_1,

$$\Sigma F_{x_1} = T - m_1 g \sin\theta = m_1 a_{x_1}$$
$$\Sigma F_{y_1} = T - m_1 g \cos\theta = m_1 a_{y_1} = 0$$
$$(a_y = 0, \text{ no net force, so the forces cancel})$$

And for m_2,

$$\Sigma F_{y_2} = m_2 g - T = m_2 a_{y_2}$$

where the masses of the string and pulley have been neglected. Since the two masses are connected by a string, the accelerations of m_1 and m_2 have the same magnitudes:

$$a_{x_1} = a_{y_2} = a$$

Then adding the first and last equations to eliminate *T*,

$$m_2 g - m_1 g \sin\theta = (m_1 + m_2)a$$
$$(\text{net force} = \textit{total} \text{ mass} \times \text{acceleration})$$

(Note that this is the equation that would be obtained by applying Newton's second law to the system as a whole, because in the system of both blocks, the *T* forces are internal forces and cancel.)

Then, solving for *a*:

$$a = \frac{m_2 g - m_1 g \sin 20°}{m_1 + m_2}$$
$$= \frac{(1.5 \text{kg})(9.8 \text{m/s}^2) - (5.0 \text{kg})(9.8 \text{m/s}^2)(0.342)}{5.0 \text{kg} + 1.5 \text{kg}}$$
$$= -0.32 \text{m/s}^2$$

The negative sign indicates that the acceleration is opposite to the assumed direction. That is, m_1 actually accelerates down the plane, and m_2 accelerates upward. As this example shows, if you assume the acceleration to be in the wrong direction, the sign on the result will give the correct direction anyway.

Could you find the tension force *T* in the string if asked to do so? How this task could be done should be quite evident from the free-body diagram.

FOLLOW-UP EXERCISE. (a) In this Example, what is the minimum amount of mass for m_2 that would cause m_1 not to accelerate up or down the plane? (b) Keeping the masses the same as in the Example, how should the angle of incline be adjusted so that m_1 would not accelerate up or down the plane?

EXAMPLE 4.7: COMPONENTS OF FORCE AND FREE-BODY DIAGRAMS

A force of 10.0 N is applied at an angle of 30° to the horizontal on a 1.25-kg block initially at rest on a frictionless surface, as illustrated in ▶ **Figure 4.14**. (a) What is the magnitude of the block's acceleration? (b) What is the magnitude of the normal force?

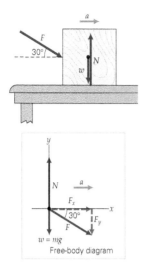

▲ FIGURE 4.14 Finding force from motional effects See Example 4.7 for description.

THINKING IT THROUGH. The applied force may be resolved into components. The horizontal component accelerates the block. The vertical component affects the normal force. (Review Figure 4.11 if necessary.) Drawing a free-body diagram for the block, as in Figure 4.14, is helpful.

SOLUTION

Given:

$F = 10.0$ N
$m = 1.25$ kg
$\theta = 30°$
$v_{\text{o}} = 0$

Find:

(a) a (acceleration)
(b) N (normal force)

(a) The acceleration of the block can be calculated using Newton's second law, and the axes are chosen so that acceleration is in the $+x$-direction. As the free-body diagram shows, only a component (F_x) of the applied force F acts in this direction. The component of F in the direction of motion is $F_x = F \cos \theta$. Applying Newton's second law in the x-direction to calculate the acceleration:

$$F_x = F \cos 30° = ma_x$$

and

$$a_x = \frac{F \cos 30°}{m} = \frac{(10.0\,\text{N})(0.866)}{1.25\,\text{kg}} = 6.93\,\text{m/s}^2$$

(b) The acceleration found in part (a) is the acceleration of the block, since the block accelerates only in the x-direction,

$a_y = 0$, thus the sum of the forces in the y-direction must be zero. That is, the downward component of F acting on the block, F_y, and its downward weight force, w, must be balanced by the upward normal force N that the surface exerts on the block. If this were not the case, then there would be a net force and an acceleration in the y-direction.

Summing the forces in the y-direction with upward taken as positive,

$$\Sigma F_y = N - F_y - w = 0$$

or

$$N - F \sin 30° - mg = 0$$

and

$$N = F \sin 30° + mg = (10.0\,\text{N})(0.500) + (1.25\,\text{kg})(9.80\,\text{m/s}^2)$$
$$= 17.3\,\text{N}$$

The surface then exerts a force of 17.3 N upward on the block, which balances the sum of the downward forces acting on it.

FOLLOW-UP EXERCISE. (a) Suppose the applied force on the block is applied for only a short time. What is the magnitude of the normal force after the applied force is removed? (b) If the block slides off the edge of the table, what would be the net force on the block just after it leaves the table (with the applied force removed)?

4.5.2 Problem-Solving Hint

There is no single fixed way to go about solving a problem. However, some general strategies or procedures are helpful in solving problems involving Newton's second law. When using the suggested problem-solving procedures introduced in Section 1.7, you might include the following steps when solving problems involving force applications:

- Draw a free-body diagram for each individual body, showing all of the forces acting on that body.
- Depending on what is to be found, apply Newton's second law either to the system as a whole (in which case internal forces cancel) or to a part of the system. Basically, *you want to obtain an equation (or set of equations) containing the quantity for which you want to solve.* Review Example 4.3. (If there are two unknown quantities, application of Newton's second law to two parts of the system may give you two equations and two unknowns. See Example 4.6.)
- Keep in mind that Newton's second law may be applied to components of acceleration and that forces may have to be resolved into components to do this. Review Example 4.7.

4.5.3 Translational Equilibrium

Several forces may act on an object without producing an acceleration. In such a case, with $\vec{a} = 0$, from Newton's second law,

$$\Sigma\vec{F}_i = 0 \quad \text{(translational equilibrium only)} \qquad (4.4)$$

That is, the vector sum of the forces, or the net force, is zero, so the object either remains at rest (as in ▼ **Figure 4.15**) *or* moves with a constant velocity (Newton's first law). In such cases, objects are said to be in **translational equilibrium**. When remaining at rest, an object is said to be in *static translational equilibrium*.

▲ **FIGURE 4.15 Several forces, no acceleration** Several external forces act on this gymnast. Nevertheless, he experiences no acceleration (static translational equilibrium). Why?

It follows that the sums of the rectangular components of the forces for an object in translational equilibrium are also zero (why?):

$$\begin{aligned} \Sigma F_x &= 0 \\ \Sigma F_y &= 0 \end{aligned} \quad \text{(translational equilibrium only)} \qquad (4.5)$$

Equations 4.5 give what is often referred to as the **condition for translational equilibrium.*** Let's apply this translational equilibrium condition to a case involving static equilibrium.

EXAMPLE 4.8: KEEP IT STRAIGHT – IN STATIC EQUILIBRIUM

Keeping a broken leg bone straight while it is healing sometimes requires *traction*, which is the procedure in which the bone is held under stretching tension forces at both ends to keep it aligned. Consider a leg under tractional tension as shown in ▼ **Figure 4.16**. The cord is attached to a suspended mass of 5.0 kg and runs over a pulley. The attached cord above the pulley makes an angle of $\theta = 40°$ with the vertical. Neglecting the mass of the lower leg and the pulley and assuming all the strings are ideal, determine the magnitude of the tension in the horizontal cord.

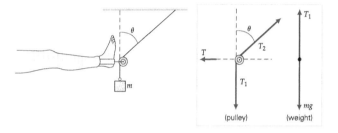

▲**FIGURE 4.16 Static translational equilibrium** See Example 4.8 for description.

THINKING IT THROUGH. The pulley is in a static equilibrium and thus has no net force on it. If the forces are summed both vertically and horizontally, they independently should add to zero. This should allow the tension in the horizontal string to be found.

SOLUTION

Given:

Listing the data:
$m = 5.0$ kg
$\theta = 40°$

Find:

T (tension) in the horizontal cord

Draw free-body diagrams for the pulley and suspended mass (shown in Figure 4.16). It should be clear that the horizontal string must exert a force to the left on the pulley as shown. Summing the vertical forces on m, it can be seen that $T_1 = mg$. Then, summing the vertical forces on the pulley,

$$\Sigma F_y = +T_2 \cos\theta - T_1 = 0$$

and summing the horizontal forces:

$$\Sigma F_x = +T_2 \sin\theta - T = 0$$

---

* For three-dimensional problems, $\Sigma F_z = 0$ also applies. However, our discussion will be restricted to forces in two dimensions.

Solving the latter equation for T, and substituting T_2 from the first:

$$T = T_2 \sin\theta = \frac{T_1}{\cos\theta}\sin\theta = mg\tan\theta$$

where $T_1 = mg$. Putting in the numbers,

$$T = mg\tan\theta = (5.0\,\text{kg})(9.8\,\text{m/s}^2)\tan 40° = 41\,\text{N}$$

FOLLOW-UP EXERCISE. Suppose the attending physician requires a tractional force on the bottom of the foot of 55 N. If the suspended mass was kept the same, would you increase or decrease the angle of the upper string? Prove your answer by calculating the required angle.

EXAMPLE 4.9: ON YOUR TOES – IN STATIC EQUILIBRIUM

An 80-kg person stands on one foot with the heel elevated (▶ **Figure 4.17a**). This gives rise to a tibia force F_1 and an Achilles tendon "pull" force F_2 as illustrated in Figure 4.17b. Typical angles are $\theta_1 = 15°$ and $\theta_2 = 21°$, respectively. (a) Find general equations for F_1 and F_2, and show that θ_2 must be greater than θ_1 to prevent damage to the Achilles tendon. (b) Compare the force applied by the Achilles tendon with the weight of the person.

THINKING IT THROUGH. This is a case of static translational equilibrium, so the x- and y-components can be summed to get equations for F_1 and F_2.

SOLUTION

Listing what is given and what is to be found:

Given:

$m = 80$ kg
$F_1 = $ tibia force
$F_2 = $ tendon "pull"
$\theta_1 = 15°, \theta_2 = 21°$

Find:

(a) general equations for F_1 and
(b) comparison of tendon force F_2 and the person's weight

(a) It is assumed that the person of mass m is at rest, standing on one foot. Then, summing the force components on the foot (Figure 4.17b),

$$\Sigma F_x = +F_1\sin\theta_1 - F_2\sin\theta_2 = 0$$
$$\Sigma F_y = +N - F_1\cos\theta_1 + F_2\cos\theta_2 - m_f g = 0$$

where m_f is the mass of the foot. From the ΣF_x equation,

$$F_1 = F_2\left(\frac{\sin\theta_2}{\sin\theta_1}\right) \qquad (1)$$

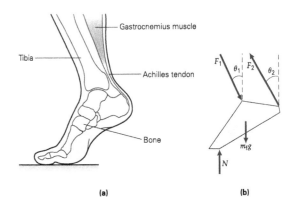

▲ **FIGURE 4.17 On your toes (a)** A person stands on one foot with the heel elevated. **(b)** The foot forces involved for this position (not to scale).

Substituting F_1 into the ΣF_y equation,

$$N - F_2\left(\frac{\sin\theta_2}{\sin\theta_1}\right)\cos\theta_1 + F_2\cos\theta_2 - m_f g = 0$$

Solving for F_2 with $N = mg$ (why?) yields,

$$F_2 = \frac{N - m_f g}{(\sin\theta_2/\tan\theta_1) - \cos\theta_2} = \frac{mg - m_f g}{\cos\theta_2\left[(\tan\theta_2/\tan\theta_1) - 1\right]} \qquad (2)$$

Examining the F_2 in Equation 2, we see that if $\theta_2 = \theta_1$, or $\tan\theta_2 = \tan\theta_1$ then F_2 is very large (why?). So to have a finite force, we must have $\tan\theta_2 > \tan\theta_1$ or $\theta_2 > \theta_1$, and $21° > 15°$, so Nature obviously knows her physics.

Then, substituting F_2 into Equation 1 to find F_1,

$$F_1 = F_2\left(\frac{\sin\theta_2}{\sin\theta_1}\right) = \frac{(m - m_f)g}{\cos\theta_2[(\tan\theta_2/\tan\theta_1)-1]}\left(\frac{\sin\theta_2}{\sin\theta_1}\right)$$
$$= \frac{(m - m_f)g\tan\theta_2}{[(\tan\theta_2/\tan\theta_1)-1]\sin\theta_1} = \frac{\tan\theta_2(m - m_f)g}{\cos\theta_1\tan\theta_2 - \sin\theta_1}$$

(Check the trig manipulation on this last step.)

(b) The person's weight is $w = mg$, where m is the mass of the person's body. This is to be compared with F_2. Then, with $m \gg m_f$ (total body mass much greater than mass of the foot), to a good approximation, m_f may be assumed negligible compared to m, that is, $w - m_f g = mg - m_f g \approx w$. So for F_2,

$$F_2 = \frac{w - m_f g}{\cos\theta_2[(\tan\theta_2/\tan\theta_1)-1]}$$
$$\approx \frac{w}{\cos 21°\left[(\tan 21°/\tan 15°)-1\right]} = 2.5w$$

The Achilles tendon force is thus approximately 2.5 times the person's weight. No wonder folks stretch or tear this tendon, even without jumping.

FOLLOW-UP EXERCISE. (a) Compare the tibia force with the weight of the person. (b) Suppose the person jumped upward from the one-foot toe position (as in taking a running jump shot in basketball). How would this jump affect F_1 and F_2?

4.6 Friction

Friction refers to the ever-present resistance to motion that occurs whenever two materials, or media, are in contact with each other. This resistance occurs for all types of media – solids, liquids, and gases – and is characterized as the force of friction ($\vec{f}$). For simplicity, up to now various kinds of friction (including air resistance) have been generally ignored in examples and exercises. Now knowing how to describe motion, you are ready to consider situations that are more realistic, in that the effects of friction are included.

In some situations, an increase in friction is desired – for example, when putting sand on an icy road or sidewalk to improve traction. Consider the forces involved in walking, as illustrated in ▼ **Figure 4.18**. Your foot pushes back on the

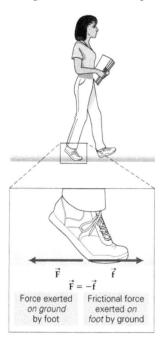

▲ FIGURE 4.18 **Friction and walking** The force of friction, $\vec{f}$, is shown in the direction of the walking motion. If you walk on a deep-pile rug, the force exerted on the rug is evident in that the pile will be bent backward.

sidewalk and the sidewalk pushes forward on you so you walk forward. This is a Newton's third law force pair. However, what if there were no friction? Your foot would simply slide backward so you cannot even push back on the sidewalk. Here the direction of the friction force is against the slipping of the foot so it is in the forward direction. Thus, friction enables us to walk!

As another example, consider a worker standing in the center of the bed of a flatbed truck that is accelerating in the forward direction. If there were no friction between the worker's shoes and the truck bed, the truck would slide out from under him. Obviously, there is a frictional force between the shoes and the bed, and it is in the forward direction. This is necessary for the worker to accelerate with the truck.

So there are situations where friction is desired (▼ **Figure 4.19a**), and situations where reduced friction is needed (Figure 4.19b). Another situation where reduced friction is promoted is in the lubrication of moving machine parts. This allows the parts to move more freely, thereby lessening wear and reducing the expenditure of energy. Automobiles would not run without friction-reducing oils and greases.

This section is concerned chiefly with friction between solid surfaces. All surfaces are microscopically rough, no matter how smooth they appear or feel. It was originally thought that friction was due primarily to the mechanical interlocking of surface irregularities, or *asperities* (high spots). However, research has shown that friction between the contacting surfaces of ordinary solids (metals in particular) is due mostly to local adhesion. When surfaces are pressed together, local welding or bonding occurs in a few small patches where the largest asperities make contact. To overcome this local adhesion, a force great enough to pull apart the bonded regions must be applied.

Friction between solids is generally classified into three types: static, kinetic (sliding), and rolling. Static friction includes all cases in which the frictional force is sufficient to prevent relative motion between surfaces. Suppose you want to move a large desk. You push on it, but the desk doesn't move. The force of static friction between the desk's legs and the floor opposes and equals the horizontal force you are applying, so there is no motion – a static condition.

(a)   (b)

▲ FIGURE 4.19 **Increasing and decreasing friction (a)** To get a fast start, drag racers need to make sure that their wheels don't slip when the starting light goes on. Just before the start of the race, they floor the accelerator to maximize the friction between their tires and the track by "burning in" the tires. This "burn-in" is done by spinning the wheels with the brakes on until the tires are extremely hot. The rubber becomes so sticky that it almost welds itself to the surface of the road. **(b)** Water serves as a good lubricant to reduce friction in rides such as this one.

Kinetic (or sliding) friction occurs when there is relative (sliding) motion at the interface of the surfaces in contact. When pushing on the desk, you can eventually get it sliding, but there is still a great deal of resistance between the desk's legs and the floor – kinetic friction.

Rolling friction occurs when one surface rotates as it moves over another surface but does not slip or slide at the point or area of contact. Rolling friction, such as that occurring between a train wheel and a rail, is attributed to small, local deformations in the contact region. This type of friction is difficult to analyze and will not be considered.

4.6.1 Frictional Forces and Coefficients of Friction

Now let's look at the forces of friction on stationary and sliding objects. These forces are called the *force of static friction* and the *force of kinetic (sliding) friction*, respectively. Experimentally, it has been found that the force of friction (f) depends on both the nature of the two surfaces, and, to a good approximation, the normal force (N) that a surface exerts on an object, that is, $f \propto N$. For an object on a horizontal surface, and with no other vertical forces, this normal force is equal in magnitude to the object's weight. (Why?) However, as was shown in Figure 4.13, on an inclined plane the normal force is in response to only a component of the weight force.

The force of static friction f_s between surfaces in contact acts in the direction that opposes the initiation of relative motion between the surfaces. The magnitude takes on a range of values given by

$$f_s \leq \mu_s N \quad \text{(static conditions)} \tag{4.6}$$

where μ_s is a constant of proportionality called the **coefficient of static friction**. ("μ_s" is the Greek letter mu. Note that it is dimensionless. How do you know this from the equation?)

The less-than-or-equal-to sign ($\leq$) indicates that the force of static friction may have different values from zero up to some maximum value. To understand this concept, look at ▼ **Figure 4.20**. In Figure 4.20a, one person pushes on a file cabinet, but it doesn't move. With no acceleration, the net force on the cabinet is zero, and $F - f_s = 0$, or $F = f_s$. Suppose that a second person also pushes, and the file cabinet still doesn't budge. Then f_s must now be larger, since the applied force has been increased. Finally, if the applied force is made large enough to overcome the static friction, motion occurs (Figure 4.20c). The greatest, or maximum, force of static friction is exerted just before the cabinet starts to slide (Figure 4.20b), and for this case, Equation 4.7 gives the maximum value of static friction:

$$f_{s_{max}} = \mu_s N \quad \text{(maximum value of static friction)} \tag{4.7}$$

Once an object is sliding, the force of friction changes to kinetic friction (f_k) This force acts in the direction opposite to the direction of the object's motion and has a magnitude of

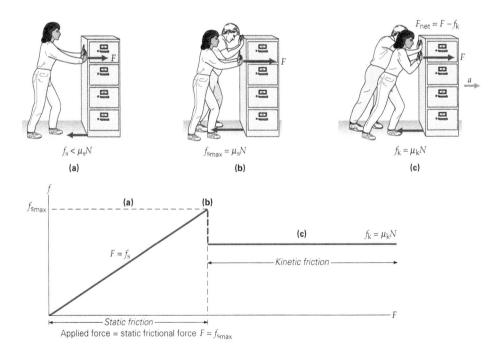

▲ **FIGURE 4.20** **Force of friction versus applied force (a)** In the static region of the graph, as the applied force F increases, so does f_s; that is $f_s = F$ and $f_s < \mu_s N$ **(b)** When the applied force F exceeds $f_{s_{max}} = \mu_s N$ the heavy file cabinet is set into motion. **(c)** Once the cabinet is moving, the frictional force decreases, since kinetic friction is less than static friction $\left(f_k < f_{s_{max}} \right)$. Thus if the applied force is maintained, there is a net force, and the cabinet is accelerated. For the cabinet to move with constant velocity, the applied force must be reduced to equal the kinetic friction force: $f_k = \mu_k N$.

$$f_k = \mu_k N \quad \text{(sliding conditions)} \qquad (4.8)$$

where μ_k is the **coefficient of kinetic friction** (sometimes called the *coefficient of sliding friction*). Note that Equations 4.7 and 4.8 are **not** vector equations, since f and N are in different directions. Generally, the coefficient of kinetic friction is less than the coefficient of static friction ($\mu_k < \mu_s$), which means that the force of kinetic friction is less than $f_{s_{max}}$. The coefficients of friction between some common materials are listed in ▼ **Table 4.1**.

TABLE 4.1 Approximate Values for Coefficients of Static and Kinetic Friction between Certain Surfaces

Friction between Materials	μ_s	μ_k
Aluminum on aluminum	1.90	1.40
Glass on glass	0.94	0.35
Rubber on concrete		
Dry	1.20	0.85
Wet	0.80	0.60
Steel on aluminum	0.61	0.47
Steel on steel		
Dry	0.75	0.48
Lubricated	0.12	0.07
Teflon on steel	0.04	0.04
Teflon on Teflon	0.04	0.04
Waxed wood on snow	0.05	0.03
Wood on wood	0.58	0.40
Lubricated ball bearings	<0.01	<0.01
Synovial joints (at the ends of most long bones – for example, elbows and hips)	0.01	0.01

Note that the force of static friction (f_s) exists in response to an applied force. The magnitude of f_s and its direction depend on the magnitude and direction of the applied force. Up to its maximum value, the force of static friction is equal in magnitude and opposite in direction to the applied force (F), since there is no acceleration ($F - f_s = ma = 0$). Thus, if the person in Figure 4.20a were to push on the cabinet in the opposite direction, f_s would also change direction to oppose the new push. If there were no applied force F, then f_s would be zero. When the magnitude of F exceeds that of $f_{s_{max}}$, the cabinet begins moving (accelerates), and kinetic friction comes into play, with $f_k = \mu_k N$. If the magnitude of F is reduced to that of f_k, the cabinet will slide with a constant velocity; if the magnitude of F is maintained greater than that of f_k, the cabinet will continue to accelerate.

It has been experimentally determined that the coefficients of friction (and therefore the forces of friction) are nearly independent of the contact area between metal surfaces. This means that the force of friction between a brick-shaped metal block and a metal surface is the same regardless of whether the block is lying on a larger side or a smaller side.

Finally, keep in mind that although the equation $f = \mu N$ holds in general for frictional forces, it may not remain linear. That is, μ is not always constant. For example, the coefficient of kinetic friction varies somewhat with the relative speed of the surfaces. However, for speeds up to several meters per second, the coefficients are relatively constant. For simplicity, our discussion will neglect any variations due to speed (or area), and the forces of static and kinetic friction will be assumed to depend only on the load (N) and the nature of the two surfaces as expressed by the given coefficients of friction.

EXAMPLE 4.10: PULLING A CRATE – STATIC AND KINETIC FORCES OF FRICTION

(a) In ▼ **Figure 4.21**, if the coefficient of static friction between the 40.0-kg crate and the floor is 0.650, what is the magnitude of the minimum horizontal force the worker must pull to get the crate moving? (b) If the worker maintains that force once the crate starts to move and the coefficient of kinetic friction between the surfaces is 0.500, what is the magnitude of the acceleration of the crate?

THINKING IT THROUGH. This situation involves applications of the forces of friction. In (a), the maximum force of static friction must be calculated. In (b), if the worker maintains an applied force of this magnitude after the crate is in motion, there will be an acceleration, since $f_k < f_{s_{max}}$.

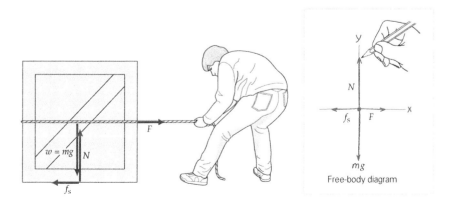

▲ FIGURE 4.21 **Forces of static and kinetic friction** See Example 4.10 for description.

SOLUTION

As usual, listing the given data and what is to be found:

Given:

$m = 40.0$ kg
$\mu_s = 0.650$
$\mu_k = 0.500$

Find:

(a) F (minimum force necessary to move crate)
(b) a (acceleration)

(a) The crate will not move until the magnitude of the applied force F slightly exceeds that of the maximum static frictional force $f_{s_{max}}$. So $f_{s_{max}}$ must be found to see what force the worker needs to apply. The weight of the crate and the normal force are equal in magnitude in this case (see the free-body diagram in Figure 4.21), so the magnitude of the maximum force of static friction is

$$f_{s_{max}} = \mu_s N = \mu_s (mg)$$
$$= (0.650)(40.0\,\text{kg})(9.80\,\text{m/s}^2) = 255\,\text{N}$$

So the crate will begin to move when the applied force F exceeds 255 N.

(b) Now with the crate in motion, the kinetic friction f_k acts on the crate. However, this force is smaller than the applied force $F = f_{s_{max}} = 255\,\text{N}$, because $\mu_k < \mu_s$. Hence, there is a net force on the crate and the acceleration of the crate can be found by using Newton's second law in the x-direction:

$$\Sigma F_x = +F - f_k = F - \mu_k N = ma_x$$

Solving for a_x,

$$a_x = \frac{F - \mu_k N}{m} = \frac{F - \mu_k (mg)}{m}$$
$$= \frac{255\,\text{N} - (0.500)(40.0\,\text{kg})(9.80\,\text{m/s}^2)}{40.0\,\text{kg}} = 1.48\,\text{m/s}^2$$

FOLLOW-UP EXERCISE. On the average, by what factor does μ_s exceed μ_k for dry (non-lubricated), metal-on-metal surfaces? (See Table 4.1.)

Let's look at another worker with the same crate, but this time with the worker applying the force at an angle (▼ **Figure 4.22**).

EXAMPLE 4.11: PULLING AT AN ANGLE – A CLOSER LOOK AT THE NORMAL FORCE

A worker pulling a crate applies a force at an angle of 30° to the horizontal, as shown in Figure 4.22. What is the magnitude of the minimum force he must apply to move the crate? (Before looking at the solution, would you expect that the force needed in this case would be greater or less than that in Example 4.10?)

THINKING IT THROUGH. Since the applied force is at an angle to the horizontal surface, the vertical component will affect the normal force. (See Figure 4.11.) This change in the normal force will, in turn, affect the maximum force of static friction.

SOLUTION

The data are the same as in Example 4.10, except that the force is applied at an angle.

Given:

$\theta = 30$

Find:

F (minimum force necessary to move the crate)

In this case, the crate will begin to move when the *horizontal component* of the applied force, $F\cos 30°$, slightly exceeds the maximum static friction force. So for the maximum friction:

$$F\cos 30° = f_{s_{max}} = \mu_s N$$

However, the magnitude of the normal force is not equal to the weight of the crate here, because of the upward component of the applied force. (See the free-body diagram in Figure 4.22.) Then by Newton's second law, since $a_y = 0$.

$$\Sigma F_y = +N + F\sin 30° - mg = 0$$

and

$$N = mg - F\sin 30°$$

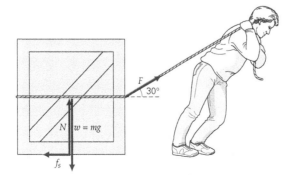
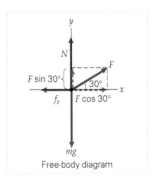

▲ FIGURE 4.22 **Pulling at an angle: a closer look at the normal force** See Example 4.11 for description.

In effect, the applied force here partially supports the weight of the crate. Substituting this expression for N into the first equation gives

$$F\cos 30° = \mu_s(mg - F\sin 30°)$$

Solving for F,

$$F = \frac{mg}{(\cos 30°/\mu_s) + \sin 30°} = \frac{(40.0\,\text{kg})(9.80\,\text{m/s}^2)}{(0.866/0.650) + (0.500)} = 214\,\text{N}$$

Thus, less applied force is needed in this case, reflecting the fact that the frictional force is less, because of the reduced normal force.

FOLLOW-UP EXERCISE. Note that in this Example, applying the force at an angle produces two effects. As the angle between the applied force and the horizontal increases, the horizontal component of the applied force is reduced. However, the normal force also gets smaller, resulting in a lower $f_{s_{max}}$. Does one effect always outweigh the other? That is, does the applied force F necessary to move the crate always decrease with increasing angle? [*Hint:* Investigate F for different angles. For example, compute F for 20° and 50°. You already have a value for 30°. What do the results tell you?]

EXAMPLE 4.12: NO SLIP, NO SLIDE – STATIC FRICTION

A crate sits in the middle of the bed on a flatbed truck that is traveling at 80 km/h on a straight, level road. The coefficient of static friction between the crate and the truck bed is 0.40. When the truck comes uniformly to a stop, the crate does not slide, but remains stationary on the truck. What is the minimum stopping distance for the truck so the crate does not slide on the truck bed?

THINKING IT THROUGH. There are three forces on the crate, as shown in the free-body diagram in ▼ **Figure 4.23** (assuming that the truck is initially traveling in the $+x$-direction). But wait.

There is a net force in the $-x$-direction, and hence there should be an acceleration in that direction. What does this mean? It means that relative to the ground, the crate is decelerating at the same rate as the truck, which is necessary for the crate not to slide – the crate and the truck slow down uniformly together.

The force creating this acceleration for the crate is the static force of friction. The acceleration is found using Newton's second law, and then is used in one of the kinematic equations to find the distance.

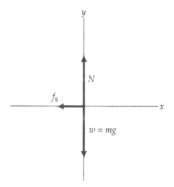

▲ FIGURE 4.23 **Free-body diagram** See Example 4.12 for description.

SOLUTION

Given:

$$v_{x_0} = 80\,\text{km/h} = 22\,\text{m/s}$$
$$\mu_s = 0.40$$

Find:

x (minimum stopping distance)

Applying Newton's second law to the crate using the maximum f_s to find the minimum stopping distance,

$$\Sigma F_x = -f_{s_{max}} = -\mu_s N = -\mu_s mg = ma_x$$

Solving for a_x,

$$a_x = -\mu_s g = -(0.40)(9.8\,\text{m/s}^2) = -3.9\,\text{m/s}^2$$

which is the maximum deceleration of the truck so the crate does not slide.

Hence, the minimum stopping distance (x) for the truck is based on this acceleration and given by Equation 2.12, where $v_x = 0$ and x_0 is taken to be zero. So,

$$v_x^2 = 0 = v_{x_0}^2 + 2(a_x)x$$

Solving for x,

$$x = \frac{v_{x_0}^2}{-2a_x} = \frac{(22\,\text{m/s})^2}{-2(-3.9\,\text{m/s}^2)} = 62\,\text{m}$$

Is the answer reasonable? This distance is about two-thirds the length of a football field.

FOLLOW-UP EXERCISE: Draw a free-body diagram and describe what happens in terms of accelerations and coefficients of friction if the crate starts to slide forward on the truck bed when the truck is braking to a stop (in other words, if a_x exceeds -3.9 m/s²).

4.6.2 Air Resistance

Air resistance refers to the resistance force acting on an object as it moves through air. In other words, air resistance is a type of frictional force. In analyses of falling objects, you can usually ignore the effect of air resistance and still get good approximations for those falling relatively short distances. However, for longer distances, air resistance cannot be ignored.

Air resistance occurs when a moving object collides with air molecules. Therefore, air resistance depends on the object's shape and size (which determine the area of the object that is exposed to collisions) as well as its speed. The larger the object and the faster it moves, the more collisions there will be with air molecules. (Air density is also a factor, but this quantity can be assumed to be constant near the Earth's surface.) To reduce air resistance (and fuel consumption), automobiles are made more "streamlined," and airfoils are used on trucks and campers (▶ **Figure 4.24**).

▲ FIGURE 4.24　**Airfoil** The airfoil at the top of the truck's cab makes the truck more streamlined and therefore reduces air resistance.

Consider a falling object. Since air resistance depends on speed, as a falling object accelerates under the influence of gravity, the retarding force of air resistance increases (▼ **Figure 4.25a**). Air resistance for human-sized objects as a general rule is proportional to the square of the speed, v^2, so the resistance builds up rather rapidly. Thus when the speed doubles, the air resistance increases by a factor of 4. Eventually, the magnitude of the retarding force equals that of the object's weight force (Figure 4.25b), so the net force on it is zero. The object then falls with a maximum constant velocity, which is called the **terminal velocity**, with magnitude v_t.

This can be easily seen from Newton's second law. For the falling object,

$$F_{net} = ma$$

or

$$mg - f = ma$$

where f is the air resistance (friction) and downward has been taken as positive for convenience. Solving for a,

$$a = g - \frac{f}{m}$$

where a is the magnitude of the instantaneous downward acceleration.

Notice that the acceleration for a falling object when air resistance is included is less than g; that is, $a < g$. As the object continues to fall, its speed increases, and the force of air resistance, f, increases (since it is speed dependent) until $a = 0$ when $f = mg$ and $f - mg = 0$. The object then falls at its constant terminal velocity.

For a skydiver with an unopened parachute, terminal velocity is about 200 km/h (about 125 mi/h). To reduce the terminal velocity so that it can be reached sooner and the time of fall extended, a skydiver will try to increase exposed body area to a maximum by assuming a spread-eagle position (▼ **Figure 4.26**).

▲ FIGURE 4.26　**Terminal velocity** Skydivers assume a spread-eagle position to maximize air resistance. This causes them to reach terminal velocity more quickly and prolongs the time of fall. Shown here is a formation of skydivers viewed from above.

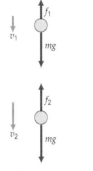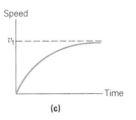

(a) As v increases, so does f.　　**(b) When $f = mg$, the object falls with a constant (terminal) velocity.**　　**(c)**

▲ FIGURE 4.25　**Air resistance and terminal velocity (a)** As the speed of a falling object increases, so does the frictional force of air resistance. **(b)** When this force of friction equals the weight of the object, the net force is zero, and the object falls with a constant (terminal) velocity. **(c)** A plot of speed versus time, showing these relationships.

This position takes advantage of the dependence of air resistance on the size and shape of the falling object. Once the parachute is open (giving a larger exposed area and a shape that catches the air), the additional air resistance slows the diver down to about 40 km/h (25 mi/h), which is more preferable for landing.

CONCEPTUAL EXAMPLE 4.13: RACE YOU DOWN – AIR RESISTANCE AND TERMINAL VELOCITY

From a high altitude, a balloonist simultaneously drops two balls of identical size, but appreciably different in mass. Assuming that both balls reach terminal velocity during the fall, which of the following is true? (a) The heavier ball reaches terminal velocity first; (b) the balls reach terminal velocity at the same time; (c) the heavier ball hits the ground first; (d) the balls hit the ground at the same time. *Clearly establish the reasoning and physical principle(s) used in determining your answer before checking it next. That is, why did you select your answer?*

REASONING AND ANSWER. Terminal velocity is reached when the weight of a ball is balanced by the frictional air resistance. Both balls initially experience the same acceleration, *g*, and their speeds and the retarding forces of air resistance increase at the same rate. The weight of the lighter ball will be balanced first, so (a) and (b) are incorrect with the lighter ball reaching terminal velocity ($a = 0$) first, the heavier ball continues to accelerate and pulls ahead of the lighter ball. Hence, the heavier ball hits the ground first, and the answer is (c), and (d) is incorrect.

FOLLOW-UP EXERCISE. Suppose the heavier ball were much larger in size than the lighter ball. How might this difference affect the outcome?

You see an example of terminal velocity quite often. Why do clouds stay seemingly suspended in the sky? Certainly the water droplets or ice crystals (high clouds) should fall – and they do. However, they are so small that their terminal velocity is reached quickly, and the very slow rate of their descent goes unnoticed. In addition, there may be some helpful updrafts that keep the water droplets and ice crystals from reaching the ground.

Chapter 4 Review

- A force is something that is capable of changing an object's state of motion. To produce a change in motion, there must be a nonzero net, or unbalanced, force:

$$\vec{\mathbf{F}}_{net} = \Sigma \vec{\mathbf{F}}_i$$

- **Newton's first law of motion** is also called the *law of inertia*, where inertia is the natural tendency of an object to maintain its state of motion. It states that in the absence of a net applied force, a body at rest remains at rest, and a body in motion remains in motion with constant velocity.

- **Newton's second law** relates the net force acting on an object or system to the (total) mass and the resulting acceleration. It defines the cause-and-effect relationship between force and acceleration:

$$\Sigma \vec{\mathbf{F}}_i = \vec{\mathbf{F}}_{net} = m\vec{\mathbf{a}} \tag{4.1}$$

The equation for weight in terms of mass is a form of Newton's second law:

$$w = mg \tag{4.2}$$

The component form of Newton's second law:

$$\Sigma(F_x\hat{\mathbf{x}} + F_y\hat{\mathbf{y}}) = m(a_x\hat{\mathbf{x}} + a_y\hat{\mathbf{y}}) = ma_x\hat{\mathbf{x}} + ma_y\hat{\mathbf{y}} \tag{4.3a}$$

and

$$\Sigma F_x = ma_x \quad \text{and} \quad \Sigma F_y = ma_y \tag{4.3b}$$

- **Newton's third law** states that for every force, there is an equal and opposite reaction force. The opposing forces of a third law force pair always act on different objects.

- An object is said to be in **translational equilibrium** when it either is at rest or moves with a constant velocity. When remaining at rest, an object is said to be in *static translational equilibrium*. The condition for translational equilibrium is represented as

$$\Sigma \vec{\mathbf{F}}_i = 0 \quad \text{(translational equilibrium)} \tag{4.4}$$

or

$$\Sigma F_x = 0 \quad \text{and} \quad \Sigma F_y = 0 \quad \text{(translational equilibrium)} \tag{4.5}$$

- **Friction** is the resistance to motion that occurs between contacting surfaces. (In general, friction occurs for all types of media – solids, liquids, and gases.)

- The frictional force between surfaces is characterized by coefficients of friction (m), one for the static case and one for the kinetic (moving) case. In many cases, $f = \mu N$ where N is the normal force – the force perpendicular to the surface (i.e., the force exerted *by* the surface *on* the object). As a ratio of forces (f/N), μ is unitless.

  Force of Static Friction:

$$f_s \leq \mu_s N \tag{4.6}$$

$$f_{s_{max}} = \mu_s N \quad \text{(maximum value of static friction)} \tag{4.7}$$

  Force of Kinetic (Sliding) Friction:

$$f_k = \mu_k N \tag{4.8}$$

- The force of air resistance on a falling object increases with increasing speed. It eventually attains a constant velocity, called the *terminal velocity*.

End of Chapter Questions and Exercises

Multiple Choice Questions

4.1 The Concepts of Force and Net Force
and
4.2 Inertia and Newton's First Law of Motion

1. Mass is related to an object's (a) weight, (b) inertia, (c) density, (d) all of the preceding.

2. A force (a) always produces motion, (b) is a scalar quantity, (c) is capable of producing a change in motion, (d) both (a) and (b).

3. If an object is moving at constant velocity, (a) there must be a force in the direction of the velocity, (b) there must be no force in the direction of the velocity, (c) there must be no net force, (d) there must be a net force in the direction of the velocity.

4. If the net force on an object is zero, the object could (a) be at rest, (b) be in motion at a constant velocity, (c) have zero acceleration, (d) all of the preceding.

5. The force required to keep a rocket ship moving at a constant velocity in deep space is (a) equal to the weight of the ship, (b) dependent on how fast the ship is moving, (c) equal to that generated by the rocket's engines at half-power, (d) zero.

4.3 Newton's Second Law of Motion

6. The newton unit of force is equivalent to (a) kg·m/s, (b) kg·m/s^2, (c) kg·m^2/s, (d) none of the preceding.

7. The acceleration of an object is (a) inversely proportional to the acting net force, (b) directly proportional to its mass, (c) directly proportional to the net force and inversely proportional to its mass, (d) none of these.

8. The weight of an object is directly proportional to (a) its mass, (b) its inertia, (c) the acceleration due to gravity, (d) all of the preceding.

4.4 Newton's Third Law of Motion

9. The action and reaction forces of Newton's third law (a) are in the same direction, (b) have different magnitudes, (c) act on different objects, (d) are the same force.

10. A brick hits a glass window. The brick breaks the glass, so (a) the magnitude of the force of the brick on the glass is greater than the magnitude of the force of the glass on the brick, (b) the magnitude of the force of the brick on the glass is smaller than the magnitude of the force of the glass on the brick, (c) the magnitude of the force of the brick on the glass is equal to the magnitude of the force of the glass on the brick, (d) none of the preceding.

11. A semi-truck collides head-on with a passenger car, causing a lot more damage to the car than to the truck. From this condition, we can say that (a) the magnitude of the force of the truck on the car is greater than the magnitude of the force of the car on the truck, (b) the magnitude of the force of the truck on the car is smaller than the magnitude of the force of the car on the truck, (c) the magnitude of the force of the truck on the car is equal to the magnitude of the force of the car on the truck, (d) none of the preceding.

4.5 More on Newton's Laws: Free-Body Diagrams and Translational Equilibrium

12. The kinematic equations of Chapter 2 cannot be used with (a) constant accelerations, (b) constant velocities, (c) variable velocities, (d) variable accelerations.

13. The condition(s) for translational equilibrium is (are) $\Sigma F_x = 0$, (b) $\Sigma F_y = 0$, (c) $\Sigma \vec{F}_i = 0$, (d) all of the preceding.

4.6 Friction

14. In general, the frictional force (a) is greater for smooth than rough surfaces, (b) depends significantly on sliding speeds, (c) is proportional to the normal force, (d) depends significantly on the surface area of contact.

15. The coefficient of kinetic friction, μ_k, (a) is usually greater than the coefficient of static friction, μ_s, (b) usually equals μ_s, (c) is usually smaller than μ_s, (d) equals the applied force that exceeds the maximum static force.

16. A crate sits in the middle of the bed of a flatbed truck. The driver accelerates the truck gradually from rest to a normal speed, but then has to make a sudden stop to avoid hitting a car. If the crate slides as the truck stops, the frictional force would be (a) in the forward direction, (b) in the backward direction, (c) zero.

Conceptual Questions

4.1 The Concepts of Force and Net Force
and
4.2 Inertia and Newton's First Law of Motion

1. (a) If an object is at rest, there must be no forces acting on it. Is this statement correct? Explain. (b) If the net force on an object is zero, can you conclude that the object is at rest? Explain.

2. When on a jet airliner that is taking off, you feel that you are being "pushed" back into the seat. Use Newton's first law to explain why.

3. An object weighs 300 N on Earth and 50 N on the Moon. Does the object also have less inertia on the Moon?

4. Consider an air-bubble level that is sitting on a horizontal surface (▶ **Figure 4.27**). Initially, the air bubble is in the middle of the horizontal glass tube. (a) If the level is pushed and a force is applied to accelerate it, which way would the bubble move? Which way would the bubble move if the force is then removed and the level slows down, due to friction? (b) Such a level is sometimes used as an "accelerometer" to indicate the direction of the acceleration. Explain the principle involved. [*Hint:* Think about pushing a pan of water.]

5. As a follow-up to Conceptual Question 4, consider a child holding a helium balloon in a closed car at rest. What would the child observe when the car (a) accelerates from rest and (b) brakes to a stop? (The balloon does not touch the roof of the car.)

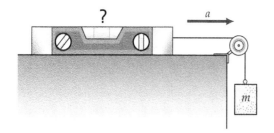

▲FIGURE 4.27 **An air-bubble level/accelerometer** See Conceptual Question 4.

6. The following is an old trick (▼ **Figure 4.28**). When the cardboard is pulled quickly, the coil falls into the glass. Why?

▲FIGURE 4.28 **Magic or physics?** See Conceptual Question 6.

7. Another old one: Referring to ▼ **Figure 4.29**, (a) how would you pull to get the upper string to break? (b) How would you pull to get the lower string to break?

▲FIGURE 4.29 **Give it a pull** See Conceptual Question 7.

8. A student weighing 600 N crouches on a scale and suddenly springs vertically upward. Will the scale read more or less than 600 N just before the student leaves the scale?

4.3 Newton's Second Law of Motion

9. An astronaut has a mass of 70 kg when measured on Earth. What is her weight in deep space, far from any celestial body? What is her mass there?

10. In general, this chapter has considered forces that are applied to objects of constant mass. What would be the situation if mass were added to or lost from a system while a constant force was being applied to the system? Give examples of situations in which this set of events might happen.

11. The engines of most rockets produce a constant thrust (forward force). However, when a rocket is fired, its acceleration increases with time as the engine continues to operate. Is this situation a violation of Newton's second law? Explain.

12. In football, good wide receivers usually have "soft" hands for catching balls (▼ **Figure 4.30**). How would you interpret this description on the basis of Newton's second law?

▲FIGURE 4.30 **Soft hands** See Conceptual Question 12.

4.4 Newton's Third Law of Motion

13. Here is a story of a horse and a farmer: One day, the farmer attaches a heavy cart to the horse and demands that the horse pull the cart. "Well," says the horse, "I cannot pull the cart, because, according to Newton's third law, if I apply a force to the cart, the cart will apply an equal and opposite force on me. The net result will be that I cannot pull the cart, since all the forces will cancel. Therefore, it is impossible for me to pull this cart." The farmer was very upset! What could he say to persuade the horse to move?

14. Is something wrong with the following statement? When a baseball is hit with a bat, there are equal and opposite forces on the bat and baseball. The forces then cancel, and there is no motion.

4.5 More on Newton's Laws: Free-Body Diagrams and Translational Equilibrium

15. Draw the free-body diagram for a person sitting in the seat of an aircraft (a) that is accelerating along the runway for takeoff, and (b) after takeoff at a 20° angle to the ground.

16. A person pushes perpendicularly on a block of wood that has been placed against a wall. Draw a free-body diagram of the block and identify the reaction forces to all the forces on the block.

17. A person on a bathroom scale (not the digital type) stands on the scale with his arms at his sides. He then quickly raises his arms over his head, and notices that the scale reading increases as he brings his arms upward. Similarly, there is a decrease as he brings his arms downward. Why does the scale reading change? (Try this yourself.)

4.6 Friction

18. Identify the direction of the friction force in the following cases: (a) a book sitting on a table; (b) a box sliding on a horizontal surface; (c) a car making a turn on a flat road; (d) the initial motion of a machine part delivered on a conveyor belt in an assembly line.

19. The purpose of a car's antilock brakes is to prevent the wheels from locking up so as to keep the car rolling rather than sliding. Why would rolling decrease the stopping distance as compared with sliding?

20. Shown in ▼ **Figure 4.31** are the front and rear wings of an Indy racing car. These wings generate **down force**, which is the vertical downward force produced by the air moving over the car. Why is such a down force desired? An Indy car can create a down force equal to twice its weight. Why not simply make the cars heavier?

▲ FIGURE 4.31   **Down force** See Conceptual Question 20.

Exercises*,†

Integrated Exercises (IEs) are two-part exercises. The first part typically requires a conceptual answer choice based on physical thinking and basic principles. The following part is quantitative calculations associated with the conceptual choice made in the first part of the exercise.

4.1 The Concepts of Force and Net Force
and
4.2 Inertia and Newton's First Law of Motion

1. • Which has more inertia, 20 cm³ of water or 10 cm³ of aluminum, and how many times more? (See Table 9.2.)

2. • Two forces act on a 5.0-kg object sitting on a frictionless horizontal surface. One force is 30 N in the $+x$-direction, and the other is 35 N in the $-x$-direction. What is the acceleration of the object?

3. • In Exercise 2, if the 35-N force acted downward at an angle of 40° relative to the horizontal, what would be the acceleration in this case?

4. • A net force of 4.0 N gives an object an acceleration of 10 m/s². What is the mass of the object?

5. • Consider a 2.0-kg ball and a 6.0-kg ball in free fall. (a) What is the net force acting on each? (b) What is the acceleration of each?

6. IE •• A hockey puck with a weight of 0.50 lb is sliding freely across a section of very smooth (frictionless) horizontal ice. (a) When it is sliding freely, how does the upward force of the ice on the puck (the normal force) compare with the upward force when the puck is sitting permanently at rest: (1) the upward force is greater when the puck is sliding; (2) the upward force is less when it is sliding; (3) the upward force is the same in both situations? (b) Calculate the upward force on the puck in both situations.

7. •• A 5.0-kg block at rest on a frictionless surface is acted on by forces $F_1 = 5.5$ N and $F_2 = 3.5$ N as illustrated in ▼ **Figure 4.32**. What additional force will keep the block at rest?

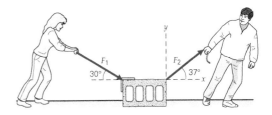

▲ FIGURE 4.32   **Two applied forces** See Exercise 7.

8. IE •• (a) You are told that an object has zero acceleration. Which of the following is true: (1) the object is at rest; (2) the object is moving with constant velocity; (3) either 1 or 2 is possible; or (4) neither 1 nor 2 is possible. (b) Two forces on the object are $F_1 = 3.6$ N at 74° below the $+x$-axis and $F_2 = 3.6$ N at 34° above the $-x$-axis. Is there a third force on the object? Why or why not? If there is a third force, what is it?

9. IE •• A fish weighing 25 lb is caught and hauled onto the boat. (a) Compare the tension in the fishing line when the fish is brought up vertically at a constant speed to the tension when the fish is held vertically at rest for the picture-taking ceremony on the wharf. In which case is the tension largest: (1) When the fish is moving up; (2) when the fish is being held steady; or (3) the tension is the same in both situations? (b) Calculate the tension in the fishing line.

---

* Unless otherwise stated, all objects are located near the Earth's surface, where $g = 9.80$ m/s².

† The bullets denote the degree of difficulty of the exercises: •, simple; ••, medium; and •••, more difficult.

10. ••• A1.5-kg object moves up the *y*-axis at a constant speed. When it reaches the origin, the forces $F_1 = 5.0$ N at 37° above the +*x*-axis, $F_2 = 2.5$ N in the +*x*-direction, $F_3 = 3.5$ N at 45° below the −*x*-axis, and $F_4 = 1.5$ N in the −*y*-direction are applied to it. (a) Will the object continue to move along the *y*-axis? (b) If not, what simultaneously applied force will keep it moving along the *y*-axis at a constant speed?

11. IE ••• Three horizontal forces (the only horizontal ones) act on a box sitting on a floor. One (call it F_1) acts due east and has a magnitude of 150 lb. A second force (call it F_2) has an easterly component of 30.0 lb and a southerly component of 40.0 lb. The box remains at rest. (Neglect friction.) (a) Sketch the two known forces on the box. In which quadrant is the unknown third force: (1) the first quadrant; (2) the second quadrant; (3) the third quadrant; or (4) the fourth quadrant? (b) Find the unknown third force in newtons and compare your answer to the sketched estimate.

4.3 Newton's Second Law of Motion

12. • A 6.0-N net force is applied to a 1.5-kg mass. What is the object's acceleration?

13. • A force acts on a 1.5-kg, mass, giving it an acceleration of 3.0 m/s². (a) If the same force acts on a 2.5-kg mass, what acceleration would be produced? (b) What is the magnitude of the force?

14. • A loaded Airbus 380 jumbo jet has a mass close to 6.0×10^5 kg. What net force is required to give the plane an acceleration of 3.5 m/s² down the runway for takeoffs?

15. IE • A 6.0-kg object is brought to the Moon, where the acceleration due to gravity is only one-sixth of that on the Earth. (a) The mass of the object on the Moon is (1) zero, (2) 1.0 kg, (3) 6.0 kg, (4) 36 kg. Why? (b) What is the weight of the object on the Moon?

16. •• A gun is fired and a 50-g bullet is accelerated to a muzzle speed of 100 m/s. If the length of the gun barrel is 0.90 m, what is the magnitude of the accelerating force? (Assume the acceleration to be constant.)

17. IE ••• ▶ **Figure 4.33** shows a product label. (a) This label is correct (1) on the Earth; (2) on the Moon, where the acceleration due to gravity is only one-sixth of that on the Earth; (3) in deep space, where there is little gravity; (4) all of the preceding. (b) What mass would a label show for an amount that weighs 2 lb on the Moon?

18. •• In a college homecoming competition, eighteen students lift a sports car. While holding the car off the ground, each student exerts an upward force of 400 N. (a) What is the mass of the car in kilograms? (b) What is its weight in pounds?

19. IE •• (a) A horizontal force acts on an object on a frictionless horizontal surface. If the force is halved and the mass of the object is doubled, the acceleration will be (1) four times, (2) two times, (3) one-half, (4) one-fourth as

▲ FIGURE 4.33 **Correct label?** See Exercise 17.

great. (b) If the acceleration of the object is 1.0 m/s², and the force on it is doubled and its mass is halved, what is the new acceleration?

20. •• A force of 50 N acts on a mass giving it an acceleration of 4.0 m/s². The same force acts on a mass m_2 and produces an acceleration of 12 m/s². What acceleration will this force produce if the total system is $m_1 + m_2$?

21. •• A student weighing 800 N crouches on a scale and suddenly springs vertically upward. His roommate notices that the scale reads 900 N momentarily just as he leaves the scale. With what acceleration does he leave the scale?

22. •• The engine of a 1.0-kg toy plane exerts a 15-N forward force. If the air exerts an 8.0-N resistive force on the plane, what is the magnitude of the acceleration of the plane?

23. •• When a horizontal force of 300 N is applied to a 75.0-kg box, the box slides on a level floor, opposed by a force of kinetic friction of 120 N. What is the magnitude of the acceleration of the box?

24. IE •• A rocket is far away from all planets and stars, so gravity is not a consideration. It is using its rocket engines to accelerate upward with an acceleration $a = 9.80$ m/s². On the floor of the main deck is a crate (object with brick pattern) with a mass of 75.0 kg (▶ **Figure 4.34**). (a) How many forces are acting on the crate: (1) zero; (2) one; (3) two; (4) three? (b) Determine the normal force on the crate and compare it to the normal force the crate would experience if it were at rest on the surface of the Earth.

25. •• An object (mass 10.0 kg) slides *upward* on a slippery vertical wall. A force *F* of 60 N acts at an angle of 60° as shown in ▶ **Figure 4.35**. (a) Determine the normal force exerted on the object by the wall. (b) Determine the object's acceleration.

▲ FIGURE 4.34   **Away we go** See Exercise 24.

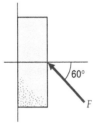

▲ FIGURE 4.35   **Up a wall** See Exercise 25. (Drawing not to scale.)

26. •• In an emergency stop to avoid an accident, a shoulder-strap seatbelt holds a 60-kg passenger in place. If the car was initially traveling at 90 km/h and came to a stop in 5.5 s along a straight, level road, what was the average force applied to the passenger by the seatbelt?

27. IE •• A student is assigned the task of measuring the startup acceleration of a large RV (recreational vehicle) using an iron ball suspended from the ceiling by a long string. In accelerating from rest, the ball no longer hangs vertically, but at an angle to the vertical. (a) Is the angle of the ball forward or backward from the vertical? (b) If the string makes an angle of 3.0° from the vertical, what is the initial acceleration of the RV?

28. •• A force of 10 N acts on two blocks on a frictionless surface (▼ **Figure 4.36**). (a) What is the acceleration of the system? (b) What force does block A exert on block B? (c) What force does block B exert on block A?

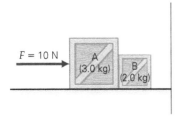

▲ FIGURE 4.36   **Forces: inside and out** See Exercise 28.

4.4 Newton's Third Law of Motion

29. IE • A book is sitting on a horizontal surface. (a) There is (are) (1) one, (2) two, or (3) three force(s) acting on the book. (b) Identify the reaction force to each force on the book.

30. •• In an Olympic figure-skating event, a 65-kg male skater pushes a 45-kg female skater, causing her to accelerate at a rate of 2.0 m/s². At what rate will the male skater accelerate? What is the direction of his acceleration?

31. IE •• A sprinter of mass 65.0 kg starts his race by pushing horizontally backward on the starting blocks with a force of 200 N. (a) What force causes him to accelerate out of the blocks: (1) his push on the blocks, (2) the downward force of gravity, or (3) the force the blocks exert forward on him? (b) Determine his initial acceleration as he leaves the blocks.

32. •• Jane and John, with masses of 50 and 60 kg, respectively, stand on a frictionless surface 10 m apart. John pulls on a rope that connects him to Jane, giving Jane an acceleration of 0.92 m/s² toward him. (a) What is John's acceleration? (b) If the pulling force is applied constantly, where will Jane and John meet?

4.5 More on Newton's Laws: Free-Body Diagrams and Translational Equilibrium

33. •• A 75.0-kg person is standing on a scale in an elevator. What is the reading of the scale in newtons if the elevator is (a) at rest, (b) moving up at a constant velocity of 2.0 m/s, (c) accelerating up at 2.00 m/s², and (d) accelerating down at 2.00 m/s²?

34. IE • (a) When an object is on an inclined plane, the normal force exerted by the inclined plane on the object is (1) less than, (2) equal to, (3) more than the weight of the object. Why? (b) For a 10-kg object on a 30° inclined plane, what are the object's weight and the normal force exerted on the object by the inclined plane?

35. IE •• The weight of a 500-kg object is 4900 N. (a) When the object is on a moving elevator, its measured weight could be (1) zero, (2) between zero and 4900 N, (3) more than 4900 N, (4) all of the preceding. Why? (b) Describe the motion if the object's measured weight is only 4000 N in a moving elevator.

36. •• A boy pulls a box of mass 30 kg with a force of 25 N in the direction shown in ▼ **Figure 4.37**. (a) Ignoring friction, what is the acceleration of the box? (b) What is the normal force exerted on the box by the ground?

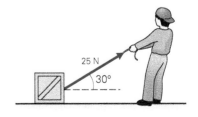

▲ FIGURE 4.37   **Pulling a box** See Exercise 36.

37. •• A girl pushes a 25-kg lawn mower as shown in ▼ **Figure 4.38**. If $F = 30$ N and $\theta = 37°$ (a) what is the acceleration of the mower, and (b) what is the normal force exerted on the mower by the lawn? Ignore friction.

▲ FIGURE 4.38   **Mowing the lawn** See Exercise 37.

38. •• A 3000-kg truck tows a 1500-kg car by a chain. If the net forward force on the truck by the ground is 3200 N, what is the acceleration of the car, and (b) what is the tension in the connecting chain?

39. •• A block of mass 25.0 kg slides down a frictionless surface inclined at 30°. To ensure that the block does not accelerate, what is the smallest force that you must exert on it and what is its direction?

40. IE •• (a) An Olympic skier coasts down a slope with an angle of inclination of 37°. Neglecting friction, there is (are), one, (2) two, (3) three force(s) acting on the skier. What is the acceleration of the skier? (c) If the skier has a speed of 5.0 m/s at the top of the slope, what is his speed when he reaches the bottom of the 35-m-long slope?

41. •• A car coasts (engine off) up a 30° grade. If the speed of the car is 25 m/s at the bottom of the grade, what is the distance traveled by the car before it comes to rest?

42. •• Assuming ideal frictionless conditions for the apparatus shown in ▼ **Figure 4.39**, what is the acceleration of the system if (a) $m_1 = 0.25$ kg, $m_2 = 0.50$ kg, and $m_3 = 0.25$ kg, and (b) $m_1 = 0.35$ kg, $m_2 = 0.15$ kg, and $m_3 = 0.50$ kg?

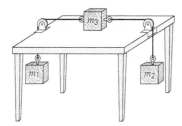

▲ FIGURE 4.39   **Which way will they accelerate?** See Exercise 42.

43. IE •• A rope is fixed at both ends on two trees and a bag is hung in the middle of the rope (causing the rope to sag vertically). (a) The tension in the rope depends on (1) only the tree separation, (2) only the sag, (3) both the tree separation and sag, (4) neither the tree separation nor the sag. (b) If the tree separation is 10 m, the mass of the bag is 5.0 kg, and the sag is 0.20 m, what is the tension in the line?

44. •• A 55-kg gymnast hangs vertically from a pair of parallel rings. (a) If the ropes supporting the rings are attached to the ceiling directly above, what is the tension in each rope? (b) If the ropes are supported so that they make an angle of 45° with the ceiling, what is the tension in each rope?

45. •• A physicist's car has a small lead weight suspended from a string attached to the interior ceiling. Starting from rest, after a fraction of a second the car accelerates at a steady rate for about 10 s. During that time, the string (with the weight on the end of it) makes a backward (opposite the acceleration) angle of 15.0° from the vertical. Determine the car's (and the weight's) acceleration during the 10-s interval.

46. •• At the end of most landing runways in airports, an extension of the runway is constructed using a special substance called formcrete. Formcrete can support the weight of cars, but crumbles under the weight of airplanes to slow them down if they run off the end of a runway. If a plane of mass 2.00×10^5 kg is to stop from a speed of 25.0 m/s on a 100-m-long stretch of formcrete, what is the average force exerted on the plane by the formcrete?

47. •• A rifle weighs 50.0 N and its barrel is 0.750 m long. It shoots a 25.0-g bullet, which leaves the barrel at a speed (muzzle velocity) of 300 m/s after being uniformly accelerated. What is the magnitude of the force exerted on the rifle by the bullet?

48. •• A horizontal force of 40 N acting on a block on a frictionless, level surface produces an acceleration of 2.5 m/s². A second block, with a mass of 4.0 kg, is dropped onto the first. What is the magnitude of the acceleration of the combination of blocks if the same force continues to act? (Assume that the second block does not slide on the first block.)

49. •• The Atwood machine consists of two masses suspended from a fixed pulley, as shown in ▶ **Figure 4.40**. It is named after the British scientist George Atwood (1746–1807), who used it to study motion and to measure the value of g. If $m_1 = 0.55$ kg and $m_2 = 0.80$ kg, what is the acceleration of the system, and (b) what is the magnitude of the tension in the string?

50. ••• In the frictionless apparatus shown in ▶ **Figure 4.41**, $m_1 = 2.0$ kg. What is m_2 if both masses are at rest? How about if both masses are moving at constant velocity?

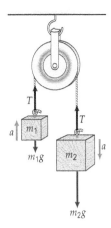

▲ **FIGURE 4.40** **Atwood machine** See Exercise 49.

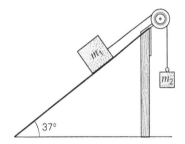

▲ **FIGURE 4.41** **Inclined Atwood machine** See Exercises 50, 51, and 66.

51. ••• In the ideal setup shown in Figure 4.41, $m_1 = 3.0$ kg and $m_2 = 2.5$ kg. (a) What is the acceleration of the masses? (b) What is the tension in the string?

4.6 Friction

52. IE • A 20-kg box sits on a rough horizontal surface. When a horizontal force of 120 N is applied, the object accelerates at 1.0 m/s². (a) If the applied force is doubled, the acceleration will (1) increase, but less than double; (2) also double; (3) increase, but more than double. Why? (b) Calculate the acceleration to prove your answer to part (a).

53. • The coefficients of static and kinetic friction between a 50.0-kg box and a horizontal surface are 0.500 and 0.400, respectively. (a) What is the acceleration of the object if a 250-N horizontal force is applied to the box? (b) What is the acceleration if the applied force is 235 N?

54. • In moving a 35.0-kg desk from one side of a classroom to the other, a professor finds that a horizontal force of 275 N is necessary to set the desk in motion, and a force of 195 N is necessary to keep it in motion at a constant speed. What are the coefficients of static and kinetic friction between the desk and the floor?

55. • A 40-kg crate is at rest on a level surface. If the coefficient of static friction between the crate and the surface is 0.69, what horizontal force is required to get the crate moving?

56. •• A packing crate is placed on a 20° inclined plane. If the coefficient of static friction between the crate and the plane is 0.65, will the crate slide down the plane if released from rest? Justify your answer.

57. •• A 1500-kg automobile travels at 90 km/h along a straight concrete highway. Faced with an emergency situation, the driver jams on the brakes, and the car skids to a stop. What is the car's stopping distance for (a) dry pavement and (b) wet pavement? [*Hint*: Refer to Table 4.1.]

58. •• A hockey player hits a puck with his stick, giving the puck an initial speed of 5.0 m/s. If the puck slows uniformly and comes to rest in a distance of 20 m, what is the coefficient of kinetic friction between the ice and the puck?

59. •• A crate sits on a flatbed truck that is traveling with a speed of 50 km/h on a straight, level road. If the coefficient of static friction between the crate and the truck bed is 0.30, in how short a distance can the truck stop with a constant acceleration without the crate sliding?

60. •• A block is projected with a speed of 2.5 m/s on a horizontal surface. If the block comes to rest in 1.5 m, what is the coefficient of kinetic friction between the block and the surface?

61. •• A block is projected with a speed of 3.0 m/s on a horizontal surface. If the coefficient of kinetic friction between the block and the surface is 0.60, how far does the block slide before coming to rest?

62. •• Suppose the slope conditions for the skier shown in ▼ **Figure 4.42** are such that the skier travels at a constant velocity. From the photo, could you find the coefficient of kinetic friction between the snowy surface and the skis? If so, describe how this would be done.

▲ **FIGURE 4.42** **A down slope run** See Exercise 62.

63. •• A block that has a mass of 2.0 kg and is 10 cm wide on each side just begins to slide down an inclined plane with a 30° angle of incline (▼ **Figure 4.43**). Another block of the same height and same material has base dimensions of 20 cm × 10 cm and thus a mass of 4.0 kg. (a) At what critical angle will the more massive block start to slide down the plane? Why? (b) Estimate the coefficient of static friction between the block and the plane.

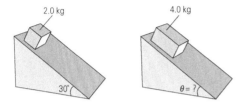

▲ FIGURE 4.43 **At what angle will it begin to slide?** See Exercise 63.

64. •• In the apparatus shown in ▼ **Figure 4.44**, $m_1 = 10$ kg and the coefficients of static and kinetic friction between m_1 and the table are 0.60 and 0.40, respectively. (a) What mass of m_2 will just barely set the system in motion? (b) After the system begins to move, what is the acceleration?

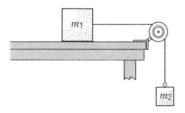

▲ FIGURE 4.44 **Friction and motion** See Exercise 64.

65. •• In loading a fish delivery truck, a person pushes a block of ice up a 20° incline at constant speed. The push is 150 N in magnitude and parallel to the incline. The block has a mass of 35.0 kg. (a) Is the incline frictionless? (b) If not, what is the force of kinetic friction on the block of ice?

66. ••• In the apparatus shown in Figure 4.41, $m_1 = 2.0$ kg and the coefficients of static and kinetic friction between m_1 and the inclined plane are 0.30 and 0.20, respectively. (a) What is m_2 if both masses are at rest? (b) What is m_2 if both masses are moving at constant velocity?

5

Work and Energy*

The pole-vaulter uses the potential energy stored in the pole to reach new heights.

A description of pole vaulting, as shown in the chapter-opening photo, might be as follows: The athlete runs with a pole, plants it into the ground, and tries to vault her body over a bar set at a certain height. However, a physicist might give a different description: The athlete has chemical potential energy stored in her body. She uses this potential energy to do work in running down the path to gain speed, or kinetic energy. When she plants the pole, most of her kinetic energy goes into elastic potential energy of the bent pole. This potential energy is used to lift the vaulter in doing work against gravity and is partially converted into gravitational potential energy. At the top, there is just enough kinetic energy left to carry the vaulter over the bar. On the way down, the gravitational potential energy is converted back to kinetic energy, which is absorbed by the mat in doing work to stop the fall. The pole vaulter participates in a game of work-energy, a game of give and take.

This chapter centers on two concepts that are important in both science and everyday life – *work* and *energy*. We commonly think of work as being associated with doing or accomplishing something. Because work makes us physically (and sometimes

---

* The mathematics in this chapter involves trigonometric functions. You may want to review these in Appendix I.

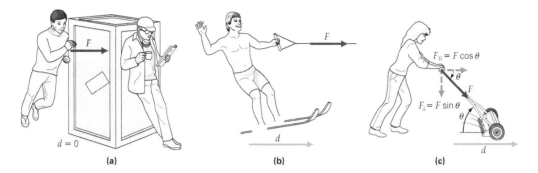

▲ **FIGURE 5.1 Work done by a constant force** The product of the magnitudes of the parallel component of force and the displacement. **(a)** If there is no displacement, no work is done: $W = 0$. **(b)** For a constant force in the same direction as the displacement, $W = Fd$. **(c)** For a constant force at an angle to the displacement, $W = (F\cos\theta)d$.

mentally) tired, machines have been invented to decrease the amount of effort expended personally. Thinking about energy tends to bring to mind the cost of fuel for transportation and heating, or perhaps the food that supplies the energy our bodies need to sustain life processes and to do work.

Although these notions do not really define work and energy, they point in the right direction. As you may have surmised, work and energy are closely related. In physics, as in everyday life, when something possesses energy, it has the ability to do work. For example, water rushing through the sluices of a dam has energy of motion, and this energy allows the water to do the work of driving a turbine or dynamo to generate electricity. Conversely, no work can be performed without energy.

Energy exists in various forms: mechanical energy, chemical energy, electrical energy, heat energy, nuclear energy, and so on. A transformation from one form to another may take place, but the total amount of energy is *conserved*, meaning there is always the same amount. This point makes the concept of energy very useful. When a physically measurable quantity is conserved, it not only gives us an insight that leads to a better understanding of nature, but also usually provides another approach to practical problems. (You will be introduced to other conserved quantities and conservation laws during the course of our study of physics.)

5.1 Work Done by a Constant Force

The word *work* is commonly used in a variety of ways: We go to work; work on projects; work at our desks or on computers; work on problems. In physics, however, *work* has a very specific meaning. Mechanically, work involves force and displacement, and the word *work* is used to describe quantitatively what is accomplished when a force acts on an object as it moves through a distance. In the simplest case of a *constant* force acting on an object, work that the force does is defined as follows:

The **work** done by a constant force acting on an object is equal to the product of the magnitudes of the displacement and the force, or component of the force, parallel to that displacement.

Work involves a force acting on an object as it moves through a distance. A force may be applied, as in ▲ **Figure 5.1a**, but *if there is no motion (no displacement), then no work is done*. However, when there is motion, a constant force F acting *in the same direction* as the displacement d does work (Figure 5.1b). The work (W) done in this case is defined as the product of their magnitudes:

$$W = Fd \qquad (5.1)$$

and work is a scalar quantity. When work is done as in Figure 5.1b, energy is expended. The relationship between work and energy is discussed in Section 5.3.

In general, work is done *on* an object *by* a force, or force *component*, parallel to the line of motion or displacement of the object (Figure 5.1c). That is, if the force acts at an angle θ to the object's displacement, then $F_{\parallel} = F\cos\theta$ is the component of the force parallel to the displacement. Thus, a more general equation for work done by a constant force is*

$$W = F_{\parallel}d = (F\cos\theta)d \quad \text{(work done by a constant force)} \qquad (5.2)$$

Notice that θ is the angle *between* the force and the displacement vectors. As a reminder of this factor, $\cos\theta$ may be written between the magnitudes of the force and displacement, $W = F(\cos\theta)d$. If $\theta = 0°$ (i.e., force and displacement are parallel or in the same direction, as in Figure 5.1b), then $W = F(\cos 0°)d = Fd$, so Equation 5.2 reduces to Equation 5.1. The perpendicular component of the force, $F_{\perp} = F\sin\theta$, does no work, since there is no displacement in this direction.

The units of work can be determined from the equation $W = Fd$. With force in newtons and displacement in meters, work has the SI unit of newton-meter (N·m). This unit is called a **joule (J)**:[†]

$$Fd = W$$

$$1\,\text{N·m} = 1\,\text{J}$$

---

* The product of two vectors (force and displacement) is a special type of vector multiplication and yields a scalar quantity equal to $(F\cos\theta)d$. Thus, work is a scalar – it does not have direction. It can, however, be positive, zero, or negative, depending on the angle.

[†] The joule (J), pronounced "jool," was named in honor of James Prescott Joule (1818–1889), a British scientist who investigated work and energy.

for example, the work done by a force of 25 N on an object as the object moves through a parallel displacement of 2.0 m is $W = Fd = (25 \text{ N})(2.0 \text{ m}) = 50 \text{ N·m}$, or 50 J.

From the previous displayed equation, it can also be seen that in the British system, work would have the unit pound-foot. However, this name is commonly written in reverse. The British standard unit of work is the **foot-pound (ft·lb)**. One ft·lb is equal to 1.36 J.

Work can be analyzed graphically. Suppose a constant force F in the x-direction acts on an object as it moves a distance x. Then $W = Fx$ and if F versus x is plotted, a horizontal straight-line graph is obtained such as shown in ▼ **Figure 5.2**. The area under the line is Fx, so this area is equal to the work done by the force over the given distance. Work done by a non-constant, or variable, force will be considered later.*

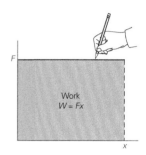

▲ FIGURE 5.2 **Graphical determination of work** Work is equal to the area under the F-versus-x curve.

Remember that *work is a scalar quantity* and may have a positive or negative value. In Figure 5.1b, the work is positive, because the force acts in the same direction as the displacement (and $\cos 0° = 1$ is positive). The work is also positive in Figure 5.1c, because a force component acts in the direction of the displacement (and $\cos\theta$ is positive).

However, if the force, or a force component, acts in the opposite direction of the displacement, the work is negative, since the cosine term is negative. For example, for $\theta = 180°$ (force opposite to the displacement), $\cos 180° = -1$, so the work is negative: $F_\parallel d = F(\cos 180°)d = -Fd$. An example is a braking force that slows down or decelerates an object. ▶ **Figure 5.3** lists all possible arrangements for the sign of work.

EXAMPLE 5.1: APPLIED PSYCHOLOGY – MECHANICAL WORK

A student holds her 1.5-kg psychology textbook out a second-story dormitory window until her arm is tired; then she releases it (▶ **Figure 5.4**). (a) How much work is done on the book by the student in simply holding it out the window? (b) How much work is done by the force of gravity during the time in which the book falls 3.0 m?

---

* Work is the area under the F-versus-x curve even if the curve is not a straight line. Finding the work in such cases generally requires advanced mathematics.

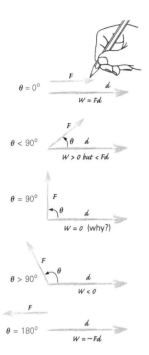

▲ FIGURE 5.3 **Determining the sign of work** The sign of work is determined by the angle between the fore and displacement vector.

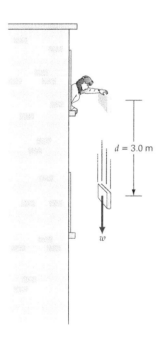

▲ FIGURE 5.4 **Mechanical work requires motion** See Example 5.1 for description.

THINKING IT THROUGH. Analyze the situations in terms of the definition of work, keeping in mind that force and displacement are the key factors.

SOLUTION

Listing the data:

Given:

$v_o = 0$ (initially at rest)
$m = 1.5$ kg
$d = 3.0$ m

Find:

(a) *W* (done by student in holding)
(b) *W* (done by gravity in falling)

(a) Even though the student gets tired (because work is performed within the body to maintain muscles in a state of tension), she does *no work on the book* in merely holding it stationary. She exerts an upward force on the book (equal in magnitude to the weight of the book), but the displacement is zero in this case ($d = 0$). Thus, $W = Fd = F \times 0 = 0$ J.

(b) While the book is falling, the only force acting on it is the force of gravity (neglecting air resistance), which is equal in magnitude to the weight of the book: $F = w = mg$. The displacement is in the same direction as the force ($\theta = 0°$) and has a magnitude of $d = 3.0$ m so the work done by gravity is

$$W = F(\cos 0°)d = (mg)d = (1.5 \text{ kg})(9.8 \text{ m/s}^2)(3.0 \text{ m})$$
$$= +44 \text{ J}$$

(The sign of work is + because the force and displacement are in the same direction.)

FOLLOW-UP EXERCISE. A 0.20-kg ball is thrown upward. How much work is done on the ball by gravity as the ball rises between heights of 2.0 and 3.0 m? (*Answers to all Follow-Up Exercises are given in Appendix V at the back of the book.*)

EXAMPLE 5.2: HARD WORK

A worker pulls a 40.0-kg crate with a rope, as illustrated in ▼ **Figure 5.5**. The coefficient of kinetic (sliding) friction between the crate and the floor is 0.550. If he moves the crate with a constant velocity a distance of 7.00 m, how much is the work done by the worker on the crate?

THINKING IT THROUGH. A good thing to do first in problems such as this is to draw a free-body diagram. This is shown in the figure. (Frictional forces were covered in Section 4.6 and in general, $f_k = \mu_k N$, where *N* is the normal force.) To find the work, the force *F* must be known. As usual in such cases, this is done by summing the forces.

Given:

$m = 40.0$ kg
$\mu_k = 0.550$
$d = 7.00$ m
$\theta = 30°$ (from figure)
$v = $ constant

Find:

W (work done by worker on crate)

SOLUTION

Then, summing the forces in the *x*- and *y*-directions and setting these equal to zero (with a constant velocity $F_{net} = 0$):

$$\Sigma F_x = F\cos 30° - f_k = F\cos 30° - \mu_k N = ma_x = 0$$
$$\Sigma F_y = N + F\sin 30° - mg = ma_y = 0$$

To find *F*, the second equation may be solved for *N*, which is then substituted in the first equation.

$$N = mg - F\sin 30°$$

(Notice that *N* is not equal to the weight of the crate. Why?) And, substituting *N* into the first equation,

$$F\cos 30° - \mu_k(mg - F\sin 30°) = 0$$

Solving for *F* and putting in values:

$$F = \frac{\mu_k mg}{(\cos 30° + \mu_k \sin 30°)} = \frac{(0.550)(40.0\text{kg})(9.80\text{m/s}^2)}{(0.866)+(0.550)(0.500)} = 189\text{N}$$

Then,

$$W = F(\cos 30°)d = (189 \text{ N})(0.866)(7.00\text{m}) = 1.15 \times 10^3 \text{ J}$$

FOLLOW-UP EXERCISE. It takes about 3.80×10^4 J of work to lose 1.00 g of body fat. What distance would the worker have to pull the crate to lose 1 g of fat? (Assume all the work goes into fat reduction.) Make an estimate before solving and see how close you come.

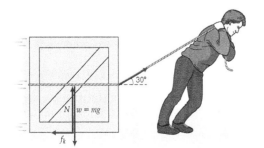
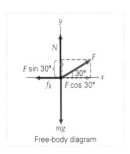

▲ FIGURE 5.5 **Doing some work** See Example 5.2 for description.

It has been said that a force does work *on* an object. For example, the force of gravity does work on a falling object, such as the book in Example 5.1. Also, when you lift an object, *you* do work *on* the object. This is sometimes described as doing work *against* gravity, because the force of gravity acts in the direction opposite that of the applied lift force and opposes it. For example, an average-sized apple has a weight of about 1 N. So, when lifting such an apple a distance of 1 m with a force equal to its weight, 1 J of work is done against gravity [$W = Fd = (1\text{ N})(1\text{ m}) = 1\text{ J}$].

If more than one force acts on an object, the work done by each can be calculated separately and added to find the **net work**. That is:

The *total* or *net work* is defined as the work done by all the forces acting on an object, or the scalar sum of the work done by each force.

This concept is illustrated in Example 5.3.

EXAMPLE 5.3: TOTAL OR NET WORK

A 0.75-kg block slides with a uniform velocity down a 20° inclined plane (▼ **Figure 5.6**). (a) How much work is done by the force of friction on the block as it slides the total length of the plane? (b) What is the net work done on the block? (c) Discuss the net work done if the angle of incline is adjusted so that the block accelerates down the plane.

THINKING IT THROUGH. (a) The length of the plane can be found using trigonometry, so this part boils down to finding the force of friction. (b) The net work is the sum of all the work done by the individual forces. (*Note*: Since the block has a uniform, or constant, velocity, the net force on it is zero. This observation should tell you the answer, but it will be shown explicitly in the solution.) (c) If there is acceleration, Newton's second law applies, which involves a net force, so there may be net work.

SOLUTION

Listing the data given, and specifically what is to be found:

Given:

$m = 0.75$ kg
$\theta = 20°$
$L = 1.2$ m (from Figure 5.6)

Find:

(a) W_f (work done on the block by friction)
(b) W_{net} (net work on the block)
(c) W_{net} (discuss with block accelerating)

(a) Note from the Figure 5.6 free-body diagram that only two forces do work, because there are only two forces parallel to the motion: f_k, the force of kinetic friction, and $mg \sin\theta$, the component of the block's weight acting down the plane. The normal force N and $mg\cos\theta$, the component of the block's weight, act perpendicular to the plane and do no work. (Why?)

First finding the work done by the frictional force:

$$W_f = f_k(\cos 180°)d = -f_k d = -\mu_k N d$$

The angle 180° indicates that the force and displacement are in opposite directions. (It is common in such cases to write $W_f = -f_k d$ directly, since kinetic friction typically opposes motion.) The distance d the block slides down the plane can be found by using trigonometry. Note that $\cos\theta = L/d$, so

$$d = \frac{L}{\cos\theta}$$

We know that $N = mg\cos\theta$, but what is μ_k? It would appear that some information is lacking. When this situation occurs, look for another approach to solve the problem. As noted earlier, there are only two forces parallel to the motion, and they are opposite, so with a constant velocity their magnitudes are equal, $f_k = mg\sin\theta$. Thus,

$$W_f = f_k d = -(mg \sin\theta)\left(\frac{L}{\cos\theta}\right) = -mg\,L\tan 20°$$
$$= -(0.75\text{kg})(9.8\text{ m/s}^2)(1.2\text{ m})(0.364) = -3.2\text{ J}$$

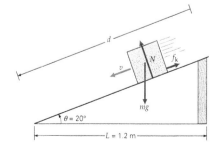

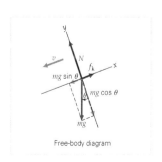

Free-body diagram

▲ FIGURE 5.6 **Total or net work** See Example 5.3 for description.

(b) To find the net work, the work done by gravity needs to be calculated and then added to the result in part (a). Since $F_\parallel$ for gravity is just $mg \sin\theta$,

$$W_g = F_\parallel d = \left(mg\sin\theta\right)\left(\frac{L}{\cos\theta}\right) = mgL\tan 20° = +3.2\,\text{J}$$

where the calculation is the same as in part (a) except for the sign. Then, the net work is

$$W_{\text{net}} = W_g + W_f = +3.2\,\text{J} + (-3.2\,\text{J}) = 0$$

(constant velocity, zero net force, zero net work). Remember that work is a scalar quantity, so scalar addition is used to find net work.

(c) If the block accelerates down the plane, then from Newton's second law, $F_{\text{net}} = mg\sin\theta - f_k = ma$. The component of the gravitational force $(mg\sin\theta)$ is greater than the opposing frictional force (f_k), so net work is done on the block, because now $|W_g| > |W_f|$. You may be wondering what the effect of nonzero net work is. As will be shown shortly, nonzero net work causes a change in the amount of kinetic energy an object has.

FOLLOW-UP EXERCISE. In part (c) of this Example, is it possible for the frictional work to be greater in magnitude than the gravitational work? What would this condition mean in terms of the block's speed?

5.1.1 Problem-Solving Hint

Note that in part (a) of Example 5.3, the equation for W_f was simplified by using algebraic expressions for N and d instead of by computing these quantities initially. It is a good rule of thumb not to plug numbers into an equation until you have to. Simplifying an equation through cancellation is easier with symbols and saves computation time.

5.2 Work Done by a Variable Force

The discussion in the preceding section was limited to work done by constant forces. In general, however, forces are variable; that is, they change in magnitude and/or angle with time and/or position.

An example of a variable force that does work is illustrated in ▶ **Figure 5.7**, which depicts a spring being stretched by an applied force F_a. As the spring is stretched (or compressed) farther and farther, its restoring force (the spring force that opposes the stretching or compression) becomes greater, and an increased applied force is required. For most springs, the spring force (F_s) is directly proportional to the change in length of the spring from its unstretched length. In equation form, this relationship is expressed

$$F_s = -k\Delta x = -k(x - x_0)$$

or, if $x_0 = 0$,

$$F_s = -kx \quad \text{(ideal spring force)} \tag{5.3}$$

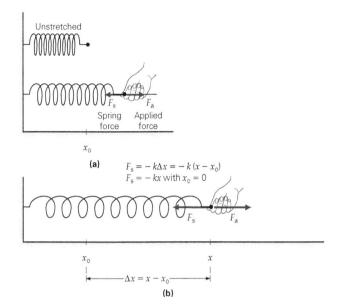

▲ FIGURE 5.7   **Spring force (a)** An applied force F_a stretches the spring, and the spring exerts an equal and opposite force F_s on the hand. **(b)** The magnitude of the force depends on the change Δx in the spring's length. This change is measured from to the end of the unstretched spring at x_0.

where x now represents the distance the spring is stretched (or compressed) from its unstretched position. As can be seen, the force varies with x. This is described by saying that the *force is a function of position*.

The k in this equation is a constant of proportionality and is commonly called the **spring constant**, or **force constant**. (See Example 5.4 to see how it is measured.) The greater the value of k, the stiffer or stronger the spring. As you should be able to prove to yourself, the SI unit of k is newtons per meter (N/m). The negative sign in Equation 5.3 indicates that the spring force acts in the direction opposite to the displacement when the spring is either stretched or compressed. Equation 5.3 is a form of what is known as *Hooke's law*, named after Robert Hooke (1635–1703), a contemporary of Isaac Newton.

The relationship expressed by the spring force equation holds only for ideal springs. Real springs approximate this linear relationship between force and displacement within certain limits. If a spring is stretched beyond a certain point, called its *elastic limit*, the spring will be permanently deformed, and the linear relationship will no longer apply.

Computing the work done by variable forces generally requires calculus. But it is fortunate that the spring force is a special case that can be computed graphically. A plot of F (the applied force) versus x is shown in ▶ **Figure 5.8**. The graph has a straight-line slope of k, with $F = kx$, where F is the applied force doing work in stretching the spring.

As described earlier, work is the area under an F-versus-x curve, and here it is in the form of a triangle, as indicated by the shaded area in the figure. Then, computing this area,

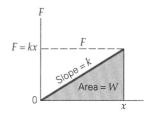

▲ FIGURE 5.8 **Work done by a uniformly variable spring force** A graph of F versus x, where F is the applied force doing work in stretching a spring, is a straight line with a slope of k. The work is equal to the area under the line, which is that of a triangle with area $= \frac{1}{2}$(altitude $\times$ base). Then $W = \frac{1}{2}Fx = \frac{1}{2}(kx)x = \frac{1}{2}kx^2$.

$$\text{area} = W = \tfrac{1}{2}(\text{altitude} \times \text{base})$$

or

$$W = \tfrac{1}{2}Fx = \tfrac{1}{2}(kx)x = \tfrac{1}{2}kx^2$$

where $F = kx$. Thus,

$$W = \tfrac{1}{2}kx^2 \quad \begin{array}{l}\text{(work done in stretching or}\\ \text{compressing a spring)}\end{array} \quad (5.4)$$

EXAMPLE 5.4: DETERMINING THE SPRING CONSTANT

A 0.15-kg mass is attached to a vertical spring and hangs at rest a distance of 4.6 cm below its original position (▼ **Figure 5.9**). An additional 0.50-kg mass is then suspended from the first mass and

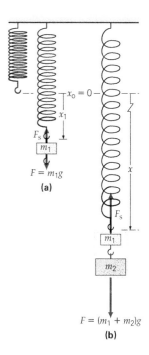

▲ FIGURE 5.9 **Determining the spring constant and the work done in stretching a spring** See Example 5.4 for description.

the system is allowed to descend to a new equilibrium position. What is the total extension of the spring? (Neglect the mass of the spring.)

THINKING IT THROUGH. The spring constant k appears in Equation 5.3. Therefore, to find the value of k for a particular instance, the spring force and distance the spring is stretched (or compressed) must be known.

SOLUTION

The data given are as follows:

Given:

$$m_1 = 0.15 \text{ kg}$$
$$x_1 = 4.6 \text{ cm} = 0.046 \text{ m}$$
$$m_2 = 0.50 \text{ kg}$$

Find:

x (total stretch distance)

The total stretch distance is given by $x = F/k$, where F is the applied force, which in this case is the weight of the mass suspended on the spring. (The negative sign in Equation 5.3 is ignored here for convenience.) However, the spring constant k is not given. But, k may be found from the data pertaining to the suspension of m_1 and resulting displacement x_1. (This method is commonly used to determine spring constants.) As seen in Figure 5.9a, the magnitudes of the weight force and the restoring spring force are equal, since $a = 0$, their magnitudes may be equated:

$$F_s = kx_1 = m_1 g$$

Solving for k,

$$k = \frac{m_1 g}{x_1} = \frac{(0.15 \text{ kg})(9.8 \text{ m/s}^2)}{0.046 \text{ m}} = 32 \text{ N/m}$$

Then, knowing k, the total extension of the spring can be found from the balanced-force situation shown in Figure 5.9b:

$$F_s = (m_1 + m_2)g = kx$$

Thus,

$$x = \frac{(m_1 + m_2)g}{k} = \frac{(0.15 \text{ kg} + 0.50 \text{ kg})(9.8 \text{ m/s}^2)}{32 \text{ N/m}} = 0.20 \text{ m (or 20 cm)}$$

FOLLOW-UP EXERCISE. How much work is done by gravity on the system in stretching the spring through both displacements in Example 5.4?

5.2.1 Problem-Solving Hint

The reference position x_o used to determine the change in length of a spring is arbitrary but is usually chosen as $x_o = 0$ for convenience. *The important quantity in computing work is the difference in position, Δx, or the net change in the length of the* *spring from its unstretched length.* As shown in ▼ **Figure 5.10** for a mass suspended on a spring, x_o can be referenced to the unloaded length of the spring or to the loaded position, which may be taken as the zero position for convenience. In Example 5.4, x_o was referenced to the end of the unloaded spring.

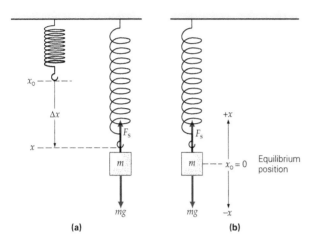

▲ **FIGURE 5.10**   **Displacement reference** The reference position x_o is arbitrary and is usually chosen for convenience. It may be **(a)** at the end of the spring at its unloaded position or **(b)** at the equilibrium position when a mass is suspended on the spring. The latter is particularly convenient in cases in which the mass oscillates up and down on the spring.

When the net force on the suspended mass is zero, the mass is said to be at its *equilibrium position* (as in Figure 5.9a with m_1 suspended). This position, rather than the unloaded length, may be taken as a zero reference ($x_o = 0$; see Figure 5.10b). The equilibrium position is a convenient reference point for cases in which the mass oscillates up and down on the spring. Also, since the displacement is in the vertical direction, the x's are often replaced by y's.

5.3 The Work-Energy Theorem: Kinetic Energy

Now that we have an operational definition of work, let's take a look at how work is related to energy. Energy is one of the most important concepts in science. It is described as something that objects or systems possess. Basically, work is something that is *done on* objects, whereas energy is something that objects *have*, which is the ability to do work.

One form of energy that is closely associated with work is *kinetic energy*. (Another basic form of energy, *potential energy*, will be discussed in Section 5.4.) Consider an object at rest on a frictionless surface. Let a horizontal force act on the object and set it in motion. Work is done *on* the object, but where does the work "go," so to speak? It goes into setting the object into motion, or changing its *kinetic* conditions. Because of its motion, we say the object has gained energy – kinetic energy, which gives it the capability to do work.

For a constant force doing work on a moving object parallel to the direction of motion, as illustrated in ▼ **Figure 5.11**, the force does an amount of work $W = Fx$. But what are the kinematic effects? The force gives the object a constant acceleration, and from Equation 2.12, $v^2 = v_o^2 + 2ax$ (with $x_o = 0$),

$$a = \frac{v^2 - v_o^2}{2x}$$

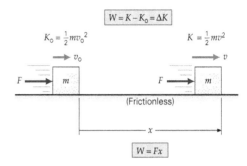

▲ **FIGURE 5.11**   **The relationship of work and kinetic energy** The work done on a block by a constant force in moving it along a horizontal frictionless surface is equal to the change in the block's kinetic energy: $W = \Delta K$.

where v_o may or may not be zero. Writing the magnitude of the force in the form of Newton's second law and substituting in the expression for a from the previous equation gives

$$F = ma = m\left(\frac{v^2 - v_o^2}{2x}\right)$$

Using this expression in the equation for work,

$$W = Fx = m\left(\frac{v^2 - v_o^2}{2x}\right)x$$
$$= \tfrac{1}{2}mv^2 - \tfrac{1}{2}mv_o^2$$

The term $\tfrac{1}{2}mv^2$ is defined as the **kinetic energy (K)** of the moving object:

$$K = \tfrac{1}{2}mv^2 \quad \text{(kinetic energy)} \tag{5.5}$$

SI unit of energy: joule (J)

Kinetic energy is often called the *energy of motion*. Note that it is directly proportional to the square of the (instantaneous) speed of a moving object, and therefore cannot be negative.

Then, in terms of kinetic energy, the previous expression for work may be written as

$$K = \tfrac{1}{2}mv^2 - \tfrac{1}{2}mv_o^2 = K - K_o = \Delta K$$

or

$$W_{\text{net}} = \Delta K \tag{5.6}$$

where it can be shown that W_{net} *is the net work if more than one force acts on the object*, as shown in Example 5.3. This equation is called the **work-energy theorem**, and it relates the work done on an object to the change in the object's kinetic energy. That is, *the net work done on a body by all the forces acting on it is equal to the change in kinetic energy of the body*. Work done on an object is a way of changing the object's energy. Both work and energy have units of joules, and both are *scalar* quantities. The work-energy theorem is true in general for variable forces and not just for the special case considered in deriving Equation 5.6. Work done on an object is a way of changing the object's energy

As a simple example of the work-energy theorem, recall that in Example 5.1 the force of gravity did +44 J of work on a book that fell from rest through a distance of $y = 3.0$ m. At that position and instant, the falling book had 44 J of kinetic energy. Since in this case the net work is that due solely to gravity, we have

$$W_{net} = Fd = mgy = \frac{mv^2}{2} = K = \Delta K$$

where $K_o = 0$. As an exercise, confirm this fact by calculating the speed of the book and computing its kinetic energy.

The work-energy theorem tells us that when work is done on an object, there is a change in or a transfer of energy. For example, a force doing work on an object that causes the object to speed up gives rise to an increase in the object's kinetic energy. Conversely, (negative) work done by the force of kinetic friction may cause a moving object to slow down and decrease its kinetic energy. So for an object to have a change in its kinetic energy, net work must be done on the object, as Equation 5.6 indicates.

When an object is in motion, it possesses kinetic energy and thus has the capability to do work. For example, a moving automobile has kinetic energy and can do work in crumpling a fender in a fender bender – not *useful* work in that case, but still work. Another example of work done by kinetic energy is shown in ▶ **Figure 5.12**.

EXAMPLE 5.5: A GAME OF SHUFFLEBOARD – THE WORK-ENERGY THEOREM

A shuffleboard player (▶ **Figure 5.13**) pushes a 0.25-kg puck that is initially at rest such that a constant horizontal force of 6.0 N acts on it through a distance of 0.50 m. (Neglect friction.) (a) What are the kinetic energy and the speed of the puck when the force is removed? (b) How much work would be required to bring the puck to rest?

THINKING IT THROUGH. Apply the work-energy theorem. If the amount of work done can be found, then this gives the change in kinetic energy.

SOLUTION

Listing the given data as usual:

▲ **FIGURE 5.12   Kinetic energy and work** A moving object, such as a wrecking ball, processes kinetic energy and can do work. A massive ball is used in demolishing this chimney.

▲ **FIGURE 5.13   Work and kinetic energy** See Example 5.5 for description.

Given:

$m = 0.25$ kg
$F = 6.0$ N
$d = 0.50$ m
$v_o = 0$

Find:

(a) K (kinetic energy) and v (speed)
(b) W (work done in stopping puck)

(a) Since the speed is not known, the kinetic energy $K = \frac{1}{2}mv^2$ cannot be computed directly. However, kinetic energy is related to work by the work-energy theorem. The work done on the puck by the player's applied force F is

$$W = Fd = (6.0 \text{ N})(0.50 \text{ m}) = 3.0 \text{ J}$$

Then, by the work-energy theorem,

$$W = \Delta K = K - K_o = 3.0 \text{ J}$$

But $K_o = \frac{1}{2}mv_o^2 = 0$, because $v_o = 0$, so

$$K = 3.0 \text{ J}$$

The speed can be found from the kinetic energy. Since $K = \frac{1}{2}mv^2$,

$$v = \sqrt{\frac{2K}{m}} = \sqrt{\frac{2(3.0 \text{ J})}{0.25 \text{ kg}}} = 4.9 \text{ m/s}$$

(b) As you might guess, the work required to bring the puck to rest is equal to the puck's kinetic energy (i.e., the amount of energy that the puck must lose to come to a stop). To confirm this equality, the previous calculation is essentially performed in reverse, with $v_o = 4.9$ m/s and $v = 0$:

$$W = K - K_o = 0 - K_o = \frac{1}{2}mv_o^2 = \frac{1}{2}(0.25 \text{ kg})(4.9 \text{ m/s})^2$$
$$= -3.0 \text{ J}$$

The negative sign indicates that the puck loses energy as it slows down. The work is done *against* the motion of the puck; that is, the opposing force is in a direction opposite that of the motion. (In the real-life situation, the opposing force could be friction.)

FOLLOW-UP EXERCISE. Suppose the puck in this Example had twice the final speed when released. Would it then take twice as much work to stop the puck? Justify your answer numerically.

5.3.1 Problem-Solving Hint

Notice how work-energy considerations were used to find speed in Example 5.5. This operation can be done in another way as well. First, the acceleration could be found from $a = F/m$, and then the kinematic equation $v^2 = v_o^2 + 2ax$ could be used to find v (where $x = d = 0.50$ m). The point is that many problems can be solved in different ways, and finding the fastest and most efficient way is often the key to success. As our discussion of energy progresses, it will be seen how useful and powerful the notions of work and energy are, both as theoretical concepts and as practical tools for solving many kinds of problems.

CONCEPTUAL EXAMPLE 5.6: KINETIC ENERGY – MASS VERSUS SPEED

In a football game, a 140-kg guard runs at a speed of 4.0 m/s, and a 70-kg free safety moves at 8.0 m/s. Which of the following is a correct statement? (a) The players have the same kinetic energy. (b) The safety has twice as much kinetic energy as the guard. (c) The guard has twice as much kinetic energy as the safety. (d) The safety has four times as much kinetic energy as the guard.

REASONING AND ANSWER. The kinetic energy of a body depends on both its mass and speed. You might think that, with half

the mass but twice the speed, the safety would have the same kinetic energy as the guard, but this is not the case. As observed from the relationship $K = \frac{1}{2}mv^2$, kinetic energy is directly proportional to the mass, but is proportional to the *square* of the speed. Thus, having half the mass decreases the kinetic energy by a factor of 2. So if the two athletes had equal speeds, the safety would have half as much kinetic energy as the guard.

However, doubling the speed increases the kinetic energy, not by a factor of 2 but by a factor of 2^2, or 4. Thus, the safety, with half the mass but twice the speed, would have $\frac{1}{2} \times 4 = 2$ (twice as much kinetic energy as the guard), and so the answer is (b).

Note that to answer this question, it was not necessary to calculate the kinetic energy of each player. But this can be done to verify the answer:

$$K_{\text{safety}} = \frac{1}{2}m_s v_s^2 = \frac{1}{2}(70 \text{ kg})(8.0 \text{ m/s})^2 = 2.2 \times 10^3 \text{ J}$$
$$K_{\text{guard}} = \frac{1}{2}m_g v_g^2 = \frac{1}{2}(140 \text{ kg})(4.0 \text{ m/s})^2 = 1.1 \times 10^3 \text{ J}$$

which explicitly shows the answer to be correct.

FOLLOW-UP EXERCISE. Suppose that the safety's speed were only 50% greater than the guard's, or 6.0 m/s. Which athlete would then have the greater kinetic energy, and how much greater?

5.3.2 Problem-Solving Hint

Note that the work-energy theorem relates the work done to the *change* in the kinetic energy. Sometimes, $v_o = 0$ and $K_o = 0$, so $W_{\text{net}} = \Delta K = K$. But take care! You *cannot* simply use the square of the change or difference in speed, $(v - v_o)^2 = (\Delta v)^2$, to calculate ΔK, as you might at first think. In terms of speed,

$$W_{\text{net}} = \Delta K = K - K_o = \frac{1}{2}mv^2 - \frac{1}{2}mv_o^2 = \frac{1}{2}m\left(v^2 - v_o^2\right)$$

Note that $(v^2 - v_o^2)$ is not the same as $(v - v_o)^2 = (\Delta v)^2$ because $(v - v_o)^2 = v^2 - 2vv_o + v_o^2$. Hence, the change in kinetic energy is *not* equal to $\frac{1}{2}m(v - v_o)^2 = \frac{1}{2}m(\Delta v)^2 \neq \Delta K$.

This observation means that to calculate work, or the change in kinetic energy, you must compute the kinetic energy of an object at one point or time (using the instantaneous speed to get the instantaneous kinetic energy) and also at another location or time. Then subtract the quantities to find the change in kinetic energy, or the work. Alternatively, you can find the difference of the *squares* of the speeds $(v^2 - v_o^2)$ first in computing the change, but remember never to use the square of the difference of the speeds. To see this hint in action, look at Conceptual Example 5.7.

CONCEPTUAL EXAMPLE 5.7: AN ACCELERATING CAR – SPEED AND KINETIC ENERGY

A car traveling at 5.0 m/s speeds up to 10 m/s, with an increase in kinetic energy that requires work W_1. Then the car's speed increases from 10 to 15 m/s, requiring additional work W_2. Which of the following relationships accurately compares the two amounts of work: (a) $W_1 > W_2$, (b) $W_1 = W_2$, or (c) $W_2 > W_1$?

REASONING AND ANSWER. As noted previously, the work-energy theorem relates the work done on the car to the *change* in its kinetic energy. Since the speeds have the same increment in each case ($\Delta v = 5.0$ m/s), it might appear that (b) would be the answer. However, keep in mind that the work is equal to the *change* in kinetic energy and involves $v_2^2 - v_1^2$, *not* $(\Delta v)^2 = (v_2 - v_1)^2$.

So the greater the speed of an object, the greater its kinetic energy. The *difference* in kinetic energy in changing speeds (or the work required to change speed) would then be greater for higher speeds for the same Δv. Therefore, (c) is the answer.

The main point is that the Δv values are the same, but more work is required to increase the kinetic energy of an object at higher speeds.

FOLLOW-UP EXERCISE. Suppose the car speeds up a third time, from 15 to 20 m/s, a change requiring work W_3. How does the work done in this increment compare with W_2? Justify your answer numerically. [*Hint:* Use a ratio.]

5.4 Potential Energy

An object in motion has kinetic energy. However, whether an object is in motion or not, it may have another form of energy – potential energy. As the name implies, an object having potential energy has the *potential* to do work. You can probably think of many examples: a compressed spring, a drawn bow, and water held back by a dam. In all such cases, the potential to do work derives from the *position* or *configuration* of bodies. A spring has energy because it is compressed, a bow because it is drawn, and the water because it has been lifted above the surface of the Earth (▼ **Figure 5.14**). Consequently, **potential energy (U)**, is often called *the energy of position* (and/or configuration).

Unlike kinetic energy, which is associated with motion, potential energy is a form of mechanical energy associated with the position of an object within a system (or configuration).

Potential energy is a property of the system, rather than the object. If the configuration of a system of objects changes, so does the potential energy of a particular object within that system.

In a sense, potential energy can be thought of as stored work. You have already seen an example of potential energy in Section 5.2 when work was done in stretching a spring from its equilibrium position. Recall that the work done in such a case is $W = \frac{1}{2}kx^2$ (with $x_0 = 0$). Note that the amount of work done depends on the amount of stretching (x). Because work is done, there is a *change* in the spring's potential energy (ΔU), which is equal to the work done *by the applied force* in stretching (or compressing) the spring:

$$W = \Delta U = U - U_0 = \tfrac{1}{2}kx^2 - \tfrac{1}{2}kx_0^2$$

Thus, with $x_0 = 0$ and $U_0 = 0$, as they are commonly taken for convenience, the *potential energy of a spring* is

$$U = \tfrac{1}{2}kx^2 \quad \text{(Potential energy of a spring)} \qquad (5.7)$$

SI unit of energy: joule (J)

[*Note*: Since the potential energy varies as x^2, the previous Problem-Solving Hint also applies, and when $x_0 \neq 0$, then $x^2 - x_0^2 \neq (x - x_0)^2$. That is, the potential energy of a spring must be calculated at different positions and then subtracted to find ΔU.]

Perhaps the most well-known type of potential energy is gravitational potential energy. In this case, position refers to the height of an object above some reference point, such as the floor or the ground. Suppose that an object of mass m is lifted a distance Δy (▶ **Figure 5.15**). Work is done against the force of gravity, and an applied force at least equal to the object's weight is necessary to lift the object: $F = w = mg$. The work done in lifting is then equal to the change in potential energy. Expressing this relationship in equation form, since there is no overall change in kinetic energy,

work done by external force = change in gravitational
potential energy

(a)   (b)

▲ **FIGURE 5.14** **Potential energy** Potential energy has many forms. **(a)** Work must be done to bend the bow, giving it potential energy. That energy is converted into kinetic energy when the arrow is released. **(b)** Gravitational potential energy is converted into kinetic energy when something falls. (Where did the gravitational potential energy of the water and the diver come from?)

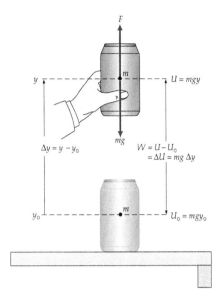

▲ FIGURE 5.15 **Gravitational potential energy** The work done in lifting an object is equal to the change in gravitational potential energy: $W = F\Delta y = mg(y - y_o)$.

or

$$W = F\Delta y = mg(y - y_o) = mgy - mgy_o = \Delta U = U - U_o$$

where y is used as the vertical coordinate. With the common choice of $y_o = 0$, such that $U_o = 0$, the **gravitational potential energy** is

$$U = mgy \quad \text{(Gravitational potential energy)} \qquad (5.8)$$

$$\text{SI unit of energy: joule (J)}$$

(Equation 5.8 represents the gravitational potential energy on or near the Earth's surface, where g is considered to be constant. A more general form of gravitational potential energy will be given in Chapter 7.5.)

EXAMPLE 5.8: MORE ENERGY NEEDED

To walk 1000 m on level ground, a 60-kg person requires an expenditure of about 1.0×10^5 J of energy. What is the total energy required if the walk is extended another 1000 m along a 5.0° incline as shown in ▼ **Figure 5.16**? (Neglect any frictional changes.)

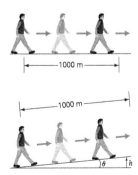

▲ FIGURE 5.16 **Adding potential energy** See Example 5.8 for description.

THINKING IT THROUGH. To walk an additional 1000 m would require another 1.0×10^5 J *plus* the additional energy for doing work against gravity in walking up the incline. From the figure, the increase in height can be seen to be $h = d\sin\theta$, where d is 1000 m.

SOLUTION

Listing the given data:

Given:

> $m = 60$ kg
> $E_o = 1.0 \times 10^5$ J (for 1000 m)
> $\theta = 5.0°$
> $d = 1000$ m (for each part of walk)

Find:

> E (total expended energy)

The additional expended energy in going up the incline is equal to gravitational potential energy gained. So,

$$\Delta U = mgh = (60 \text{ kg})(9.8 \text{ m/s}^2)(1000 \text{ m})\sin 5.0° = 5.1 \times 10^4 \text{ J}$$

Then, the total energy expended for the 2000-m walk is

$$\text{Total } E = 2E_o + \Delta U = 2(1.0 \times 10^5 \text{ J}) + 0.51 \times 10^5 \text{ J} = 2.5 \times 10^5 \text{ J}$$

Notice that the value of ΔU was expressed as a multiple of 10^5 in the last equation so it could be added to the E_o term, and the result was rounded to two significant figures per the rules given in Chapter 1.6.

FOLLOW-UP EXERCISE. If the angle of incline were doubled and the walk *just* up the incline is repeated, will the additional energy expended by the person in doing work against gravity be doubled? Justify your answer.

EXAMPLE 5.9: A THROWN BALL – KINETIC ENERGY AND GRAVITATIONAL POTENTIAL ENERGY

A 0.50-kg ball is thrown vertically upward with an initial velocity of 10 m/s (▼ **Figure 5.17**). (a) What is the change in the ball's kinetic energy between the starting point and the ball's maximum height? (b) What is the change in the ball's potential energy between the starting point and the ball's maximum height? (Neglect air resistance.)

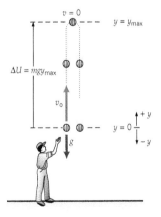

▲ FIGURE 5.17 **Kinetic and potential energies** (The ball is displaced sideways for clarity.) See Example 5.9 for description.

THINKING IT THROUGH. Kinetic energy is lost and gravitational potential energy is gained as the ball travels upward.

SOLUTION

Studying Figure 5.17 and listing the given data:

Given:

$m = 0.50$ kg
$v_o = 10$ m/s

Find:

(a) ΔK (change in kinetic energy from y_o to y_{max})
(b) ΔU (change in potential energy from y_o to y_{max})

(a) To find the *change* in kinetic energy, the kinetic energy is computed at each point. The initial velocity is v_o and at the maximum height $v = 0$, so $K = 0$. Thus,

$$\Delta K = K - K_o = 0 - K_o = -\tfrac{1}{2}mv_o^2 = -\tfrac{1}{2}(0.50 \text{ kg})(10 \text{ m/s})^2$$
$$= -25 \text{ J}$$

That is, the ball loses 25 J of kinetic energy as negative work is done on it by the force of gravity. (The gravitational force and the ball's displacement are in opposite directions.)

(b) To find the change in potential energy, we need to know the ball's height above its starting point when $v = 0$. Using Equation 2.11′, $v^2 = v_o^2 - 2gy$ (with $y_o = 0$ and $v = 0$), to find y_{max},

$$y_{max} = \frac{v_o^2}{2g} = \frac{(10 \text{ m/s})^2}{2(9.8 \text{ m/s}^2)} = 5.1 \text{ m}$$

Then, with $y_o = 0$ and $U_o = 0$

$$\Delta U = U = mgy_{max} = (0.50 \text{ kg})(9.8 \text{ m/s}^2)(5.1 \text{ m}) = +25 \text{ J}$$

The potential energy increases by 25 J, as might be expected. This is an example of the conservation of energy, as will be discussed shortly.

FOLLOW-UP EXERCISE. In this Example, what are the overall changes in the ball's kinetic and potential energies when the ball returns to the starting point?

5.4.1 Zero Reference Point

An important point is illustrated in Example 5.9; namely, the choice of a zero reference point. Potential energy is the energy of *position*, and the potential energy at a particular position (U) is meaningful only when referenced to the potential energy at some other position (U_o). The reference position or point is arbitrary, as is the origin of a set of coordinate axes for analyzing a system. Reference points are usually chosen with convenience in mind – for example, $y_o = 0$. The value of the potential energy at a particular position depends on the reference point used. However, *the difference, or change, in potential energy associated with two positions is the same regardless of the reference position*.

If, in Example 5.9, ground level had been taken as the zero reference point, then U_o at the release point would not have been zero. However, U at the maximum height would have been greater, and $\Delta U = U - U_o$ would have been the same. This concept is illustrated in ▼ **Figure 5.18**. Note in Figure 5.18a that the potential energy can be negative. When an object has a negative potential energy, it is said to be in a *potential energy well*, which is analogous to being in an actual well. Work is needed to raise the object to a higher position in the well or to get it out of the well.

It is also said that gravitational potential energy is *independent of path*. This means that only the change in height (Δy) is the consideration, not the path that leads to the change in height. An object could travel many paths leading to the same Δy.

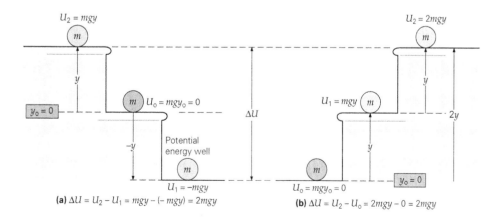

(a) $\Delta U = U_2 - U_1 = mgy - (-mgy) = 2mgy$

(b) $\Delta U = U_2 - U_o = 2mgy - 0 = 2mgy$

▲ **FIGURE 5.18** **Reference point and change in potential energy (a)** The choice of a reference point (zero height) is arbitrary and may give rise to a negative potential energy. An object is said to be in a potential energy well in this case. **(b)** The well may be avoided by selecting a new zero reference. Note that the difference, or change, in potential energy (ΔU) associated with the two positions is the same, regardless of the reference point. There is no physical difference, even though there are two coordinate systems and two different zero reference points.

5.5 Conservation of Energy

Conservation laws are the cornerstones of physics, both theoretically and practically. Most scientists would probably name conservation of energy as the most profound and far-reaching of these important laws. Saying that a physical quantity is *conserved* means it is constant, or has a constant value. Because so many things continually change in physical processes, conserved quantities are extremely helpful in our attempts to understand and describe a situation. Keep in mind, though, that many quantities are conserved only under special conditions.

One of the most important conservation laws is that concerning conservation of energy. (You have seen this topic in Example 5.9.) A familiar statement is that *the total energy of the universe is conserved.* This statement is true, because the whole universe is taken to be a system. A system is defined as a definite quantity of matter enclosed by boundaries, either real or imaginary. In effect, the universe is the largest possible closed, or isolated, system we can imagine. On a smaller scale, a classroom might be considered a system, and so might an arbitrary cubic meter of air.

Within a *closed system*, particles can interact with each other, but have absolutely no interaction with anything outside. In general, then, the amount of energy in a system remains constant when no work is done on or by the isolated system and no energy is transferred to or from the system (including thermal energy and radiation). Thus, the law of conservation of total energy may be stated as follows:

The total energy of an isolated system is always conserved.

Within such a system, energy may be converted from one form to another, but the total amount of all forms of energy is constant, or unchanged. Energy can never be created or destroyed.

CONCEPTUAL EXAMPLE 5.10: VIOLATION
OF THE CONSERVATION OF ENERGY?

A static, uniform liquid is in one side of a double container as shown in ▼ **Figure 5.19a**. If the valve is open, the level will fall, because the liquid has (gravitational) potential energy. This may be computed by assuming all the mass of the liquid to be concentrated at its center of mass, which is at a height $h/2$. (More

on the center of mass in Chapter 6.5.) When the valve is open, the liquid flows into the container on the right, and when static equilibrium is reached, each container has liquid to a height of $h/2$, with centers of mass at $h/4$. This being the case, the potential energy of the liquid before opening the valve was $U_o = (mg)h/2$, and afterward, with half the total mass in each container (Figure 5.19b), $U = (m/2)g(h/4) + (m/2)g(h/4) = 2(m/2)g(h/4) = (mg)h/4$. Whoa. Was half of the energy lost?

REASONING AND ANSWER. No; by the conservation of total energy, it must be around somewhere. Where might it have gone? When the liquid flows from one container to the other, because of internal friction and friction against the walls, half of the potential energy is first converted to kinetic energy (flow of liquid), then to heat (thermal energy), which is transferred to the surroundings as the liquid comes to equilibrium. (This means a constant temperature and no internal fluctuations.)

FOLLOW-UP EXERCISE. What would happen in this Example in the absence of friction?

5.5.1 Conservative and Nonconservative Forces

A general distinction can be made among systems by considering two categories of forces that may act within or on them: conservative and nonconservative forces. You have already been introduced to a couple of conservative forces: the force due to gravity and the spring force. A classic nonconservative force, friction, was considered in Chapter 4.6.

A **conservative force** is defined as follows:

A force is said to be conservative if the work done by it in moving an object is independent of the object's path.

This definition means that the work done by a conservative force depends only on the initial and final positions of an object.

The concept of conservative and nonconservative forces is sometimes difficult to comprehend at first. Because this concept is so important in the conservation of energy, let's consider some illustrative examples to increase understanding.

First, what does *independent of path* mean? As an example of path independence, consider picking an object up from the floor and placing it on a table. This is doing work against the *conservative force of gravity.* The work done is equal to the

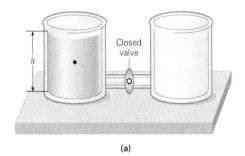
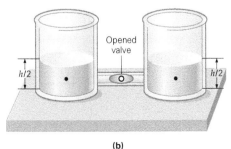

▲ FIGURE 5.19 **Is energy lost?** See Conceptual Example 5.10 for description.

potential energy gained, $mg\Delta y$, where Δy is the *vertical* distance between the object's position on the floor and its position on the table. This is the important point. You may have carried the object over to the sink before putting it on the table, or walked around to the other side of the table. But only the vertical displacement makes a difference in the work done because that is in the direction of the vertical force. (Note that it was said in the last section that gravitational potential energy is independent of path. Now you know why.)

For any horizontal displacement no work is done, since the displacement and force are at right angles. The magnitude of the work done is equal to the change in potential energy (under frictionless conditions only), and in fact, *the concept of potential energy is associated only with conservative forces.* A change in potential energy can be defined in terms of the work done by a conservative force.

Conversely, a **nonconservative force** *does depend on path.*

A force is said to be nonconservative if the work done by it in moving an object depends on the object's path.

Friction is a nonconservative force. A longer path would produce more work done by friction than a shorter one, and more energy would be lost to heat on the longer path. So the work done by friction certainly depends on the path. Hence, in a sense, a conservative force allows you to conserve or store energy as potential energy, whereas a nonconservative force does not.

Another approach to explain the distinction between conservative and nonconservative forces is through an equivalent statement of the previous definition of conservative force:

A force is conservative if the work done by it in moving an object through a round trip is zero.

Notice that for the *conservative* gravitational force, the force and displacement are sometimes in the same direction (in which case positive work is done by the force) and sometimes in opposite directions (in which case negative work is done by the force) during a round trip. Think of the simple case of the book falling to the floor and being placed back on the table. With positive and negative work, the total work done by gravity is zero.

However, for a *nonconservative* force like that of kinetic friction, which opposes motion or is in the opposite direction to the displacement, the total work done by such a force in a round trip can *never* be zero and is always negative (i.e., energy is lost). But don't get the idea that nonconservative forces only take energy away from a system. On the contrary, nonconservative pushes and pulls (forces) that add to the energy of a system are often supplied, such as when you push a stalled car and get it moving.

5.5.2 Conservation of Total Mechanical Energy

The idea of a conservative force allows us to extend the conservation of energy to the special case of mechanical energy, which greatly helps to better analyze many physical situations. The sum of the kinetic and potential energies is called the **total mechanical energy**:

$$
\begin{array}{ccccc}
E & = & K & + & U \\
\text{total} & & \text{kinetic} & & \text{potential} \\
\text{mechanical} & & \text{energy} & & \text{energy} \\
\text{energy} & & & &
\end{array}
\qquad (5.9)
$$

For a **conservative system** (i.e., a system in which only conservative forces do work), the total mechanical energy is constant, or conserved:

$$E = E_0$$

Substituting for E and E_0 from Equation 5.9,

$$K + U = K_0 + U_0 \qquad (5.10a)$$

Equation 5.10a is a mathematical statement of the law of the conservation of mechanical energy:

In a conservative system, the sum of all types of kinetic energy and potential energy is constant and equals the total mechanical energy of the system at any time.

In special cases when there is only one object involved, Equation 5.10a could be written as:

$$\tfrac{1}{2}mv^2 + U = \tfrac{1}{2}mv_0^2 + U_0 \qquad (5.10b)$$

While the kinetic and potential energies in a conservative system may change, their sum is always constant. For a conservative system when work is done and energy is transferred within a system, Equation 5.10a can also be written as

$$(K - K_0) + (U - U_0) = 0 \qquad (5.11a)$$

or

$$\Delta K + \Delta U = 0 \quad \text{(for a conservative system)} \qquad (5.11b)$$

This expression indicates that these quantities are related in a seesaw fashion: If there is a decrease in potential energy, then the kinetic energy must increase by an equal amount to keep the sum of the changes equal to zero. However, if nonconservative forces are present, mechanical energy can be lost (as with friction), or gained, or stays the same (if the nonconservative works balance out).

EXAMPLE 5.11: LOOK OUT BELOW!
CONSERVATION OF MECHANICAL ENERGY

A painter on a scaffold drops a 1.50-kg can of paint from a height of 6.00 m. (a) What is the kinetic energy of the can when the can is at a height of 4.00 m? (b) With what speed will the can hit the ground? (Neglect air resistance.) (c) Show that the expression for speed from energy considerations is the same as that from kinematics (Section 2.5).

THINKING IT THROUGH. Total mechanical energy is conserved, since only the conservative force of gravity acts on the system (the can). The initial total mechanical energy can be found, and the potential energy decreases as the kinetic energy (as well as speed) increases.

SOLUTION

Listing the given data and what is to be found:

Given:

$m = 1.50$ kg
$y_\text{o} = 6.00$ m
$y = 4.00$ m

Find:

(a) K (kinetic energy at $y = 4.00$ m)
(b) v (speed just before hitting the ground)
(c) Compare speeds

(a) First, it is convenient to find the can's initial total mechanical energy, since this quantity is conserved while the can is falling. With $v_\text{o} = 0$, the can's total mechanical energy is initially all potential energy. Taking the ground as the zero reference point,

$$E = K_\text{o} + U_\text{o} = 0 + mgy_\text{o} = (1.50\,\text{kg})(9.80\,\text{m/s}^2)(6.00\,\text{m}) = 88.2\,\text{J}$$

The relation $E = K + U$ continues to hold while the can is falling, and now E is known. Rearranging the equation, $K = E - U$ and K can be found at $y = 4.00$ m

$$K = E - U = E - mgy = 88.2\,\text{J} - (1.50\,\text{kg})(9.80\,\text{m/s}^2)(4.00\,\text{m})$$
$$= 29.4\,\text{J}$$

Alternatively, the change in (in this case, the loss of) potential energy, ΔU, could have been computed. Whatever potential energy was lost must have been gained as kinetic energy (Equation 5.11). Then,

$$\Delta K + \Delta U = 0$$
$$(K - K_\text{o}) + (U - U_\text{o}) = (K - K_\text{o}) + (mgy - mgy_\text{o}) = 0$$

with $K_\text{o} = 0$ (because $v_\text{o} = 0$),

$$K = mg(y_\text{o} - y) = (1.50\,\text{kg})(9.8\,\text{m/s}^2)(6.00\,\text{m} - 4.00\,\text{m})$$
$$= 29.4\,\text{J}$$

(b) Just before the can strikes the ground ($y = 0$, $U = 0$), the total mechanical energy is all kinetic energy,

$$E = K = \tfrac{1}{2}mv^2$$

Thus, the speed is,

$$v = \sqrt{\frac{2E}{m}} = \sqrt{\frac{2(88.2\,\text{J})}{1.50\,\text{kg}}} = 10.8\,\text{m/s}$$

(c) Basically, all of the potential energy of a free-falling object released from some height y is converted into kinetic energy just before the object hits the ground, so

$$|\Delta K| = |\Delta U|$$

(Why absolute values?) Thus,

$$\tfrac{1}{2}mv^2 = mgy$$

and

$$v = \sqrt{2gy} = \sqrt{2(9.8\,\text{m/s}^2)(6.00\,\text{m})} = 10.8\,\text{m/s}$$

Note that the mass cancels and is not a consideration. This result is also obtained from a kinematic equation (Equation 2.12): $v^2 = v_\text{o}^2 - 2g(y - y_\text{o})$. With $v_\text{o} = 0$, $y_\text{o} = 0$, and $-y$ (downward),

$$v = \sqrt{2gy}$$

FOLLOW-UP EXERCISE. A painter on the ground wishes to toss a paintbrush vertically upward a distance of 5.0 m to her partner on the scaffold. Use methods of conservation of mechanical energy to determine the minimum speed that she must give to the brush.

CONCEPTUAL EXAMPLE 5.12: A MATTER OF DIRECTION? SPEED AND CONSERVATION OF ENERGY

Three balls of equal mass m are projected with the same speed in different directions, as shown in ▼ **Figure 5.20**. If air resistance is neglected, which ball would you expect to strike the ground with the greatest speed: (a) ball 1, (b) ball 2, (c) ball 3, or (d) all balls strike with the same speed?

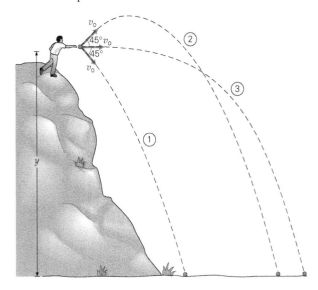

▲ FIGURE 5.20 **Speed and energy** See Example 5.12 for description.

REASONING AND ANSWER. All of the balls have the same initial kinetic energy, $K_o = \frac{1}{2}mv_o^2$. (Recall that energy is a scalar quantity, and the different directions of projection do not produce any difference in the kinetic energies.) Regardless of their trajectories, all of the balls ultimately descend a distance y relative to their common starting point, so they all lose the same amount of potential energy. (Recall that U is energy of *position* and is *independent* of path.)

By the law of conservation of mechanical energy, the amount of potential energy each ball loses is equal to the amount of kinetic energy it gains. Since all of the balls start with the same amount of kinetic energy and gain the same amount of kinetic energy, all three will have equal kinetic energies just before striking the ground. This means that their speeds must be equal, so the answer is (d).

Although balls 1 and 2 are projected at 45° angles, this factor is not relevant. Since the change in potential energy is independent of path, it is independent of the projection angle. The vertical distance between the starting point and the ground is the same (y) for projectiles at any angle. (*Note*: Although the strike speeds are equal, the *times* the balls take to reach the ground are different. Refer to Chapter 3, Conceptual Example 3.11 for another approach.)

FOLLOW-UP EXERCISE. Would the balls strike the ground with different speeds if their masses were different? (Neglect air resistance.)

**EXAMPLE 5.13: CONSERVATIVE FORCES –
MECHANICAL ENERGY OF A SPRING**

A 0.30-kg block sliding on a horizontal frictionless surface with a speed of 2.5 m/s, as depicted in ▼ **Figure 5.21**, strikes a light spring that has a spring constant of 3.0×10^3 N/m. (a) What is the total mechanical energy of the system? (b) What is the kinetic energy K_1 of the block when the spring is compressed a distance $x_1 = 1.0$ cm? (Assume that no energy is lost in the collision.)

THINKING IT THROUGH. (a) Initially, the total mechanical energy is all kinetic energy. (a) The total energy is the same as in part (a), but it is now divided between kinetic energy and spring potential energy (assuming the spring is not fully compressed).

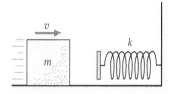

▲ FIGURE 5.21   **Conservative force and the mechanical energy of a spring** See Example 5.13 for description.

SOLUTION

Given:

$m = 0.30$ kg
$v_o = 2.5$ m/s
$k = 3.0 \times 10^3$ N/m
$x_1 = 1.0$ cm $= 0.010$ m

Find:

(a) E (total mechanical energy)
(b) K_1 (kinetic energy)
(a) Just before the block makes contact with the spring, the total mechanical energy of the system is all in the form of kinetic energy,

$$E = K_o = \tfrac{1}{2}mv_o^2 = \tfrac{1}{2}(0.30 \text{ kg})(2.5 \text{ m/s})^2 = 0.94 \text{ J}$$

Since the system is conservative (i.e., no mechanical energy is lost), this quantity is the total mechanical energy at any time.

(b) When the spring is compressed a distance x_1, it has gained potential energy $U_1 = \tfrac{1}{2}kx_1^2$, and the block has kinetic energy K_1, so

$$E = K_1 + U_1 = K_1 + \tfrac{1}{2}kx_1^2$$

Solving for K_1,

$$\begin{aligned} K_1 &= E - \tfrac{1}{2}kx_1^2 \\ &= 0.94 \text{ J} - \tfrac{1}{2}(3.0 \times 10^3 \text{ N/m})(0.010 \text{ m})^2 \\ &= 0.94 \text{ J} - 0.15 \text{ J} = 0.79 \text{ J} \end{aligned}$$

FOLLOW-UP EXERCISE. How far will the spring in this Example be compressed when the block comes to a stop? (Solve using energy principles.)

5.5.3 Total Energy and Nonconservative Forces

In the preceding examples, the force of friction was ignored; however, friction is probably the most common nonconservative force. In general, both conservative and nonconservative forces can do work on objects. But when nonconservative forces do work, the total mechanical energy is not conserved. Mechanical energy is "lost" through the work done by nonconservative forces, such as friction. Mechanical energy could also be converted to other non-mechanical types of energy such as heat and sound.

You might think that an energy approach can no longer be used to analyze problems involving such nonconservative forces, since mechanical energy can be gained or lost (▶ **Figure 5.22**). However, in some instances, the total energy can be used to find out how much energy was lost to the work done by a nonconservative force. Suppose an object initially has mechanical energy and that nonconservative forces do an amount of work W_{nc} on it. Starting with the work-energy theorem,

$$W_{net} = \Delta K = K - K_o$$

▲ **FIGURE 5.22** **Nonconservative force and energy loss** Friction is a nonconservative force – when friction is present and does work, mechanical energy is not conserved. Can you tell from the photo what is happening to the work being done by the motor on the grinding wheel after the work is converted into rotational kinetic energy?

In general, the net work (W_{net}) may be done by both conservative forces (W_c) and nonconservative forces (W_{nc}), so

$$W_c + W_{nc} = K - K_o \qquad (5.12)$$

But from Equation 5.10a, the work done by conservative forces is equal to $W_c = -\Delta U = -(U - U_o)$, so Equation 5.12 then becomes

$$W_{nc} = K - K_o + (U - U_o)$$
$$= (K + U) - (K_o + U_o)$$

Therefore,

$$W_{nc} = E - E_o = \Delta E \qquad (5.13)$$

Hence, the work done by the nonconservative forces acting on a system is equal to the change in mechanical energy. Notice that for dissipative forces, $E_o > E$. Thus, the change is negative, indicating a decrease in mechanical energy. This condition agrees in sign with W_{nc}, which, for friction, would also be negative. Example 5.14 illustrates this concept.

EXAMPLE 5.14: NONCONSERVATIVE FORCE – DOWNHILL RACER

A skier with a mass of 80 kg starts from rest at the top of a slope and skis down from an elevation of 110 m (▶ **Figure 5.23**). The speed of the skier at the bottom of the slope is 20 m/s. (a) Show that the system is nonconservative. (b) How much work is done by the nonconservative force of friction?

▲ FIGURE 5.23 **Work done by a nonconservative force** See Example 5.14 for description.

THINKING IT THROUGH. (a) If the system is nonconservative, then $E_o \neq E$, (actually here $E < E_o$), and these quantities can be computed. (b) The work cannot be determined from force-distance considerations, but W_{nc} is equal to the difference in total energies (Equation 5.13).

SOLUTION

Given:

$m = 80$ kg
$v_o = 0$
$v = 20$ m/s
$y_o = 110$ m

Find:

(a) Show that E is not conserved.
(b) W_{nc} (work done by friction)

(a) If the system is conservative, the total mechanical energy is constant. Taking $U_o = 0$ at the bottom of the hill, the initial energy at the top of the hill is

$$E_o = U = mgy_o = (80 \text{ kg})(9.8 \text{ m/s}^2)(110 \text{ m}) = 8.6 \times 10^4 \text{ J}$$

And the energy at the bottom of the slope is all kinetic, thus

$$E = K = \tfrac{1}{2}mv^2 = \tfrac{1}{2}(80 \text{ kg})(20 \text{ m/s})^2 = 1.6 \times 10^4 \text{ J}$$

Therefore, $E_o \neq E$, i.e., $E < E_o$, so this system is not conservative.

(b) The amount of work done by the nonconservative force of friction is equal to the change in the mechanical energy, or to the amount of mechanical energy lost (Equation 5.13):

$$W_{nc} = E - E_o = 1.6 \times 10^4 \text{ J} - 8.6 \times 10^4 \text{ J} = -7.0 \times 10^4 \text{ J}$$

This quantity is more than 80% of the initial energy. (Where did this energy actually go?)

FOLLOW-UP EXERCISE. In free fall, air resistance is sometimes negligible, but for skydivers, air resistance has a very practical effect. Typically, a skydiver descends about 450 m before reaching a terminal velocity (Chapter 4.6) of 60 m/s. (a) What is the percentage of energy loss to nonconservative forces during this descent? (b) Show that after terminal velocity is reached, the rate of energy loss in J/s is given by (60 mg), where m is the mass of the skydiver and both m and g must be in standard SI units.

EXAMPLE 5.15: NONCONSERVATIVE FORCE – ONE MORE TIME

A 0.75-kg block slides on a frictionless surface with a speed of 20 m/s. It then slides over a rough area 1.0 m in length and onto another frictionless surface. The coefficient of kinetic friction between the block and the rough surface is 0.17. What is the speed of the block after it passes across the rough surface?

THINKING IT THROUGH. The task of finding the final speed implies that equations involving kinetic energy can be used, where the final kinetic energy can be found by using the conservation of *total* energy. Note that the initial and final energies are kinetic energies, since there is no change in gravitational potential energy. It is always good to make a sketch of the situation for clarity and understanding (▼ **Figure 5.24**).

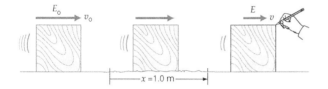

▲ FIGURE 5.24   **A nonconservative rough spot** See Example 5.15 for description.

SOLUTION

Listing the data,

Given:

$m = 0.75$ kg
$x = 1.0$ m
$\mu_{\text{k}} = 0.17$
$v_{\text{o}} = 2.0$ m/s

Find:

v (final speed of block)

For this nonconservative system, from Equation 5.13

$$W_{\text{nc}} = E - E_{\text{o}} = K - K_{\text{o}}$$

In the rough area, the block loses energy, because of the work done by friction (W_{nc}) and thus

$$W_{\text{nc}} = f_{\text{k}}x = -\mu_{\text{k}}Nx = -\mu_{\text{k}}mgx$$

[The result is negative because f_{k} and the displacement x are in opposite directions; that is, $f_{\text{k}}(\cos 180°)x = -f_{\text{k}}x$.]

Then, rearranging the energy equation and writing the terms out in detail,

$$K = K_{\text{o}} + W_{\text{nc}}$$

or

$$\tfrac{1}{2}mv^2 = \tfrac{1}{2}mv_{\text{o}}^2 - \mu_{\text{k}}mgx$$

Solving for v yields,

$$\begin{aligned} v &= \sqrt{v_{\text{o}}^2 - 2\mu_{\text{k}}gx} \\ &= \sqrt{(2.0\text{ m/s})^2 - 2(0.17)(9.8\text{ m/s}^2)(1.0\text{ m})} \\ &= 0.82\text{ m/s} \end{aligned}$$

Note that the mass of the block was not needed. Also, it can be easily shown that the block lost more than 80% of its energy to friction.

FOLLOW-UP EXERCISE. Suppose the coefficient of kinetic friction between the block and the rough surface were 0.25. What would happen to the block in this case?

Note that in a closed nonconservative system, the *total energy* (*not* the total mechanical energy) is conserved (including non-mechanical forms of energy, such as thermal energy). But not all of the energy is available for mechanical work. For a conservative system, you get back what you put in, so to speak. That is, if you do work on the system, the transferred energy is available to do work. Conservative systems are idealizations, because all real systems are nonconservative to some degree. However, working with ideal conservative systems gives an insight into the conservation of energy.

Total energy is always conserved in a closed or isolated system. During the course of study, you will learn about other forms of energy, such as thermal, electrical, and nuclear energies. In general, on the microscopic and submicroscopic levels, these forms of energy can be described in terms of kinetic energy and potential energy. Also, you will learn that mass is a form of energy and that the law of conservation of energy must take this form into account in order to be applied to the analysis of nuclear reactions.

5.6 Power

A particular task may require a certain amount of work, but that work might be done over different lengths of time or at different rates. For example, suppose that you have to mow a lawn. This task takes a certain amount of work, but you might do the job in a half-hour, or you might take an hour. There's a practical distinction to be made here. That is, there is usually not only an interest in the amount of work done, but also an interest in how fast it is done – that is, the rate at which it is done. *The time rate of doing work* is called **power**.

The average power ($\bar{P}$) is the work done divided by the time it takes to do the work, or work per unit of time:

$$\bar{P} = \frac{W}{t} \tag{5.14}$$

The work (and power) done by a constant force of magnitude F acting while an object moves through a parallel displacement of magnitude d is

$$\bar{P} = \frac{W}{t} = \frac{Wd}{t} = F\left(\frac{d}{t}\right) = F\,\bar{v} \qquad (5.15)$$

SI unit of power: J/s or watt (W)

where it is assumed that the force is in the direction of the displacement. Here, $\bar{v}$ is the magnitude of the average velocity. If the velocity is constant, then $\bar{P} = P = Fv$. If the force and displacement are not in the same direction, then we can write

$$\begin{aligned}\bar{P} &= \frac{F(\cos\theta)d}{t} \\ &= F\,\bar{v}\cos\theta\end{aligned} \qquad (5.16)$$

where θ is the angle between the force and the displacement.

As can be seen from Equation 5.15, the SI unit of power is joules per second (J/s), but this unit is given another name, the **watt (W)**:

$$1\,\text{J/s} = 1\,\text{watt (W)}$$

The SI unit of power is named in honor of James Watt (1736–1819), a Scottish engineer who developed one of the first practical steam engines. A common unit of electrical power is the *kilowatt* (kW). The British unit of power is foot-pound per second (ft·lb/s). However, a larger unit coined by Watt, the **horsepower (hp)**, is more commonly used:*

$$1\,\text{hp} = 550\,\text{ft·lb/s} = 746\,\text{W}$$

Power tells how fast work is being done *or* how fast energy is transferred. For example, motors have power ratings commonly given in horsepower. A 2-hp motor can do a given amount of work in half the time that a 1-hp motor would take, or twice the work in the same amount of time. That is, a 2-hp motor is twice as "powerful" as a 1-hp motor.

EXAMPLE 5.16: A CRANE HOIST – WORK AND POWER

A crane hoist like the one shown in ▶ **Figure 5.25** lifts a load of 1.0 metric ton a vertical distance of 25 m in 9.0 s at a constant velocity. How much useful work is done on the load by the hoist each second?

---

* In Watt's time, steam engines were replacing horses for work in mines and mills. To characterize the performance of his new engine, which was more efficient than existing ones, Watt used the average rate at which a horse could do work as a unit – a horsepower.

▲ FIGURE 5.25 **Power delivery** See Example 5.16 for description.

THINKING IT THROUGH. By definition, the useful work done each second (i.e., per second) is the power output, so this is what needs to be found (Equation 5.15).

SOLUTION

Given:

$m = 1.0$ metric ton $= 1.0 \times 10^3$ kg
$y = 25$ m
$t = 9.0$ s

Find:

P (power, work per second)

Since the load moves with a constant velocity, $\bar{P} = P$. (Why?) The work is done against gravity, so $F = mg$, and

$$\begin{aligned}P &= \frac{W}{t} = \frac{Fd}{t} = \frac{mgy}{t} \\ &= \frac{(1.0 \times 10^3\,\text{kg})(9.8\,\text{m/s}^2)(25\,\text{m})}{9.0\,\text{s}} \\ &= 2.7 \times 10^4\,\text{W (or 27 kW)}\end{aligned}$$

Thus, since a watt (W) is a joule per second (J/s), the hoist did 2.7×10^4 J of work each second. Note that the velocity has a magnitude of $v = d/t = (25\,\text{m})/(9.0\,\text{s}) = 2.8$ m/s, and the displacement is parallel to the applied force, therefore the power could also be found using

$$P = Fv = mgv = (1.0 \times 10^3\,\text{kg})(9.80\,\text{m/s}^2)(2.8\,\text{m/s}) = 2.7 \times 10^4\,\text{W}$$

FOLLOW-UP EXERCISE. If the hoist motor of the crane in this Example is rated at 70 hp, what percentage of this power output goes into useful work?

EXAMPLE 5.17: CLEANING UP – WORK AND TIME

The motors of two vacuum cleaners have net power outputs of 1.00 hp and 0.500 hp, respectively. (a) How much work in joules can each motor do in 3.00 min? (b) How long does each motor take to do 97.0 kJ of useful work?

THINKING IT THROUGH. (a) Since power is work/time ($P = W/t$), the work can be computed. Note that power is given in horsepower units, which is converted to watts. (b) This part of the problem is another application of Equation 5.15.

SOLUTION

Given:

$P_1 = 1.00$ hp $= 746$ W
$P_2 = 0.500$ hp $= 373$ W
(a) $t = 3.00$ min $= 180$ s
(b) $W = 97.0$ kJ $= 97.0 \times 10^3$ J

Find:

(a) W (work for each)
(b) t (time for each)

(a) Since $P = W/t$, for the 1.00-hp motor:

$$W_1 = P_1 t = (746 \text{ W})(180 \text{ s}) = 1.34 \times 10^5 \text{ J}$$

And for the 0.500-hp motor:

$$W_2 = P_2 t = (373 \text{ W})(180 \text{ s}) = 0.67 \times 10^5 \text{ J}$$

Note that in the same amount of time, the smaller motor does half as much work as the larger one, as would be expected.

(b) The times are given by $t = W/P$, and for the same amount of work,

$$t_1 = \frac{W}{P_1} = \frac{97.0 \times 10^3 \text{ J}}{746 \text{ W}} = 130 \text{ s}$$

and

$$t_2 = \frac{W}{P_2} = \frac{97.0 \times 10^3 \text{ J}}{373 \text{ W}} = 260 \text{ s}$$

So, the smaller motor takes twice as long as the larger one to do the same amount of work.

FOLLOW-UP EXERCISE. (a) A 10-hp motor breaks down and is temporarily replaced with a 5-hp motor. What can you say about the rate of work output? (b) Suppose the situation were reversed – a 5-hp motor is replaced with a 10-hp motor. What can you say about the rate of work output for this case?

5.6.1 Efficiency

Machines and motors are commonly used items in our daily lives, and comments are made about their efficiencies – for example, one machine is more efficient than another. Efficiency involves work, energy, and/or power. Both simple and complex machines that do work have mechanical parts that move, so some input energy is always lost because of friction or some other cause (perhaps in the form of sound). Thus, not all of the input energy goes into doing useful work.

Mechanical efficiency is essentially a measure of what you get out for what you put in – that is, the *useful* work output compared with the energy input. **Efficiency**, ε, is given as a fraction (or percentage):

$$\varepsilon = \frac{\text{work output}}{\text{energy input}} (\times 100\%) = \frac{W_{out}}{E_{in}} (\times 100\%) \qquad (5.17)$$

Efficiency is a unitless quantity

For example, if a machine has a 100-J (electric energy) input and a 40-J (useful work) output, then its efficiency is

$$\varepsilon = \frac{W_{out}}{E_{in}} = \frac{40 \text{ J}}{100 \text{ J}} = 0.40 (\times 100\%) = 40\%$$

An efficiency of 0.40, or 40%, means that 60% of the energy input is lost because of friction or some other cause and doesn't serve its intended purpose. Note that if both terms of the ratio in Equation 5.17 are divided by time t, we obtain $W_{out}/t = P_{out}$ and $E_{in}/t = P_{in}$. So efficiency can be written in terms of power, P:

$$\varepsilon = \frac{P_{out}}{P_{in}} (\times 100\%) \qquad (5.18)$$

**EXAMPLE 5.18: HOME IMPROVEMENT –
MECHANICAL EFFICIENCY AND WORK OUTPUT**

The motor of an electric drill with an efficiency of 80% has a power input of 600 W. How much useful work is done by the drill in 30 s?

THINKING IT THROUGH. Given the efficiency and power input, the power output P_{out} can readily be found from Equation 5.18. This quantity is related to the work output ($P_{out} = W_{out}/t$), from which W_{out} may be found.

SOLUTION

Given:

$\varepsilon = 80\% = 0.80$
$P_{in} = 600$ W
$t = 30$ s

Find:

W_{out} (work output)

First, rearranging Equation 5.18 to find the power output:

$$P_{out} = \varepsilon P_{in} = (0.80)(600 \text{ W}) = 4.8 \times 10^2 \text{ W}$$

Then, substituting this value into the equation relating power output and work output,

$$W_{out} = P_{out} t = (4.8 \times 10^2 \text{ W})(30 \text{ s}) = 1.4 \times 10^4 \text{ J}$$

FOLLOW-UP EXERCISE. (a) Is it possible to have a mechanical efficiency of 100%? (b) What would an efficiency of greater than 100% imply?

▼ **Table 5.1** lists the typical efficiencies of some machines. You may be surprised by the relatively low efficiency of the automobile. Much of the energy input (from gasoline combustion) is lost as exhaust heat and through the cooling system (more than 60%), and friction accounts for a good deal more. About 20%–25% of the input energy is converted to useful work that goes into propelling the vehicle. Air conditioning, power steering, and a high-powered audio system are nice, but they also use energy and contribute to the car's decrease in efficiency.

TABLE 5.1   Typical Efficiencies of Some Machines

Machine	Efficiency (approximate %)
Compressor	85–90
Electric motor	70–95
Automobile (gasoline)	20–25
Automobile (diesel)	25–30
Human muscle[a]	20–25
Steam engine	5–10

[a] Technically not a machine but used to perform work.

Chapter 5 Review

- **Work done by a constant force** is the product of the magnitude of the displacement and the component of the force parallel to the displacement:

$$W = F_\| d = (F \cos\theta)d \quad \text{(work done by a constant force)} \quad (5.2)$$

- Calculating work done by a variable force requires advanced mathematics. An example of a variable force is the **spring force**, given by *Hooke's law*:

$$F_s = -kx \quad \text{(ideal spring force)} \quad (5.3)$$

- The **work done on a spring** is given by

$$W = \tfrac{1}{2}kx^2 \quad (5.4)$$

- **Kinetic energy** is the energy of motion and is given by

$$K = \tfrac{1}{2}mv^2 \quad (5.5)$$

- By the **work-energy theorem**, the net work done on an object is equal to the change in the kinetic energy of the object:

$$W_{net} = K - K_o = \Delta K \quad (5.6)$$

- **Potential energy** is the energy of position and/or configuration.

- The elastic **potential energy of a spring** is given by

$$U = \tfrac{1}{2}kx^2 \quad \text{(relative to unstretched position } x_o = 0) \quad (5.7)$$

- The most common type of potential energy is **gravitational potential energy**, associated with the gravitational attraction near the Earth's surface.

$$U = mgy \quad \text{(relative to a chosen zero level } y_o = 0) \quad (5.8)$$

- **Conservation of energy:** The total energy of the universe or of an isolated system is always conserved.

- **Conservation of mechanical energy:** The total mechanical energy (kinetic plus potential) is constant in a conservative system:

$$K + U = K_o + U_o \quad (5.10a)$$

- In systems with **nonconservative forces**, where mechanical energy is gained or lost, the work done by a nonconservative force is given by

$$W_{nc} = E - E_o = \Delta E \quad (5.13)$$

- **Power** is the time rate of doing work (or expending energy). **Average power** is given by

$$\bar{P} = \frac{W}{t} = \frac{Fd}{t} = F\bar{v}$$
$$\text{(constant force in direction of } d \text{ and } v) \quad (5.15)$$

$$\bar{P} = \frac{F(\cos\theta)d}{t} = F\bar{v}\cos\theta$$
$$\text{(constant force acts at an angle } \theta \text{ between } d \text{ and } v) \quad (5.16)$$

- **Efficiency** relates work output to energy (work) input as a fraction or percent:

$$\varepsilon = \frac{W_{out}}{E_{in}}(\times 100\%) \quad (5.17)$$

$$\varepsilon = \frac{P_{out}}{P_{in}}(\times 100\%) \quad (5.18)$$

End of Chapter Questions and Exercises

Multiple Choice Questions

5.1 Work Done by a Constant Force

1. The units of work are (a) N·m, (b) kg·m²/s², (c) J, (d) all of the preceding.

2. For a particular force and displacement, the most work is done when the angle between them is (a) 30°, (b) 60°, (c) 90°, (d) 180°.

3. A pitcher throws a fastball. When the catcher catches it, (a) positive work is done, (b) negative work is done, (c) the net work is zero.

4. Work done in free fall (a) is only positive, (b) is only negative, or (c) can be either positive or negative.

5. Which one of the following has units of work: (a) N, (b) N/s, (c) J·s, or (d) N·m?

5.2 Work Done by a Variable Force

6. The work done by a variable force of the form $F = kx$ is equal to (a) kx^2, (b) kx, (c) $\frac{1}{2}kx^2$, (d) none of the preceding.

5.3 The Work-Energy Theorem: Kinetic Energy

7. Which of the following is a scalar quantity: (a) work, (b) force, (c) kinetic energy, or (d) both a and c?

8. If the angle between the net force and the displacement of an object is greater than 90°, (a) kinetic energy increases, (b) kinetic energy decreases, (c) kinetic energy remains the same, (d) the object stops.

9. Two identical cars, A and B, traveling at 55 mi/h collide head-on. A third identical car, C, crashes into a brick wall going 55 mi/h. Which car has the least damage: (a) car A, (b) car B, (c) car C, or (d) all the same?

5.4 Potential Energy

10. A change in gravitational potential energy (a) is always positive, (b) depends on the reference point, (c) depends on the path, (d) depends only on the initial and final positions.

11. The reference point for gravitational potential energy may be (a) zero, (b) negative, (c) positive, (d) all of the preceding.

5.5 Conservation of Energy

12. Energy cannot be (a) transferred, (b) conserved, (c) created, (d) in different forms.

13. If a nonconservative force acts on an object, and does work, then (a) the object's kinetic energy is conserved, (b) the object's potential energy is conserved, (c) the mechanical energy is conserved, (d) the mechanical energy is not conserved.

14. The speed of a pendulum is greatest (a) when the pendulum's kinetic energy is a minimum, (b) when the pendulum's acceleration is a maximum, (c) when the pendulum's potential energy is a minimum, (d) none of the preceding.

15. Two springs are identical except for their spring constants, $k_2 > k_1$. If the same force is used to stretch the springs, (a) spring 1 will be stretched farther than spring 2, (b) spring 2 will be stretched farther than spring 1, (c) both will be stretched the same distance.

16. Two identical stones are thrown from the top of a tall building. Stone 1 is thrown vertically downward with an initial speed v, and stone 2 is thrown vertically upward with the same initial speed. Neglecting air resistance, which stone hits the ground with a greater speed: (a) stone 1, (b) stone 2, or (c) both have the same speed?

5.6 Power

17. Which of the following is not a unit of power: (a) J/s, (b) W·s, (c) W, or (d) hp?

18. Consider a 2.0-hp motor and a 1.0-hp motor. Compared to the 2.0-hp motor, for a given amount of work, the 1.0 hp motor can (a) do twice as much work in half the time, (b) half the work in the same time, (c) one-quarter of the work in three quarters of the time, (d) none of the preceding.

Conceptual Questions

5.1 Work Done by a Constant Force

1. (a) As a weightlifter lifts a barbell from the floor in the "clean" procedure (▼ **Figure 5.26a**), has she done work? Why or why not? (b) In raising the barbell above her head in the "jerk" procedure, is she doing work? Explain. (c) In holding the barbell above her head (Figure 5.26b), is she doing more work, less work, or the same amount of work as in lifting the barbell? Explain. (d) If the weightlifter drops the barbell, is work done on the barbell? Explain what happens in this situation.

2. You are carrying a backpack across campus. What is the work done by your vertical carrying force on the backpack? Explain.

3. A jet plane flies in a vertical circular loop. In what regions of the loop is the work done by the force of gravity on the plane positive and/or negative? Is the work constant? If not, are there maximum and minimum instantaneous values? Explain.

(a)   (b)

▲ FIGURE 5.26 **Woman at work in the clean and jerk?** During the clean, she moves the barbell from the floor to a racked position across the deltoids. During the jerk she raises the barbell to a stationary position above her head. See Conceptual Question 1.

5.2 Work Done by a Variable Force

4. Does it take twice the work to stretch a spring 2 cm from its equilibrium position as it does to stretch it 1 cm from its equilibrium position?

5. If a spring is compressed 2.0 cm from its equilibrium position and then compressed an additional 2.0 cm, how much more work is done in the second compression than in the first? Explain.

5.3 The Work-Energy Theorem: Kinetic Energy

6. You want to decrease the kinetic energy of an object as much as you can. You can do so by either reducing the mass by half or reducing the speed by half. Which option should you pick, and why?

7. A certain amount of work W is required to accelerate a car from rest to a speed v. How much work is required to accelerate the car from rest to a speed of $v/2$?

8. A certain amount of work W is required to accelerate a car from rest to a speed v. If instead an amount of work equal to $2W$ is done on the car, what is the car's speed?

9. Car B is traveling twice as fast as car A, but car A has four times the mass of car B. Which car has the greater kinetic energy?

5.4 Potential Energy

10. If a spring is stretched from its position from x_0 to x, what is the change in potential energy then proportional to? (Express the answer in terms of x_0 and x.)

11. A lab notebook sits on a table 0.75 m above the floor. Your lab partner tells you the book has zero potential energy, and another student says it has 8.0 J of potential energy. Who is correct?

5.5 Conservation of Energy

12. When you throw an object into the air, is its initial speed the same as its speed just before it returns to your hand? Explain by applying the concept of conservation of mechanical energy.

13. A student throws a ball vertically upward so it just reaches the height of a window on the second floor of a dormitory. At the same time that the ball is thrown upward, a student at the window drops an identical ball. Are the mechanical energies of the balls the same at half the height of the window? Explain.

5.6 Power

14. If you check your electricity bill, you will note that you are paying the power company for so many kilowatt-hours (kwh). Are you really paying for power? Explain.

15. (a) Does efficiency describe how fast work is done? Explain. (b) Does a more powerful machine always perform more work than a less powerful one? Explain.

16. Two students who weigh the same start at the same ground floor location at the same time to go to the same classroom on the third floor by different routes. If they arrive at different times, which student will have expended more power? Explain.

Exercises*

Integrated Exercises (IEs) are two-part exercises. The first part typically requires a conceptual answer choice based on physical thinking and basic principles. The following part is quantitative calculations associated with the conceptual choice made in the first part of the exercise.

5.1 Work Done by a Constant Force

1. • If a person does 50 J of work in moving a 30-kg box over a 10-m distance on a horizontal surface, what is the minimum force required?

2. • A 5.0-kg box slides a 10-m distance on ice. If the coefficient of kinetic friction is 0.20, what is the work done by the friction force?

3. • A passenger at an airport pulls a rolling suitcase by its handle. If the force used is 10 N and the handle makes an angle of 25° to the horizontal, what is the work done by the pulling force while the passenger walks 200 m?

4. •• A 3.00-kg block slides down a frictionless plane inclined 20° to the horizontal. If the length of the plane's surface is 1.50 m, how much work is done, and by what force?

5. •• Suppose the coefficient of kinetic friction between the block and the plane in Exercise 4 is 0.275. What would be the net work done in this case?

6. •• A father pulls his young daughter on a sled with a constant velocity on a level surface a distance of 10 m, as illustrated in ▼ **Figure 5.27a**. If the total mass of the sled and the girl is 35 kg and the coefficient of kinetic friction between the sled runners and the snow is 0.20, how much work does the father do?

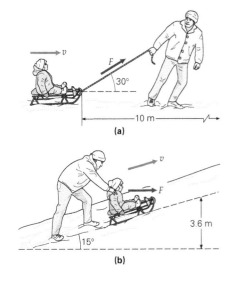

▲ FIGURE 5.27 **Fun and work** See Exercises 6 and 7.

* The bullets denote the degree of difficulty of the exercises: •, simple; ••, medium; and •••, more difficult.

7. •• A father pushes horizontally on his daughter's sled to move it up a snowy incline, as illustrated in Figure 5.27b. If the sled moves up the hill with a constant velocity, how much work is done by the father in moving it from the bottom to the top of the hill? (Some necessary data are given in Exercise 6.)

8. •• A block on a level frictionless surface has two forces applied, as shown in ▼ **Figure 5.28**. (a) What force F_2 would cause the block to move in a straight line to the right? (b) If the block moves horizontally by 50 cm, how much work is done by each force? (c) What is the total work done by the two forces?

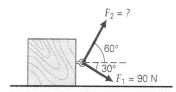

▲ FIGURE 5.28   **Make it go straight** See Exercise 8.

9. •• A 0.50-kg shuffleboard puck slides a distance of 3.0 m on the board. If the coefficient of kinetic friction between the puck and the board is 0.15, what work is done by the force of friction?

10. •• A crate is dragged 3.0 m along a rough floor with a constant velocity by a worker applying a force of 500 N to a rope at an angle of 30° to the horizontal. (a) How many forces are acting on the crate? (b) How much work does each of these forces do? (c) What is the total work done on the crate?

11. IE •• A hot air balloon ascends at a constant rate. (a) The weight of the balloon does (1) positive work, (2) negative work, (3) no work. Why? (b) A hot air balloon with a mass of 500 kg ascends at a constant rate of 1.50 m/s for 20.0 s. How much work is done by the upward buoyant force? (Neglect air resistance.)

12. IE •• A hockey puck with a mass of 200 g and an initial speed of 25.0 m/s slides freely to rest in the space of 100 m on a sheet of horizontal ice. How many forces do nonzero work on it as it slows: (a) (1) none, (2) one, (3) two, or (4) three? Explain. (b) Determine the work done by all the individual forces on the puck as it slows.

5.2   Work Done by a Variable Force

13. • To measure the spring constant of a certain spring, a student applies a 4.0-N force, and the spring stretches by 5.0 cm. What is the spring constant?

14. • A spring has a spring constant of 30 N/m. How much work is required to stretch the spring 2.0 cm from its equilibrium position?

15. • If it takes 400 J of work to stretch a spring 8.00 cm, what is the spring constant?

16. • If a 10-N force is used to compress a spring with a spring constant of 4.0×10^2 N/m, what is the resulting spring compression?

17. IE • A certain amount of work is required to stretch a spring from its equilibrium position. (a) If twice the work is performed on the spring, the spring will stretch more by a factor of (1) $\sqrt{2}$, (2) 2, (3) $1/\sqrt{2}$, (4) 1/2. Why? (b) If 100 J of work is done to pull a spring 1.0 cm, what work is required to stretch it 3.0 cm?

18. • Compute the work done by the variable force in the graph of F versus x in ▼ **Figure 5.29**. [*Hint*: The area of a triangle is $A = 1/2 \times$ altitude $\times$ base]

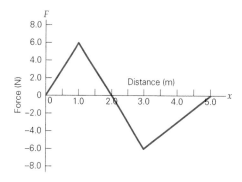

▲ FIGURE 5.29   **How much work is done?** See Exercise 18.

19. IE •• A spring with a force constant of 50 N/m is to be stretched from 0 to 20 cm. (a) The work required to stretch the spring from 10 to 20 cm is (1) more than, (2) the same as, (3) less than that required to stretch it from 0 to 10 cm. (b) Compare the two work values to prove your answer to part (a).

20. •• A particular spring has a force constant of 2.5×10^3 N/m. (a) How much work is done in stretching the relaxed spring by 6.0 cm? (b) How much more work is done in stretching the spring an additional 2.0 cm?

5.3   The Work-Energy Theorem: Kinetic Energy

21. IE • A 0.20-kg object with a horizontal speed of 10 m/s hits a wall and bounces directly back with only half the original speed. (a) What percentage of the object's initial kinetic energy is lost: (1) 25%, (2) 50%, or (3) 75%? (b) How much kinetic energy is lost in the ball's collision with the wall?

22. • A 1200-kg automobile travels at 90 km/h. (a) What is its kinetic energy? (b) What net work would be required to bring it to a stop?

23. • A constant net force of 75 N acts on an object initially at rest as it moves through a parallel distance of 0.60 m. (a) What is the final kinetic energy of the object? (b) If the object has a mass of 0.20 kg, what is its final speed?

24. IE •• A 2.00-kg mass is attached to a vertical spring with a spring constant of 250 N/m. A student pushes on the mass vertically upward with her hand while

slowly lowering it to its equilibrium position. (a) How many forces do nonzero work on the object: (1) one, (2) two, or (3) three? Explain your reasoning. (b) Calculate the work done on the object by each of the forces acting on it as it is lowered into position.

25. •• The stopping distance of a vehicle is an important safety factor. Assuming a constant braking force, use the work-energy theorem to show that a vehicle's stopping distance is proportional to the square of its initial speed. If an automobile traveling at 45 km/h is brought to a stop in 50 m, what would be the stopping distance for an initial speed of 90 km/h?

26. IE •• A large car of mass $2m$ travels at speed v. A small car of mass m travels with a speed $2v$. Both skid to a stop with the same coefficient of friction. (a) The small car will have (1) a longer, (2) the same, (3) a shorter stopping distance. (b) Calculate the ratio of the stopping distance of the small car to that of the large car. (Use the work-energy theorem, not Newton's laws.)

5.4 Potential Energy

27. • How much more gravitational potential energy does a 1.0-kg hammer have when it is on a shelf 1.2 m high than when it is on a shelf 0.90 m high?

28. IE • You are told that the gravitational potential energy of a 2.0-kg object has decreased by 10 J. (a) With this information, you can determine (1) the object's initial height, (2) the object's final height, (3) both the initial and the final height, (4) only the difference between the two heights. Why? (b) What can you say has physically happened to the object?

29. •• Six identical books, 4.0 cm thick and each with a mass of 0.80 kg, lie individually on a flat table. How much work would be needed to stack the books one on top of the other?

30. IE •• The floor of the basement of a house is 3.0 m below ground level, and the floor of the attic is 4.5 m above ground level. (a) If an object in the attic were brought to the basement, the change in potential energy will be greatest relative to which floor: (1) attic, (2) ground, (3) basement, or (4) all the same? Why? (b) What are the respective potential energies of 1.5-kg objects in the basement and attic, relative to ground level? (c) What is the change in potential energy if the object in the attic is brought to the basement?

31. •• A 0.50-kg mass is placed on the end of a vertical spring that has a spring constant of 75 N/m and eased down into its equilibrium position. (a) Determine the change in spring (elastic) potential energy of the system. (b) Determine the system's change in gravitational potential energy.

32. •• A horizontal spring, resting on a frictionless tabletop, is stretched 15 cm from its unstretched configuration and a 1.00-kg mass is attached to it. The system is released from rest. A fraction of a second later, the spring finds itself compressed 3.0 cm from its unstretched configuration. How does its final potential energy compare to its initial potential energy? (Give your answer as a ratio, final to initial.)

5.5 Conservation of Energy

33. • A 0.300-kg ball is thrown vertically upward with an initial speed of 10.0 m/s. If the initial potential energy is taken as zero, find the ball's kinetic, potential, and mechanical energies (a) at its initial position, (b) at 2.50 m above the initial position, and (c) at its maximum height.

34. • What is the maximum height reached by the ball in Exercise 33?

35. IE •• A girl swings back and forth on a swing with ropes that are 4.00 m long. The maximum height she reaches is 2.00 m above the ground. At the lowest point of the swing, she is 0.500 m above the ground. (a) The girl attains the maximum speed (1) at the top, (2) in the middle, (3) at the bottom of the swing. Why? (b) What is the girl's maximum speed?

36. IE •• A 500-g (small) mass on the end of a 1.50-m-long string is pulled aside 15° from the vertical and shoved downward (toward the bottom of its motion) with a speed of 2.00 m/s. (a) Is the angle on the other side (1) greater than, (2) less than, or (3) the same as the angle on the initial side (15°)? Explain in terms of energy. (b) Calculate the angle it goes to on the other side, neglecting air resistance.

37. •• A 0.20-kg rubber ball is dropped from a height of 1.0 m above the floor and it bounces back to a height of 0.70 m. (a) What is the ball's speed just before hitting the floor? (b) What is the speed of the ball just as it leaves the ground? (c) How much energy was lost and where did it go?

38. •• A skier coasts down a very smooth, 10-m-high slope similar to the one shown in Figure 5.23. If the speed of the skier on the top of the slope is 5.0 m/s, what is his speed at the bottom of the slope?

39. •• A roller coaster travels on a frictionless track as shown in ▼ **Figure 5.30**. (a) If the speed of the roller coaster at point A is 5.0 m/s, what is its speed at point B? (b) Will it reach point C? (c) What minimum speed at point A is required for the roller coaster to reach point C?

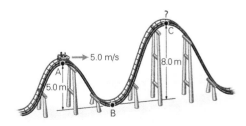

▲ **FIGURE 5.30   Energy conversion(s)** See Exercise 39.

40. •• A simple pendulum has a length of 0.75 m and a bob with a mass of 0.15 kg. The bob is released from an angle of 25° relative to a vertical reference line (▼ **Figure 5.31**). (a) Show that the vertical height of the bob when it is released is $h = L(1 - \cos 25°)$. (b) What is the kinetic energy of the bob when the string is at an angle of 9.0°? (c) What is the speed of the bob at the bottom of the swing? (Neglect friction and the mass of the string.)

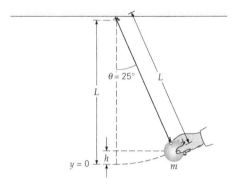

▲ FIGURE 5.31 **A pendulum swings** See Exercise 40.

41. •• Suppose the simple pendulum in Exercise 40 were released from an angle of 60°. (a) What would be the speed of the bob at the bottom of the swing? (b) To what height would the bob swing on the other side? (c) What angle of release would give half the speed of that for the 60° release angle at the bottom of the swing?

42. •• A 1.5-kg box that is sliding on a frictionless surface with a speed of 12 m/s approaches a horizontal spring. (See Figure 5.21.) The spring has a spring constant of 2000 N/m. If one end of the spring is fixed and the other end changes its position, (a) how far will the spring be compressed in stopping the box? (b) How far will the spring be compressed when the box's speed is reduced to half of its initial speed?

43. •• A 0.50-kg mass is suspended on a spring that stretches 3.0 cm. (a) What is the spring constant? (b) What added mass would stretch the spring an additional 2.0 cm? (c) What is the change in potential energy when the mass is added?

44. •• A vertical spring with a force constant of 300 N/m is compressed 6.0 cm and a 0.25-kg ball placed on top. The spring is released and the ball flies vertically upward. How high does the ball go?

45. •• A block with a mass $m_1 = 6.0$ kg sitting on a frictionless table is connected to a suspended mass $m_2 = 2.0$ kg by a light string passing over a frictionless pulley. Using energy considerations, find the speed at which m² hits the floor after descending 0.75 m. (*Note*: A similar problem in Example 4.6 was solved using Newton's laws.)

46. ••• A hiker plans to swing on a rope across a ravine in the mountains, as illustrated in ▼ **Figure 5.32**, and to drop when she is just above the far edge. (a) At what horizontal speed should she be moving when she starts to swing? (b) Below what speed would she be in danger of falling into the ravine? Explain.

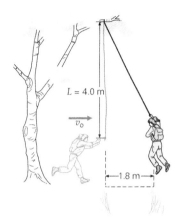

▲ FIGURE 5.32 **Can she make it?** See Exercise 46.

47. ••• In Exercise 38, if the skier has a mass of 60 kg and the force of friction retards his motion by doing 2500 J of work, what is his speed at the bottom of the slope?

5.6 Power

48. • A 1500-kg race car can go from 0 to 90 km/h in 5.0 s. What average power is required to do this?

49. • The two 0.50-kg weights of a cuckoo clock descend 1.5 m in a three-day period. At what rate is their total gravitational potential energy decreased?

50. •• A pump lifts 200 kg of water per hour a height of 5.0 m. What is the minimum necessary power output rating of the water pump in watts and horsepower?

51. •• A race car is driven at a constant velocity of 200 km/h on a straight, level track. The power delivered to the wheels is 150 kW. What is the total resistive force on the car?

52. •• An electric motor with a 2.0-hp output drives a machine with an efficiency of 40%. What is the energy output of the machine per second?

53. •• Water is lifted out of a well 30.0 m deep by a motor rated at 1.00 hp. Assuming 90% efficiency, how many kilograms of water can be lifted in 1 min?

54. •• How much power must you exert to horizontally drag a 25.0-kg table 10.0 m across a brick floor in 30.0 s at constant velocity, assuming the coefficient of kinetic friction between the table and floor is 0.550?

55. ••• A 3250-kg aircraft takes 12.5 min to achieve its cruising altitude of 10.0 km and cruising speed of 850 km/h. If the plane's engines deliver, on average, 1500 hp during this time, what is the efficiency of the engines?

6

Linear Momentum and Collisions

The collision between the racquet and the ball changed the momentum of the ball.

Tomorrow, sportscasters may say that the momentum of the entire match changed as a result of the clutch shot shown in the photograph. This player is said to have gained momentum and went on to win the match. But regardless of the effect on the player, it's clear that the momentum of the *ball* in the chapter-opening photograph must have changed dramatically. The ball was traveling toward the player at a good rate of speed – with a lot of momentum. But a collision with a racquet – with plenty of momentum of its own – changed the ball's direction in a fraction of a second. A fan might say that the player turned the ball around. After studying Newton's second law in Section 4.4.3, you might say that the force the racquet applied to the ball gave it a large acceleration, reversing its velocity vector. Yet if you summed

the momenta (plural of momentum) of the ball and racquet just before the collision and just afterward, you'd discover that although both the ball and the racquet had momentum changes, the total momentum didn't change.

If you were bowling and the ball bounced off the pins and rolled back toward you, you would probably be very surprised. But why? What leads us to expect that the ball will send the pins flying and continue on its way, rather than rebounding? You might say that the momentum of the ball carries it forward even after the collision (and you would be right) – but what does that really mean? In this chapter, the concept of *momentum* will be studied, and you will learn how it is particularly useful in analyzing motion and collisions.

6.1 Linear Momentum

The term *momentum* may bring to mind a football player running down the field, knocking down players who are trying to stop him. Or you might have heard someone say that a team lost its momentum (and so lost the game). Such everyday usage gives some insight into the meaning of momentum. They suggest the idea of mass in motion and therefore inertia. We tend to think of heavy or massive objects in motion as having a great deal of momentum, even if they move very slowly. However, according to the technical definition of momentum, a light object can have just as much momentum as a heavier one, and sometimes more.

Newton referred to what modern physicists term **linear momentum ($\vec{\mathbf{p}}$)** as "the quantity of motion arising from velocity and the quantity of matter conjointly." In other words, the momentum of a body is proportional to *both* its mass and velocity. By definition, the linear momentum of an object is the product of its mass and velocity:

$$\vec{\mathbf{p}} = m\vec{\mathbf{v}} \qquad (6.1)$$

SI unit of momentum: kilogram-meter per second (kg·m/s)

It is common to refer to linear momentum as simply *momentum*. Momentum is a vector quantity that has the same direction as the velocity, and *x*- and *y*-components with magnitudes of $p_x = mv_x$ and $p_y = mv_y$, respectively.

Equation 6.1 expresses the momentum of a single object or particle. For a system of more than one particle, the **total linear momentum ($\vec{\mathbf{P}}$)** of the system is the vector sum of the *momenta* of the individual particles:

$$\vec{\mathbf{P}} = \vec{\mathbf{p}}_1 + \vec{\mathbf{p}}_2 + \vec{\mathbf{p}}_3 + \cdots = \sum \vec{\mathbf{p}}_i \qquad (6.2)$$

Note: $\vec{\mathbf{P}}$ (upper case) signifies the *total* momentum, while $\vec{\mathbf{p}}$ (lower case) signifies an *individual* momentum.

EXAMPLE 6.1: MOMENTUM – MASS AND VELOCITY

A 100-kg football player runs with a velocity of 4.0 m/s straight down the field. A 1.0-kg artillery shell leaves the barrel of a gun with a muzzle velocity of 500 m/s. Which has the greater momentum (magnitude), the football player or the shell?

THINKING IT THROUGH. Given the mass and velocity of an object, the magnitude of its momentum can be calculated from Equation 6.1.

SOLUTION

As usual, first listing the given data and using the subscripts p and s to refer to the player and shell, respectively:

Given:

$m_p = 100$ kg
$v_p = 4.0$ m/s
$m_s = 1.0$ kg
$v_s = 500$ m/s

Find: p_p and p_s (magnitudes of the momenta)

The magnitude of the momentum of the football player is

$$p_p = m_p v_p = (100 \, \text{kg})(4.0 \, \text{m/s}) = 4.0 \times 10^{22} \, \text{kg·m/s}$$

and that of the shell is

$$p_s = m_s v_s = (1.0 \, \text{kg})(500 \, \text{m/s}) = 5.0 \times 10^2 \, \text{kg·m/s}$$

Thus, the less massive shell has the greater momentum. Remember, the magnitude of momentum depends on *both* the mass *and* the magnitude of the velocity.

FOLLOW-UP EXERCISE. What would the football player's speed have to be for his momentum to have the same magnitude as the artillery shell's momentum? Would this speed be realistic? (*Answers to all Follow-Up Exercises are given in Appendix V at the back of the book.*)

EXAMPLE 6.2: LINEAR MOMENTUM – SOME BALLPARK COMPARISONS

Consider the three objects shown in ▼ **Figure 6.1**: a 0.22-caliber bullet (mass 10^{-3} kg and velocity 10^2 m/s), a cruise ship (mass 10^8 kg and velocity 10^1 m/s), and a glacier (mass 10^{12} kg and velocity 10^{-5} m/s). Compute order-of-magnitude values of the linear momentum of the objects.

THINKING IT THROUGH. Given the mass and velocity of each object, the magnitude of its momentum can be estimated with order of magnitude using Equation 6.1.

 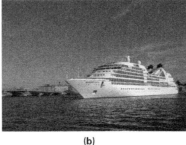 

(a)   (b)   (c)

▲ FIGURE 6.1   **Three moving objects: a comparison of momenta and kinetic energies (a)** A 0.22-caliber bullet shattering a glass; **(b)** a cruise ship; **(c)** a glacier, Glacier Bay, Alaska. See Example 6.2 for description.

SOLUTION

Listing the given data and using the subscripts b, s, and g to refer to the bullet, ship, and glacier, respectively:

Given:

$$m_b \approx 10^{-3} \text{ kg}; v \approx 10^2 \text{ m/s}$$
$$m_s \approx 10^8 \text{ kg}; v \approx 10^1 \text{ m/s}$$
$$m_g \approx 10^{12} \text{ kg}; v \approx 10^{-5} \text{ m/s}$$

Find: p for each object

Bullet:

$$p_b = m_b v_b \approx (10^{-3} \text{ kg})(10^2 \text{ m/s}) = 10^{-1} \text{ kg·m/s}$$

Ship:

$$p_s = m_s v_s \approx (10^8 \text{ kg})(10^1 \text{ m/s}) = 10^9 \text{ kg·m/s}$$

Glacier:

$$p_g = m_g v_g \approx (10^{12} \text{ kg})(10^{-5} \text{ m/s}) = 10^7 \text{ kg·m/s}$$

So the ship has the largest momentum, and the bullet has the smallest according to the estimates.

FOLLOW-UP EXERCISE. Which of the objects in this Example has (1) the greatest kinetic energy and (2) the least kinetic energy? Justify your choices using order-of-magnitude calculations. [Notice here that the dependence is on the square of the speed, $K = 1/2(mv^2)$.]

EXAMPLE 6.3: TOTAL MOMENTUM – A VECTOR SUM

What is the total momentum for each of the systems of particles illustrated in ▶ **Figure 6.2a and b**?

THINKING IT THROUGH. The total momentum is the vector sum of the individual momenta (Equation 6.2). This quantity can be computed using the components of each vector.

SOLUTION

Given: Magnitudes and directions of momenta from Figure 6.2

Find:

(a) Total momentum ($\vec{\mathbf{P}}$) for Figure 6.2a.

(b) Total momentum ($\vec{\mathbf{P}}$) for Figure 6.2b.

(a) The total momentum of a system is the vector sum of the momenta of the individual particles, so

$$\vec{\mathbf{P}} = \vec{\mathbf{p}}_1 + \vec{\mathbf{p}}_2$$
$$= (2.0 \text{ kg·m/s})\hat{\mathbf{x}} + (3.0 \text{ kg·m/s})\hat{\mathbf{x}}$$
$$= (5.0 \text{ kg·m/s})\hat{\mathbf{x}} \quad (+x\text{-direction})$$

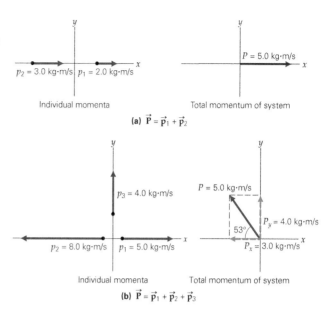

▲ FIGURE 6.2 **Total momentum** The total momentum of a system of particles is the vector sum of the particles' individual momenta. See Example 6.3 for description.

(b) Computing the total momenta in the x- and y-directions gives

$$P_x = p_{1x} + p_{2x} = 5.0 \text{ kg·m/s} + (-8.0 \text{ kg·m/s})$$
$$= -(3.0 \text{ kg·m/s}) \quad (-x\text{-direction})$$
$$P_y = p_{3x} = 4.0 \text{ kg·m/s} \quad (+y\text{-direction})$$

Then

$$\vec{\mathbf{P}} = \vec{\mathbf{P}}_x + \vec{\mathbf{P}}_y = (-3.0 \text{ kg·m/s})\hat{\mathbf{x}} + (4.0 \text{ kg·m/s})\hat{\mathbf{y}}$$

or

$$P = 5.0 \text{ kg·m/s at } 53° \text{ relative to the } -x\text{-axis}$$

FOLLOW-UP EXERCISE. In this Example, if $\vec{\mathbf{p}}_1$ and $\vec{\mathbf{p}}_2$ in part (a) were added to $\vec{\mathbf{p}}_2$ and $\vec{\mathbf{p}}_3$ in part (b), what would be the total momentum?

In Example 6.3a, each of the momenta was along one of the coordinate axes and thus was added straightforwardly. If the motion of one (or more) of the particles is not along an axis, its momentum vector may be broken up, or resolved, into rectangular components, and then individual components can be added to find the components of the total momentum, just as you learned to do with force components in Section 6.4.3.

Since momentum is a vector, a change in momentum can result from a change in magnitude and/or direction. Examples of changes in the momenta of particles because of changes of direction on collision are illustrated in ▶ **Figure 6.3**. In the figure, the magnitude of a particle's momentum is taken to be the same both before and after collision (as indicated by the arrows of equal length). Figure 6.3a illustrates a direct rebound – a 180°

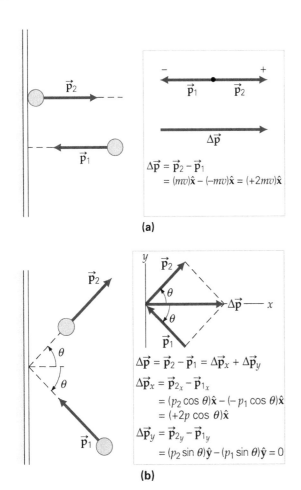

$$\Delta\vec{\mathbf{p}} = \vec{\mathbf{p}}_2 - \vec{\mathbf{p}}_1$$
$$= (mv)\hat{\mathbf{x}} - (-mv)\hat{\mathbf{x}} = (+2mv)\hat{\mathbf{x}}$$

(a)

$$\Delta\vec{\mathbf{p}} = \vec{\mathbf{p}}_2 - \vec{\mathbf{p}}_1 = \Delta\vec{\mathbf{p}}_x + \Delta\vec{\mathbf{p}}_y$$
$$\Delta\vec{\mathbf{p}}_x = \vec{\mathbf{p}}_{2_x} - \vec{\mathbf{p}}_{1_x}$$
$$= (p_2\cos\theta)\hat{\mathbf{x}} - (-p_1\cos\theta)\hat{\mathbf{x}}$$
$$= (+2p\cos\theta)\hat{\mathbf{x}}$$
$$\Delta\vec{\mathbf{p}}_y = \vec{\mathbf{p}}_{2_y} - \vec{\mathbf{p}}_{1_y}$$
$$= (p_2\sin\theta)\hat{\mathbf{y}} - (p_1\sin\theta)\hat{\mathbf{y}} = 0$$

(b)

▲ **FIGURE 6.3 Change in momentum** The change in momentum is given by the *difference* in the momentum vectors. **(a)** Here, the vector sum is zero, but the vector *difference*, or change in momentum, is not. (The particles are displaced for convenience.) **(b)** The change in momentum is found by computing the change in the components.

change in direction. Note that the change in momentum ($\Delta\vec{\mathbf{p}}$) is the *vector* difference and that directional signs for the vectors are important. Figure 6.3b shows a glancing collision, for which the change in momentum is given by analyzing changes in the *x*- and *y*-components.

6.1.1 Force and Momentum

As you know from Section 6.4.3, if an object has a change in velocity, a net force must be acting on it. Similarly, since momentum is directly related to velocity, a change in momentum also requires a net force. In fact, Newton originally expressed his second law of motion in terms of momentum rather than acceleration. The force-momentum relationship may be seen by starting with $\vec{\mathbf{F}}_{net} = m\vec{\mathbf{a}}$ and using $\vec{\mathbf{a}} = (\vec{\mathbf{v}} - \vec{\mathbf{v}}_o)/\Delta t$, where the mass is assumed to be constant. Thus,

$$\vec{\mathbf{F}}_{net} = m\vec{\mathbf{a}} = \frac{m(\vec{\mathbf{v}} - \vec{\mathbf{v}}_o)}{\Delta t} = \frac{m\vec{\mathbf{v}} - m\vec{\mathbf{v}}_o}{\Delta t} = \frac{\vec{\mathbf{p}} - \vec{\mathbf{p}}_o}{\Delta t} = \frac{\Delta\vec{\mathbf{p}}}{\Delta t}$$

or

$$\vec{\mathbf{F}}_{net} = \frac{\Delta\vec{\mathbf{p}}}{\Delta t} \qquad (6.3)$$

where $\vec{\mathbf{F}}_{net}$ is the *average* net force on the object if the acceleration is not constant (or the *instantaneous* net force if Δt goes to zero).

Expressed in this form, Newton's second law states that *the net external force acting on an object is equal to the time rate of change of the object's momentum*. It is easily seen from the development of Equation 6.3 that the equations $\vec{\mathbf{F}}_{net} = m\vec{\mathbf{a}}$ and $\vec{\mathbf{F}}_{net} = \Delta\vec{\mathbf{p}}/\Delta t$ are equivalent if the mass is constant. In some situations, however, the mass may vary. This factor will not be a consideration here in the discussion of particle collisions, but a special case will be given later in the chapter. The more general form of Newton's second law, Equation 6.3, is true even if the mass varies.

Just as the equation $\vec{\mathbf{F}}_{net} = m\vec{\mathbf{a}}$ indicates that an acceleration is evidence of a net force, the equation $\vec{\mathbf{F}}_{net} = \Delta\vec{\mathbf{p}}/\Delta t$ indicates that *a change in momentum is evidence of a net force*. For example, as illustrated in ▼ **Figure 6.4**, the momentum of a projectile is tangential to the projectile's parabolic path and changes in both magnitude and direction. The change in momentum indicates that there is a net force acting on the projectile, which of course is the force of gravity. Changes in momentum were illustrated in Figure 6.3. Can you identify the forces in these two cases? (Think in terms of Newton's third law.)

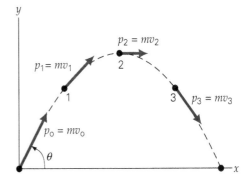

▲ **FIGURE 6.4 Change in the momentum of a projectile** The total momentum vector of a projectile is tangential to the projectile's path (as is its velocity); this vector changes in magnitude and direction, because of the action of an external force (gravity). The *x*-component of the momentum is constant. (Why?)

6.2 Impulse

When two objects – such as a hammer and a nail, a golf club and a golf ball, or even two cars – collide, they can exert a large force on one another for a short period of time, or an *impulse*, as shown in the chapter-opening photograph and ▶ **Figure 6.5a**. The force is not constant in this situation. However, Newton's second law in momentum form is still useful for analyzing such situations by using average values. Written in this form, the law states that the *average* force is equal to the time rate of change of momentum: $\vec{\mathbf{F}}_{avg} = \Delta\vec{\mathbf{p}}/\Delta t$ (Equation 6.3). Rewriting the equation to express the change in momentum (with only one force acting on the object),

$$\vec{\mathbf{F}}_{avg}\Delta t = \Delta\vec{\mathbf{p}} = \vec{\mathbf{p}} - \vec{\mathbf{p}}_o \qquad (6.4)$$

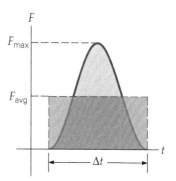

▲ **FIGURE 6.6   Average impulse force** The area under the average force curve ($F_{avg}\,\Delta t$, within the dashed red lines) is the same as the area under the F-versus-t curve, which is usually difficult to evaluate.

TABLE 6.1   Some Typical Contact Times (Δt)

	Δt (milliseconds)
Golf ball (hit by a driver)	1.0
Baseball (hit off tee)	1.3
Tennis (forehand)	5.0
Football (kick)	8.0
Soccer (header)	23.0

(a)

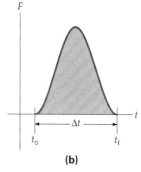

(b)

▲**FIGURE 6.5   Collision impulse (a)** A collision impulse causes the tennis ball to be deformed. **(b)** The impulse is the area under the curve of an F-versus-t graph. Note that the impulse force on the ball is not constant, but rises to a maximum.

The term $\vec{\mathbf{F}}_{avg}\Delta t$ is known as the **impulse ($\vec{\mathbf{I}}$)** of the force:

$$\vec{\mathbf{I}} = \vec{\mathbf{F}}_{avg}\,\Delta t = \Delta\vec{\mathbf{p}} = m\vec{\mathbf{v}} - m\vec{\mathbf{v}}_o \qquad (6.5)$$

SI unit of impulse and momentum: newton-second (N·s)

Thus, *the impulse exerted on an object is equal to the change in the object's momentum.* This statement is referred to as the **impulse–momentum theorem.** Impulse has units of newton-second (N·s), which are also units of momentum ($1 \text{ N·s} = 1 \text{ kg·m/s}^2\text{·s} = 1 \text{ kg·m/s}$).

In Section 5.3, it was learned that by the work-energy theorem ($W_{net} = F_{net}\Delta x = \Delta K$), the area under an F_{net}-versus-x curve is equal to the net work, or change in kinetic energy. Similarly, the area under an F_{net}-versus-t curve is equal to the impulse, or the change in momentum (Figure 6.5b). Forces between interacting objects usually vary with time and are therefore not constant forces. However, in general, it is convenient to talk about the equivalent *constant* average force $\vec{\mathbf{F}}_{avg}$ acting over a time interval Δt to give the same impulse (same area under the force-versus-time curve), as shown in ▶ **Figure 6.6.** Some typical contact times in sports are given in ▶ **Table 6.1.**

EXAMPLE 6.4: TEEING OFF – THE IMPULSE-MOMENTUM THEOREM

A golfer drives a 0.046-kg ball from an elevated tee, giving the ball an initial horizontal speed of 40 m/s (about 90 mi/h). What is the magnitude of the average force exerted by the club on the ball during this time?

THINKING IT THROUGH. The average force on the ball is equal to the time rate of change of its momentum, and this can be computed (Equation 6.5). The Δt of the collision is obtained from Table 6.1.

SOLUTION

Given:
$$m = 0.046 \text{ kg}$$
$$v = 40 \text{ m/s}$$
$$v_o = 0$$
$$\Delta t = 1.0 \text{ ms} = 1.0 \times 10^{-3} \text{ s (Table 6.1)}$$

Find: F_{avg} (average force)

The mass and the initial and final velocities are given, so the change in momentum can be easily found. Then the magnitude of the average force can be computed from the impulse-momentum theorem:

$$F_{avg}\,\Delta t = p = p_o = mv - mv_o$$

and

$$F_{avg} = \frac{mv - mv_o}{\Delta t} = \frac{(0.046\,\text{kg})(40\,\text{m/s}) - 0}{1.0 \times 10^{-3}\,\text{s}}$$
$$= 1.8 \times 10^3 \text{ N (or about 410 lb)}$$

[This is a very large force compared with the weight of the ball, $w = mg = (0.046\ \text{kg})(9.8\ \text{m/s}^2) = 0.45\ \text{N}$ (or about 0.22 lb).] The force is in the direction of the acceleration and is the *average* force. The instantaneous force is even greater than this value near the midpoint of the time interval of the collision (Δt in Figure 6.6).

FOLLOW-UP EXERCISE. Suppose the golfer in this Example drives the ball with the same average force, but "follows through" on the swing so as to increase the contact time to 1.5 ms. What effect would this change have on the initial horizontal speed of the drive?

Example 6.4 illustrates the large forces that colliding objects can exert on one another during short contact times. In some cases, the contact time may be shortened to maximize the impulse – for example, in a karate chop. However, in other instances, the Δt may be manipulated to reduce the force. Suppose there is a fixed change in momentum in a given situation. Then, with $\Delta p = F_{\text{avg}}\ \Delta t$, if Δt could be made longer, the average impulse F_{avg} would be reduced.

You have probably tried to minimize the impulse on occasion. For example, in catching a hard, fast-moving ball, you quickly learn not to catch it with your arms rigid, but rather to move your hands with the ball. This movement increases the contact time and reduces the impulse and the "sting" (▼ **Figure 6.7**).

| (a) | (b) |

▲ **FIGURE 6.7   Adjust the impulse (a)** The change in momentum in catching the ball is a constant mv_0. If the ball is stopped quickly (small Δt), the impulse force is large (big F_{avg}) and stings the catcher's bare hands. **(b)** Increasing the contact time (large Δt) by moving the hands with the ball reduces the impulse force and makes catching more enjoyable.

When jumping from a height onto a hard surface, you should not land stiff-legged. The abrupt stop (small Δt) would apply a large force to your leg bones and joints and could cause injury. If you bend your knees as you land, the time interval Δt increases and this makes the force smaller. The impulse is vertically upward, opposite your velocity ($F_{\text{avg}}\ \Delta t = \Delta p = -mv_0$ with the final velocity being zero) and its value is a constant. The automobile airbag uses the same principle to reduce injuries.

EXAMPLE 6.5: IMPULSE AND BODY INJURY

A 70.0-kg worker jumps stiff-legged from a height of 1.00 m onto a concrete floor. (a) What is the magnitude of the impulse he feels on landing, assuming a sudden stop in 8.00 ms? (b) What is the average force?

THINKING IT THROUGH. The impulse is $F_{\text{avg}}\ \Delta t$, which cannot be calculated directly from the given data. But impulse is equal to the change in momentum, $F_{\text{avg}}\ \Delta t = \Delta p = mv - mv_0$. So the impulse can be calculated from the difference in momenta if v_0 is found.

SOLUTION

Given:

$$m = 70.0\ \text{kg}$$
$$h = 1.00\ \text{m}$$
$$\Delta t = 8.00\ \text{ms} = 8.00 \times 10^{-3}\ \text{s}$$

Find:

(a) I (impulse on worker)
(b) F_{avg} (average force)

There are two different parts here: (1) the worker descending after jumping and (2) the sudden stop after hitting the floor. So we must be careful with notation and distinguish between the two parts (a and b) with the subscripts of 1 and 2, respectively. (b) Knowing Δt, the average force can be calculated.

(a) Here, the initial velocity of the worker is $v_{1o} = 0$, and the final velocity just before hitting the floor may be found using $v^2 = v_0^2 - 2gh$ (Equation 2.12'), with the result of

$$v_{1f} = -\sqrt{2gh}$$

where the negative sign indicates direction (downward). The v_{1f} in (a) is then the initial velocity with which the stiff-legged worker hits the floor; that is, $v_{2o} = v_{1f} = -\sqrt{2gh}$, and the final velocity at the second phase is $v_{2f} = 0$. Then,

$$I = F_{\text{avg}}\Delta t = mv_{2f} - mv_{2o} = 0 - m\left(-\sqrt{2gh}\right) = +m\sqrt{2gh}$$
$$= (70.0\,\text{kg})\sqrt{2\left(9.80\,\text{m/s}^2\right)(1.00\,\text{m})} = 310\,\text{kg·m/s}$$

where the impulse is in the upward direction.

(b) With a Δt of 8.00×10^{-3} s for the sudden stop on impact, this would give a force of

$$F_{\text{avg}} = \frac{\Delta p}{\Delta t} = \frac{310\,\text{kg·m/s}}{8.00 \times 10^{-3}\,\text{s}} = 3.88 \times 10^4\,\text{N}$$

(about 8730 lb of force!)

and the force is upward on the stiff legs.

FOLLOW-UP EXERCISE. Suppose the worker bent his knees and increased the contact time to 0.60 s on landing. What would be the average impulse force on him in this case?

In some instances, the impulse force may be relatively constant and the contact time (Δt) deliberately increased to produce a greater impulse, and thus a greater change in momentum ($F_{\text{avg}}\ \Delta t = \Delta p$). This is the principle of "following through" in sports, for example, when hitting a ball with a bat or racquet, or a golf club. In the latter case (▶ **Figure 6.8a**), assuming that the golfer supplies the same average force with each swing, the longer the contact time, the greater the impulse or change in momentum the ball receives. That is, with $F_{avg} \Delta t = mv$ (since $v_o = 0$), the greater the value of Δt, the greater the final speed of the ball. (This principle was illustrated in the Follow-Up Exercise in Example 6.4.) In some instances, a long follow-through may primarily be used to improve control of the ball's direction (Figure 6.8b).

(a)

(b)

▲ FIGURE 6.8 **Increasing the contact time (a)** A golfer follows through on a drive. One reason he does so is to increase the contact time so that the ball receives greater impulse and momentum. **(b)** The follow-through on a long putt increases the contact time for greater momentum, but the main reason here is for directional control.

The word *impulse* implies that the impulse force acts only briefly (like an "impulsive" action), and this is true in many instances. However, the definition of *impulse* places no limit on the time interval of a collision over which the force may act. Technically, a comet at its closest approach to the Sun is involved in a collision, because in physics, collision forces do *not* have to be contact forces. Basically, a collision is any interaction between objects in which there is an exchange of momentum and/or energy.

As you might expect from the work-energy theorem and the impulse-momentum theorem, momentum and kinetic energy are directly related. A little algebraic manipulation of the equation for kinetic energy (Equation 5.5) allows us to express kinetic energy (K) in terms of the *magnitude* of momentum (p):

$$K = \frac{1}{2}mv^2 = \frac{(mv)^2}{2m} = \frac{p^2}{2m} \qquad (6.6)$$

Thus, kinetic energy and momentum are intimately related, but they are different quantities.

6.3 Conservation of Linear Momentum

Like total mechanical energy, the total momentum of a system is a conserved quantity under certain conditions. This fact allows us to analyze a wide range of situations and to readily solve many problems. Conservation of momentum is one of the most important principles in physics. In particular, it is used to analyze collisions of objects ranging from subatomic particles to automobiles in traffic accidents.

For the linear momentum of a single object to be conserved (i.e., to remain constant with time), one condition must hold that is apparent from the momentum form of Newton's second law (Equation 6.3). If the net force acting on a particle is zero; that is,

$$\vec{F}_{net} = \frac{\Delta \vec{p}}{\Delta t} = 0$$

then

$$\Delta \vec{p} = 0 = \vec{p} - \vec{p}_o$$

where $\vec{p}_o$ is the initial momentum and $\vec{p}$ is the momentum at some later time. Since these two values are equal, the momentum is conserved, and

$$\vec{p} = \vec{p}_o \text{ or } m\vec{v} = m\vec{v}_o \quad \text{(final momentum = initial momentum)}$$

Note that this conservation is consistent with Newton's first law: An object remains at rest ($\vec{p} = 0$), or in motion with a *uniform* velocity (constant $\vec{p} \neq 0$), unless acted on by a net external force.

The conservation of momentum can be extended to a system of particles if Newton's second law is written in terms of the net force acting on the system and of the momenta of the particles: $\vec{F}_{net} = \Sigma \vec{F}_i$ and $\vec{P} = \Sigma \vec{p}_i = \Sigma m_i \vec{v}_i$.

Because $\vec{F}_{net} = \Delta \vec{P}/\Delta t$, and if there is no net external force acting *on the system*, then $\vec{F}_{net} = 0$, and $\Delta \vec{F} = 0$; so $\vec{P} = \vec{P}_o$, and the *total* momentum is conserved. This generalized condition is referred to as the law of **conservation of linear momentum**:

$$\vec{P} = \vec{P}_o \qquad (6.7)$$

Thus, the total linear momentum of a system, $\vec{P} = \Sigma \vec{p}_i$, is conserved if the net external force acting on the system is zero.

There are various ways to achieve this condition. For example, recall from Section 5.5 that a *closed*, or *isolated*, system is one on which no net external force acts, so the total linear momentum of an isolated system is conserved.

Within a system, internal forces may act – for example, when particles collide. These are force pairs of Newton's third law, and there is a good reason that such forces are not explicitly referred to in the condition for the conservation of momentum. By Newton's third law, these internal forces are equal and opposite and vectorially cancel each other. Thus, *the net internal force of a system is always zero.*

An important point to understand, however, is that the momenta of *individual* particles or objects within a system may change. But in the absence of a net external force, the *vector sum* of all the momenta (the total system momentum $\vec{\mathbf{P}}$) remains the same. If the objects are initially at rest (i.e., the total momentum is zero) and then are set in motion as the result of internal forces, the total momentum must still add to zero. This principle is illustrated in ▼ **Figure 6.9** and analyzed in Example 6.6. Objects in an isolated system may transfer momentum among themselves, but the total momentum after the changes must add up to the initial value, assuming the net external force on the system is zero.

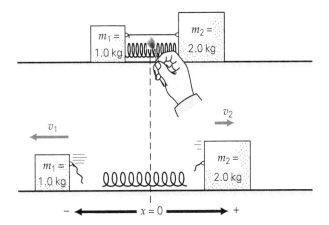

▲ **FIGURE 6.9   An internal force and the conservation of momentum** The spring force is an internal force, so the momentum of the system is conserved. See Example 6.6.

The conservation of momentum is often a powerful and convenient tool for analyzing situations involving motion and collisions. Its application is illustrated in the following Examples. (Notice that conservation of momentum, in many cases, bypasses the need to know the forces involved.)

EXAMPLE 6.6: BEFORE AND AFTER – CONSERVATION OF MOMENTUM

Two masses, $m_1 = 1.0$ kg and $m_2 = 2.0$ kg, are held on either side of a light compressed spring by a light string joining them, as shown in Figure 6.9. The string is burned (negligible external force), and the masses move apart on the frictionless surface, with m_1 having a velocity of 1.8 m/s to the left. What is the velocity of m_2?

THINKING IT THROUGH. With no net external force (the weights are each canceled by a normal force), the total momentum of the system is conserved. It is initially zero, so after the string is burned, the momentum of m_2 must be *equal to and opposite* that of m_1. (Vector addition gives zero total momentum. Also, note that the term *light* indicates that the masses of the spring and string can be ignored.)

SOLUTION

Listing the data:

Given:

$m_1 = 1.0$ kg
$m_2 = 2.0$ kg
$v_1 = -1.8$ m/s (left)

Find: v_2 (velocity – speed and direction)

Here, the system consists of the two masses and the spring. Since the spring force is internal to the system, the momentum of the system is conserved. It should be apparent that the initial total momentum of the system $\left(\vec{\mathbf{P}}_\mathrm{o}\right)$ is zero, and therefore the final momentum must also be zero. Thus,

$$\vec{\mathbf{P}}_\mathrm{o} = \vec{\mathbf{P}} = 0 \text{ and } \vec{\mathbf{P}} = \vec{\mathbf{p}}_1 + \vec{\mathbf{p}}_2 = 0$$

(The momentum of the "light" spring does not come into the equations, because its mass is negligible.) Then,

$$\vec{\mathbf{p}}_2 = -\vec{\mathbf{p}}_1$$

which means that the momenta of m_1 and m_2 are equal and opposite. Using directional signs (with $+$ indicating the direction to the right in the figure),

$$m_2 v_2 = -m_1 v_1$$

and

$$v_2 = -\left(\frac{m_1}{m_2}\right)v_1 = -\left(\frac{1.0\,\mathrm{kg}}{2.0\,\mathrm{kg}}\right)(-1.8\,\mathrm{m/s}) = +0.90\,\mathrm{m/s}$$

Thus, the velocity of m_2 is 0.90 m/s in the $+x$-direction, or to the right in the figure. This value is half that of v_1 as you might have expected, since m_2 has twice the mass of m_1.

FOLLOW-UP EXERCISE. (a) Suppose that the large block in Figure 6.9 were attached to the Earth's surface so that the block could not move when the string was burned. Would momentum be conserved in this case? Explain. (b) Two girls, each having a mass of 50 kg, stand at rest on skateboards with negligible friction. The first girl tosses a 2.5-kg ball to the second. If the speed of the ball is 10 m/s, what is the speed of each girl after the ball is caught, and what is the momentum of the ball before it is tossed, while it is in the air, and after it is caught?

INTEGRATED EXAMPLE 6.7: CONSERVATION OF LINEAR MOMENTUM – FRAGMENTS AND COMPONENTS

A 30-g bullet with a speed of 400 m/s strikes a glancing blow to a target brick of mass 1.0 kg. The brick breaks into two fragments. The bullet deflects at an angle of 30° above the $+x$-axis and has a reduced speed of 100 m/s. One piece of the brick (with mass 0.75 kg) goes off to the right, or in the initial direction of the bullet, with a speed of 5.0 m/s. (a) Taking the $+x$-axis to the right, will the other piece of the brick move in the (1) second quadrant, (2) third quadrant, or (3) fourth quadrant? (b) Determine the speed and direction of the other piece of the brick immediately after collision (where gravity can be neglected).

(a) **CONCEPTUAL REASONING.** The conservation of linear momentum can be applied because there is no net external force on the system – the bullet and brick. Initially, all of the momentum is in the forward $+x$-direction (▼ **Figure 6.10**). Afterward, one piece of the brick flies off in the $+x$-direction, and the bullet at an angle of 30° to the x-axis. The bullet's momentum has a positive y-component, so the other piece of the brick must have a negative y-component because there was no initial momentum in the y-direction. Hence, with the total momentum in the $+x$-direction (before and after), the answer is (3) fourth quadrant.

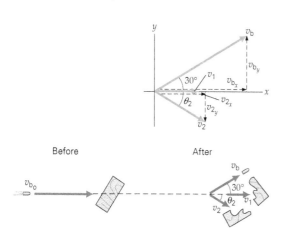

▲ **FIGURE 6.10 A glancing collision** Momentum is conserved in an isolated system. The motion in two dimensions may be analyzed in terms of the components of momentum, which are also conserved.

(b) **QUANTITATIVE REASONING AND SOLUTION.** There is one object with momentum before collision (the bullet), and three with momenta afterward (the bullet and the two fragments). By the conservation of linear momentum, the total (vector) momentum after collision equals that before collision. As is often the case, a sketch of the situation is helpful, with the vectors resolved in component form (Figure 6.10). Applying the conservation of linear momentum should allow the velocity (speed and direction) of the second fragment to be determined.

Given:

$m_b = 30$ g $= 0.030$ kg
$v_{b_o} = 400$ m/s (initial bullet speed)
$v_b = 100$ m/s (final bullet speed)
$\theta_b = 30°$ (final bullet angle)
$M = 1.0$ kg (brick mass)
$m_1 = 0.75$ kg (mass of large fragment)
$\theta_1 = 0°$ and $v_1 = 5.0$ m/s
$m_2 = 0.25$ kg (mass of small fragment)

Find: v_2 (speed of the smaller brick fragment)
θ_2 (direction of the fragment relative to the original direction of the bullet).

With no external forces (gravity neglected), the total linear momentum is conserved. Therefore, both the x- and y-components of the total momentum can be equated before and after (see Figure 6.10):

before		after
$x:$ $m_b v_{b_o}$	$=$	$m_b v_b \cos\theta_b + m_1 v_1 + m_2 v_2 \cos\theta_2$
$y:$ 0	$=$	$m_b v_b \sin\theta_b - m_2 v_2 \sin\theta_2$

The x-equation can be rearranged to solve for the magnitude of the x-velocity of the smaller fragment:

$$v_2 \cos\theta_2 = \frac{m_b v_{b_o} - m_b v_b \cos\theta - m_1 v_1}{m_2}$$

$$= \frac{(0.030\,\text{kg})(400\,\text{m/s}) - (0.030\,\text{kg})(100\,\text{m/s})(0.886) - (0.75\,\text{kg})(5.0\,\text{m/s})}{0.25\,\text{kg}}$$

$$= 22\,\text{m/s}$$

Similarly, the y-equation can be solved for the magnitude of the y-velocity component of the smaller fragment:

$$v_2 \sin\theta_2 = \frac{m_b v_{b_o} \sin\theta_b}{m_2} = \frac{(0.030\,\text{kg})(100\,\text{m/s})(0.50)}{0.25\,\text{kg}} = 6.0\,\text{m/s}$$

Forming a ratio,

$$\frac{v_2 \sin\theta_2}{v_2 \cos\theta_2} = \frac{6.0\,\text{m/s}}{22\,\text{m/s}} = 0.27 = \tan\theta_2$$

(where the v_2 terms cancel, and $(\sin\theta_2 / \cos\theta_2) = \tan\theta_2$). Then,

$$\theta_2 = \tan^{-1}(0.27) = 15°$$

and from the x-equation,

$$v_2 = \frac{22\,\text{m/s}}{\cos 15°} = \frac{22\,\text{m/s}}{0.97} = 23\,\text{m/s}$$

FOLLOW-UP EXERCISE. Is the kinetic energy conserved for the collision in this Example? If not, where did the energy go?

EXAMPLE 6.8: PHYSICS ON ICE

A physicist is lowered from a helicopter to the middle of a smooth, level, frozen lake, the surface of which has negligible friction, and challenged to make her way off the ice. Walking is out of the question. (Why?) As she stands there pondering her predicament, she decides to use the conservation of momentum by throwing her heavy, identical mittens, which will provide her with the momentum to get herself to shore. To get to the shore more quickly, which should this sly physicist do: throw both mittens at once or throw them separately, one after the other with the same speed?

THINKING IT THROUGH. The initial momentum of the system (physicist and mittens) is zero. With no net external force, by the conservation of momentum, the total momentum *remains* zero. So if the physicist throws the mittens in one direction, she will go in the opposite direction (because momenta vectors in opposite directions add to zero). Then which way of throwing gives greater speed? If both the mittens were thrown together, the magnitude of their momentum would be $2mv$, where v is relative to the ice and m is the mass of one mitten.

When thrown separately, the first mitten would have a momentum of mv. The physicist and the second mitten would then be in motion, and throwing the second mitten would add some more momentum to the physicist and increase her speed, but would the speed now be greater than that if both mittens were thrown simultaneously?

Let's analyze the conditions of the second throw. After throwing the first mitten, the physicist "system" would have less mass. With less mass, the second throw would produce a greater acceleration and speed things up. But on the other hand, after the first throw, the second mitten is moving with the person, and when thrown in the opposite direction, the mitten would have a velocity less than v relative to the ice (or to a stationary observer). So which effect would be greater? What do you think? Sometimes situations are difficult to analyze intuitively, and you must apply scientific principles to figure them out.

SOLUTION

Given:

> m = mass of single mitten
> M = mass of physicist
> $-v$ = velocity of thrown mitten(s), in the negative direction
> V_p = velocity of physicist in the positive direction

Find: Which method of mitten throwing gives the physicist the greater speed

When the mittens are *thrown together*, by the conservation of momentum,

$$\begin{array}{cc} \text{before} & \text{after} \end{array}$$
$$0 = 2m(-v) + MV_p, \text{ so } V_p = \frac{2mv}{M} \qquad (1)$$

when they are thrown separately,

$$\begin{array}{cc} \text{before} & \text{after} \end{array}$$
$$\text{1st throw:} \quad 0 = 2m(-v) + (M+m)V_{p1}, \text{ so } V_{p1} = \frac{mv}{M+m} \qquad (2)$$

$$\text{2nd throw:} \quad (M+m)V_{p1} = m(V_{p1} - v) + MV_{p2}.$$

Note that in the last m term of the second throw, the quantities in the parentheses represent the velocity of the mitten relative to the ice. With an initial velocity of $+V_{p_1}$ when the first mitten is thrown in the negative direction, then $V_{p1} - v$. (Recall relative velocities from Section 3.4.)

Solving for V_{p2}:

$$V_{p2} = V_{p1} + \left(\frac{m}{M}\right)v = \frac{mv}{M+m} + \left(\frac{m}{M}\right)v = \left(\frac{m}{M+m} + \frac{m}{M}\right)v \qquad (3)$$

where Equation 6.2 was substituted for V_{p1} after the first throw.

Now, when the mittens are thrown together (Equation 6.1), $V_p = (2m/M)v$, so the question is whether the result of Equation 6.3

is greater or less than that of Equation 1. Notice that with a greater denominator for the $m/(M+m)$ term in Equation 6.3 it is less than the m/M term. So,

$$\left(\frac{m}{M+m} + \frac{m}{M}\right) < \frac{2m}{M}$$

and therefore, $V_p > V_{P2}$, or (thrown together) > (thrown separately).

FOLLOW-UP EXERCISE. Suppose the second throw were in the direction of the physicist's velocity from the first throw, i.e., the second throw is opposite to the first throw. Would this throw bring her to a stop?

As mentioned previously, the conservation of momentum is used to analyze the collisions of objects ranging from subatomic particles to automobiles in traffic accidents. In many instances, however, external forces may be acting on the objects, which means that the momentum is not conserved.

But, as will be learned in the next section, the conservation of momentum often allows a good approximation *over the short time of a collision,* during which the internal forces (which conserve system momentum) are much greater than the external forces. For example, external forces such as gravity and friction also act on colliding objects but are often relatively small compared with the internal forces of the collision. (This concept was implied in Example 6.7.) Therefore, if the objects interact for only a brief time, the effects of the external forces may be negligible compared with those of the large internal forces during that time and the conservation of linear momentum may be used.

6.4 Elastic and Inelastic Collisions

In general, a *collision* may be defined as a meeting or interaction of particles or objects that causes an exchange of energy and/or momentum. Taking a closer look at collisions in terms of the conservation of momentum is simpler for an isolated system, such as a system of particles (or balls) involved in head-on collisions. For simplicity, only collisions in one dimension will be considered, which can be analyzed in terms of the conservation of energy. On the basis of what happens to the total kinetic energy, two types of collisions are defined: *elastic* and *inelastic.*

In an **elastic collision**, the total kinetic energy is conserved. That is, the *total* kinetic energy of all the objects of the system after the collision is the same as the *total* kinetic energy before the collision (▶ **Figure 6.11a**). Kinetic energy may be traded between objects of a system, but the total kinetic energy in the system remains constant. That is,

$$\begin{array}{ll} \text{total } K \text{ final} = \text{total } K \text{ initial} & \text{(condition for} \\ K_f = K_i & \text{an elastic collision)} \quad (6.8) \end{array}$$

(a)

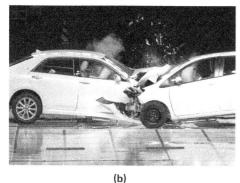

(b)

▲ **FIGURE 6.11 Collisions (a)** Approximate elastic collisions. **(b)** An inelastic collision.

During such a collision, some or all of the initial kinetic energy is temporarily converted to potential energy as the objects are deformed. But after the maximum deformations occur, the objects *elastically* "spring" back to their original shapes, and the system regains all of its original kinetic energy. For example, two steel balls or two billiard balls may have a nearly elastic collision, with each ball having the same shape afterward as before; that is, there is no permanent deformation.

In an **inelastic collision**, total kinetic energy is *not* conserved (Figure 6.11b). For example, one or more of the colliding objects may not regain the original shapes, and/or sound or frictional heat may be generated and some kinetic energy is lost. Then,

$$\text{total } K \text{ final} < \text{total } K \text{ initial} \quad \text{(condition for}$$
$$K_f < K_i \qquad\qquad \text{an inelastic collision)} \quad (6.9)$$

For example, a hollow aluminum ball that collides with a solid steel ball may be dented. Permanent deformation of the ball takes work, and that work is done at the expense of the original kinetic energy of the system. Everyday collisions are inelastic.

But, **for isolated systems, momentum is conserved in both elastic and inelastic collisions**. *For an inelastic collision, only an amount of kinetic energy consistent with the conservation of momentum may be lost.* It may seem strange that kinetic energy can be lost and momentum still conserved, but this fact provides insight into the difference between scalar and vector quantities and the differences in their conservation requirements.

6.4.1 Momentum and Energy in Inelastic Collisions

To see how momentum can remain constant while the kinetic energy changes (decreases) in inelastic collisions, consider the two examples illustrated in ▼ **Figure 6.12**. In Figure 6.12a, two balls of equal mass ($m_1 = m_2$) approach each other with equal and opposite velocities ($v_{1o} = -v_{2o}$). Hence, the total momentum before the collision is (vectorially) zero, but the (scalar) total kinetic energy is *not* zero. After the collision, the balls are stuck together and stationary, so the total momentum is unchanged – still zero.

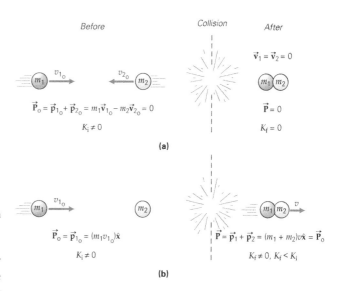

▲ **FIGURE 6.12 Inelastic collisions (a, b)** In inelastic collisions, momentum is conserved, but kinetic energy is not. Collisions like the ones shown here, in which the objects stick together, are called *completely* or *totally inelastic collisions*. The maximum amount of kinetic energy lost is consistent with the law of conservation of momentum.

Momentum is conserved because the forces of collision are internal to the system of the two balls – there is no net external force on the system. The total kinetic energy, however, has decreased to zero. In this case, some of the kinetic energy went into the work done in permanently deforming the balls. Some energy may also have gone into doing work against friction (producing heat) or may have been lost in some other way (e.g., in producing sound).

It should be noted that the balls need not stick together after collision. In a less inelastic collision, the balls may recoil in opposite directions at reduced, but equal speeds. The momentum would still be conserved (still equal to zero – why?), but the kinetic energy would again not be conserved. Under all conditions, the amount of kinetic energy that can be lost must be consistent with the conservation of momentum.

In Figure 6.12b, one ball is initially at rest as the other approaches. The balls stick together after collision but are still in motion. Both cases in Figure 6.12 are examples of a **completely inelastic collision**, in which the objects stick together, and hence both objects have the same velocity after colliding. The coupling of colliding railroad cars is a practical example of a completely (or totally) inelastic collision.

Assume that the balls in Figure 6.12b have different masses. Since the momentum is conserved even in inelastic collisions,

$$\begin{matrix} \text{before} & & \text{after} \\ m_1 v_{1o} & = & (m_1 + m_2)v \end{matrix}$$

and

$$v = \left(\frac{m_1}{m_1 + m_2}\right)v_{1o} \quad \begin{array}{l}(m_2 \text{ initially at rest,} \\ \text{completely inelastic collision only})\end{array} \quad (6.10)$$

Thus, v is less than v_{1o}, since $m_1/(m_1 + m_2)$ must be less than 1. Now consider how much kinetic energy has been lost. Initially, $K_i = 1/2(m_1 v_o^2)$, and after collision the final kinetic energy is:

$$K_f = \frac{1}{2}(m_1 + m_2)v^2$$

Substituting for v from Equation 6.10 and simplifying the result,

$$K_f = \tfrac{1}{2}(m_1 + m_2)\left(\frac{m_1 v_{1o}}{m_1 + m_2}\right)^2 = \frac{(1/2)m_1^2 v_{1o}^2}{m_1 + m_2}$$

$$= \left(\frac{m_1}{m_1 + m_2}\right)\tfrac{1}{2}m_1 v_{1o}^2 = \left(\frac{m_1}{m_1 + m_2}\right)K_i$$

and

$$\frac{K_f}{K_i} = \frac{m_1}{m_1 + m_2} \quad \begin{array}{l}(m_2 \text{ initially at rest,} \\ \text{completely inelastic collision only})\end{array} \quad (6.11)$$

Equation 6.11 gives the fractional amount of the initial kinetic energy that remains with the system after a completely inelastic collision. For example, if the masses of the balls are equal ($m_1 = m_2$), then $m_1/(m_1 + m_2) = 1/2$, and $K_f/K_i = 1/2$ or $K_f = K_i/2$. That is, half of the initial kinetic energy is lost.

Note that not all of the kinetic energy can be lost in this case, no matter what the masses of the balls are. The total momentum after collision cannot be zero, since it was not zero initially. Thus, after the collision, the balls must be moving and must have some kinetic energy ($K_f \neq 0$). *In a completely inelastic collision, the maximum amount of kinetic energy lost must be consistent with the conservation of momentum.*

EXAMPLE 6.9: STUCK TOGETHER – COMPLETELY INELASTIC COLLISION

A 1.0-kg ball with a speed of 4.5 m/s strikes a 2.0-kg stationary ball. If the collision is completely inelastic, (a) what are the speeds of the balls after the collision? (b) What percentage of the initial kinetic energy do the balls have after the collision? (c) What is the total momentum after the collision?

THINKING IT THROUGH. Recall the definition of a completely inelastic collision. The balls stick together after collision; kinetic energy is not conserved, but total momentum is.

SOLUTION

Listing the data:

Given:

$m_1 = 1.0$ kg
$m_2 = 2.0$ kg
$v_o = 4.5$ m/s

Find:

(a) v (speed after collision)
(b) K_f/K_i ($\times 100\%$)
(c) P_f (total momentum after collision)

(a) The momentum is conserved and

$$\vec{P}_f = \vec{P}_o \text{ or } (m_1 + m_2)v = m_1 v_o$$

The balls stick together and have the same speed after collision. This speed is then

$$v = \left(\frac{m_1}{m_1 + m_2}\right)v_o = \left(\frac{1.0\,\text{kg}}{1.0\,\text{kg} + 2.0\,\text{kg}}\right)(4.5\,\text{m/s}) = 1.5\,\text{m/s}$$

(b) The fractional part of the initial kinetic energy that the balls have after the completely inelastic collision is given by Equation 6.11. Notice that this fraction, as given by the masses, is the same as that for the speeds (Equation 6.11) in this special case. By inspection,

$$\frac{K_f}{K_i} = \frac{m_1}{m_1 + m_2} = \frac{1.0\,\text{kg}}{1.0\,\text{kg} + 2.0\,\text{kg}} = \frac{1}{3} = 0.33 \ (\times 100\%) = 33\%$$

Let's show this relationship explicitly:

$$\frac{K_f}{K_i} = \frac{1/2(m_1 + m_2)v^2}{1/2(m_1 v_o^2)} = \frac{1/2(1.0\,\text{kg} + 2.0\,\text{kg})(1.5\,\text{m/s})^2}{1/2(1.0\,\text{kg})(4.5\,\text{m/s})^2}$$

$$= 0.33 \ (= 33\%)$$

Keep in mind that Equation 6.11 applies *only* to *completely* inelastic collisions in which m_2 is initially at rest. For other types of collisions, the initial and final values of the kinetic energy must be computed explicitly.

(c) The total momentum is conserved in all collisions (in the absence of external forces), so the total momentum after collision is the same as before collision. That value is the momentum of the incident ball, with a magnitude of

$$P_f = p_{1o} = m_1 v_o = (1.0\,\text{kg})(4.5\,\text{m/s}) = 4.5\,\text{kg·m/s}$$

and the same direction as that of the incoming ball. Also, as a double check,

$$P_f = (m_1 + m_2)v = 4.5\,\text{kg·m/s}$$

FOLLOW-UP EXERCISE. A small hard-metal ball of mass m collides with a larger, stationary, soft-metal ball of mass M. A *minimum* amount of work W is required to make a dent in the larger ball. If the smaller ball initially has kinetic energy $K = W$, will the larger ball be dented in a completely inelastic collision between the two balls?

6.4.2 Momentum and Energy in Elastic Collisions

For *elastic* collisions, there are two conservation criteria: conservation of momentum (which holds for both elastic and inelastic collisions) and conservation of kinetic energy (for elastic collisions only). That is, for the elastic collision of two objects:

$$\text{Momentum } \vec{\mathbf{P}}: \quad m_1\vec{\mathbf{v}}_{1o} + m_2\vec{\mathbf{v}}_{2o} = m_1\vec{\mathbf{v}}_1 + m_2\vec{\mathbf{v}}_2 \tag{6.12}$$

before   after

$$\text{Kinetic energy } K: \quad \tfrac{1}{2}m_1v_{1o}^2 + \tfrac{1}{2}m_2v_{2o}^2 = \tfrac{1}{2}m_1v_1^2 + \tfrac{1}{2}m_2v_2^2 \tag{6.13}$$

▼ **Figure 6.13** illustrates two objects traveling prior to a one-dimensional, head-on collision with $v_{1o} > v_{2o}$ (both in $+x$-direction). For this two-object situation,

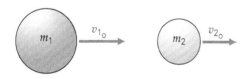

▲ **FIGURE 6.13 Elastic collision coming up** Two objects traveling prior to collision with $v_{1o} > v_{2o}$. See text for description.

before   after

$$\text{Total momentum: } m_1v_{1o} + m_2v_{2o} = m_1v_1 + m_2v_2 \tag{1}$$

(where signs are used to indicate directions and the v's indicate magnitudes).

$$\text{Kinetic energy: } \tfrac{1}{2}m_1v_{1o}^2 + \tfrac{1}{2}m_2v_{2o}^2 = \tfrac{1}{2}m_1v_1^2 + \tfrac{1}{2}m_2v_2^2 \tag{2}$$

If the masses and the initial velocities of the objects are known (which they usually are), then there are two unknown quantities, the final velocities after collision. To find them, Equations 1 and 2 are solved simultaneously. First the equation for momentum conservation is written as follows:

$$m_1(v_{1o} - v_1) = m_2(v_{2o} - v_2) \tag{3}$$

Then, canceling the 1/2 terms in Equation 2, rearranging, and factoring $[a^2 - b^2 = (a - b)(a + b)]$:

$$m_1(v_{1o} - v_1)(v_{1o} + v_1) = m_2(v_{2o} - v_2)(v_{2o} + v_2) \tag{4}$$

Dividing Equation 4 by 3 and rearranging yields

$$v_{1o} - v_{2o} = -(v_1 - v_2) \tag{5}$$

This equation shows that the magnitudes of the relative velocities before and after collision are equal. That is, the relative speed of approach of object m_1 to object m_2 before collision is the same as their relative speed of separation after collision. (See Section 3.4.) Notice that this relation is independent of the values of the masses of the objects, and holds for any mass combination as long as the collision is elastic and *one-dimensional*.

Then, combining Equation 5 with 3 to eliminate v_2 and get v_1 in terms of the two initial velocities,

$$v_1 = \left(\frac{m_1 - m_2}{m_1 + m_2}\right)v_{1o} + \left(\frac{2m_2}{m_1 + m_2}\right)v_{2o} \tag{6.14}$$

Similarly, eliminating v_1 to find v_2,

$$v_2 = \left(\frac{2m_1}{m_1 + m_2}\right)v_{1o} - \left(\frac{m_1 - m_2}{m_1 + m_2}\right)v_{2o} \tag{6.15}$$

6.4.3 One Object Initially at Rest

For this common, special case, say with $v_{2o} = 0$, there are only the first terms in Equations 6.14 and 6.15. In addition, if $m_1 = m_2$, then $v_1 = 0$ and $v_2 = v_{1o}$. That is, the objects *completely* exchange momentum and kinetic energy. The incoming object is stopped on collision, and the originally stationary object moves off with the same velocity as the incoming ball, obviously conserving the system's momentum and kinetic energy. (A real-world example that comes close to these conditions is the head-on collision of billiard balls.)

You can also get some approximates for special cases from the equations for one object initially at rest (taken to be m_2):

For $m_1 \gg m_2$ (massive incoming ball): $v_1 \approx v_{1o}$ and $v_2 \approx 2v_{1o}$

That is, the massive incoming object is slowed down only slightly and the light (less massive) object is knocked away with a velocity almost twice that of the initial velocity of the massive object. (Think of a bowling ball hitting a pin.)

For $m_1 \ll m_2$ (light incoming ball): $v_1 \approx -v_{1o}$ and $v_2 \approx 0$

That is, if a light (small mass) object elastically collides with a massive stationary one, the massive object remains *almost* stationary and the light object recoils backward with approximately the same speed that it had before collision. (Think of a ping-pong ball hitting a bowling ball.)

EXAMPLE 6.10: ELASTIC COLLISION – CONSERVATION OF MOMENTUM AND KINETIC ENERGY

A 0.30-kg ball with a speed of 2.0 m/s in +x-direction has a head-on elastic collision with a stationary 0.70-kg ball. What are the velocities of the balls after collision?

THINKING IT THROUGH. The incoming ball is less massive than the stationary one, so it might be expected that the objects separate in opposite directions after collision, with the less massive ball recoiling from the more massive one. Equations 6.14 and 6.15 can be used to find the velocities with $v_{2_0} = 0$.

SOLUTION

Using the previous notation in listing the data,

Given:

$m_1 = 0.30$ kg and $v_{1_0} = 2.0$ m/s
$m_2 = 0.70$ kg and $v_{2_0} = 0$

Find: v_1 and v_2

Directly from Equations 6.13 and 6.14, the velocities after collision are

$$v_1 = \left(\frac{m_1 - m_2}{m_1 + m_2}\right)v_{1_0} = \left(\frac{0.30\,\text{kg} - 0.70\,\text{kg}}{0.30\,\text{kg} + 0.70\,\text{kg}}\right)(2.0\,\text{m/s}) = -0.80\,\text{m/s}$$

$$v_2 = \left(\frac{2m_1}{m_1 + m_2}\right)v_{1_0} = \left[\frac{2(0.30\,\text{kg})}{0.30\,\text{kg} + 0.70\,\text{kg}}\right](2.0\,\text{m/s}) = 1.2\,\text{m/s}$$

FOLLOW-UP EXERCISE. What would be the separation distance of the two objects 2.5 s after collision?

6.4.4 Two Colliding Objects, Both Initially Moving

Now let's look at some examples where both terms in Equations 6.14 and 6.15 are needed.

EXAMPLE 6.11: COLLISIONS – OVERTAKING AND COMING TOGETHER

The precollision conditions for two elastic collisions are shown in ▶ **Figure 6.14**. What are the final velocities in each case?

THINKING IT THROUGH. These collisions are direct applications of Equations 6.14 and 6.15. Notice that in (a) the 4.0-kg object will overtake and collide with the 1.0-kg object.

SOLUTION

Listing the data from the figure with the +x-direction taken to the right:

Given:

(a) $m_1 = 4.0$ kg $v_{1_0} = 10$ m/s
    $m_2 = 1.0$ kg $v_{2_0} = 5.0$ m/s

(b) $m_1 = 2.0$ kg $v_{1_0} = 6.0$ m/s
    $m_2 = 4.0$ kg $v_{2_0} = -6.0$ m/s

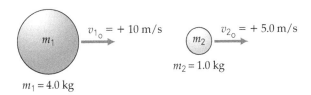

(a)

(b)

▲ **FIGURE 6.14**   **Collisions** (a) Overtaking and (b) coming together. See Example 6.11 for description.

Find: v_1 and v_2 (velocities after collision)

Then, substituting into the collision equations,

(a) Equation 6.14:

$$v_1 = \left(\frac{m_1 - m_2}{m_1 + m_2}\right)v_{1_0} + \left(\frac{2m_2}{m_1 + m_2}\right)v_{2_0}$$

$$= \left(\frac{4.0\,\text{kg} - 1.0\,\text{kg}}{4.0\,\text{kg} + 1.0\,\text{kg}}\right)(10\,\text{m/s}) + \left(\frac{2(1.0\,\text{kg})}{4.0\,\text{kg} + 1.0\,\text{kg}}\right)(5.0\,\text{m/s})$$

$$= 3/5(10\,\text{m/s}) + 2/5(5.0\,\text{m/s}) = 8.0\,\text{m/s}$$

Similarly, Equation 6.15 gives:

$$v_2 = 13\,\text{m/s}$$

So the more massive object overtakes and collides with the less massive object, transferring momentum (increasing velocity).

(b) Applying the collision equations for this situation (Equation 6.14):

$$v_1 = \left(\frac{2.0\,\text{kg} - 4.0\,\text{kg}}{2.0\,\text{kg} + 4.0\,\text{kg}}\right)(6.0\,\text{m/s}) + \left(\frac{2(4.0\,\text{kg})}{2.0\,\text{kg} + 4.0\,\text{kg}}\right)(-6.0\,\text{m/s})$$

$$= -1/3(6.0\,\text{m/s}) + 4/3(-6.0\,\text{m/s}) = -10\,\text{m/s}$$

Similarly, Equation 6.15 gives

$$v_2 = 2.0 \text{ m/s}$$

Here, the less massive object goes in the opposite (negative) direction after collision, with a greater momentum obtained from the more massive object.

FOLLOW-UP EXERCISE. Show that in parts (a) and (b) of this Example, the magnitude of momentum gained by one object is the same as that lost by the other.

INTEGRATED EXAMPLE 6.12: EQUAL AND OPPOSITE

Two balls of equal mass, with equal but opposite velocities, approach each other for a head-on elastic collision. (a) After collision, the balls will (1) move off stuck together, (2) both be at rest, (3) move off in the same direction, or (4) recoil in opposite directions. (b) Prove your answer explicitly.

(a) **CONCEPTUAL REASONING.** Make a sketch of the situation. Then, looking at the choices, (1) is eliminated because if they stuck together it would be an inelastic collision. If they come to rest after collision, momentum would be conserved (why?), but not kinetic energy, so (2) is not applicable for an elastic collision. If they both moved off in the same direction after collision, (3), the momentum would not be conserved (zero before, nonzero after). The answer must be (4). This is the only option by which momentum and kinetic energy could be conserved. To maintain the zero momentum before collision, the objects would have to recoil in opposite directions with the same speeds as before collision.

(b) **QUANTITATIVE REASONING AND SOLUTION.** To explicitly show that (4) is correct, Equations 6.13 and 6.14 may be used. Since no numerical values are given, we work with symbols.

Given: $m_1 = m_2 = m$ (taking m_1 to be initially traveling in $+x$-direction) v_{1o} and $-v_{2o}$ (with equal speeds)

Find: v_1 and v_2

Then, substituting into Equations 6.14 and 6.15, without writing the equations out [see part (a) in Example 6.11],

$$v_1 = \left(\frac{0}{2m}\right)v_{1o} + \left(\frac{2m}{2m}\right)(-v_{2o}) = -v_{2o}$$

and

$$v_2 = \left(\frac{2m}{2m}\right)v_{1o} + \left(\frac{0}{2m}\right)(-v_{2o}) = v_{1o}$$

From the results, it can be seen that after collision the balls recoil in opposite directions.

FOLLOW-UP EXERCISE. Show that momentum and kinetic energy are conserved in this Example.

CONCEPTUAL EXAMPLE 6.13: TWO IN, ONE OUT?

A novelty collision device, as shown in ▶ **Figure 6.15**, consists of five identical metal balls. When one ball swings in, after multiple collisions, one ball swings out at the other end of the row of balls. When two balls swing in, two swing out; when three swing in, three swing out, and so on – always the same number out as in.

Suppose that two balls, each of mass m, swing in at velocity v and collide with the next ball. Why doesn't one ball swing out at the other end with a velocity $2v$?

(a)

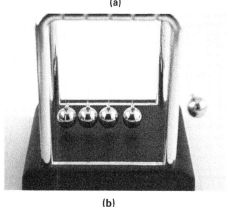

(b)

▲ FIGURE 6.15   **One in, one out** See Example 6.13 for description.

REASONING AND ANSWER. The collisions along the horizontal row of balls are approximately elastic. The case of two balls swinging in and one ball swinging out with twice the velocity wouldn't violate the conservation of momentum: $(2m)v = m(2v)$. However, another condition applies if we assume elastic collisions – the conservation of kinetic energy. Let's check to see if this condition is upheld for this case:

$$\begin{array}{cc} \text{before} & \text{after} \\ \multicolumn{2}{c}{K_i = K_f} \end{array}$$

$$\frac{1}{2}(2m)v^2 \overset{?}{=} \frac{1}{2}m(2v)^2$$

$$mv^2 \neq 2mv^2$$

Hence, the kinetic energy would *not* be conserved if this happened, and the equation is telling us that this situation violates established physical principles and does not occur. Note that there's a big violation – more energy out than in.

FOLLOW-UP EXERCISE. Suppose the first ball of mass m were replaced with a ball of mass $2m$. When this ball is pulled back and allowed to swing in, how many balls will swing out? [*Hint:* Think about the analogous situation for the first two balls as in Figure 6.14a, and remember that the balls in the row are actually colliding. It may help to think of them as being separated.]

6.5 Center of Mass

The conservation of total momentum gives a method of analyzing a "system of particles." Such a system may be virtually anything – for example, a volume of gas, water in a container, or a baseball. Another important concept, the center of mass, allows us to analyze the overall motion of a system of particles. It involves representing the whole system as a single particle or point mass. This concept will be introduced here and applied in more detail in the upcoming chapters.

It has been seen that if no net external force acts on a particle, the particle's linear momentum is constant. Similarly, if no net external force acts on a *system* of particles, the linear momentum of the *system* is constant. This similarity implies that a system of particles might be represented by an *equivalent* single particle. Moving rigid objects, such as balls, automobiles, and so forth, are essentially systems of particles and can be effectively represented by equivalent single particles when analyzing motion. Such representation is done through the concept of the **center of mass (CM)**:

The center of mass is the point at which all of the mass of an object or system may be considered to be concentrated, for the purposes of describing its linear or translational motion only.

Even if a rigid object is rotating, an important result (beyond the scope of this text to derive) is that the center of mass still moves as though it were a particle. The center of mass is sometimes described as the *balance point* of a solid object. For example, if you balance a meterstick on your finger, the center of mass of the stick is located directly above your finger, and all of the mass (or weight) seems to be concentrated there. This also applies to the gymnast in ▼ **Figure 6.16**.

▲ FIGURE 6.16 **Center of mass** The center of mass of this gymnast is on a line directly above the hand on the balance beam.

An expression similar to Newton's second law for a single particle applies to a *system* when the center of mass is used:

$$\vec{F}_{net} = M\vec{A}_{CM} \quad \text{(Newton's second law for a system)} \quad (6.16)$$

Here, $\vec{F}_{net}$ is the net *external* force on the system, M is the total mass of the system which the sum of the masses of the particles of the system ($M = m_1 + m_2 + m_3 + \cdots + m_n$, where the system has n particles), and $\vec{A}_{CM}$ is the acceleration of the center of mass of the system. In words, Equation 6.16 says that the *center of mass* of a system of particles moves as though all the mass of the system were concentrated there and acted on by the resultant of the external forces. Note that the movement of the individual parts of the system is *not* predicted by Equation 6.16.

It follows that *if the net external force on a system is zero,* the total linear momentum of the center of mass is conserved (i.e., it stays constant), because

$$\vec{F}_{net} = M\vec{A}_{CM} = M\left(\frac{\Delta\vec{V}_{CM}}{\Delta t}\right) = \frac{\Delta(M\vec{V}_{CM})}{\Delta t} = \frac{\Delta\vec{P}_{CM}}{\Delta t} = 0 \quad (6.17)$$

Then, $\Delta\vec{P}_{CM}/\Delta t = 0$, which means that there is no change in $\vec{P}_{CM}$ during a time Δt, or the total momentum of the system, $\vec{P}_{CM} = M\vec{V}_{CM}$ is constant (but not necessarily zero). Since M is constant (why?), $\vec{V}_{CM}$ is a constant in this case. Thus, the center of mass either moves with a constant velocity or remains at rest.

Although you may more readily visualize the center of mass of a solid object, the concept of the center of mass applies to any system of particles or objects, even a quantity of gas. For a system of n particles arranged in one dimension, along the x-axis (▼ **Figure 6.17**), the location of the center of mass is given by

$$X_{CM} = \frac{m_1X_1 + m_2X_2 + m_3X_3 + \cdots + m_nX_n}{m_1 + m_2 + m_3 + \cdots + m_n} \quad (6.18)$$

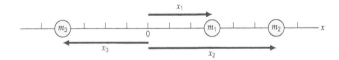

▲ FIGURE 6.17 **System of particles in one dimension** Where is the system's center of mass? See Example 6.14 for description.

That is, X_{CM} is the x-coordinate of the center of mass (CM) of a system of particles. In shorthand notation (using signs to indicate vector directions in one dimension), this relationship is expressed as

$$X_{CM} = \frac{\Sigma m_i x_i}{M} \quad (6.19)$$

where Σ stands for the summation of the products $m_i\,x_i$ for n particles ($i = 1, 2, 3\ldots, n$). If $\Sigma m_i\,x_i = 0$, then $X_{CM} = 0$, and the center of mass of the one-dimensional system is located at the origin.

Other coordinates of the center of mass for systems of particles are similarly defined. For a two-dimensional distribution of masses, the coordinates of the center of mass are $(X_{\text{CM}}, Y_{\text{CM}})$.

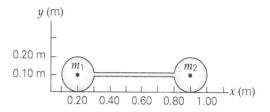

▲ FIGURE 6.18   **Location of the center of mass** See Example 6.15 for description.

EXAMPLE 6.14: FINDING THE CENTER OF MASS – A SUMMATION PROCESS

Three masses – 2.0 kg, 3.0 kg, and 6.0 kg – are located at positions (3.0, 0), (6.0, 0), and (−4.0, 0), respectively, in meters from the origin (Figure 6.17). Where is the center of mass of this system?

THINKING IT THROUGH. Since all $y_i = 0$, obviously $Y_{\text{CM}} = 0$, and the CM lies somewhere on the x-axis. The masses and the positions are given, so we can use Equation 6.19 to calculate X_{CM} directly. However, keep in mind that the positions are located by vector displacements from the origin and are indicated in one dimension by the appropriate signs (+ or −).

SOLUTION

Listing the data:

Given:

$m_1 = 2.0$ kg, $x_1 = 3.0$ m
$m_2 = 3.0$ kg, $x_2 = 6.0$ m
$m_3 = 6.0$ kg, $x_3 = -4.0$ m

Find: X_{CM} (CM coordinate)

Then, performing the summation as indicated in Equation 6.19 yields,

$$X_{\text{CM}} = \frac{\Sigma m_i x_i}{M}$$
$$= \frac{(2.0\,\text{kg})(3.0\,\text{m}) + (3.0\,\text{kg})(6.0\,\text{m}) + (6.0\,\text{kg})(-4.0\,\text{m})}{2.0\,\text{kg} + 3.0\,\text{kg} + 6.0\,\text{kg}} = 0$$

The center of mass is at the origin.

FOLLOW-UP EXERCISE. At what position should a fourth mass of 8.0 kg be added so the CM is at $X = +1.0$ m?

EXAMPLE 6.15: A DUMBBELL –CENTER OF MASS REVISITED

A dumbbell (▶ **Figure 6.18**) has a connecting bar of negligible mass. Find the location of the center of mass (a) if m_1 and m_2 are each 5.0 kg and (b) if m_1 is 5.0 kg and m_2 is 10.0 kg.

THINKING IT THROUGH. This Example shows how the location of the center of mass depends on the distribution of mass. In part (b), you might expect the center of mass to be located closer to the more massive end of the dumbbell.

SOLUTION

Listing the data, with the coordinates to be used in Equation 6.19,

Given:

$x_1 = 0.20$ m, $x_2 = 0.90$ m
$y_1 = y_2 = 0.10$ m

(a) $m_1 = m_2 = 5.0$ kg
(b) $m_1 = 5.0$ kg, $m_2 = 10.0$ kg

Find:

(a) $(X_{\text{CM}}, Y_{\text{CM}})$, with $m_1 = m_2$
(b) $(X_{\text{CM}}, Y_{\text{CM}})$, with $m_1 \neq m_2$

Note that each mass is considered to be a particle located at the center of the sphere (its center of mass).

(a) X_{CM} is given by a two-term sum.

$$X_{\text{CM}} = \frac{m_1 x_1 + m_2 x_2}{m_1 + m_2} = \frac{(5.0\,\text{kg})(0.20\,\text{m}) + (5.0\,\text{kg})(0.90\,\text{m})}{5.0\,\text{kg} + 5.0\,\text{kg}} = 0.55\,\text{m}$$

Similarly, $Y_{\text{CM}} = 0.10$ m, as you can prove for yourself. (You might have seen this right away, since each center of mass is at this height.) The center of mass of the dumbbell is then located at $(X_{\text{CM}}, Y_{\text{CM}}) = (0.55$ m, 0.10 m), or midway between the end masses.

(b) With $m_2 = 10.0$ kg,

$$X_{\text{CM}} = \frac{m_1 x_1 + m_2 x_2}{m_1 + m_2}$$
$$= \frac{(5.0\,\text{kg})(0.20\,\text{m}) + (10.0\,\text{kg})(0.90\,\text{m})}{5.0\,\text{kg} + 10.0\,\text{kg}} = 0.67\,\text{m}$$

which is two-thirds of the way between the masses. (Note that the distance of the CM from the center of m_1 is $\Delta x = 0.67$ m − 0.20 m = 0.47 m. With the distance $L = 0.70$ m between the centers of the masses, $\Delta x/L = 0.47$ m/0.70 m = 0.67, or 2/3.) You might expect the balance point of the dumbbell in this case to be closer to m_2, and it is. The y-coordinate of the center of mass is again $Y_{\text{CM}} = 0.10$ m.

FOLLOW-UP EXERCISE. In part (b) of this Example, take the origin of the coordinate axes to be at the point where m_1 touches the x-axis. What are the coordinates of the CM in this case, and how does its location compare with that found in the Example?

In Example 6.15, when the value of one of the masses changed, the x-coordinate of the center of mass changed. However, the centers of the end masses were still at the same height, and Y_{CM} remained the same. To increase Y_{CM}, one or both of the end masses would have to be in a higher position.

Now let's see how the concept of the center of mass can be applied to a realistic situation.

INTEGRATED EXAMPLE 6.16: INTERNAL MOTION – WHERE'S THE CENTER OF MASS AND THE MAN?

A 75.0-kg man stands in the far end of a 50.0-kg boat 100 m from the shore, as illustrated in ▼ **Figure 6.19**. If he walks to the other end of the 6.00-m-long boat, (a) does the CM (1) move to the right, (2) move to the left, or (3) remain stationary? Neglect friction and assume the CM of the boat is at its midpoint. (b) After walking to the other end of the boat, how far is he from the shore?

(a) **CONCEPTUAL REASONING.** With no net external force, the acceleration of the center of mass of the man-boat system is zero (Equation 6.18), and so is the total momentum by Equation 6.17 $\left(\vec{\mathbf{P}}_{CM} = M\vec{\mathbf{V}}_{CM} = 0\right)$. Hence, the velocity of the center of mass of the system is zero, or the center of mass is stationary and remains so to conserve system momentum; that is, X_{CM_i} (initial) $= X_{CM_f}$ (final), so the answer is (3).

(b) **QUANTITATIVE REASONING AND SOLUTION.** The answer is *not* 100 m − 6.00 m = 94.0 m, because the boat moves as the man walks. Why? The positions of the masses of the man and the boat determine the location of the CM of the system, both before and after the man walks. Since the CM does not move, $X_{CM_i} = X_{CM_f}$. Using this fact and finding the value of X_{CM_i}, this value can be used in the calculation of X_{CM_f}, which will contain the unknown we are looking for.

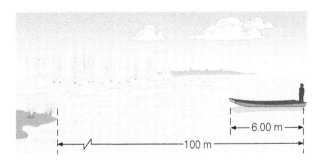

▲ **FIGURE 6.19 Walking toward shore** See Example 6.16 for description.

Taking the shore as the origin ($x = 0$),

Given:

$m_m = 75.0$ kg
$x_{m_i} = 100$ m
$m_b = 50.0$ kg
$x_{b_i} = 94.0$ m + 3.00 m = 97.0 m (CM position of the boat)

Find: x_{m_f} (distance of man from shore)

Note that if we take the man's final position to be a distance x_{m_f} from the shore, then the final position of the boat's center of mass will be $x_{b_f} = x_{m_f} + 3.00$ m, since the man will be at the front of the boat, 3.00 m from its CM, but on the other side. Then initially,

$$X_{CM_i} = \frac{m_m x_{m_i} + m_b x_{b_i}}{m_m + m_b}$$

$$= \frac{(75.0\,\text{kg})(100\,\text{m}) + (50.0\,\text{kg})(97.0\,\text{m})}{75.0\,\text{kg} + 50.0\,\text{kg}} = 98.8\,\text{m}$$

Finally, the CM must be at the same location, since $V_{CM} = 0$. Then from Equation 6.19,

$$X_{CM_f} = \frac{m_m x_{m_f} + m_b x_{b_f}}{m_m + m_b}$$

$$= \frac{(75.0\,\text{kg})x_{m_f} + (50.0\,\text{kg})(x_{m_f} + 3.00\,\text{m})}{75.0\,\text{kg} + 50.0\,\text{kg}} = 98.8\,\text{m}$$

Here, $X_{CM_f} = 98.8$ m $= X_{CM_i}$, since the CM does not move. Then, solving for x_{m_f},

$$(125\,\text{kg})(98.8\,\text{m}) = (125\,\text{kg})x_{m_f} + (50.0\,\text{kg})(3.00\,\text{m})$$

and

$$x_{m_f} = 97.6\,\text{m}$$

from the shore.

FOLLOW-UP EXERCISE. Suppose the man then walks back to his original position at the opposite end of the boat. Would he then be 100 m from shore again?

6.5.1 Center of Gravity

As you know, mass and weight are related. Closely associated with the concept of the center of mass is the concept of the **center of gravity (CG)**, the point where all of the *weight* of an object may be considered to be concentrated when the object is represented as a particle. If the acceleration due to gravity is constant in both magnitude and direction over the extent of the object, Equation 6.20 can be rewritten as (with all $g_i = g$),

$$M_g X_{CM} = \Sigma m_i g x_i \qquad (6.20)$$

Then, the object's weight M_g acts as though its mass were concentrated at X_{CM}, and the center of mass and the center of gravity coincide. As you may have noticed, the location of the center of gravity was implied in some previous figures in Chapter 4, where the vector arrows for weight ($w = mg$) were drawn from a point at or near the center of an object.

For practical purposes, the center of gravity is usually considered to coincide with the center of mass. That is, the acceleration due to gravity is constant for all parts of the object. (Note the constant g in Equation 6.20.) There would be a difference in the

locations of the two points if an object were so large that the acceleration due to gravity was different at different parts of the object.

In some cases, the center of mass or the center of gravity of an object may be located by symmetry. For example, for a spherical object that is homogeneous (i.e., the mass is distributed evenly throughout), the center of mass is at the geometrical center (-or center of symmetry). In Example 6.15a, where the end masses of the dumbbell were equal, it was probably apparent that the center of mass was midway between them.

The location of the center of mass or center of gravity of an irregularly shaped object is not so evident and is usually difficult to calculate (even with advanced mathematical methods that are beyond the scope of this book). In some instances, the center of mass may be located experimentally. For example, the center of mass of a flat, irregularly shaped object can be determined experimentally by suspending it freely from different points (▼ **Figure 6.20**). A moment's thought should convince you that the center of mass (or center of gravity) always lies vertically below the point of suspension. Since the center of mass is defined as the point at which all the mass of a body can be considered to be concentrated, this is analogous to a particle of mass suspended from a string. Suspending the object from two or more points and marking the vertical lines on which the center of mass must lie locates the center of mass as the intersection of the lines.

The center of mass (or center of gravity) of an object may not lie in the body of the object (▼ **Figure 6.21**). For example, the center of mass of a homogeneous ring is at the ring's center. The mass in any section of the ring is compensated for by the mass in an equivalent section directly across the ring, and by symmetry, the center of mass is at the center of the ring. For an L-shaped object with uniform legs, equal in mass and length, the center of mass lies on a line that makes a 45° angle with both legs. Its location can easily be determined by suspending the L from a point on one of the legs and noting where a vertical line from that point intersects the diagonal line.

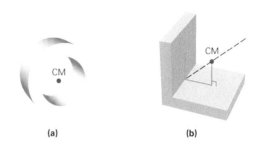

(a) **(b)**

▲ **FIGURE 6.21** **The center of mass may be located outside a body** The center of mass (and center of gravity) may lie either inside or outside a body, depending on the distribution of that object's mass. **(a)** For a uniform ring, the center of mass is at the center of the ring. **(b)** For an L-shaped object, if the mass distribution is uniform and the legs are of equal length, the center of mass lies on the diagonal between the legs.

In the high jump, the location of CG is very important. Jumping raises the CG. It takes energy to do this, and the higher the jump, the more energy it takes. Therefore, a high jumper wants to clear the bar while keeping her CG low. A jumper will try to keep her CG as close to the bar as possible when passing over it. In the "Fosbury flop" style (made famous by Dick Fosbury in the 1968 Olympics), the jumper arches her body backward over the bar (▼ **Figure 6.22**). With the legs, head, and arms below the bar, the

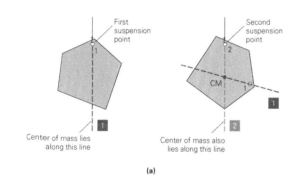

(a)

(b)

▲ **FIGURE 6.20** **Location of the center of mass by suspension** **(a)** The center of mass of a flat, irregularly shaped object can be found by suspending the object from two or more points. The CM (and CG) lies on a vertical line under any point of suspension, so the intersection of two such lines marks its location midway through the thickness of the body. The sheet could be balanced horizontally at this point. Why? **(b)** The process is illustrated with a cutout map of the continental United States. Note that a plumb line dropped from any other point (third photo) does in fact pass through the CM as located in the first two photos.

▲ **FIGURE 6.22** **Center of gravity** By arching her body backward over the bar, the high jumper lowers her center of gravity. See text for description.

CG is lower than in the "layout" style, where the body is nearly parallel to the ground when going over the bar. With the "flop," a jumper may be able to make her CG (which is outside the body) pass underneath the bar while successfully clearing the bar.

6.6 Jet Propulsion and Rockets

The word *jet* is sometimes used to refer to a stream of liquid or gas emitted at a high speed – for example, a jet of water from a fountain or a jet of air from an automobile tire. **Jet propulsion** is the application of such jets to the production of motion. This concept usually brings to mind jet planes and rockets, but squid and octopi propel themselves by squirting jets of water (▼ **Figure 6.23**).

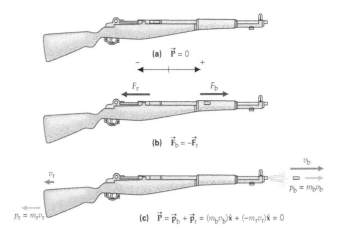

▲ **FIGURE 6.24 Conservation of momentum (a)** Before the rifle is fired, the total momentum of the rifle and bullet (as an isolated system) is zero. **(b)** During firing, there are equal and opposite internal forces, and the instantaneous total momentum of the rifle-bullet system remains zero (neglecting external forces, such as those that arise when a rifle is being held). **(c)** When the bullet leaves the barrel, the total momentum of the system is still zero. (The vector equation is written in boldface [vector] notation and then in sign-magnitude notation so as to indicate directions.)

▲ **FIGURE 6.23 Jet propulsion** Squid and octopi propel themselves by squirting jets of water. Shown here is a Giant Octopus jetting away.

You have probably tried the simple application of blowing up a balloon and releasing it. Lacking any guidance or rigid exhaust system, the balloon zigzags around, driven by the escaping air. In terms of Newton's third law, the air is forced out by the contraction of the stretched balloon – that is, the balloon exerts a force on the air. Thus, there must be an equal and opposite reaction force exerted by the air on the balloon. It is this force that propels the balloon on its erratic path.

Jet propulsion is explained by Newton's third law, and in the absence of external forces, the conservation of momentum also applies. You may understand this concept better by considering the recoil of a rifle, taking the rifle and the bullet as an isolated system (▶ **Figure 6.24**).

Initially, the total momentum of this system is zero. When the rifle is fired (by remote control to avoid external forces), the expansion of the gases from the exploding charge accelerates the bullet down the barrel. These gases push backward on the rifle as well, producing a recoil force (the "kick" experienced by a person firing a weapon). Since the initial momentum of the system is zero and the force of the expanding gas is an internal force, the momenta of the bullet and of the rifle must be equal and opposite at any instant. After the bullet leaves the barrel, there is no propelling force, so the

bullet and the rifle move with constant velocities (unless acted on by a net external force such as gravity or air resistance).

Similarly, the thrust of a rocket is created by exhausting the gas from burning fuel out the rear of the rocket. The expanding gas exerts a force on the rocket that propels the rocket in the forward direction (▶ **Figure 6.25**). The rocket exerts a reaction force on the gas, so the gas is directed out the exhaust nozzle. If the rocket is at rest when the engines are turned on and there are no external forces (as in deep space, where friction is zero and gravitational forces are negligible), then the instantaneous momentum of the exhaust gas is equal and opposite to that of the rocket. The numerous exhaust gas molecules have small masses and high velocities, and the rocket has a much larger mass and a smaller velocity.

Unlike a rifle firing a single shot, a rocket continuously loses mass when burning fuel. (The rocket is more like a machine gun.) Thus, the rocket is a system for which the mass is not constant. As the mass of the rocket decreases, it accelerates more easily. Multistage rockets take advantage of this fact. The hull of a burnt-out stage is jettisoned to give a further in-flight reduction in mass (Figure 6.25c). The payload (cargo) is typically a very small part of the initial mass of rockets for space flights.

Suppose that the purpose of a space flight is to land a payload on the Moon. At some point on the journey, the gravitational attraction of the Moon will become greater than that of the Earth, and the spacecraft will accelerate toward the Moon. A soft landing is desirable, so the spacecraft must be slowed down enough to go into orbit around the Moon or land on it. This slowing down is accomplished by using the rocket engines to apply a *reverse thrust*, or braking thrust. The spacecraft is maneuvered through a 180° angle, or turned around, which is quite easy to do

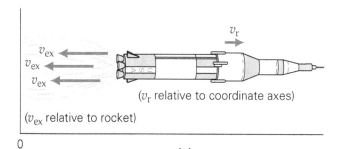

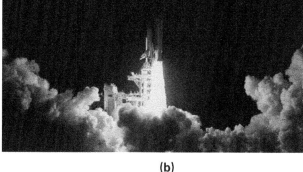

(a)

(b)

(c)

▲ FIGURE 6.25 **Jet propulsion and mass reduction (a)** A rocket burning fuel is continuously losing mass and thus becomes easier to accelerate. The resulting force on the rocket (the thrust) depends on the product of the rate of change of its mass with time and the velocity of the exhaust gases: $(\Delta m / \Delta t)\vec{\mathbf{v}}_{ex}$. Since the mass is decreasing, $\Delta m/\Delta t$ is negative, and the thrust is opposite $\vec{\mathbf{v}}_{ex}$. **(b)** The space shuttle uses a multistage rocket. Both of the two booster rockets and the huge external fuel tank are jettisoned in flight. **(c)** The multistage rockets of *Saturn V*.

in space. The rocket engines are then fired, expelling the exhaust gas toward the Moon and supplying a braking action. That is, the force on the rocket is opposite its velocity.

You have experienced a reverse thrust effect if you have flown in a commercial jet. In this instance, however, the craft is not turned around. Instead, after touchdown, the jet engines are revved up, and a braking action can be felt. Ordinarily, revving up the engines accelerates the plane forward. The reverse thrust is accomplished by activating thrust reversers in the engines that deflect the exhaust gases forward (▼ **Figure 6.26**). The gas experiences an impulse force and a change in momentum in the forward direction (see Figure 6.3b), and the engine and the aircraft have an equal and opposite momentum change, thus experiencing a braking impulse force.

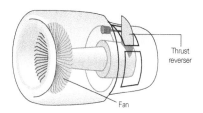

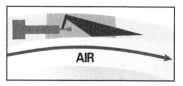

Normal operation

Thrust reverser activated

▲ FIGURE 6.26 **Reverse thrust** Thrust reversers are activated on jet engines during landing to help slow the plane. The gas experiences an impulse force and a change in momentum in the forward direction, and the plane experiences an equal and opposite momentum change and a braking impulse force.

Chapter 6 Review

- The **linear momentum ($\vec{\mathbf{p}}$)** of a particle is a vector and is defined as the product of mass and velocity:

$$\vec{\mathbf{p}} = m\vec{\mathbf{v}} \qquad (6.1)$$

- The **total linear momentum ($\vec{\mathbf{P}}$)** of a system is the vector sum of the momenta of the individual particles:

$$\vec{\mathbf{P}} = \vec{\mathbf{p}}_1 + \vec{\mathbf{p}}_2 + \vec{\mathbf{p}}_3 + \cdots + = \sum \vec{\mathbf{p}}_i \qquad (6.2)$$

- **Newton's second law in terms of momentum (for a particle):**

$$\vec{\mathbf{F}}_{net} = \frac{\Delta \vec{\mathbf{p}}}{\Delta t} \qquad (6.3)$$

- The **impulse-momentum theorem** relates the impulse acting on an object to its change in momentum:

$$\text{Impulse} = \vec{F}_{\text{avg}} \Delta t = \Delta \vec{p} = m\vec{v} - m\vec{v}_{\text{o}} \qquad (6.5)$$

- **Conservation of linear momentum:** In the absence of a net external force, the total linear momentum of a system is conserved:

$$\vec{P} = \vec{P}_{\text{o}} \qquad (6.7)$$

- **In an elastic collision, the total kinetic energy of the system is conserved.**
- **Momentum is conserved in both elastic and inelastic collisions.** In a completely inelastic collision, objects stick together after impact.
- **Conditions for an elastic collision:**

$$\vec{P}_{\text{f}} = \vec{P}_{\text{i}} \qquad (6.8)$$
$$K_{\text{f}} = K_{\text{i}}$$

- **Conditions for an inelastic collision:**

$$\vec{P}_{\text{f}} = \vec{P}_{\text{i}} \qquad (6.9)$$
$$K_{\text{f}} < K_{\text{i}}$$

- **Final velocity in a head-on, two-body completely inelastic collision** ($v_{2_{\text{o}}} = 0$):

$$v = \left(\frac{m_1}{m_1 + m_2} \right) v_{1_{\text{o}}} \qquad (6.10)$$

- **Ratio of kinetic energies in a head-on, two-body completely inelastic collision** ($v_{2_{\text{o}}} = 0$):

$$\frac{K_{\text{f}}}{K_{\text{i}}} = \frac{m_1}{m_1 + m_2} \qquad (6.11)$$

- **Final velocities in a head-on, two-body elastic collision:**

$$v_1 = \left(\frac{m_1 - m_2}{m_1 + m_2} \right) v_{1_{\text{o}}} + \left(\frac{2m_2}{m_1 + m_2} \right) v_{2_{\text{o}}} \qquad (6.14)$$

$$v_2 = \left(\frac{2m_1}{m_1 + m_2} \right) v_{1_{\text{o}}} - \left(\frac{m_1 - m_2}{m_1 + m_2} \right) v_{2_{\text{o}}} \qquad (6.15)$$

- The **center of mass** is the point at which all of the mass of an object or system may be considered to be concentrated. The center of mass does not necessarily lie within an object. (The **center of gravity** is the point at which all the weight may be considered to be concentrated.)

- **Coordinates of the center of mass** (using signs for directions):

$$X_{\text{CM}} = \frac{\Sigma m_i x_i}{M} \qquad (6.19)$$

End of Chapter Questions and Exercises

Multiple Choice Questions

6.1 Linear Momentum

1. Linear momentum has units of (a) N/m, (b) kg·m/s, (c) N/s, (d) all of the preceding.
2. Linear momentum is (a) unrelated to mass, (b) a scalar quantity, (c) a vector quantity, (d) unrelated to force.
3. A net force on an object can cause (a) an acceleration, (b) a change in momentum, (c) a change in velocity, (d) all of the preceding.
4. A change in momentum requires which of the following: (a) an unbalanced force, (b) a change in velocity, (c) an acceleration, or (d) any of these?

6.2 Impulse

5. Impulse has units (a) of kg·m/s, (b) of N·s, (c) the same as momentum, (d) all of the preceding.
6. Impulse is equal to (a) $F\Delta x$, (b) the change in kinetic energy, (c) the change in momentum, (d) $\Delta p/\Delta t$.
7. Impulse (a) is the time rate of change of momentum, (b) is the force per unit time, (c) has the same units as momentum, (d) none of these.

6.3 Conservation of Linear Momentum

8. The conservation of linear momentum is described by (a) the momentum-impulse theorem, (b) the work-energy theorem, (c) Newton's first law, (d) conservation of energy.
9. The linear momentum of an object is conserved if (a) the force acting on the object is conservative, (b) a single, unbalanced internal force is acting on the object, (c) the mechanical energy is conserved, (d) none of the preceding.
10. Internal forces do not affect the conservation of momentum because (a) they cancel each other, (b) their effects are canceled by external forces, (c) they can never produce a change in velocity, (d) Newton's second law is not applicable to them.

6.4 Elastic and Inelastic Collisions

11. Which of the following is *not* conserved in an inelastic collision: (a) momentum, (b) mass, (c) kinetic energy, or (d) total energy?
12. A rubber ball of mass m traveling horizontally with a speed v hits a wall and bounces back with the same speed. The magnitude of the change in momentum is (a) 0, (b) mv, (c) $mv/2$, (d) $2mv$.

13. In a head-on elastic collision, a mass m_1 strikes a stationary mass m_2. There is a complete transfer of energy if (a) $m_1 = m_2$, (b) $m_1 \gg m_2$, (c) $m_1 \ll m_2$, (d) the masses stick together.

14. The condition for a two-object inelastic collision is (a) $K_f < K_i$, (b) $p_i \neq p_f$, (c) $m_1 > m_2$, (d) $v_1 < v_2$.

6.5 Center of Mass

15. The center of mass of an object (a) always lies at the center of the object, (b) is at the location of the most massive particle in the object, (c) always lies within the object, (d) none of the preceding.

16. The center of mass and center of gravity coincide (a) if the acceleration due to gravity is constant, (b) if momentum is conserved, (c) if momentum is not conserved, (d) only for irregularly shaped objects.

Conceptual Questions

6.1 Linear Momentum

1. In a football game, does a fast-running running back always have more linear momentum than a slow-moving, more massive lineman? Explain.

2. Two objects have the same momentum. Do they necessarily have the same kinetic energy? Explain.

3. Two objects have the same kinetic energy. Do they necessarily have the same momentum? Explain.

6.2 Impulse

4. "Follow-through" is very important in many sports, such as in serving a tennis ball. Explain how follow-through can increase the speed of the tennis ball when it is served.

5. A karate student tries *not* to follow through in order to break a board, as shown in ▼ **Figure 6.27**. How can the abrupt stop of the hand (with no follow-through) generate so much force?

▲ FIGURE 6.27 **A karate chop** See Conceptual Question 5 and Exercise 18.

6. Explain the difference for each of the following pairs of actions in terms of impulse: (a) a golfer's long drive and a short chip shot; (b) a boxer's jab and a knockout

punch; (c) a baseball player's bunting action and a home-run swing.

7. When jumping from a height to the ground, it is advised to land with the legs bent rather than stiff-legged. Why is this?

6.3 Conservation of Linear Momentum

8. Imagine yourself standing in the middle of a frozen lake. The ice is so smooth that it is frictionless. How could you get to shore? (You couldn't walk. Why?)

9. A stationary object receives a direct hit by another object moving toward it. Is it possible for both objects to be at rest after the collision? Explain.

10. Does the conservation of momentum follow from Newton's third law?

6.4 Elastic and Inelastic Collisions

11. Since $K = p^2/2m$, how can kinetic energy be lost in an inelastic collision while the total momentum is still conserved? Explain.

12. Can all of the kinetic energy be lost in the collision of two objects? Explain.

13. Automobiles used to have firm steel bumpers for safety. Today, auto bumpers are made out of materials that crumple or collapse on sufficient impact. Why is this?

14. Two balls of equal mass collide head on in a completely inelastic collision and come to rest. (a) Is the kinetic energy conserved? (b) Is the momentum conserved? Explain.

6.5 Center of Mass

15. ▼ **Figure 6.28** shows a flamingo standing on one of its two legs, with its other leg lifted. What can you say about the location of the flamingo's center of mass?

▲ FIGURE 6.28 **Delicate balance** See Conceptual Question 15.

16. Two identical objects are located a distance d apart. If one of the objects remains at rest and the other moves away with a constant velocity, what is the effect on the center of mass of the system?

Exercises*

Integrated Exercises (IEs) are two-part exercises. The first part typically requires a conceptual answer choice based on physical thinking and basic principles. The following part is quantitative calculations associated with the conceptual choice made in the first part of the exercise.

6.1 Linear Momentum

1. • If a 60-kg woman is riding in a car traveling at 90 km/h, what is her linear momentum relative to (a) the ground and (b) the car?

2. • The linear momentum of a runner in a 100-m dash is 7.5×10^2 kg·m/s. If the runner's speed is 10 m/s, what is his mass?

3. • Find the magnitude of the linear momentum of (a) a 7.1-kg bowling ball traveling at 12 m/s and (b) a 1200-kg automobile traveling at 90 km/h.

4. • In a football game, a lineman usually has more mass than a running back. (a) Will a lineman always have greater linear momentum than a running back? Why? (b) Who has greater linear momentum, a 75-kg running back running at 8.5 m/s or a 120-kg lineman moving at m/s?

5. •• A 0.150-kg baseball traveling with a horizontal speed of 4.50 m/s is hit by a bat and then moves with a speed of 34.7 m/s in the opposite direction. What is the change in the ball's momentum?

6. •• A 15.0-g rubber bullet hits a wall with a speed of 150 m/s. If the bullet bounces straight back with a speed of 120 m/s, what is the change in momentum of the bullet?

7. •• A 5.0-g bullet with a speed of 200 m/s is fired horizontally into a 0.75-kg wooden block at rest on a figure. If the block containing the bullet slides a distance of 0.20 m before coming to rest, (a) what is the coefficient of kinetic friction between the block and the figure? (b) What fraction of the bullet's energy is dissipated in the collision?

8. •• Two runners of mass 70 kg and 60 kg, respectively, have a total linear momentum of 350 kg·m/s. The heavier runner is running at 2.0 m/s. Determine the possible velocities of the lighter runner.

9. •• A 0.20-kg billiard ball traveling at a speed of 15 m/s strikes the side rail of a pool figure at an angle of 60° (▶ **Figure 6.29**). If the ball rebounds at the same speed and angle, what is the change in its momentum?

10. •• Suppose the billiard ball in Figure 6.29 approaches the rail at a speed of 15 m/s and an angle of 60°, as shown, but rebounds at a speed of 10 m/s and an angle of 50°. What is the change in momentum in this case? [*Hint:* Use components.]

11. •• A loaded tractor-trailer with a total mass of 5000 kg traveling at 3.0 km/h hits a loading dock and comes to

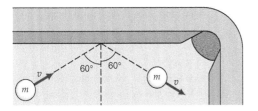

▲ **FIGURE 6.29 Glancing collision** See Exercises 9, 10, and 27.

a stop in 0.64 s. What is the magnitude of the average force exerted on the truck by the dock?

12. •• A 2.0-kg mud ball drops from rest at a height of 15 m. If the impact between the ball and the ground lasts 0.50 s, what is the average net force exerted by the ball on the ground?

13. **IE** •• In football practice, two wide receivers run different pass receiving patterns. One with a mass of 80.0 kg runs at 45° northeast at a speed of 5.00 m/s. The second receiver (mass of 90.0 kg) runs straight down the field (due east) at 6.00 m/s. (a) What is the direction of their total momentum: (1) exactly northeast, (2) to the north of northeast, (3) exactly east, or (4) to the east of northeast? (b) Justify your answer in part (a) by actually computing their total momentum.

14. •• A major league catcher catches a fastball moving at 95.0 mi/h and his hand and glove recoil 10.0 cm in bringing the ball to rest. If it took 0.00470 s to bring the ball (with a mass of 250 g) to rest in the glove, (a) what are the magnitude and direction of the change in momentum of the ball? (b) Find the average force the ball exerts on the hand and glove.

6.2 Impulse

15. • When tossed upward and hit horizontally by a batter, a 0.20-kg softball receives an impulse of 3.0 N·s. With what horizontal speed does the ball move away from the bat?

16. • An automobile with a linear momentum of 3.0×10^4 kg·m/s is brought to a stop in 5.0 s. What is the magnitude of the average braking force?

17. • A pool player imparts an impulse of 3.2 N·s to a stationary 0.25-kg cue ball with a cue stick. What is the speed of the ball just after impact?

18. •• For the karate chop in Figure 6.27, assume that the hand has a mass of 0.35 kg and that the speeds of the hand just before and just after hitting the board are 10 m/s and 0, respectively. What is the average force exerted by the fist on the board if (a) the fist follows through, so the contact time is 3.0 ms, and (b) the fist stops abruptly, so the contact time is only 0.30 ms?

19. **IE** •• A car with a mass of 1500 kg is rolling on a level road at 30.0 m/s. It receives an impulse with a magnitude of 2000 N·s and its speed is reduced. (a) Was this impulse caused by (1) the driver hitting the accelerator,

---

* The bullets denote the degree of difficulty of the exercises: •, simple; ••, medium; and •••, more difficult.

(2) the driver putting on the brakes, or (3) the driver turning the steering wheel? (b) What was the car's speed after the impulse was applied?

20. •• A volleyball is traveling toward you. (a) Which action will require a greater force on the volleyball, your catching the ball or your hitting the ball back? Why? (b) A 0.45-kg volleyball travels with a horizontal velocity of 4.0 m/s over the net. You jump up and hit the ball back with a horizontal velocity of 7.0 m/s. If the contact time is 0.040 s, what was the average force on the ball?

21. •• A boy catches – with bare hands and his arms rigidly extended – a 0.16-kg baseball coming directly toward him at a speed of 25 m/s. He emits an audible "Ouch!" because the ball stings his hands. He learns quickly to move his hands with the ball as he catches it. If the contact time of the collision is increased from 3.5 ms to 8.5 ms in this way, how do the magnitudes of the average impulse forces compare?

22. •• A one-dimensional impulse force acts on a 3.0-kg object as diagrammed in ▼ **Figure 6.30**. Find (a) the magnitude of the impulse given to the object, (b) the magnitude of the average force, and (c) the final speed if the object had an initial speed of 6.0 m/s.

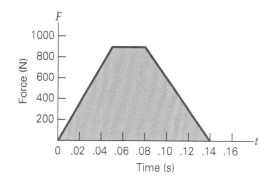

▲ FIGURE 6.30 **Force-versus-time graph** See Exercise 22.

23. •• A 0.45-kg piece of putty is dropped from a height of 2.5 m above a flat surface. When it hits the surface, the putty comes to rest in 0.30 s. What is the average force exerted on the putty by the surface?

24. •• A 50-kg driver sits in her car waiting for the traffic light to change. Another car hits her from behind in a head-on, rear-end collision and her car suddenly receives an acceleration of 16 m/s². If all of this takes place in 0.25 s, (a) what is the impulse on the driver? (b) What is the average force exerted on the driver, and what exerts this force?

25. •• An incoming 0.14-kg baseball has a speed of 45 m/s. The batter hits the ball, giving it a speed of 60 m/s. If the contact time is 0.040 s, what is the average force of the bat on the ball?

26. •• At a shooting competition, a contestant fires and a 12.0-g bullet leaves the rifle with a muzzle speed of 130 m/s. The bullet hits the thick target backing and

stops after traveling 4.00 cm. Assuming a uniform acceleration, (a) what is the impulse on the target? (b) What is the average force on the target?

27. •• If the billiard ball in Figure 6.29 is in contact with the rail for 0.010 s, what is the magnitude of the average force exerted on the ball? (See Exercise 9.)

28. •• A 15 000-N automobile travels at a speed of 45 km/h northward along a street, and a 7 500-N sports car travels at a speed of 60 km/h eastward along an intersecting street. (a) If neither driver brakes and the cars collide at the intersection and lock bumpers, what will the velocity of the cars be immediately after the collision? (b) What percentage of the initial kinetic energy will be lost in the collision?

6.3 Conservation of Linear Momentum

29. • A 60-kg astronaut floating at rest in space outside a space capsule throws his 0.50-kg hammer such that it moves with a speed of 10 m/s relative to the capsule. What happens to the astronaut?

30. • In a pairs figure-skating competition, a 65-kg man and his 45-kg female partner stand facing each other on skates on the ice. If they push apart and the woman has a velocity of 1.5 m/s eastward, what is the velocity of her partner? (Neglect friction.)

31. •• To get off a frozen, frictionless lake, a 65.0-kg person takes off a 0.150-kg shoe and throws it horizontally, directly away from the shore with a speed of 2.00 m/s. If the person is 5.00 m from the shore, how long does he take to reach it?

32. IE •• An object initially at rest explodes and splits into three fragments. The first fragment flies off to the west, and the second fragment flies off to the south. The third fragment will fly off toward a general direction of (1) southwest, (2) north of east, (3) either due north or due east. Why? (b) If the object has a mass of 3.0 kg, the first fragment has a mass of 0.50 kg and a speed of 2.8 m/s, and the second fragment has a mass of 1.3 kg and a speed of 1.5 m/s, what are the speed and direction of the third fragment?

33. •• Consider two string-suspended balls, both with a mass of 0.15 kg. (Similar to the arrangement in Figure 6.15, but with only two balls.) One ball is pulled back in line with the other so it has a vertical height of 10 cm, and is then released. (a) What is the speed of the ball just before hitting the stationary one? (b) If the collision is completely inelastic, to what height do the balls swing?

34. •• A stationary cherry bomb explodes into three pieces of equal mass. The first piece has an initial velocity of $(10 \text{ m/s})\hat{x}$. The second piece has an initial velocity of $(6.0 \text{ m/s})\hat{x} - (3.0 \text{ m/s})\hat{y}$. What is the velocity of the third piece?

35. •• Two ice skaters not paying attention collide in a completely inelastic collision. Prior to the collision, skater 1, with a mass of 60 kg, has a velocity of 5.0 km/h eastward, and moves at a right angle to skater 2, who has

a mass of 75 kg and a velocity of 7.5 km/h southward. What is the velocity of the skaters after collision?

36. •• Two balls of equal mass (0.50 kg) approach the origin along the positive *x*- and *y*-axes at the same speed (3.3 m/s). (a) What is the total momentum of the system? (b) Will the balls necessarily collide at the origin? What is the total momentum of the system after both balls have passed through the origin?

37. •• A 1200-kg car moving to the right with a speed of 25 m/s collides with a 1500-kg truck and locks bumpers with the truck. Calculate the velocity of the combination after the collision if the truck is initially (a) at rest, (b) moving to the right with a speed of 20 m/s, and (c) moving to the left with a speed of 20 m/s.

38. •• A 10-g bullet moving horizontally at 400 m/s penetrates a 3.0-kg wood block resting on a horizontal surface. If the bullet slows down to 300 m/s after emerging from the block, what is the speed of the block immediately after the bullet emerges (▼ **Figure 6.31**)?

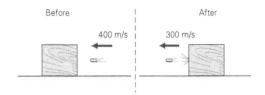

Before   After

400 m/s   300 m/s

▲ FIGURE 6.31 **Momentum transfer?** See Exercise 38.

39. •• An explosion of a 10.0-kg bomb releases only two separate pieces. The bomb was initially at rest and a 4.00 kg piece travels westward at 100 m/s immediately after the explosion. (a) What are the speed and direction of the other piece immediately after the explosion? (b) How much kinetic energy was released in this explosion?

40. •• A 1600-kg (empty) truck rolls with a speed of 2.5 m/s under a loading bin, and a mass of 3500 kg is deposited into the truck. What is the truck's speed immediately after loading?

41. IE •• A crowd control method utilizes "rubber" bullets instead of real ones. Suppose that, in a test, one of these "bullets" with a mass of 5.00 g is traveling at 250 m/s to the right. It hits a stationary target head-on. The target's mass is 2.50 kg and it rests on a smooth surface. The bullet bounces backward (to the left) off the target at 100 m/s. (a) Which way must the target move after the collision: (1) right, (2) left, (3) it could be stationary, or (4) you can't tell from the data given? (b) Determine the recoil speed of the target after the collision.

42. •• For a movie scene, a 75-kg stuntman drops from a tree onto a 50-kg sled that is moving on a frozen lake with a velocity of 10 m/s toward the shore. (a) What is

the speed of the sled after the stuntman is on board? (b) If the sled hits the bank and stops, but the stuntman keeps on going, with what speed does he leave the sled? (Neglect friction.)

43. •• A 90-kg astronaut is stranded in space at a point 6.0 m from his spaceship, and he needs to get back in 4.0 min to control the spaceship. To get back, he throws a 0.50-kg piece of equipment so that it moves at a speed of 4.0 m/s directly away from the spaceship. (a) Does he get back in time? (b) How fast must he throw the piece of equipment so he gets back in time?

6.4 Elastic and Inelastic Collisions

44. •• For the apparatus in Figure 6.15, one ball swinging in at a speed of $2v_o$ will not cause two balls to swing out with speeds v_o. (a) Which law of physics precludes this situation from happening: the law of conservation of momentum or the law of conservation of mechanical energy? (b) Prove this law mathematically.

45. •• A proton of mass *m* moving with a speed of 3.0×10^6 m/s undergoes a head-on elastic collision with an alpha particle of mass 4*m*, which is initially at rest. What are the velocities of the two particles after the collision?

46. •• A 4.0-kg ball with a velocity of 4.0 m/s in the +*x*-direction collides head-on elastically with a stationary 2.0-kg ball. What are the velocities of the balls after the collision?

47. •• A dropped rubber ball hits the floor with a speed of 8.0 m/s and rebounds to a height of 0.25 m. What fraction of the initial kinetic energy was lost in the collision?

48. •• At a county fair, two children ram each other head-on while riding on the bumper cars. Jill and her car, traveling left at 3.50 m/s, have a total mass of 325 kg. Jack and his car, traveling to the right at 2.00 m/s, have a total mass of 290 kg. Assuming the collision to be elastic, determine their velocities after the collision.

49. •• In a high-speed chase, a policeman's car bumps a criminal's car directly from behind to get his attention. The policeman's car is moving at 40.0 m/s to the right and has a total mass of 1800 kg. The criminal's car is initially moving in the same direction at 38.0 m/s. His car has a total mass of 1500 kg. Assuming an elastic collision, determine their two velocities immediately after the bump.

50. IE •• ▶ **Figure 6.32** shows a bird catching a fish. Assume that initially the fish jumps up and that the bird coasts horizontally and does not touch the water with its feet or flap its wings. (a) Is this kind of collision (1) elastic, (2) inelastic, or (3) completely inelastic? Why? (b) If the mass of the bird is 5.0 kg, the mass of the fish is 0.80 kg, and the bird coasts with a speed of 6.5 m/s before grabbing, what is the speed of the bird after grabbing the fish?

▲ FIGURE 6.32 **Elastic or inelastic?** See Exercise 50.

51. •• A 1.0-kg object moving at 10 m/s collides with a stationary 2.0-kg object as shown in ▼ **Figure 6.33**. If the collision is perfectly inelastic, how far along the inclined plane will the combined system travel? (Neglect friction.)

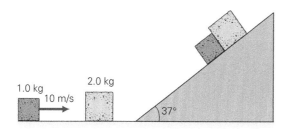

▲ FIGURE 6.33 **How far is up?** See Exercises 51 and 55.

52. •• In a pool game, a cue ball traveling at 0.75 m/s hits the stationary eight ball. The eight ball moves off with a velocity of 0.25 m/s at an angle of 37° relative to the cue ball's initial direction. Assuming that the collision is inelastic, at what angle will the cue ball be deflected, and what will be its speed?

53. •• Two balls approach each other as shown in ▶ **Figure 6.34**, where $m = 2.0$ kg, $v = 3.0$ m/s, $M = 4.0$ kg, and $V = 5.0$ m/s. If the balls collide and stick together at the origin, (a) what are the components of the velocity v of the balls after collision, and (b) what is the angle θ?

54. IE •• A car traveling east and a minivan traveling south collide in a completely inelastic collision at a perpendicular intersection. (a) Right after the collision, will the car and minivan move toward a general direction (1) south of east, (2) north of west, or (3) either due south or due east? Why? (b) If the initial speed of the 1500-kg car was 90.0 km/h and the initial speed of the 3000-kg minivan was 60.0 km/h, what is the velocity of the vehicles immediately after collision?

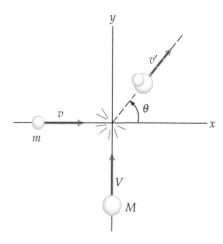

▲ FIGURE 6.34 **A completely inelastic collision** See Exercise 53.

55. •• A 1.0-kg object moving at 2.0 m/s collides elastically with a stationary 1.0-kg object, similar to the situation shown in Figure 6.33. How far will the initially stationary object travel along a 37° inclined plane? (Neglect friction.)

56. •• A fellow student states that the total momentum of a three-particle system ($m_1 = 0.25$ kg, $m_2 = 0.20$ kg, and $m_3 = 0.33$ kg) is initially zero. He calculates that after an inelastic triple collision the particles have velocities of 4.0 m/s at 0°, 6.0 m at 120°, and 2.5 m/s at 230°, respectively, with angles measured from the +x-axis. Do you agree with his calculations? If not, assuming the first two answers to be correct, what should be the momentum of the third particle so the total momentum is zero?

57. •• A freight car with a mass of 25 000 kg rolls down an inclined track through a vertical distance of 1.5 m. At the bottom of the incline, on a level track, the car collides and couples with an identical freight car that was at rest. What percentage of the initial kinetic energy is lost in the collision?

6.5 Center of Mass

58. (a) The center of mass of a system consisting of two 0.10-kg particles is located at the origin. If one of the particles is at (0, 0.45 m), where is the other? (b) If the masses are moved so their center of mass is located at (0.25 m, 0.15 m), can you tell where the particles are located?

59. •• Find the center of mass of a system composed of three spherical objects with masses of 3.0, 2.0, and 4.0 kg and centers located at (−6.0 m, 0), (1.0 m, 0), and (3.0 m, 0), respectively.

60. IE •• A 3.0-kg rod of length 5.0 m has at opposite ends point masses of 4.0 kg and 6.0 kg. (a) Will the center of mass of this system be (1) nearer to the 4.0-kg mass, (2) nearer to the 6.0-kg mass, or (3) at the center of the rod? Why? (b) Where is the center of mass of the system?

61. •• A piece of uniform sheet metal measures 25 cm by 25 cm. If a circular piece with a radius of 5.0 cm is cut from the center of the sheet, where is the sheet's center of mass now?

62. •• Locate the center of mass of the system shown in ▼ **Figure 6.35** (a) if all of the masses are equal, (b) if $m_2 = m_4 = 2m_1 = 2m_3$, (c) if $m_1 = 1.0$ kg, $m_2 = 2.0$ kg, $m_3 = 3.0$ kg, and $m_4 = 4.0$ kg.

(0, 4.0 m)   (4.0 m, 4.0 m)

m_2 m_3

m_1 m_4

(0, 0)   (4.0 m, 0)

▲ FIGURE 6.35   **Where's the center of mass?** See Exercise 62.

63. •• Two skaters with masses of 65 and 45 kg, respectively, stand 8.0 m apart, each holding one end of a piece of rope. (a) If they pull themselves along the rope until they meet, how far does each skater travel? (Neglect friction.) (b) If only the 45-kg skater pulls along the rope until she meets her friend (who just holds onto the rope), how far does each skater travel?

7

Circular Motion and Gravitation

Sufficient centripetal force explains why these riders won't fall from their seats even when upside down.

People often say that rides like the circular one in the chapter-opening photograph "defy gravity." Of course, you know that in reality, gravity cannot be defied; it commands respect. There is nothing that will shield you from it and no place in the universe where you can go to be entirely free of gravity.

Circular motion is everywhere, from atoms to galaxies, from flagella of bacteria to Ferris wheels. Two terms are frequently used to describe such motion. In general, we say that an object *rotates* when the axis of rotation lies within the body and that it *revolves* when the axis is outside the body. Thus, the Earth rotates on its axis and revolves about the Sun.

Such motion is in two dimensions, and so can be described by rectangular components as used in Chapter 3. However, it is usually more convenient to describe circular motion in terms of angular quantities that will be introduced in this chapter. Being familiar with the description of circular motion will make the study of rotating rigid bodies in Chapter 8 much easier.

Gravity plays a major role in determining the motions of the planets, since it supplies the force necessary to maintain their orbits. Newton's law of gravitation will be considered in this chapter. This law describes the fundamental force of gravity, and will be used to analyze planetary motion. The same considerations will help you understand the motions of Earth satellites, which include one natural satellite (the Moon) and many artificial ones.

7.1 Angular Measure*

Motion is described as a change of position with time (Section 2.1). As you might guess, *angular speed* and *angular velocity* also involve a time rate of change of position, which is expressed by an *angular change*. Consider a particle traveling in a circular path, as shown in ▶ **Figure 7.1**. At a particular instant, the particle's position (P) may be designated by the Cartesian coordinates x and y. However, the position may also be designated by the *polar coordinates* r and θ. The distance r extends from the origin, and the angle θ is commonly measured counterclockwise from the positive x-axis. The transformation Equations that relate one set of coordinates to the other are

$$x = r\cos\theta \tag{7.1a}$$

$$y = r\sin\theta \tag{7.1b}$$

as can be seen from the x- and y-coordinates of point P in Figure 7.1.

Note that r is the same for any point on a given circle. As a particle travels in a circle, the value of r is constant, and only θ changes with time. Thus, circular motion can be described by

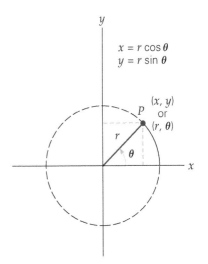

▲ FIGURE 7.1   **Polar coordinates** A point (P) may be described by polar coordinates instead of Cartesian coordinates – that is, by (r, θ) instead of (x, y). For a circle, θ is the angular distance and r is the radial distance. The two types of coordinates are related by the transformation equations $x = r\cos\theta$ and $y = r\sin\theta$.

using one polar coordinate (θ) that changes with time, instead of two Cartesian coordinates (x) and (y), both of which change with time.

Analogous to linear displacement is **angular displacement**, the magnitude of which is given by

$$\Delta\theta = \theta - \theta_0 \tag{7.2}$$

or simply $\Delta\theta = \theta$ when $\theta_0 = 0$. (The direction of the angular displacement will be explained in the next section on angular velocity.) A unit commonly used to express angular displacement is the degree (°); there are 360° in one complete circle.[†]

It is important to be able to relate the angular description of circular motion to the orbital or tangential description – that is, to relate the angular displacement to the arc length s. The *arc length* is the distance traveled along the circular path, and the angle θ is said to *subtend* (define) the arc length. A quantity that is very convenient for relating angle to arc length is the **radian (rad)**. The angle in radians is given by the ratio of the arc length (s) and the radius (r) – that is, θ (in radians) equals s/r. When $s = r$, the angle is equal to one radian, $\theta = s/r = r/r = 1$ rad (▶ **Figure 7.2**). Thus, (with the angle in radians),

$$s = r\theta \tag{7.3}$$

[†] A degree may be divided into the smaller units of minutes (1° = 60 min) and seconds (1 min = 60 s). These divisions have nothing to do with time units.

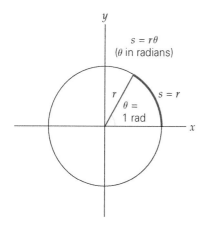

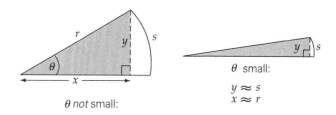

▲ FIGURE 7.3 **The small-angle approximation** See text for description.

EXAMPLE 7.1: FINDING ARC LENGTH – USING RADIAN MEASURE

A spectator standing at the center of a circular running track observes a runner start a practice race 256 m due east of her own position (▼ **Figure 7.4**). The runner runs on the track to the finish line, which is located due north of the observer's position. What is the distance of the run?

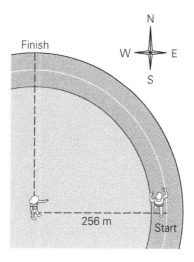

▲ FIGURE 7.4 **Arc length – found by means of radians** See Example 7.1 for description.

THINKING IT THROUGH. Note that the subtending angle of the section of circular track is $\theta = 90°$. The arc length (s) can be found, since the radius r of the circle is known.

SOLUTION

Listing what is given and what is to be found,

Given:

$r = 256$ m
$\theta = 90° = \pi/2$ rad

Find: s (arc length)

Simply using Equation 7.3 to find the arc length,

$$s = r\theta = (256\,\text{m})\left(\frac{\pi}{2}\right) = 402\,\text{m}$$

▲ FIGURE 7.2 **Radian measure** Angular displacement may be measured either in degrees or in radians (rad). An angle θ is subtended by an arc length s. When $s = r$, the angle subtending s is defined to be 1 rad. More generally, $\theta = s/r$, where θ is in radians. One radian is equal to 57.3°.

which is an important relationship between the circular arc length s and the radius of the circle r. (Notice that since $\theta = s/r$, the angle in radians is the ratio of two lengths. This means that a radian measure is a pure number – that is, it is dimensionless and has no units.)

To get a general relationship between radians and degrees, consider the distance around a complete circle (360°). For one full circle, with $s = 2\pi r$ (the circumference), there are a total of $\theta = s/r = 2\pi r/r = 2\pi$ rad in 360°; that is,

$$2\pi\,\text{rad} = 360°$$

This relationship can be used to obtain convenient conversions of common angles (▼ **Table 7.1**). Also, dividing both sides of this relationship by 2π, the degree value of 1 rad is obtained:

$$1\,\text{rad} = 360°/2\pi = 57.3°$$

TABLE 7.1   Equivalent Degree and Radian Measures

Degrees	Radians
360°	2π
180°	π
90°	$\pi/2$
60°	$\pi/3$
57.3°	1
45°	$\pi/4$
30°	$\pi/6$

Notice in Table 7.1 that the angles in radians are expressed in terms of π explicitly, for convenience.

As shown in ▶ **Figure 7.3**, $\sin\theta = y/r$ and $\tan\theta = y/x$. However, when the angle θ is small, θ (in rad) $= s/r \approx y/r \approx y/x$. Thus θ (in rad) $\approx \sin\theta \approx \tan\theta$.

Note that the unitless "rad" is omitted, and the equation is dimensionally correct.

FOLLOW-UP EXERCISE. What path length would the runner have traveled when going an angular distance of 210° around the track? (*Answers to all Follow-Up Exercises are given in Appendix V at the back of the book.*)

EXAMPLE 7.2: HOW FAR AWAY?
A USEFUL APPROXIMATION

A sailor sights a distant tanker ship and finds that it subtends an angle of 1.15° as illustrated in ▼ **Figure 7.5a**. He knows from the shipping charts that the tanker is 150 m in length. Approximately how far away is the tanker?

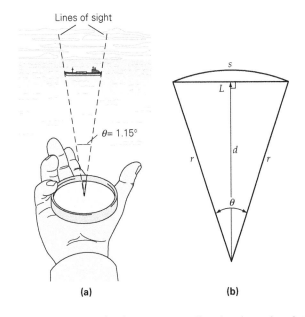

▲ **FIGURE 7.5**   **Angular distance** For small angles, the arc length is approximately a straight line, or the chord length. Knowing the length of the tanker, how far away it can be found by measuring its angular size. See Example 7.2 for description. (Drawing not to scale for clarity.)

THINKING IT THROUGH. Note that in Figure 7.3, for small angles, the arc length approximates the *y*-length of the triangle (the opposite side from θ), or $s \approx y$. Hence, if the length and the angle are known, the radial distance can be found, which is approximately equal to the tanker's distance from the sailor.

THINKING IT THROUGH. To approximate the distance, the ship's length is taken to be nearly equal to the arc length subtended by the measured angle. This approximation is good for small angles.

SOLUTION

The data are as follows:

Given:

$\theta = 1.15°(1 \text{ rad}/57.3°) = 0.0201 \text{ rad}$
$L = 150 \text{ m}$

Find: *r* (radial distance)

Knowing the arc length and angle, Equation 7.3 can be used to find *r*.

$$r = \frac{s}{\theta} \approx \frac{150\,\text{m}}{0.0201} = 7.46 \times 10^3\,\text{m} = 7.46\,\text{km}$$

(Note that the unitless rad is omitted.)

The distance *r* is an approximation, obtained by assuming that for small angles, the arc length *s* and the straight-line chord length *L* are very nearly the same length (Figure 7.5b). How good is this approximation? To check, let's compute the actual distance *d* to the middle of the ship. From the geometry, $\tan(\theta/2) = (L/2)/d$, so

$$d = \frac{L}{2\tan(\theta/2)} = \frac{150\,\text{m}}{2\tan(1.15°/2)} = 7.47 \times 10^3\,\text{m} = 7.47\,\text{km}$$

The first calculation is a pretty good approximation – the values derived by the two methods are nearly equal.

FOLLOW-UP EXERCISE. As pointed out, the approximation used in this Example is for *small* angles. You might wonder what is small. To investigate this question, what would be the percentage error of the approximated distance to the tanker for angles of 10° and 20°?

7.1.1 Problem-Solving Hint

In computing trigonometric functions such as $\tan \theta$ or $\sin \theta$, the angle may be expressed in degrees or radians; for example, $\sin 30° = \sin [(\pi/6) \text{ rad}] = \sin (0.524 \text{ rad}) = 0.500$. When finding trig functions with a calculator, note that there is usually a way to change the angle entry between *deg* and *rad* modes. Hand calculators commonly are set in *deg* (degree) mode, so if you want to find the value of, say, $\sin (1.22 \text{ rad})$, first change to *rad* mode and enter sin 1.22, and $\sin (1.22 \text{ rad}) = 0.939$. (Or you could convert rads to degrees first and use *deg* mode on your calculator.) Some calculators may have a third mode, *grad*. The grad is a little-used angular unit. One grad is 1/100 of a right (90°) angle; that is, there are 100 grads in a right angle.

7.2 Angular Speed and Velocity

The description of circular motion in angular form is analogous to the description of linear motion. In fact, you'll notice that the equations are almost identical mathematically, with different symbols being used to indicate that the quantities have different meanings. The lowercase Greek letter omega with a bar over it is used to represent average angular speed ($\bar{\omega}$), the magnitude of the angular displacement divided by the total time to travel the angular distance:

$$\bar{\omega} = \frac{\Delta\theta}{\Delta t} = \frac{\theta - \theta_\text{o}}{t - t_\text{o}} \quad \text{(average angular speed)} \tag{7.4}$$

It is commonly said that the units of angular speed are radians per second. Technically, the unit is 1/s or s^{-1} since the radian is unitless. But it is useful to keep the rad to indicate that the quantity is angular speed. The **instantaneous angular speed (ω)** is given by considering a very small time interval – that is, as Δt approaches zero.

As in the linear case, if the angular speed is *constant,* then $\bar{\omega} = \omega$. Taking θ_0 and t_0 to be zero in Equation 7.4,

$$\omega = \frac{\theta}{t} \quad \text{or} \quad \theta = \omega t \quad \text{(constant angular speed)} \quad (7.5)$$

SI unit of angular speed: radians per second (rad/s or s^{-1})

Another common descriptive unit for angular speed is revolutions per minute (rpm); for example, a DVD (*d*igital *v*ideo *d*isc) rotates at a speed of 570–1600 rpm (depending on the location of the track). This nonstandard unit of revolutions per minute is readily converted to radians per second, since 1 revolution = 2π rad. For example,

(600 rev/min) (2π rad/rev) (1 min/60 s) = 20π rad/s ($= 63$ rad/s).*

The **average angular velocity** and the **instantaneous angular velocity** are analogous to their linear counterparts. Angular velocity is associated with angular displacement. Both are vectors and thus have direction; however, this directionality is, by convention, specified in a special way. In one-dimensional, or linear, motion, a particle can go only in one direction or the other, positive (+) or negative (−), so the displacement and velocity vectors can have only these two directions. In the angular case, a particle moves one way or the other, but the motion is along its *circular path.*

Thus, the angular displacement and angular velocity vectors of a particle in circular motion can have only two directions, which correspond to going around the circular path with either increasing or decreasing angular displacement from θ_0 – that is, counterclockwise or clockwise. Let's focus on the angular velocity vector $\vec{\omega}$. (The direction of the angular displacement will be the same as that of the angular velocity. Why?)

The *direction* of the angular velocity vector is given by a *right-hand rule* as illustrated in ▶ **Figure 7.6a**. When the fingers of your right hand are curled in the direction of the circular motion, your extended thumb points in the direction of $\vec{\omega}$. Note that since circular motion can be in only one of two circular *senses,* clockwise or counterclockwise, then (+) and (−) signs can be used to distinguish circular rotation directions. It is customary to take a counterclockwise rotation as (+), since positive angular displacement is conventionally measured counterclockwise from the +x-axis.

Why not just designate the direction of the angular velocity vector to be either clockwise or counterclockwise? This designation is not used because clockwise (cw) and counterclockwise (ccw) are directional senses or indications rather than actual directions.

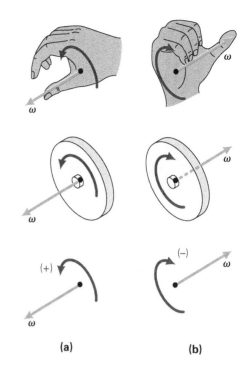

▲ **FIGURE 7.6 Angular velocity** The direction of the angular velocity vector for an object in rotational motion is given by the right-hand rule: When the fingers of the right hand are curled in the direction of the rotation, the extended thumb points in the direction of the angular velocity vector. Circular senses or directions are commonly indicated by **(a)** positive and **(b)** negative signs.

These rotational senses are like right and left. If you faced another person and each of you were asked whether something was on the right or left, your answers would disagree. Similarly, if you held this book up toward a person facing you and rotated it, would it be rotating cw or ccw for both of you? Check it out. We can use cw and ccw to indicate rotational "directions" when they are specified relative to a reference – for example, the +x-axis.

Referring to Figure 7.6, imagine yourself being first on one side of one of the rotating disks and then on the other. Then apply the right-hand rule on both sides. You should find that the direction of the angular velocity vector is the same for both locations – there is no ambiguity in using + and − to indicate rotational senses or directions.

7.2.1 Relationship between Tangential and Angular Speeds

A particle moving in a circle has an instantaneous velocity tangential to its circular path. For a constant angular velocity, the particle's *orbital speed,* or tangential speed (v_t), the magnitude of the tangential velocity, is also constant. How the angular and tangential speeds are related is revealed by starting with Equation 7.3 ($s = r\theta$) and Equation 7.5 ($\theta = \omega t$):

$$s = r\theta = r(\omega t)$$

---

* It is often convenient to leave the angular speed with π in symbol form; in this case, 20π rad/s.

The arc length, or distance, is also given by

$$s = v_t t$$

Combining the equations for *s* gives the relationship between the tangential speed (v_t) and the angular speed (ω),

$$v_t = r\omega \quad \text{(tangential speed relation} \atop \text{to angular speed for circular motion)}} \quad (7.6)$$

where ω is in radians per second. Equation 7.6 holds in general for instantaneous tangential and angular speeds for solid- or rigid-body rotation about a fixed axis, even when ω might vary with time.

Note that all the particles of a solid object rotating with constant angular velocity have the same angular speed, but the tangential speeds are different at different distances from the axis of rotation (▼ **Figure 7.7a**).

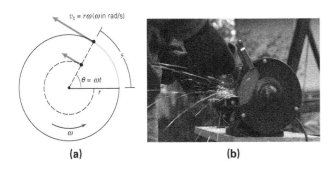

(a)   **(b)**

▲ **FIGURE 7.7** **Tangential and angular speeds (a)** Tangential and angular speeds are related by $v_t = r\omega$, where ω is in radians per second. Note that all of the particles of an object rotating about a fixed axis travel in circles. All the particles have the same angular speed ω, but particles at different distances from the axis of rotation have different tangential speeds. **(b)** Sparks from a grinding wheel provide a graphic illustration of instantaneous tangential velocity. (Why do the paths curve slightly?)

EXAMPLE 7.3: MERRY-GO-ROUNDS – DO SOME GO FASTER THAN OTHERS?

An amusement park merry-go-round at its constant operational speed makes one complete rotation in 45 s. Two children are on horses, one at 3.0 m from the center of the ride and the other farther out, 6.0 m from the center. What are (a) the angular speed and (b) the tangential speed of each child?

THINKING IT THROUGH. The angular speed of each child is constant and the same, since both children make a complete rotation in the same time. However, the tangential speeds will be different, because the radii are different. That is, the child at the greater radius travels in a larger circle during the rotation time and thus must travel faster.

SOLUTION

Given:
$\theta = 2\pi$ rad (one rotation)
$t = 45$ s
$r_1 = 3.0$ m
$r_2 = 6.0$ m

Find:

(a) ω_1 and ω_2 (angular speeds)
(b) v_1 and v_2 (tangential speeds)

(a) As noted, $\omega_1 = \omega_2 =$ constant, i.e., both riders rotate at the same constant angular speed. All points on the merry-go-round travel through 2π rad in the time it takes to make one rotation. The angular speed can be found from Equation 7.5 (constant ω) as

$$\omega_1 = \omega_2 = \frac{\theta}{t} = \frac{2\pi \text{ rad}}{45 \text{ s}} = 0.14 \text{ rad/s}$$

(b) The tangential speed is different at different radial locations on the merry-go-round. All of the "particles" making up the merry-go-round go through one rotation in the same amount of time. Therefore, the farther a particle is from the center, the longer its circular path, and the greater its tangential speed, as Equation 7.6 indicates. (See also Figure 7.7a.) Thus,

$$v_{t1} = r_1\omega = (3.0\,\text{m})(0.14\,\text{rad/s}) = 0.42\,\text{m/s}$$

and

$$v_{t2} = r_2\omega = (6.0\,\text{m})(0.14\,\text{rad/s}) = 0.84\,\text{m/s}$$

(Note that the rad has been dropped from the answer. Why?)

Therefore, a rider on the outer part of the ride has a greater tangential speed than a rider closer to the center. Here, rider 2 has a radius twice that of rider 1 and therefore goes twice as fast.

FOLLOW-UP EXERCISE. (a) On an old 45-rpm record, the beginning track is 8.0 cm from the center, and the end track is 5.0 cm from the center. What are the angular speeds and the tangential speeds at these distances when the record is spinning at 45 rpm? (b) For races on oval tracks, why do inside and outside runners have different starting points (called a "staggered" start), such that some runners start "ahead" of others?

7.2.2 Period and Frequency

Some other quantities commonly used to describe circular motion are period and frequency. The time it takes an object in circular motion to make one complete revolution (or rotation), or *cycle*, is called the **period** (*T*). For example, the period of revolution of the Earth about the Sun is one year, and the period of the Earth's axial rotation is 24 h.* The standard unit of period is the second (s). The period is sometimes given in seconds per cycle (s/cycle).

Closely related to the period is the **frequency (*f*)**, which is the number of revolutions, or cycles, made in a given time, generally a second. For example, if a particle traveling uniformly in a circular orbit makes 5.0 revolutions in 2.0 s, the frequency (of revolution) is $f = (5.0\text{ rev})/(2.0\text{ s}) = 2.5$ rev/s, or 2.5 cycles/s (cps, or cycles per second).

Revolution and *cycle* are merely descriptive terms used for convenience and are *not* units. Without these descriptive terms, it can be seen that the unit of frequency is inverse seconds (1/s, or s⁻¹), which is called the **hertz (Hz)** in the SI.† The two quantities are inversely related by

$$f = \frac{1}{T} \quad \text{(relationship of frequency and period)} \quad (7.7)$$

SI unit of frequency: hertz (Hz, 1/s or s⁻¹)

where the period is in seconds and the frequency is in hertz, or inverse seconds.

For uniform circular motion, the tangential orbital speed is related to the period *T* by $v = 2\pi/T$ – that is, the distance traveled in one revolution divided by the time for one revolution (one period). The frequency can also be related to the angular speed.

Since an angular distance of 2π rad is traveled in one period (by definition of the period), then

$$\omega = \frac{2\pi}{T} = 2\pi f \quad \begin{array}{l}\text{(angular speed in terms}\\\text{of period and frequency)}\end{array} \quad (7.8)$$

Notice that ω and *f* have the same units of 1/s. This notation can easily cause confusion, which is why the unitless radian (rad) term is often used in angular speed (rad/s) and cycles in frequency (cycles/s).

EXAMPLE 7.4: FREQUENCY AND PERIOD – AN INVERSE RELATIONSHIP

A DVD rotates in a player at a constant speed of 800 rpm. What are the DVD's (a) frequency and (b) period of revolution?

THINKING IT THROUGH. The relationships for the frequency (*f*), the period (*T*), and the angular frequency (ω), are expressed in Equations 7.7 and 7.8, so these equations can be used.

SOLUTION

The angular speed is not in standard units and so must be converted. Revolutions per minute (rpm) is converted to radians per second (rad/s).

Given:

$$\omega = \left(\frac{800\text{ rev}}{\text{min}}\right)\left(\frac{1\text{ min}}{60\text{ s}}\right)\left(\frac{2\pi\text{ rad}}{\text{rev}}\right) = 83.7\text{ rad/s}$$

Find:

(a) *f* (frequency)
(b) *T* (period)

(a) Rearranging Equation 7.8 and solving for *f*,

$$f = \frac{\omega}{2\pi} = \frac{83.7\text{ rad/s}}{2\pi\text{ rad/cycle}} = 13.3\text{ Hz}$$

The units of 2π are rad/cycle or revolution, so the result is in cycles/second or inverse seconds, which is the hertz.

(b) Equation 7.8 could be used to find *T*, but Equation 7.7 is a bit simpler:

$$T = \frac{1}{f} = \frac{1}{13.3\text{ Hz}} = 0.0752\text{ s}$$

Thus, the DVD takes 0.0752 s to make one revolution. (Notice that since Hz = 1/s, the equation is dimensionally correct.)

FOLLOW-UP EXERCISE. If the period of a particular DVD is 0.0500 s, what is the DVD's angular speed in revolutions per minute?

7.3 Uniform Circular Motion and Centripetal Acceleration

A simple but important type of circular motion is **uniform circular motion**, which occurs when an object moves at *a constant speed in a circular path*. An example of this motion may be a car going around a circular track at a constant speed. The motion of the Moon around the Earth can be approximated by uniform circular motion. Such motion is curvilinear, and as discussed in Section 3.1 there must be an acceleration. But what are its magnitude and direction?

7.3.1 Centripetal Acceleration

The acceleration of uniform circular motion is not in the same direction as the instantaneous velocity (which is tangent to the circular path at any point). If it were, the object would speed up, and the circular motion wouldn't be uniform. Recall that acceleration is the time rate of change of velocity and that velocity has both *magnitude* and *direction*. In uniform circular motion, the direction of the velocity is continuously

---

* The discussion applies to rotations as well as revolutions. Revolutions will be used as a general term, as is commonly done; for example, a DVD rotates at so many revolutions per minute.

† Named for Heinrich Hertz (1857–1894), a German physicist and pioneering investigator of electromagnetic waves, which also are characterized by frequency.

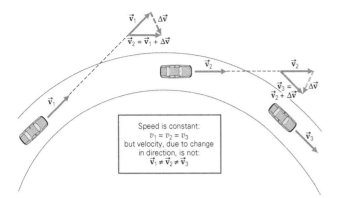

▲ FIGURE 7.8 **Uniform circular motion** The speed of an object in uniform circular motion is constant, but the object's velocity changes in the direction of motion. Thus, there is an acceleration.

changing, which is a clue to the direction of the acceleration (▲ **Figure 7.8**).

The velocity vectors at the beginning and end of a time interval give the change in velocity, or $\Delta \vec{v}$, via vector subtraction. All of the instantaneous velocity vectors have the same magnitude or length (constant speed), but they differ in direction. Note that because $\Delta \vec{v}$ is not zero, there must be an acceleration ($\vec{a} = \Delta \vec{v}/\Delta t$).

As illustrated in ▶ **Figure 7.9**, as Δt (or $\Delta \theta$) becomes smaller, $\Delta \vec{v}$ points more toward the center of the circular path. As Δt approaches zero, the instantaneous change in the velocity, and therefore the acceleration, points directly toward the center of the circle. As a result, the acceleration in uniform circular motion is called **centripetal acceleration (a_c)**, which means "center-seeking" acceleration (from the Latin *centri*, "center," and *petere*, "to fall toward" or "to seek").

The centripetal acceleration must be directed radially inward; that is, with no component in the direction of the perpendicular (tangential) velocity, or else the magnitude of that velocity would change (▶ **Figure 7.10**). Note that for an object in uniform circular motion, the direction of the centripetal acceleration is continuously changing. In terms of x- and y-components, a_x and a_y are not constant. (Can you describe how this differs from the acceleration in projectile motion?)

The magnitude of the centripetal acceleration can be deduced from the small shaded triangles in Figure 7.9. (For very short time intervals, the arc length Δs is almost a straight line – the chord.) These two triangles are similar triangles, because each has a pair of equal sides surrounding the same angle $\Delta \theta$. (Note that the velocity vectors have the same magnitude.) Thus, Δv is to v as Δs is to r, which can be written as*

$$\frac{\Delta v}{v} \approx \frac{\Delta s}{r}$$

---
* The subscript t will be dropped with the understanding that v is tangential speed in uniform circular motion.

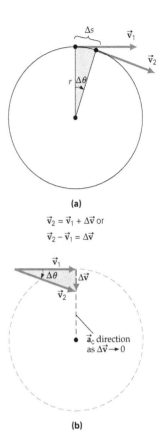

(a)

$$\vec{v}_2 = \vec{v}_1 + \Delta \vec{v} \text{ or}$$
$$\vec{v}_2 - \vec{v}_1 = \Delta \vec{v}$$

(b)

▲ FIGURE 7.9 **Analysis of centripetal acceleration** (a) The velocity vector of an object in uniform circular motion is constantly changing direction. (b) As Δt, the time interval for $\Delta \theta$, is taken to be smaller and smaller and approaches zero, $\Delta \vec{v}$ (the change in the velocity, and therefore an acceleration) is directed toward the center of the circle. The result is a centripetal, or center-seeking, acceleration that has a magnitude of $a_c = v^2/r$.

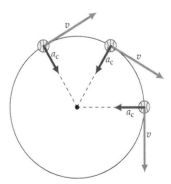

▲ FIGURE 7.10 **Centripetal acceleration** For an object in uniform circular motion, the centripetal acceleration is directed radially inward. There is no acceleration component in the tangential direction; if there were, the magnitude of the velocity (tangential *speed*) would change.

The arc length Δs is the distance traveled in time Δt; thus, $\Delta s = v\Delta t$ so

$$\frac{\Delta v}{v} \approx \frac{\Delta s}{r} = \frac{v\Delta t}{r}$$

and

$$\frac{\Delta v}{\Delta t} \approx \frac{v^2}{r}$$

Then, as Δt approaches zero, this approximation becomes exact. The instantaneous centripetal acceleration, $a_c = \Delta v/\Delta t$, thus has a magnitude of

$$a_c = \frac{v^2}{r} \quad \text{(magnitude of centripetal acceleration in terms of tangential speed)} \tag{7.9}$$

Using Equation 7.6 ($v = r\omega$), the equation for centripetal acceleration can also be written in terms of the angular speed:

$$a_c = \frac{v^2}{r} = \frac{(r\omega)^2}{r} = r\omega^2 \quad \text{(magnitude of centripetal acceleration in terms of angular speed)} \tag{7.10}$$

EXAMPLE 7.5: A CENTRIFUGE – CENTRIPETAL ACCELERATION

A laboratory centrifuge like that shown in ▼ **Figure 7.11** operates at a rotational speed of 12 000 rpm. (a) What is the magnitude of the centripetal acceleration of a red blood cell at a radial distance of 8.00 cm from the centrifuge's axis of rotation? (b) How does this acceleration compare with g (the gravitational acceleration on the surface of the Earth)?

▲ FIGURE 7.11 **Centrifuge** Centrifuges are used to separate particles of different sizes and densities suspended in liquids. For example, red and white blood cells can be separated from each other and from the plasma that makes up the liquid portion of the blood in the centrifuge tube. When spinning, the tubes are horizontal.

THINKING IT THROUGH. Here, the angular speed and the radius are given, so the magnitude of the centripetal acceleration can be computed directly from Equation 7.10. The result can be compared with g by using $g = 9.80$ m/s².

SOLUTION

The data are as follows:

Given:

$$\omega = (1.20\times10^4 \text{ rpm})\left[\frac{(\pi/30) \text{ rad/s}}{\text{rpm}}\right] = 1.26\times10^3 \text{ rad/s}$$
$$r = 8.00 \text{ cm} = 0.0800 \text{ m}$$

Find:

(a) a_c
(b) How a_c compares with g

(a) The centripetal acceleration is found from Equation 7.10:

$$a_c = r\omega^2 = (0.0800 \text{ m})(1.26\times10^3 \text{ rad/s})^2 = 1.27\times10^5 \text{ m/s}^2$$

(b) Using the relationship $1\,g = 9.80$ m/s² to express a_c in terms of g,

$$a_c = (1.27\times10^5 \text{ m/s}^2)(1g/9.80 \text{ m/s}^2) = 1.30\times10^4\,g \; (= 13\,000\,g!)$$

FOLLOW-UP EXERCISE. What angular speed in revolutions per minute would give a centripetal acceleration of 1 g at the radial distance in this Example, and, taking gravity into account, what would be the resultant acceleration?

7.3.2 Centripetal Force

For an acceleration to exist, there must be a net force. Thus, for a centripetal (inward) acceleration to exist, there must be a **centripetal force** (net inward force). Expressing the magnitude of this force in terms of Newton's second law ($\vec{F}_{net} = m\vec{a}$) and inserting the expression for centripetal acceleration from Equation 7.9 for magnitude,

$$F_c = ma_c = \frac{mv^2}{r} \tag{7.11}$$

The centripetal force, like the centripetal acceleration, is directed radially toward the center of the circular path.

CONCEPTUAL EXAMPLE 7.6: BREAKING AWAY

A ball attached to a string is swung with uniform motion in a horizontal circle above a person's head (▶ **Figure 7.12a**). If the string breaks, which of the trajectories shown in Figure 7.12b (viewed from above) would the ball follow?

REASONING AND ANSWER. When the string breaks, the centripetal force provided by the string tension goes to zero. There is no force in the outward direction, so the ball could not follow trajectory *a*. Newton's first law states that if no force acts on an object in motion, the object will continue to move in a straight line. This factor rules out trajectories *b*, *d*, and *e*.

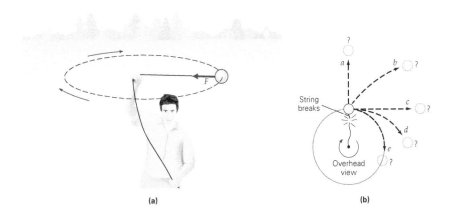

▲ FIGURE 7.12   **Centripetal force (a)** A ball is swung in a horizontal circle. **(b)** If the string breaks and the centripetal force goes to zero, what happens to the ball? See Example 7.6 for description.

It should be evident from the previous discussion that at any instant (including the instant when the string breaks), the isolated ball has a horizontal, tangential velocity. The downward force of gravity acts on it, but this force affects only its vertical motion, which is not visible in Figure 7.12b. The ball thus flies off tangentially and is essentially a horizontal projectile (with $v_{x_0} = v$, $v_{y_0} = 0$ and $a_y = -g$). Viewed from above, the ball would appear to follow the path labeled *c*.

FOLLOW-UP EXERCISE. If you swing a ball in a horizontal circle about your head, can the string be exactly horizontal? (See Figure 7.12a.) Explain your answer. [*Hint:* Analyze the forces acting on the ball.]

Keep in mind that, in general, a net force applied at an angle to the direction of motion of an object produces changes in the magnitude *and* direction of the velocity. However, when a net force of constant magnitude is continuously applied at an angle of 90° to the direction of motion (as is centripetal force), only the direction of the velocity changes. This is because there is no force component parallel to the velocity. Also notice that because the centripetal force is always perpendicular to the direction of motion, this force does no work. (Why?) Therefore, *a centripetal force does not change the kinetic energy or speed of the object.*

Note that the centripetal force in the form $F_c = mv^2/r$ is *not* a new individual force, but rather the cause of the centripetal acceleration, and is supplied by either a real force or the vector sum of several forces.

The force supplying the centripetal acceleration for satellites is gravity. In Conceptual Example 7.6, it was the tension in the string. Another force that often supplies centripetal acceleration is friction. Suppose that an automobile moves into a level, circular curve. To negotiate the curve, the car must have a centripetal acceleration, which is supplied by the force of friction between the tires and the road.

However, this static friction (why static?) has a maximum limiting value. If the speed of the car is high enough or the curve is sharp enough, the friction will not be sufficient to supply the necessary centripetal acceleration, and the car will skid outward from the center of the curve. If the car moves onto a wet or icy spot, the friction between the tires and the road may be reduced, allowing the car to skid at an even lower speed. (Banking a curve helps vehicles negotiate the curve without slipping.) To see the details of a car turning, consider the next example.

EXAMPLE 7.7: WHERE THE RUBBER MEETS THE ROAD – FRICTION AND CENTRIPETAL FORCE

A car approaches a level, circular curve with a radius of 45.0 m. If the concrete pavement is dry, what is the maximum speed at which the car can negotiate the curve at a constant speed?

THINKING IT THROUGH. The car is in uniform circular motion on the curve, so there must be a centripetal force. This force is supplied by static friction, so the maximum static frictional force provides the centripetal force when the car is at its maximum tangential speed.

SOLUTION

Given:

   $r = 45.0$ m
   $\mu_s = 1.20$ (from Table 4.1)

Find: v (maximum speed)

To go around the curve at a particular speed, the car must have a centripetal acceleration, and therefore a centripetal force must act on it. In the vertical direction, the weight of the car (downward) and the normal force on the car by the road (upward) are equal and opposite so they cancel. Static friction between the tires and the road is the only other force (in the horizontal direction) acting on the car and thus is the net force on it. (The tires are not slipping or skidding relative to the road.)

Recall from Section 4.6 that the maximum frictional force is given by $f_{s_{max}} = \mu_s N$ (Equation 4.7), where N is the magnitude of the normal force on the car and is equal in magnitude to the weight of the car, mg, on the level road (why?). Thus the magnitude of the maximum static frictional force is equal to the magnitude of the centripetal force ($F_c = mv^2/r$). From this the maximum speed can be found. To find $f_{s_{max}}$, the coefficient of friction between rubber and dry concrete is needed, and from Table 4.1, $\mu_s = 1.20$. Then,

$$f_{s_{max}} = F_c = \mu_s N = \mu_s mg = \frac{mv^2}{r}$$

So $\mu_s g = (v^2/r)$, thus

$$v = \sqrt{\mu_s rg} = \sqrt{(1.20)(45.0\,\text{m})(9.80\,\text{m/s}^2)} = 23.0\,\text{m/s}$$

FOLLOW-UP EXERCISE. Would the centripetal force be the same for all types of vehicles as in this Example? Explain.

The proper safe speed for driving on a highway curve is an important consideration. The coefficient of friction between tires and the road may vary, depending on weather, road conditions, the design of the tires, the amount of tread wear, and so on. When a curved road is designed, safety may be promoted by banking, or inclining, the roadway. This design reduces the chances of skidding because the normal force exerted on the car by the road then has a component toward the center of the curve that reduces the need for friction. In fact, for a circular curve with a given banking angle and radius, there is one speed for which no friction is required at all. This condition is used in banking design. (See Conceptual Question 12 at the end of the chapter.)

Let's look at one more example of centripetal force, this time with two objects in uniform circular motion.

EXAMPLE 7.8: STRUNG OUT – CENTRIPETAL FORCE AND NEWTON'S SECOND LAW

Suppose that two masses, $m_1 = 2.5$ kg and $m_2 = 3.5$ kg, are connected by light strings and are in uniform circular motion on a horizontal frictionless surface as illustrated in ▼ **Figure 7.13**, where $r_1 = 1.0$ m and $r_2 = 1.3$ m. The tension forces acting on the masses are $T_1 = 4.5$ N and $T_2 = 2.9$ N, which are the respective tensions in the strings. Find the magnitude of the centripetal acceleration and the tangential speed of (a) mass m_2 and (b) mass m_1.

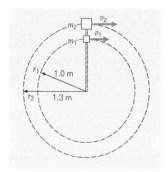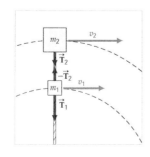

▲ FIGURE 7.13 **Centripetal force and Newton's second law** See Example 7.8 for description.

THINKING IT THROUGH. The centripetal forces on the masses are supplied by the tensions (T_1 and T_2) in the strings. By isolating the masses, a_c for each mass can be found, because the net force on a mass is equal to the mass's centripetal force ($F_c = ma_c$). The tangential speeds can then be found, since the radii are known ($a_c = v^2/r$).

SOLUTION

Given:

$\quad r_1 = 1.0$ m and $r_2 = 1.3$ m
$\quad m_1 = 2.5$ kg and $m_2 = 3.5$ kg
$\quad T_1 = 4.5$ N and $T_2 = 2.9$ N

Find: a_c (centripetal acceleration) and v
$\qquad$ (tangential speed) (a) m_2 and (b) m_1

(a) By isolating m_2 in Figure 7.13, it can seen that the centripetal force is provided by the tension in the string. (T_2 is the only force acting on m_2 toward the center of its circular path.) Thus,

$$T_2 = m_2 a_{c2}$$

and

$$a_{c2} = \frac{T_2}{m_2} = \frac{2.9\,\text{N}}{3.5\,\text{kg}} = 0.83\,\text{m/s}^2$$

where the acceleration is toward the center of the circle. The tangential speed of m_2 can be found from $a_c = v^2/r$:

$$v_2 = \sqrt{a_{c2}r_2} = \sqrt{(0.83\,\text{m/s}^2)(1.3\,\text{m})} = 1.0\,\text{m/s}$$

(b) The situation is a bit different for m_1. In this case, two radial forces are acting on m_1: the string tensions T_1 (inward) and $-T_2$ (outward). By Newton's second law, in order to have a centripetal acceleration, there must be a net force, which is given by the difference in the two tensions, so we expect $T_1 > T_2$, and

$$F_{net1} = +T_1 + (-T_2) = m_1 a_{c1} = \frac{m_1 v_1^2}{r_1}$$

where the radial direction (toward the center of the circular path) is taken to be positive. Then

$$a_{c1} = \frac{T_1 - T_2}{m_1} = \frac{4.5\,\text{N} - 2.9\,\text{N}}{2.5\,\text{kg}} = 0.64\,\text{m/s}^2$$

and

$$v_1 = \sqrt{a_{c1}r_1} = \sqrt{(0.64\,\text{m/s}^2)(1.0\,\text{m})} = 0.80\,\text{m/s}$$

FOLLOW-UP EXERCISE. Notice in this Example that the centripetal acceleration of m_2 is greater than that of m_1 yet $r_2 > r_1$, and $a_c \propto 1/r$. Is something wrong here? Explain.

INTEGRATED EXAMPLE 7.9: CENTER-SEEKING FORCE – ONE MORE TIME

A 1.0-m cord is used to suspend a 0.50-kg tetherball from the top of the pole. After being hit several times, the ball goes around the pole in uniform circular motion with a tangential speed of 1.1 m/s at an angle of 20° relative to the pole. (a) The force that supplies the centripetal acceleration is (1) the weight of the ball, (2) a component of the tension force in the string, (3) the total tension in the string. (b) What is the magnitude of the centripetal force?

(a) CONCEPTUAL REASONING. The centripetal force, being a "center-seeking" force, is directed perpendicularly toward the pole, about which the ball is in circular motion. As suggested in the problem-solving procedures provided in Section 1.7, it is almost always helpful to sketch a diagram, such as that in ▼ **Figure 7.14**.

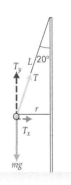

▲ FIGURE 7.14   **Ball on a string** See Example 7.9 for description.

Immediately, it can be seen that (1) and (3) are not correct, as these forces are not directly toward the circle's center located on the pole. (m_g and T_y are equal and opposite, because there is no acceleration in the y-direction.) The answer is obviously (2), with a component of the tension force, T_x, supplying the centripetal force.

(b) QUANTITATIVE REASONING AND SOLUTION. T_x supplies the centripetal force, and the given data are for the dynamical form of the centripetal force, i.e., $T_x = F_c = mv^2/r$ (Equation 7.11).

Given:

$L = 1.0$ m
$v = 1.1$ m/s
$m = 0.50$ kg and $\theta = 20°$

Find: F_c (magnitude of the centripetal force)

As pointed out previously, the magnitude of the centripetal force may be found using Equation 7.11:

$$F_c = T_x = \frac{mv^2}{r}$$

But the radial distance r is needed. From the figure, this quantity can be seen to be $r = L \sin 20°$, so

$$F_c = \frac{mv^2}{L\sin 20°} = \frac{(0.50\,\text{kg})(1.1\,\text{m/s})^2}{(1.0\,\text{m})(0.342)} = 1.8\,\text{N}$$

FOLLOW-UP EXERCISE. (a) What is the magnitude of the tension T in the string? (b) What is the period of the ball's rotation?

7.4 Angular Acceleration

As you might have guessed, there is another type of acceleration besides linear, and that is *angular acceleration*. This quantity is the time rate of change of angular velocity. In circular motion, if there is an angular acceleration, the motion is not uniform, because the speed and/or direction would be changing. Analogous to the linear case, the magnitude of the **average angular acceleration** ($\bar{\alpha}$) is given by

$$\bar{\alpha} = \frac{\Delta \omega}{\Delta t}$$

where the bar over the alpha indicates that it is an average value, as usual. Taking $t_0 = 0$, and if the angular acceleration is constant, so that $\bar{\alpha} = \alpha$, then

$$\alpha = \frac{\omega - \omega_0}{t} \quad \text{(constant angular acceleration)}$$

SI unit of angular acceleration: radians per second squared (rad/s²)

and rearranging,

$$\omega = \omega_0 + \alpha t \quad \text{(constant angular acceleration only)} \quad (7.12)$$

No boldface vector symbols with over arrows are used in Equation 7.12, because positive and negative signs will be used to indicate angular directions, as described earlier. As in the case of linear motion, if the angular acceleration increases the angular velocity, both quantities have the same sign, meaning that their vector directions are the same (i.e., α is in the same direction as ω as given by the right-hand rule). If the angular acceleration decreases the angular velocity, then the two quantities have opposite signs, meaning that their vectors are opposed (i.e., α is in the direction opposite to ω as given by the right-hand rule, or is an angular deceleration, so to speak).

EXAMPLE 7.10: A ROTATING DVD – ANGULAR ACCELERATION

A DVD accelerates uniformly from rest to its operational speed of 570 rpm in 3.50 s. (a) What is the angular acceleration of the DVD during this time? (b) What is the angular acceleration of the DVD after this time? (c) If the DVD comes uniformly to a stop in 4.50 s, what is its angular acceleration during this part of the motion?

THINKING IT THROUGH. (a) Once given the initial and final angular velocities, the constant (uniform) angular acceleration can be calculated (Equation 7.12), since the amount of time during which the DVD accelerates is known. (b) Keep in mind that the operational angular speed is constant. (c) Everything is given for Equation 7.12, but a negative result should be expected. Why?

SOLUTION

Given:

$$\omega_o = 0$$

$$\omega = (570\,\text{rpm})\left[\frac{(\pi/30)\,\text{rad/s}}{\text{rpm}}\right] = 59.7\,\text{rad/s}$$

$t_1 = 3.50$ s (starting up)
$t_2 = 4.50$ s (in coming to a stop)

Find:

(a) α (during startup)
(b) α (in operation)
(c) α (in coming to a stop)

(a) Using Equation 7.12, the acceleration during startup is

$$\alpha = \frac{\omega - \omega_o}{t_1} = \frac{59.7\,\text{rad/s} - 0}{3.50\,\text{s}} = 17.1\,\text{rad/s}^2$$

in the direction of the angular velocity.

(b) After the DVD reaches its operational speed, the angular velocity remains constant, so $\alpha = 0$.

(c) Again using Equation 7.12, but this time with $\omega_o = 570$ rpm and $\omega = 0$.

$$\alpha = \frac{\omega - \omega_o}{t_2} = \frac{0 - 59.7\,\text{rad/s}}{4.50\,\text{s}} = -13.3\,\text{rad/s}^2$$

where the − sign indicates that the angular acceleration is in the direction opposite that of the angular velocity (which is taken as +).

FOLLOW-UP EXERCISE. (a) What are the directions of the angular velocity and angular acceleration vectors in part (a) of this Example if the DVD rotates clockwise when viewed from above? (b) Do the directions of these vectors change in part (c)? Explain.

As with arc length and angle ($s = r\theta$) and tangential and angular speeds ($v = r\omega$), there is a relationship between the magnitudes of the tangential acceleration and the angular acceleration. The **tangential acceleration** (a_t) is associated with changes in tangential speed and hence continuously changes direction. The magnitudes of the tangential and angular accelerations are related by a factor of r. For circular motion with a constant radius r,

$$a_t = \frac{\Delta v}{\Delta t} = \frac{\Delta(r\omega)}{\Delta t} = \frac{r\Delta\omega}{\Delta t} = r\alpha$$

so

$$a_t = r\alpha \quad \left(\text{magnitude of tangential acceleration}\right) \qquad (7.13)$$

The tangential acceleration (a_t) is written with a subscript t to distinguish it from the radial, or centripetal, acceleration (a_c). Centripetal acceleration is necessary for circular motion, but tangential acceleration is not. For uniform circular motion, there is no angular acceleration ($\alpha = 0$) or tangential acceleration, as can be seen from Equation 7.13. There is only centripetal acceleration (▼ **Figure 7.15a**).

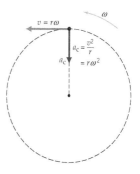

(a) Uniform circular motion
($\alpha = 0$)

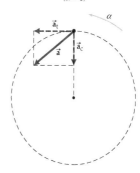

(b) Nonuniform circular motion
($\vec{a} = \vec{a}_t + \vec{a}_c$)

▲ **FIGURE 7.15   Acceleration and circular motion (a)** In uniform circular motion, there is centripetal acceleration, but no angular acceleration ($\alpha = 0$) or tangential acceleration ($a_t = r\alpha = 0$). **(b)** In nonuniform circular motion, there are angular and tangential accelerations, and the total acceleration is the vector sum of the tangential and centripetal accelerations.

However, when there is an angular acceleration a (and therefore a tangential acceleration of magnitude $a_t = r\alpha$), there is a change in *both* the angular *and* tangential velocities. As a result, the centripetal acceleration $a_c = v^2/r = r\omega^2$ must increase or decrease if the object is to maintain the same circular orbit (i.e., if r is to stay the same). When there are both tangential and centripetal accelerations, the instantaneous acceleration is their vector sum (Figure 7.15b).

The tangential acceleration vector and the centripetal acceleration vector are perpendicular to each other at any instant, and the acceleration is $\vec{a} = a_t\hat{t} + a_c\hat{r}$ where $\hat{t}$ and $\hat{r}$ are unit vectors directed tangentially and radially inward, respectively. You should be able to find the magnitude of $\vec{a}$ and the angle it makes relative to $\vec{a}_t$ by using trigonometry (Figure 7.15b).

Other equations for angular kinematics can be derived, as was done for the linear equations in Section 2.4. That development will not be shown here; the set of angular equations with their linear counterparts for constant accelerations is listed in ▼ **Table 7.2**. A quick review of Section 2.4 (with a change of symbols) will show you how the angular equations are derived.

TABLE 7.2   Equations for Linear and Angular Motion with Constant Acceleration[a]

Linear	Angular	
$x = \bar{v}t$	$\theta = \bar{\omega}t$	(1)
$\vec{v} = \dfrac{v + v_o}{2}$	$\bar{\omega} = \dfrac{\omega + \omega_o}{2}$	(2)
$v = v_o + at$	$\omega = \omega_o + \alpha t$	(3)
$x = x_o + v_o t + \frac{1}{2}at^2$	$\theta = \theta_o + \omega_o t + \frac{1}{2}\alpha t^2$	(4)
$v^2 = v_o^2 + 2a(x - x_o)$	$\omega^2 = \omega_o^2 + 2\alpha(\theta - \theta_o)$	(5)

[a] The first equation in each column is general; that is, not limited to situations where the acceleration is constant.

EXAMPLE 7.11: EVEN COOKING – ROTATIONAL KINEMATICS

A microwave oven has a 30-cm-diameter rotating plate for even cooking. The plate accelerates from rest at a uniform rate of 0.87 rad/s² for 0.50 s before reaching its constant operational speed. (a) How many revolutions does the plate make before reaching its operational speed? (b) What are the operational angular speed of the plate and the operational tangential speed at its rim?

THINKING IT THROUGH. This Example involves the use of the angular kinematic equations (Table 7.2). In (a), the angular distance θ will give the number of revolutions. For (b), first find v and then $v = r\omega$.

SOLUTION

Listing the given data and what is to be found:

Given:

$d = 30$ cm
$r = 15$ cm $= 0.15$ m (radius)
$\omega_o = 0$ (at rest)
$\alpha = 0.87$ rad/s² and $t = 0.50$ s

Find:

(a) θ (in revolutions)
(b) ω and v (angular and tangential speeds, respectively)

(a) To find the angular distance θ *in radians*, use Equation 4 from Table 7.2 with $\theta_o = 0$:

$$\theta = \omega_o t + \tfrac{1}{2}\alpha t^2 = 0 + \tfrac{1}{2}(0.87 \text{ rad/s}^2)(0.50 \text{ s})^2 = 0.11 \text{ rad}$$

Since 1 rev $= 2\pi$ rad,

$$\theta = (0.11 \text{ rad})\left(\frac{1 \text{ rev}}{2\pi \text{ rad}}\right) = 0.018 \text{ rev}$$

so the plate reaches its operational speed in only a small fraction of a revolution.

(b) From Table 7.2, it can be seen that Equation 3 (same as Equation 7.12) gives the angular speed, and

$$\omega = \omega_o + \alpha t = 0 + (0.87 \text{ rad/s}^2)(0.50 \text{ s}) = 0.44 \text{ rad/s}$$

Then, Equation 7.6 gives the tangential speed at the rim radius:

$$v = r\omega = (0.15 \text{ m})(0.44 \text{ rad/s}) = 0.66 \text{ m/s}$$

FOLLOW-UP EXERCISE. (a) When the oven is turned off, the plate makes half a revolution before stopping. What is the plate's angular acceleration during this period? (b) How long does it take to stop?

7.5 Newton's Law of Gravitation

Another of Isaac Newton's many accomplishments was the formulation of the **universal law of gravitation.** This law is very powerful and fundamental. Without it, for example, we would not understand the cause of tides or know how to put satellites into particular orbits around the Earth. This law allows us to analyze the motions of planets, comets, stars, and even galaxies. The word *universal* in the name indicates that it is believed to apply everywhere in the universe. (This term highlights the importance of the law, but for brevity, it is common to refer simply to *Newton's law of gravitation* or the *law of gravitation.*)

Newton's law of gravitation in mathematical form gives a simple relationship for the gravitational interaction between two particles, or point masses, m_1 and m_2 separated by a distance r

(▼ **Figure 7.16a**). Basically, every particle in the universe has an attractive gravitational interaction with every other particle because of their masses. The forces of mutual interaction are equal and opposite, forming a force pair as described by Newton's third law (Section 4.4); that is, $\vec{F}_{12} = -\vec{F}_{21}$ in Figure 7.16a.

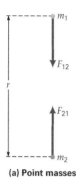

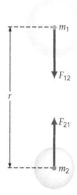

(a) Point masses

(b) Homogeneous spheres

▲ FIGURE 7.16 **Universal law of gravitation (a)** Any two particles, or point masses, are gravitationally attracted to each other with a force that has a magnitude given by Newton's universal law of gravitation. **(b)** For homogeneous spheres, the masses may be considered to be concentrated at their centers.

The gravitational attraction, or force (F_g), decreases as the square of the distance (r^2) between two point masses increases; that is, the magnitude of the gravitational force and the distance separating the two particles are related as follows:

$$F_g \propto \frac{1}{r^2}$$

(This type of relationship is called an *inverse-square*; that is, F_g is inversely proportional to r^2.)

Newton's law also correctly postulates that the gravitational force, or attraction of a body, depends on the body's mass – the greater the mass, the greater the attraction. However, because gravity is a mutual interaction between masses, it should be directly proportional to both masses; that is, to their product ($F_g \propto m_1 m_2$).

Hence, **Newton's law of gravitation** has the form $F_g \propto m_1 m_2 / r^2$. Expressed as an equation with a constant of proportionality, the magnitude of the mutually attractive gravitational force (F_g) between two masses is given by

$$F_g = \frac{Gm_1 m_2}{r^2} \quad \text{(Newton's law of gravitation)} \quad (7.14)$$

$$G = 6.67 \times 10^{-11} \, \text{N·m}^2 / \text{kg}^2$$

where G is a constant called the **universal gravitational constant**. This constant is often referred to as "big G" to distinguish it from "little g," the acceleration due to gravity. Note from Equation 7.14 that F_g approaches zero only when r becomes infinitely large. That is, the gravitational force has, or acts over, an *infinite range*.

How did Newton come to his conclusions about the force of gravity? Legend has it that his insight came after he observed an apple fall from a tree to the ground. Newton had been wondering what supplied the centripetal force to keep the Moon in orbit and might have had this thought: "If gravity attracts an apple toward the Earth, perhaps it also attracts the Moon, and the Moon is 'falling', or accelerating toward the Earth, under the influence of gravity" (▼ **Figure 7.17**).

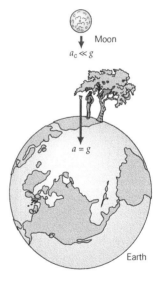

▲ FIGURE 7.17 **Gravitational insight?** Newton developed his law of gravitation while studying the orbital motion of the Moon. According to legend, his thinking was spurred when he observed an apple falling from a tree. He supposedly wondered whether the force causing the apple to accelerate toward the ground could extend to the Moon and cause it to "fall" or accelerate toward Earth – that is, supply its orbital centripetal acceleration.

Whether or not the legendary apple did the trick, Newton assumed that the Moon and the Earth were attracted to each other and could be treated as point masses, with their total masses concentrated at their centers (Figure 7.17b). The inverse-square relationship had been speculated on by some of his contemporaries. Newton's achievement was demonstrating that the relationship could be deduced from one of Johannes Kepler's laws of planetary motion (Section 7.6).

Newton expressed Equation 7.14 as a proportion ($F_g \propto m_1 m_2/r^2$) because he did not know the value of G. It was not until 1798 (seventy-one years after Newton's death) that the value of the universal gravitational constant was experimentally determined by an English physicist, Henry Cavendish. Cavendish used a sensitive balance to measure the gravitational force between separated spherical masses (as illustrated in Figure 7.16b). If F, r, and the m's are known, G can be computed from Equation 7.14.

As mentioned earlier, Newton considered the nearly spherical Earth and Moon to be point masses located at their respective centers. It took him some years, using mathematical methods he developed, to prove that this is the case only for spherical, *homogeneous* objects.* The concept is illustrated in ▼ **Figure 7.18**.

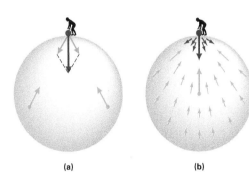

(a)   (b)

▲ **FIGURE 7.18 Uniform spherical masses** (a) Gravity acts between any two particles. The resultant gravitational force exerted on an object outside a homogeneous sphere by two particles at symmetric locations within the sphere is directed toward the center of the sphere. (b) Because of the sphere's symmetry and uniform distribution of mass, the net effect is as though all the mass of the sphere were concentrated as a particle at its center. For this special case, the gravitational center of force and center of mass coincide, but this is generally not true for other objects. (Only a few of the red force arrows are shown because of space considerations.)

EXAMPLE 7.12: GREATER GRAVITATIONAL ATTRACTION?

The gravitational attractions of the Sun and the Moon give rise to ocean tides. It is sometimes said that since the Moon is closer to the Earth than the Sun, the Moon's gravitational attraction is much stronger and therefore has a greater influence on ocean tides. Is this true?

---

* For a homogeneous sphere, the equivalent point mass is located at the center of mass. However, this is a special case. The center of gravitational force and the center of mass of a configuration of particles or an object do not generally coincide.

THINKING IT THROUGH. To see if this is true, the gravitational attractions of the Moon and the Sun on the Earth can be easily calculated using Newton's law of gravitation. The masses and distances are given on the inside back cover of the book. (Assume that the bodies are solid, homogenous spheres.)

SOLUTION

No data are given, so this must be available from references:

Given:

From tables on the inside back cover:

$m_E = 6.0 \times 10^{24}$ kg (mass of the Earth)
$m_M = 7.4 \times 10^{22}$ kg (mass of the Moon)
$m_S = 2.0 \times 10^{30}$ kg (mass of the Sun)
$r_{EM} = 3.8 \times 10^8$ m (average distance between)
$r_{ES} = 1.5 \times 10^8$ km (average distance between)

Find:

F_{EM} (Earth–Moon force)
F_{ES} (Earth–Sun force)

The average distances are taken to be the distance from the center of one to the center of the other. Using Equation 7.14, remembering to change kilometers to meters:

$$F_{EM} = \frac{Gm_1 m_2}{r^2} = \frac{GM_E m_M}{r_{EM}^2}$$
$$= \frac{(6.67 \times 10^{-11}\,\text{N·m}^2/\text{kg}^2)(6.0 \times 10^{24}\,\text{kg})(7.4 \times 10^{22}\,\text{kg})}{(3.8 \times 10^8\,\text{m})^2}$$
$$= 2.1 \times 10^{20}\,\text{N} \quad \text{(Earth–Moon)}$$
$$F_{ES} = \frac{Gm_E m_s}{r_{ES}^2} = \frac{(6.67 \times 10^{-11}\,\text{m}^2/\text{kg}^2)(6.0 \times 10^{24}\,\text{kg})(2.0 \times 10^{30}\,\text{kg})}{(1.5 \times 10^{11}\,\text{m})^2}$$
$$= 3.6 \times 10^{22}\,\text{N} \quad \text{(Earth–Sun)}$$

So, the gravitational attraction of the Sun on the Earth is much greater than that of the Moon on the Earth, on the order of 100 times greater. But it is well known that the Moon has the major influence on tides. How is this with less gravitational attraction? Basically, it is because the gravitational *differential* of the Moon with less gravitational attraction is greater. That is, the ocean water on the side of the Earth toward the Moon is closer and the gravitational attraction forms a tidal bulge. The Earth is less attracted toward the Moon, but it is somewhat displaced, leaving the least attracted water on the opposite side of the Earth where another tidal bulge is formed. As the Moon revolves about the Earth, the tidal bulges tag along, and there are two high tides (bulges) and two low tides daily. (Actually, the two high tides are 12 h, 25 min apart.)

Even though the Sun has greater gravitational attraction, the differential distances from the Sun to the water-Earth-water are miniscule, and so the Sun has little effect on the daily tides.

FOLLOW-UP EXERCISE. The gravitational attraction of the Earth on the Moon provides the centripetal force that keeps the Moon revolving in its orbit. It is sometimes said the Moon is "falling" (accelerating) toward the Earth. What is the magnitude of the Moon's acceleration in "falling" toward the Earth? And with this acceleration, why doesn't the Moon get closer to the Earth?

The acceleration due to gravity at a particular distance from a planet can also be investigated by using Newton's second law of motion and the law of gravitation. The magnitude of the acceleration due to gravity, which will generally be written as a_g at a distance r (r is greater than the radius of the planet) from the center of a spherical mass M, is found by setting the force of gravitational attraction due to that spherical mass equal to ma_g. This is the net force on an object of mass m at a distance r:

$$ma_g = \frac{GmM}{r^2}$$

Then, the acceleration due to gravity at any distance r from the planet's center is

$$a_g = \frac{GM}{r^2} \tag{7.15}$$

Notice that a_g is proportional to $1/r^2$, so the farther away an object is from the planet, the smaller its acceleration due to gravity and the smaller the attractive force (ma_g) on the object. The force is directed toward the center of the planet.

Equation 7.15 can be applied to the Moon or any planet. For example, taking the Earth to be a point mass M_E located at its center and R_E as its radius, we obtain the acceleration due to gravity at the Earth's surface ($a_{gE} = g$) by setting the distance r to be equal to R_E:

$$a_{gE} = g = \frac{GM_E}{R_E^2} \tag{7.16}$$

This equation has several interesting implications. First, it reveals that taking g to be constant everywhere on the surface of the Earth involves the assumption that the Earth has a homogeneous distribution of mass and that the distance from the center of the Earth to any location on its surface is the same. These two assumptions are not exactly true. Therefore, taking g to be a constant is an approximation, but one that works pretty well for most situations.

Also, you can see why the acceleration due to gravity is the same for all free-falling objects – that is, independent of the mass of the object. The mass of the object doesn't appear in Equation 7.16, so all objects in free fall accelerate at the same rate.

Finally, if you're observant, you'll notice that Equation 7.16 can be used to compute the mass of the Earth. All of the other quantities in the equation are measurable and their values are known, so M_E can readily be calculated. This is what Cavendish did after he determined the value of G experimentally.

The acceleration due to gravity does vary with altitude. At a distance h above the Earth's surface, $r = R_e + h$. The acceleration is then given by

$$a_g = \frac{GM_E}{(R_E + h)^2} \quad \text{(at a distance } h \text{ above the Earth's surface)} \tag{7.17}$$

7.5.1 Problem-Solving Hint

When comparing accelerations due to gravity or gravitational forces, you will often find it convenient to work with ratios. For example, comparing a_g with g (Equations 7.15 and 7.16) for the Earth gives

$$\frac{a_g}{g} = \frac{GM_E / r^2}{GM_E / R_E^2} = \frac{R_E^2}{r^2} = \left(\frac{R_E}{r}\right)^2 \text{ or } \frac{a_g}{g} = \left(\frac{R_E}{r}\right)^2$$

Note how the constants cancel out. Taking $r = R_E + h$ you can easily compute a_g/g, or the acceleration due to gravity at some altitude h above the Earth compared with g on the Earth's surface (9.80 m/s²).

Because R_E is very large compared with everyday altitudes above the Earth's surface, the acceleration due to gravity does not decrease very rapidly with height. At an altitude of 16 km (10 mi, about twice as high as modern jet airliners fly), $a_g/g = 0.99$, and thus a_g, is still 99% of the value of g at the Earth's surface. At an altitude of 320 km (200 mi), a_g is 91% of g. This is the approximate altitude of typical near-Earth orbits.

EXAMPLE 7.13: GEOSYNCHRONOUS SATELLITE ORBIT

Some communication and weather satellites are launched into circular orbits above the Earth's equator so they are *synchronous* (from the Greek *syn-*, same, and *chronos*, time) with the Earth's rotation. That is, they remain "fixed" or "hover" over one point on the equator. At what altitude are these geosynchronous satellites?

THINKING IT THROUGH. To remain above one location at the equator, the period of the satellite's revolution must be the same as the Earth's period of rotation, that is, 24 h. Also, the centripetal force keeping the satellite in orbit is supplied by the gravitational force of the Earth, $F_g = F_c$. The distance between the center of the Earth and the satellite is $r = R_E + h$. (R_E is the radius of the Earth and h is the height or altitude of the satellite above the Earth's surface.)

SOLUTION

Listing the known data:

Given:

T (period) $= 24$ h $= 8.64 \times 10^4$ s
$r = R_E + h$
From solar system data inside back cover:
$R_E = 6.4 \times 10^3$ km $= 6.4 \times 10^6$ m, $M_E = 6.0 \times 10^{24}$ kg

Find: h (altitude)

Setting the magnitudes of gravitational force and the motional centripetal force equal ($F_g = F_c$), where m is the mass of the satellite, and putting the values in terms of angular speed,

$$F_g = F_c$$
$$\frac{GmM_E}{r^2} = \frac{mv^2}{r} = \frac{m(r\omega)^2}{r} = mr\omega^2$$

using the relationship $\omega = 2\pi/T$,

$$r^3 = \frac{GM_E}{\omega^2} = GM_E \left(\frac{T}{2\pi}\right)^2 = \left(\frac{GM_E}{4\pi^2}\right)T^2$$
$$= \frac{(6.67 \times 10^{-11}\,\text{N·m}^2\,/\,\text{kg}^2)(6.0 \times 10^{24}\,\text{kg})(8.64 \times 10^4\,\text{s})^2}{4\pi^2}$$
$$= 76 \times 10^{21}\,\text{m}^3$$

taking the cube root:

$$r = \sqrt[3]{76 \times 10^{21}\,\text{m}^3} = 4.2 \times 10^7\,\text{m}$$

So,

$$h = r - R_E = 4.2 \times 10^7\,\text{m} - 0.64 \times 10^7\,\text{m} = 3.6 \times 10^7\,\text{m}$$
$$= 3.6 \times 10^4\,\text{km}\ (= 22\,000\,\text{mi})$$

FOLLOW-UP EXERCISE. Show that the period of a satellite in orbit close to the Earth's surface ($h \ll R_E$) may be approximated by $T^2 \approx 4R_E$ and compute T. (Neglect air resistance.)

Another aspect of the decrease of g with altitude concerns potential energy. In Section 5.5, it was learned that $U = mgh$ for an object at a height h above some zero reference point, since g is essentially constant near the Earth's surface. This potential energy is equal to the work done in raising the object a distance h above the Earth's surface in a *uniform* gravitational field.

But what if the change in altitude is so large that g cannot be considered constant while work is done in moving an object, such as a satellite? In this case, the equation $U = mgh$ doesn't apply. In general, it can be shown (using mathematical methods beyond the scope of this book) that the **gravitational potential energy** (U) of two point masses separated by a distance r is given by

$$U = -\frac{Gm_1m_2}{r} \tag{7.18}$$

The negative sign in Equation 7.18 arises from the choice of the zero reference point (the point where $U = 0$), which is $r = \infty$ (infinity).

In terms of the Earth and a mass m at an altitude h above the Earth's surface,

$$U = -\frac{Gm_1m_2}{r} = -\frac{GmM_E}{R_E + h} \tag{7.19}$$

where r is the distance separating the Earth's center and the mass. This means that on the Earth we can visualize ourselves as being in a negative gravitational potential energy well (▼ **Figure 7.19**) that extends to infinity, because the force of gravity has an infinite range. As h increases, so does U. That is, U becomes *less negative*, or gets closer to zero (i.e., more positive), corresponding to a higher position in the potential energy well.

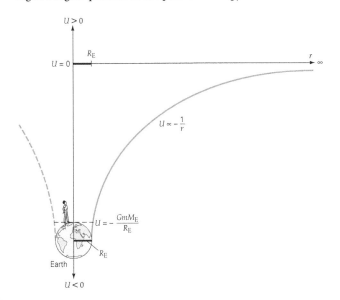

▲ **FIGURE 7.19   Gravitational potential energy well** On the Earth, we can visualize ourselves as being in a negative gravitational potential energy well. As with an actual well or hole in the ground, work must be done against gravity to get higher in the well. The potential energy of an object increases as the object moves higher in the well. This means that the value of U becomes less negative. The top of the Earth's gravitational well is at infinity, where the gravitational potential energy is, by choice, zero.

Thus, when gravity does negative work (an object moves higher in the well) or gravity does positive work (an object falls lower in the well), there is a *change* in potential energy. As with finite potential energy wells, this *change* in energy is one of the most important things in analyzing situations such as these.

EXAMPLE 7.14: DIFFERENT ORBITS – CHANGE IN GRAVITATIONAL POTENTIAL ENERGY

Two 50-kg satellites move in circular orbits about the Earth at altitudes of 1000 km (about 620 mi) and 36 000 km (about 22 000 mi), respectively. The lower one monitors particles about to enter the atmosphere, and the higher, geosynchronous one takes weather pictures from its stationary position with respect to the Earth's surface over the equator (see Example 7.13). What is the difference in the gravitational potential energies of the two satellites in their respective orbits?

THINKING IT THROUGH. The potential energies of the satellites are given by Equation 7.19. Since an increase in altitude (h) results in a less negative value of U, the satellite with the greater h is higher in the gravitational-potential energy well and has more gravitational potential energy.

SOLUTION

Listing the data so we can better see what's given (with two significant figures):

Given:

$m = 50$ kg
$h_1 = 1000$ km $= 1.0 \times 10^6$ m
$h_2 = 36\ 000$ km $= 36 \times 10^6$ m
$M_E = 6.0 \times 10^{24}$ kg (from the inside the back cover of the book)
$R_E = 6.4 \times 10^6$ m

Find: ΔU (difference in potential energy)

The difference in the gravitational potential energy can be computed directly from Equation 7.19. Keep in mind that the potential energy is the energy of position, so we compute the potential energies for each position or altitude and subtract one from the other. Thus,

$$\Delta U = U_2 - U_1 = -\frac{GmM_E}{R_E + h_2} - \left(-\frac{GmM_E}{R_E + h_1}\right)$$

$$= GmM_E\left(\frac{1}{R_E + h_1} - \frac{1}{R_E + h_2}\right)$$

$$= (6.67 \times 10^{-11}\,\text{N·m}^2/\text{kg}^2)(50\,\text{kg})(6.0 \times 10^{24}\,\text{kg})$$

$$\times \left[\frac{1}{6.4 \times 10^6\,\text{m} + 1.0 \times 10^6\,\text{m}} - \frac{1}{6.4 \times 10^6\,\text{m} + 36 \times 10^6\,\text{m}}\right]$$

$$= +2.2 \times 10^9\,\text{J}$$

Because ΔU is positive, m_2 is higher in the gravitational potential energy well than m_1. Note that even though both U_1 and U_2 *are* negative, U_2 is "more positive," or "less negative," and closer to zero. Thus, it takes more energy to get a satellite farther from the Earth.

FOLLOW-UP EXERCISE. Suppose that the altitude of the higher satellite in this Example were doubled, to 72 000 km. Would the difference in the gravitational potential energies of the two satellites then be twice as great? Justify your answer.

Substituting the gravitational potential energy (Equation 7.18) into the equation for the total mechanical energy gives the equation a different form than it had in Chapter 5. For example, the total mechanical energy of a mass m_1 moving at a distance r from mass m_2 is

$$E = K + U = \frac{1}{2}m_1v^2 - \frac{Gm_1m_2}{r} \qquad (7.20)$$

This equation and the principle of the conservation of energy can be applied to the Earth's motion about the Sun by neglecting

other gravitational forces. The Earth's orbit is not quite circular, but slightly elliptical. At *perihelion* (the point of the Earth's closest approach to the Sun), the mutual gravitational potential energy is less (a larger negative value) than it is at *aphelion* (the point farthest from the Sun). Therefore, as can be seen from Equation 7.20 in the form $1/2(m_1v^2) = E + Gm_1m_2/r$ where E is constant, the Earth's kinetic energy and orbital speed are greatest at perihelion (the smallest value of r) and least at aphelion (the greatest value of r). Or, in general, the Earth's orbital speed is greater when it is nearer the Sun than when it is farther away.

Mutual gravitational potential energy also applies to a group, or *configuration*, of more than two masses. That is, there is gravitational potential energy due to the several masses in a configuration, because work was needed to be done in bringing the masses together. Suppose that there is a single fixed mass m_1, and another mass m_2 is brought close to m_1 from an infinite distance (where $U = 0$). The work done against the attractive force of gravity is negative (why?) and equal to the change in the mutual potential energy of the masses, which are now separated by a distance r_{12}; that is, $U_{12} = -Gm_1m_2/r_{12}$.

If a third mass m_3 is brought close to the other two fixed masses, there are then two forces of gravity acting on m_3, so $U_{13} = -Gm_1m_3/r_{13}$ and $U_{23} = -Gm_2m_3/r_{23}$. The total gravitational potential energy of the configuration is therefore

$$U = U_{12} + U_{13} + U_{23} = -\frac{Gm_1m_2}{r_{12}} - \frac{Gm_1m_3}{r_{13}} - \frac{Gm_2m_3}{r_{23}} \quad (7.21)$$

A fourth mass could be brought in to further prove the point, but this development should be sufficient to suggest that the total gravitational potential energy of a configuration of particles is equal to the sum of the individual potential energies for all pairs of particles.

EXAMPLE 7.15: TOTAL GRAVITATIONAL POTENTIAL ENERGY – ENERGY OF CONFIGURATION

Three masses are in a configuration as shown in ▼ **Figure 7.20**. What is their total gravitational potential energy?

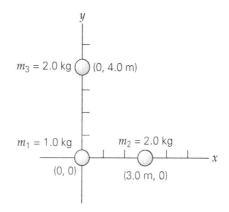

▲ FIGURE 7.20 **Total gravitational potential energy** See Example 7.15 for description.

THINKING IT THROUGH. Equation 7.21 applies, but be sure to keep your masses and their distances distinct.

SOLUTION

From the figure, the data are:

Given:

$m_1 = 1.0\ \text{kg}$
$m_2 = 2.0\ \text{kg}$
$m_3 = 2.0\ \text{kg}$
$r_{12} = 3.0\ \text{m};\ r_{13} = 4.0\ \text{m};\ r_{23} = 5.0\ \text{m}$ (3-4-5 right-angle triangle)

Find: U (total gravitational potential energy)

Equation 7.21 can be used directly, since only three masses are used in this Example. (Note that Equation 7.21 can be extended to any number of masses.) Then,

$$U = U_{12} + U_{13} + U_{23} = -\frac{Gm_1m_2}{r_{12}} - \frac{Gm_1m_3}{r_{13}} - \frac{Gm_2m_3}{r_{23}}$$

$$= (6.67 \times 10^{-11}\ \text{N·m}^2/\text{kg}^2)$$

$$\times \left[-\frac{(1.0\ \text{kg})(2.0\ \text{kg})}{3.0\ \text{m}} - \frac{(1.0\ \text{kg})(2.0\ \text{kg})}{4.0\ \text{m}} - \frac{(2.0\ \text{kg})(2.0\ \text{kg})}{5.0\ \text{m}} \right]$$

$$= -1.3 \times 10^{-10}\ \text{J}$$

FOLLOW-UP EXERCISE. Explain what the *negative* potential energy in this Example means in physical terms.

7.6 Kepler's Laws and Earth Satellites

The force of gravity determines the motions of the planets and satellites and holds the solar system (and galaxy) together. A general description of planetary motion had been set forth shortly before Newton's time by the German astronomer and mathematician Johannes Kepler (1571–1630). Kepler formulated three *empirical* laws from observational data gathered during a twenty-year period by the Danish astronomer Tycho Brahe (1546–1601).

Kepler went to Prague to assist Brahe, who was the official mathematician at the court of the Holy Roman Emperor. Brahe died the next year, and Kepler succeeded him, inheriting his records of the positions of the planets. Analyzing these data, Kepler announced the first two of his three laws in 1609 (the year Galileo built his first telescope). These laws were applied initially only to Mars. Kepler's third law came ten years later.

Interestingly enough, Kepler's laws of planetary motion, which took him about fifteen years to deduce from observed data, can now be derived theoretically with a page or two of calculations. These three laws apply not only to planets, but also to any system composed of a body revolving about a much more massive body to which the inverse-square law of gravitation applies (such as the Moon, artificial Earth satellites, and solar-bound comets).

7.6.1 Kepler's First Law (the Law of Orbits)

Planets move in elliptical orbits, with the Sun at one of the focal points.

An ellipse, shown in ▼ **Figure 7.21a**, has, in general, an oval shape, resembling a flattened circle. In fact, a circle is a special case of an ellipse in which the focal points, or *foci* (plural of *focus*), are at the same point (the center of the circle). Although the orbits of the planets are elliptical, most do not deviate very much from circles (Mercury and the dwarf planet Pluto are notable exceptions; see "Eccentricity," Appendix II). For example, the difference between the perihelion and aphelion of the Earth (its closest and farthest distances from the Sun, respectively) is about 5 million km. This distance may sound like a lot, but it is only a little more than 3% of 150 million km, which is the average distance between the Earth and the Sun.

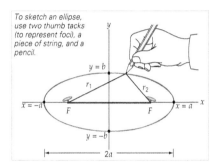

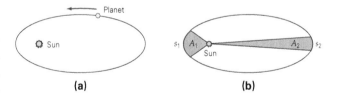

(a)   **(b)**

▲ **FIGURE 7.21   Kepler's first and second laws of planetary motion** **(a)** In general, an ellipse has an oval shape. The sum of the distances from the focal points *F* to any point on the ellipse is constant: $r_1 + r_2 = 2a$. Here, $2a$ is the length of the line joining the two points on the ellipse at the greatest distance from its center, called the *major axis*. (The line joining the two points closest to the center is *b*, the *minor axis*.) Planets revolve about the Sun in elliptical orbits for which the Sun is at one of the focal points and nothing is at the other. **(b)** A line joining the Sun and a planet sweeps out equal areas in equal times. Since $A_1 = A_2$, a planet travels faster along s_1 than along s_2. (See text for description.)

7.6.2 Kepler's Second Law (the Law of Areas)

A line from the Sun to a planet sweeps out equal areas in equal lengths of time.

This law is illustrated in Figure 7.21b. Since the time to travel the different orbital distances (s_1 and s_2) is the same such that the areas swept out (A_1 and A_2) are equal, this law tells you that the orbital speed of a planet varies in different parts of its orbit. Because a planet's orbit is elliptical, its orbital speed is greater when it is closer to the Sun than when it is farther away. The conservation of energy was used in Section 7.5 (Equation 7.20) to deduce this relationship for the Earth.

7.6.3 Kepler's Third Law (the Law of Periods)

The square of the orbital period of a planet is directly proportional to the cube of the average distance of the planet from the Sun; that is, $T^2 \propto r^3$.

Kepler's third law is easily derived for the special case of a planet with a circular orbit, using Newton's law of gravitation. Since the centripetal force is supplied by the force of gravity, the expressions for these forces can be set equal:

$$\underset{\text{centripetal force}}{\frac{m_p v^2}{r}} = \underset{\text{gravitational force}}{\frac{Gm_p M_S}{r^2}}$$

and

$$v = \sqrt{\frac{GM_S}{r}}$$

In these equations, m_p and M_S are the masses of the planet and the Sun, respectively, and v is the planet's orbital speed. But $v = 2\pi r/T$ (circumference/period = distance/time), so

$$\frac{2\pi r}{T} = \sqrt{\frac{GM_S}{r}}$$

Squaring both sides and solving for T^2 gives

$$T^2 = \left(\frac{4\pi^2}{GM_S}\right)r^3$$

or

$$T^2 = Kr^3 \qquad (7.22)$$

The constant K for solar system planetary orbits is easily evaluated from orbital data (for T and r) for the Earth: $K = 2.97 \times 10^{-19}$ s^2/m^3. As an exercise, you might wish to convert K to the more useful units of y^2/km^3. (*Note:* This value of K applies to all the planets in our solar system but does *not* apply to planet satellites, as Example 7.16 will show.)

If you look inside the back cover and in Appendix II, you will find the masses of the Sun and the planets of the solar system. How were these masses determined? The following Example shows how Kepler's third law can be used to do this.

EXAMPLE 7.16: BY JOVE!

The planet Jupiter is the largest in the solar system, both in volume and mass. Jupiter has 79 known moons, the four largest having been discovered by Galileo in 1610. Two of these moons, Io and Europa, are shown in ▼ **Figure 7.22**. Given that Io is an average distance of 4.22×10^5 km from the enter of Jupiter and has an orbital period of 1.77 days, compute the mass of Jupiter.

▲ FIGURE 7.22 **Jupiter and moons** Two of Jupiter's moons discovered by Galileo, Europa and Io, are shown here. Europa is on the left, and Io on the right below the Great Red Spot. (Its shadow is seen on the planet.)

THINKING IT THROUGH. Given the values for Io's distance from the planet (r) and period (T), this would appear to be an application of Kepler's third law, and it is. However, keep in mind that the M_S in Equation 7.22 is the mass of the Sun, which the planets orbit. The third law can be applied to any satellite, as long as the M is that of the body being orbited by the satellite. In this case, it will be M_J, the mass of Jupiter.

SOLUTION

Given:

$r = 4.22 \times 10^5$ km $= 4.22 \times 10^8$ m
$T = (1.77$ days$) (18.64 \times 10^4$ s/day$) = 1.53 \times 10^5$ s

Find: M_J (mass of Jupiter)

With r and T known, K can be found in Equation 7.22 (written K_I, indicating it is for Io-Jupiter)*

$$K_I = \frac{T^2}{r^3} = \frac{(1.53 \times 10^5 \text{ s})^2}{(4.22 \times 10^8 \text{ m})^3} = 3.11 \times 10^{-16} \text{ s}^2/\text{m}^3$$

---

* Note that this is different from the K for planets orbiting about the Sun because it is about the mass of Jupiter, not the mass of the Sun.

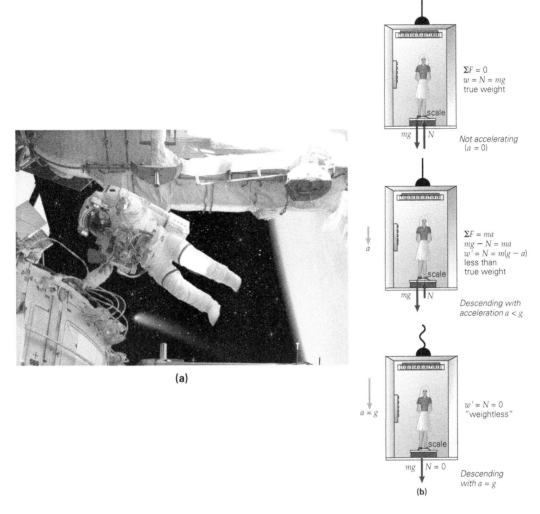

$\Sigma F = 0$
$w = N = mg$
true weight

Not accelerating
(a = 0)

$\Sigma F = ma$
$mg - N = ma$
$w' = N = m(g - a)$
less than
true weight

Descending with
acceleration a < g

$w' = N = 0$
"weightless"

Descending
with a = g

(b)

▲ **FIGURE 7.23**   **Apparent weightlessness (a)** An astronaut "floats" in space, seemingly in a weightless condition. (He is not being held up.) **(b)** In a stationary elevator (top), a scale reads the passenger's true weight. The weight reading is the reaction force N of the scale on the person. If the elevator is descending with an acceleration $a < g$ (middle), the reaction force and apparent weight are less than the true weight. If the elevator were in free fall ($a = g$; bottom), the reaction force and indicated weight would be zero, since the scale would be falling as fast as the person.

Then, writing K_I explicitly, $K_I = (4\pi^2 / GM_J)$, and

$$M_J = \frac{4\pi^2}{GK_I} = \frac{4\pi^2}{(6.67 \times 10^{-11}\,\text{N·m}^2\,/\,\text{kg}^2)(3.11 \times 10^{-16}\,\text{s}^2\,/\,\text{m}^3)}$$
$$= 1.90 \times 10^{27}\,\text{kg}$$

FOLLOW-UP EXERCISE. Compute the mass of the Sun from Earth's orbital data.

7.6.4 Earth's Satellites and Weightlessness

The advent of the space age and the use of orbiting satellites have brought us the terms *weightlessness* and *zero gravity*, because astronauts appear to "float" about in orbiting space-craft (▲ **Figure 7.23a**). However, these terms are misnomers.

As mentioned earlier in the chapter, gravity is an infinite-range force, and the Earth's gravity acts on a spacecraft and astronauts, supplying the centripetal force necessary to keep them in orbit. Gravity there is not zero, so there must be weight.

A better term to describe the floating effect of astronauts in orbiting spacecraft would be *apparent weightlessness.** The astronauts "float" because both the astronauts and the spacecraft are centripetally accelerating (or "falling") toward the Earth at the same rate. To help you understand this effect, consider the analogous situation of a person standing on a scale in an elevator (Figure 7.23b). The "weight" measurement that the scale registers is actually the normal force N of the scale on the person. In a nonaccelerating elevator ($a = 0$), $N = mg = w$, and N is equal to the true weight of the individual. However, suppose the elevator

---

* This term will be used here for description, with the understanding that it is *apparent* zero-*g*.

is descending with an acceleration a, where $a < g$. As the vector diagram in the figure shows,

$$mg - N = ma$$

and the *apparent* weight w' is

$$w' = N = m(g - a) < mg$$

where the downward direction is taken as positive in this instance. With a downward acceleration a, we see that N is less than mg, hence the scale indicates that the person weighs less than his or her true weight. Note that the *apparent acceleration* due to gravity is $g' = g - a$.

Now suppose the elevator were in free fall, with $a = g$. As you can see, N (and thus the apparent weight w') would be zero. Essentially, the scale is accelerating, or falling, at the same rate as the person. The scale may indicate a "weightless" condition ($N = 0$), but gravity still acts, as would be noted by the sudden stop of the elevator.

Space has been called the *final frontier.* Someday, instead of brief stays in Earth-orbiting spacecraft, there may be permanent space colonies with "artificial" gravity in the future. One proposal is to have a huge, rotating space colony in the form of a wheel – somewhat like an automobile tire, with the inhabitants living inside the tire. As you know, centripetal force is necessary to keep an object in rotational circular motion. On the rotating Earth, that force is supplied by gravity, and we refer to it as *weight.* We exert a force on the ground, and the normal force (by Newton's third law) exerted upward on our feet is what is actually sensed and gives the feeling of "having our feet on solid ground."

In a rotating space colony, the situation is somewhat reversed. The rotating colony would supply the centripetal force on the inhabitants, and the centripetal force would be perceived as a normal force acting on the soles of the feet, providing artificial gravity. Rotation at the proper speed would produce a simulation of "normal" gravity ($a_c \approx g = 9.80$ m/s²) within the colony wheel. Note that in the colonists' world, "down" would be outward, toward the periphery of the space station, and "up" would always be inward, toward the axis of rotation (▶ **Figure 7.24**).

Chapter 7 Review

- The **radian (rad)** is a measure of angle; 1 rad is the angle of a circle subtended by an arc length (s) equal to the radius (r). Arc length (angle in radians):

$$s = r\theta \tag{7.3}$$

- *Angular kinematic equations* for $\theta_o = 0$ and $t_o = 0$ (see Table 7.2 for linear analogues):

$$\theta = \bar{\omega}t \quad \text{(in general, not limited to constant acceleration)} \tag{7.5}$$

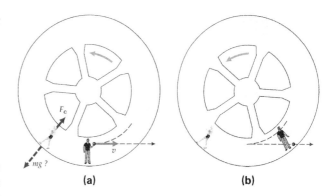

▲ **FIGURE 7.24** **Space colony and artificial gravity** It has been suggested that a space colony could be housed in a huge, rotating wheel. The rotation would supply the "artificial gravity" for the colonists. **(a)** In the frame of reference of someone in a rotating space colony, centripetal force, coming from the normal force N of the floor, would be perceived as weight sensation or artificial gravity. We are used to feeling N upward on our feet to balance gravity. Rotation at the proper speed would simulate normal gravity. To an outside observer, a dropped ball would follow a tangential straight-line path, as shown. **(b)** A colonist on board the space colony would observe the ball to fall downward as in a normal gravitational situation.

$$\bar{\omega} = \frac{\omega + \omega_o}{2} \qquad\qquad (2, \text{Table } 7.2)$$
$$\omega = \omega_o + \alpha t \qquad\qquad\quad (3, \text{Table } 7.2)$$
$$\theta = \theta_o + \omega_o t + \tfrac{1}{2}\alpha t^2 \qquad (4, \text{Table } 7.2)$$
$$\omega^2 = \omega_o^2 + 2\alpha(\theta - \theta_o) \qquad (5, \text{Table } 7.2)$$

constant acceleration only

- **Tangential speed** (v_t) and **angular speed** (ω) for circular motion are directly proportional, with the radius r being the constant of proportionality:

$$v_t = r\omega \tag{7.6}$$

- The **frequency** (f) and **period** (T) are inversely related:

$$f = \frac{1}{T} \tag{7.7}$$

- **Angular speed** (*with uniform circular motion*) in terms of period (T) and frequency (f):

$$\omega = \frac{2\pi}{T} = 2\pi f \tag{7.8}$$

- In uniform circular motion, a **centripetal acceleration** (a_c) is required and is always directed toward the center of the circular path, and its magnitude is given by:

$$a_c = \frac{v^2}{r} = r\omega^2 \tag{7.10}$$

- A **centripetal force,** F_c, (the net force directed toward the center of a circle) is a requirement for circular motion, the magnitude of which is

$$F_c = ma_c = \frac{mv^2}{r} \qquad (7.11)$$

- **Angular acceleration** (α) is the time rate of change of angular velocity and is related to the **tangential acceleration** (a_t) in magnitude by

$$a_t = r\alpha \qquad (7.13)$$

- According to **Newton's law of gravitation**, every particle attracts every other particle in the universe with a force that is proportional to the masses of both particles and inversely proportional to the square of the distance between them:

$$F_g = \frac{Gm_1m_2}{r^2} \qquad (7.14)$$
$$G = 6.67 \times 10^{-11}\,\text{N·m}^2/\text{kg}^2$$

- **Acceleration due to gravity at an altitude *h* above the Earth's surface:**

$$a_g = \frac{GM_E}{(R_E + h)^2} \qquad (7.17)$$

- **Gravitational potential energy of a system of two masses:**

$$U = -\frac{Gm_1m_2}{r} \qquad (7.18)$$

- **Kepler's first law (law of orbits):** Planets move in elliptical orbits, with the Sun at one of the focal points.
- **Kepler's second law (law of areas):** A line from the Sun to a planet sweeps out equal areas in equal lengths of time.
- **Kepler's third law (law of periods):**

$$T^2 = Kr^3 \qquad (7.22)$$

(*K* depends on the mass of the object orbited; for objects orbiting the Sun, $K = 2.97 \times 10^{-19}\,\text{s}^2/\text{m}^3$.)

End of Chapter Questions and Exercises

Multiple Choice Questions

7.1 Angular Measure

1. The radian unit is a ratio of (a) degree/time, (b) length, (c) length/length, (d) length/time.
2. For the polar coordinates of a particle traveling in a circle, the variables are (a) both *r* and θ, (b) only *r*, (c) only θ, (d) none of the preceding.

3. Which of the following is the greatest angle: (a) $3\pi/2$ rad, (b) $5\pi/8$ rad, or (c) 220°?

7.2 Angular Speed and Velocity

4. Viewed from above, a turntable rotates counterclockwise. The angular velocity vector is then (a) tangential to the turntable's rim, (b) out of the plane of the turntable, (c) counterclockwise, (d) none of the preceding.
5. The frequency unit of hertz is equivalent to (a) that of the period, (b) that of the cycle, (c) radian/s, (d) s^{-1}.
6. The unit of angular speed is (a) rad, (b) s^{-1} (c) s, (d) rad/rpm.
7. The particles in a uniformly rotating object all have the same (a) angular acceleration, (b) angular speed, (c) tangential velocity, (d) both (a) and (b).

7.3 Uniform Circular Motion and Centripetal Acceleration

8. Uniform circular motion requires (a) centripetal acceleration, (b) angular speed, (c) tangential velocity, (d) all of the preceding.
9. In uniform circular motion, there is a (a) constant velocity, (b) constant angular velocity, (c) zero acceleration, (d) nonzero tangential acceleration.
10. If the centripetal force on a particle in uniform circular motion is increased, (a) the tangential speed will remain constant, (b) the tangential speed will decrease, (c) the radius of the circular path will increase, (d) the tangential speed will increase and/or the radius will decrease.

7.4 Angular Acceleration

11. The unit of angular acceleration is (a) s^{-2}, (b) rpm, (c) rad^2/s, (d) s^2.
12. The angular acceleration in circular motion (a) is equal in magnitude to the tangential acceleration divided by the radius, (b) increases the angular velocity if both angular velocity and angular acceleration are in the same direction, (c) has units of s^{-2}, (d) all of the preceding.
13. In circular motion, the tangential acceleration (a) does not depend on the angular acceleration, (b) is constant, (c) has units of s^{-2}, (d) none of these.
14. For uniform circular motion, (a) $\alpha = 0$, (b) $\omega = 0$, (c) $v = 0$, (d) none of the preceding.

7.5 Newton's Law of Gravitation

15. The gravitational force is (a) a linear function of distance, (b) an inverse function of distance, (c) an inverse function of distance squared, (d) sometimes repulsive.
16. The acceleration due to gravity of an object on the Earth's surface (a) is a universal constant, like *G*, (b) does not depend on the Earth's mass, (c) is directly proportional to the Earth's radius, (d) does not depend on the object's mass.
17. Compared with its value on the Earth's surface, the value of the acceleration due to gravity at an altitude of one Earth radius is (a) the same, (b) two times as great, (c) one-half as great, (d) one-fourth as great.

7.6 Kepler's Laws and Earth Satellites

18. A new planet is discovered and its period determined. The new planet's distance from the Sun could then be found by using Kepler's (a) first law, (b) second law, (c) third law.

19. As a planet moves in its elliptical orbit, (a) its speed is constant, (b) its distance from the Sun is constant, (c) it moves faster when it is closer to the Sun, (d) it moves slower when it is closer to the Sun.

Conceptual Questions

7.1 Angular Measure

1. Why does 1 rad equal 57.3°? Wouldn't it be more convenient to have an even number of degrees?

2. A wheel rotates about a rigid axis through its center. Do all points on the wheel travel the same distance? How about the same *angular* distance?

7.2 Angular Speed and Velocity

3. Do all points on a wheel rotating about a fixed axis through its center have the same angular velocity? The same tangential speed? Explain.

4. When "clockwise" or "counterclockwise" is used to describe rotational motion, why is a phrase such as "viewed from above" added?

5. Imagine yourself standing on the edge of an operating merry-go-round. How would your tangential speed be affected if you walked toward the center? (Watch out for the horses going up and down.)

6. A car's speedometer is set to read in relationship to the angular speed of the rear wheels. If for winter the tires are changed to larger diameter all-weather tires, would this affect the speedometer reading? Explain. How about the odometer?

7.3 Uniform Circular Motion and Centripetal Acceleration

7. The spin cycle of a washing machine is used to extract water from recently washed clothes. Explain the physical principle(s) involved.

8. Can a car be moving with a constant speed of 100 km/h and still be accelerating? Explain.

9. On the rotating Earth, at what location(s) would a person have (a) the greatest and (b) the least centripetal acceleration? (c) How does the centripetal acceleration for a person at 40° N latitude compare to that of a person at 40° S latitude? (What supplies the centripetal acceleration?)

10. When rounding a curve in a fast-moving car, we experience a feeling of being thrown outward (▶ **Figure 7.25**). It is sometimes said that this effect occurs because of an outward centrifugal (center-fleeing) force. However, in terms of Newton's laws for a ground-based observer, this pseudo, or false, force doesn't really exist. Analyze the situation in the figure to show that this is the case (i.e., that the force does not exist). [*Hint:* Start with Newton's first law.]

▲ FIGURE 7.25 **A center-fleeing force?** See Conceptual Question 10.

11. Many curves have banked turns, which allow the cars to travel faster around the curves than if the road were flat. Actually, cars could also make turns on these banked curves if there were no friction at all. Explain this statement using the free-body diagram shown in ▼ **Figure 7.26**.

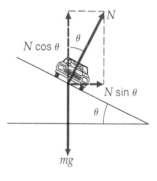

▲ FIGURE 7.26 **Banking safety** See Conceptual Question 11.

7.4 Angular Acceleration

12. A car increases its speed when it is on a circular track. Does the car have centripetal acceleration? How about angular acceleration? Explain.

13. Is it possible for a car traveling on a circular track to have angular acceleration, but not centripetal acceleration? Explain.

14. Is it possible for a car traveling on a circular track to have a change in tangential acceleration and no change in centripetal acceleration?

7.5 Newton's Law of Gravitation

15. If the mass of the Moon were doubled, how would this affect its orbit?

16. Weighing yourself at a park in Ecuador through which the equator runs, you would find that you weigh slightly less than normal. Why is this?

17. Can you determine the mass of the Earth simply by measuring the gravitational acceleration near the Earth's surface? If yes, give the details.

Exercises*

Integrated Exercises (IEs) *are two-part exercises. The first part typically requires a conceptual answer choice based on physical thinking and basic principles. The following part is quantitative calculations associated with the conceptual choice made in the first part of the exercise.*

7.1 Angular Measure

1. • The polar coordinates of a point are (5.3 m, 32°). What are the point's Cartesian coordinates?

2. • Convert the following angles from degrees to radians, to two significant figures: (a) 15°, (b) 45°, (c) 90°, and (d) 120°.

3. • Convert the following angles from radians to degrees: (a) $\pi/6$ rad, (b) $5\pi/12$ rad, (c) $3\pi/4$ rad, and (d) π rad.

4. • Express the following angles in degrees, radians, and/or revolutions (rev) as appropriate: (a) 105°, (b) 1.8 rad, and (c) 5/7 rev.

5. • You measure the length of a distant car to be subtended by an angular distance of 1.5°. If the car is actually 5.0 m long, approximately how far away is the car?

6. • How large an angle in radians and degrees does the diameter of the Moon subtend to a person on the Earth?

7. •• The hour, minute, and second hands on a clock are 0.25, 0.30, and 0.35 m long, respectively. What are the distances traveled by the tips of the hands in a 30-min interval?

8. •• A car with a 65-cm-diameter wheel travels 3.0 km. How many revolutions does the wheel make in this distance?

9. •• Two gear wheels with radii of 25 and 60 cm have interlocking teeth. How many radians does the smaller wheel turn when the larger wheel turns 4.0 rev?

10. •• At the end of her routine, an ice skater spins through 7.50 revolutions with her arms always fully outstretched at right angles to her body. If her arms are 60.0 cm long, through what arc length distance do the tips of her fingers move during her finish?

11. ••• (a) Could a circular pie be cut such that all of the wedge-shaped pieces have an arc length along the outer crust equal to the pie's radius? (b) If not, how many such pieces could you cut, and what would be the angular dimension of the final piece?

7.2 Angular Speed and Velocity

12. • A computer DVD-ROM has a variable angular speed from 200 rpm to 450 rpm. Express this range of angular speed in radians per second.

13. • A race car makes two and a half laps around a circular track in 3.0 min. What is the car's average angular speed?

14. • What are the angular speeds of the (a) second hand, (b) minute hand, and (c) hour hand of a clock? Are the speeds constant?

15. • What is the period of revolution for (a) a 12 500-rpm centrifuge and (b) a 9 500-rpm computer hard disk drive?

16. •• Determine which has the greater angular speed: particle A, which travels 160° in 2.00 s, or particle B, which travels 4π rad in 8.00 s.

17. •• The tangential speed of a particle on a rotating wheel is 3.0 m/s. If the particle is 0.20 m from the axis of rotation, how long will the particle take to make one revolution?

18. •• A merry-go-round makes 24 revolutions in a 3.0-min ride. (a) What is its average angular speed in rad/s? (b) What are the tangential speeds of two people 4.0 m and 5.0 m from the center, or axis of rotation?

19. •• A little boy jumps onto a small merry-go-round (radius of 2.00 m) in a park and rotates for 2.30 s through an arc length distance of 2.55 m before coming to rest. If he landed (and stayed) at a distance of 1.75 m from the central axis of rotation of the merry-go-round, what was his average angular speed and average tangential speed?

7.3 Uniform Circular Motion and Centripetal Acceleration

20. • An Indy car with a speed of 120 km/h goes around a level, circular track with a radius of 1.00 km. What is the centripetal acceleration of the car?

21. • A wheel of radius 1.5 m rotates at a uniform speed. If a point on the rim of the wheel has a centripetal acceleration of 1.2 m/s², what is the point's tangential speed?

22. • A rotating cylinder about 16 km long and 7.0 km in diameter is designed to be used as a space colony. With what angular speed must it rotate so that the residents on it will experience the same acceleration due to gravity as on Earth?

23. •• An airplane pilot is going to demonstrate flying in a tight vertical circle. To ensure that she doesn't black out at the bottom of the circle, the acceleration must not exceed 4.0 *g*. If the speed of the plane is 50 m/s at the bottom of the circle, what is the minimum radius of the circle so that the 4.0 *g* limit is not exceeded?

24. •• Imagine that you swing about your head a ball attached to the end of a string. The ball moves at a constant speed in a horizontal circle. (a) Can the string be exactly horizontal? Why or why not? (b) If the mass of the ball is 0.250 kg, the radius of the circle is 1.50 m, and it takes 1.20 s for the ball to make one revolution, what is the ball's tangential speed? (c) What centripetal force are you imparting to the ball via the string?

25. •• In Exercise 24, if you supplied a tension force of 12.5 N to the string, what angle would the string make relative to the horizontal?

26. •• A car with a constant speed of 83.0 km/h enters a circular flat curve with a radius of curvature of 0.400 km.

If the friction between the road and the car's tires can supply a centripetal acceleration of 1.25 m/s², does the car negotiate the curve safely? Justify your answer.

27. **IE ••** A student is to swing a bucket of water in a vertical circle without spilling any (▼ **Figure 7.27**). (a) Explain how this task is possible. (b) If the distance from his shoulder to the center of mass of the bucket of water is 1.0 m, what is the minimum speed required to keep the water from coming out of the bucket at the top of the swing?

▲ FIGURE 7.27   **Weightless water?** See Exercise 27.

28. **••** In performing a "figure 8" maneuver, a figure skater wants to make the top part of the 8 approximately a circle of radius 2.20 m. He needs to glide through this part of the figure at approximately a constant speed, taking 4.50 s. His skates digging into the ice are capable of providing a maximum centripetal acceleration of 3.25 m/s². Will he be able to do this as planned? If not, what adjustment can he make if he wants this part of the figure to remain the same size (assume the ice conditions and skates don't change)?

29. **••** A light string of length of 56.0 cm connects two small square blocks, each with a mass of 1.50 kg. The system is placed on a slippery (frictionless) sheet of horizontal ice and spun so that the two blocks rotate uniformly about their common center of mass, which itself does not move. They are supposed to rotate with a period of 0.750 s. If the string can exert a force of only 100 N before it breaks, determine whether this string will work.

30. **IE ••** A jet pilot puts an aircraft with a constant speed into a vertical circular loop. (a) Which is greater, the normal force exerted on the seat by the pilot at the bottom of the loop or that at the top of the loop? Why? (b) If the speed of the aircraft is 700 km/h and the radius of the circle is 2.0 km, calculate the normal forces exerted on the seat by the pilot at the bottom and top of the loop. Express your answer in terms of the pilot's weight.

31. **•••** A block of mass m slides down an inclined plane into a loop-the-loop of radius r (▼ **Figure 7.28**). (a) Neglecting friction, what is the minimum speed the block must have at the highest point of the loop in order to stay in the loop? [*Hint:* What force must act on the block at the top of the loop to keep the block on a circular path?] (b) At what vertical height on the inclined plane (in terms of the radius of the loop) must the block be released if it is to have the required minimum speed at the top of the loop?

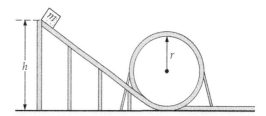

▲ FIGURE 7.28   **Loop-the-loop** See Exercise 31.

32. **•••** For a scene in a movie, a stunt driver drives a 150×10^3 kg SUV with a length of 4.25 m around a circular curve with a radius of curvature of 0.333 km (▼ **Figure 7.29**). The vehicle is to be driven off the edge of a gully 10.0 m wide, and land on the other side 2.96 m below the initial side. What is the minimum centripetal acceleration the SUV must have in going around the circular curve to clear the gully and land on the other side?

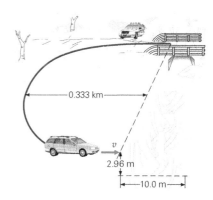

▲ FIGURE 7.29   **Over the gully** See Exercise 32.

7.4 Angular Acceleration

33. **•** A merry-go-round accelerating uniformly from rest achieves its operating speed of 2.5 rpm in 5 revolutions. What is the magnitude of its angular acceleration?

34. **••** A flywheel rotates with an angular speed of 25 rev/s. As it is brought to rest with a constant acceleration, it turns 50 rev. (a) What is the magnitude of the angular acceleration? (b) How much time does it take to stop?

35. **IE ••** A car on a circular track accelerates from rest. (a) The car experiences (1) only angular acceleration, (2) only centripetal acceleration, (3) both angular and centripetal accelerations. Why? (b) If the radius of the track is 0.30 km and the magnitude of the constant angular acceleration is 4.5×10^{-3} rad/s², how long does the car take to make one lap around the track? (c) What is the total (vector) acceleration of the car when it has completed half of a lap?

36. **••** Show that for a constant acceleration,

$$\theta = \theta_\circ + \frac{(\omega^2 - \omega_\circ^2)}{2\alpha}$$

37. **••** The blades of a fan running at low speed turn at 250 rpm. When the fan is switched to high speed, the rotation rate increases uniformly to 350 rpm in 5.75 s. (a) What is the magnitude of the angular acceleration of the blades? (b) How many revolutions do the blades go through while the fan is accelerating?

38. **••** In the spin-dry cycle of a modern washing machine, a wet towel with a mass of 1.50 kg is "stuck" to the inside surface of the perforated (to allow the water out) washing cylinder. To have decent removal of water, damp/wet clothes need to experience a centripetal acceleration of at least 10 *g*. Assuming this value, and that the cylinder has a radius of 35.0 cm, determine the constant angular acceleration of the towel required if the washing machine takes 2.50 s to achieve its final angular speed.

39. **•••** A pendulum swinging in a circular arc under the influence of gravity, as shown in ▼ **Figure 7.30**, has both centripetal and tangential components of acceleration. (a) If the pendulum bob has a speed of 2.7 m/s when the cord makes an angle of $\theta = 15°$ with the vertical, what are the magnitudes of the components at this time? (b) Where is the centripetal acceleration a maximum? What is the value of the tangential acceleration at that location?

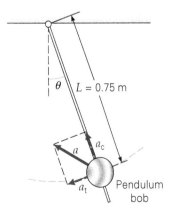

▲ FIGURE 7.30   **A swinging pendulum** See Exercise 39.

40. **•••** A simple pendulum of length 2.00 m is released from a horizontal position. When it makes an angle of 30° from the vertical, determine (a) its angular acceleration, (b) its centripetal acceleration, and (c) the tension in the string. Assume the bob's mass is 1.50 kg.

7.5 Newton's Law of Gravitation

41. **•** From the known mass and radius of the Moon (see the tables inside the back cover of the book), compute the value of the acceleration due to gravity, g_M, at the surface of the Moon.

42. **•** The gravitational forces of the Earth and the Moon are attractive, so there must be a point on a line joining their centers where the gravitational forces on an object cancel. How far is this distance from the Earth's center?

43. **••** Four identical masses of 2.5 kg each are located at the corners of a square with 1.0-m sides. What is the net force on any one of the masses?

44. **••** The average density of the Earth is 5.52 g/cm³. Assuming this is a uniform density, compute the value of *G*.

45. **••** A 100-kg object is taken to a height of 300 km above the Earth's surface. (a) What is the object's mass at this height? (b) What is the object's weight at this height?

46. **••** A man has a mass of 75 kg on the Earth's surface. How far above the surface of the Earth would he have to go to "lose" 10% of his body weight?

47. **••** It takes 27 days for the Moon to orbit the Earth in a nearly circular orbit of radius 3.80×10^5 km. (a) Show in symbol notation that the mass of the Earth can be found using these data. (b) Compute the Earth's mass and compare with the value given inside the back cover of the book.

48. **IE ••** Two objects are attracting each other with a certain gravitational force. (a) If the distance between the objects is halved, the new gravitational force will (1) increase by a factor of 2, (2) increase by a factor of 4, (3) decrease by a factor of 2, (4) decrease by a factor of 4. Why? (b) If the original force between the two objects is 0.90 N, and the distance is tripled, what is the new gravitational force between the objects?

49. **••** During the Apollo lunar explorations of the late 1960s and early 1970s, the main section of the spaceship remained in orbit about the Moon with one astronaut in it while the other two astronauts descended to the surface in the landing module. If the main section orbited about 50 mi above the lunar surface, determine that section's centripetal acceleration.

7.6 Kepler's Laws and Earth Satellites

50. **•** An instrument package is projected vertically upward to collect data near the top of the Earth's atmosphere (at an altitude of about 900 km). (a) What initial speed is required at the Earth's surface for the package to reach this height? (b) What percentage of the escape speed is this initial speed?

51. • What is the orbital speed of a geosynchronous satellite? (See Example 7.13.)

52. •• In the year 2056, Martian Colony I wants to put a Mars-synchronous communication satellite in orbit about Mars to facilitate communications with the new bases being planned on the Red Planet. At what distance above the Martian equator would this satellite be placed? (To a good approximation, the Martian day is the same length as that of the Earth's.)

53. •• The asteroid belt that lies between Mars and Jupiter may be the debris of a planet that broke apart or that was not able to form as a result of Jupiter's strong gravitation. An average asteroid has a period of about 5.0 y. Approximately how far from the Sun would this "fifth" planet have been?

54. •• Using a development similar to Kepler's law of periods for planets orbiting the Sun, find the required altitude of geosynchronous satellites above the Earth. [*Hint:* The period of such satellites is the same as that of the Earth.]

55. •• Venus has a rotational period of 243 days. What would be the altitude of a synchronous satellite for this planet (similar to geosynchronous satellite on the Earth)?

56. ••• A small space probe is put into circular orbit about a newly discovered moon of Saturn. The moon's radius is known to be 550 km. If the probe orbits at a height of 1500 km above the moon's surface and takes 2.00 Earth days to make one orbit, determine the moon's mass.

Rotational Motion and Equilibrium

The long pole plays a very important role in maintaining the balance and equilibrium of the girl walking on a tightrope.

It's always a good idea to keep your equilibrium – but it's more important in some situations than in others. When looking at the chapter-opening photo, your first reaction is probably to wonder how this young girl keeps from falling. Presumably, the pole must help – but in what way? You'll find out in this chapter.

It might be said that the tightrope walker is in equilibrium. Translational equilibrium ($\Sigma \vec{F}_i = 0$) was discussed in Section 4.5, but here there is another consideration; namely, rotation. Should the walker start to fall (and it is hoped she doesn't), there would

be a sideways rotation about the rope. To avoid this calamity, another condition must be met: rotational equilibrium, which will be considered in this chapter.

The tightrope walker is striving to avoid rotational motion. But rotational motion is very important in physics, because rotating objects are all around us: wheels on vehicles, gears and pulleys in machinery, planets in our solar system, and even many bones in the human body. (Can you think of bones that rotate in sockets?)

Fortunately, the equations describing rotational motion can be written as almost direct analogues of those for translational (linear) motion. In Section 7.4, this similarity was pointed out with respect to the linear and angular kinematic equations. With the addition of equations describing rotational dynamics, you will be able to analyze the general motions of real objects that can rotate, as well as translate.

8.1 Rigid Bodies, Translations, and Rotations

In previous chapters, it was convenient to consider motion with the understanding that an object can be represented by a particle located at the center of mass of the object. Rotation, or spinning, was not a consideration, because a particle, or point mass, has no physical dimensions. Rotational motion becomes relevant when analyzing the motion of a solid, extended object or *rigid body*, which is the focus of this chapter.

> A **rigid body** is an object or a system of particles in which the distances between particles are fixed (remain constant).

A quantity of liquid water is not a rigid body, but the ice that would form if the water were frozen would be. The discussion of rigid body rotation is therefore restricted to solids. Actually, the concept of a rigid body is an idealization. In reality, the particles (atoms and molecules) of a solid vibrate constantly. Also, solids can undergo elastic (and inelastic) deformations in collisions (Section 6.4). Even so, most solids can be considered rigid bodies for purposes of analyzing rotational motion.

A rigid body may be subject to either or both of two types of motions: *translational* and *rotational*. Translational motion is basically the linear motion studied in previous chapters. If an object has only (pure) **translational motion**, *every particle in it has the same instantaneous velocity*, which means that the object is not rotating (▶ **Figure 8.1a**).

An object may have only (pure) **rotational motion** (motion about a fixed axis), and *all of the particles of the object have the same instantaneous angular velocity* and travel in circles about the axis of rotation (Figure 8.1b).*

General rigid body motion is a combination of both translational and rotational motions. When you throw a ball, the translational motion is described by the motion of its center of mass (as in projectile motion). But the ball may also spin, or rotate, and it usually does. A common example of rigid body motion involving both translation and rotation is rolling, as illustrated in Figure 8.1c. The combined motion of any point or particle is given by the vector sum of the particle's instantaneous velocity

---
* The words *rotation* and *revolution* are commonly used synonymously. In general, this book uses *rotation* when the axis of rotation goes through the body (e.g., the Earth's rotation on its axis, in a period of 24 h) and *revolution* when the axis is outside the body (e.g., the revolution of the Earth about the Sun, in a period of 365 days).

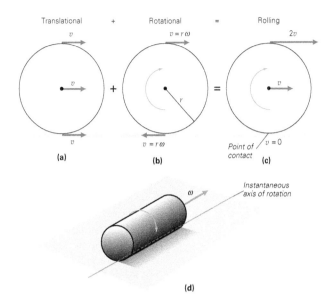

▲ **FIGURE 8.1** **Rolling – a combination of translational and rotational motions** (a) In pure translational motion, all the particles of an object have the same instantaneous velocity. (b) In pure rotational motion, all the particles of an object have the same instantaneous angular velocity. (c) Rolling is a combination of translational and rotational motions. Summing the velocity vectors for these two motions shows that the point of contact (for a sphere) or the line of contact (for a cylinder) is instantaneously at rest. (d) The line of contact for a cylinder is called the *instantaneous axis of rotation*. Note that the center of mass of a rolling object on a level surface moves linearly and remains over the point or line of contact. (Is the ω vector in the right direction?)

vectors. (Three points or particles are shown in the figure – one at the top, one in the middle, and one at the bottom of the object.)

At each instant, a rolling object rotates about an **instantaneous axis of rotation** through the point of contact of the object with the surface it is rolling on (for a sphere) or along the line of contact of the object with the surface (for a cylinder; Figure 8.1d). The location of this axis changes with time. However, note in Figure 8.1c that the point or line of contact of the body with the surface is instantaneously at rest (and thus has zero velocity), as can be seen from the vector addition of the combined motions at that point. Also, the point on the top has twice the tangential speed (2*v*) of the middle (center-of-mass) point (*v*), because the top point is twice as far away from the instantaneous axis of rotation as the middle point. (With a radius *r*, for the middle point, $r\omega = v$, and for the top point, $2r\omega = 2v$.)

When an object rolls without slipping – for example, when a ball (or cylinder) rolls in a straight line on a flat surface – it turns through an angle θ, and a point (or line) on the object that was initially in contact with the surface moves through an arc distance *s* (▶ **Figure 8.2**). And from Section 7.1, $s = r\theta$ (Equation 7.3). The center of mass of the ball is directly over the point of contact and moves a linear distance *s*. Then

$$v_{CM} = \frac{s}{t} = \frac{r\theta}{t} = r\omega$$

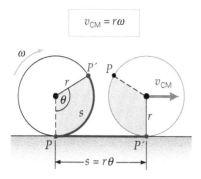

$$v_{CM} = r\omega$$

▲ FIGURE 8.2 **Rolling without slipping** As an object rolls without slipping, the length of the arc between two points of contact on the circumference is equal to the linear distance traveled. (Think of paint coming off a roller.) This distance is $s = r\theta$. The speed of the center of mass is $v_{CM} = r\omega$.

where $\omega = \theta/t$. In terms of the speed of the center of mass and the angular speed ω **the condition for rolling without slipping** is

$$v_{CM} = r\omega \quad \text{(rolling, no slipping)} \qquad (8.1)$$

The condition for rolling without slipping is also expressed by

$$s = r\theta \quad \text{(rolling, no slipping)} \qquad (8.1a)$$

where s is the distance the object rolls (the distance the center of mass moves).

By carrying Equation 8.1 one step further, an expression for the time rate of change of the velocity can be obtained.

$$a_{CM} = \frac{\Delta v_{CM}}{\Delta t} = \frac{\Delta(r\omega)}{\Delta t} = \frac{r\Delta\omega}{\Delta t} = r\alpha$$

which yields an equation for accelerated rolling without slipping:

$$a_{CM} = r\alpha \quad \text{(accelerated rolling without slipping)} \qquad (8.1b)$$

where $\alpha = \Delta\omega/\Delta t$ (for a constant α).

Essentially, an object will roll without slipping if the coefficient of static friction between the object and surface is great enough to prevent slippage. There may be a combination of rolling and slipping motions – for example, the slipping of a car's wheels when traveling through mud or ice. If there is rolling and slipping, then there is no clear relationship between the translational and rotational motions, and $v_{CM} = r\omega$ does *not* hold.

INTEGRATED EXAMPLE 8.1: ROLLING WITHOUT SLIPPING

A cylinder rolls on a horizontal surface without slipping with a constant speed of v. (a) At any point in time, the tangential speed of the top of the cylinder is (1) v, (2) $r\omega$, (3) $v + r\omega$, or (4) zero. (b) The cylinder has a radius of 12 cm and a center-of-mass speed of 0.10 m/s

as it rolls without slipping. If it continues to travel at this speed for 2.0 s, through what angle does the cylinder rotate during this time?

(a) CONCEPTUAL REASONING. Since the cylinder rolls without slipping, the relationship $v_{CM} = r\omega$ applies. As was shown in Figure 8.1, the speed at the point of contact is zero, v for the center (of mass), and $2v$ at the top. With $v = 0$ at point of contact, and $v = r\omega$, the answer is (3), $v + r\omega = v + v = 2v$.

(b) QUANTITATIVE REASONING AND SOLUTION. Since the radius and the translational speed are known, the angular speed can be calculated from the nonslipping condition, $v_{CM} = r\omega$. With this relationship and the time, the angle of rotation may be calculated. Listing the data:

Given:

$$r = 12 \text{ cm} = 0.12 \text{ m}$$
$$v_{CM} = 0.10 \text{ m/s}$$
$$t = 2.0 \text{ s}$$

Find: θ (angle of rotation)

Using $v_{CM} = r\omega$ to find the angular speed,

$$\omega = \frac{v_{CM}}{r} = \frac{0.10 \text{ m/s}}{0.12 \text{ m}} = 0.83 \text{ rad/s}$$

Then,

$$\theta = \omega t = (0.83 \text{ rad/s})(2.0 \text{ s}) = 1.7 \text{ rad}$$

The cylinder makes a little over one-quarter of a rotation. (Right? Check it yourself.)

FOLLOW-UP EXERCISE. How far does the CM of the cylinder travel linearly in part (b) of this Example? Find the distance by using two different methods: translational and rotational. (*Answers to all Follow-Up Exercises are given in Appendix V at the back of the book.*)

8.2 Torque, Equilibrium, and Stability

8.2.1 Torque

As with translational motion, a force is necessary to produce a change in rotational motion. However, the rate of change of rotational motion depends not only on the magnitude of the force, but also on the perpendicular distance of its line of action from the axis of rotation, $r_\perp$ (▶ **Figure 8.3a,b**). The line of action of a force is an imaginary line extending through the force vector arrow – that is, an extended line along which the force acts. (Note that if force is applied at the axis of rotation, $r_\perp$ is zero and there is no change in rotational motion about that axis.)

Figure 8.3 shows that $r_\perp = r\sin\theta$, where r is the straight-line distance between the axis of rotation and the force line of action and θ is the angle between the line of r (or radial vector $\vec{r}$) and the force vector $\vec{F}$. The perpendicular distance $r_\perp$ is called the **moment arm** or **lever arm**.

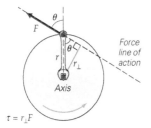

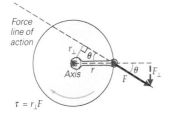

(a) Counterclockwise torque

$\tau = r_\perp F$

(b) Smaller clockwise torque

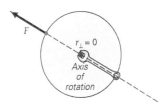

$\tau = r_\perp F$

(c) Zero torque

▲ **FIGURE 8.3** **Torque and moment arm** (a) The perpendicular distance $r_\perp$ from the axis of rotation to the line of action of a force is called the *moment arm* (or *lever arm*) and is equal to $r\sin\theta$ (where θ is the angle between the line of r, or radial vector $\vec{\mathbf{r}}$, and the force vector $\vec{\mathbf{F}}$). The magnitude of the torque (τ), or twisting force, that produces rotational motion is $\tau = r_\perp F$. (b) The same force in the opposite direction with a smaller moment arm produces a smaller torque in the opposite direction. Note that $r_\perp F = rF_\perp$ or $(r\sin\theta)F = r(F\sin\theta)$. (c) When a force acts through the axis of rotation, $r_\perp = 0$ and $\tau = 0$.

The product of the force and the lever arm is called **torque** ($\vec{\tau}$), from the Latin *torquere*, meaning "to twist." The magnitude of the torque provided by the force is

$$\tau = r_\perp F = rF\sin\theta \qquad (8.2)$$

SI unit of torque: meter × newton (m·N)

(The symbolism $r_\perp F$ is commonly used to denote torque, but also note from Figure 8.3b that $r_\perp F = rF_\perp$.) The SI units of torque are meter × newton (m·N) the same as the units of work, $W = Fd$ (N·m or J). However, the units of torque are usually written in reverse order as m·N to avoid confusion. But keep in mind that torque is *not* work, and its unit is *not* the joule.

Rotational acceleration is not *always* produced when a force acts on a stationary rigid body. From Equation 8.2, it can be seen that when the force acts through the axis of rotation such that $\theta = 0$, then $\tau = 0$ (Figure 8.3c). Also, when $\theta = 90°$, the torque is at a maximum and the force acts perpendicularly to r. The

angular acceleration depends on *where* a perpendicular force is applied (and therefore on the length of the lever arm). As a practical example, think of applying a force to a heavy glass door that swings in and out. Where you apply the force makes a great difference in how easily the door opens or rotates (through the hinge axis). Have you ever tried to open such a door and inadvertently pushed on the side near the hinges? This force produces a small torque and thus little or no rotational acceleration.

Torque in rotational motion can be thought of as the analogue of force in translational motion. An unbalanced or net force changes translational motion, whereas an unbalanced or net torque changes rotational motion. Torque is a vector. Its direction is always perpendicular to the plane of the force and moment arm and is given by a *right-hand rule* similar to that for angular velocity given in Section 7.2. If the fingers of the right hand are curled around the axis of rotation in the direction that the torque would produce a rotational (angular) acceleration, the extended thumb points in the direction of the torque. A sign convention, as in the case of linear motion, can be used to represent torque directions, as will be discussed shortly.

EXAMPLE 8.2: LIFTING AND HOLDING – MUSCLE TORQUE AT WORK

In the human body torques produced by the contraction of muscles cause some bones to rotate at joints. For example, when you lift something with your forearm, a torque is applied on the lower arm by the biceps muscle (▼ **Figure 8.4**). With the axis of rotation through the elbow joint and the muscle attached 4.0 cm from the joint, what are the magnitudes of the muscle torques for cases (a) and (b) in Figure 8.4 if the muscle exerts a force of 600 N?

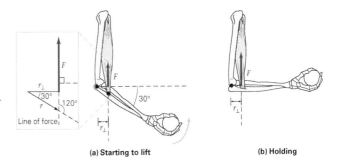

(a) Starting to lift **(b) Holding**

▲ **FIGURE 8.4** **Human torque** In our bodies, torques produced by the contraction of muscles cause bones to rotate at joints. Here the bicep muscles supply the force. See Example 8.2 for description.

THINKING IT THROUGH. As in many rotational situations, it is important to know the orientations of the $\vec{\mathbf{r}}$ and $\vec{\mathbf{F}}$ vectors so that the angle *between* them can be found to determine the lever arm. Note in the inset in Figure 8.4a that if the tails of the $\vec{\mathbf{r}}$ and $\vec{\mathbf{F}}$ vectors were put together, the angle between them would be greater than 90°; that is, $30° + 90° = 120°$. In Figure 8.4b, the angle is 90°. This Example demonstrates an important point, namely that θ is the angle *between* the radial vector $\vec{\mathbf{r}}$ and the force vector $\vec{\mathbf{F}}$.

SOLUTION

First listing the data given here and in the figure:

Given:

$r = 4.0 \text{ cm} = 0.040 \text{ m}$
$F = 600 \text{ N}$
$\theta_a = 30° + 90° = 120°$
$\theta_b = 90°$

Find:

(a) τ_a (muscle torque) for Figure 8.4a
(b) τ_b (muscle torque) for Figure 8.4b

(a) In this case, $\vec{r}$ is directed along the forearm, so the angle between the $\vec{r}$ and $\vec{F}$ vectors is $\theta_a = 120°$. Using Equation 8.2,

$$\tau_a = rF\sin(120°) = (0.040\text{m})(600\text{N})(0.866) = 21\text{m·N}.$$

(b) Here, the distance r and the line of action of the force are perpendicular ($\theta_b = 90°$), and $r_\perp = r\sin 90° = r$. Then,

$$\tau_b = r_\perp F = rF = (0.040\text{m})(600\text{N}) = 24\text{m·N}$$

The torque is greater in (b). This is to be expected because the maximum value of the torque (τ_{max}) occurs when $\theta = 90°$.

FOLLOW-UP EXERCISE. In part (a) of this Example, there must have been a net torque, since the ball was accelerated upward by a rotation of the forearm. In part (b), the ball is just being held and there is no rotational acceleration, so there is no net torque on the system. Identify the other torque(s) in each case.

CONCEPTUAL EXAMPLE 8.3: MY ACHING BACK

A person bends over as shown in ▶ **Figure 8.5**. For most of us, the center of gravity of the human body is in or near the chest region. When bending over, the force of gravity on the person's upper torso, acting through its center of gravity, gives rise to a torque that tends to produce rotation about an axis at the base of the spine that could cause us to fall over – but this doesn't usually happen. So why don't we fall when bending over like this? (Consider only the upper torso.)

REASONING AND ANSWER. If this were the only torque acting, we would indeed fall when bending forward. But since there is no rotation and fall, another force must be producing a torque such that the net torque is zero. Where does this other torque come from? Obviously from inside the body through a complicated combination of back muscles.

Representing the vector sum of all the back muscle forces as the net force F_b (as shown in Figure 8.5), it can be seen that the back muscles exert a force that counterbalances the torque on the torso's center of gravity.

FOLLOW-UP EXERCISE. Suppose the person was bent over holding a heavy object he had just picked up. How would this affect the back muscle force?

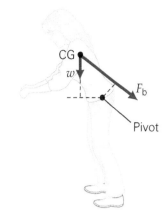

▲ FIGURE 8.5 **Torque but no rotation** When bending over, a person's weight, acting through the upper torso's center of gravity, gives rise to a counterclockwise torque that tends to produce rotation about an axis at the base of the spine. However, the back muscles attached between the shoulders combine to produce a force, F_b, and the resulting clockwise torque counterbalances that of gravity, such that the net torque is zero.

Before considering rotational dynamics with net torques and rotational motions, let's look at a situation in which the forces and torques acting on an object are balanced, and the object is in equilibrium.

8.2.2 Equilibrium

In general, equilibrium means that forces and torques are in balance. Unbalanced forces produce translational accelerations, but *balanced* forces produce the condition called *translational equilibrium*. Similarly, unbalanced torques produce rotational accelerations, but *balanced* torques produce *rotational equilibrium*.

According to Newton's first law of motion, when the sum of the forces acting on a body is zero, the body remains either at rest (static) or in motion with a constant velocity. In either case, the body is said to be in **translational equilibrium** (Section 4.5). Stated another way, the *condition for translational equilibrium* is that the net force on a body is zero; that is, $\vec{F}_{net} = \Sigma\vec{F}_i = 0$. It should be apparent that this condition is satisfied for the situations illustrated in ▶ **Figure 8.6a** and **b**. Forces with lines of action through the same point are called concurrent forces. When these forces vectorially add to zero, as in Figure 8.6a and b, the body is in translational equilibrium.

But what about the situation pictured in Figure 8.6c? Here, $\Sigma\vec{F} = 0$, but the opposing forces will cause the object to rotate, and it will clearly not be in a state of static equilibrium. (Such a pair of equal and opposite forces that do not have the same line of action is called a *couple*.) Thus, the condition $\Sigma\vec{F} = 0$ is a necessary, but *not sufficient*, condition for static equilibrium.

Since $\vec{F}_{net} = \Sigma\vec{F}_i = 0$ is the condition for translational equilibrium, you might predict (and correctly so) that $\vec{\tau}_{net} = \Sigma\vec{\tau}_i = 0$ is the *condition for rotational equilibrium*. That is, if the sum of the torques acting on an object is zero, then the object is in **rotational equilibrium** – it remains rotationally at rest or rotates with a constant angular velocity.

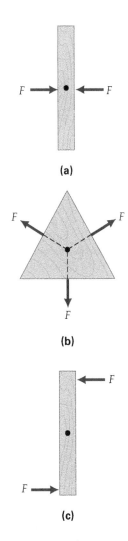

(a)

(b)

(c)

▲ FIGURE 8.6   **Equilibrium and forces** Forces with lines of action through the same point are said to be *concurrent*. The resultants of the concurrent forces acting on the objects in **(a)** and **(b)** are zero $(\vec{F}_{net} = \Sigma \vec{F}_i = 0)$, and the objects are in static equilibrium, because the net torque *and* net force are zero. In **(c)**, the object is in *translational* equilibrium, but it will undergo angular acceleration; thus, the object is *not* in rotational equilibrium.

Thus, there are actually *two* equilibrium conditions. Taken together, they define **mechanical equilibrium**. *A body is said to be in mechanical equilibrium when the conditions for both translational and rotational equilibrium are satisfied:*

$$\vec{F}_{net} = \Sigma \vec{F}_i = 0 \quad \text{(for translational equilibrium)}$$
$$\vec{\tau}_{net} = \Sigma \vec{\tau}_i = 0 \quad \text{(for rotational equilibrium)} \tag{8.3}$$

A rigid body in mechanical equilibrium may be either at rest or moving with a constant linear and/or angular velocity. An example of the latter is an object rolling (rotating) without slipping on a level surface, with the center of mass of the object having a constant velocity. Of greater practical interest is **static equilibrium**, the condition that exists when a rigid body remains

at rest – that is, a body for which $v = 0$ and $\omega = 0$. There are many instances in which we do not want things to move, and this absence of motion can occur only if the equilibrium conditions are satisfied. It is particularly comforting to know, for example, that a bridge over which cars are crossing is in static equilibrium and not subject to translational or rotational motions.

Let's consider examples of static translational equilibrium and static rotational equilibrium separately and then an example in which both apply.

EXAMPLE 8.4: TRANSLATIONAL STATIC EQUILIBRIUM – NO TRANSLATIONAL ACCELERATION OR MOTION

A picture hangs motionless on a wall as shown in ▼ **Figure 8.7a**. If the picture has a mass of 3.0 kg, what are the magnitudes of the tension forces in the wires?

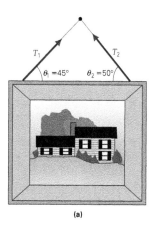
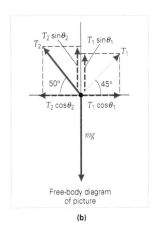

(a)   **(b)**

▲ FIGURE 8.7   **Translational static equilibrium** **(a)** Since the picture hangs motionless on the wall, the sum of the forces acting on it must be zero. The forces are concurrent, with their lines of action passing through a common point at the nail. **(b)** In the free-body diagram, all the forces are represented as acting at the common point. T_1 and T_2 have been moved to this point for convenience. Note, however, that the forces shown are acting on the *picture*. See Example 8.4 for description.

THINKING IT THROUGH. Since the picture remains motionless, it must be in static equilibrium, so applying the conditions for mechanical equilibrium should give equations that yield the tensions. Note that all the forces (tension and weight forces) are concurrent; that is, their lines of action pass through a common point, the nail. Because of this, the condition for rotational equilibrium $\left(\Sigma \vec{\tau}_i = 0\right)$ is automatically satisfied. With respect to the axis of rotation, the moment arms $(r_\perp)$ of the forces are zero, and therefore the torques are zero. Thus, only translational equilibrium needs to be considered.

SOLUTION
Given:

$\theta_1 = 45°, \theta_2 = 50°$
$m = 3.00$ kg

Find: T_1 and T_2

It is helpful to isolate the forces acting on the picture in a free-body diagram, as was done in Section 4.5 for force problems (Figure 8.7b). The diagram shows the concurrent forces acting through their common point. Note that all the force vectors have been moved to that point, which is taken as the origin of the coordinate axes. The weight force mg acts downward.

With the system in static equilibrium, the net force on the picture is zero; that is, $\Sigma\vec{F}_i = 0$. Thus, the sums of the rectangular components are also zero: $\Sigma F_x = 0$ and $\Sigma F_y = 0$. Then (using positive and negative signs for directions),

$$\Sigma F_x: + T_1\cos\theta_1 - T_2\cos\theta_2 = 0 \qquad (1)$$
$$\Sigma F_y: + T_1\sin\theta_1 - T_2\sin\theta_2 - mg = 0 \qquad (2)$$

Then, solving for T_2 in Equation 1 (or T_1 if you like),

$$T_2 = T_1\left(\frac{\cos\theta_1}{\cos\theta_2}\right) \qquad (3)$$

Then substituting Equation 3 into Equation 2 so as to eliminate T_2 and solving for T_1 with a little algebra,

$$T_1\left[\sin 45° + \left(\frac{\cos 45°}{\cos 50°}\right)\sin 50°\right] - mg =$$

$$T_1\left[0.707 + \left(\frac{0.707}{0.643}\right)(0.766)\right] - (3.00\,\text{kg})(9.80\,\text{m/s}^2) = 0$$

and

$$T_1 = \frac{29.4\,\text{N}}{1.55} = 19.0\,\text{N}$$

Then, using Equation 2 to find T_2,

$$T_2 = T_1\left(\frac{\cos\theta_1}{\cos\theta_2}\right) = 19.0\,\text{N}\left(\frac{0.707}{0.643}\right) = 20.9\,\text{N}$$

FOLLOW-UP EXERCISE. Analyze the situation in Figure 8.7 that would result if the wires were at equal angles and were shortened such that the angles were decreased but kept equal. Carry your analysis to the limit where the angles approach zero. Is the answer realistic?

As pointed out earlier, torque is a vector and therefore has direction. Similar to linear motion (Section 2.2), in which plus and minus signs were used to express opposite directions (e.g., $+x$ and $-x$), torque directions can be designated as being plus or minus, depending on the rotational acceleration they tend to produce. The rotational "directions" are taken as clockwise or counterclockwise around the axis of rotation. A torque that tends to produce a counterclockwise rotation will be taken as positive ($+$), and a torque that tends to produce a clockwise rotation will be taken as negative ($-$). (See the right-hand rule in Section 7.2.) To illustrate, let's apply this convention to the situation in Example 8.5.

EXAMPLE 8.5: ROTATIONAL STATIC EQUILIBRIUM – NO ROTATIONAL MOTION

Three masses are suspended from a meterstick as shown in ▼ **Figure 8.8a**. How much mass must be suspended on the right side for the system to be in static equilibrium? (Neglect the mass of the meterstick.)

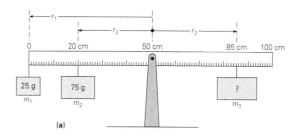

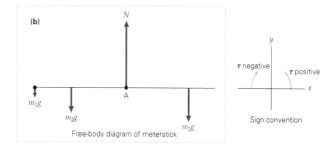

▲ **FIGURE 8.8   Rotational static equilibrium (a)** For the meter-stick to be in rotational equilibrium, the sum of the torques acting about any selected axis must be zero. **(b)** Here the axis is taken to be through point A, the 50-cm position. (The mass of the meterstick is considered negligible.) See Example 8.5 for description.

THINKING IT THROUGH. As the free-body diagram (Figure 8.8b) shows, the translational equilibrium condition will be satisfied with the upward normal force $\vec{N}$ balancing the downward forces caused by the objects weights, so long as the stick remains horizontal. But $\vec{N}$ is not known if m_3 is unknown, so applying the condition for rotational equilibrium should give the required value of m_3. (Note that the lever arms are measured from the pivot point, the center of the meterstick.)

SOLUTION

From the figure the data are (using cgs units for convenience):

Given:

$$m_1 = 25\text{ g},\ r_1 = 50\text{ cm}$$
$$m_2 = 75\text{ g},\ r_2 = 30\text{ cm}$$
$$r_3 = 35\text{ cm}$$

Find: m_3 (unknown mass)

Because the condition for translational equilibrium ($\Sigma\vec{F}_i = 0$) is satisfied (there is no $\vec{F}_{net}$ in the y-direction), $N - Mg = 0$, or $N = Mg$, where M is the total mass. This is true no matter what the total mass may be – that is, regardless of how much mass is added for m_3. However, unless the proper mass for m_3 is placed on the right side, the stick will experience a net torque and begin to rotate.

Notice that the masses on the left side produce torques that would tend to rotate the stick counterclockwise, and a mass on the right side produces a torque that would tend to rotate the stick clockwise. The condition for rotational equilibrium is applied by summing the torques about an axis. Here, this axis is conveniently taken to be through the center of the stick at the 50-cm position, or point A in Figure 8.8b. Then, noting that the force N passes through the axis of rotation ($r_\perp = 0$) and produces no torque,

(using sign convention for torque vectors)

$$\Sigma\tau_i: \quad \tau_1 + \tau_2 + \tau_3 = +r_1F_1 + r_2F_2 - r_3F_3$$
$$= r_1(m_1g) + r_2(m_2g) - r_3(m_3g) = 0$$

Noting that the g's cancel and solving for m_3,

$$m_3 = \frac{m_1r_1 + m_2r_2}{r_3} = \frac{(25\,g)(50\,cm) + (75\,g)(30\,cm)}{35\,cm} = 100\,g$$

(The mass of the stick was neglected. If the stick is uniform, however, its mass will not affect the equilibrium, as long as the pivot point is at the 50-cm mark. Why?)

FOLLOW-UP EXERCISE. The axis of rotation could have been taken through any point along the stick. That is, if a system is in static rotational equilibrium, the condition $\Sigma\vec{\tau}_i = 0$ holds for *any* axis of rotation. Show that the preceding statement is true for the system in this Example by taking the axis of rotation through the left end of the stick ($x = 0$).

In general, the conditions for both translational and rotational equilibrium need to be written explicitly to solve a statics problem. Example 8.6 is one such case.

EXAMPLE 8.6: STATIC EQUILIBRIUM – NO TRANSLATION, NO ROTATION

A ladder with a mass of 15 kg rests against a smooth wall (▶ **Figure 8.9a**). A painter who has a mass of 78 kg stands on the ladder as shown in the figure. What is the magnitude of frictional force that must act on the bottom of the ladder to keep it from slipping?

THINKING IT THROUGH. Here there are a variety of forces and torques. However, the ladder will not slip as long as the conditions for static equilibrium are satisfied. Summing both the forces and torques to zero should enable us to solve for the necessary frictional force. Also, as will be seen, choosing a convenient axis of rotation, such that one or more τ's are zero in the summation of the torques, can simplify the torque equation.

SOLUTION

Given:

$m_1 = 15$ kg
$m_m = 78$ kg
Distances given in figure

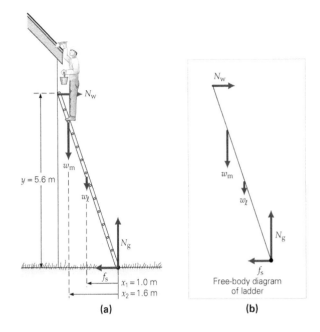

▲ **FIGURE 8.9** **Static equilibrium** For the painter's sake, the ladder has to be in static equilibrium; that is, both the sum of the forces and the sum of the torques must be zero. See Example 8.6 for description.

Find: f_s (force of static friction)

Because the wall is smooth, there is negligible friction between it and the ladder, and only the normal reaction force of the wall (N_w) acts on the ladder at this point (Figure 8.9b).

In applying the conditions for static equilibrium, any axis of rotation may be chosen. (The conditions must hold for all parts of a system that is in static equilibrium; that is, there can't be motion in any part of the system.) Note that choosing an axis at the end of the ladder where it touches the ground eliminates the torques due to f_s and N_g, since the moment arms are zero. Then writing the equations for force components and torque (using mg for w):

$$\Sigma F_x: \quad N_w - f_s = 0$$
$$\Sigma F_y: \quad N_g - m_m g - m_\ell g = 0$$

and

$$\Sigma\tau_i: \quad (m_\ell g)x_1 + (m_m g)x_2 + (-N_w y) = 0$$

The weight of the ladder is considered to be concentrated at its center of gravity. Solving the third equation for N_w and substituting the given values for the masses and distances yields

$$N_w = \frac{(m_\ell g)x_1 + (m_m g)x_2}{y}$$
$$= \frac{(15\,kg)(9.8\,m/s^2)(1.0\,m) + (78\,kg)(9.8\,m/s^2)(1.6\,m)}{5.6\,m}$$
$$= 2.4 \times 10^2\,N$$

Then from the ΣF_x equation,

$$f_s = N_w = 2.4 \times 10^2\,N$$

FOLLOW-UP EXERCISE. In this Example, would the frictional force between the ladder and the ground (call it f_{s1}) remain the same if there were friction between the wall and the ladder (call it f_{s2})? Justify your answer.

8.2.3 Problem-Solving Hint

As the preceding Examples have shown, a good procedure to follow in working problems involving static equilibrium is as follows:

1. Sketch a space diagram of the problem.
2. Draw a free-body diagram, showing and labeling all external forces and, if necessary, resolving the forces into x- and y-components.
3. Apply the equilibrium conditions. Sum the forces: $\Sigma\vec{\mathbf{F}}_i = 0$, usually in component form; $\Sigma F_x = 0$ and $\Sigma F_y = 0$. Sum the torques: $\Sigma\vec{\boldsymbol{\tau}}_i = 0$. Remember to select an appropriate axis of rotation to reduce the number of terms as much as possible. Use $+$ or $-$ sign conventions for both $\vec{\mathbf{F}}$ and $\vec{\boldsymbol{\tau}}$ vectors.
4. Solve for the unknown quantities.

8.2.4 Stability and Center of Gravity

The equilibrium of a particle or a rigid body can be either stable or unstable in a gravitational field. For rigid bodies, these categories of equilibria are conveniently analyzed in terms of the center of gravity of the body. Recall from Section 6.5 that the **center of gravity** is the point at which all the weight of an object may be considered to be acting as if the object were a particle. When the acceleration due to gravity is constant, the center of gravity and the center of mass coincide.

If an object is in stable equilibrium, *any small displacement results in a restoring force or torque, which tends to return the object to its original equilibrium position.* As illustrated in ▶ **Figure 8.10a**, a ball in a bowl is in stable equilibrium. Analogously, the center of gravity of an extended body on the right is in stable equilibrium. Any slight displacement raises its center of gravity (CG), and a restoring gravitational force tends to return it to the position of minimum potential energy. This force actually produces a restoring torque that is due to a component of the weight force and that tends to rotate the object about a pivot point back to its original position.

For an object in unstable equilibrium, *any small displacement from equilibrium results in a torque that tends to rotate the object farther away from its equilibrium position.* This situation is illustrated in Figure 8.10b. Note that the center of gravity of the object is at the top of an overturned, or inverted, potential energy bowl; that is, the potential energy is at a maximum in this case. Small displacements or slight disturbances have profound effects on objects that are in unstable equilibrium. It doesn't take much to cause such an object to change its position.

Yet even if the angular displacement of an object in stable equilibrium is quite substantial, the object will still be restored

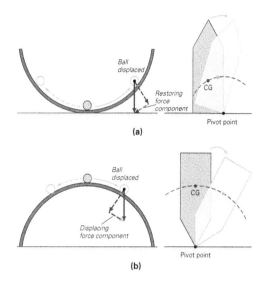

▲ FIGURE 8.10 **Stable and unstable equilibria (a)** When an object is in stable equilibrium, any small displacement from an equilibrium position results in a force or torque that tends to return the object to that position. A ball in a bowl (left) returns to the bottom after being displaced. Analogously, the center of gravity (CG) of an extended object (right) can be thought of as being on an inverted potential energy "bowl:" A small displacement raises the CG, increasing the object's potential energy. When released, the object will rotate about the pivot point back to its stable equilibrium position. **(b)** For an object in unstable equilibrium, any small displacement from its equilibrium position results in a force or torque that tends to take the object farther away from that position. The ball on top of an overturned bowl (left) is in unstable equilibrium. For an extended object (right), the CG can be thought of as being on an inverted potential energy bowl. A small displacement lowers the CG, decreasing the object's potential energy.

to its equilibrium position. As you might have surmised, the condition for stable equilibrium is:

An object is in stable equilibrium as long as its center of gravity after a small displacement still lies above and inside the object's original base of support. That is, the line of action of the weight force through the center of gravity intersects the original base of support.

When this is the case, there will always be a restoring gravitational torque (▶ **Figure 8.11a**). However, when the center of gravity or center of mass falls outside the base of support, over goes the object – because of a gravitational torque that rotates it away from its equilibrium position (Figure 8.11b).

Rigid bodies with wide bases and low centers of gravity are therefore most stable and least likely to tip over. This relationship is evident in the design of high-speed race cars, which have wide wheel bases and centers of gravity close to the ground (▶ **Figure 8.12a**). SUVs, on the other hand, can roll over more easily. Why? And how about the yoga couple in Figure 8.12b?

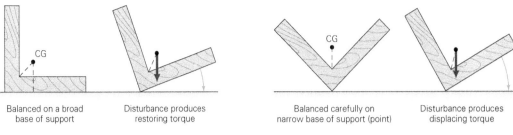

(a) Stable Equilibrium   **(b) Unstable Equilibrium**

▲ **FIGURE 8.11   Examples of stable and unstable equilibria (a)** When the center of gravity is above and inside an object's base of support, the object is in stable equilibrium. There is a restoring torque when the object is displaced. Note how the line of action of the weight intersects the original base of support after the displacement. **(b)** When the center of gravity lies outside the base of support, the object is unstable. (There is a displacing torque.)

(a)   (b)

▲ FIGURE 8.12   **Stable and unstable (a)** Race cars are very stable because of their wide wheel bases and low center of gravity. **(b)** The yoga man's base of support is very narrow: the small area of hands-to-ground contact. As long as the couple's center of gravity remains above the hand area, they are in equilibrium, but a displacement of only a few centimeters would probably be enough to topple them.

Another classic example of equilibrium is the Leaning Tower of Pisa (▶ **Figure 8.13**), from which Galileo allegedly performed his "free-fall" experiments. The tower started leaning before its completion in 1350 because of the soft subsoil beneath it. In 1990, the lean was about 5.5° from the vertical (about 5 m, or 17 ft, at the top) with an average increase in the lean of about 1.2 mm a year.

Attempts have been made to stop the lean increase. In 1930, cement was injected under the base, but the lean continued to increase. In the 1990s, major actions were taken. The tower was cabled back and counterweights were added to the high side. The tower settled back to about a 5° lean, or a shift of about 40 cm at the top. Moral of the story: Keep that center of gravity above the base of support.

EXAMPLE 8.7: STACK THEM UP – CENTER OF GRAVITY

Uniform, identical bricks 20 cm long are stacked so that 4.0 cm of each brick extends beyond the brick beneath, as shown in ▶ **Figure 8.14a**. How many bricks can be stacked in this way before the stack falls over?

THINKING IT THROUGH. As each brick is added, the center of mass (or center of gravity) of the stack moves to the right. The stack will be stable as long as the combined center of mass (CM) is over the base of support – the bottom brick. All of the bricks have the

▲ FIGURE 8.13   **Hold it stable!** The Leaning Tower of Pisa. Although leaning, it is still in stable equilibrium. Why?

same mass, and the CM of each is located at its midpoint. So the horizontal location of the stack's CM must be computed as bricks are added, until the CM extends beyond the base. The location of the CM was discussed in Section 6.5 (see Equation 6.19).

SOLUTION

Given: Brick length $= 20$ cm

Find: Maximum number of bricks that yields stability displacement of each brick $= 4.0$ cm

Taking the origin to be at the center of the bottom brick, the horizontal coordinate of the center of mass (or center of gravity) for the first two bricks in the stack is given by Equation 6.19, where $m_1 = m_2 = m$ and x_2 is the displacement of the second brick:

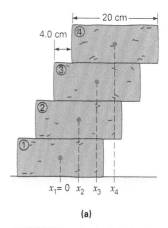

▲ FIGURE 8.14 **Stack them up!** (a) How many bricks can be stacked like this before the stack falls? See Example 8.7. (b) Try a similar experiment with books.

$$X_{CM2} = \frac{mx_1 + mx_2}{m+m} = \frac{m(x_1 + x_2)}{2m} = \frac{x_1 + x_2}{2}$$
$$= \frac{0 + 4.0\,cm}{2} = 2.0\,cm$$

The masses of the bricks cancel out (since they are all the same). For three bricks,

$$X_{CM3} = \frac{m(x_1 + x_2 + x_3)}{3m} = \frac{0 + 4.0\,cm + 8.0\,cm}{3}$$
$$= 4.0\,cm$$

For four bricks,

$$X_{CM4} = \frac{m(x_1 + x_2 + x_3 + x_4)}{4m}$$
$$= \frac{0 + 4.0\,cm + 8.0\,cm + 12.0\,cm}{4} = 6.0\,cm$$

and so on.

This series of results shows that the center of mass of the stack moves horizontally 2.0 cm for each brick added to the bottom one. For a stack of six bricks, the center of mass is 10 cm from the origin and directly over the edge of the bottom brick (2.0 cm × 5 added bricks = 10 cm, which is half the length of the bottom brick), so the stack is just at unstable equilibrium. The stack may not topple if the sixth brick is positioned very carefully, but it is doubtful that this could be done in practice. A seventh brick would definitely cause the stack to fall off the bottom brick. Why? (As shown in Figure 8.14b, you can try it yourself and stack them up using books. Don't let the librarian catch you.)

FOLLOW-UP EXERCISE. If the bricks in this Example were stacked so that, alternately, 4.0 cm and 6.0 cm extended beyond the brick beneath, how many bricks could be stacked before the stack toppled?

8.3 Rotational Dynamics

8.3.1 Moment of Inertia

Torque is the rotational analogue of force in linear motion, and a net torque produces rotational acceleration. To analyze this relationship, consider a constant net force acting on a particle of mass m about the given axis (▼ Figure 8.15). The magnitude of the net torque on the particle is

$$\tau_{net} = r_\perp F = rF_\perp = rma_\perp$$
$$= mr^2\alpha \quad \text{(net torque on a particle)} \tag{8.4}$$

where $a_\perp = a_t = r\alpha$ is the tangential acceleration (a_t, Equation 7.13). For the rotation of a rigid body about a fixed axis, this equation can be applied to each particle in the object and the results summed over the entire body (n particles) to find the total torque acting on the body. Since all the particles of a rotating rigid body have the same angular acceleration, we can simply add the individual torque magnitudes:

$$\tau_{net} = \Sigma\tau_i = \tau_1 + \tau_2 + \tau_3 + \cdots + \tau_n$$
$$= m_1 r_1^2\alpha + m_2^2 r_2\alpha + m_2 r_3^2\alpha + \cdots + m_n r_n^2\alpha \tag{8.5}$$
$$= (m_1 r_1^2 + m_2 r_2^2 + m_3 r_3^2 + \cdots + m_n r_n^2) = (\Sigma m_i r_i^2)\alpha$$

$$\tau_{net} = r_\perp F = rF_\perp = mr^2\alpha$$

▲ FIGURE 8.15 **Torque on a particle** The magnitude of the torque on a particle of mass m is $\tau = mr^2\alpha$. See text for description.

But for a rigid body, the masses (m_i) and the distances from the axis of rotation (r_i) do not change. Therefore, the quantity in the parentheses in Equation 8.5 is constant, and it is called the **moment of inertia**, I. For rotation about a given axis, it is defined as:

$$I = \Sigma m_i r_i^2 \quad \text{(moment of inertia)} \tag{8.6}$$

SI unit of moment of inertia: kilogram-meters squared (kg·m²)
The magnitude of the net torque can be conveniently written as

$$\tau_{net} = I\alpha \quad \text{(net torque on a rigid body)} \tag{8.7}$$

This is the *rotational form of Newton's second law* ($\vec{\tau}_{net} = I\vec{\alpha}$, in vector form). Keep in mind that, as a *net* force is necessary to produce a translational acceleration, a *net* torque (τ_{net}) is necessary to produce an angular acceleration.

By comparing the rotational form of Newton's second law ($\vec{\tau}_{net} = I\vec{\alpha}$) with the translational form ($\vec{F}_{net} = m\vec{a}$), where m is a measure of translational inertia, it can be seen that the moment of inertia I is a measure of **rotational inertia**, *or a body's tendency to resist change in its rotational motion*. Although I is constant for a rigid body and is the rotational analogue of mass, unlike the mass of a particle, the moment of inertia of a body is referenced to a particular axis and can have different values for different axes.

The moment of inertia also depends on the mass distribution of the body *relative* to its axis of rotation. It is easier (i.e., it takes less torque) to give an object an angular acceleration about some axes than about others. Example 8.8 illustrates this point.

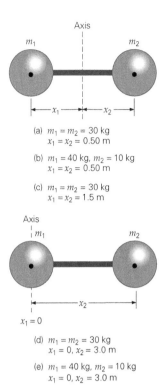

(a) $m_1 = m_2 = 30$ kg
   $x_1 = x_2 = 0.50$ m

(b) $m_1 = 40$ kg, $m_2 = 10$ kg
   $x_1 = x_2 = 0.50$ m

(c) $m_1 = m_2 = 30$ kg
   $x_1 = x_2 = 1.5$ m

(d) $m_1 = m_2 = 30$ kg
   $x_1 = 0$, $x_2 = 3.0$ m

(e) $m_1 = 40$ kg, $m_2 = 10$ kg
   $x_1 = 0$, $x_2 = 3.0$ m

▲ **FIGURE 8.16 Moment of inertia** The moment of inertia depends on the distribution of mass relative to a particular axis of rotation and, in general, has a different value for each axis. This difference reflects the fact that objects are easier or more difficult to rotate about certain axes. See Example 8.8 for description.

EXAMPLE 8.8: ROTATIONAL INERTIA – MASS DISTRIBUTION AND AXIS OF ROTATION

Find the moment of inertia about the axis indicated for each of the one-dimensional dumbbell configurations in ▶ **Figure 8.16**. (Neglect the mass of the connecting bar and give your answers to three significant figures for comparison.)

THINKING IT THROUGH. This is a direct application of Equation 8.6 for cases with different masses and distances. It will show that the moment of inertia of an object depends on the axis of rotation and on the mass distribution relative to the axis of rotation. The sum for I will include only two terms (two masses).

SOLUTION

Given: Values of m and r in the figure.

Find: I (moment of inertia)

With $I = m_1 r_1^2 + m_2 r_2^2$:

(a) $I = (30 \text{ kg})(0.50 \text{ m})^2 + (30 \text{ kg})(0.50 \text{ m})^2 = 15.0 \text{ kg·m}^2$
(b) $I = (40 \text{ kg})(0.50 \text{ m})^2 + (10 \text{ kg})(0.50 \text{ m})^2 = 12.5 \text{ kg·m}^2$
(c) $I = (30 \text{ kg})(1.5 \text{ m})^2 + (30 \text{ kg})(1.5 \text{ m})^2 = 135 \text{ kg·m}^2$
(d) $I = (30 \text{ kg})(0 \text{ m})^2 + (30 \text{ kg})(3.0 \text{ m})^2 = 270 \text{ kg·m}^2$
(e) $I = (40 \text{ kg})(0 \text{ m})^2 + (10 \text{ kg})(3.0 \text{ m})^2 = 90.0 \text{ kg·m}^2$

This Example clearly shows how the moment of inertia depends on mass *and* its distribution relative to a particular axis of rotation. In general, the moment of inertia is larger the farther the mass is from the axis of rotation. This principle is important in the design of flywheels, which are used in automobiles to keep the engine running smoothly between cylinder firings. The mass of a flywheel is concentrated near the rim, giving a large moment of inertia, which resists changes in motion.

FOLLOW-UP EXERCISE. In parts (d) and (e) of this Example, would the moments of inertia be different if the axis of rotation went through m_2? Explain.

The moment of inertia is an important consideration in rotational motion. By changing the axis of rotation and the relative mass distribution, the value of I can be changed and the motion affected. You were probably told to do this when playing softball or baseball as a child. When at bat, children are often instructed to "choke up" on the bat – to move their hands farther up on the handle.

Now you know why. In doing so, the child moves the axis of rotation of the bat closer to the more massive end of the bat (or its center of mass). Hence, the moment of inertia of the bat is decreased (smaller r in the mr^2 term). Then, when a swing is taken, the angular acceleration is greater. The bat gets around quicker, and the chance of hitting the ball before it goes past is greater. A batter has only a fraction of a second to swing, and with $\theta = 1/2(\alpha t^2)$, a larger α allows the bat to rotate more quickly (swing faster).

8.3.2 Parallel Axis Theorem

Calculations of the moments of inertia of most extended rigid bodies require math that is beyond the scope of this book. The results for some common shapes are given in ▼ **Figure 8.17**. The rotational axes are generally taken along axes of symmetry – that is, axes running through the center of mass that give a symmetrical mass distribution. However, there are examples, such as a rod with an axis of rotation through one end (Figure 8.17c). This axis is parallel to an axis of rotation through the center of mass of the rod (Figure 8.17b). The moment of inertia about such a parallel axis is given by a useful theorem called the **parallel axis theorem**; namely,

$$I = I_{CM} + Md^2$$

where I is the moment of inertia about an axis that is parallel to one through the center of mass and at a distance d from it, I_{CM} is the moment of inertia about an axis through the center of mass, and M is the total mass of the body (▶ **Figure 8.18**). For the axis through the end of the rod (Figure 8.17c), the moment of inertia is obtained by applying the parallel axis theorem to the thin rod in Figure 8.17b:

$$I = I_{CM} + Md^2 = \frac{1}{12}ML^2 + M\left(\frac{L}{2}\right)^2 = \frac{1}{12}ML^2$$
$$= \frac{1}{4}ML^2 = \frac{1}{3}ML^2$$

8.3.3 Applications of Rotational Dynamics

The rotational form of Newton's second law allows us to analyze dynamic rotational situations. Examples 8.9 and 8.10 illustrate

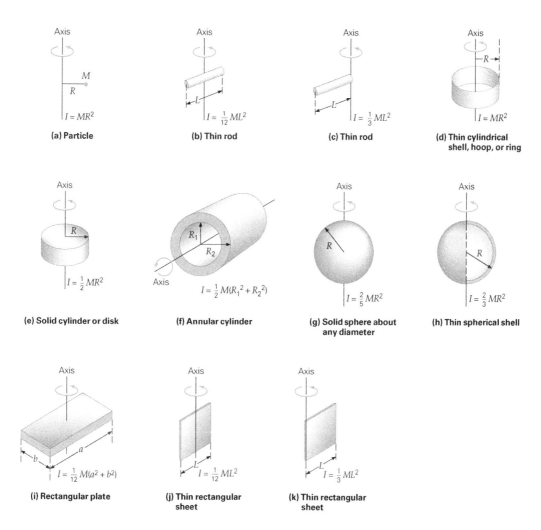

(a) Particle $I = MR^2$

(b) Thin rod $I = \frac{1}{12}ML^2$

(c) Thin rod $I = \frac{1}{3}ML^2$

(d) Thin cylindrical shell, hoop, or ring $I = MR^2$

(e) Solid cylinder or disk $I = \frac{1}{2}MR^2$

(f) Annular cylinder $I = \frac{1}{2}M(R_1^2 + R_2^2)$

(g) Solid sphere about any diameter $I = \frac{2}{5}MR^2$

(h) Thin spherical shell $I = \frac{2}{3}MR^2$

(i) Rectangular plate $I = \frac{1}{12}M(a^2 + b^2)$

(j) Thin rectangular sheet $I = \frac{1}{12}ML^2$

(k) Thin rectangular sheet $I = \frac{1}{3}ML^2$

▲ FIGURE 8.17 **Moments of inertia of some uniform density objects with common shapes**

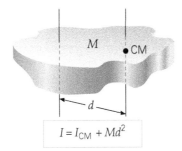

$$I = I_{CM} + Md^2$$

▲ **FIGURE 8.18 Parallel axis theorem** The moment of inertia about an axis parallel to another through the center of mass of a body is $I = I_{CM} + Md^2$, where M is the total mass of the body and d is the distance between the two axes.

how this is done. In such situations, it is very important to make certain that all the data are properly listed to help with the increasing number of variables.

EXAMPLE 8.9: OPENING THE DOOR – TORQUE IN ACTION

A student opens a uniform 12-kg door by applying a constant force of 40 N at a perpendicular distance of 0.90 m from the hinges (▼ **Figure 8.19**). If the door is 2.0 m in height and 1.0 m wide, what is the magnitude of its angular acceleration? (Assume that the door rotates freely on its hinges.)

THINKING IT THROUGH. From the given information, the applied net torque can be calculated. To find the angular acceleration of the door, the moment of inertia is needed. This can be calculated, since the door's mass and dimensions are known.

SOLUTION

From the information given in the problem:

▲ **FIGURE 8.19 Torque in action** See Example 8.9 for description.

Given:

$M = 12$ kg, $F = 40$ N
$r_\perp = r = 0.90$ m
$h = 2.0$ m (door height)

Find: α (magnitude of angular acceleration)

The rotational form of Newton's second law can be applied, $\tau_{net} = I\alpha$, where I is about the hinge axis. τ_{net} can be found from the given data, so the problem boils down to determining the moment of inertia of the door.

Looking at Figure 8.17, it can be seen that case (*k*) applies to a door (treated as a uniform rectangle) rotating on hinges, so $I = 1/3(ML^2)$, where $L = w$, the width of the door. Then,

$$\tau_{net} = I\alpha$$

or

$$\alpha = \frac{\tau_{net}}{I} = \frac{r_\perp}{1/3(ML^2)} = \frac{3rF}{Mw^2} = \frac{3(0.90\,\text{m})(40\,\text{N})}{(12\,\text{kg})(1.0\,\text{m})^2}$$
$$= 9.0\,\text{rad/s}^2$$

FOLLOW-UP EXERCISE. In this Example, if the constant torque were applied through an angular distance of 45° and then removed, how long would the door take to swing completely open (90°)? (Neglect friction.)

In problems involving pulleys in Section 4.5, the mass (and hence the inertia) of the pulley was neglected in order to simplify things. Now you know how to include those quantities and can treat pulleys more realistically, as seen in Example 8.10.

EXAMPLE 8.10: PULLEYS HAVE MASS, TOO – TAKING PULLEY INERTIA INTO ACCOUNT

A block of mass *m* hangs from a string wrapped around a frictionless, disk-shaped pulley of mass *M* and radius *R*, as shown in ▶ **Figure 8.20**. If the block descends from rest under the influence of gravity, what is the magnitude of its linear acceleration? (Neglect the mass of the string.)

THINKING IT THROUGH. Real pulleys have mass and rotational inertia, which affect their motion. The suspended mass (via the string) applies a torque to the pulley. Here the rotational form of Newton's second law can be used to find the angular acceleration of the pulley and then its tangential acceleration, which is the same in magnitude as the linear acceleration of the block. (Why?) No numerical values are given so the answer will be in symbol form.

SOLUTION

The linear acceleration of the block depends on the angular acceleration of the pulley, so we look at the pulley system first. The pulley

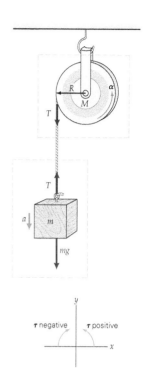

▲ FIGURE 8.20 Pulley with inertia Taking the mass, or rotational inertia, of a pulley into account allows a more realistic description of the motion. The directional sign convention for torque is shown. See Example 8.10.

is treated as a disk and thus has a moment of inertia $I = 1/2(MR^2)$ (Figure 8.17e). A torque due to the tension force in the string (T) acts on the pulley. With $\tau = I\alpha$ (considering only the upper dashed box in Figure 8.20),

$$\tau_{\text{net}} = r_{\perp}F = RT = I\alpha = \left(\frac{1}{2}MR^2\right)\alpha$$

such that

$$\alpha = \frac{2T}{MR}$$

The linear acceleration of the block and the angular acceleration of the pulley are related by $a = R\alpha$, where a is the tangential acceleration, and

$$a = R\alpha = \frac{2T}{M} \qquad (1)$$

But T is unknown. Looking at the descending mass (the lower dashed box) and summing the forces in the vertical direction (choosing down as positive) gives

$$mg - T = ma$$

or

$$T = mg - ma \qquad (2)$$

Using Equation 2 to eliminate T from Equation 1 yields

$$a = \frac{2T}{M} = \frac{2(mg - ma)}{M}$$

And solving for a,

$$a = \frac{2mg}{(2m + M)} \qquad (3)$$

Note that if $M \to 0$ (as in the case of ideal, massless pulleys in previous chapters), then $I \to 0$ and $a \to g$ (from Equation 3). Here, however, $M \neq 0$, so $a < g$. (Why?)

FOLLOW-UP EXERCISE. Pulleys can be analyzed even more realistically. In this Example, friction at the axle was neglected, but practically, a frictional torque (τ_f) exists and should be included. What would be the form, as in Equation 3, of the angular acceleration in this case? Show that your result is dimensionally correct.

In pulley problems, as before, the mass of the string will be neglected – an approach that still gives a good approximation if the string is relatively light. Taking the mass of the string into account would give a continuously varying mass hanging on the pulley, thus producing a variable torque. Such problems are beyond the scope of this book.

Suppose you had masses suspended from each side of a pulley. Here, you would have to compute the net torque. If the values of the masses are unknown, so which way the pulley would rotate cannot be determined, then simply assume a direction. As in the linear case, if the result came out with the opposite sign, it would indicate that you had assumed the wrong direction.

8.3.4 Problem-Solving Hint

For problems such as Example 8.10, dealing with coupled rotational and translational motions, keep in mind that with no string slippage, the magnitudes of the accelerations are usually related by $a = r\alpha$, while $v = r\omega$ relates the magnitudes of the velocities at any instant of time. Applying Newton's second law (in rotational or linear form) to different parts of the system gives equations that can be combined by using such relationships. Also, for rolling without slipping, $a = r\alpha$ and $v = r\omega$, relate the angular quantities to the linear motion of the center of mass.

Another application of rotational dynamics is the analysis of motion of objects that can roll.

The string of a yo-yo sitting on a level surface is pulled as shown in
▼ Figure 8.21. Will the yo-yo roll (a) toward the person or (b) away
from the person?

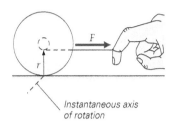

▲ FIGURE 8.21   **Pulling the yo-yo's string** See Conceptual Example
8.11 for description.

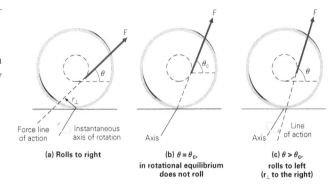

▲ FIGURE 8.22   **The angle makes a difference** (a) With the line of
action to the left of the instantaneous axis, the yo-yo rolls to the right.
(b) At a critical angle θ_c the line of action passes through the axis, and
the yo-yo is in equilibrium. (c) When the line of action is to the right of
the axis, the yo-yo rolls to the left.

REASONING AND ANSWER. Apply the physics just studied to
the situation. Note that the instantaneous axis of rotation is along
the line of contact of the yo-yo with the surface. If you had a stick
standing vertically in place of the $\vec{r}$ vector and pulled on a string
attached to the top of the stick in the direction of $\vec{F}$, which way
would the stick rotate? Of course, it would rotate clockwise (about
its instantaneous axis of rotation). The yo-yo reacts similarly; that
is, it rolls in the direction of the pull, so the answer is (a). (Get a
yo-yo and try it if you're a nonbeliever.)

There is more interesting physics in our yo-yo situation. The
pull force is not the only force acting on the yo-yo; there are three
others. Do they contribute torques? Let's identify these forces.
There's the weight of the yo-yo and the normal force from the surface. Also, there is a horizontal force of static friction between the
yo-yo and the surface. (Otherwise the yo-yo would slide rather
than roll.) But these three forces act through the line of contact
or through the instantaneous axis of rotation, so they produce no
torques here. (Why?)

What would happen if the angle of the string or pull force were
increased (relative to the horizontal), as illustrated in ▶ Figure 8.22a?
The yo-yo would still roll to the right. As can be seen in Figure 8.22b,
at some critical angle θ_c the line of force goes through the axis of
rotation, and the net torque on the yo-yo becomes zero, so the yo-yo
does not roll.

If this critical angle is exceeded (Figure 8.22c), the yo-yo will
begin to roll counterclockwise, or to the left. Note that the line of
action of the force is on the other side of the axis of rotation from
that in Figure 8.22a and that the lever arm ($r_\perp$) has changed directions, resulting in a reversed net torque direction.

FOLLOW-UP EXERCISE: Suppose you set the yo-yo string at
the critical angle, with the string over a round, horizontal bar at
the appropriate height, and you suspend a weight on the end of the
string to supply the force for the equilibrium condition. What will
happen if you then pull the yo-yo toward you, away from its equilibrium position, and release it?

8.4 Rotational Work and Kinetic Energy

This section gives the rotational analogues of various equations
of linear motion associated with work and kinetic energy for
constant torques. Because their development is similar to that
given for their linear counterparts, detailed discussion is not
needed. As in Section 5.1, it is understood that W is the net work
if more than one force or torque acts on an object.

Rotational Work: We can go directly from work done by a force
to work done by a torque, since the two are related ($\tau = r_\perp F$). For
rotational motion, the **rotational work**, $W = Fs$, done by a single
force F acting tangentially along an arc length s is

$$W = Fs = F(r_\perp\theta) = \tau\theta$$

where θ is in radians. Thus, for a single torque acting through an
angle of rotation θ,

$$W = \tau\theta \quad \text{(rotational work for a single force)} \qquad (8.9)$$

In this book, both the torque (τ) and angular displacement (θ)
vectors are almost always along the fixed axis of rotation, so you
will not need to be concerned about parallel components, as you
were for translational work. The torque and angular displacement may be in opposite directions, in which case the torque
does negative work and slows the rotation of the body. Negative
rotational work is analogous to F and d being in opposite directions for translational motion.

Rotational Power: An expression for the instantaneous
rotational power, the rotational analogue of power (the time
rate of doing work), P, is easily obtained from Equation 8.9:

$$P = \frac{W}{t} = \tau\left(\frac{\theta}{t}\right) = \tau\omega \quad \text{(rotational power)} \qquad (8.10)$$

8.4.1 The Work: Energy Theorem and Kinetic Energy

The relationship between the net rotational work done on a rigid body (more than one force acting) and the change in rotational kinetic energy of the body can be derived as follows, starting with the equation for rotational work:

$$W_{net} = \tau\theta = I\alpha\theta$$

Since we assume the torques are due only to constant forces, α is constant. But from rotational kinematics in Chapter 7, it is known that for a constant angular acceleration, $\omega^2 = \omega_o^2 + 2\alpha\theta$, and

$$W_{net} = I\left(\frac{\omega^2 - \omega_o^2}{2}\right) = \frac{1}{2}I\omega^2 - \frac{1}{2}I\omega_o^2$$

From Equation 5.6 (work-energy), $W_{net} = \Delta K$. Therefore,

$$W_{net} = \frac{1}{2}I\omega^2 - \frac{1}{2}I\omega_o^2 = K - K_o = \Delta K \qquad (8.11)$$

Then the expression for rotational kinetic energy is

$$K = \frac{1}{2}I\omega^2 \quad \text{(rotational kinetic energy)} \qquad (8.12)$$

Thus, *the net rotational work done on an object is equal to the change in the rotational kinetic energy of the object*. Consequently, to change the rotational kinetic energy of an object, a net torque must be applied.

It is possible to derive the expression for the kinetic energy of a rotating rigid body (about a fixed axis) directly. Summing the instantaneous kinetic energies of the body's individual particles relative to the fixed axis gives

$$K = \frac{1}{2}\Sigma m_i v_i^2 = \frac{1}{2}\left(\Sigma m_i r_i^2\right)\omega^2 = \frac{1}{2}I\omega^2$$

where, for each particle of the body, $v_i = r_i\omega$. So, Equation 8.12 doesn't represent a new form of energy; rather, it is simply another expression for kinetic energy, in a form that is more convenient for rigid body rotation.

A summary of translational and rotational analogues is given in ▼ **Table 8.1**. (The table also contains angular momentum, which will be discussed in Section 8.5.)

When an object has both translational and rotational motion, its total kinetic energy may be divided into parts to reflect the two kinds of motion. For example, for a cylinder rolling without slipping on a level surface, the motion is a combination of the translational motion of the center of mass and the rotational motion about an axis through the center of mass.

The rotational kinetic energy and the translational kinetic are, respectively

$$K_r = \frac{1}{2}I_{CM}\omega^2 \quad \text{and} \quad K_t = \frac{1}{2}Mv_{CM}^2$$

The total K is

$$K = \frac{1}{2}I_{CM}\omega^2 + \frac{1}{2}Mv_{CM}^2 \quad \text{(rolling, no slipping)} \qquad (8.13)$$

$$\underset{\text{total}}{K} = \underset{\text{rotational}}{K_r} + \underset{\text{translational}}{K_t}$$

Note that although a cylinder was used as an example here, this is a general result and applies to any object that is rolling without slipping.

Thus, *the total kinetic energy of such an object is the sum of two contributions: the translational kinetic energy of the object's center of mass and the rotational kinetic energy of the object relative to a horizontal axis through its center of mass.*

EXAMPLE 8.12: DIVISION OF ENERGY – ROTATIONAL AND TRANSLATIONAL

A uniform, solid 1.0-kg cylinder rolls without slipping at a speed of 1.8 m/s on a flat surface. (a) What is the total kinetic energy of the cylinder? (b) What percentage of this total is rotational kinetic energy?

THINKING IT THROUGH. The cylinder has both rotational and translational kinetic energies, so Equation 8.13 applies, and its terms are related by the condition of rolling without slipping.

Given:

$$M = 1.0 \text{ kg}$$
$$v_{CM} = 1.8 \text{ m/s}$$
$$I_{CM} = \frac{1}{2}(MR^2) \text{ (from Figure 8.17e)}$$

TABLE 8.1   Translational and Rotational Quantities and Equations

Translational		Rotational	
Force:	$\vec{F}$	Torque (magnitude):	$\tau = rF\sin\theta$
Mass (inertia):	m	Moment of inertia:	$I = \Sigma m_i r_i^2$
Newton's second law:	$\vec{F}_{net} = m\vec{a}$	Newton's second law:	$\vec{\tau}_{net} = I\vec{\alpha}$
Work:	$W = Fd$	Work:	$W = \tau\theta$
Power:	$P = Fv$	Power:	$P = \tau\omega$
Kinetic energy:	$K = \frac{1}{2}mv^2$	Kinetic energy:	$K = \frac{1}{2}I\omega^2$
Work-energy theorem:	$W_{net} = \frac{1}{2}mv^2 - \frac{1}{2}mv_o^2 = \Delta K$	Work-energy theorem:	$W_{net} = \frac{1}{2}I\omega^2 - \frac{1}{2}I\omega_o^2 = \Delta K$
Linear momentum:	$\vec{p} = m\vec{v}$	Angular momentum:	$\vec{L} = I\vec{\omega}$

Find:

(a) *K* (total kinetic energy)

(b) K_r/K (×100%) (percentage of rotational energy)

(a) The cylinder rolls without slipping, so the condition $v_{CM} = R\omega$ applies. Then the total kinetic energy (K) is the sum of the rotational kinetic energy K_r and the translational kinetic energy K_t of the center of mass, K_{CM} (Equation 8.13):

$$K = \frac{1}{2}I_{CM}\omega^2 + \frac{1}{2}Mv_{CM}^2 = \frac{1}{2}\left(\frac{1}{2}MR^2\right)\left(\frac{v_{CM}}{R}\right)^2 + \frac{1}{2}Mv_{CM}^2$$

$$= \frac{1}{4}Mv_{CM}^2 + \frac{1}{2}Mv_{CM}^2$$

$$= \frac{3}{4}Mv_{CM}^2 = \frac{3}{4}(1.0\,\text{kg})(1.8\,\text{m/s})^2 = 2.4\,\text{J}$$

(b) The rotational kinetic energy K_r of the cylinder is the first term of the preceding equation, so, forming a ratio in symbol form,

$$\frac{K_r}{K} = \frac{1/4(Mv_{CM}^2)}{3/4(Mv_{CM}^2)} = \frac{1}{3}(\times 100\%) = 33\%$$

Thus, the total kinetic energy of the cylinder is made up of rotational and translational parts, with one-third being rotational.

Note that in part (b) the radius of the cylinder was not needed, nor was the mass. Because a ratio was used, these quantities canceled. However, *don't* think that this exact division of energy is a general result. It is easy to show that the percentage is different for objects with different moments of inertia. For example, you should expect a rolling sphere to have a smaller percentage of rotational kinetic energy than a cylinder has, because the sphere has a smaller moment of inertia ($I = (2/5)MR^2$).

FOLLOW-UP EXERCISE. Potential energy can be brought into the act by applying the conservation of energy to an object rolling up or down an inclined plane. In this Example, suppose that the cylinder rolled up a 20° inclined plane without slipping. (a) At what vertical height (measured by the vertical distance of its CM) on the plane does the cylinder stop? (b) To find the height in part (a), you probably equated the initial total kinetic energy to the final gravitational potential energy. That is, the total kinetic energy was reduced by the work done by gravity. However, a frictional force also acts (to prevent slipping). Is there not work done here, too?

EXAMPLE 8.13: ROLLING DOWN

A uniform cylindrical hoop is released from rest at a height of 0.25 m near the top of an inclined plane (▶ **Figure 8.23**). If the hoop rolls down the plane without slipping and no energy is lost due to friction, what is the linear speed of the cylinder's center of mass at the bottom of the incline?

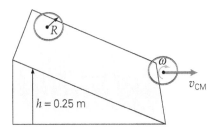

▲FIGURE 8.23 **Rolling motion and energy** When an object rolls down an inclined plane, potential energy is converted to translational *and* rotational kinetic energy. This makes the rolling slower than frictionless sliding. See Example 8.13 for a description.

THINKING IT THROUGH. Here, gravitational potential energy is converted into kinetic energy – both rotational and translational. The conservation of (mechanical) energy applies, since no energy is lost to friction. The initial kinetic energy is zero since the hoop is released from rest.

SOLUTION

Given:

$$h = 0.25\ \text{m}$$
$$(v_{CM})_o = \omega_o = 0$$
$$I_{CM} = MR^2 \ \text{(from Figure 8.17d)}$$

Find: v_{CM} (speed of CM)

The total mechanical energy of the cylinder is conserved:

$$E_o = E \quad \text{or} \quad K_o + U_o = K + U$$

Since $(v_{CM})_o = \omega_o = 0$ at the top of the incline and assuming that $U = 0$ at the bottom, this becomes

$$Mgh = \frac{1}{2}I_{CM}\omega^2 + \frac{1}{2}Mv_{CM}^2$$

Using the rolling without slipping condition, $v_{CM} = R\omega$, gives

$$Mgh = \frac{1}{2}(MR^2)\left(\frac{v_{CM}}{R}\right)^2 + \frac{1}{2}Mv_{CM}^2 = Mv_{CM}^2$$

Solving for v_{CM},

$$v_{CM} = \sqrt{gh} = \sqrt{(9.8\,\text{m/s}^2)(0.25\,\text{m})} = 1.6\,\text{m/s}$$

FOLLOW-UP EXERCISE. Suppose the inclined plane in this Example were frictionless and the hoop slid down the plane instead of rolling. How would the speed at the bottom compare in this case? Why are the speeds different?

8.4.2 A Fixed Race

As Example 8.13 shows for an object rolling down an incline without slipping, v_{CM} is independent of *M* and *R*. The masses and radii cancel, so all objects of a particular shape (with the same equation for the moment of inertia) roll with the same speed, regardless of their size or density. But the rolling speed does vary with the moment of inertia, which varies with an object's shape. Therefore, rigid bodies with different shapes roll with different speeds. For example, if you released a cylindrical hoop, a solid cylinder, and a uniform sphere at the same time from the top of an inclined plane, the sphere would win the race to the bottom, followed by the cylinder, with the hoop coming in last – every time!

You can try this as an experiment with a couple of cans of food or other cylindrical containers – one full of some solid material (in effect, a rigid body) and one empty and with the ends cut out – and a smooth, solid ball. Remember that the masses and the radii make no difference. You might think that an annular cylinder (a hollow cylinder with inner and outer radii that are appreciably different – Figure 8.17f) would be a possible front-runner, or "front-roller," in such a race, but it wouldn't be. The rolling race down an incline is fixed (or rigged by the laws of physics!) even when you vary the masses and the radii.

8.5 Angular Momentum

Another important quantity in rotational motion is angular momentum. Recall from Section 6.1 how the linear momentum of an object is changed by a force. Analogously, changes in angular momentum are associated with torques. As has been learned, torque is the product of a moment arm and a force. In a similar manner, **angular momentum ($\vec{L}$)** is the product of a moment arm ($r_\perp$) and a linear momentum (p). For a particle of mass m rotating with an angular velocity of ω about an axis with a radius of r, the magnitude of the linear momentum is $p = mv$, where $v = r\omega$. (For circular motion, $r_\perp = r$, since $\vec{v}$ is perpendicular to $\vec{r}$.) The magnitude of the angular momentum is

$$L = r_\perp p = mr_\perp v = mr_\perp^2 \omega \qquad (8.14)$$

(single-particle angular momentum)

SI unit of angular momentum: kilogram-meters squared per second (kg·m²/s).

For a system of particles making up a rigid body, all the particles travel in circles, and the magnitude of the total angular momentum is

$$L = (\Sigma m_i r_i^2)\omega = I\omega \quad \text{(rigid body angular momentum)} \qquad (8.15)$$

which, for rotation about a fixed axis, is (in vector notation)

$$\vec{L} = I\vec{\omega} \qquad (8.16)$$

Thus, $\vec{L}$ is in the direction of the angular velocity vector ($\vec{\omega}$). This direction is given by the right-hand rule (Section 8.2).

Since the moment of inertia, I, depends on the axis rotation, the angular momentum of a rigid body also varies with different axes of rotation.

For linear motion, the magnitude of the change in the total linear momentum of a system is related to the net external force by $F_{net} = \Delta P/\Delta t$. Angular momentum is analogously related to net torque (in magnitude form):

$$\tau_{net} = I\alpha = \frac{I\Delta\omega}{\Delta t} = \frac{\Delta(I\omega)}{\Delta t} = \frac{\Delta L}{\Delta t}$$

That is,

$$\tau_{net} = \frac{\Delta L}{\Delta t} \qquad (8.17)$$

Thus, the net torque is equal to *the time rate of change of angular momentum*. In other words, a net torque results in a *change* in angular momentum, just as a net force results in a change in linear momentum.

8.5.1 Conservation of Angular Momentum

Equation 8.17 was derived using $\tau_{net} = I\alpha$, which applies to a rigid system of particles or a rigid body having a constant moment of inertia. However, Equation 8.17 is a general equation that also applies to even nonrigid systems of particles. In such a system, there may be a change in the internal mass distribution and a change in the moment of inertia.

If the net torque on a system is zero, then, by Equation 8.17, $\tau_{net} = (\Delta L/\Delta t) = 0$, and

$$\Delta L = L - L_o = I\omega - I_o\omega_o = 0$$

or

$$I\omega = I_o\omega_o \quad \text{(conservation of angular momentum)} \qquad (8.18)$$

Thus, the condition for the conservation of angular momentum is as follows:

In the absence of an external, unbalanced torque, the total (vector) angular momentum of a system is conserved (remains constant).

Just as the internal forces cannot change a system's linear momentum, neither can internal torques change a system's angular momentum.

For a rigid body with a constant moment of inertia (i.e., $I = I_o$), the angular speed remains constant ($\omega = \omega_o$) in the absence of a net torque. But it is possible for the moment of inertia to change in some systems, giving rise to a change in the angular speed, as the following Example illustrates.

EXAMPLE 8.14: PULL IT DOWN – CONSERVATION OF ANGULAR MOMENTUM

A small ball at the end of a string that passes through a tube is swung in a circle, as illustrated in ▼ **Figure 8.24**. When the string is pulled downward through the tube, the angular speed of the ball increases. (a) Is the increase in angular speed caused by a torque due to the pulling force? (b) If the ball is initially moving at 2.8 m/s in a circle with a radius of 0.30 m, what will be its tangential speed if the string is pulled down to reduce the radius of the circle to 0.15 m? (Neglect the mass of the string.)

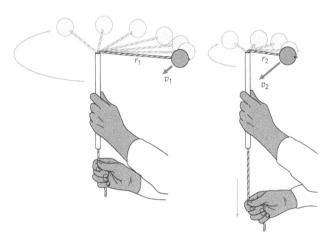

▲ FIGURE 8.24 **Conservation of angular momentum** When the string is pulled downward through the tube, the revolving ball speeds up. See Example 8.14 for description.

THINKING IT THROUGH. (a) A force is applied to the ball via the string, but consider the axis of rotation. (b) In the absence of a net torque, the angular momentum is conserved (Equation 8.18), and the tangential speed is related to the angular speed by $v = r\omega$.

SOLUTION

Given:

$r_1 = 0.30$ m
$r_2 = 0.15$ m
$v_1 = 2.8$ m/s

Find:

(a) Cause of the increase in angular speed
(b) v_2 (final tangential speed)

(a) The change in the angular velocity, or an angular acceleration, is not caused by a torque due to the pulling force. The force on the ball, as transmitted by the string (tension), acts through the axis of rotation, and therefore the torque is zero. Because the rotating portion of the string is shortened, the moment of inertia of the ball ($I = mr^2$, from Figure 8.17a) decreases. Because of the absence of an external torque, the angular momentum ($I\omega$) of the ball is conserved, and if I is reduced, ω must increase.

(b) Because the angular momentum is conserved, we can equate the magnitudes of the angular momenta:

$$I_o\omega_o = I\omega$$

Then, using $I = mr^2$ and $\omega = v/r$ gives

$$mr_1v_1 = mr_2v_2$$

and

$$v_2 = \left(\frac{r_1}{r_2}\right)v_1 = \left(\frac{0.30\,\text{m}}{0.15\,\text{m}}\right)(2.8\,\text{m/s}) = 5.6\,\text{m/s}$$

When the radial distance is shortened, the ball speeds up.

FOLLOW-UP EXERCISE. Let's look at the situation in this Example in terms of work and energy. If the initial speed is the same and the vertical pulling force is 7.8 N, recalculate the final speed of the 0.10-kg ball.

Example 8.14 should help you understand Kepler's law of equal areas (Section 7.6) from another viewpoint. A planet's angular momentum is conserved to a good approximation by neglecting the weak gravitational torques from other planets. (The Sun's gravitational force on a planet produces little or no torque. Why?) When a planet is closer to the Sun in its elliptical orbit and so has a shorter moment arm, its speed is greater, by the conservation of angular momentum. (Hence the law of equal areas really is a result of angular momentum conservation.) Similarly, when an orbiting satellite's altitude varies during the course of an elliptical orbit about a planet, the satellite speeds up or slows down in accordance with the same principle.

8.5.2 Real-Life Angular Momentum

If L is constant, what happens to ω when I is made smaller by reducing r? The angular speed must increase to compensate and keep L constant. Ice skaters and ballerinas perform dizzying spins by pulling in and raising their arms to reduce their moment of inertia (▶ **Figure 8.25a,b**). Similarly, a diver spins during a high dive by tucking in the body and limbs, greatly decreasing his or her moment of inertia. The enormous wind speeds of tornadoes and hurricanes represent another example of the same effect (Figure 8.25c).

Angular momentum also plays a role in figure-skating jumps in which the skater spins in the air, such as a triple axel or triple lutz. A torque applied on the jump gives the skater angular momentum, and the arms and legs are drawn into the body, which, as in spinning on one's toes, decreases the moment of

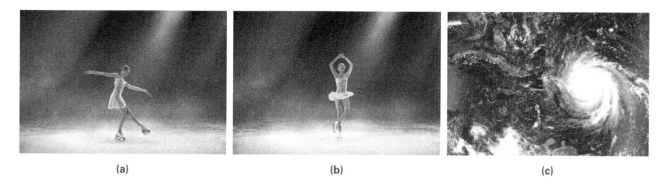

(a)   (b)   (c)

▲ **FIGURE 8.25** **Change in moment of inertia (a)** When the skater spins slowly with outstretched arms and legs, her moment of inertia is relatively large. (The masses are farther from the axis of rotation.) Note that she is isolated, with no external torques (neglecting friction) acting on her, so her angular momentum, $L = I\omega$ is conserved. **(b)** Pulling her arms and legs inward decreases her moment of inertia. (Why?) Consequently, ω must increase, and she goes into a dizzying spin. **(c)** The same principle helps explain the violence of the winds that spiral around the center of a hurricane. As air rushes in toward the low-pressure center of the storm, the air's rotational speed must increase for angular momentum to be conserved.

inertia and increases the angular speed so that multiple spins can be made during the jump. To land with a smaller rate of spin, the skater opens the arms and projects the nonlanding leg. You may have noticed that most jump landings proceed in a curved arc, which allows the skater to gain control.

EXAMPLE 8.15: A SKATER MODEL

Real-life situations are generally complicated, but some can be approximately analyzed by using simple models. Such a model for a skater's spin is shown in ▶ **Figure 8.26**, with a cylinder and rods representing the skater. In (a), the skater goes into the spin with the "arms" out, and in (b) the "arms" are over the head to achieve a faster spin by the conservation of angular momentum. If the initial spin rate is 1 revolution per 1.5 s, what is the angular speed when the arms are tucked in?

THINKING IT THROUGH. The body and arms of a skater are approximated by the cylinder and rods, for which the moments of inertia are known (Figure 8.17). Special attention must be given to finding the moment of inertia of the arms around the axis of rotation (through the cylinder). This can be done by applying the parallel axis theorem (Equation 8.8).

With the angular momentum conserved, $L = L_0$ or $I\omega = I_0\omega_0$. Knowing the initial angular speed, and given quantities to evaluate the moments of inertia (Figure 8.26), the final angular speed can be found.

SOLUTION

Listing the given data (see Figure 8.26):

Given:

$\omega_0 = (1 \text{ rev}/1.5 \text{ s})(2\pi \text{ rad/rev}) = 4.2 \text{ rad/s}$
$M_c = 75 \text{ kg (cylinder or body)}$
$M_r = 5.0 \text{ kg (one rod or arm)}$
$R = 20 \text{ cm} = 0.20 \text{ m}$
$L = 80 \text{ cm} = 0.80 \text{ m}$
Momenta of inertia (from Figure 8.17).
Cylinder: $I_c = 1/2(M_cR^2)$ rod: $I_r = 1/12(M_rL^2)$

(a) Arms extended (not to scale)

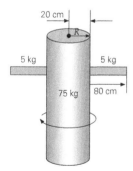

(b) Arms overhead

▲ **FIGURE 8.26** **Skater model** Change in moment of inertia and spin. See Example 8.15.

212   *College Physics Essentials*

Find: ω (final angular speed)

Let's first compute the moments of inertia of the system using the parallel axis theorem, $I = I_{CM} + Md^2$ (Equation 8.8).

Before: The I_c of the cylinder is straightforward (Figure 8.17e):

$$I_c = \frac{1}{2}M_cR^2 = \frac{1}{2}(75\,\text{kg})(0.20\,\text{m})^2 = 1.5\,\text{kg·m}^2$$

Referencing the moment of inertia of a horizontal rod (Figure 8.26a) to the cylinder's axis of rotation using the parallel axis theorem:

$$
\begin{aligned}
I_r &= I_{CM(rod)} + Md^2 \\
&= \frac{1}{12}M_rL^2 + M_r[R+(L/2)]^2 \\
&= \frac{1}{12}(5.0\,\text{kg})(0.80\,\text{m})^2 + (5.0\,\text{kg})(0.20\,\text{m}+0.40\,\text{m})^2 \\
&= 2.1\,\text{kg·m}^2
\end{aligned}
$$

(where the parallel axis through the CM of the rod is a distance of $R + (L/2)$ from the axis of rotation)

And, $I_o = I_c + 2I_r = 1.5\text{ kg·m}^2 + 2(2.1\text{ kg·m}^2) = 5.7\text{ kg·m}^2$

After: In Figure 8.26b, treating an arm mass as if its center of mass is now only about 20 cm from the axis of rotation, the moment of inertia of each arm is $I = M_rR^2$ (Figure 8.17b), and,

$$
\begin{aligned}
I &= I_c + 2(M_rR^2) = 1.5\,\text{kg·m}^2 + 2(5.0\,\text{kg·m}^2)(0.20\,\text{m})^2 \\
&= 1.9\,\text{kg·m}^2
\end{aligned}
$$

Then with the conservation of angular momentum, $L = L_o$ or $I\omega = I_o\omega_o$ and

$$\omega = \left(\frac{I_o}{I}\right)\omega_o = \left(\frac{5.7\,\text{kg·m}^2}{1.9\,\text{kg·m}^2}\right)(4.2\,\text{rad/s}) = 13\,\text{rad/s}$$

So the angular speed increases by a factor of 3.

FOLLOW-UP EXERCISE. Suppose a skater with 75% of the mass of the skater in the Exercise did a spin. What would be the spin rate ω in this case? (Consider all masses to be reduced by 75%.)

Angular momentum, $\vec{L}$, is a vector, and when it is conserved or constant, its magnitude *and* direction must remain unchanged. Thus, when no external torques act, the direction of $\vec{L}$ is fixed in space. This is the principle behind passing a football accurately, as well as that behind the movement of a gyrocompass (▶ **Figure 8.27**). A football is normally passed with a spiraling rotation. This spin, or gyroscopic action, stabilizes the ball's spin

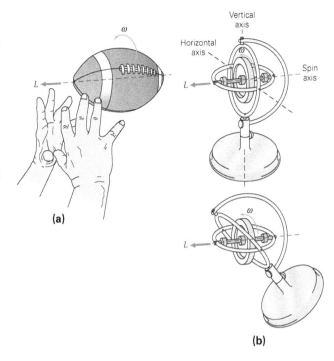

(a)

(b)

▲ **FIGURE 8.27** **Constant direction of angular momentum** When angular momentum is conserved, its direction is constant in space. **(a)** This principle can be demonstrated by a passed football. **(b)** Gyroscopic action also occurs in a gyroscope, a rotating wheel that is universally mounted on gimbals (rings) so that it is free to turn about any axis. When the frame moves, the wheel maintains its direction. This is the principle of the gyrocompass.

axis in the direction of motion. Similarly, rifle bullets are set spinning by the rifling in the barrel for directional stability.

The $\vec{L}$ vector of a spinning gyroscope in the compass is set in a particular direction (usually north). In the absence of external torques, the compass direction remains fixed, even though its carrier (an airplane or ship, for example) changes directions. You may have played with a toy gyroscope that is set spinning and placed on a pedestal. In a "sleeping" condition, the gyro stands straight up with its angular momentum vector fixed in space for some time. The gyro's center of gravity is on the axis of rotation, so there is no net torque due to its weight.

However, the gyroscope eventually slows down because of friction, causing $\vec{L}$ to tilt. In watching this motion, you may have noticed that the spin axis revolves, or *precesses*, about the vertical axis. It revolves tilted over, so to speak (Figure 8.27b). Since the gyroscope precesses, the angular momentum vector $\vec{L}$ is no longer constant in direction, indicating that a torque must be acting to produce a change ($\Delta\vec{L}$) with time.

As can be seen from the figure, the torque arises from the weight force, since the center of gravity no longer lies directly above the point of support or on the vertical axis of rotation. The instantaneous torque is such that the gyroscope's axis moves or precesses about the vertical axis.

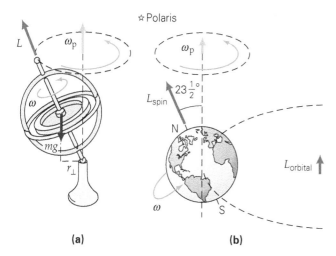

(a) **(b)**

▲ FIGURE 8.28 **Precession** An external torque causes a change in angular momentum. **(a)** For a spinning gyroscope, this change is directional, and the axis of rotation precesses at angular acceleration ω_p about a vertical line. (The torque due to the weight force would point out of the page as drawn here, as would $\Delta\vec{L}$.) Note that although there is a torque that would topple a nonspinning gyroscope, a spinning gyroscope doesn't fall. **(b)** Similarly, the Earth's axis precesses because of gravitational torques caused by the Sun and the Moon. We don't notice this motion because the period of precession is about 26 000 years.

In a similar manner, the Earth's rotational axis precesses. The Earth's spin axis is tilted $23\frac{1}{2}°$ with respect to a line perpendicular to the plane of its revolution about the Sun; the axis precesses about this line (▲ **Figure 8.28**). The precession is due to slight gravitational torques exerted on the Earth by the Sun and the Moon.

The period of the precession of the Earth's axis is about 26 000 years, so the precession has little day-to-day effect. However, it does have an interesting long-term effect. Polaris will not always be (nor has it always been) the North Star – that is, the star toward which the Earth's axis of rotation points. About 5000 years ago, Alpha Draconis was the North Star, and 5000 years from now it will be Alpha Cephei, which is at an angular distance of about 68° away from Polaris on the circle described by the precession of the Earth's axis.

Finally, a common example in which angular momentum is an important consideration is the helicopter. What would happen if a helicopter had a single rotor? Since the motor supplying the torque is internal, the angular momentum would be conserved. Initially, $L = 0$; hence, to conserve the total angular momentum of the system (rotor plus body), the separate angular momenta of the rotor and body would have to be in opposite directions to cancel. Thus, on takeoff, the rotor would rotate one way and the helicopter body the other, which is not a desirable situation.

To prevent this, helicopters have two rotors. Large helicopters have two overlapping rotors (▼ **Figure 8.29a**). The oppositely rotating rotors cancel each other's angular momenta, so the helicopter body does not have to rotate to provide canceling angular momentum. The rotors are offset at different heights so that they do not collide.

Small helicopters with a single overhead rotor have small "antitorque" tail rotors (Figure 8.29b). The tail rotor produces a thrust like a propeller and supplies the torque to counterbalance the torque produced by the overhead rotor. The tail rotor also helps in steering the craft. By increasing or decreasing the tail rotor's thrust, the helicopter turns (rotates) one way or the other.

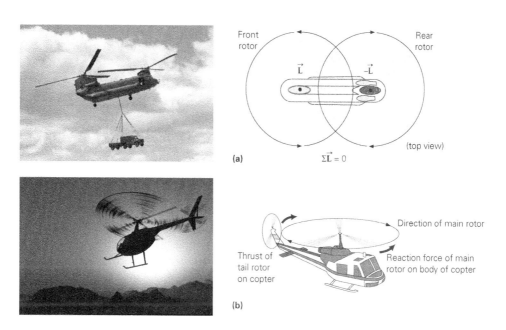

▲ FIGURE 8.29 **Different rotors** See text for description.

Chapter 8 Review

- In **pure translational motion**, all of the particles that make up a rigid object have the same instantaneous velocity.
- In **pure rotational motion (about a fixed axis)**, all of the particles that make up a rigid object have the same instantaneous angular velocity.
- Condition for rolling without slipping:

$$v_{CM} = r\omega \quad \text{(rolling, no slipping)} \tag{8.1}$$

- **Torque** ($\vec{\tau}$), the rotational analogue of force, is the product of a force and a moment arm, or lever arm. Its magnitude is

$$\tau = r_\perp F = rF\sin\theta \tag{8.2}$$

- Mechanical **equilibrium** requires that the net force, or summation of the forces, be zero (translational equilibrium) and that the net torque, or summation of the torques, be zero (rotational equilibrium).
- Conditions for translational and rotational mechanical equilibrium, respectively:

$$\vec{F}_{net} = \Sigma\vec{F}_i = 0 \quad \text{and} \quad \vec{\tau}_{net} = \Sigma\vec{\tau}_i = 0 \tag{8.3}$$

- An object is in **stable equilibrium** as long as its center of gravity, upon small displacement, lies above and inside the object's original base of support.
- **Moment of inertia (*I*)** is the rotational analogue of mass and is given by

$$I = \Sigma m_i r_i^2 \tag{8.6}$$

- *Rotational form of Newton's second law:*

$$\vec{\tau}_{net} = I\vec{\alpha} \tag{8.7}$$

- *Parallel axis theorem:*

$$I = I_{CM} + Md^2 \tag{8.8}$$

- *Rotational work:*

$$W = \tau\theta \tag{8.9}$$

- *Rotational power:*

$$P = \tau\omega \tag{8.10}$$

- *Work-energy theorem (rotational):*

$$W_{net} = \frac{1}{2}I\omega^2 - \frac{1}{2}I\omega_o^2 = \Delta K \tag{8.11}$$

- *Rotational kinetic energy:*

$$K = \frac{1}{2}I\omega^2 \tag{8.12}$$

- *Kinetic energy of a rolling object (no slipping):*

$$K = \frac{1}{2}I_{CM}\omega^2 + \frac{1}{2}Mv_{CM}^2 \tag{8.13}$$

- **Angular *momentum*:** The product of a moment arm and linear momentum or the product of a moment of inertia and angular velocity.
- *Angular momentum of a particle in circular motion (magnitude):*

$$L = r_\perp p = mr_\perp v = mr_\perp^2\omega \tag{8.14}$$

- *Angular momentum of a rigid body:*

$$\vec{L} = I\vec{\omega} \tag{8.16}$$

- *Torque as change in angular momentum (magnitude form):*

$$\tau_{net} = \frac{\Delta L}{\Delta t} \tag{8.17}$$

- *Conservation of angular momentum (with $\tau_{net} = 0$):*

$$L = L_o \quad \text{or} \quad I\omega = I_o\omega_o \tag{8.18}$$

End of Chapter Questions and Exercises

Multiple Choice Questions

8.1 Rigid Bodies, Translations, and Rotations

1. In pure rotational motion of a rigid body, (a) all the particles of the body have the same angular velocity, (b) all the particles of the body have the same tangential velocity, (c) acceleration is always zero, (d) there are always two simultaneous axes of rotation.

2. For an object with only rotational motion, all particles of the object have the same (a) instantaneous velocity, (b) average velocity, (c) distance from the axis of rotation, (d) instantaneous angular velocity.

3. The condition for rolling without slipping is (a) $a_c = r\omega^2$, (b) $v_{CM} = r\omega$, (c) $F = ma$, (d) $a_c = v^2/r$.

4. A rolling object (a) has an axis of rotation through the axis of symmetry, (b) has a zero velocity at the point or line of contact, (c) will slip if $s = r\theta$, (d) all of the preceding.

5. For the tires on your rolling, but skidding car, (a) $v_{CM} = r\omega$, (b) $v_{CM} > r\omega$, (c) $v_{CM} < r\omega$, (d) none of the preceding.

8.2 Torque, Equilibrium, and Stability

6. It is possible to have a net torque when (a) all forces act through the axis of rotation, (b) $\Sigma \vec{F}_i = 0$, (c) an object is in rotational equilibrium, (d) an object remains in unstable equilibrium.

7. If an object in stable equilibrium is displaced slightly, (a) there will be a restoring force or torque, (b) the object returns to its original equilibrium position, (c) its center of gravity still lies above and inside the object's original base of support, (d) all of the preceding.

8. Torque has the same units as (a) work, (b) force, (c) angular velocity, (d) angular acceleration.

8.3 Rotational Dynamics

9. The moment of inertia of a rigid body (a) depends on the axis of rotation, (b) cannot be zero, (c) depends on mass distribution, (d) all of the preceding.

10. Which of the following best describes the physical quantity called torque: (a) rotational analogue of force, (b) energy due to rotation, (c) rate of change of linear momentum, or (d) force that is tangent to a circle?

11. In general, the moment of inertia is greater when (a) more mass is farther from the axis of rotation, (b) more mass is closer to the axis of rotation, (c) it makes no difference.

12. A solid sphere (radius R) and a solid cylinder (radius R) with equal masses are released simultaneously from the top of a frictionless inclined plane. Then, (a) the sphere reaches the bottom first, (b) the cylinder reaches the bottom first, (c) they reach the bottom together.

13. The moment of inertia about an axis parallel to the axis through the center of mass depends on (a) the mass of the rigid body, (b) the distance between the axes, (c) the moment of inertia about the axis through the center of mass, (d) all of the preceding.

8.4 Rotational Work and Kinetic Energy

14. From $W = \tau\theta$, the unit of rotational work is the (a) watt, (b) N·m, (c) kg·rad/s², (d) N·rad.

15. A bowling ball rolls without slipping on a flat surface. The ball has (a) rotational kinetic energy, (b) translational kinetic energy, (c) both translational and rotational kinetic energy, (d) neither translational nor rotational kinetic energy.

8.5 Angular Momentum

16. The angular momentum may be increased by (a) increasing the moment of inertia, (b) increasing the angular velocity, (c) increasing both the moment of inertia and the angular velocity.

17. The units of angular momentum are (a) N·m, (b) kg·m/s², (c) kg·m²/s, (d) J·m.

Conceptual Questions

8.1 Rigid Bodies, Translations, and Rotations

1. Suppose someone in your physics class says that it is possible for a rigid body to have translational motion and rotational motion at the same time. Would you agree? If so, give an example.

2. For a rolling cylinder, what would happen if the tangential speed v were less than $r\omega$? Is it possible for v to be greater than $r\omega$? Explain.

3. If the top of your automobile tire is moving with a speed of v, what is the reading of your speedometer?

8.2 Torque, Equilibrium, and Stability

4. A force can produce torque. Will a large force always produce a larger torque? Explain.

5. Can an object in motion be in translational equilibrium? Explain.

6. If an object is in mechanical equilibrium, what conditions must be satisfied?

7. What is the difference between stable and unstable equilibrium? Give an example of each.

8.3 Rotational Dynamics

8. (a) Does the moment of inertia of a rigid body depend in any way on the center of mass of the body? Explain. (b) Can a moment of inertia have a negative value? If so, explain what this would mean.

9. Why does the moment of inertia of a rigid body have different values for different axes of rotation? What does this mean physically?

10. Two cylinders of equal mass sitting on a horizontal surface are made from materials with different densities. (a) Which cylinder will have the greater moment of inertia about an axis passing horizontally through the center? (b) Which cylinder will have the greater moment of inertia about an axis along the surface of contact?

11. Here is an interesting experiment you can try for yourself at home. Prepare a hard-boiled egg and have a raw egg available. Set them both spinning on the kitchen table. Stop both eggs quickly, and the release both. You will notice the hard-boiled one remains at rest, whereas the raw one starts spinning again. Explain.

12. Why does jerking a paper towel from a roll cause the paper to tear more easily than pulling it smoothly? Will the amount of paper on the roll affect the results?

13. Tightrope walkers are continually in danger of falling (unstable equilibrium). Commonly, a performer carries a long pole while walking the tight rope, as shown in the chapter-opening photo. What is the purpose of the pole? (In walking along a narrow board or rail, you probably extend your arms for the same reason.)

14. A solid cylinder and an annular cylinder of equal mass are rolling on the floor with the same speed. (a) If the solid cylinder's radius is equal to the annular cylinder's inner radius, which cylinder would be harder to stop? Explain. (b) Would it make any stopping difference if the solid cylinder's radius were equal to the annular cylinder's outer radius? Justify your answer explicitly.

8.4 Rotational Work and Kinetic Energy

15. Can you increase the rotational kinetic energy of a wheel without changing its translational kinetic energy? Explain.

16. In order to produce fuel-efficient vehicles, automobile manufacturers want to minimize rotational kinetic energy and maximize translational kinetic energy when a car is traveling. If you were the designer of wheels of a certain diameter, how would you design them?

17. What is required to produce a change in rotational kinetic energy?

8.5 Angular Momentum

18. A child stands on the edge of a rotating playground merry-go-round (the hand-driven type). He then starts to walk toward the center of the merry-go-round. This can result in a dangerous situation. Why?

19. The release of vast amounts of carbon dioxide may result in an increase in the Earth's average temperature through the so-called greenhouse effect and cause melting of the polar ice caps. If this occurred and the ocean level rose substantially, what effect would it have on the Earth's rotation?

20. In the classroom demonstration illustrated in ▼ **Figure 8.30**, a person on a rotating stool holds a rotating bicycle wheel by handles attached to the wheel. When the wheel is held horizontally, she rotates one way (clockwise as viewed from above). When the wheel is turned over, she rotates in the opposite direction. Explain why this occurs. [*Hint*: Consider angular momentum vectors.]

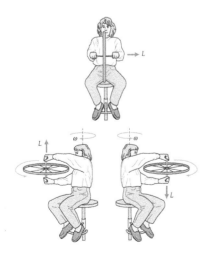

▲ **FIGURE 8.30 Faster rotation** See Conceptual Question 20.

21. Two ice skaters that weigh the same skate toward each other with the same mass and same speed on parallel paths. As they pass each other, they link arms. (a) What is the velocity of their center of mass after they link arms? (b) What happens to their initial, translational kinetic energies?

Exercises*

Integrated Exercises (IEs) *are two-part exercises. The first part typically requires a conceptual answer choice based on physical thinking and basic principles. The following part is quantitative calculations associated with the conceptual choice made in the first part of the exercise.*

8.1 Rigid Bodies, Translations, and Rotations

1. • A wheel rolls uniformly on level ground without slipping. A piece of mud on the wheel flies off when it is at the 9 o'clock position (rear of wheel). Describe the subsequent motion of the mud.

2. • A rope goes over a circular pulley with a radius of 6.5 cm. If the pulley makes 4 revolutions without the rope slipping, what length of rope passes over the pulley?

3. • A wheel rolls 5 revolutions on a horizontal surface without slipping. If the center of the wheel moves 3.2 m, what is the radius of the wheel?

4. •• A bowling ball with a radius of 15.0 cm travels down the lane so that its center of mass is moving at 3.60 m/s. The bowler estimates that it makes about 7.50 complete revolutions in 2.00 s. Is it rolling without slipping? Prove your answer, assuming that the bowler's quick observation limits answers to two significant figures.

5. •• A ball with a radius of 15 cm rolls on a level surface, and the translational speed of the center of mass is 0.25 m/s. What is the angular speed about the center of mass if the ball rolls without slipping?

6. IE •• (a) When a disk rolls without slipping, should the product $r\omega$ be (1) greater than, (2) equal to, or (3) less than v_{CM}? (b) A disk with a radius of 0.15 m rotates through 270° as it travels 0.71 m. Does the disk roll without slipping? Prove your answer.

8.2 Torque, Equilibrium, and Stability

7. • A torque of 18 m·N is required to loosen a bolt. If a force of 50 N is applied, what should be the minimum length of wrench?

8. • The drain plug on a car's engine has been tightened to a torque of 25 m·N. If a 0.15-m-long wrench is used to change the oil, what is the minimum force needed to loosen the plug?

---

* The bullets denote the degree of difficulty of the exercises: •, simple; ••, medium; and •••, more difficult.

9. • In Exercise 8, due to limited work space, you must crawl under the car. The force thus cannot be applied perpendicularly to the length of the wrench. If the applied force makes a 30° angle with the length of the wrench, what is the force required to loosen the drain plug?

10. • How many different positions of stable equilibrium and unstable equilibrium are there for a cube? Consider each surface, edge, and corner to be a different position.

11. IE •• Two children are sitting on opposite ends of a uniform seesaw of negligible mass. (a) Can the seesaw be balanced if the masses of the children are different? How? (b) If a 35-kg child is 2.0 m from the pivot point (or fulcrum), how far from the pivot point will her 30-kg playmate have to sit on the other side for the seesaw to be in equilibrium?

12. • A uniform meterstick pivoted at its center, as in Example 8.5, has a 100-g mass suspended at the 25.0-cm position. (a) At what position should a 75.0-g mass be suspended to put the system in equilibrium? (b) What mass would have to be suspended at the 90.0-cm position for the system to be in equilibrium?

13. IE •• Telephone and electrical lines are allowed to sag between poles so that the tension will not be too great when something hits or sits on the line. (a) Is it possible to have the lines perfectly horizontal? Why or why not? (b) Suppose that a line were stretched almost perfectly horizontally between two poles that are 30 m apart. If a 0.25-kg bird perches on the wire midway between the poles and the wire sags 1.0 cm, what would be the tension in the wire? (Neglect the mass of the wire.)

14. •• In ▼ **Figure 8.31**, what is the force F_m supplied by the deltoid muscle so as to hold up the outstretched arm if the mass of the arm is 3.0 kg? (F_j is the joint force on the bone of the upper arm – the humerus.)

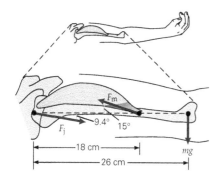

▲ FIGURE 8.31 **Arm in static equilibrium** See Exercise 14.

15. •• A bowling ball (mass 7.00 kg and radius 17.0 cm) is released so fast that it skids without rotating down the lane (at least for a while). Assume the ball skids to the right and the coefficient of sliding friction between the ball and the lane surface is 0.400. (a) What is the

direction of the torque exerted by the friction on the ball about the center of mass of the ball? (b) Determine the magnitude of this torque (again about the ball's center of mass).

16. •• In doing physical therapy for an injured knee joint, a person raises a 5.0-kg weighted boot as shown in ▼ **Figure 8.32**. Compute the torque due to the boot for each position shown.

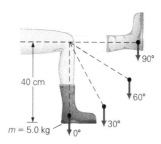

▲ FIGURE 8.32 **Torque in physical therapy** See Exercise 16.

17. •• An artist wishes to construct a birds and bees mobile, as shown in ▼ **Figure 8.33**. If the mass of the bee on the lower left is 0.10 kg and each vertical support string has a length of 30 cm, what are the masses of the other birds and bees? (Neglect the masses of the bars and strings.)

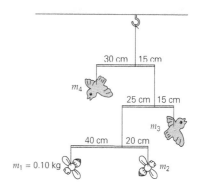

▲ FIGURE 8.33 **Birds and bees** See Exercise 17.

18. IE •• The location of a person's center of gravity relative to his or her height can be found by using the arrangement shown in ▼ **Figure 8.34**. The scales are initially adjusted to zero with the board alone. (a) Would you expect the location of the center of gravity to be (1) midway between the scales, (2) toward the scale at the person's head, or (3) toward the scale at the person's feet? Why? (b) Locate the center of gravity of the person relative to the horizontal dimension.

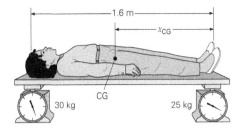

▲ FIGURE 8.34 **Locating the center of gravity** See Exercise 18.

19. •• (a) How many uniform, identical textbooks of width 25.0 cm can be stacked on top of each other on a level surface without the stack falling over if each successive book is displaced 3.00 cm in width relative to the book below it? (b) If the books are 5.00 cm thick, what will be the height of the center of mass of the stack above the level surface?

20. •• If four metersticks were stacked on a table with 10 cm, 15 cm, 30 cm, and 50 cm, respectively, hanging over the edge, as shown in ▼ **Figure 8.35**, would the top meterstick remain on the table?

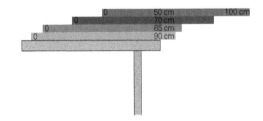

▲ FIGURE 8.35   **Will they fall off?** See Exercise 20.

21. •• A 10.0-kg solid uniform cube with 0.500-m sides rests on a level surface. What is the minimum amount of work necessary to put the cube into an unstable equilibrium position?

22. •• While standing on a long board resting on a scaffold, a 70-kg painter paints the side of a house, as shown in ▼ **Figure 8.36**. If the mass of the board is 15 kg, how close to the end can the painter stand without tipping the board over?

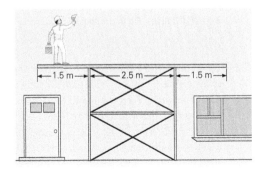

▲ FIGURE 8.36   **Not too far!** See Exercise 22.

23. •• A mass is suspended by two cords as shown in ▼ **Figure 8.37**. What are the tensions in the cords?

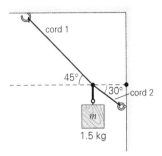

▲ FIGURE 8.37   **A lot of tension** See Exercises 23 and 24.

24. •• If the cord attached to the vertical wall in Figure 8.37 were horizontal (instead of at a 30° angle), what would the tensions in the cords be?

25. •• A force is applied to a cord wrapped around a solid 2.0-kg cylinder as shown in ▼ **Figure 8.38**. Assuming the cylinder rolls without slipping, what is the force of friction acting on the cylinder?

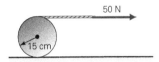

▲ FIGURE 8.38   **No slipping** See Exercise 25.

8.3 Rotational Dynamics

26. • A fixed 0.15-kg solid-disk pulley with a radius of 0.075 m is acted on by a net torque of 6.4 m·N. What is the angular acceleration of the pulley?

27. • What net torque is required to give a uniform 20-kg solid ball with a radius of 0.20 m an angular acceleration of 20 rad/s²?

28. • For the system of masses shown in ▶ **Figure 8.39**, find the moment of inertia about (a) the x-axis, (b) the y-axis, and (c) an axis through the origin and perpendicular to the page (z-axis). (Neglect the masses of the connecting rods.)

29. •• A 2000-kg Ferris wheel accelerates from rest to an angular speed of 20 rad/s in 12 s. Approximate the Ferris wheel as a circular disk with a radius of 30 m. What is the net torque on the wheel?

30. IE •• Two objects of different masses are joined by a light rod. (a) Is the moment of inertia about the center of mass the minimum or the maximum? Why? (b) If the two masses are 3.0 and 5.0 kg and the length of the rod

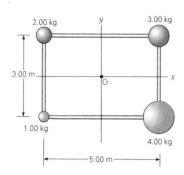

▲ FIGURE 8.39   Moments of inertia about different axes
See Exercise 28.

is 2.0 m, find the moments of inertia of the system about an axis perpendicular to the rod, through the center of the rod and the center of mass.

31. •• Two masses are suspended from a pulley as shown in ▼ **Figure 8.40** (the Atwood machine revisited; see Chapter 4, Exercise 49). The pulley itself has a mass of 0.20 kg, a radius of 0.15 m, and a constant torque of 0.35 m·N due to the friction between the rotating pulley and its axle. What is the magnitude of the acceleration of the suspended masses if $m_1 = 0.40$ kg and $m_2 = 0.80$ kg? (Neglect the mass of the string.)

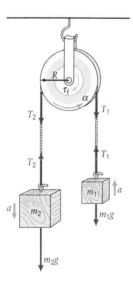

▲ FIGURE 8.40   The Atwood machine revisited See Exercise 31.

32. •• To start her lawn mower, Julie pulls on a cord that is wrapped around a pulley. The pulley has a moment of inertia about its central axis of $I = 0.550$ kg·m² and a radius of 5.00 cm. There is an equivalent frictional torque impeding her pull of $\tau_f = 0.430$ m·N. To accelerate the pulley at $\alpha = 4.55$ rad/s², (a) how much torque does Julie need to apply to the pulley? (b) How much tension must the rope exert?

33. •• For the system shown in ▼ **Figure 8.41**, $m_1 = 8.0$ kg, $m_2 = 3.0$ kg, $\theta = 30°$, and the radius and mass of the pulley are 0.10 m and 0.10 kg, respectively. (a) What is the acceleration of the masses? (Neglect friction and the string's mass.) (b) If the pulley has a constant frictional torque of 0.050 m·N when the system is in motion, what is the acceleration of the masses? [*Hint:* Isolate the forces. The tensions in the strings are different. Why?]

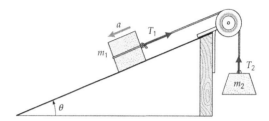

▲ FIGURE   8.41   Inclined   plane   and   pulley See Exercise 33.

34. •• A meterstick pivoted about a horizontal axis through the 0-cm end is held in a horizontal position and let go. (a) What is the initial tangential acceleration of the 100 cm position? Are you surprised by this result? (b) Which position has a tangential acceleration equal to the acceleration due to gravity?

35. •• Pennies are placed every 10 cm on a meterstick. One end of the stick is put on a table and the other end is held horizontally with a finger, as shown in ▼ **Figure 8.42**. If the finger is pulled away, what are the accelerations of the pennies?

▲ FIGURE 8.42   Money left behind? See Exercise 35.

36. ••• A uniform 2.0-kg cylinder of radius 0.15 m is suspended by two strings wrapped around it (▶ **Figure 8.43**). As the cylinder descends, the strings unwind from it. What is the acceleration of the center of mass of the cylinder? (Neglect the mass of the string.)

8.4   Rotational Work and Kinetic Energy

37. • A constant retarding torque of 12 m·N stops a rolling wheel of diameter 0.80 m in a distance of 15 m. How much work is done by the torque?

38. • A person opens a door by applying a 15-N force perpendicular to it at a distance 0.90 m from the hinges. The door is pushed wide open (to 120°) in 2.0 s. (a) How much work was done? (b) What was the average power delivered?

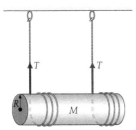

▲ FIGURE 8.43   **Unwinding with gravity** See Exercise 36.

39. IE • In Figure 8.20, a mass m descends a vertical distance from rest. (Neglect friction and the mass of the string.) (a) From the conservation of mechanical energy, will the linear speed of the descending mass be (1) greater than, (2) equal to, or (3) less than $\sqrt{2gh}$? Why? (b) If $m = 1.0$ kg, $M = 0.30$ kg, and $R = 0.15$ m, what is the linear speed of the mass after it has descended a vertical distance of 2.0 m from rest?

40. • A constant torque of 10 m·N is applied to the rim of a 10-kg uniform disk of radius 0.20 m. What is the angular speed of the disk about an axis through its center after it rotates 2.0 revolutions from rest?

41. • A 2.5-kg pulley of radius 0.15 m is pivoted about an axis through its center. What constant torque is required for the pulley to reach an angular speed of 25 rad/s after rotating 3.0 revolutions, starting from rest?

42. •• A solid ball of mass m rolls along a horizontal surface with a translational speed of v. What percent of its total kinetic energy is translational?

43. •• A pencil 18 cm long stands vertically on its point end on a horizontal table. If it falls over without slipping, with what tangential speed does the eraser end strike the table?

44. •• A uniform sphere and a uniform cylinder with the same mass and radius roll at the same velocity side by side on a level surface without slipping. If the sphere and the cylinder approach an inclined plane and roll up it without slipping, will they be at the same height on the plane when they come to a stop? If not, what will be the percentage difference of the heights?

45. •• A hoop starts from rest at a height 1.2 m above the base of an inclined plane and rolls down under the influence of gravity. What is the linear speed of the hoop's center of mass just as the hoop leaves the incline and rolls onto a horizontal surface? (Neglect friction.)

46. •• A cylindrical hoop, a cylinder, and a sphere of equal radius and mass are released at the same time from the top of an inclined plane. Using the conservation of mechanical energy, show that the sphere always gets to the bottom of the incline first with the fastest speed and that the hoop always arrives last with the slowest speed.

47. •• For the following objects, which all roll without slipping, determine the rotational kinetic energy about the center of mass as a percentage of the total kinetic energy: (a) a solid sphere, (b) a thin spherical shell, and (c) a thin cylindrical shell.

48. •• An industrial flywheel with a moment of inertia of 4.25×10^2 kg·m^2 rotates with a speed of 7500 rpm. (a) How much work is required to bring the flywheel to rest? (b) If this work is done uniformly in 1.5 min, how much power is required?

49. ••• A steel ball rolls down an incline into a loop-the-loop of radius R (▼ **Figure 8.44a**). (a) What minimum speed must the ball have at the top of the loop in order to stay on the track? (b) At what vertical height (h) on the incline, in terms of the radius of the loop, must the ball be released in order for it to have the required minimum speed at the top of the loop? (Neglect frictional losses.) (c) Figure 8.44b shows the loop-the-loop of a roller coaster. What are the sensations of the riders if the roller coaster has the minimum speed or a greater speed at the top of the loop? [*Hint:* In case the speed is below the minimum, seat and shoulder straps hold the riders in.]

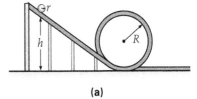

(a)

(b)

▲ FIGURE 8.44   **Loop-the-loop and rotational speed** See Exercise 49.

8.5 Angular Momentum

50. • What is the angular momentum of a 2.0-g particle moving counterclockwise (as viewed from above) with an angular speed of 5π rad/s in a horizontal circle of radius 15 cm? (Give the magnitude and direction.)

51. • A 10-kg rotating disk of radius 0.25 m has an angular momentum of 0.45 kg·m^2/s. What is the angular speed of the disk?

52. •• Compute the ratio of the magnitudes of the Earth's orbital angular momentum and its rotational angular momentum. Are these momenta in the same direction?

53. •• The Earth revolves about the Sun and spins on its axis, which is tilted $23\frac{1}{2}°$ to its orbital plane. (a) Assuming a circular orbit, what is the magnitude of the angular momentum associated with the Earth's orbital motion about the Sun? (b) What is the magnitude of the angular momentum associated with the Earth's rotation on its axis?

54. •• The period of the Moon's rotation is the same as the period of its revolution: 27.3 days (sidereal). What is the angular momentum for each rotation and revolution? (Because the periods are equal, we see only one side of the Moon from Earth.)

55. IE •• Circular disks are used in automobile clutches and transmissions. When a rotating disk couples to a stationary one through frictional force, the energy from the rotating disk can transfer to the stationary one. (a) Is the angular speed of the coupled disks (1) greater than, (2) less than, or (3) the same as the angular speed of the original rotating disk? Why? (b) If a disk rotating at 800 rpm couples to a stationary disk with three times the moment of inertia, what is the angular speed of the combination?

56. •• An ice skater has a moment of inertia of 100 kg·m^2 when his arms are outstretched and a moment of inertia of 75 kg·m^2 when his arms are tucked in close to his chest. If he starts to spin at an angular speed of 2.0 rps (revolutions per second) with his arms outstretched, what will his angular speed be when they are tucked in?

57. •• An ice skater spinning with outstretched arms has an angular speed of 4.0 rad/s. She tucks in her arms, decreasing her moment of inertia by 7.5%. (a) What is the resulting angular speed? (b) By what factor does the skater's kinetic energy change? (Neglect any frictional effects.) (c) Where does the extra kinetic energy come from?

58. ••• While repairing his bicycle, a student turns it upside down and sets the front wheel spinning at 2.00 rev/s. Assume the wheel has a mass of 3.25 kg and all of the mass is located on the rim, which has a radius of 41.0 cm. To slow the wheel, he places his hand on the tire, thereby exerting a tangential force of friction on the wheel. It takes 3.50 s to come to rest. Use the change in angular momentum to determine the force he exerts on the wheel. Assume the frictional force of the axle is negligible.

59. ••• A comet approaches the Sun as illustrated in ▼ Figure 8.45 and is deflected by the Sun's gravitational attraction. This event is considered a collision, and b is called the *impact parameter*. Find the distance of closest approach (d) in terms of the impact parameter and the velocities (v_0 at large distances and v at closest approach). Assume that the radius of the Sun is negligible compared to d. (As the figure shows, the tail of a comet always "points" away from the Sun.)

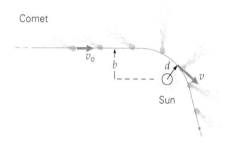

▲ FIGURE 8.45   **A comet "collision"** See Exercise 59.

9

Solids and Fluids

The hang glider's lift and subsequent motion can be explained by fluid buoyancy and fluid dynamics.

Shown in the chapter-opening photo are white clouds, blue sky, and unseen air that makes gliding possible. We walk on the solid surface of the Earth and in our daily lives use solid objects of all sorts, from scissors to computers. But we are surrounded by fluids – liquids and gases – some of which are indispensable. Without water, survival would be for only a few days at most. By far the most abundant substance in our bodies is water, and it is in the watery environment of our cells that all chemical processes on which life depends take place. Also, we are surrounded by gases and the gaseous oxygen in the air is essential for life processes.

On the basis of general physical distinctions, matter is commonly divided into three phases: solid, liquid, and gas. A *solid* has a definite shape and volume. A *liquid* has a fairly definite volume but assumes the shape of its container. A *gas* takes on the shape and volume of its container. Solids and liquids are

223

sometimes called *condensed matter*. In this chapter, a different classification scheme will be used and matter will be considered in terms of solids and fluids. Liquids and gases are referred to collectively as fluids. A **fluid** is a substance that can flow; liquids and gases qualify, but solids do not.

A simplistic description of solids is that they are made up of particles called atoms that are held rigidly together by interatomic forces. In Section 8.1, the concept of an ideal rigid body was used to describe rotational motion. Real solid bodies are not absolutely rigid and can be elastically deformed by external forces. Elasticity usually brings to mind a rubber band or spring that will resume its original dimensions even after being greatly deformed. In fact, all materials – even very hard steel – are elastic to some degree. But, as will be learned, such deformation has an *elastic limit*.

Fluids, however, have little or no elastic response to a force. Instead, the force merely causes an unconfined fluid to flow. This chapter pays particular attention to the behavior of fluids, shedding light on such questions as how hydraulic lifts work, why icebergs and ocean liners float, and what "10W−30" on a bottle of motor oil means. You'll also discover why the person in the chapter-opening photo can soar, with the aid of suitably shaped plastic materials.

Because of their fluidity, liquids and gases have many properties in common, and it is convenient to study them together. But there are important differences as well. For example, liquids are not very compressible, whereas gases are easily compressed.

9.1 Solids and Elastic Moduli

As stated previously, all solid materials are elastic to some degree. That is, a body that is slightly deformed by an applied force will return to its original dimensions or shape when the force is removed. The deformation may not be noticeable for many materials, but it's there.

You may be able to visualize why materials are elastic if you think in terms of the simplistic model of a solid in ▼ **Figure 9.1**. The atoms of the solid substance are imagined to be held together by springs. The elasticity of the springs represents the resilient nature of the interatomic forces. The springs resist

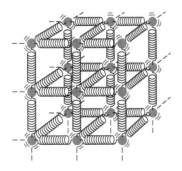

▲ FIGURE 9.1 **A springy solid** The elastic nature of interatomic forces is indicated by simplistically representing them as springs, which, like the forces, resist deformation.

permanent deformation, as do the forces between atoms. The elastic properties of solids are commonly discussed in terms of stress and strain. **Stress** is a measure of the force causing a deformation. **Strain** is a relative measure of the deformation a stress causes. Quantitatively, *stress is the applied force per unit cross-sectional area*:

$$\text{stress} = \frac{F}{A} \tag{9.1}$$

SI unit of stress: newton per square meter (N/m²)

Here, F is the magnitude of the applied force normal (perpendicular) to the cross-sectional area. Equation 9.1 shows that the SI units for stress are newtons per square meter (N/m²).

As illustrated in ▼ **Figure 9.2**, a force applied to the ends of a rod gives rise to either a *tensile stress* (an elongating tension, $\Delta L > 0$) or a *compressional stress* (a shortening tension, $\Delta L < 0$), depending on the direction of the force. In both these cases, the *tensile strain* is the ratio of the change in length ($\Delta L = L - L_o$) to the original length (L_o), without regard to the sign, so the absolute value, $|\Delta L|$, is used. Then,

$$\text{strain} = \frac{|\text{change in length}|}{\text{original length}} = \frac{|\Delta L|}{L_o} = \frac{|L - L_o|}{L_o} \tag{9.2}$$

Strain is a positive unitless quantity

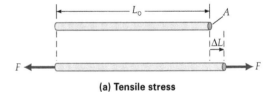

(a) Tensile stress

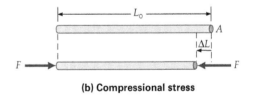

(b) Compressional stress

▲ FIGURE 9.2 **Tensile and compressional stresses** Tensile and compressional stresses are due to forces applied normally to the surface area of the ends of bodies. **(a)** A tension, or tensile stress, tends to increase the length of an object. **(b)** A compressional stress tends to shorten the length. ($\Delta L = L - L_o$) can be positive, as in (a), or negative, as in (b). The sign is not needed in Equation 9.2, so the absolute value, $|\Delta L|$, is used.

Thus the strain is the *fractional change* in length. For example, if the strain is 0.05, the length of the material has changed by 5% of the original length.

As might be expected, the resulting strain depends on the applied stress. For relatively small stresses, this is a direct proportion; that is, stress is proportional to ($\propto$) strain. The constant of proportionality, which depends on the nature of the material, is called the **elastic modulus**; that is,

$$\text{stress} = \text{elastic modulus} \times \text{strain}$$

or

$$\text{elastic modulus} = \frac{\text{stress}}{\text{strain}} \qquad (9.3)$$

SI unit of elastic modulus: newton per square meter (N/m^2)

The elastic modulus is the stress divided by the strain, and the elastic modulus has the same units as stress. (Why?)

Three general types of elastic moduli (plural of *modulus*) are associated with stresses that produce changes in length, shape, and volume. These are called *Young's modulus,* the *shear modulus,* and the *bulk modulus,* respectively.

9.1.1 Change in Length: Young's Modulus

▼ **Figure 9.3** is a typical graph of the tensile stress versus the strain for a metal rod. The curve is a straight line up to a point where true elastic behavior ends. Beyond this point, the strain begins to increase more rapidly to the **elastic limit**. If the tension is removed at this point, the material will return to its original length. If the tension is applied beyond the elastic limit and then removed, the material will recover somewhat, but will retain some permanent deformation.

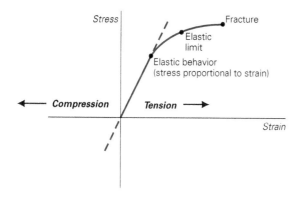

▲ **FIGURE 9.3** **Stress versus strain** A plot of stress versus strain for a typical metal rod is a straight line up to a point where true elastic behavior ends. Then elastic deformation continues until the elastic limit is reached. Beyond that, the rod will be permanently deformed and will eventually fracture or break.

The straight-line part of the graph shows a direct proportionality between stress and strain. This relationship, first formalized by the English physicist Robert Hooke in 1678, is known as *Hooke's law.* (It is the same general relationship as that given for a spring in Section 5.2 – see Figure 5.7.) The elastic modulus for a tension or a compression is called **Young's modulus (Y):**[*]

---

[*] Thomas Young (1773–1829) was an English physician and physicist who also demonstrated the wave nature of light. See Young's double-slit experiment, Section 24.1.

$$\frac{F}{A} = Y\left(\frac{\Delta L}{L_\text{o}}\right) \quad \text{or} \quad Y = \frac{F/A}{\Delta L/L_\text{o}} \quad \text{(Young's modulus)} \quad (9.4)$$

SI unit of Young's modulus: newton per square meter (N/m^2)

The units of Young's modulus are the same as those of stress, newtons per square meter (N/m^2), since the strain is unitless. Some typical values of Young's modulus are given in ▼ **Table 9.1**.

To obtain a conceptual or physical understanding of Young's modulus, let's solve Equation 9.4 for ΔL:

$$\Delta L = \left(\frac{FL_\text{o}}{A}\right)\frac{1}{Y} \quad \text{or} \quad \Delta L \propto \frac{1}{Y}$$

Hence, the larger the Young's modulus of a material, the smaller its change in length (with other parameters being equal).

TABLE 9.1   Elastic Moduli for Various Materials (in N/m^2)

Substance	Young's Modulus (Y)	Shear Modulus (S)	Bulk Modulus (B)
Solids			
Aluminum	7.0×10^{10}	2.5×10^{10}	7.0×10^{10}
Bone (limb)	1.5×10^{10} (tension)	1.2×10^{10}	
	9.3×10^{9} (compression)		
Brass	9.0×10^{10}	3.5×10^{10}	7.5×10^{10}
Copper	11×10^{10}	3.8×10^{10}	12×10^{10}
Glass	5.7×10^{10}	2.4×10^{10}	4.0×10^{10}
Iron	15×10^{10}	6.0×10^{10}	12×10^{10}
Nylon	5.0×10^{9}	8.0×10^{8}	
Steel	20×10^{10}	8.2×10^{10}	15×10^{10}
Liquids			
Alcohol, ethyl			1.0×10^{9}
Glycerin			4.5×10^{9}
Mercury			26×10^{9}
Water			2.2×10^{9}

EXAMPLE 9.1: PULLING MY LEG – UNDER A LOT OF STRESS

The femur (upper leg bone) is the longest and strongest bone in the body. Taking a typical femur to be approximately circular in cross-section with a radius of 2.0 cm, how much force would be required to extend a patient's femur by 0.010% while in horizontal traction?

THINKING IT THROUGH. Equation 9.4 should apply, but where does the percentage increase fit in? This question can be answered as soon as it is recognized that the $\Delta L/L_\text{o}$ term is the *fractional* increase in length. For example, if you had a spring with a length of 10 cm (L_o) and you stretched it 1.0 cm (ΔL), then $\Delta L/L_\text{o} = 1.0$ cm/10 cm $= 0.10$. This ratio can readily be changed to a percentage, and the spring's length was increased by 10%. So the percentage increase is really just the value of the $\Delta L/L_\text{o}$ term (multiplied by 100%).

SOLUTION

Listing the data:

Given:

$$r = 2.0 \text{ cm} = 0.020 \text{ m}$$
$$\Delta L/L_{\text{o}} = 0.010\% = 1.0 \times 10^{-4}$$
$$Y = 1.5 \times 10^{10} \text{ N/m}^2 \text{ (for bone, from Table 9.1)}$$

Find: F (tensile force)

Using Equation 9.4,

$$F = Y(\Delta L/L_{\text{o}})A = Y(\Delta L/L_{\text{o}})\pi r^2$$
$$= (1.5 \times 10^{10} \text{ N/m}^2)(1.0 \times 10^{-4})\pi(0.020\,\text{m})^2 = 1.9 \times 10^3 \text{ N}$$

How much force is this? Quite a bit – in fact, more than 400 lb. The femur is a pretty strong bone.

FOLLOW-UP EXERCISE. A total mass of 16 kg is suspended from a 0.10-cm-diameter steel wire. (a) By what percentage does the length of the wire increase? (b) The tensile or ultimate strength of a material is the maximum stress the material can support before breaking or fracturing. If the tensile strength of the steel wire in (a) is $4.9 \times 10^8 \text{ N/m}^2$, how much mass could be suspended before the wire would break? *(Answers to all Follow-Up Exercises are given in Appendix V at the back of the book.)*

9.1.2 Change in Shape: Shear Modulus

Another way an elastic body can be deformed is by a *shear stress*. In this case, the deformation is due to an applied force that is *tangential* to the surface area (▶ **Figure 9.4a**). A change in shape results without a change in volume. The *shear strain* is given by x/h, where x is the relative displacement of the faces and h is the distance between them.

The shear strain may be defined in terms of the *shear angle* ϕ. As Figure 9.4b shows, $\tan \phi = x/h$. But the shear angle is usually quite small, so a good approximation is $\tan \phi \approx \phi \approx x/h$, where ϕ is in radians.* (If $\phi = 10°$, for example, there is only 1.0% difference between ϕ and $\tan \phi$.) The shear **modulus (S)**, sometimes called the *modulus of rigidity*, is then

$$S = \frac{F/A}{x/h} \approx \frac{F/A}{\phi} \quad \text{(Shear modulus)} \quad (9.5)$$

SI unit of shear modulus: newton per square meter (N/m^2)

Note in Table 9.1 that the shear modulus is generally less than Young's modulus. In fact, S is approximately $Y/3$ for many materials, which indicates a greater response to a shear stress than to a tensile stress. Note also that the inverse relationship $\phi \approx 1/S$ is similar to that pointed out previously for Young's modulus.

Liquids do not have shear moduli (or Young's moduli) – hence the gaps in Table 9.1. A shear stress cannot be effectively applied to a liquid or a gas because fluids deform continuously in response. That is, *fluids cannot support a shear.*

_____

* See Section 7.1 and Figure 7.3, The Small-Angle Approximation.

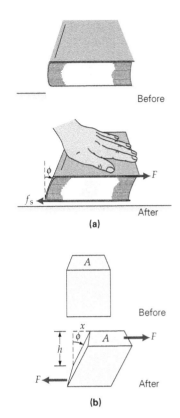

▲ **FIGURE 9.4   Shear stress and strain (a)** A shear stress is produced when a force is applied tangentially to a surface area. **(b)** The strain is measured in terms of the relative displacement of the object's faces, or the shear angle ϕ.

9.1.3 Change in Volume: Bulk Modulus

Suppose that a force directed inward acts over the entire surface of a body (▼ **Figure 9.5**). Such a *volume stress* is often applied by pressure transmitted by a fluid. An elastic material will be compressed by a volume stress; that is, the material will show a change in volume, but not in general shape, in response to a pressure change Δp. [Pressure (p) is force per

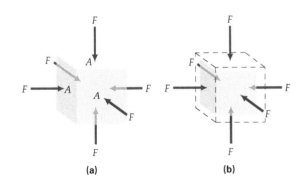

▲ **FIGURE 9.5   Volume stress and strain (a)** A volume stress is applied when a normal force acts over an entire surface area, as shown here for a cube. This type of stress most commonly occurs in gases. **(b)** The resulting strain is a change in volume.

unit area; Section 9.2.] The change in pressure is equal to the volume stress, or $\Delta p = F/A$. The *volume strain* is the ratio of the volume change (ΔV) to the original volume (V_o). The **bulk modulus (B)** is then

$$B = \frac{F/A}{-\Delta V/V_o} = -\frac{\Delta p}{\Delta V/V_o} \quad \text{(Bulk modulus)} \quad (9.6)$$

SI unit of bulk modulus: newton per square meter (N/m²)

The minus sign is introduced to make B a positive quantity, since $\Delta V = V - V_o$ is negative for an increase in external pressure (when Δp is positive). Similarly to the previous moduli relationships, $\Delta V \propto 1/B$.

Bulk moduli of selected solids and liquids are listed in Table 9.1. Gases also have bulk moduli, since they can be compressed. For a gas, it is common to talk about the reciprocal of the bulk modulus, which is called the **compressibility (k)**:

$$k = \frac{1}{B} \quad \text{(compressiblility for gases)} \quad (9.7)$$

The change in volume ΔV is thus directly proportional to the compressibility k.

Solids and liquids are relatively incompressible and thus have small values of compressibility. Conversely, gases are easily compressed and have large compressibilities, which vary with pressure and temperature.

EXAMPLE 9.2: COMPRESSING A LIQUID – VOLUME STRESS AND BULK MODULUS

By how much should the pressure on a liter of water be changed to compress it by 0.10%?

THINKING IT THROUGH. Similarly to the fractional change in length, $\Delta L/L_o$, the fractional change in volume is given by $-\Delta V/V_o$, which may be expressed as a percentage. The pressure change can then be found from Equation 9.6. Compression implies a negative ΔV.

SOLUTION

Given:

$$-\Delta V/V_o = 0.0010 \text{ (or 0.10\%)}$$
$$V_o = 1.0 \text{ L} = 1000 \text{ cm}^3$$
$$B_{H_2O} = 2.2 \times 10^9 \text{ N/m}^2 \text{ (from Table 9.1)}$$

Find: Δp

Note that $-\Delta V/V_o$ is the *fractional* change in the volume. With $V_o = 1000$ cm³, the change (reduction) in volume is

$$-\Delta V = 0.0010 \, V_o = 0.0010(1000 \text{ cm}^3) = 1.0 \text{ cm}^3$$

However, the change in volume is not needed. The fractional change, as listed in the given data, can be used directly in Equation 9.6 to find the increase in pressure:

$$\Delta p = B\left(\frac{-\Delta V}{V_o}\right) = (2.2 \times 10^9 \text{ N/m}^2)(0.0010) = 2.2 \times 10^6 \text{ N/m}^2$$

(This increase is about 22 times normal atmospheric pressure, which, as will be shown shortly, is the pressure at about a depth 220 m or about 700 ft under the ocean!)

FOLLOW-UP EXERCISE. If an extra 1.0×10^6 N/m² of pressure above normal atmospheric pressure is applied to a half-liter of water, what is the change in the water's volume?

9.2 Fluids: Pressure and Pascal's Principle

When a force is applied to an area, **pressure**, or the *force per unit area*, is defined as

$$p = \frac{F}{A} \quad \text{(Pressure)} \quad (9.8a)$$

SI unit of pressure: newton per square meter (N/m²), or pascal (Pa)

Pressure has SI units of newton per square meter (N/m²), or **pascal (Pa)**, in honor of the French scientist and philosopher Blaise Pascal (1623–1662), who studied fluids and pressure. By definition, 1 Pa = 1 N/m².

The force in Equation 9.8a is understood to be acting normally (perpendicularly) to the surface area. F may be the perpendicular component of a force that acts at an angle to a surface (▼ Figure 9.6).

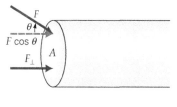

$$p = \frac{F_\perp}{A} = \frac{F \cos\theta}{A}$$

▲ FIGURE 9.6 **Pressure** Pressure is usually written $p = F/A$, where it is understood that F is the force or component of force normal to the surface. In general, then, $p = F(\cos\theta)/A$.

As Figure 9.6 shows, in the more general case,

$$p = \frac{F_\perp}{A} = \frac{F \cos\theta}{A} \quad (9.8b)$$

In the British system, a common unit of pressure is pound per square inch (lb/in², or psi). Other units, some of which will be introduced later, are used in special applications. Before continuing, here's a "solid" example of the relationship between force and pressure.

CONCEPTUAL EXAMPLE 9.3: FORCE AND PRESSURE – TAKING A NAP ON A BED OF NAILS

Suppose you are getting ready to take a nap, and you have a choice of lying stretched out on your back on (a) a bed of nails, (b) a hardwood floor, or (c) a couch. Which one would you choose for the most comfort, and *why?*

REASONING AND ANSWER. The comfortable choice is quite apparent – the couch. But here, the conceptual question is *why*.

First let's look at the prospect of lying or standing on a bed of nails, an old trick that originated in India and used to be demonstrated in carnival sideshows (see Figure 9.25 later in this chapter). There is really no trick here, just physics – namely, force and pressure. It is the force per unit area, or pressure ($p = F/A$), that determines whether a nail will pierce the skin. The force is determined by the weight of the person lying on the nails. The area is determined by the *effective* area of the nails in contact with the skin (neglecting one's clothes).

If there were only one nail, the pressure would be very great – a situation in which the lone nail would pierce the skin. However, when a bed of nails is used, the pressure is then reduced to a level at which the nails do not pierce the skin, due to a relatively large effective area of contact.

When you are lying on a hardwood floor, the area in contact with your body is appreciable and the pressure is reduced, but it still may be a bit uncomfortable. Parts of your body, such as your neck and the curved portion of your back, are *not* in contact with a surface, but they would be on a couch. On a soft couch, the body sinks into it and the contact surface is greater, therefore reduced pressure and more comfort. So (c) is the answer.

FOLLOW-UP EXERCISE. What are a couple of important considerations in constructing a bed of nails to lie on?

Now, let's take a quick review of density, which is an important consideration in the study of fluids. Recall from Section 1.4 that the density (ρ) of a substance is defined as mass per *unit* volume (Equation 1.1):

$$\text{density} = \frac{\text{mass}}{\text{volume}}$$

$$\rho = \frac{m}{V}$$

SI unit of density: kilogram per cubic meter (kg/m³)
(common cgs unit: gram per cubic centimeter, or g/cm³)

The densities of some common substances are given in ▶ **Table 9.2**.

Water has a density of 1.00×10^3 kg/m³ (or 1.00 g/cm³) from the original definition of the kilogram (Section 1.2). Mercury has a density of 13.6×10^3 kg/m³ (or 13.6 g/cm³). Hence, mercury is 13.6 times as dense as water. Gasoline, however, is less dense than water. See Table 9.2. (*Note:* Be careful not to confuse the symbol for density, ρ [Greek rho], with that for pressure, p.)

9.2.1 Pressure and Depth

If you have gone scuba diving, you well know that pressure increases with depth, having felt the increased pressure on your

TABLE 9.2 Densities of Some Common Substances (in kg/m³)

Solids	Density (ρ)	Liquids	Density (ρ)	Gases[a]	Density (ρ)
Aluminum	2.7×10^3	Alcohol, ethyl	0.79×10^3	Air	1.29
Brass	8.7×10^3	Alcohol, methyl	0.82×10^3	Helium	0.18
Copper	8.9×10^3	Blood, whole	1.05×10^3	Hydrogen	0.090
Glass	2.6×10^3	Blood plasma	1.03×10^3	Oxygen	1.43
Gold	19.3×10^3	Gasoline	0.68×10^3	Water vapor (100°C)	0.63
Ice	0.92×10^3	Kerosene	0.82×10^3		
Iron, steel	7.8×10^3	Mercury	13.6×10^3		
Lead	11.4×10^3	Seawater (4°C)	1.03×10^3		
Silver	10.5×10^3	Water, fresh (4°C)	1.00×10^3		
Wood, oak	0.81×10^3				

[a] At 0°C and 1 atm, unless otherwise specified.

eardrums. An opposite effect is commonly felt when you fly in a plane. With increasing altitude, your ears may "pop" because of *reduced* external air pressure.

How the pressure in a fluid varies with depth can be demonstrated by considering a container of liquid at rest. Imagine that you can isolate an imaginary rectangular column of liquid, as shown in ▼ **Figure 9.7**. Then the force on the bottom of the container below the column (or the hand) is equal to the weight of the liquid making up the column: $F = w = mg$. Since density is $\rho = m/V$, the mass in the column is equal to the density times the volume; that is, $m = \rho V$. (The liquid is assumed incompressible, so ρ is constant.)

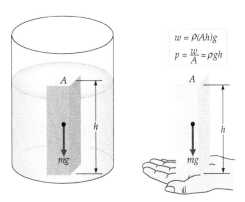

▲ **FIGURE 9.7 Pressure and depth** The extra pressure at a depth h in a liquid is due to the weight of the liquid above: $p = \rho g h$, where ρ is the density of the liquid (assumed to be constant). This is shown for an imaginary rectangular column of liquid.

The volume of the isolated liquid column is equal to the height of the column times the area of its base, or $V = hA$. Thus,

$$F = w = mg = \rho V g = \rho g h A$$

With $p = F/A$, the pressure at a depth h due to the weight of the column is

$$p = \rho g h \quad \text{(Fluid pressure at depth } h\text{)} \qquad (9.9)$$

This is a general result for incompressible liquids. The pressure is the same everywhere on a horizontal plane at a depth h (with ρ and g constant). Note that Equation 9.9 is independent of the base area of the rectangular column. The whole cylindrical column of the liquid in the container in Figure 9.7 could have been taken with the same result.

The derivation of Equation 9.9 did not take into account pressure being applied to the open surface of the liquid. This factor adds to the pressure at a depth h to give a *total* pressure of

$$p = p_{\text{o}} + \rho g h \quad \text{(Total fluid pressure at depth } h\text{)} \qquad (9.10)$$

where p_{o} is the pressure applied to the liquid surface (that is, the pressure applied at $h = 0$). For an open container, $p_{\text{o}} = p_{\text{a}}$, atmospheric pressure, or the weight (force) per unit area due to the gases in the atmosphere above the liquid's surface. The average atmospheric pressure at sea level is sometimes used as a unit, called an **atmosphere (atm)**:

$$1\,\text{atm} = 101.325\,\text{kPa} = 1.01325 \times 10^5\,\text{N/m}^2 = 14.7\,\text{lb/in}^2$$

The measurement of atmospheric pressure will be described shortly.

EXAMPLE 9.4: A SCUBA DIVER – PRESSURE AND FORCE

(a) What is the total pressure on the back of a scuba diver in a lake at a depth of 8.00 m? (b) What is the force on the diver's back due to the water alone? (Take the surface of the back to be a rectangle 60.0 cm by 50.0 cm.)

THINKING IT THROUGH. (a) This is a direct application of Equation 9.10 in which p_{o} is taken as the atmospheric pressure p_{a}. (b) Knowing the area and the pressure due to the water, the force can be found from the definition of pressure, $p = F/A$.

SOLUTION

Given:

$h = 8.00$ m
$A = 60.0$ cm $\times$ 50.0 cm
  $= 0.600$ m $\times$ 0.500 m $= 0.300$ m^2
$p_{\text{a}} = 1.01 \times 10^5$ N/m^2

Find:

(a) p (total pressure)
(b) F (force due to water)

(a) The total pressure is the sum of the pressure due to the water and the atmospheric pressure (p_{a}). By Equation 9.10, this is

$p = p_{\text{a}} + \rho g h$

$\quad = (1.01 \times 10^5\,\text{N/m}^2) + (1.00 \times 10^3\,\text{kg/m}^3)(9.80\,\text{m/s}^2)(8.00\,\text{m})$

$\quad = (1.01 \times 10^5\,\text{N/m}^2) + (0.784 \times 10^5\,\text{N/m}^2) = 1.79 \times 10^5\,\text{N/m}^2\,(\text{or Pa})$

$\quad \text{(expressed in atmospheres)} \approx 1.8\,\text{atm}$

This is also the inward pressure on the diver's eardrums.

(b) The pressure $p_{\text{H}_2\text{O}}$ due to the water alone is the $\rho g h$ portion of the preceding equation, so $p_{\text{H}_2\text{O}} = 0.784 \times 10^5\,\text{N/m}^2$.

Then, $p_{\text{H}_2\text{O}} = F/A$, and

$F = p_{\text{H}_2\text{O}} A = (0.784 \times 10^5\,\text{N/m}^2)(0.300\,\text{m}^2)$

$\quad = 2.35 \times 10^4\,\text{N}\,(\text{or } 5.29 \times 10^3\,\text{lb, about 2.6 tons!})$

FOLLOW-UP EXERCISE. You might question the answer to part (b) of this Example – how could the diver support such a force? To get a better idea of the forces our bodies can support, what would be the force on the diver's back at the water surface from atmospheric pressure alone? How do you suppose our bodies can support such forces or pressures?

9.2.2 Pascal's Principle

When the pressure (e.g., air pressure) is increased on the entire open surface of an incompressible liquid at rest, the pressure at any point in the liquid or on the boundary surfaces increases by the same amount. The effect is the same if pressure is applied to any surface of an enclosed fluid by means of a piston (▼ **Figure 9.8**). The transmission of pressure in fluids was studied by Pascal, and the observed effect is called **Pascal's principle**:

Pressure applied to an enclosed fluid is transmitted undiminished to every point in the fluid and to the walls of the container.

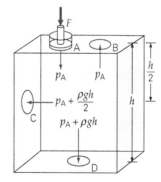

▲ **FIGURE 9.8 Pascal's principle.** The pressure applied at point A is fully transmitted to all parts of the fluid and to the walls of the container. There is also pressure due to the weight of the fluid above at different depths (for instance, $\rho g h/2$ at C and $\rho g h$ at D).

For an incompressible liquid, the change in pressure is transmitted essentially instantaneously. For a gas, a change in pressure

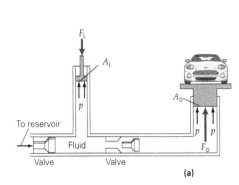
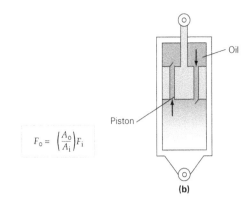

$$F_o = \left(\frac{A_o}{A_i}\right)F_i$$

▲ FIGURE 9.9   **The hydraulic lift and shock absorbers (a)** Because the input and output pressures are equal (Pascal's principle), a small input force on the small piston gives a large output force on the large piston, in proportion to the ratio of the piston areas. **(b)** A simplified exposed view of one type of shock absorber. (See Example 9.5 for description.)

will generally be accompanied by a change in volume or temperature (or both), but after equilibrium has been reestablished, Pascal's principle remains valid.

Common practical applications of Pascal's principle include the hydraulic braking systems used on automobiles. Through tubes filled with brake fluid, a force on the brake pedal transmits a force to the wheel brake cylinder. Similarly, hydraulic lifts and jacks are used to raise automobiles and other heavy objects (▲ **Figure 9.9**).

Using Pascal's principle, it can be shown how such systems allow us not only to transmit force from one place to another, but also to multiply that force. The input pressure p_i supplied by compressed air for a garage lift, for example, gives an input force F_i on a small input piston area A_i (Figure 9.9). The full magnitude of the pressure is transmitted to the large output piston, which has an area A_o. Since $p_i = p_o$, it follows that

$$\frac{F_i}{A_i} = \frac{F_o}{A_o}$$

and

$$F_o = \left(\frac{A_o}{A_i}\right)F_i \quad \text{(hydraulic force multiplication)} \quad (9.11)$$

With A_o larger than A_i, then F_o will be larger than F_i. The input force is greatly multiplied if the input piston has a relatively small area.

EXAMPLE 9.5: THE HYDRAULIC LIFT – PASCAL'S PRINCIPLE

A garage lift has input and lift (output) pistons with diameters of 10 and 30 cm, respectively. The lift is used to hold up a car with a weight of 1.4×10^4 N (3150 lb). (a) What is the magnitude of the force on the input piston? (b) What pressure is applied to the input piston?

THINKING IT THROUGH. (a) Pascal's principle, as expressed in the hydraulic Equation 9.11, has four variables, and three are given (areas via diameters). (b) The pressure is simply $p = F/A$.

SOLUTION

Given:

$$d_i = 10 \text{ cm} = 0.10 \text{ m}$$
$$d_o = 30 \text{ cm} = 0.30 \text{ m}$$
$$F_o = 1.4 \times 10^4 \text{ N}$$

Find:

(a) F_i (input force)
(b) p_i (input pressure)

(a) Rearranging Equation 9.11 and using $A = \pi r^2 = \pi d^2/4$ for the circular piston $(r = d/2)$ gives

$$F_i = \left(\frac{A_i}{A_o}\right)F_o = \left(\frac{\pi d_i^2/4}{\pi d_o^2/4}\right)F_o = \left(\frac{d_i}{d_o}\right)^2 F_o$$

or

$$F_i = \left(\frac{0.10 \text{ m}}{0.30 \text{ m}}\right)^2 F_o = \frac{F_o}{9} = \frac{1.4 \times 10^4 \text{ N}}{9} = 1.6 \times 10^3 \text{ N}$$

The input force is one-ninth of the output force; in other words, the force was multiplied by 9 (i.e., $F_o = 9F_i$).

(Note that we didn't really need to write the complete expressions for the areas. The area of a circle is proportional to the square of the diameter of the circle. If the ratio of the piston diameters is 3 to 1, the ratio of their areas must therefore be 9 to 1, and this ratio is used directly in Equation 9.11.)

(b) Then applying Equation 9.8a:

$$p_i = \frac{F_i}{A_i} = \frac{F_i}{\pi r_i^2} = \frac{F_i}{\pi(d/2)^2} = \frac{1.6 \times 10^3 \text{ N}}{\pi(0.10 \text{ m})^2/4}$$
$$= 2.0 \times 10^5 \text{ N/m}^2 \quad (= 200 \text{ kPa})$$

This pressure is about 30 lb/in^2, a common pressure used in automobile tires and about twice atmospheric pressure (which is approximately 100 kPa, or 15 lb/in^2).

FOLLOW-UP EXERCISE. Pascal's principle is used in shock absorbers on automobiles and on the landing gear of airplanes. Suppose that the larger input piston of a shock absorber on a jet plane has a diameter of 8.0 cm. What would be the diameter of a smaller output piston (channel) that would reduce the force by a factor of 10?

As Example 9.5 shows, forces produced by pistons relate directly to their diameters: $F_i = (d_i/d_o)^2 F_o$ or $F_o = (d_o/d_i)^2 F_i$. By making $d_o \gg d_i$, huge factors of force multiplication can be obtained, as is typical for hydraulic presses, jacks, and earth-moving equipment. Inversely, force reductions may be obtained by making $d_i > d_o$, as in Follow-Up Exercise 9.5.

However, don't think that you are getting something for nothing with large force multiplications. Energy is still a factor, and it can never be multiplied by a machine. (Why not?) Looking at the work involved and assuming that the work output is equal to the work input, $W_o = W_i$ (an ideal condition – why?). Then, with $W = Fx$ (Equation 5.1),

$$F_o x_o = F_i x_i$$

or

$$F_o = \left(\frac{x_i}{x_o}\right) F_i$$

where x_o and x_i are the output and input distances moved by the respective pistons.

Thus, the output force can be much greater than the input force only if the input distance is much greater than the output distance. For example, if $F_o = 10F_i$, then $x_i = 10x_o$, and the input piston must travel 10 times the distance of the output piston. *Force is multiplied at the expense of distance.*

9.2.3 Pressure Measurement

Pressure can be measured by mechanical devices that are often spring loaded (such as a tire gauge). Another type of instrument, called a manometer, uses a liquid – usually mercury – to measure pressure. An *open-tube manometer* is illustrated in ▼ Figure 9.10a. One end of the U-shaped tube is open to the atmosphere, and the other is connected to the container of gas whose pressure is to be measured. The liquid in the U-tube acts as a reservoir through which pressure is transmitted according to Pascal's principle.

The pressure of the gas (p) is balanced by the pressure of the column of liquid (of height h, the difference in the heights of the columns) and the atmospheric pressure (p_a) on the open liquid surface:

$$p = p_a + \rho gh \quad \text{(Absolute pressure)} \tag{9.12}$$

The pressure p is called the **absolute pressure**.

You may have measured pressure using pressure gauges; a tire gauge used to measure air pressure in automobile tires is a common example (Figure 9.10b). Such gauges, quite appropriately, measure **gauge pressure**. A pressure gauge registers only the pressure *above* (or *below*) atmospheric pressure. Hence, to get the absolute pressure (p), you have to add the atmospheric pressure (p_a) to the gauge pressure (p_g): $p = p_a + p_g$, where $p_a = 101$ kPa $= 14.7$ lb/in^2, as will be shown shortly.

For example, suppose your tire gauge reads a pressure of 200 kPa (≈ 30 lb/in^2). The absolute pressure within the tire is

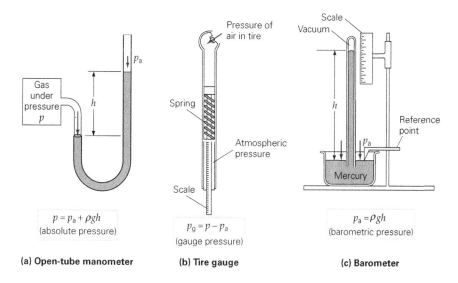

(a) **Open-tube manometer**
$p = p_a + \rho gh$ (absolute pressure)

(b) **Tire gauge**
$p_g = p - p_a$ (gauge pressure)

(c) **Barometer**
$p_a = \rho gh$ (barometric pressure)

▲ FIGURE 9.10   **Pressure measurement (a)** For an open-tube manometer, the pressure of the gas in the container is balanced by the pressure of the liquid column and atmospheric pressure acting on the open surface of the liquid. The absolute pressure of the gas equals the sum of the atmospheric pressure (p_a) and ρgh, the gauge pressure. **(b)** A tire gauge measures gauge pressure, the difference between the pressure in the tire and atmospheric pressure: $p_{gauge} = p - p_a$. Thus, if a tire gauge reads 200 kPa (30 lb/in^2), the actual pressure within the tire is 1 atm higher, or 300 kPa. **(c)** A barometer is a closed-tube manometer that is exposed to the atmosphere and thus reads only atmospheric pressure.

then $p = p_a + p_g = 101\,kPa + 200\,kPa = 301\,kPa$ ($\approx$ about 45 lb/in²). Hence, the absolute pressure on the inside of the tire is 45 psi, and that on the outside is 15 psi. If you open the valve or get a puncture, the internal and external pressures equalize and you have a flat!

Atmospheric pressure can be measured with a *barometer*. The principle of a mercury barometer is illustrated in Figure 9.10c. The device was invented by Evangelista Torricelli (1608–1647), Galileo's successor as professor of mathematics at an academy in Florence. A simple barometer consists of a tube filled with mercury that is inverted into a reservoir. Some mercury runs from the tube into the reservoir, but a column supported by the air pressure on the surface of the reservoir remains in the tube. This device can be considered a *closed-tube manometer,* and the pressure it measures is just the atmospheric pressure, since the gauge pressure (the pressure *above* atmospheric pressure) is zero.

The atmospheric pressure is then equal to the pressure due to the weight of the column of mercury, or

$$p = \rho g h \tag{9.13}$$

A *standard atmosphere* is defined as the pressure supporting a column of mercury exactly 76 cm in height at sea level and at 0°C.

Changes in atmospheric pressure can be observed as changes in the height of a column of mercury. These changes are due primarily to high- and low-pressure air masses that travel across the country. Atmospheric pressure is commonly reported in terms of the height of the barometer column, and weather forecasters say that the barometer is rising or falling. That is,

1 atm (about 101 kPa) = 76 cm Hg = 760 mm Hg

= 29.92 in. Hg (about 30 in. Hg)

In honor of Torricelli, a pressure supporting 1 mm of mercury is given the name *torr*:

1 mm Hg = 1 torr

and

1 atm = 760 torr*

Because mercury is highly toxic, it is sealed inside a barometer. A safer and less expensive device that is widely used to measure atmospheric pressure is the *aneroid* ("without fluid") *barometer*. In an aneroid barometer, a sensitive metal diaphragm on an evacuated container (something like a drumhead) responds to pressure changes, which are indicated on a dial. This is the

kind of barometer you frequently find in homes in decorative wall mountings (▼ **Figure 9.11**). Modern digital barometers may use various capacitive, electromagnetic, piezoelectric, or optical pressure sensors.

Since air is compressible, the atmospheric density and pressure are greatest at the Earth's surface and decrease with altitude. We live at the bottom of the atmosphere, but don't notice its pressure very much in our daily activities because our bodies are composed largely of fluids, which exert a matching outward pressure. Indeed, the external pressure of the atmosphere is so important to our normal functioning that it is taken with us wherever we can. For example, the pressurized suits worn by astronauts in space or on the Moon are needed to provide an external pressure similar to that on the Earth's surface.

▲ FIGURE 9.11   **Aneroid barometer** Changes in atmospheric pressure on a sensitive metal diaphragm are reflected on the dial face of the barometer. A prediction of fair weather is generally associated with high barometric pressures, and rainy weather with low barometric pressures.

EXAMPLE 9.6: AN IV – A GRAVITY ASSIST

Consider a hospital patient who receives an IV (intravenous) injection under gravity flow, as shown in ▶ **Figure 9.12**. If the blood gauge pressure in the vein is 20.0 mm Hg, above what height should the bottle be placed for the IV blood transfusion to function properly?

THINKING IT THROUGH. The fluid gauge pressure at the bottom of the IV tube must be greater than the pressure in the vein and can be computed from Equation 9.9. (The liquid is assumed to be incompressible.)

SOLUTION

Given:

$p_v = 20.0$ mm Hg (vein gauge pressure)
$\rho = 1.05 \times 10^3$ kg/m³
(whole blood density from Table 9.2)

▲ FIGURE 9.12 **What height is needed?** See Example 9.6 for description.

Find: h (height for $p_v > 20$ mm Hg)

First, the common medical unit of mm Hg (or torr) needs to be changed to the SI unit of pascal (Pa, or N/m²):

$$p_v = (20.0 \, \text{mm Hg})[133 \, \text{Pa/(mm Hg)}] = 2.66 \times 10^3 \, \text{Pa}$$

Then, for $p = \rho g h > p_v$,

$$h > \frac{p_v}{\rho g} = \frac{2.66 \times 10^3 \, \text{Pa}}{(1.05 \times 10^3 \, \text{kg/m}^3)(9.80 \, \text{m/s}^2)} = 0.259 \, \text{m} \, (\approx 26 \, \text{cm})$$

The IV bottle needs to be at least 26 cm above the injection site.

FOLLOW-UP EXERCISE. The normal (gauge) blood pressure range is commonly reported as 120/80 (in millimeters Hg). Why is the blood pressure of 20 mm Hg in this Example so low?

9.3 Buoyancy and Archimedes' Principle

When placed in a fluid, an object will either sink or float. This is most commonly observed with liquids; for example, objects float or sink in water. But the same effect occurs in gases: A falling object sinks in the atmosphere, while other objects float (▼ **Figure 9.13**).

▲ FIGURE 9.13 **Fluid buoyancy** The air is a fluid in which objects such as this dirigible float. The helium inside the blimp is less dense than the surrounding air, and displacing its volume of air, the blimp is supported by the resulting buoyant force.

Things float because they are buoyant, or are buoyed up. For example, if you immerse a cork in water and release it, the cork will be buoyed up to the surface and float there. From your knowledge of forces, you know that such motion requires an upward net force on an object. That is, there must be an upward force acting on the object that is greater than the downward force of its weight. The forces are equal when the object floats in equilibrium. The upward force resulting from an object being wholly or partially immersed in a fluid is called the **buoyant force**.

How the buoyant force comes about can be seen by considering a buoyant object being held under the surface of a fluid (▼ **Figure 9.14a**). The pressures on the upper and lower surfaces of the block are $p_1 = \rho_f g h_1$ *and* $p_2 = \rho_f g h_2$, respectively, where ρ_f

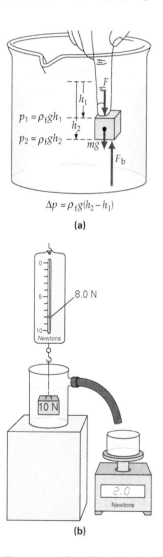

▲ FIGURE 9.14 **Buoyancy and Archimedes' principle (a)** A buoyant force arises from the difference in pressure at different depths. The pressure on the bottom of the submerged block (p_2) is greater than that on the top (p_1), so there is a (buoyant) force directed upward. (Shifted for clarity.) **(b)** Archimedes' principle: The buoyant force on the object is equal to the weight of the volume of fluid displaced. (The scale is set to read zero when the container is empty.)

is the density of the fluid. Thus, there is a pressure difference $\Delta p = p_2 - p_1 = \rho_f g(h_2 - h_1)$ between the top and bottom of the block, which gives an upward force (the buoyant force) F_b. This force is balanced by the applied force and the weight of the block.

It is not difficult to derive an expression for the magnitude of the buoyant force. Pressure is force per unit area. Thus, if both the top and bottom areas of the block are A, the magnitude of the net buoyant force in terms of the pressure difference is

$$F_b = p_2 A - p_1 A = (\Delta p)A = \rho_f g(h_2 - h_1)A$$

Since $(h_2 - h_1)A$ is the volume of the block and hence the volume of fluid displaced by the block, V_f, the expression for F_b may be written as

$$F_b = \rho_f g V_f$$

But $\rho_f V_f$ is simply the mass of the fluid displaced by the block, m_f. Thus, the expression for the buoyant force becomes $F_b = m_f g$: The magnitude of the buoyant force is equal to the weight of the fluid displaced by the block (Figure 9.14b). This general result is known as **Archimedes' principle**:*

A body immersed wholly or partially in a fluid experiences a buoyant force equal in magnitude to the weight of the volume of fluid that is displaced:

$$F_b = m_f g = \rho_f g V_f \qquad (9.14)$$

INTEGRATED EXAMPLE 9.7: LIGHTER THAN AIR – BUOYANT FORCE

A spherical helium-filled weather balloon has a radius of 1.10 m. (a) Does the buoyant force on the balloon depend on (1) the density of helium, (2) density of air, or (3) the weight of the rubber "skin"? [$\rho_{air} = 1.29$ kg/m³ and $\rho_{He} = 0.180$ kg/m³.] (b) Compute the magnitude of the buoyant force on the balloon. (c) The balloon's rubber skin has a mass of 1.20 kg. When released, what is the magnitude of the balloon's initial acceleration if it carries a payload with a mass of 3.52 kg?

(a) **CONCEPTUAL REASONING.** The upward buoyant force has nothing to do with the helium or rubber skin and is equal to the weight of the displaced air, which can be found from the balloon's volume and the density of air. So the answer is (2).

(b, c) **QUANTITATIVE REASONING AND SOLUTION**

---

* Archimedes (287–212 BCE), a Greek scientist, was given the task of determining whether a gold crown made for the king was pure gold or contained a quantity of silver. Legend has it that the solution came to him upon immersing himself in a full bath tub. On doing so, he noticed that water overflowed the tub. It is said that Archimedes was so excited that he jumped out and ran home through the streets of the city (unclothed) shouting "Eureka! Eureka!" (Greek for "I have found it").

Given:

$\rho_{air} = 1.29$ kg/m³
$\rho_{He} = 0.180$ kg/m³
$m_s = 1.20$ kg, $m_p = 3.52$ kg
$r = 1.10$ m

Find:

(b) F_b (buoyant force)
(c) a (initial acceleration)

(b) The volume of the balloon is

$$V = (4/3)\pi r^3 = (4/3)\pi(1.10\,\text{m})^3 = 5.58\,\text{m}^3$$

Then the buoyant force is equal to the weight of the air displaced:

$$F_b = m_{air} g = (\rho_{air} V)g = (1.29\,\text{kg/m}^3)(5.58\,\text{m}^3)(9.80\,\text{m/s}^2) = 70.5\,\text{N}$$

(c) Draw a free-body diagram. There are three weight forces downward – those of the helium, the rubber skin, and the payload – and the upward buoyant force. Sum these forces to find the net force, and then use Newton's second law to find the acceleration. The weights of the helium, rubber skin, and payload are as follows:

$$w_{He} = m_{He} g = (\rho_{He} V)g = (0.180\,\text{kg/m}^3)(5.58\,\text{m}^3)(9.80\,\text{m/s}^2) = 9.84\,\text{N}$$
$$w_s = m_s g = (1.20\,\text{kg})(9.80\,\text{m/s}^2) = 11.8\,\text{N}$$
$$w_p = m_p g = (3.52\,\text{kg})(9.80\,\text{m/s}^2) = 35.5\,\text{N}$$

Summing the forces (taking upward as positive),

$$F_{net} = F_b - w_{He} - w_s - w_p = 70.5\,\text{N} - 11.8\,\text{N} - 35.5\,\text{N} = 13.4\,\text{N}$$

and with the masses found from the weights:

$$a = \frac{F_{net}}{m_{total}} = \frac{F_{net}}{m_{He} + m_s + m_p} = \frac{13.4\,\text{N}}{1.00\,\text{kg} + 1.20\,\text{kg} + 3.52\,\text{kg}}$$
$$= 2.34\,\text{m/s}^2$$

FOLLOW-UP EXERCISE. As the balloon rises, it eventually stops accelerating and rises at a constant velocity for a short time, then starts sinking toward the ground. Explain this behavior in terms of atmospheric density and temperature. [*Hint:* Temperature and air density generally decrease with altitude. The pressure of a quantity of gas is directly proportional to temperature.]

EXAMPLE 9.8: YOUR BUOYANCY IN AIR

Air is a fluid and our bodies displace air. And so, a buoyant force is acting on each of us. Estimate the magnitude of the buoyant force on a 75-kg person due to the air displaced.

THINKING IT THROUGH. The key word here is *estimate*, because not much data are given. We know that the buoyant force is $F_b = \rho_a g V$, where ρ_a is the density of air (which can be found in

Table 9.2), and V is the volume of the air displaced, which is the same as the volume of the person. The question is, how do we find the volume of the person?

The mass is given, and if the density of the person were known, the volume could be found ($\rho = m/V$) or $V = m/\rho$. Here is where the estimate comes in. Most people can barely float in water, so the density of the human body is about that of water, $\rho = 1000$ kg/m³. Using this estimate, the buoyant force can also be estimated.

SOLUTION

Given:

> $m = 75$ kg
> $\rho_a = 1.29$ kg/m³ (Table 9.2)
> $\rho_p = 1000$ kg/m³ (estimated density of person)

Find: F_b (buoyant force)

First, let's find the volume of the person:

$$V_p = \frac{m}{\rho_p} = \frac{75\,\text{kg}}{1000\,\text{kg/m}^3} = 0.075\,\text{m}^3$$

Then,

$$F_b = \rho_a g V_p = (1.29\,\text{kg/m}^3)(9.8\,\text{m/s}^2)(0.075\,\text{m}^3)$$
$$= 0.95\,\text{N}\,(\approx 1.0\,\text{N or }0.225\,\text{lb})$$

This amount is not much when you weigh yourself. But it does mean that your weight is ≈ 0.2 lb more than the scale reading.

FOLLOW-UP EXERCISE. Estimate the buoyant force on a helium-filled weather balloon that has a diameter on the order of a meteorologist's arm span (arms held horizontally), and compare with the result in the Example.

INTEGRATED EXAMPLE 9.9: WEIGHT AND BUOYANT FORCE – ARCHIMEDES' PRINCIPLE

A container of water with an overflow tube, similar to that shown in Figure 9.14b, sits on a scale that reads 40 N. The water level is just below the exit tube in the side of the container. (a) An 8.0-N cube of wood is placed in the container. The water displaced by the floating cube runs out the exit tube into another container that is not on the scale. Will the scale reading then be (1) exactly 48 N, (2) between 40 N and 48 N, (3) exactly 40 N, or (4) less than 40 N? (b) Suppose you pushed down on the wooden cube with your finger such that the top surface of the cube was even with the water level. How much force would have to be applied if the wooden cube measured 10 cm on a side?

(a) **CONCEPTUAL REASONING.** By Archimedes' principle, the block is buoyed upward with a force equal in magnitude to the weight of the water displaced. Since the block floats, the upward buoyant force must balance the weight of the cube and so has a magnitude of 8.0 N. Thus, a volume of water weighing 8.0 N is displaced from the container as 8.0 N of weight is added to the container. The scale still reads 40 N, so the answer is (3).

Note that the upward buoyant force and the block's weight act *on the block*. The reaction force (pressure) of the block *on the water* is transmitted to the bottom of the container (Pascal's principle) and is registered on the scale. (Make a sketch showing the forces on the cube.)

(b) **QUANTITATIVE REASONING AND SOLUTION.** Here three forces are acting on the stationary cube: the buoyant force upward and the weight and the force applied by the finger downward. The weight of the cube is known, so to find the applied finger force, we need to determine the buoyant force on the cube.

Given:

> $\ell = 10$ cm $= 0.10$ m (cube length)
> $w = 8.0$ N (cube weight)

Find: F_f (downward applied force necessary to put cube even with water level

The summation of the forces acting on the wood cube is $\Sigma F_y = +F_b - w - F_f = 0$, where F_b is the upward buoyant force and F_f is the downward force applied by the finger. Hence, $F_f = F_b - w$. As we know, the magnitude of the buoyant force is equal to the weight of the water the cube displaces, which is given by $F_b = \rho_f g V_f$ (Equation 9.14). The density of the fluid is that of water, which is known (1.0×10^3 kg/m³, Table 9.2), so

$$F_b = \rho_f g V_f = (1.0 \times 10^3\,\text{kg/m}^3)(9.8\,\text{m/s}^2)(0.10\,\text{m})^3 = 9.8\,\text{N}$$

Thus,

$$F_f = F_b - w = 9.8\,\text{N} - 8.0\,\text{N} = 1.8\,\text{N}$$

FOLLOW-UP EXERCISE. In part (a), would the scale still read 40 N if the object had a density greater than that of water? In part (b), what would the scale read?

9.3.1 Buoyancy and Density

It is commonly said that helium and hot air balloons float because they are lighter than air. To be technically correct, it should be said that the balloons are *less dense than air*. An object's density will tell you whether it will sink or float in a fluid, as long as you also know the density of the fluid. Consider a solid uniform object that is totally immersed in a fluid. The weight of the object is

$$w_o = m_o g = \rho_o V_o g$$

The weight of the volume of fluid displaced, or the magnitude of the buoyant force, is

$$F_b = w_f = m_f g = \rho_f V_f g$$

If the object is *completely submerged*, $V_f = V_o$. Dividing the second equation by the first gives

$$\frac{F_b}{w_o} = \frac{\rho_f}{\rho_o} \quad \text{or} \quad F_b = \left(\frac{\rho_f}{\rho_o}\right) w_o \quad \text{(object completely submerged)}$$

$$\text{(9.15)}$$

Thus, if ρ_o is less than ρ_f, then F_b will be greater than w_o, and the object will be buoyed to the surface and float. If ρ_o is greater than ρ_f, then F_b will be less than w_o, and the object will sink. If ρ_o equals ρ_f, then F_b will be equal to w_o, and the object will remain in equilibrium at any submerged depth (as long as the density of the fluid is constant). If the object is not uniform, so that its density varies over its volume, then the density of the object in Equation 9.15 is the average density.

Expressed in words, these three conditions are as follows:

- An object will float in a fluid if the average density of the object is less than the density of the fluid ($\rho_o < \rho_f$)
- An object will sink in a fluid if the average density of the object is greater than the density of the fluid ($\rho_o > \rho_f$)
- An object will be in equilibrium at any submerged depth in a fluid if the average density of the object and the density of the fluid are equal ($\rho_o = \rho_f$)

A quick look at Table 9.2 will tell you whether an object will float in a fluid, regardless of the shape or volume of the object. The three conditions just stated also apply to a fluid in a fluid, provided that the two are immiscible (do not mix). For example, you might think that cream is "heavier" than skim milk, but that's not so: Since cream floats on milk, it is less dense than milk.

In general, the densities of objects or fluids will be assumed to be uniform and constant in this book. (The density of the atmosphere does vary with altitude but is relatively constant near the surface of the Earth.) In any event, in practical applications it is the *average* density of an object that often matters with regard to floating and sinking. For example, an ocean liner is, on average, less dense than water, even though it is made of steel. Most of its volume is occupied by air, so the liner's average density is less than that of water. Similarly, the human body has air-filled spaces, so most of us float in water. The surface depth at which a person floats depends on his or her density. (Why?)

In some instances, the overall density of an object is purposefully varied. For example, a submarine submerges by flooding its tanks with seawater (called "*taking on ballast*"), which increases its average density. When the sub is ready to surface, the water is pumped out of the tanks, so the average density of the sub becomes less than that of the surrounding seawater.

Similarly, many fish control their depths by using their *swim bladders* or *gas bladders*. A fish changes or maintains buoyancy by regulating the volume of gas in the gas bladder. Maintaining neutral buoyancy (neither rising nor sinking) is important because it allows the fish to stay at a particular depth for feeding. Some fish may move up and down in the water in search of food. Instead of using up energy to swim up and down, the fish alters its buoyancy to rise and sink.

This is accomplished by adjusting the quantities of gas in the gas bladder. Gas is transferred from the gas bladder to the adjoining blood vessels and back again. Deflating the bladder decreases the volume and increases the average density, and the fish sinks. Gas is forced into the surrounding blood vessels and carried away.

Conversely, to inflate the bladder, gases are forced into the bladder from the blood vessels, thereby increasing the volume

and decreasing the average density, and the fish rises. These processes are complex, but Archimedes' principle is being applied in a biological setting.

EXAMPLE 9.10: FLOAT OR SINK?
COMPARISON OF DENSITIES

A uniform solid cube of material 10.0 cm on each side has a mass of 700 g. (a) Will the cube float in water? (b) If so, how much of its volume would be submerged?

THINKING IT THROUGH. (a) The question is whether the density of the material the cube is made of is greater or less than that of water, so we compute the cube's density. (b) If the cube floats, then the buoyant force and the cube's weight are equal. Both of these forces are related to the cube's volume, so we can write them in terms of that volume and equate them.

SOLUTION

It is sometimes convenient to work in cgs units in comparing small quantities. For densities in grams per cubic centimeter, divide the values in Table 9.2 by 10^3, or drop the " $\times 10^3$" from the values given for solids and liquids, and replace with " $\times 10^{-3}$" for gases.

Given:

$$m = 700 \text{ g}$$
$$L = 10.0 \text{ cm}$$
$$\rho_{H_2O} = 1.00 \times 10^3 \text{ kg/m}^3$$
$$= 1.00 \text{ g/cm}^3 \text{ (Table 9.2)}$$

Find:

(a) Whether the cube will float in water
(b) The percentage of the volume submerged if the cube does float

(a) The density of the cube is

$$\rho_c = \frac{m}{V_c} = \frac{m}{L^3} = \frac{700 \text{ g}}{(10.0 \text{ cm})^3} = 0.700 \text{ g/cm}^3 < \rho_{H_2O} = 1.00 \text{ g/cm}^3$$

Since ρ_c is less than ρ_{H_2O} the cube will float.

(b) The weight of the cube is $w_c = \rho_c g V_c$. When the cube is floating, it is in equilibrium, which means that its weight is balanced by the buoyant force. That is, $F_b = \rho_{H_2O} g V_{H_2O}$, where V_{H_2O} is the volume of water the submerged part of the cube displaces. Equating the expressions for weight and buoyant force gives

$$\rho_{H_2O} g V_{H_2O} = \rho_c g V_c$$

or

$$\frac{V_{H_2O}}{V_c} = \frac{\rho_c}{\rho_{H_2O}} = \frac{0.700 \text{ g/cm}^3}{1.00 \text{ g/cm}^2} = 0.700$$

Thus, $V_{H_2O} = 0.700 \, V_c$, and 70.0% of the cube is submerged.

▲ FIGURE 9.15 **The tip of the iceberg** The vast majority of an iceberg's bulk is underneath the water, as illustrated here in a false photo. See Example 9.10 Follow-Up Exercise.

FOLLOW-UP EXERCISE. Most of an iceberg floating in the ocean is submerged (▲ **Figure 9.15**). The visible portion is the proverbial "tip of the iceberg." What percentage of an iceberg's volume is seen above the surface? (*Note:* Icebergs are frozen *fresh* water floating in salty sea water.)

A quantity called specific gravity is related to density. It is commonly used for liquids, but also applies to solids. The **specific gravity (sp. gr.)** of a substance is equal to the ratio of the density of the substance (ρ_s) to the density of water (ρ_{H_2O}) at 4°C, the temperature for maximum density:

$$\text{sp. gr.} = \frac{\rho_s}{\rho_{H_2O}}$$

Because it is a ratio of densities, specific gravity has no units. In cgs units, ρ_{H_2O}, so

$$\text{sp. gr.} = \frac{\rho_s}{1.00} = \rho_s \quad (\rho_s \text{ in g/cm}^3 \text{ only})$$

That is, the specific gravity of a substance is equal to the numerical value of its density *in cgs units.* For example, if a liquid has a density of 1.5 g/cm³, its specific gravity is 1.5, which tells you that it is 1.5 times as dense as water. (As pointed out earlier, to get density values for solids and liquids in grams per cubic centimeter, divide the value in Table 9.2 by 10³.)

9.4 Fluid Dynamics and Bernoulli's Equation

In general, fluid motion is difficult to analyze. For example, think of trying to describe the motion of a particle (a molecule, as an approximation) of water in a rushing stream. The overall motion of the stream may be apparent, but a mathematical description of

the motion of any one particle of it may be virtually impossible because of eddy currents (small whirlpool motions), the gushing of water over rocks, frictional drag on the stream bottom, and so on. A basic description of fluid flow is conveniently obtained by ignoring such complications and considering an ideal fluid. Actual fluid flow can then be approximated with reference to this simpler theoretical model.

In this simplified approach to fluid dynamics, it is customary to consider four characteristics of an **ideal fluid**. In such a fluid, flow is (1) *steady*, (2) *irrotational*, (3) *nonviscous*, and (4) *incompressible*.

Condition 1: **Steady flow** means that all the particles of a fluid have the same velocity as they pass a given point.

Steady flow might be called smooth or regular flow. The path of steady flow can be depicted in the form of **streamlines** (▼ **Figure 9.16a**). Every particle that passes a particular point moves along a streamline. That is, every particle moves along the same path (streamline) as particles that passed by earlier. Streamlines never cross; if they did, a particle would have alternative paths and abrupt changes in its velocity, in which case the flow would not be steady.

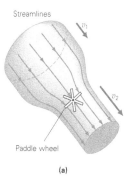

(a)

(b)

▲ FIGURE 9.16 **Streamline flow (a)** Streamlines never cross and are closer together in regions of greater fluid velocity. The stationary paddle wheel indicates that the flow is irrotational, or without whirlpools and eddy currents. **(b)** The smoke from an extinguished candle begins to rise in nearly streamline flow, but quickly becomes rotational and turbulent.

Steady flow requires low velocities. For example, steady flow is approximated by the flow relative to a canoe that is gliding slowly through still water. When the flow velocity is high, eddies tend to appear, especially near boundaries, and the flow becomes turbulent, as in Figure 9.16b.

Streamlines also indicate the relative magnitude of the velocity of a fluid. The velocity is greater where the streamlines are closer together. Notice this effect in Figure 9.16a. The reason for it will be explained shortly.

Condition 2: **Irrotational flow** means that a fluid element (a small volume of the fluid) has no net angular velocity. This condition eliminates the possibility of whirlpools and eddy currents (nonturbulent flow).

Consider the small paddle wheel in Figure 9.16a. With a zero net torque, the wheel does not rotate. Thus, the flow is irrotational.

Condition 3: **Nonviscous flow** means that viscosity is negligible.

Viscosity refers to a fluid's internal friction, or resistance to flow (e.g., honey has a much greater viscosity than water). A truly nonviscous fluid would flow freely with no internal energy loss. Also, there would be no frictional drag between the fluid and the walls containing it. In reality, when a liquid flows through a pipe, the speed is lower near the walls because of frictional drag and is higher toward the center of the pipe. (Viscosity is discussed in more detail in Section 9.5.)

Condition 4: **Incompressible flow** means that the fluid's density is constant.

Liquids can usually be considered incompressible. Gases, by contrast, are quite compressible. Sometimes, however, gases approximate incompressible flow – for example, air flowing relative to the wings of an airplane traveling at low speeds. Theoretical or ideal fluid flow is not characteristic of most real situations, but the analysis of ideal flow provides results that approximate, or generally describe, a variety of applications. Usually, this analysis is derived, not from Newton's laws, but instead from two basic principles: conservation of mass and conservation of energy.

9.4.1 Equation of Continuity

If there are no losses of fluid within a tube, the mass of fluid flowing into the tube in a given time must be equal to the mass flowing out of the tube in the same time (by the conservation of mass). For example, in ▶ **Figure 9.17a**, the mass (Δm_1) entering the tube during a short time (Δt) is

$$\Delta m_1 = \rho_1 \Delta V_1 = \rho_1(A_1 \Delta x_1) = \rho_1(A_1 v_1 \Delta t)$$

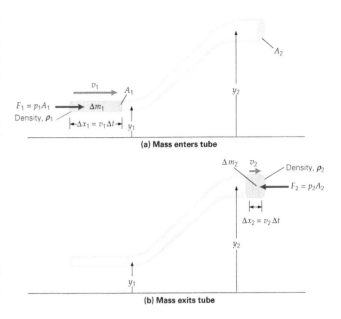

▲ **FIGURE 9.17**   **Flow continuity** Ideal fluid flow can be described in terms of the conservation of mass by the equation of continuity. See text for description.

where A_1 is the cross-sectional area of the tube at the entrance and, in a time Δt, a fluid particle moves a distance equal to $v_1 \Delta t$. Similarly, the mass leaving the tube in the same interval is (Figure 9.17b)

$$\Delta m_2 = \rho_2 \Delta V_2 = \rho_2(A_2 \Delta x_2) = \rho_2(A_2 v_2 \Delta t)$$

Since the mass is conserved, $\Delta m_1 = \Delta m_2$, and it follows that

$$\rho_1 A_1 v_1 = \rho_2 A_2 v_2 \quad \text{or} \quad \rho A v = \text{constant}$$
$$\text{(Equation of continuity)} \quad (9.16)$$

This general result is called the equation of continuity.

For an incompressible fluid, the density ρ is constant, so

$$A_1 v_1 = A_2 v_2 \quad \text{or} \quad A v = \text{constant}$$
$$\text{(for incompressible fluid)} \quad (9.17)$$

This is sometimes called the **flow rate equation**. Av is called the *volume rate of flow,* and is the volume of fluid that passes by a point in the tube per unit time. (The units of Av are m²·m/s = m³/s, volume per time.)

Note that the flow rate equation shows that the fluid speed is greater where the cross-sectional area of the tube is smaller. That is,

$$v_2 = \left(\frac{A_1}{A_2}\right) v_1$$

and v_2 is greater than v_1 if A_2 is less than A_1. This effect is evident in the common experience that the speed of water is greater from a hose fitted with a nozzle than that from the same hose without a nozzle (▼ **Figure 9.18**).

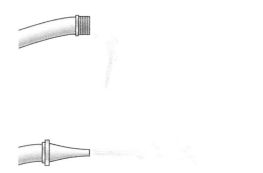

▲ FIGURE 9.18 **Flow rate** By the flow rate equation, the speed of a fluid is greater when the cross-sectional area of the tube through which the fluid is flowing is smaller. Think of a hose that is equipped with a nozzle such that the cross-sectional area of the hose is made smaller.

The flow rate equation can be applied to the flow of blood in your body. Blood flows from the heart into the aorta. It then makes a circuit through the circulatory system, passing through arteries, arterioles (small arteries), capillaries, and venules (small veins) and back to the heart through veins. The speed is lowest in the capillaries. Is this a contradiction? No: The *total* area of the capillaries is much larger than that of the arteries or veins, so the flow rate equation is still valid.

EXAMPLE 9.11: BLOOD FLOW – CHOLESTEROL AND PLAQUE

High cholesterol in the blood can cause fatty deposits called plaques to form on the walls of blood vessels. Suppose a plaque reduces the effective radius of an artery by 25%. How does this partial blockage affect the speed of blood through the artery?

THINKING IT THROUGH. The flow rate equation (Equation 9.17) applies, but note that no values of area or speed are given. This indicates that we should use ratios.

SOLUTION

Taking the unclogged artery to have a radius r_1, that the plaque then reduces the effective radius to r_2.

Given:

$r_2 = 0.75 r_1$ (for a 25% reduction)

Find: v_2

Writing the flow rate equation in terms of the radii,

$$A_1 v_1 = A_2 v_2$$
$$\left(\pi r_1^2\right) v_1 = \left(\pi r_2^2\right) v_2$$

Rearranging and canceling,

$$v_2 = \left(\frac{r_1}{r_2}\right)^2 v_1$$

From the given information, $r_1/r_2 = 1/0.75$, so

$$v_2 = (1/0.75)^2 v_1 = 1.8 v_1$$

Hence, the speed through the clogged artery increases by 80%.

FOLLOW-UP EXERCISE. By how much would the effective radius of an artery have to be reduced to have a 50% increase in the speed of the blood flowing through it?

EXAMPLE 9.12: SPEED OF BLOOD IN THE AORTA

Blood flows at a rate of 5.00 L/min through an aorta with a radius of 1.00 cm. What is the speed of blood flow in the aorta?

THINKING IT THROUGH. It is noted that the flow rate is a volume flow rate, which implies the use of the flow rate equation (Equation 9.17), Av = constant. Since the constant is in terms of volume/time, the given flow rate is the constant.

SOLUTION

Listing the data:

Given:

Flow rate = 5.00 L/min
$r = 1.00$ cm $= 1.00 \times 10^{-2}$ m

Find: v (blood speed)

Let's first find the cross-sectional area of the circular aorta.

$$A = \pi r^2 = (3.14)(1.00 \times 10^{-2}\,\text{m})^2 = 3.14 \times 10^{-4}\,\text{m}^2$$

Then the (volume) flow rate needs to be put into standard units.

$$5.00\,\text{L/min} = (5.00\,\text{L/min})(10^{-3}\,\text{m}^3/\text{L})(1\,\text{min}/60\,\text{s})$$
$$= 8.33 \times 10^{-5}\,\text{m}^3/\text{s (constant)}$$

Using the flow rate equation,

$$v = \frac{\text{constant}}{A} = \frac{8.33 \times 10^{-5}\,\text{m}^3/\text{s}}{3.14 \times 10^{-4}\,\text{m}^2} = 0.265\,\text{m/s}$$

FOLLOW-UP EXERCISE. Constrictions of the arteries occur with hardening of the arteries. If the radius of the aorta in this Example were constricted to 0.900 cm, what would be the percentage change in blood flow speed assuming the blood flow rate is constant?

9.4.2 Bernoulli's Equation

The conservation of energy or the general work-energy theorem leads to another relationship that has great generality for fluid flow. This relationship was first derived in 1738 by the Swiss mathematician Daniel Bernoulli (1700–1782) and is named for him. Bernoulli's result was based on the work-energy theorem: $W_{net} = \Delta K + \Delta U$.

In working with a fluid, $W_{net} = F_{net}\Delta x = \Delta p(A\Delta x) = \Delta p \Delta V$. With $\rho = m/V$, $\Delta V = \Delta m/\rho$, $W_{net} = (\Delta m/\rho)(p_1 - p_2)$, where Δm is a mass increment as in the derivation of the continuity equation. Thus,

$$\frac{\Delta m}{\rho}(p_1 - p_2) = \tfrac{1}{2}\Delta m\left(v_2^2 - v_1^2\right) + \Delta m g(y_2 - y_1)$$

Canceling each Δm and rearranging gives the common form of **Bernoulli's equation**:

$$p_1 + \tfrac{1}{2}\rho v_1^2 + \rho g y_1 = p_2 + \tfrac{1}{2}\rho v_2^2 + \rho g y_2 \quad \text{(Bernoulli's equation)} \quad (9.18)$$

or

$$p + \tfrac{1}{2}\rho v^2 + \rho g y = \text{constant}$$

Bernoulli's equation, or principle, can be applied to many situations. For example, for a fluid at rest ($v_2 = v_1 = 0$). Bernoulli's equation becomes

$$p_2 - p_1 = \rho g(y_1 - y_2)$$

This is the pressure-depth relationship derived earlier (Equation 9.10). Also, if there is horizontal flow ($y_1 = y_2$), then $p + \tfrac{1}{2}\rho v^2 = \text{constant}$, which indicates that the pressure decreases if the speed of the fluid increases (and vice versa). This effect is illustrated in ▼ **Figure 9.19**, where the difference in flow heights through the pipe is considered negligible (so the $\rho g y$ term drops out).

Chimneys and smokestacks are tall in order to take advantage of the more consistent and higher wind speeds at greater heights. The faster the wind blows over the top of a chimney, the lower the pressure there, and the greater the pressure difference between the bottom and top of the chimney. Thus, the chimney "draws" exhaust out more efficiently. Bernoulli's equation and the continuity equation ($Av = \text{constant}$) also tell you that if the cross-sectional area of a pipe is reduced so that the speed of the fluid passing through it is increased, then the pressure is reduced.

The Bernoulli effect (as it is sometimes called) gives a *simplistic* explanation for the lift of an airplane. Ideal airflow over an airfoil or wing is shown in ▶ **Figure 9.20**. (Turbulence is neglected.) The wing is curved on the top side and is angled relative to the incident streamlines. As a result, the streamlines above the wing are closer together than those below, which causes a higher air speed and lower pressure above the wing. With a higher pressure on the bottom of the wing, there is a net upward force, or *lift*.

This common explanation of lift is simplistic because Bernoulli's effect does not apply to the situation. Bernoulli's principle requires the conditions of both ideal fluid flow and energy conservation within the system, neither of which is satisfied in aircraft flying conditions. It is perhaps better to rely on Newton's laws. Basically, the wing deflects the airflow downward, giving rise to a downward change in the airflow momentum and a downward force (Newton's second law). This results in an upward reaction force on the wing (Newton's third law). When this upward force exceeds the weight of the plane, there is enough lift for takeoff and flight.

EXAMPLE 9.13: FLOW RATE FROM A TANK – BERNOULLI'S EQUATION

A cylindrical tank containing water has a small hole punched in its side below the water level, and water runs out (▶ **Figure 9.21**). What is the approximate initial flow rate of water out of the tank in terms of the heights shown?

THINKING IT THROUGH. Equation 9.17 ($A_1 v_2 = A_2 v_2$) is the flow rate equation, where Av has units of m³/s, or volume/time. The v terms can be related by Bernoulli's equation, which also contains y, and can be used to find differences in height. The areas are not given, so relating the v terms might require some sort of approximation, as will be seen. (Note that the *approximate* initial flow rate is wanted.)

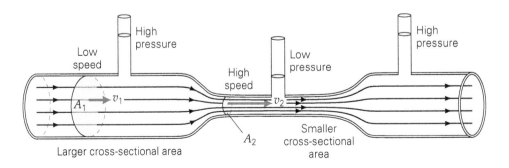

▲ **FIGURE 9.19** **Flow rate and pressure** Taking the horizontal difference in flow heights to be negligible in a constricted pipe, we obtain, for Bernoulli's equation, $p + \tfrac{1}{2}\rho v^2 = \text{constant}$. In a region of smaller cross-sectional area, the flow speed is greater (see flow rate equation); from Bernoulli's equation, the pressure in that region is lower than in other regions.

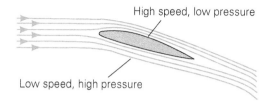

▲ **FIGURE 9.20 Airplane lift – Bernoulli's principle in action** Because of the shape and orientation of an airfoil or airplane wing, the air streamlines are closer together, and the air speed is greater above the wing than below it. By Bernoulli's principle, the resulting pressure difference supplies part of the upward force called the lift. (However, Bernoulli's principle is not applicable; see text.)

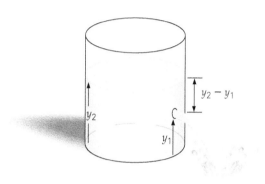

▲ **FIGURE 9.21 Fluid flow from a tank** The flow rate is given by Bernoulli's equation. See Example 9.13 for description.

SOLUTION

Given:

No specific values are given, so symbols will be used.

Find: An expression for the approximate initial water flow rate from the hole

Bernoulli's equation,

$$p_1 + \tfrac{1}{2}\rho v_1^2 + \rho g y_1 = p_2 + \tfrac{1}{2}\rho v_2^2 + \rho g y_2$$

can be used. Note that $y_2 - y_1$ is just the height of the surface of the liquid above the hole. The atmospheric pressures acting on the open surface and at the hole, p_1 and p_2, respectively, are essentially equal and cancel from the equation, as does the density, so

$$v_1^2 - v_2^2 = 2g(y_2 - y_1)$$

By the equation of continuity (the flow rate equation, Equation 9.17), $A_1 v_1 = A_2 v_2$, where A_2 is the cross-sectional area of the tank and A_1 is that of the hole. Since A_2 is much greater than A_1, then v_1 is much greater than v_2 (initially, $v_2 \approx 0$). So, to a good approximation,

$$v_1^2 = 2g(y_2 - y_1) \quad \text{or} \quad v_1 = \sqrt{2g(y_2 - y_1)}$$

The flow rate (volume/time) is then

$$\text{flow rate} = A_1 v_1 = A_1\sqrt{2g(y_2 - y_1)}$$

Given the area of the hole and the height of the liquid above it, the initial speed of the water coming from the hole and the flow rate can be found. (What happens as the water level falls?)

FOLLOW-UP EXERCISE. What would be the percentage change in the initial flow rate from the tank in this Example if the diameter of the small circular hole were increased by 30.0%?

CONCEPTUAL EXAMPLE 9.14: A STREAM OF WATER – SMALLER AND SMALLER

You have probably observed that a steady stream of water flowing out of a kitchen faucet gets smaller the farther the water falls from the faucet. Why does that happen?

REASONING AND ANSWER. This effect can be explained by Bernoulli's principle. As the water falls, it accelerates and its speed increases. Then, by Bernoulli's principle, the liquid pressure inside the stream decreases. (See Figure 9.19.) A pressure difference between that inside stream and the atmospheric pressure on the outside is thus created. As a result, there is an increasing inward force as the stream falls, so it becomes smaller. Eventually, the stream may get so thin that it breaks up into individual droplets.

FOLLOW-UP EXERCISE. The equation of continuity can also be used to explain this stream effect. Give this explanation.

9.5 Surface Tension, Viscosity, and Poiseuille's Law (Optional)

9.5.1 Surface Tension

The molecules of a liquid exert small attractive forces on each other. Even though molecules are electrically neutral overall, there is often some slight asymmetry of charge that gives rise to attractive forces between them (called *van der Waals forces*).* Within a liquid, any molecule is completely surrounded by other molecules, and the net force is zero (▶ **Figure 9.22a**).

However, for molecules at the surface of the liquid, there is no attractive force acting from above the surface. (The effect of air molecules is small and considered negligible.) As a result, net forces act upon the molecules of the surface layer, due to the attraction of neighboring molecules just below the surface. This inward pull on the surface molecules causes the surface of the liquid to contract and to resist being stretched or broken, a property called **surface tension**.

---

* After Johannes van der Waals (1837–1923), a Dutch scientist who first postulated an intermolecular force.

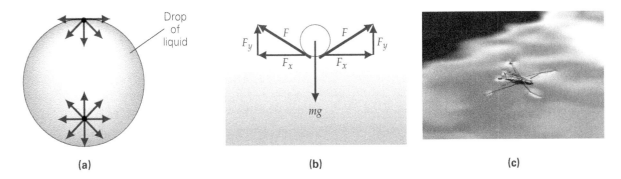

(a)   **(b)**   **(c)**

▲ FIGURE 9.22 **Surface tension (a)** The net force on a molecule in the interior of a liquid is zero, because the molecule is surrounded by other molecules. However, a nonzero fluid force acts on a molecule at the surface, due to the attractive forces of the neighboring molecules just below the surface. **(b)** For an object such as a needle to form a depression on the surface, work must be done, since more interior molecules must be brought to the surface to increase its area. As a result, the surface area acts like a stretched elastic membrane, and the weight of the object is supported by the upward components of the surface tension. **(c)** Insects such as this water strider can walk on water because of the upward components of the surface tension, much as you might walk on a large trampoline. Note the depressions in the surface of the liquid where the legs touch it.

(a)   **(b)**

▲ FIGURE 9.23 **Surface tension at work** Because of surface tension, **(a)** water droplets and **(b)** soap bubbles tend to assume the shape that minimizes their surface area – that of a sphere.

If a sewing needle is carefully placed on the surface of a bowl of water, the surface acts like an elastic membrane under tension. There is a slight depression in the surface, and molecular forces along the depression act at an angle to the surface (Figure 9.22b). The vertical components of these forces balance the weight (*mg*) of the needle, and the needle "floats" on the surface. Similarly, surface tension supports the weight of a water strider (Figure 9.22c).

The net effect of surface tension is to make the surface area of a liquid as small as possible. That is, a given volume of liquid tends to assume the shape that has the least surface area. As a result, drops of water and soap bubbles have spherical shapes, because a sphere has the smallest surface area for a given volume (▲ **Figure 9.23**). In forming a drop or bubble, surface tension pulls the molecules together to minimize the surface area.

9.5.2 Viscosity

All real fluids have an internal resistance to flow, or viscosity, which can be considered to be friction between the molecules of a fluid. In liquids, viscosity is caused by short-range cohesive forces, and in gases, it is caused by collisions between molecules. (See the discussion of air resistance in Section 4.6.) The viscous drag for both liquids and gases depends on their speeds and may be directly proportional to it in some cases. However, the relationship varies with the conditions; for example, the drag is approximately proportional to either v^2 or v^3 in turbulent flow.

Internal friction causes the layers of a fluid to move relative to each other in response to a shear stress. This layered motion, called *laminar flow*, is characteristic of steady flow for viscous liquids at low velocities (▶ **Figure 9.24a**). At higher velocities, the flow becomes rotational, or *turbulent*, and difficult to analyze.

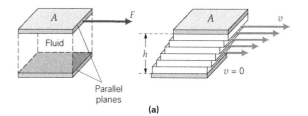

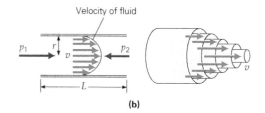

▲ FIGURE 9.24 **Laminar flow (a)** A shear stress causes layers of a fluid to move over each other in laminar flow. The shear force and the flow rate depend on the viscosity of the fluid. **(b)** For laminar flow through a pipe, the speed of the fluid is less near the walls of the pipe than near the center because of frictional drag between the walls and the fluid.

Since there are shear stresses and shear strains (deformation) in laminar flow, the viscous property of a fluid can be described by a coefficient. Viscosity is characterized by a *coefficient of viscosity*, η (the Greek letter eta), commonly referred to as simply the viscosity.

The coefficient of viscosity is, in effect, the ratio of the shear stress to the rate of change of the shear strain (since motion is involved). Unit analysis shows that the SI unit of viscosity is the pascal-second (Pa·s). This combined unit is called the *poiseuille* (Pl), in honor of the French scientist Jean Poiseuille (1797–1869), who studied the flow of liquids, particularly blood. (Poiseuille's law on flow rate will be presented shortly.) The cgs unit of viscosity is the *poise* (P). A smaller multiple, the *centipoise* (cP), is widely used because of its convenient size; $1\ \text{P} = 10^2\ \text{cP}$.

The viscosities of some fluids are listed in ▶ **Table 9.3**. The greater the viscosity of a liquid, which is easier to visualize than that of a gas, the greater the shear stress required to get the layers of the liquid to slide along each other. Note, for example, the large viscosity of glycerin compared to that of water.

As you might expect, viscosity, and thus fluid flow, varies with temperature, which is evident from the old saying, "slow as molasses in January." A familiar application is the viscosity grading of motor oil used in automobiles. In winter, a low-viscosity, or relatively thin, oil should be used (such as SAE grade 0W or 5W), because it will flow more readily, particularly when the engine is cold at startup. In summer, a higher viscosity, or thicker, oil is used (SAE 20 or 30).*

---

* *SAE* stands for *Society of Automotive Engineers*, an organization that designates the grades of motor oils based on their viscosity.

TABLE 9.3 Viscosities of Various Fluids[a]

Fluid	Viscosity (η) [Poiseuille (Pl)]
Liquids	
Alcohol, ethyl	1.2×10^{-3}
Blood, whole (37°C)	1.7×10^{-3}
Blood plasma (37°C)	2.5×10^{-3}
Glycerin	1.5×10^{-3}
Mercury	1.55×10^{-3}
Oil, light machine	1.1
Water	1.00×10^{-3}
Gases	
Air	1.9×10^{-5}
Oxygen	2.2×10^{-5}

[a] At 20°C unless otherwise indicated.

Seasonal changes in the grade of motor oil are not necessary if you use the multigrade, year-round oils. These oils contain additives called viscosity improvers, which are polymers whose molecules are long, coiled chains. An increase in temperature causes the molecules to uncoil and intertwine. Thus, the normal decrease in viscosity is counteracted. The action is reversed on cooling, and the oil maintains a relatively small viscosity range over a large temperature range. Such motor oils are graded, for example, as SAE 5W-20 ("five-W-twenty").

9.5.3 Poiseuille's Law

Viscosity makes analyzing fluid flow difficult. For example, when a fluid flows through a pipe, there is frictional drag between the liquid and the walls, and the fluid speed is greater toward the center of the pipe (Figure 9.24b). In practice, this effect makes a difference in a fluid's *average flow rate* $Q = A\bar{v} = \Delta V/\Delta t$ (see Equation 9.17), which describes the volume (ΔV) of fluid flowing past a given point during a time Δt. The SI unit of flow rate is cubic meters per second (m³/s). The flow rate depends on the properties of the fluid and the dimensions of the pipe, as well as on the pressure difference (Δp) between the ends of the pipe.

Jean Poiseuille studied flow in pipes and tubes, assuming a constant viscosity and steady or laminar flow. He derived the following relationship, known as **Poiseuille's law**, for the flow rate:

$$Q = \frac{\Delta V}{\Delta t} = \frac{\pi r^4 \Delta p}{8 \eta L} \quad \text{(Poiseuille's law)} \qquad (9.19)$$

Here, r is the radius of the pipe and L is its length.

As expected, the flow rate is inversely proportional to the viscosity (η) and the length of the pipe. Also as expected, the flow rate is directly proportional to the pressure difference Δp between the ends of the pipe. Somewhat surprisingly, however, the flow rate is proportional to r^4, which makes it more highly dependent on the radius of the tube than might have been thought. That is the reason why large pipes should be used to ease the requirement for large pressure difference (stronger pumps) in fluid transfer applications.

Chapter 9 Review

- In the deformation of elastic solids, **stress** is a measure of the force causing the deformation:

$$\text{stress} = \frac{F}{A} \qquad (9.1)$$

Strain is a relative measure of the deformation a stress causes:

$$\text{strain} = \frac{\text{change in length}}{\text{original length}} = \frac{\Delta L}{L_\text{o}} = \frac{|L - L_\text{o}|}{L_\text{o}} \qquad (9.2)$$

- An **elastic modulus** is the ratio of stress to strain.

Young's modulus:

$$Y = \frac{F/A}{\Delta L/L_\text{o}} \qquad (9.4)$$

Shear modulus:

$$S = \frac{F/A}{x/h} \approx \frac{F/A}{\phi} \qquad (9.5)$$

Bulk modulus:

$$B = \frac{F/A}{-\Delta V/V_\text{o}} = -\frac{\Delta p}{\Delta V/V_\text{o}} \qquad (9.6)$$

- Pressure is the force per unit area:

$$p = \frac{F}{A} \qquad (9.8\text{a})$$

Pressure-depth relationship (for an incompressible fluid at constant density):

$$p = p_\text{o} + \rho g h \qquad (9.10)$$

- **Pascal's principle.** Pressure applied to an enclosed fluid is transmitted undiminished to every point in the fluid and to the walls of the container.
- **Archimedes' principle.** A body immersed wholly or partially in a fluid is buoyed up by a force equal in magnitude to the weight of the volume of fluid displaced.

Buoyant force:

$$F_\text{b} = m_\text{f} g = \rho_\text{f} g V_\text{f} \qquad (9.14)$$

- An object will float in a fluid if the average density of the object is less than the density of the fluid. If the average density of the object is greater than the density of the fluid, the object will sink.
- For an ideal fluid, the flow is (1) steady, (2) irrotational, (3) nonviscous, and (4) incompressible. The following equations describe such a flow:

Equation of continuity:

$$\rho_1 A_1 v_1 = \rho_2 A_2 v_2 \quad \text{or} \quad \rho A v = \text{constant} \qquad (9.16)$$

Flow rate equation (for an incompressible fluid):

$$A_1 v_1 = A_2 v_2 \quad \text{or} \quad A v = \text{constant} \qquad (9.17)$$

Bernoulli's equation (for an incompressible fluid):

$$p_1 + \tfrac{1}{2} \rho v_1^2 + \rho g y_1 = p_2 + \tfrac{1}{2} \rho v_2^2 + \rho g y_2$$

or

$$p + \tfrac{1}{2} \rho v^2 + \rho g y = \text{constant} \qquad (9.18)$$

- Bernoulli's equation is a statement of the conservation of energy for a fluid.
- **Surface tension:** The inward pull on the surface molecules of a liquid that causes the surface to contract and resist being stretched or broken.
- **Viscosity:** A fluid's internal resistance to flow. All real fluids have a nonzero viscosity.
- **Poiseuille's law** (flow rate in pipes and tubes for fluids with constant viscosity and steady or laminar flow):

$$Q = \frac{\Delta V}{\Delta t} = \frac{\pi r^4 \Delta p}{8 \eta L} \qquad (9.19)$$

End of Chapter Questions and Exercises

Multiple Choice Questions

9.1 Solids and Elastic Moduli

1. The pressure on an elastic body is described by (a) a modulus, (b) work, (c) stress, (d) strain.
2. Shear moduli are not zero for (a) solids, (b) liquids, (c) gases, (d) all of these.
3. A relative measure of deformation is (a) a modulus, (b) work, (c) stress, (d) strain.
4. The volume stress for the bulk modulus is (a) Δp, (b) ΔV, (c) V_o, (d) $\Delta V/V_\text{o}$.

9.2 Fluids: Pressure and Pascal's Principle

5. For a liquid in an open container, the absolute pressure at any depth depends on (a) atmospheric pressure,

(b) liquid density, (c) acceleration due to gravity, (d) all of the preceding.

6. For the pressure-depth relationship for a fluid ($p = \rho g h$) it is assumed that (a) the pressure decreases with depth, (b) a pressure difference depends on the reference point, (c) the fluid density is constant, (d) the relationship applies only to liquids.

7. When measuring automobile tire pressure, what type of pressure is this: (a) gauge, (b) absolute, (c) atmospheric, or (d) all of the preceding?

9.3 Buoyancy and Archimedes' Principle

8. A wood block floats in a swimming pool. The buoyant force exerted on the block by water depends on (a) the volume of water in the pool, (b) the volume of the wood block, (c) the volume of the wood block under water, (d) all of the preceding.

9. If a submerged object displaces an amount of liquid of greater weight than its own and is then released, the object will (a) rise to the surface and float, (b) sink, (c) remain in equilibrium at its submerged position, (d) none of the preceding.

10. A rock is thrown into a lake. While sinking, the buoyant force (a) is zero, (b) decreases, (c) increases, (d) remains constant.

11. A glass containing an ice cube is filled to the brim and the cube floats on the surface. When the ice cube melts, (a) water will spill over the sides of the glass, (b) the water level decreases, (c) the water level is at the top of the glass without any spill.

12. Comparing an object's average density (ρ_o) to that of a fluid (ρ_f). What is the condition for the object to sink: (a) $\rho_o < \rho_f$, (b) $\rho_f < \rho_o$, or (c) $\rho_f = \rho_o$?

9.4 Fluid Dynamics and Bernoulli's Equation

13. An ideal fluid is not (a) steady, (b) compressible, (c) irrotational, (d) nonviscous.

14. Bernoulli's equation is based primarily on (a) Newton's laws, (b) conservation of momentum, (c) conservation of mass, (d) conservation of energy.

15. According to Bernoulli's equation, if the pressure on the liquid in Figure 9.19 is increased, (a) the flow speed always increases, (b) the height of the liquid always increases, (c) both the flow speed and the height of the liquid may increase, (d) none of the preceding.

9.5 Surface Tension, Viscosity, and Poiseuille's Law (Optional)

16. Water droplets and soap bubbles tend to assume the shape of a sphere. This effect is due to (a) viscosity, (b) surface tension, (c) laminar flow, (d) none of the preceding.

17. Some insects can walk on water because (a) the density of water is greater than that of the insect, (b) water is viscous, (c) water has surface tension, (d) none of the preceding.

18. The viscosity of a fluid is due to (a) forces causing friction between the molecules, (b) surface tension, (c) density, (d) none of the preceding.

Conceptual Questions

9.1 Solids and Elastic Moduli

1. Which has a greater Young's modulus, a steel wire or a rubber band? Explain.

2. Why are scissors sometimes called shears? Is this a descriptive name in the physical sense?

3. Ancient stonemasons sometimes split huge blocks of rock by inserting wooden pegs into holes drilled in the rock and then pouring water on the pegs. Can you explain the physics that underlies this technique? [*Hint:* Think about sponges and paper towels.]

9.2 Fluids: Pressure and Pascal's Principle

4. ▼ **Figure 9.25** shows a famous "bed of nails" trick. The woman stands on a bed of nails and the nails do not pierce her feet. Explain why.

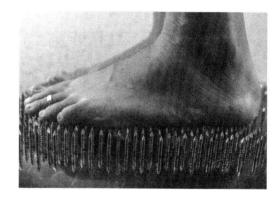

▲ FIGURE 9.25 **A bed of nails** See Conceptual Question 4.

5. Automobile tires are inflated to about 30 lb/in², whereas thin bicycle tires are inflated to 90 to 115 lb/in² – at least three times as much pressure! Why?

6. What is the principle of drinking through a straw? (Liquids aren't "sucked" up.)

7. What is the absolute pressure inside a flat tire?

8. (a) Two dams form artificial lakes of equal depth. However, one lake backs up 15 km behind the dam, and the other backs up 50 km behind. What effect does the difference in length have on the pressures on the dams? (b) Dams are usually thicker at the bottom. Why?

9. Water towers (storage tanks) are generally bulb shaped, as shown in ▶ **Figure 9.26**. Wouldn't it be better to have a cylindrical storage tank of the same height? Explain.

▲ FIGURE 9.26 **Why a bulb-shaped water tower?** See Conceptual Question 9.

10. A type of water dispenser for pets contains an inverted plastic bottle, as shown in ▼ **Figure 9.27**. When a certain amount of water is drunk from the bowl, more water flows automatically from the bottle into the bowl. The bowl never overflows. Explain the operation of the dispenser.

▲ FIGURE 9.27 **Pet barometers** See Conceptual Question 10.

9.3 Buoyancy and Archimedes' Principle

11. (a) What is the most important factor in constructing a life jacket that will keep a person afloat? (b) Why is it so easy to float in Utah's Great Salt Lake?

12. An ice cube floats in a glass of water. As the ice melts, how does the level of the water in the glass change? Would it make any difference if the ice cube were hollow? Explain.

13. A heavy object is dropped into a lake. As it descends below the surface, does the pressure on it increase? Does the buoyant force on the object increase?

14. Ocean liners weigh thousands of tons. How are they made to float?

15. Two blocks of equal volume, one iron and one aluminum, are dropped into a body of water. Which block will experience the greater buoyant force? Why?

9.4 Fluid Dynamics and Bernoulli's Equation

16. The speed of blood flow is greater in arteries than in capillaries. However, the flow rate equation ($Av = $ constant) seems to predict that the speed should be greater in the smaller capillaries. Can you resolve this apparent inconsistency?

17. When driving your car on an interstate at the posted speed limit (of course) and an 18-wheeler quickly passes you going in the opposite direction, you feel a force toward the truck. Why is this?

18. A pump spray bottle or "atomizer" operates by the Bernoulli principle. Explain how this works.

9.5 Surface Tension, Viscosity, and Poiseuille's Law (Optional)

19. A motor oil is labeled 5W-20. What do the numbers 5 and 20 measure? How about the W?

Exercises*

Integrated Exercises (IEs) are two-part exercises. The first part typically requires a conceptual answer choice based on physical thinking and basic principles. The following part is quantitative calculations associated with the conceptual choice made in the first part of the exercise.

9.1 Solids and Elastic Moduli

1. • A 5.0-m-long rod is stretched 0.10 m by a force. What is the strain in the rod?

2. • Suppose you use the tip of one finger to support a 1.0-kg object. If your finger has a diameter of 2.0 cm, what is the stress on your finger?

3. • A 2.5-m nylon fishing line used to hold up a 8.0-kg fish has a diameter of 1.6 mm. How much is the line elongated?

4. • A tennis racket has nylon strings. If one of the strings with a diameter of 1.0 mm is under a tension of 15 N, how much is it lengthened from its original length of 40 cm?

5. •• A copper wire has a length of 5.0 m and a diameter of 3.0 mm. Under what load will its length increase by 0.30 mm?

6. •• A metal wire 1.0 mm in diameter and 2.0 m long hangs vertically with a 6.0-kg object suspended from it. If the wire stretches 1.4 mm under the tension, what is the value of Young's modulus for the metal?

7. IE •• When railroad tracks are installed, gaps are left between the rails. (a) Should a greater gap be used if the rails are installed on (1) a cold day or (2) a hot day? Or (3) does the temperature not make any difference? Why? (b) Each steel rail is 8.0 m long and has a cross-sectional area of 0.0025 m². On a hot day, each rail thermally expands as much as 3.0×10^{-3} m. If there were no

---

* The bullets denote the degree of difficulty of the exercises: •, simple; ••, medium; and •••, more difficult.

gaps between the rails, what would be the force on the ends of each rail?

8. •• A rectangular steel column (20.0 cm × 15.0 cm) supports a load of 12.0 metric tons. If the column is 2.00 m in length before being stressed, what is the decrease in length?

9. IE •• A bimetallic rod as illustrated in ▼ **Figure 9.28** is composed of brass and copper. (a) If the rod is subjected to a compressive force, will the rod bend toward the brass or the copper? Why? (b) Justify your answer mathematically if the compressive force is 5.00×10^4 N.

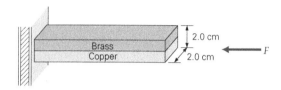

▲FIGURE 9.28 **Bimetallic rod and mechanical stress** See Exercise 9.

10. IE •• Two same-size metal posts, one aluminum and one copper, are subjected to equal shear stresses. (a) Which post will show the larger deformation angle, (1) the copper post or (2) the aluminum post? Or (3) Is the angle the same for both? Why? (b) By what factor is the deformation angle of one post greater than the other?

11. •• An 85.0-kg person stands on one leg and 90% of the weight is supported by the upper leg connecting the knee and hip joint – the femur. Assuming the femur is 0.650 m long and has a radius of 2.00 cm, by how much is the bone compressed?

12. •• Two metal plates are held together by two steel rivets, each of diameter 0.20 cm and length 1.0 cm. How much force must be applied parallel to the plates to shear off both rivets?

13. IE •• (a) Which of the liquids in Table 9.1 has the greatest compressibility? Why? (b) For equal volumes of ethyl alcohol and water, which would require more pressure to be compressed by 0.10%, and how many times more?

14. •• How much pressure would be required to compress a quantity of mercury by 0.010%?

15. ••• A brass cube 6.0 cm on each side is placed in a pressure chamber and subjected to a pressure of 1.2×10^7 N/m² on all of its surfaces. By how much will each side be compressed under this pressure?

9.2 Fluids: Pressure and Pascal's Principle

16. IE • In his original barometer, Pascal used water instead of mercury. (a) Water is less dense than mercury, so the water barometer would have (1) a higher height than, (2) a lower height than, or (3) the same height as the mercury barometer. Why? (b) How high would the water column have been?

17. • If you dive to a depth of 10 m below the surface of a lake, (a) what is the pressure due to the water alone? (b) What is the absolute pressure at that depth?

18. IE • In an open U-tube, the pressure of a water column on one side is balanced by the pressure of a column of gasoline on the other side. (a) Compared to the height of the water column, the gasoline column will have (1) a higher height, (2) a lower height, or (3) the same height. Why? (b) If the height of the water column is 15 cm, what is the height of the gasoline column?

19. • A 75.0-kg athlete performs a single-hand handstand. If the area of the hand in contact with the floor is 125 cm², what pressure is exerted on the floor?

20. • A rectangular fish tank measuring 0.75 m × 0.50 m is filled with water to a height of 65 cm. What is the gauge pressure on the bottom of the tank?

21. • (a) What is the absolute pressure at a depth of 10 m in a lake? (b) What is the gauge pressure?

22. •• The gauge pressure in both tires of a bicycle is 690 kPa. If the bicycle and the rider have a combined mass of 90.0 kg, what is the area of contact of *each* tire with the ground? (Assume that each tire supports half the total weight of the bicycle.)

23. •• In a sample of seawater taken from an oil spill, an oil layer 4.0 cm thick floats on 55 cm of water. If the density of the oil is 0.75×10^3 kg/m³, what is the absolute pressure on the bottom of the container?

24. •• The door and the seals on an aircraft are subject to a tremendous amount of force during flight. At an altitude of 10 000 m (about 33 000 ft), the air pressure outside the airplane is only 2.7×10^4 N/m² while the inside is still at normal atmospheric pressure, due to pressurization of the cabin. Calculate the force due to the air pressure on a door of area 3.0 m².

25. •• The pressure exerted by a person's lungs can be measured by having the person blow as hard as possible into one side of a manometer. If a person blowing into one side of an open tube manometer produces an 80-cm difference between the heights of the columns of water in the manometer arms, what is the gauge pressure of the lungs?

26. •• To drink a soda (assume same density as water) through a straw requires that you lower the pressure at the top of the straw. What does the pressure need to be at the top of a straw that is 15.0 cm above the surface of the soda in order for the soda to reach your lips?

27. •• During a plane flight, a passenger experiences ear pain due to clogged Eustachian tubes. Assuming the pressure in his tubes remained at 1.00 atm (from sea level) and the cabin pressure is maintained at 0.900 atm, determine the air pressure force (including its direction) on one eardrum, assuming it has a diameter of 0.800 cm.

28. •• Here is a demonstration Pascal used to show the importance of a fluid's pressure on the fluid's depth (▼ **Figure 9.29**): An oak barrel with a lid of area 0.20 m² is filled with water. A long, thin tube of cross-sectional area 5.0×10^{-5} m² is inserted into a hole at the center of the lid, and water is poured into the tube. When the water reaches 12 m high, the barrel bursts. (a) What was the weight of the water in the tube? (b) What was the pressure of the water on the lid of the barrel? (c) What was the net force on the lid due to the water pressure?

▲ FIGURE 9.29 **Pascal and the bursting barrel** See Exercise 28.

29. •• In a head-on auto collision, the driver, who had his air bags disconnected, hits his head on the windshield, fracturing his skull. Assuming the driver's head has a mass of 4.0 kg, the area of the head to hit the windshield to be 5.0 cm², and an impact time of 3.0 ms, with what speed does his head hit the windshield? (Take the compressive fracture strength of the cranial bone to be 1.0×10^8 Pa.)

30. •• A cylinder has a diameter of 15 cm (▼ **Figure 9.30**). The water level in the cylinder is maintained at a constant height of 0.45 m. If the diameter of the spout pipe is 0.50 cm, how high is h, the vertical stream of water? (Assume the water to be an ideal fluid.)

▲ FIGURE 9.30 **How high a fountain?** See Exercise 30.

31. •• In 1960, the U.S. Navy's bathyscaphe *Trieste* (a submersible) descended to a depth of 10 912 m (about 35 000 ft) into the Mariana Trench in the Pacific Ocean. (a) What was the pressure at that depth? (Assume that seawater is incompressible.) (b) What was the force on a circular observation window with a diameter of 15 cm?

32. •• The output piston of a hydraulic press has a cross-sectional area of 0.25 m². (a) How much pressure on the input piston is required for the press to generate a force of 1.5×10^6 N? (b) What force is applied to the input piston if it has a diameter of 5.0 cm?

33. •• A hydraulic lift in a garage has two pistons: a small one of cross-sectional area 4.00 cm² and a large one of cross-sectional area 250 cm². (a) If this lift is designed to raise a 3500-kg car, what minimum force must be applied to the small piston? (b) If the force is applied through compressed air, what must be the minimum air pressure applied to the small piston?

34. •• The Magdeburg water bridge is a channel bridge over the River Elbe in Germany (▼ **Figure 9.31**). Its dimensions are length 918 m, width 43.0 m, and depth 4.25 m. (a) When filled with water, what is the weight of the water? (b) What is the pressure on the bridge floor?

▲ FIGURE 9.31 **Water bridge** See Exercise 34.

9.3 Buoyancy and Archimedes' Principle

35. IE • (a) If the density of an object is exactly equal to the density of a fluid, the object will (1) float, (2) sink, (3) stay at any height in the fluid, as long as it is totally immersed. (b) A cube 8.5 cm on each side has a mass of 0.65 kg. Will the cube float or sink in water? Prove your answer.

36. • A rectangular boat, as illustrated in ▶ **Figure 9.32**, is overloaded such that the water level is just 1.0 cm below the top of the boat. What is the combined mass of the people and the boat?

37. •• An object has a weight of 8.0 N in air. However, it apparently weighs only 4.0 N when it is completely submerged in water. What is the density of the object?

38. •• When a 0.80-kg crown is submerged in water, its apparent weight is measured to be 7.3 N. Is the crown pure gold?

▲ FIGURE 9.32 **An overloaded boat** See Exercise 36.

39. •• A steel cube 0.30 m on each side is suspended from a scale and immersed in water. What will the scale read?

40. •• A solid ball has a weight of 3.0 N. When it is submerged in water, it has an apparent weight of 2.7 N. What is the density of the ball?

41. •• A wood cube 0.30 m on each side has a density of 700 kg/m³ and floats levelly in water. (a) What is the distance from the top of the wood to the water surface? (b) What mass has to be placed on top of the wood so that its top is just at the water level?

42. •• (a) Given a piece of metal with a light string attached, a scale, and a container of water in which the piece of metal can be submersed, how could you find the volume of the piece without using the variation in the water level? (b) An object has a weight of 0.882 N. It is suspended from a scale, which reads 0.735 N when the piece is submerged in water. What are the volume and density of the piece of metal?

43. •• An aquarium is filled with a liquid. A cork cube, 10.0 cm on a side, is pushed and held at rest completely submerged in the liquid. It takes a force of 7.84 N to hold it under the liquid. If the density of cork is 200 kg/m³, find the density of the liquid.

44. •• A block of iron quickly sinks in water, but ships constructed of iron float. A solid cube of iron 1.0 m on each side is made into sheets. To make these sheets into a hollow cube that will not sink, what should be the minimum length of the sides of the sheets?

45. •• Plans are being made to bring back the zeppelin, a lighter-than-air airship like the Goodyear blimp that carries passengers and cargo, but is filled with helium, not flammable hydrogen as was used in the ill-fated *Hindenburg*. One design calls for the ship to be 110 m long and to have a total mass (without helium) of 30.0 metric tons. Assuming the ship's "envelope" to be cylindrical, what would its diameter have to be so as to lift the total weight of the ship and the helium?

46. •• A girl floats in a lake with 97% of her body beneath the water. What are (a) her mass density and (b) her weight density?

9.4 Fluid Dynamics and Bernoulli's Equation

47. • An ideal fluid is moving at 3.0 m/s in a section of a pipe of radius 0.20 m. If the radius in another section is 0.35 m, what is the flow speed there?

48. IE • (a) If the radius of a pipe narrows to half of its original size, will the flow speed in the narrow section (1) increase by a factor of 2, (2) increase by a factor of 4, (3) decrease by a factor of 2, or (4) decrease by a factor of 4? Why? (b) If the radius widens to three times its original size, what is the ratio of the flow speed in the wider section to that in the narrow section?

49. •• Water flows through a horizontal tube similar to that in Figure 9.19. However, in this case, the constricted part of the tube is half the diameter of the larger part. If the water speed is 1.5 m/s in the larger parts of the tube, by how much does the pressure drop in the constricted part? Express the final answer in atmospheres.

50. •• The speed of blood in a major artery of diameter 1.0 cm is 4.5 cm/s. (a) What is the flow rate in the artery? (b) If the capillary system has a total cross-sectional area of 2500 cm², the average speed of blood through the capillaries is what percentage of that through the major artery? (c) Why must blood flow at low speed through the capillaries?

51. •• The blood flow speed through an aorta with a radius of 1.00 cm is 0.265 m/s. If hardening of the arteries causes the aorta to be constricted to a radius of 0.800 cm, by how much would the blood flow speed increase?

52. •• Using the data and result of Exercise 51, calculate the pressure difference between the two areas of the aorta. (Blood density: $\rho = 1.05 \times 10^3$ kg/m³.)

53. •• In Conceptual Example 9.12, it was explained why a stream of water from a faucet necks shrinks down into a smaller cross-sectional area as it descends. Suppose at the top of the stream it has a cross-sectional area of 2.0 cm², and a vertical distance 5.0 cm below the cross-sectional area of the stream is 0.80 cm². What is (a) the speed of the water and (b) the flow rate?

54. •• Water flows at a rate of 25 L/min through a horizontal 7.0-cm-diameter pipe under a pressure of 6.0 Pa. At one point, calcium deposits reduce the cross-sectional area of the pipe to 30 cm². What is the pressure at this point? (Consider the water to be an ideal fluid.)

9.5 Surface Tension, Viscosity, and Poiseuille's Law (Optional)

55. •• The pulmonary artery, which connects the heart to the lungs, is about 8.0 cm long and has an inside diameter of 5.0 mm. If the flow rate in it is to be 25 mL/s, what is the required pressure difference over its length?

56. •• A hospital patient receives a quick 500-cc blood transfusion through a needle with a length of 5.0 cm and an inner diameter of 1.0 mm. If the blood bag is suspended 0.85 m above the needle, how long does the transfusion take? (Neglect the viscosity of the blood flowing in the plastic tube between the bag and the needle.)

10

Temperature and Kinetic Theory

The melting of polar ice, as shown in this photo, due to a rise in our planet's average temperature.

Global warning is becoming a popular topic due to increasing evidence that human activities have played a major role in the increase in the Earth's atmospheric temperature. Increasing global temperatures will cause polar ice to melt and sea levels to rise. Satellite photos during the last forty years clearly show the shrinking of the Arctic ice caps. This chapter-opening photograph shows that melting viewed from ground level.

Temperature and heat are frequent subjects of conversation, but if you had to explain what these words mean, you might have some difficulty. For example, a temperature change can result from the addition or removal of heat. Temperature, therefore, must be related to heat. But how? And what is heat? In this chapter, you'll find that the answers to such questions which will lead to an understanding of far-reaching physical principles.

An early theory considered heat to be a fluid-like substance called *caloric* (from the Latin *calor,* meaning "heat") that could flow into and out of a body. Even though this theory has been abandoned, it is still common to speak of heat as "flowing" from one object to another. Heat is now known to be energy in transit, and temperature and thermal properties are explained by considering the molecular behavior of substances. This and the next two chapters examine temperature and heat both in terms of the microscopic (molecular) and macroscopic. You'll also encounter the ideal gas law, which, among other things, explains the quantitative relationship between gas pressure and temperature.

10.1 Temperature, Heat, and Thermal Energy

Our study of thermal physics begins with everyday definitions of **temperature** (T) and **heat** (Q). Temperature denotes a relative measure of "hotness" or "coldness." For example, a hot stove has a higher temperature than an ice cube. Note that hot and cold are relative terms, like *tall* and *short*. Clearly this is a qualitative description and does not really define what temperature is or means physically. The physical meaning of temperature will be discussed shortly.

Heat is related to temperature *differences* and the *(molecular) energy transferred between objects or systems because of temperature differences*. For example, when heat energy is transferred from a hot to a cold object, this may result in an increase in the total energy of the molecules that comprise the cold object.*

On a microscopic level, temperature is associated with molecular motion. In the kinetic theory of gases (Sections 10.5 and 10.6), the temperature of a gas sample is found to be the determining factor in the energy of its molecules. In general, a gas molecule can consist of more than one atom, such as the diatomic (two atoms) gas molecule in ▼ **Figure 10.1**. Thus, besides *translational* (or straight-line or linear motion) kinetic energy, the molecules may have energy due to *rotations* and/or *vibrations*. For a system consisting of a gas, the sum of these molecular kinetic energies is its **thermal energy**, E_{th}. More generally, internal energy includes intermolecular and intramolecular energies as well (depicted in Figure 10.1).

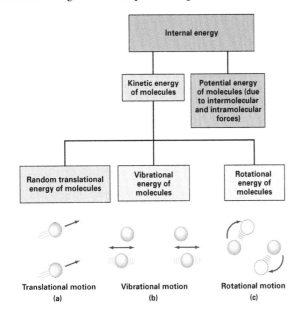

▲ **FIGURE 10.1 Internal energy** The internal energy of a system is composed of molecular kinetic energy as well as intermolecular and intramolecular potential energies. Molecular kinetic energy can come in three forms (shown here for a diatomic molecule): **(a)** translational kinetic energy, **(b)** linear vibrational kinetic energy, and **(c)** rotational kinetic energy.

Note that specifically for gas samples a higher temperature does not necessarily mean that one system has a greater thermal energy than another because thermal energy also depends on the number of molecules that comprise the system. For example, the air temperature inside a warm classroom on a cold winter day is higher than that of the outdoor air. However, all that cold outside air contains far more thermal energy than the inside air because there is so much more of it (it has many more molecules). In general, then, the thermal energy of a system depends on its temperature as well as the number of molecules in that system.

When heat is transferred between any two objects, not just gases, regardless of whether they are touching, (thermal energy can be transferred by a process called radiation that does not involved physical contact. This will be covered in Chapter 11) the objects are said to be in **thermal contact**. When there is no longer a net heat transfer between objects in thermal contact (i.e., they have come to the same temperature) it is said that they are in **thermal equilibrium**.

10.2 The Celsius and Fahrenheit Temperature Scales

A quantitative measure of temperature may be obtained by using a **thermometer**, a device constructed to make use of a property of a substance that changes with temperature. By far the most commonly used property is thermal expansion (Section 10.4), a change in the dimensions or volume of a substance that occurs when its temperature changes.

A common everyday thermometer is the liquid-in-glass type, which is based on the thermal expansion of a liquid. In these a liquid expands into a glass stem, rising into a hollow tube. Mercury and alcohol (dyed red to make it visible) are the liquids in most liquid-in-glass thermometers. These are chosen because of their relatively large thermal expansion rates and because they remain liquid over normal temperature ranges.

Thermometers are calibrated, and thus temperature scales defined, when numerical values for temperatures are assigned at two fixed (thermal) conditions, or points. The ice point and the steam point of water at standard atmospheric pressure were historically convenient fixed points. Also known as the freezing and boiling points, these are the temperatures at which pure water freezes and boils, respectively, at a pressure of 1 atm.

Two familiar temperature scales are the Fahrenheit scale[†] (used in the United States) and the Celsius scale[‡] (used in most of the rest of the world). As shown in ▶ **Figure 10.2**, the ice and steam points are chosen as 32 °F and 212 °F, respectively, on the Fahrenheit scale and 0 °C and 100 °C, respectively, on the Celsius scale. On the Fahrenheit scale, there are 180 equal intervals, or degrees (°F), between these two points and on the Celsius scale, there are 100 degrees (°C). Therefore, the Celsius degree is 1.8 times the size of the Fahrenheit degree.

---

* Note: Some of the heat may go into doing work and not into thermal energy (Section 12.2).

[†] Daniel Gabriel Fahrenheit (1686–1736), a German instrument maker, who constructed the first alcohol and mercury thermometers.

[‡] Named for Anders Celsius (1701–1744), a Swedish astronomer.

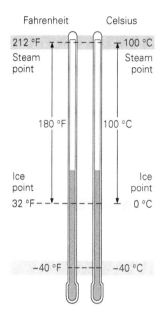

Fahrenheit   Celsius

212 °F 100 °C
Steam point Steam point
180 °F 100 °C
Ice point Ice point
32 °F 0 °C
−40 °F −40 °C

▲ **FIGURE 10.2 Celsius and Fahrenheit temperature scales** Between the ice and steam fixed points, there are 100 degrees on the Celsius scale and 180 degrees on the Fahrenheit scale. Thus, a Celsius degree is 1.8 times larger than a Fahrenheit degree.

The relationships for converting between these scales are as follows:

$$T_F = \frac{9}{5}T_C + 32 \quad \text{(Celsius to Fahrenheit conversion)} \quad (10.1a)$$

and its inverse

$$T_C = \frac{5}{9}(T_F - 32) \quad \text{(Fahrenheit to Celsius conversion)} \quad (10.1b)$$

EXAMPLE 10.1: CONVERTING TEMPERATURE SCALE READINGS – FAHRENHEIT AND CELSIUS

What are (a) the typical room temperature of 20 °C and a cold temperature of −15 °C on the Fahrenheit scale, and (b) a cold temperature of −10 °F and normal body temperature, 98.6 °F, on the Celsius scale?

THINKING IT THROUGH. These are direct applications of Equations 10.1a and 10.1b.

SOLUTION

Given:
(a) $T_C = 20$ °C and $T_C = -15$ °C
(b) $T_F = -10$ °F and $T_F = 98.6$ °F

Find: For each temperature,

(a) T_F
(b) T_C

(a) Using Equation 10.1a:

$$\text{for } T_C = 20\,°C \quad \text{or} \quad T_F = \frac{9}{5}T_C + 32 = \left[\frac{9}{5}(20) + 32\right]°F = 68\,°F$$

$$\text{for } T_C = -15\,°C \quad \text{or} \quad T_F = \frac{9}{5}T_C + 32 = \left[\frac{9}{5}(-15) + 32\right]°F = 5.0\,°F$$

(b) Using Equation 10.1b:

$$\text{for } T_F = -10\,°F \quad \text{or} \quad T_C = \frac{5}{9}(T_F - 32) = \left[\frac{5}{9}(-10 - 32)\right]°C = -23\,°C$$

$$\text{for } T_F = 98.6\,°F \quad \text{or} \quad T_C = \frac{5}{9}(T_F - 32) = \left[\frac{5}{9}(98.6 - 32)\right]°C = 37.0\,°C$$

An important medical comment – it is important to know your scales. An elevated body temperature measured as 40.0 °C represents a temperature elevation of 3.0 °C above normal. However since one Celsius degree is 1.8 times larger than one Fahrenheit degree, on the Fahrenheit scale this is an increase of 3.0 × 1.8 °F = 5.4 °F, or a body temperature 104.0 °F, which is serious.

FOLLOW-UP EXERCISE. Convert the following temperatures: (a) −40 °F to Celsius and (b) −40 °C to Fahrenheit.

10.2.1 Problem-Solving Hint

Because Equations 10.1a and 10.1b are similar-looking, it is easy to confuse them. Since they are equivalent, you need to know only one of them – say, Celsius to Fahrenheit, Equation 10.1a. Solving this for T_C gives Equation 10.1b. To check a conversion equation for correctness, you should test it using a known temperature, such as the boiling point of water. For example, converting $T_F = 212$ °F using Equation 10.1b produces

$$T_C = \frac{5}{9}(T_F - 32) = \left[\frac{5}{9}(212 - 32)\right]°C = \frac{5}{9}(180)\,°C = 100\,°C$$

and thus this equation is the correct one.

Liquid-in-glass thermometers are adequate for many temperature measurements, but for sensitive measurements and to define intermediate temperatures more precisely, other types of thermometers are used. One of these, a gas thermometer, is discussed in detail in the next section.

10.3 Gas Laws and Absolute (Kelvin) Temperature

Different liquid-in-glass thermometers show slightly different readings for temperatures other than fixed points because of the liquids' different expansion properties. A thermometer that uses a gas gives the same readings regardless of the gas used because at low densities all gases exhibit the same expansion behavior.

To describe the macroscopic (large scale) behavior of a gas it is customary to use its pressure, volume, and temperature (p, V, and T). When its temperature is constant, the pressure and volume of a gas are found to be *inversely* related meaning $p \propto \dfrac{1}{V}$

$$pV = \text{constant} \quad \text{or} \quad p_1V_1 = p_2V_2 \quad \text{(at constant temperature)} \quad (10.2)$$

This relationship is known as **Boyle's law**, after Robert Boyle (1627–1691), the English chemist who discovered it.

However, when the gas pressure is constant, its volume is found to be directly proportional to its *absolute* (or Kelvin) temperature T (defined shortly) meaning $V \propto T$

$$\frac{V}{T} = \text{constant} \quad \text{or} \quad \frac{V_1}{T_1} = \frac{V_2}{T_2} \quad \text{(at constant pressure)} \quad (10.3)$$

This relationship is known as **Charles's law**, after the French scientist Jacques Charles (1746–1823) who discovered it.

Low-density (or *ideal*) gases obey these laws, which can be combined into a single relationship. For a fixed quantity of gas, then, the expression pV/T must be constant. Combining Equations 10.2 and 10.3 yields the macroscopic (ratio) form of the **ideal gas law**:

$$\frac{pV}{T} = \text{constant} \quad \text{or} \quad \frac{p_1V_1}{T_1} = \frac{p_2V_2}{T_2}$$
$$\text{(ideal gas law, ratio form)} \quad (10.4)$$

10.3.1 Microscopic Form of the Ideal Gas Law

Equation 10.4 can be rewritten so it applies on the microscopic level. On this level, the "amount" of gas sample is designated by the number of molecules N. If volume and temperature are constant, the gas pressure is directly related to N (more molecules, higher pressure). Including this observation and at the same time rewriting the ratio form of the ideal gas law as an equation yields:

$$\frac{pV}{T} = Nk_B \quad \text{or} \quad pV = Nk_BT \quad \text{(microscopic ideal gas law)} \quad (10.5)$$

Here k_B is known as **Boltzmann's constant** and has a value of 1.38×10^{-23} J/K. It is named after the Austrian physicist Ludwig Boltzmann (1844–1906), who first determined it.

10.3.2 Macroscopic Form of the Ideal Gas Law

Equation 10.5 can be rewritten in macroscopic (macro means large) form, which recasts it in terms of quantities measured by everyday laboratory equipment. The macroscopic form is commonly written as

$$pV = nRT \quad \text{(microscopic ideal gas law)} \quad (10.6)$$

Here R is the **ideal gas constant** [$R = 8.31$ J/(mol·K)] and n represents the quantity of gas expressed in *moles*. A **mole*** (abbreviated mol) is defined as the quantity of a substance that contains **Avogadro's number** ($N_A = 6.02 \times 10^{23}$ molecules/mol) of molecules. The number of moles is related to the number of molecules at N by $n = N/N_A$ or $N = nN_A$.

One way to think of this is in analogy to finding the number of egg cartons (dozen analogous to n) by dividing the number of eggs, N, by 12 (analogous to N_A). Thus 36 eggs is equivalent to 36/12 or 3 cartons. Clearly either description – 3 dozen (macroscopic) or 36 (microscopic) – represents the same total number of eggs.

Since the microscopic and macroscopic viewpoints must lead to the same results, Equation 10.6 must be equivalent to Equation 10.5. Therefore it must be that $nR = Nk_B$. Using $n = N/N_A$, this becomes $k_B = (R/N_A)$, which leads to the following interpretation of the Boltzmann constant:

the Boltzmann constant is the microscopic version of the macroscopic ideal gas constant, expressed on a per molecule rather than a per mole basis.

Inserting the Boltzmann constant and Avogadro's number into $R = k_BN_A$, as previously stated, the value of R is

$$R = 8.31 \text{J/(mol·K)}$$

Notice the units tell the story. R's units are energy *per mole* whereas k_B is *per molecule*. Because there are a huge number of molecules per mole, the numerical value of the Boltzmann constant is much smaller than R.

To use Equation 10.6, (the number of moles, i.e., the quantity) of gas needs to be known. This is done using the concept of the **molar mass**, M, of a compound or element. Molar mass is the mass of one mole of substance, so $M = mN_A$, where m is the **molecular mass** or the mass of one molecule. Because molecular masses are so very small in relation to the SI standard kilogram, another unit, the **atomic mass unit** (u), is used:

$$1 \text{ atomic mass unit (u)} = 1.660\,54 \times 10^{-27} \text{ kg}$$

The molecular mass is determined from the chemical formula and the atomic masses of the atoms. (The latter are listed in Appendix III and IV and are commonly rounded to 2 or 3 significant figures.) For example, water, H_2O, with

---

* While in the final production stages of this book (2019), the International Committee for Weights and Measures has adopted a new definition of the mole based on fundamental physical constants, such as Avogadro's number (Section 10.3.2) among others.

two hydrogen atoms and one oxygen atom, has a molecular mass of $2(m_H) + 1(m_O) = 2(1.0\ u) + 1(16.0\ u) = 18.0\ u$ because the atomic mass of each hydrogen atom is 1.0 u and that of an oxygen atom is 16.0 u. Then, one mole of water has a *molar* mass of (18 u) $(1.660\ 54 \times 10^{-27}\ kg/u)(6.02 \times 10^{23}/mol) = 0.0180\ kg/mol = 18.0\ g/mol$. Similarly, the oxygen we breathe, O_2, has a molecular mass of $2 \times 16.0\ u = 32.0\ u$. Hence, one mole of oxygen has a mass of 32.0 g. The general rule is, then, if you know the molecular mass value (in terms of u) then that number is also the molar mass expressed in g/mol.

The reverse calculation can also be made. For example, suppose you want to know the mass of a water molecule (H_2O). If you know the molar mass of water to be 18.0 g/mol, then the molecular mass (m_{H_2O}) is

$$m_{H_2O} = \frac{M(\text{molar mass})}{N_A} = \frac{(18.0\ g/mol)}{6.02 \times 10^{23}\ \text{molecular/mol}}$$
$$= 2.99 \times 10^{-23}\ g/molecule = 2.99 \times 10^{-26}\ kg/molecule$$

10.3.3 Absolute Zero and the Kelvin Temperature Scale

The ideal gas law tells us that the product of the pressure and the volume of a sample of ideal gas is directly proportional to the temperature of the gas: $pV \propto T$. This relationship allows a gas to be used to measure temperature in a ***constant volume gas thermometer***. Holding the volume of a gas constant (as in a rigid container) means that $p \propto T$ (▼ **Figure 10.3**). Thus a constant volume gas thermometer displays temperature in terms

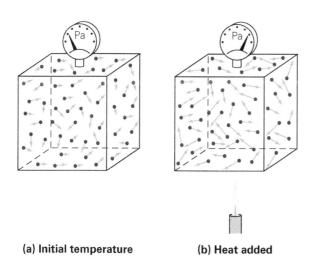

(a) Initial temperature   **(b) Heat added**

▲ **FIGURE 10.3 Constant volume gas thermometer** Such a thermometer indicates temperature as a function of pressure, since for a low-density gas at constant volume, the pressure and temperature are proportional. **(a)** At some initial temperature, the pressure reading has a certain value. **(b)** When the gas has its internal energy increased by adding heat, the pressure (and thus temperature) reading rises. On the microscopic scale this is because, on average, the molecules are now moving faster (have more kinetic energy).

of pressure, and a plot of pressure versus temperature gives a straight calibration line (▶ **Figure 10.4a**).

As shown in Figure 10.4b, p versus T measurements of real gases (plotted data points) are linear over a large temperature range but deviate from the values predicted by the ideal gas law at very low temperatures as they begin to liquefy. The absolute minimum temperature for an ideal gas is therefore inferred by extending the line to the T-axis, as in Figure 10.4b. Regardless of the type of gas, this temperature is −273.15 °C and is called **absolute zero**. Absolute zero is believed to be the lower limit of temperature, but it has never been attained. There is no known upper limit to temperature. For example, the temperatures at the centers of some stars are estimated to exceed 100 million degrees Celsius.

Absolute zero is the foundation of the **Kelvin temperature scale***, named after the British scientist Lord Kelvin [born William Thomson (1824–1907)] who proposed it in 1848. On this scale, −273.15 °C is taken as the zero point – that is, as 0 K (▶ **Figure 10.5**). The size of a single unit of Kelvin temperature is the same as that of the degree Celsius, so temperatures on these scales are related by

$$T = T_C + 273.15 \quad \text{(Celsius to Kelvin conversion)} \quad (10.7a)$$

The symbol *T with no subscript* is used to represent absolute temperature in **kelvins**. The kelvin is abbreviated as K (not degrees Kelvin, °K). For general calculations, it is common to round the 273.15 in Equation 10.8 to 273; that is,

$$T = T_C + 273 \quad \text{(for general calculations)} \quad (10.7b)$$

For example, a room temperature of 20 °C would be said to be 293 kelvins on the absolute scale. The Kelvin scale is the official SI temperature scale; however, the Celsius scale is used in most parts of the world for everyday temperature readings. The absolute temperature in kelvins is used primarily in scientific applications.

10.3.4 Problem-Solving Hints

Kelvin temperatures must be used when using the ideal gas law. It is a common mistake to use Celsius or Fahrenheit temperatures. Just suppose you mistakenly used a Celsius temperature of $T = 0\ °C$ in the gas law. This would mean $pV = 0$, which makes no sense, since neither p nor V is ever zero. Furthermore, negative temperatures make no sense in the ideal gas law, and, unlike the other two scales, there are no negative temperatures on the Kelvin scale. The bottom line for doing ideal gas calculations is to always convert the temperatures into kelvins before starting.

---

* While in the final production stages of this book (2019), the International Committee for Weights and Measures has adopted a new definition of the Kelvin temperature scale based on fundamental physical constants, such as Boltzmann's constant (Section 10.3.1) among others.

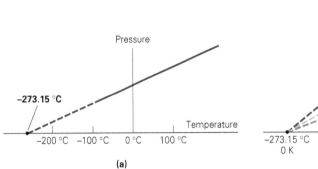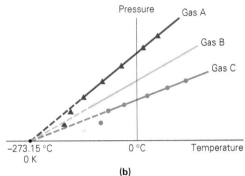

▲ FIGURE 10.4 **Pressure versus temperature** **(a)** A low-density gas kept at a constant volume gives a straight line on a graph of p versus T (in Celsius). When the line is extended to the zero pressure value, a temperature of −273.15 °C is obtained, which is taken to be absolute zero. **(b)** Extrapolation of such lines for *all* low-density gases indicates the same absolute zero temperature. Note that the actual behavior of gases deviates from a straight-line relationship at very low temperatures because the gases start to liquefy.

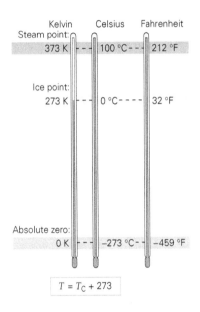

▲ FIGURE 10.5 **The Kelvin temperature scale** The lowest temperature on the Kelvin scale (corresponding to −273.15 °C) is absolute zero or 0 K. A degree interval on the Kelvin scale, called a kelvin and abbreviated K, is equivalent to a temperature change of 1 °C.

EXAMPLE 10.2: DEEPEST FREEZE – ABSOLUTE ZERO ON THE FAHRENHEIT SCALE

What is absolute zero on the Fahrenheit scale?

THINKING IT THROUGH. This requires the conversion of 0 K to the Fahrenheit scale. But first a conversion to the Celsius scale is in order. (Why?)

SOLUTION

Given:

$T = 0$ K

Find: T_F

Kelvin temperatures are related to Celsius values by Equation 10.7a. Converting 0 K to Celsius:

$$T_\text{C} = T - 273.15 = (0 - 273.15)\,°\text{C} = -273.15\,°\text{C}$$

Then to Fahrenheit (Equation 10.1) gives

$$T_\text{F} = \frac{9}{5}T_\text{C} + 32 = \left[\frac{9}{5}(-273.15) + 32\right]°\text{F} = -459.67\,°\text{F} \approx -460\,°\text{F}$$

FOLLOW-UP EXERCISE. There is an absolute temperature scale associated with the Fahrenheit temperature scale called the Rankine scale. A Rankine degree is the same size as a Fahrenheit degree, and absolute zero is taken as 0 °R (zero degrees Rankine). Write the conversion equations between (a) the Rankine and the Fahrenheit scales, (b) the Rankine and the Celsius scales, and (c) the Rankine and the Kelvin scales.

Initially, gas thermometers were calibrated using the ice and steam points of water as two fixed points. The Kelvin scale, however, uses absolute zero and a single fixed point adopted in 1954 by the International Committee on Weights and Measures. This fixed point is actually the **triple point of water**, a unique set of conditions at which water coexists simultaneously in equilibrium as a solid (ice), liquid (water), and gas (water vapor) – a temperature of 0.01 °C and a pressure of 4.58 mm Hg (611.73 Pa) – and provides a reproducible reference temperature for the Kelvin scale. The temperature of the triple point on the Kelvin scale is thus assigned a value of 273.16 K.

The Kelvin temperature scale has special significance. As will be seen in Section 10.5, absolute temperature is one of the factors that determine the thermal energy of an ideal gas. Now let's use various forms of the ideal gas law, remembering this requires the use of absolute temperatures.

EXAMPLE 10.3: THE IDEAL GAS LAW – USING ABSOLUTE TEMPERATURES

A quantity of ideal gas in a rigid container is initially at room temperature (20 °C) and a particular pressure (p_1). If the gas is heated to a temperature of 60 °C, by what factor does the pressure change?

THINKING IT THROUGH. Absolute temperatures are required here, the Celsius temperatures should be converted to kelvins. A "factor" of change implies a ratio (p_2/p_1), so the ideal gas law in ratio form should apply. The container is rigid, thus $V_1 = V_2$.

SOLUTION

Given:

$$T_1 = 20 \text{ °C} = (20 + 273)\text{K} = 293 \text{ K}$$

$$T_2 = 60 \text{ °C} = (60 + 273)\text{K} = 333 \text{ K}$$

$$V_1 = V_2$$

Find: p_2/p_1 (pressure ratio or factor)

The ideal gas law in ratio form is $(p_2 V_2/T_2) = (p_1 V_1/T_1)$ and since $V_1 = V_2$, this leads to

$$\frac{p_2}{p_1} = \frac{T_2}{T_1} = \frac{333\,\text{K}}{293\,\text{K}} = 1.14$$

Thus p_2 is 1.14 times p_1 and the pressure increases by 14%. Notice that by mistakenly using Celsius, the ratio is (60 °C)/(20 °C) = 3, or $p_2 = 3p_1$, which is wrong of course.

FOLLOW-UP EXERCISE. If the gas in this Example is heated from an initial temperature of 20 °C so the pressure increases by a factor of 1.26, what is the final Celsius temperature?

EXAMPLE 10.4: THE IDEAL GAS LAW – HOW MUCH OXYGEN?

A patient needing breathing therapy purchased a filled oxygen (O_2) tank. The tank has a volume of 2.5 L and is filled with pure oxygen at an absolute pressure of 100 atm when at room temperature (20 °C). What is the mass of the oxygen in the tank?

THINKING IT THROUGH. Since the mass of oxygen (O_2) is to be determined, the number of moles of the gas in the tank needs to be found first. That implies that the *macroscopic* form of the ideal gas law should be used. Then, knowing the molar mass of oxygen (O_2), the mass in the tank can be determined. To avoid any numerical errors, make sure the units of pressure and volume are in SI units, and temperatures, of course, are in kelvins.

SOLUTION

Given:

$$T = (20 + 273) \text{ K} = 293 \text{ K}$$

$$p = 100 \text{ atm} = (100)(1.01 \times 10^5 \text{ Pa}) = 1.01 \times 10^7 \text{ Pa}$$

$$V = 2.5 \text{ L} = 0.0025 \text{ m}^3 \text{ (because 1 m}^3 = 1000 \text{ L)}$$

Find: m_{O_2}

Using the macroscopic form of the ideal gas law the number of moles is found

$$n = \frac{pV}{RT} = \frac{(1.01 \times 10^7 \text{ Pa})(0.0025\,\text{m}^3)}{[8.31\,\text{J}/(\text{mol·K})](293\,\text{K})} = 10.4\,\text{mol}$$

Oxygen has a molecular mass of 2 × 16.0 u = 32.0 u, so its molar mass is 32.0 g/mol. Therefore, the mass of oxygen in the tank is

$$m_{O_2} = (10.4\,\text{mol})(32.0\,\text{g/mol}) = 333\,\text{g} = 0.333\,\text{kg}$$

FOLLOW-UP EXERCISE. Suppose this tank was used for a while, reducing its pressure to 65 atm as measured in a warmer room at 30 °C. What is the mass of oxygen in it now?

10.4 Thermal Expansion

Changes in the dimensions of materials are common thermal effects. For example, thermal expansion provides a means of creating a thermometer. Thermal expansion of gases is described by the ideal gas law, and this expansion is usually obvious. Less dramatic, but by no means less important, is the thermal expansion of liquids and solids. Let's consider solids first.

10.4.1 Solids

Thermal expansion in solids results from a change in the average distance between atoms of that solid. The atoms are held together by bonding forces, which can be simplistically represented as springs in a simple model of a solid. (See Figure 9.1). With increased temperature, the atoms vibrate back and forth over greater distances while their average separation also increases. With larger average atomic spacing in all directions, the solid expands as a whole.

The change in one dimension of a solid (length, width, or thickness) is called *linear* expansion. For small temperature changes ΔT, linear expansion (or contraction) is approximately proportional to ΔT (▶ **Figure 10.6a**). Experimentally the fractional change in length [$(L-L_0)/L_0$ or $\Delta L/L_0$, where L_0 is the original length] is proportional to the change in temperature and this is expressed mathematically as

$$\frac{\Delta L}{L_0} = \alpha \Delta T \quad \text{or} \quad \Delta L = \alpha L_0 \Delta T \quad \text{(linear expansion)} \tag{10.8}$$

where α is the **thermal coefficient of linear expansion** and depends on the type of material. The units of α are 1/ °C or °C^{-1}. Values of α for some materials are shown in ▶ **Table 10.1**. A solid may have different coefficients of linear expansion in different directions, but for simplicity it will be assumed that the same coefficient applies to all directions (in other words, isotropic expansion of solids is assumed). Also, these coefficients may vary slightly with temperature. However, this variation is negligible for most common applications, thus α will be taken to be constant.

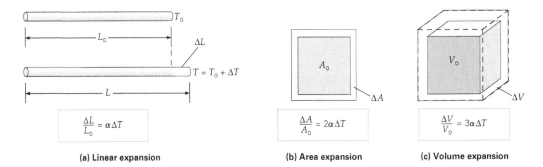

▲ FIGURE 10.6 **Thermal expansion (a)** Linear expansion is proportional to the temperature change; that is, the change in length ΔL is proportional to ΔT. Mathematically $\Delta L = \alpha L_0 \Delta T$, where α is the thermal coefficient of linear expansion. **(b)** For isotropic expansion, the thermal coefficient of area expansion is approximately 2α. **(c)** The thermal coefficient of volume expansion for solids is about 3α.

TABLE 10.1 Values of Thermal Expansion Coefficients (in °C^{-1}) for Some Materials at 20 °C

Material	Coefficient of Linear Expansion (α)	Material	Coefficient of Volume Expansion (β)
Aluminum	24×10^{-6}	Alcohol, ethyl	1.1×10^{-4}
Brass	19×10^{-6}	Gasoline	9.5×10^{-4}
Brick or concrete	12×10^{-6}	Glycerin	4.9×10^{-4}
Copper	17×10^{-6}	Mercury	1.8×10^{-4}
Glass, window	9.0×10^{-6}	Water	2.1×10^{-4}
Glass, Pyrex	3.3×10^{-6}		
Gold	14×10^{-6}	Air (and most gases at 1 atm)	3.5×10^{-3}
Ice	52×10^{-6}		
Iron and steel	12×10^{-6}		

A fractional change may also be expressed as a percent change. For example, by analogy, if you invested \$100 ($\$_0$) and made \$10 ($\Delta\$$), then the fractional change would be $\Delta\$/\$_0 = 10/100 = 0.10$, or 10% if you choose to express it as a percentage.

Equation 10.8 can be rewritten to give the length (L) as a function of temperature change:

$$\Delta L = L - L_0 = \alpha L_0 \Delta T \quad \text{or} \quad L = L_0 + \alpha L_0 \Delta T \quad (10.9)$$

This form is useful for computing the thermal expansion of areas of flat objects. For a square of initial side L_0, the initial area is $A_0 = L_0^2$. Thus the expanded area, A, is given by

$$A = L^2 = L_0^2(1 + \alpha\Delta T)^2 = A_0(1 + 2\alpha\Delta T + \alpha^2[\Delta T]^2)$$

This result, while derived for a flat square, is actually true regardless of shape. Because the values of α for solids are much less than 1 ($\sim 10^{-5}$, see Table 10.1), the quadratic or second-order term in this result (containing $\alpha^2 \approx (10^{-5})^2 = 10^{-10} \ll 10^{-5}$) can be dropped with negligible error. With that assumption, then, the change in area $\Delta A = A - A_0$ is

$$A = A_0(1 + 2\alpha\Delta T) \quad \text{or} \quad \frac{\Delta A}{A_0} = 2\alpha\Delta T$$
$$\text{(area expansion-any shape)} \quad (10.10)$$

Note that the **thermal coefficient of area expansion**, 2α (Figure 10.6b) is twice as large as the coefficient of linear expansion for a given material.

Similarly, volume expansion of a solid is given by

$$V = V_0(1 + 3\alpha\Delta T) \quad \text{or} \quad \frac{\Delta V}{V_0} = 3\alpha\Delta T$$
$$\text{(volume expansion-any shape)} \quad (10.11)$$

Thus the **thermal coefficient of volume expansion** (Figure 10.6c) is equal to 3α (for isotropic solids).

The thermal expansion of materials is an important consideration in construction; for example, seams of flexible material like asphalt between sections of concrete highways to allow room for expansion and prevent cracking. Expansion gaps in large bridges and between railroad rails are necessary to prevent damage. The Golden Gate Bridge in San Francisco varies in length by about 1 m between summer and winter, and the Eiffel Tower in Paris varies 0.36 cm for each degree Celsius change. Obviously, then, the thermal expansion of steel beams and girders can cause tremendous pressures, as the following Example shows.

**EXAMPLE 10.5: TEMPERATURE RISING –
THERMAL EXPANSION AND STRESS**

A steel beam is 5.0 m long at a temperature of 20 °C (68 °F). On a hot day, the temperature rises to 40 °C (104 °F). (a) What is the change in the beam's length due to thermal expansion? (b) Suppose that the ends of the beam are initially in contact with rigid vertical supports. How much force will the expanded beam exert on the supports if the beam has a cross-sectional area of 60 cm²?

THINKING IT THROUGH. (a) This is a direct application of Equation 10.9. (b) As the constricted beam expands, it applies a stress, and hence a force, to the supports. For linear expansion, Young's modulus (Section 9.1) comes into play.

SOLUTION

Given:

$L_0 = 5.0 \text{ m}$
$T_o = 20 \text{ °C}$
$T = 40 \text{ °C}$
$\alpha = 12 \times 10^{-6} \text{ °C}^{-1}$ (from Table 10.1)
$A = 60 \text{ cm}^2 \left(\dfrac{1 \text{ m}}{100 \text{ cm}} \right)^2 = 6.0 \times 10^{-3} \text{ m}^2$

Find:

(a) ΔL (change in length)
(b) F (force)

(a) Using Equation 10.9 to find ΔL with
$\Delta T = T - T_o = 40 \text{ °C} - 20 \text{ °C} = 20 \text{ °C}$ then:

$$\Delta L = \alpha L_0 \Delta T$$
$$= (12 \times 10^{-6} \text{ °C}^{-1})(5.0 \text{ m})(20 \text{ °C}) = 1.2 \times 10^{-3} \text{ m} = 1.2 \text{ mm}$$

This may not seem like much of an expansion, but it can give rise to a great deal of force if the beam is constrained and kept from expanding, as part (b) will show.

(b) By Newton's third law, if the beam is kept from expanding, the force the beam exerts on its constraint supports is equal to the force exerted by the supports to prevent the beam from expanding. This is the same as the force that would be required to compress the beam by that length. Using Young's modulus and Equation 9.4 with $Y = 20 \times 10^{10} \text{ N/m}^2$ (Table 9.1), the stress on the beam is

$$\frac{F}{A} = \frac{Y \Delta L}{L_0} = \frac{(20 \times 10^{10} \text{ N/m}^2)(1.2 \times 10^{-3} \text{ m})}{5.0 \text{ m}} = 4.8 \times 10^7 \text{ N/m}^2$$

Thus the force is

$$F = (4.8 \times 10^7 \text{ N/m}^2)A = (4.8 \times 10^7 \text{ N/m}^2)(6.0 \times 10^{-3} \text{ m}^2)$$
$$= 2.9 \times 10^5 \text{ N} \text{ (about 65 000 lb, or 32.5 tons!)}$$

Follow-Up Exercise. Expansion gaps between identical steel beams laid end to end are specified to be 0.060% of the length of each beam at an installation temperature of 68 °F. What is the maximum temperature at which noncontact expansion can be guaranteed?

CONCEPTUAL EXAMPLE 10.6: LARGER OR SMALLER? AREA EXPANSION

A circular piece is cut from a flat metal sheet (▶ **Figure 10.7a**). If the sheet is then heated in an oven, the size of the hole will (a) become larger, (b) become smaller, (c) remain unchanged.

REASONING AND ANSWER

It is a common misconception to think that the area of the hole will shrink because the metal expands inwardly around it. To counter this misconception, think of the piece of metal removed from the

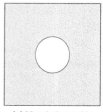 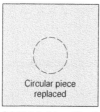

| **(a) Metal plate with hole** | **(b) Metal plate without hole** |

▲ FIGURE 10.7 **A larger or smaller hole?** See Example 10.6.

hole rather than of the hole itself. This piece would expand with increasing temperature. The metal in the heated sheet reacts as if the piece that was removed were still part of it. (Think of putting the piece of metal back into the hole after heating, as in Figure 10.12b, or consider drawing a circle on an uncut metal sheet and heating it.) So the answer is (a).

FOLLOW-UP EXERCISE. A student is trying to fit a bearing onto a shaft. The inside diameter of the bearing is just slightly smaller than the outside diameter of the shaft. Should the student heat the bearing or the shaft in order to fit the shaft inside the bearing?

Fluids (liquids and gases), like solids, normally expand with increasing temperature. Because fluids have no definite shape, only volume expansion (and not linear or area expansion) is meaningful. The analogous expression for fluid expansion is

$$\frac{\Delta V}{V_o} = \beta \Delta T \quad \text{(fluid volume expansion-any shape)} \quad (10.12)$$

where β is the **coefficient of volume expansion for fluids**. Note in Table 10.1 that the values of β for fluids are typically larger than the values of 3α for solids.

Unlike most liquids, water exhibits an anomalous expansion in volume near its ice point. The volume of a given amount of water decreases as it is cooled from room temperature, until its temperature reaches 4 °C (▶ **Figure 10.8a**). Below 4 °C, the volume increases, and therefore the density decreases (Figure 10.8b). This means that water has its maximum density at 4 °C (actually, 3.98 °C). This is why ice floats in ice water and frozen water pipes burst – water expands by about 9% on freezing.

This property has an important environmental effect: Bodies of water freeze at the top first, and the ice that forms floats. As it cool toward 4 °C, the water near the surface contracts, becomes denser, and sinks. The warmer, less dense water rises. However, once the colder water on top drops below 4 °C, it becomes less dense and remains at the surface, where it freezes. If water did not have this property, lakes and ponds would freeze from the bottom up, which would destroy much of their animal and plant life (and ice skating would be a lot less popular!). There would also be no ice caps at the polar regions since the solid ice would reside at the bottom of the ocean covered by liquid water.

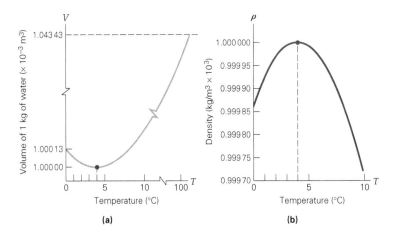

▲ FIGURE 10.8 **Thermal expansion of water** Water exhibits nonlinear expansion behavior near its ice point. **(a)** Above 4 °C (actually, 3.98 °C), water expands with increasing temperature, but from 4 °C down to 0 °C, it expands with decreasing temperature. **(b)** As a result, water has its maximum density near 4 °C.

10.5 The Kinetic Theory of Gases

One of the major accomplishments of early theoretical physics was to derive the ideal gas law from mechanical principles – modeling the molecules as "point" particles subject to Newton's laws. Among other accomplishments, this **kinetic theory of gases** led to a realization that temperature determines the translational kinetic energy of the molecules. As a starting point, the molecules were assumed to have large distances between them so molecular-molecular collisions were neglected. The only relevant molecular collisions were assumed to be those with the walls of the container or objects in the gas sample. These collisions result in what is measured as the gas's pressure.

In this section, the kinetic theory of monatomic (single atom, also called noble or inert) gases, such as helium and neon, is considered. More complicated diatomic gases, such as O_2, will be considered in (optional) Section 10.6.

According to kinetic theory, the molecules undergo perfectly elastic collisions with the walls of its container. From Newton's laws, the force on the walls of the container can be calculated from the change in momentum of the molecules when they collide with the walls (▶ **Figure 10.9**). The final result expresses the gas's affect on the wall in terms of its pressure (force/area):

$$pV = \frac{1}{3} N m v_{rms}^2 \qquad (10.13)$$

Here, V is the gas volume, N is the number of gas molecules, and m is the mass of *one* molecule. v_{rms} is a special kind of "average speed" obtained by averaging the squares of the speeds and then taking the square root of that average – that is, $\sqrt{(v^2)_{avg}} = v_{rms}$. As a result, v_{rms} is called the root-mean-square (rms) speed.

This theoretical result, if correct, must agree with the microscopic ideal gas law. To get agreement, equate the right hand side of Equation 10.13 to that of Equation 10.5.

$$\frac{1}{2} m v_{rms}^2 = \frac{3}{2} k_B T \qquad (10.14a)$$

But the average kinetic energy *per molecule* is related to the rms speed by $K_{avg} = \frac{1}{2} m v_{rms}^2$ thus

$$K_{avg} = \frac{3}{2} k_B T \qquad (10.14b)$$

and therefore the physical meaning of temperature is that it is the sole determiner of the average molecular kinetic energy of a gas sample.

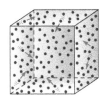

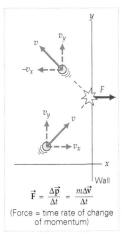

▲ FIGURE 10.9 **Kinetic theory of gases** The pressure a gas exerts on the walls of a container is due to the force resulting from the change in momentum of the molecules that collide with that wall. The wall must exert a force on each molecule to change its momentum. The molecules then exert a reaction force on the wall. The sum of the normal components of these collision forces divided by the wall area ($p = F/A$) gives the pressure exerted on the wall by the gas sample.

INTEGRATED EXAMPLE 10.7: MOLECULAR SPEED – RELATION TO ABSOLUTE TEMPERATURE

A helium-filled balloon is at 20 °C. (a) If it is heated to 40 °C, the rms speed of its molecules will (1) double, (2) increase by less than a factor of 2, (3) be half as much, (4) decrease by less than a factor of 2. Explain. (b) Calculate the rms speeds at these two temperatures. (Take the mass of the helium molecule to be $4\ u = 6.65 \times 10^{-27}$ kg.)

(a) **CONCEPTUAL REASONING.** According to Equation 10.14a, the rms speed is proportional to the square root of the absolute temperature, $v_{rms} \propto \sqrt{T}$. Therefore, a higher temperature will increase the rms speed; thus (3) and (4) are not possible. When the temperature increases from 20 °C to 40 °C, the *absolute* temperature T increases only from 293 K to 313 K, not even close to doubling. Furthermore, even if the absolute temperature were to double, the square root would *not* double (but it would increase). Thus, the correct answer is (2) – increase by less than a factor of 2.

(b) **QUANTATIVE REASONING AND SOLUTION.** All the data needed to solve for the rms speed are given. The Celsius temperatures must be changed to kelvins.

Given:

$m_{He} = 6.65 \times 10^{-27}$ kg
$T_1 = 20\ °C = (273 + 20)K = 293$ K
$T_2 = 40\ °C = (273 + 40)K = 313$ K

Find: v_{rms} (rms speed)

Rearranging Equation 10.14a:
At 20 °C

$$v_{rms} = \sqrt{\frac{3k_B T}{m}} = \sqrt{\frac{3(1.38 \times 10^{-22}\ J/K)(293K)}{6.65 \times 10^{-27}\ kg}}$$

$$= 1.35 \times 10^3\ m/s = 1.35\ km/s\ (\text{over 3000 mph})$$

and at 40 °C

$$v_{rms} = \sqrt{\frac{3(1.38 \times 10^{-23}\ J/K)(313K)}{6.65 \times 10^{-27}\ kg}}$$

$$= 1.40 \times 10^3\ m/s = 1.40\ km/s$$

FOLLOW-UP EXERCISE. In this Example, if the rms speed is to double its value at 20 °C, what would be the new Celsius temperature?

10.5.1 Thermal Energy of Monatomic Gases

Because the molecules in an ideal monatomic gas are modeled as "point masses" they cannot possess rotational or vibrational energy. Thus the thermal energy of this type of gas is just the total *translational* kinetic energy. With N molecules, a gas

sample's thermal energy of a monatomic gas is $E_{th} = N(K_{avg})$.* From $K_{avg} = (3/2)k_B T$ a microscopic expression for thermal energy is found:

$$E_{th} = \frac{3}{2} N k_B T \quad \text{(microscopic)} \quad (10.15a)$$

Using $Nk_B = nR$, this can be converted to the macroscopic form

$$E_{th} = \frac{3}{2} nRT \quad \text{(macroscopic)} \quad (10.15b)$$

Thus, the thermal (or internal) energy of an ideal monatomic gas is also directly proportional to its absolute temperature. That is, if the absolute temperature of a gas, say, doubled from 200 K to 400 K, then the thermal energy of the gas also would double.

10.6 Kinetic Theory, Diatomic Gases, and the Equipartition Theorem (Optional)

It is interesting to investigate how the kinetic theory can be adapted to work beyond monatomic gases because in the real world, most gases are not monatomic gases. For example, the mixture of gases of the air we breathe consists mainly of diatomic molecules of nitrogen (N_2, 78% by volume) and oxygen (O_2, 21% by volume). How does one deal with these and even those molecules consisting of more than two atoms, such as carbon dioxide (CO_2). The latter type are clearly of interest, but because of their molecular complexity our discussion will be limited to diatomic molecules.

10.6.1 The Equipartition Theorem

From Section 10.5, the average *translational* (or linear) kinetic energy of a monatomic molecule is determined by its gas's temperature. But this is also true for any molecule, monatomic or not. To make it clear that this refers only to translational energy, a subscript "trans" will be used. Thus $K_{avg,trans} = (3/2)k_B T$ is true for all species of gases.

Since a diatomic molecule is free to rotate and/or vibrate in addition to moving linearly, these additional forms of energy need to be accounted for when determining the total energy of a molecule. In deriving Equations 10.15a and 10.15b, scientists realized that the factor of 3 was because of the fact that the molecules had three independent linear dimensions in which to move. Thus, they reasoned, each molecule had three independent motions connected to its total linear kinetic energy. Each possible way a molecule has for possessing energy is called a *degree of freedom*.

---

* An analogy might be how the total points scored by a class on an exam can be found based on the number of students and the class average. For example, a class of 40 students and a test average of 80 means that there were a total of (40 students) (80 pts/student) = 3200 class points.

On this basis, a generalization (for ideal gases) called the **equipartition theorem** was proposed. (As the name implies, the thermal energy of a gas or molecule is "partitioned," or divided, equally for each degree of freedom.) That is,

> the thermal energy of an ideal gas sample is divided equally among each degree of freedom its molecules possess. Each degree of freedom contributes $(1/2)Nk_BT$ (or $(1/2)nRT$) to the thermal energy of the gas. On a per molecule basis each degree of freedom contributes $(1/2)k_BT$ of energy.

10.6.2 Thermal Energy of a Diatomic Gas

Let us use the equipartition theorem to compute the thermal energy of a diatomic gas. A diatomic molecule might rotate (see Figure 10.1), having rotational kinetic energies about three independent axes of rotations (i.e., three additional degrees of freedom). A diatomic gas might also vibrate, thus having both vibrational kinetic and potential energies (two additional degrees of freedom). Altogether then, a diatomic molecule could have as many as seven degrees of freedom.

Consider a symmetric diatomic molecule – for example, O_2. A classical model describes such a molecule as if the molecules were small masses connected by a rigid rod (▶ **Figure 10.10**). The rotational moments of inertia about each of two of the axes (here x and y) that pass perpendicularly through the center of the rod have the same value. However, the moment of inertia about the z-axis is negligible compared to these (see Figure 10.10 caption for a detailed explanation). Thus, only two degrees of freedom are associated with the rotational kinetic energies of diatomic molecules. Furthermore, quantum theory predicts (and experiment verifies) that at normal (room) temperatures, the vibrational kinetic energy and potential energy can be ignored. Thus, the thermal energy of a diatomic gas is composed of the molecular energies associated with three linear degrees of freedom and the two rotational degrees of freedom, for a total of five degrees of freedom.

Thus on average per molecule:

$$K_{avg} = K_{trans} + K_{rot} = 3\left[\tfrac{1}{2}k_BT\right] + 2\left[\tfrac{1}{2}k_BT\right]$$
$$= \frac{5}{2}k_BT \quad \text{(diatomic molecule)} \tag{10.16a}$$

and for the gas sample as a whole:

$$E_{th} = K_{tras} + K_{rot} = 3\left(\tfrac{1}{2}nRT\right) + 2\left(\tfrac{1}{2}nRT\right)$$
$$= \frac{5}{2}nRT = \frac{5}{2}Nk_BT \quad \text{(diatomic gas)} \tag{10.16b}$$

Thus, at a given temperature and assuming equal amounts (moles), a monatomic gas has 40% less thermal energy than a diatomic gas. Equivalently, the monatomic sample possesses only 60% (three-fifths) of the thermal energy of the diatomic sample.

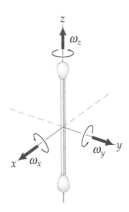

▲ **FIGURE 10.10   Model of a diatomic gas molecule** A dumbbell-like molecule can rotate about three axes. The moments of inertia, I_x and I_y, about the x- and y-axes are the same. Since the masses (molecules) on the ends of the rod are small, so the moment of inertia about the z-axis I_z is negligible compared to I_x and I_y.

EXAMPLE 10.8: MONATOMIC VERSUS DIATOMIC – ARE TWO ATOMS BETTER THAN ONE?

More than 99% of air consists of diatomic gases, mainly nitrogen and oxygen. There are also traces of other gases such as radon (Rn), a monatomic gas arising from radioactive decay of uranium in the ground. (a) Calculate the thermal energy of 1.00-mol samples oxygen and radon at room temperature (20 °C). (b) For each sample, calculate the thermal energy associated with just molecular *translational* kinetic energy.

THINKING IT THROUGH. (a) The number of degrees of freedom associated with the molecules of each type of gas must be taken into account. (b) Only three degrees of freedom contribute to the translational kinetic energy portion of the thermal energy of both gases.

SOLUTION

Listing the data and converting to kelvins because thermal energy *must* be computed using absolute temperature:

Given:

$n = 1.00$ mol
$T = (20 + 273)\text{K} = 293$ K

Find:

(a) E_{th} (O_2 and Rn at 20 °C)
(b) E_{trans} (O_2 and Rn at 20 °C)

(a) The thermal energy of the radon sample is found by using Equation 10.15b:

$$E_{th,Rn} = \frac{3}{2}nRT = \frac{3}{2}(1.00\,\text{mol})[8.31\text{J}/(\text{mol}\cdot\text{K})](293\text{K})$$
$$= 3.65 \times 10^3 \text{ J}$$

The oxygen calculation must also include energy stored as two extra degrees of freedom, due to rotation. Thus

$$E_{th,O_2} = \frac{5}{2}nRT = \frac{5}{2}(1.00\,mol)[8.31\,J/(mol\cdot K)](293\,K)$$
$$= 6.09 \times 10^3\,J$$

Thus even though each sample has the same number of molecules and the same temperature, the oxygen sample has 1.67 times more thermal energy.

(b) For radon, the thermal energy is all in the form of translational kinetic energy; hence, the answer is the same as in (a):

$$E_{trans,Rn} = 3.65 \times 10^3\,J$$

For oxygen, only $(3/2)nRT$ is in the form of translational kinetic energy, so the answer is the same as for radon; that is, $E_{trans,O_2} = 3.65 \times 10^3\,J$.

FOLLOW-UP EXERCISE. (a) In this Example, how much energy is associated with the rotational motion of the oxygen molecules? (b) Which sample has the higher rms speed? (*Note:* The mass of one radon atom is about seven times the mass of an oxygen molecule.) Explain your reasoning.

Chapter 10 Review

- Temperature scale conversions:

$$T_F = \frac{9}{5}T_C + 32 \tag{10.1a}$$

$$T_C = \frac{5}{9}(T_F - 32) \tag{10.1b}$$

$$T = T_C + 273 \tag{10.7}$$

- **Heat** is the net energy transferred from one object to another because of temperature differences. Once transferred, the energy represents the *change* (loss or gain) of the internal energy of each object.
- **Internal energy** of a system consists of molecular kinetic energies and energies due to intra- and inter-molecular forces.
- **Thermal energy** of a sample is defined as the sum of the kinetic energies of its molecules and is related to its temperature. For an ideal gas sample, the thermal energy is its internal energy.
- The **ideal gas law** relates the pressure, volume, and absolute temperature of an ideal gas.

$$pV = Nk_BT \tag{10.5}$$

$$pV = nRT \tag{10.6}$$

where T must be in Kelvin and $k_B = 1.38 \times 10^{-23}$ J/K and $R = 8.31$ J/(mol·K).

- **Thermal coefficients of expansion** relate the fractional change in dimension(s) to a change in temperature.
  Thermal expansion of solids:

Linear : $\quad \dfrac{\Delta L}{L_0} = \alpha \Delta T \quad$ or $\quad L = L_0(1 + \alpha \Delta T) \qquad$ (10.8)

Area : $\quad \dfrac{\Delta A}{A_0} = 2\alpha \Delta T \quad$ or $\quad A = A_0(1 + 2\alpha \Delta T) \qquad$ (10.10)

Volume : $\quad \dfrac{\Delta V}{V_0} = 3\alpha \Delta T \quad$ or $\quad V = V_0(1 + 3\alpha \Delta T)$ (10.11)

Thermal volume expansion of fluids:

$$\frac{\Delta V}{V_0} = \beta \Delta T \tag{10.12}$$

- According to the **kinetic theory of gases**, the energies of gas molecules as well as gas samples as a whole are proportional to the absolute temperature of the gas.

$$\frac{1}{2}mv_{rms}^2 = \frac{3}{2}k_BT \quad \text{(ideal gases)} \tag{10.14a}$$

$$E_{th} = \frac{3}{2}Nk_BT = \frac{3}{2}nRT \quad \text{(monatomic gases)} \tag{10.15a}$$

$$E_{th} = \frac{5}{2}Nk_BT = \frac{5}{2}nRT \quad \text{(diatomic gases)} \tag{10.15b}$$

End of Chapter Questions and Exercises

Multiple Choice Questions

10.1 Temperature and Heat
and
10.2 The Celsius and Fahrenheit Temperature Scales

1. Temperature differences between objects are associated with (a) energy exchanges, (b) heat, (c) molecular energy differences, or (d) all of these.
2. What types of energies can possibly contribute to the energy of a diatomic gas sample: (a) rotational kinetic energy, (b) translational kinetic energy, (c) vibrational kinetic energy, or (d) all of the preceding?
3. An object at a higher temperature (a) must, (b) may, or (c) must not have more internal energy than another object at a lower temperature.

10.3 Gas Laws, Absolute Temperature, and the Kelvin Temperature Scale

4. The temperature in the ideal gas law must be expressed on which scale: (a) Celsius, (b) Fahrenheit, (c) Kelvin, or (d) any of the preceding?

5. If an ideal gas at constant volume were to reach absolute zero, (a) its pressure would reach zero, (b) its pressure would reach infinity, (c) its mass would disappear, or (d) its mass would be infinite.

6. When the temperature of a quantity of gas is increased, (a) the pressure must increase, (b) the volume must increase, (c) both the pressure and volume must increase, (d) none of the these is necessarily true.

10.4 Thermal Expansion

7. What is the predominant cause of thermal expansion: (a) atom sizes change, (b) atom shapes change, or (c) the average distances between atoms change?

8. The units of the thermal coefficient of linear expansion are: (a) m/°C, (b) m²/°C, (c) m·°C, or (d) 1/°C.

9. Which of the following describes the behavior of water density from 0 °C to 4 °C: (a) it increases with increasing temperature, (b) it remains constant, (c) it decreases with increasing temperature, or (d) none of these?

10.5 The Kinetic Theory of Gases

10. If the average translational kinetic energy of the molecules in an ideal monatomic gas, initially at 20 °C, doubles, what is the final temperature of the gas: (a) 10 °C, (b) 40 °C, (c) 313 °C, or (d) 586 °C?

11. If the temperature of a given amount of ideal gas is raised from 300 to 600 K, is the thermal energy of the gas (a) doubled, (b) halved, (c) unchanged, or (d) none of the preceding?

12. Two different gas samples are at the same temperature. The sample with the more massive molecules will have (a) a higher, (b) a lower, or (c) the same rms speed as that of the sample with less massive molecules.

10.6 Kinetic Theory, Diatomic Gases, and the Equipartition Theorem (Optional)

13. Which of the following is a diatomic molecule: (a) He, (b) N_2, (c) CO_2, or (d) Ne?

14. A diatomic gas at room temperature has an thermal energy of (a) $(3/2)nRT$, (b) $(5/2)nRT$, (c) $(7/2)nRT$, or (d) none of the preceding.

15. On average, is the thermal energy of a gas divided equally among (a) each molecule, (b) each degree of freedom, (c) translational motion, rotational motion, and vibrational motion, or (d) none of the preceding?

Conceptual Questions

10.1 Temperature and Heat
and
10.2 The Celsius and Fahrenheit Temperature Scales

1. Heat flows spontaneously from a body at a higher temperature to one at a lower temperature that is in thermal contact with it. Does heat always flow from a body with more thermal energy to one with less thermal energy? Explain.

2. When temperature changes during the day, which scale, Celsius or Fahrenheit, will register a smaller numerical change? Explain.

3. What forms of energy make up the thermal energy of monatomic gases? How about diatomic gases?

10.3 Gas Laws, Absolute Temperature, and the Kelvin Temperature Scale

4. A type of constant volume gas thermometer is shown in ▼ **Figure 10.11**. Describe how it operates.

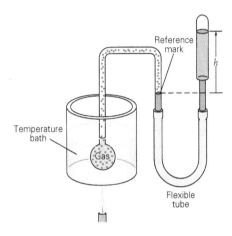

▲ FIGURE 10.11 **A type of constant volume gas thermometer** See Conceptual Question 4.

5. Describe how a constant pressure gas thermometer might be constructed.

6. In terms of the ideal gas law, what would a temperature of absolute zero imply? How about a negative absolute temperature?

7. Excited about a New Year's Eve party in Times Square, you pump up ten balloons in your warm apartment and take them to the cold square. However, you are very disappointed with your decorations. Why?

8. Which contains more molecules, 1 mole of oxygen or 1 mole of nitrogen? Explain.

9. Which contains more atoms, 1 mole of helium or 1 mole of oxygen? Explain.

10. How many moles of helium contain the same number of atoms as a mole of carbon dioxide?

10.4 Thermal Expansion

11. A cube of ice sits on a bimetallic strip at room temperature (▶ **Figure 10.12**). What will happen if (a) the upper strip is aluminum and the lower strip is brass, and (b) the upper strip is iron and the lower strip is copper? (c) If the cube is made of a hot metal rather than ice and the two strips are brass and copper, which metal should be on the top layer to keep the cube from falling off?

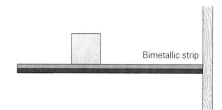

▲ FIGURE 10.12 **Which way will the cube go?** See Conceptual Question 11.

12. A solid flat circular metal disk rotates freely about a fixed axis perpendicular through its center. If the disk is heated while it is rotating, will there be any effect on the rate of rotation (the angular speed)? Explain.

13. A physics professor is demonstrating thermal expansion using a spherical ball and a ring, both of the same material. At room temperature, the ball fits through the ring. When the ball alone is heated it then does not fit through the ring. However, if both the ball and the ring are heated, the ball again fits through the ring. Explain why the ball now fits.

14. A circular ring of iron has a tight-fitting iron bar across its diameter, as illustrated in ▼ **Figure 10.13**. If the arrangement is heated in an oven to a high temperature, will the circular ring be distorted? What if the bar is made of aluminum?

▲ FIGURE 10.13 **Stress out of shape?** See Conceptual Question 14.

15. We often use hot water to loosen tightly sealed metal lids on glass jars. Explain how this works.

10.5 The Kinetic Theory of Gases

16. Gas sample A has twice as much average translational kinetic energy per molecule as gas sample B. What can be said about the absolute temperatures of the gas samples?

17. Equal quantities of helium (He) and neon (Ne) are at the same temperature. Which has more thermal energy? Which has a higher average molecular speed? Explain.

18. Two separate samples of helium (He) and neon (Ne) are at the same temperature. The helium sample has twice the thermal energy as the neon sample. Explain how this can be.

10.6 Kinetic Theory, Diatomic Gases, and the Equipartition Theorem (Optional)

19. If a monatomic gas and a diatomic gas sample are both at room temperature, will they have the same average total kinetic energy per molecule? Explain.

20. Explain clearly why a diatomic gas has more thermal energy than the same quantity of a monatomic gas at the same temperature.

21. A monatomic and diatomic gas sample both have n moles at temperature T. What is the difference in their thermal energies? Express your answer in terms of n, R, and T.

Exercises*

Integrated Exercises (IEs) are two-part exercises. The first part typically requires a conceptual answer choice based on physical thinking and basic principles. The following part is quantitative calculations associated with the conceptual choice made in the first part of the exercise.

10.1 Temperature and Heat
and
10.2 The Celsius and Fahrenheit Temperature Scales

1. • A person running a fever has a body temperature of 40 °C. What is this temperature on the Fahrenheit scale?

2. • Convert the following to Celsius readings: (a) 80 °F, (b) 0 °F, and (c) −10 °F.

3. • Convert the following to Fahrenheit readings: (a) 120 °C, (b) 12 °C, and (c) −5 °C.

4. • Which is the lower temperature: (a) 245 °C or 245 °F? (b) 200 °C or 375 °F?

5. • The coldest inhabited village in the world is Oymyakon, a town located in eastern Siberia, where it gets as cold as −94 °F. What is this temperature on the Celsius scale?

6. • The highest and lowest offically recorded air temperatures in the world are, respectively, 58 °C (Libya, 1922) and −89 °C (Antarctica, 1983). What are these temperatures on the Fahrenheit scale?

7. • The highest and lowest recorded air temperatures in the United States are, respectively, 134 °F (Death Valley, California, 1913) and –80 °F (Prospect Creek, Alaska, 1971). What are these temperatures on the Celsius scale?

8. •• During open-heart surgery it is common to cool the patient's body down to slow body processes and gain an extra margin of safety. A drop of 8.5 °C is typical in these types of operations. If a patient's normal body temperature is 98.2 °F, what is her final temperature in both Celsius and Fahrenheit?

---

* The bullets denote the degree of difficulty of the exercises: •, simple; ••, medium; and •••, more difficult.

9. •• In the troposphere (the lowest part of the atmosphere), the temperature decreases rather uniformly with altitude at a so-called "lapse" rate of about 6.5 °C/km. What are the temperatures (a) near the top of the troposphere (which has an average thickness of 11 km) and (b) outside a commercial aircraft flying at a cruising altitude of 34 000 ft? (Assume that the ground temperature is normal room temperature.)

10. IE •• The temperature drops from 60 °F during the day to 35 °F during the night. (a) The corresponding temperature drop on the Celsius scale is (1) greater than, (2) the same as, or (3) less than. Explain. (b) Compute the temperature drop on the Celsius scale.

11. IE •• There is one temperature at which the Celsius and Fahrenheit scales have the same reading. (a) To find that temperature, would you set (1) $5T_F = 9T_C$, (2) $9T_F = 5T_C$, or (3) $T_F = T_C$? Why? (b) Find the temperature.

12. •• (a) The largest temperature drop recorded in the United States in one day occurred in Browning, Montana, in 1916, when the temperature went from 7 °C to −49 °C. What is the corresponding change on the Fahrenheit scale? (b) On the Moon, the average surface temperature is 127 °C during the days and −183 °C during the nights. What is the corresponding difference on the Fahrenheit scale?

10.3 Gas Laws, Absolute Temperature, and the Kelvin Temperature Scale

13. • Convert the following temperatures to absolute temperatures in kelvins: (a) 0 °C, (b) 100 °C, (c) 20 °C, and (d) −35 °C.

14. • Convert the following temperatures to Celsius: (a) 0 K, (b) 250 K, (c) 273 K, and (d) 325 K.

15. • (a) Derive an equation for converting Fahrenheit temperatures directly to absolute temperatures in kelvins. (b) Which is the lower temperature, 300 °F or 300 K?

16. • When lightning strikes, it can heat the air around it to more than 30 000 K, five times the surface temperature of the Sun. (a) What is this temperature on the Fahrenheit and Celsius scales? (b) The temperature is sometimes reported to be 30 000 °C. Assuming that 30 000 K is correct, what is the percentage error of this Celsius value?

17. • How many moles are in (a) 40 g of water, (b) 245 g of CO_2 (carbon dioxide), (c) 138 g of N_2 (nitrogen), and (d) 56 g of O_2 (oxygen) at STP*?

18. IE • (a) In a constant volume gas thermometer, if the pressure of the gas decreases, has the temperature of the gas (1) increased, (2) decreased, or (3) remained the same? Why? (b) The initial absolute pressure of

a gas is 1000 Pa at room temperature (20 °C). If the pressure increases to 1500 Pa, what is the new Celsius temperature?

19. • If the pressure of an ideal gas is doubled while its absolute temperature is halved, what is the ratio of the final volume to the initial volume?

20. •• Show that 1.00 mol of ideal gas under STP occupies a volume of 0.0224 m³ = 22.4 L.

21. •• What volume is occupied by 160 g of oxygen under a pressure of 2.00 atm and a temperature of 300 K?

22. •• An athlete has a large lung capacity, 7.0 L. Assuming air to be an ideal gas, how many molecules of air are in the athlete's lungs when the air temperature in the lungs is 37 °C under normal atmospheric pressure?

23. •• Is there a temperature that has the same numerical value on the Kelvin and the Fahrenheit scales? Justify your answer.

24. •• A husband buys a helium-filled anniversary balloon for his wife. The balloon has a volume of 3.5 L in the warm store at 74 °F. When he takes it outside, where the temperature is 48 °F, he finds it has shrunk. By how much has the volume decreased?

25. •• An automobile tire is filled to an absolute pressure of 3.0 atm at a temperature of 30 °C. Later it is driven to a place where the temperature is only −20 °C. What is the absolute pressure of the tire at the cold place? (Assume that the air in the tire behaves as an ideal gas at constant volume.)

26. •• On a warm day (92 °F), an air-filled balloon occupies a volume of 0.200 m³ and has an absolute pressure of 20.0 lb/in². If the balloon is cooled to 32 °F in a refrigerator while its pressure is reduced to 14.7 lb/in², what is the volume of the air in the container?

27. •• A steel-belted radial automobile tire is inflated to a gauge pressure of 30.0 lb/in² when the temperature is 61 °F. Later in the day, the temperature rises to 100 °F. Assuming the volume of the tire is constant, what is the tire's pressure at the elevated temperature?

28. •• A scuba diver takes a tank of air on a deep dive. The tank's volume is 10 L and it is completely filled with air at an absolute pressure of 232 atm at the start of the dive. The air temperature at the surface is 94 °F and the diver ends up in deep water at 60 °F. Assuming thermal equilibrium and neglecting air loss, determine the absolute pressure of the air when it is cold.

29. IE •• (a) If the temperature of an ideal gas increases and its volume decreases, will the pressure of the gas (1) increase, (2) remain the same, or (3) decrease? Why? (b) The Kelvin temperature of an ideal gas is doubled and its volume is halved. How is the pressure affected?

30. •• If 2.4 m³ of a gas initially at STP is compressed to 1.6 m³ and its temperature is raised to 30 °C, what is its final pressure?

---

* Note: STP refers to Standard Temperature and Pressure. These values are 273 K and 1 atm.

31. **IE** •• The pressure on a low-density gas in a balloon is kept constant as its temperature is increased. (a) Does the volume of the gas (1) increase, (2) decrease, or (3) remain the same? Why? (b) If the temperature is increased from 10 °C to 40 °C, what is the percentage change in the volume of the gas?

32. ••• A diver releases an air bubble of volume 2.0 cm³ from a depth of 15 m below the surface of a lake, where the temperature is 7.0 °C. What is the volume of the bubble when it reaches just below the surface of the lake, where the temperature is 20 °C?

33. ••• Show that for Kelvin temperatures in the millions, (a) the Kelvin and Celsius temperatures are about the same and (b) both Kelvin and Celsius temperatures are about half the Fahrenheit temperature. (c) For a typical stellar interior temperature of 10 million K, what is the percentage error (compared to exact values) if these approximations are used to estimate the values in Celsius and Fahrenheit?

10.4 Thermal Expansion

34. • A steel beam 10 m long is installed in a structure at 20 °C. What is the beam's change in length when the temperature reaches (a) −25 °C and (b) 45 °C?

35. **IE** • An aluminum tape measure is accurate at 20 °C. (a) If the tape measure is placed in a freezer, would it read (1) high, (2) low, or (3) the same? Why? (b) If the temperature of the freezer is −5.0 °C, what would be the stick's percentage error because of thermal contraction?

36. • Concrete highway slabs are poured in lengths of 5.0 m. How wide should the expansion gaps between the slabs be at a temperature of 20 °C to ensure that there will be no contact between adjacent slabs over a temperature range of −25 °C to 45 °C?

37. • A man's gold wedding ring has an inner diameter of 2.4 cm at 20 °C. If the ring is dropped into boiling water, what will be the change in the inner diameter of the ring?

38. •• A circular steel plate of radius exactly 15.0 cm is cooled from 350 °C to 20 °C. By what percentage does the plate's area decrease?

39. •• What temperature change would cause a 0.20% increase in the volume of a quantity of water that was initially at 20 °C?

40. •• A piece of copper tubing used in plumbing has a length of 60.0 cm and an inner diameter of 1.50 cm at 20 °C. When hot water at 85 °C flows through the tube, what are (a) the tube's new length and (b) the change in its cross-sectional area?

41. •• A pie plate is filled up to the brim with pumpkin pie filling. The pie plate is made of Pyrex and its expansion can be neglected. It is a cylinder with an inside depth of 2.10 cm and an inside diameter of 30.0 cm. The pie is prepared at a room temperature of 68 °F and placed in an oven at 400 °F. When it taken out, 151 cm³ of the pie filling has flowed out and over the rim. Determine the coefficient of volume expansion of the pie filling, assuming it is a fluid.

42. **IE** •• A circular piece is cut from an aluminum sheet at room temperature. (a) When the sheet is then placed in an oven, will the hole (1) get larger, (2) get smaller, or (3) remain the same? Why? (b) If the diameter of the hole is 8.00 cm at 20 °C and the temperature of the oven is 150 °C, what will be the new area of the hole?

43. •• One morning, an employee at a rental car company fills a car's steel gas tank to the top and then parks the car a short distance away. (a) That afternoon, when the temperature increases, will any gas overflow? Why? (b) If the temperatures in the morning and afternoon are, respectively, 10 °C and 30 °C and the gas tank can hold 25 gal in the morning, how much gas will be lost? (Neglect the expansion of the tank.)

44. •• A copper block has an internal spherical cavity with a 10-cm diameter (▼ **Figure 10.14**). The block is heated in an oven from 20 °C to 500 K. (a) Does the cavity get larger or smaller? (b) What is the change in the cavity's volume?

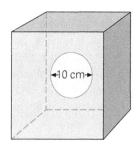

▲ FIGURE 10.14 **A hole in a block** See Exercise 44.

45. ••• A brass rod has a circular cross-section of radius 5.00 cm. The rod fits into a circular hole in a copper sheet with a clearance of 0.010 mm completely around it when both the rod and the sheet are at 20 °C. (a) At what temperature will the clearance be zero? (b) Would such a tight fit be possible if the sheet were brass and the rod were copper?

46. ••• An aluminum rod is measured with a steel tape at 20 °C, and the length of the rod is found to be 75 cm. What length will the tape indicate when both the rod and the tape are at (a) −10 °C? (b) 50 °C? [*Hint:* Both the rod and tape will either expand or shrink as temperature changes. Keep as many significant figures as needed to express the answer.]

10.5 The Kinetic Theory of Gases

47. • If the average kinetic energy per molecule of a monatomic gas sample is 7.0×10^{-21} J, what is the Celsius temperature of the gas?

48. • What is the average kinetic energy per molecule in a monatomic gas sample at (a) 10 °C and (b) 90 °C?

49. IE • If the Celsius temperature of a monatomic gas is doubled, (a) will the thermal energy of the gas (1) double, (2) increase by less than a factor of 2, (3) be half as much, or (4) decrease by less than a factor of 2? Why? (b) If the temperature is raised from 20 °C to 40 °C, what is the ratio of the final thermal energy to initial thermal energy?

50. • What is the rms speed of the molecules in an oxygen sample at 0 °C?

51. • (a) What is the average kinetic energy per molecule of a monatomic gas sample at 25 °C? (b) What is the rms speed of the molecules if the gas is helium?

52. •• (a) Estimate the total translational kinetic energy in a classroom at normal room temperature. Assume the room measures 4.00 m by 10.0 m by 3.00 m. (b) If this energy were all harnessed, how high would it be able to lift an elephant with a mass of 1200 kg?

53. •• A quantity of an ideal gas is at 0 °C. An equal quantity of another ideal gas is at twice the absolute temperature. What is its Celsius temperature?

54. IE •• A sample of oxygen and another sample of nitrogen are at the same temperature. (a) The rms speed of the nitrogen sample is (1) greater than, (2) the same as, or (3) less than the rms speed of the oxygen sample. Explain. (b) Calculate the ratio of the rms speed in the nitrogen sample to in the oxygen sample.

55. •• If 2.0 mol of oxygen is confined in a 10-L bottle under a pressure of 6.0 atm, what is the average kinetic energy of an oxygen molecule?

56. •• If the temperature of an ideal gas increases from 300 to 600 K, what happens to the rms speed of the gas molecules?

57. •• If the temperature of an ideal gas is raised from 25 °C to 100 °C, how much faster is the new rms speed of the gas molecules?

58. •• If the rms speed of the molecules in an ideal gas at 20 °C increases by a factor of 2, what is the new Celsius temperature?

59. IE ••• During the race to develop the atomic bomb in World War II, it was necessary to separate the lighter "fissionable" isotope of uranium (U-235) from the more massive "non-fissionable" isotope (U-238). The uranium was first converted into a gas, uranium hexafluoride (UF_6), and the two uranium isotopes were separated by a process called gaseous diffusion which relied on the *difference* in their rms speeds. (a) As a two-component molecular mixture at room temperature, which of the two types of molecules would be moving faster, on average: (1) $^{235}UF_6$ or (2) $^{238}UF_6$. Or (3) would they move equally fast? Explain. (b) Determine the ratio of their rms speeds, light molecule to heavy molecule. Treat the molecules as ideal gases and neglect rotations and/or vibrations of the molecules. The masses of the three atoms in atomic mass units are 238 and 235 for the two uranium isotopes and 19 for fluorine.

10.6 Kinetic Theory, Diatomic Gases, and the Equipartition Theorem (Optional)

60. • At 30 °C what are the thermal energies of 1.00 mol of helium and 1 mol of oxygen gas?

61. • If 1.0 mol of a monatomic gas has a thermal energy of 5.0×10^3 J at a certain temperature, what is the thermal energy of 1.0 mol of a diatomic gas at the same temperature?

62. •• For an average molecule of nitrogen at 10 °C, what are its (a) translational kinetic energy, (b) rotational kinetic energy, and (c) total energy? Repeat for helium gas at the same temperature.

63. •• A diatomic gas sample has a certain thermal energy at 25 °C. If a monatomic gas sample with the same number of molecules is to have the same thermal energy, what is the Celsius temperature of the monatomic gas?

11

Heat

A hot athlete cools off by transferring energy from his warm skin to cool water.

Heat exchanges are crucial to our existence. Our bodies must balance heat loss and gain to stay within a narrow temperature range necessary for life – a delicate thermal balance from which deviations can have serious consequences. Our bodies convert food energy (chemical potential) to mechanical work; but this process is not perfect – it converts less than 20%, depending on which muscles are doing the work. The rest becomes the heat energy that can be transferred to the environment via various mechanisms. After intense exercise, a particularly efficient mechanism is perspiring, and/or the evaporation of water. The triumphant athlete in the above photo is encouraging thermal energy loss from his warm body by applying cool water on his warm skin. Heat transfer from his skin to the water will warm the water and eventually cause evaporation, thus cooling his body.

On a larger scale, heat exchanges are important to our planet's ecosystem. The average temperature of the Earth, critical to the environment and the survival of the organisms that inhabit it, is maintained through a heat exchange balance. Each day, a vast quantity of solar energy reaches our atmosphere and surface. Changes in this balance could have serious consequences. For example, scientists are concerned that a buildup of atmospheric "greenhouse" gases, a product of our industrial society, could significantly raise the Earth's average temperature, which would undoubtedly have a negative effect on life on Earth as we know it.

On a more practical level, care must be taken when handling anything that has recently been in contact with a source of heat. Yet while the copper bottom of a steel pot on a stove can be very hot, the steel pot handle might only be warm. Why the difference? The answer has to do with thermal conduction, one of the several heat transfer mechanisms covered in this chapter. In addition, the definition of heat and how it is measured will be covered. Put together, these concepts can lead to the understanding of many phenomena such as the conversion of thermal energy into mechanical work – thermal engines – covered in Chapter 12.

11.1 Definition and Units of Heat

Like work, **heat** is a way to *transfer* energy. In the 1800s, it was thought that heat described the *amount* of energy an object possessed, but this is not true. Rather, heat is the name used to describe an *internal energy transfer*. "Heat," or "heat energy," is internal energy added to, or removed from, an object due to temperature differences.

Heat then is internal energy *in transit*, and is measured in the standard SI unit, the joule (J). However, there are other nonstandard, commonly used units for heat. An important one is the *kilocalorie* (kcal):

One kilocalorie (kcal) is the amount of heat needed to raise the temperature of 1 kg of water by 1 °C.

The *calorie* (cal) is also used (1 kcal = 1000 cal):

One calorie (cal) is the amount of heat needed to raise the temperature of 1 g of water by 1 °C.

A familiar use of the kilocalorie is for specifying the energy values of foods. In this context, the word is shortened to *Calorie* (Cal). Thus people on diets are really counting kilocalories (Calories), sometimes called dietary calories. This quantity refers to the food energy that is available for mechanical movement, maintaining body temperature, and/or increasing body mass. The capital C is used to distinguish the larger kilocalorie from the smaller calorie. In many countries, but not the United States, the joule is used for food energy values.

A unit of heat sometimes used in industry is the *British thermal unit* (Btu):

One Btu is the amount of heat needed to raise the temperature of 1 lb of water by 1 °F.

The conversion factor is 1 Btu = 252 cal = 0.252 kcal. If you buy an air conditioner or an electric heater, you will find it is rated in Btu per hour – in other words, a power rating. For example, window air conditioners range from 4000 to 25 000 Btu/h. These specify the rate at which the air conditioner can transfer thermal energy, say, out of a warm room.

11.1.1 The Mechanical Equivalent of Heat

The idea that heat is actually a transfer of energy is the result of work by many scientists. Early observations were made by an American, Benjamin Thompson (known as Count Rumford, 1753−1814), while he was supervising the boring of cannon barrels in Germany. Rumford noticed that water put into the bore of a cannon (to prevent overheating during drilling) boiled away and had to be replenished. The theory of heat at that time depicted it as a "caloric fluid," which flowed from hot objects to colder ones. Rumford did several experiments to try to detect "caloric fluid" by measuring changes in the weights of heated substances. No weight change was detected, thus he concluded that the mechanical work done by friction was actually responsible for the heating of the water.

This conclusion was proven quantitatively by the English scientist James Joule (1818–1889) after whom the unit of energy is named. Using the apparatus illustrated in ▼ **Figure 11.1**, Joule demonstrated that when mechanical work was done on the water, the water exhibited an increase in temperature (and thus thermal energy). He determined that there were actually *two ways* to increase the internal energy of the water (and all objects); that is, do mechanical work on them or transfer internal energy (heat) to them. For every 4186 J of work done on the water, its temperature rose 1 °C per kg. Thus, 4186 J was equivalent to 1 kcal:

$$1 \, \text{kcal} = 4186 \, \text{J} = 4.186 \, \text{kJ} \quad \text{or} \quad 1 \, \text{cal} = 4.186 \, \text{J}$$

These conversion factors are called the **mechanical equivalent of heat**. Example 11.1 illustrates an everyday use of this important discovery.

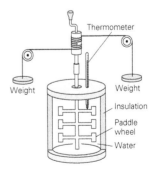

▲ FIGURE 11.1 **Joule's apparatus for determining the mechanical equivalent of heat** As the weights descend, the paddle wheels churn the water, and the mechanical energy, or work, is converted into internal energy, thus raising the water temperature. For every 4186 J of work done, the temperature of the water rose 1 °C per kilogram. Thus, 4186 J is equivalent to 1 kcal.

EXAMPLE 11.1: WORKING OFF THAT BIRTHDAY CAKE – MECHANICAL EQUIVALENT OF HEAT TO THE RESCUE

At a birthday party, a student eats a piece of cake (food energy value of 200 Cal). To prevent this energy from being stored as fat, she takes a stationary bicycle workout class right after the party. During this exercise, she does work at a rate of 200 watts. How long must the student bicycle to achieve her goal of "working off" the cake's energy at this rate (assumed constant)?

THINKING IT THROUGH. Recall that power is the rate at which work is done, and 1 W = 1 J/s. To find the time to do this work, the food energy content is expressed in joules and the definition of power, power = work/time, used.

SOLUTION

The work required to "burn off" the chemical energy content of the cake is 200 Cal. Listing the data and converting to SI units (remember that Cal means kcal):

Given:

$$W = (200\,\text{kcal})\left(\frac{4186\,\text{J}}{\text{kcal}}\right) = 8.37 \times 10^5\,\text{J}$$
$$P = 200\,\text{W} = 200\,\text{J/s}$$

Find: Δt (time to "burn off" 200 Cal)

Rearranging the definition of power ($P = W/\Delta t$) to find the time required:

$$\Delta t = \frac{W}{P} = \frac{8.37 \times 10^5\,\text{J}}{200\,\text{J/s}} = 4.19 \times 10^3\,\text{s} = 69.8\,\text{min} = 1.16\,\text{h}$$

FOLLOW-UP EXERCISE. In this Example, if the 200 Cal were to be worked off instead by doing vertical lifts of 1.0 m with a 40-kg barbell, how long would it take, assuming one lift could be done every second? Assume the relevant work is done only during the lift stage.

11.2 Specific Heat and Calorimetry

11.2.1 Specific Heats of Solids and Liquids

When heat is added to a solid or liquid, the energy can result in a *phase change without an associated temperature change*, such as ice melting into liquid water at a constant 0 °C. Phase changes such as these will be discussed later in this chapter. In this section, only heat transfers that result in temperature changes will be considered. Since different substances have different molecular configurations, if equal amounts of heat are added to equal masses of different substances, the resulting temperature changes will *not* generally be the same.

To characterize the thermal property of a substance that depends only on the type of material it is composed of, the concept of *specific heat capacity,* or **specific heat (*c*)** was created. It is defined as the heat required to raise (or lower) the temperature of 1 kg of a substance by 1 °C. If heat Q is added to, or removed from, a substance of mass m resulting in a temperature change of ΔT, then the specific heat c of that material is

$$c = \frac{Q}{m\Delta T} \quad \text{(definition of specific heat)} \qquad (11.1)$$

SI units of specific heat are J/(kg·°C) or J/(kg·K)
(since 1 K = 1 °C).

Specific heat is a characteristic of the substance and *not* the amount of it. The specific heats of some common substances are listed in ▼ **Table 11.1**. Specific heats can vary slightly with temperature, but they will be considered constant for our purposes. To find the heat (Q) required to change the temperature of an object by ΔT, Equation 11.1 can be written:

$$Q = cm\Delta T \qquad (11.2)$$

TABLE 11.1 Specific Heats of Various Substances (Solids and Liquids) at 20 °C and 1 atm

Substance	Specific Heat (*c*) (J/(kg·°C))
Solids	
Aluminum	920
Copper	390
Glass	840
Ice (−10 °C)	2100
Iron or steel	460
Lead	130
Soil (average)	1050
Wood (average)	1680
Human body (average)	3500
Liquids	
Ethyl alcohol	2450
Glycerin	2410
Mercury	139
Water (15 °C)	4186
Gas	
Steam (100 °C)	2000

The larger the specific heat of a substance, the more heat must be transferred to or taken from it (per kilogram) to change its temperature by a certain amount. That is, for a given mass, a substance with a higher specific heat requires more heat for a given temperature change than one with a lower specific heat. Table 11.1 shows that metals have specific heats considerably lower than that of water. Thus it takes only a small amount of heat to produce a relatively large temperature increase in a metal object, compared to the same mass of water.

You have been the victim of the high specific heat of water if you have ever burned your mouth on a baked potato or the hot cheese on a pizza. These foods have high water content. Due to water's high specific heat, when hot cheese, for example, contacts your tongue, a relatively large amount of heat is transferred to bring the cheese temperature down. The large specific heat of water is also responsible for the mild climate of places near large bodies of water. (See Section 11.4 for more details.)

Note from Equation 11.2 when there is a temperature increase, and therefore ΔT is positive ($T_f > T_i$), then the sign of Q is positive. This corresponds to energy being *added* to a system. Conversely, ΔT and Q are negative when energy is *removed from* a system. This sign convention will be used throughout this book.

EXAMPLE 11.2: BIRTHDAY CAKE REVISITED – HEAT FOR A WARM BATH?

At the birthday party in Example 11.1, a student ate a piece of cake (200 Cal). To get an idea of the magnitude of the amount of (chemical or food) energy in that cake, the student wants to know how much water at 20 °C could be brought to 45 °C, assuming all of the 200 Cal could be converted to thermal energy and transferred to the water. Can you help her out?

THINKING IT THROUGH. The thermal energy from the cake is used to heat water from 20 °C to 45 °C. Using the mechanical equivalent of heat and Equation 11.1, the mass of water can be found.

SOLUTION

Listing the data given and converting to SI units (remember that Cal means kcal):

Given:

$$Q = (200\ \text{kcal})\left(\frac{4186\ \text{J}}{\text{kcal}}\right) = 8.37 \times 10^5\ \text{J}$$
$$T_i = 20\ °C,\ T_f = 45\ °C$$
$$c_w = 4186\ \text{J/(kg·°C)}\ \text{(from Table 11.1)}$$

Find: m (mass of water)

Solving Equation 11.1 for m gives

$$m = \frac{Q}{c\Delta T} = \frac{8.37 \times 10^5\ \text{J}}{[4186\ \text{J/(kg·°C)}](45\ °C - 20\ °C)} = 8.00\ \text{kg}$$

FOLLOW-UP EXERCISE. In this Example, how would the answer change if the water was initially at a temperature of 5 °C rather than 20 °C?

INTEGRATED EXAMPLE 11.3: COOKING CLASS 101 – STUDYING SPECIFIC HEATS WHILE LEARNING HOW TO BOIL WATER

To prepare pasta, you bring a pot of water from room temperature (20 °C) to its boiling point (100 °C). The pot itself has a mass of 0.900 kg, is made of steel, and holds 3.00 kg of water. (a) Which of the following is true: (1) the pot requires more heat than the water, (2) the water requires more heat than the pot, or (3) they require the same amount of heat? (b) Determine the required heat for both the water and the pot, and the ratio Q_w/Q_{pot}.

(a) **CONCEPTUAL REASONING.** The temperature increase is the same for the water and the pot. Thus, the required heat is determined by the product of mass and specific heat. There is 3.00 kg of water to heat. This is more than three times the mass of the pot. From Table 11.1, the specific heat of water

is about nine times larger than that of steel. Both factors together indicate that the water will require significantly more heat than the pot, so the answer is (2).

(b) **QUANTITATIVE REASONING AND SOLUTION.** The heat needed can be found using Equation 11.1, after looking up the specific heats in Table 11.1. The temperature change is easily determined from the initial and final values.

Listing the data given:

Given:

$$m_{pot} = 0.900\ \text{kg}$$
$$m_w = 3.00\ \text{kg}$$
$$c_{pot} = 460\ \text{J/(kg·°C)}\ \text{(from Table 11.1)}$$
$$c_w = 4186\ \text{J/(kg·°C)}\ \text{(from Table 11.1)}$$
$$\Delta T = T_f - T_i = 100\ °C - 20\ °C = 80\ °C\ \text{(both)}$$

Find: Q_w, Q_{pot} and Q_w/Q_{pot} (the heat for the water, the pot, and their ratio)

From Equation 11.2, the heat for the water is

$$Q_w = c_w m_w \Delta T_w$$
$$= [4186\ \text{J/(kg·°C)}](3.00\ \text{kg})(80\ °C) = 1.00 \times 10^6\ \text{J}$$

and for the pot it is

$$Q_{pot} = c_{pot} m_{pot} \Delta T_{pot}$$
$$= [460\ \text{J/(kg·°C)}](0.900\ \text{kg})(80\ °C) = 3.31 \times 10^4\ \text{J}$$

Therefore,

$$\frac{Q_w}{Q_{pot}} = \frac{1.00 \times 10^6\ \text{J}}{3.31 \times 10^4\ \text{J}} = 30.2$$

Hence the water requires more than thirty times the heat required for the pot, because it has both more mass and a greater specific heat. Notice also that both heats are positive, since both temperature changes are positive, indicating heat energy flow *into* the pot and water.

FOLLOW-UP EXERCISE. (a) In this Example, if the pot were the same mass as the steel pot but instead made of aluminum, would the heat ratio (water to pot) be smaller or larger than the answer for the steel pot? Explain. (b) Verify your answer.

Calorimetry*

Calorimetry is a technique that uses heat exchanges coupled with energy conservation to determine thermal properties of substances. The measurements are made in a *calorimeter* that typically consists of an insulated container (with a thermometer) designed to prevent significant heat exchange with the environment (▶ Figure 11.2). Our discussions will assume there are no such exchanges.

---

* In this section, calorimetry will not involve phase changes, such as ice melting or water boiling. These effects are discussed in Section 11.3.

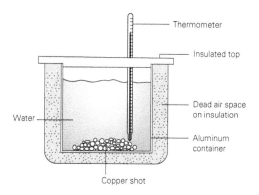

▲ FIGURE 11.2   Sketch of a calorimetry apparatus (Example 11.4) The aluminum cup fits inside a larger container which serves to create an insulation barrier to the environment. The larger container does not participate in the heat exchanges. The insulating cover prevents energy losses via evaporation and a thermometer is inserted through it. The copper shot is typically preheated in boiling water at 100 °C.

Calorimetry is frequently used to determine unknown specific heats. Typically, a substance of known mass and temperature is put into a quantity of water in a calorimeter. The water is at a different (known) temperature than the substance. After determining the final equilibrium temperature of the system, energy conservation is applied to determine the substance's specific heat, c. This procedure is called the *method of mixtures* and is illustrated in Example 11.4. Since there are no energy exchanges with the environment, the algebraic sum of all the heat losses ($Q < 0$) and gains ($Q > 0$) must add up to zero, or $\sum_i Q_i = 0$. In writing this sum, care must be taken to make sure each of the terms has the correct sign.

EXAMPLE 11.4: CALORIMETRY USING THE METHOD OF MIXTURES

Students in a physics lab are to determine the specific heat of copper experimentally. They place 0.150 kg of copper shot into boiling water and let it stay for a while, so as to reach a temperature of 100 °C. Then they carefully pour the hot shot into a calorimeter cup (see Figure 11.2) containing 0.200 kg of water at 20.0 °C. The final temperature of the mixture in the cup is measured to be 25.0 °C. If the aluminum cup has a mass of 0.0450 kg, what is the specific heat of copper? (Assume that there is no heat exchange with the surroundings.)

THINKING IT THROUGH. The conservation of heat energy is involved, thus $\sum_i Q_i = 0$ should be the starting point, making sure each term has the correct signs, as well as identifying each with the proper subscript.

SOLUTION

Here the subscripts Cu, w, and Al are used to refer to the copper, water, and aluminum calorimeter cup, respectively. The

subscripts h, i, and f refer to the temperature of the hot metal shot, the water (and cup) initially at room temperature, and the final temperature of the system, respectively. Masses must be expressed in kilograms.

Given:

$$m_{\mathrm{Cu}} = 0.150 \text{ kg}$$
$$m_{\mathrm{w}} = 0.200 \text{ kg}$$
$$c_{\mathrm{w}} = 4186 \text{ J/(kg·°C)} \text{ (from Table 11.1)}$$
$$m_{\mathrm{Al}} = 0.0450 \text{ kg}$$
$$c_{\mathrm{Al}} = 920 \text{ J/(kg·°C)} \text{ (from Table 11.1)}$$
$$T_{\mathrm{h}} = 100 \text{ °C (initial temperature of Cu shot)}$$
$$T_{\mathrm{i}} = 20.0 \text{ °C (initial temperature of water and aluminum)}$$
$$T_{\mathrm{f}} = 25.0 \text{ °C (system's final temperature)}$$

Find: c_{Cu} (specific heat)

Assuming no heat exchange with the surroundings, then $\sum_i Q_i = 0$. Equation 11.2 can be used to determine the heat exchanges of the water and aluminum since both have a temperature change of +5.0 °C:

$$Q_{\mathrm{w}} = c_{\mathrm{w}} m_{\mathrm{w}} \Delta T_{\mathrm{w}} = [4186 \text{ J/(kg·°C)}](0.200 \text{ kg})(+5.0\text{°C})$$
$$= +4.19 \times 10^3 \text{ J}$$

$$Q_{\mathrm{Al}} = c_{\mathrm{Al}} m_{\mathrm{Al}} \Delta T_{\mathrm{Al}} = [920 \text{ J/(kg·°C)}](0.045 \text{ kg})(+5.0\text{°C})$$
$$= +2.07 \times 10^2 \text{ J}$$

Writing out the sum gives:

$$\sum_i Q_i = Q_{\mathrm{w}} + Q_{\mathrm{Al}} + Q_{\mathrm{Cu}} = +4.19 \times 10^3 \text{ J} + 2.07 \times 10^2 \text{ J} + Q_{\mathrm{Cu}} = 0$$

Thus $Q_{\mathrm{Cu}} = c_{\mathrm{Cu}} m_{\mathrm{Cu}} \Delta T_{\mathrm{Cu}} = -4.40 \times 10^3 \text{ J}$ and   Equation 11.2 can be employed to determine the only unknown, copper's specific heat, since the mass and temperature change (-75 °C) of copper are known:

$$c_{\mathrm{Cu}} = \frac{Q_{\mathrm{Cu}}}{m_{\mathrm{Cu}} \Delta T_{\mathrm{Cu}}} = \frac{-4.40 \times 10^3 \text{ J}}{(0.150 \text{ kg})(-75.0\text{°C})}$$
$$= 391 \text{ J/(kg·°C)}$$

Notice that the proper use of signs resulted in no sign for the specific heat, as required. If, for example, the copper heat term had a positive (incorrect) sign, the answer would have been negative – a clue that there is likely a sign error in one or more of the heat terms in the sum.

FOLLOW-UP EXERCISE. In this Example, use the now known specific heat of copper to determine the equilibrium temperature if the calorimeter (water and cup) initially were at a temperature of 30 °C.

11.2.2 Specific Heat of Gases

When heat is added to or removed from most materials, they expand or contract. During expansion, for example, the materials do work on the environment. For most solids and liquids, this work is negligible, because the volume changes are very small. This is why this effect wasn't included in our discussion of specific heat of solids and liquids.

However, for gases, expansion and contraction *can* be significant. It is therefore important to specify the *conditions* under which heat is transferred when referring to a gas. If heat is added to a gas at constant volume (a *rigid* container), the gas does no work on the environment because although it exerts a force on the container walls, they do not move. In this case, all the heat goes into increasing the gas's thermal energy, resulting in a temperature increase. However, if the same amount of heat is added and *expansion allowed*, then a portion of that heat is converted to work done by the gas (exerting a force *and* moving the container walls). Since less of the heat goes into the gas's thermal energy, a constant pressure process results in a *smaller* temperature change. Since specific heat is inversely related to temperature change (Equation 11.1), the specific heat for a gas at constant pressure is larger than at constant volume.

To designate the physical quantities held constant while heat is added to or removed from a gas, a subscript notation is used: c_p designates the specific heat value at constant pressure (p), and c_v that value at constant volume (v). Note that specific heat for water vapor (H_2O) in Table 11.1 is at constant pressure. Specific heats of gases play an important role in adiabatic thermodynamic processes (Section 12.2.7).

11.3 Phase Changes and Latent Heat

Matter normally exists in one of three normal **phases**: solid, liquid, or gas (▼ **Figure 11.3**). The phase that a substance is in depends, in large part, on its temperature and the pressure on it. Most commonly, adding or removing heat from a substance is the way to change its phase.

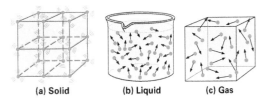

(a) Solid   **(b) Liquid**   **(c) Gas**

▲ FIGURE 11.3 **Three phases of matter (a)** The molecules of a solid are held together by bonds; consequently, a solid has a definite shape and volume. **(b)** The molecules of a liquid can move more freely, so a liquid has a definite volume and assumes the shape of its container. **(c)** The molecules of a gas interact weakly and are separated by relatively large distances; thus, a gas has no definite shape or volume, unless it is confined in a container.

In the solid phase, molecules are held together by attractive forces, or bonds (Figure 11.3a). Adding heat causes increased molecular motion and if enough heat is added to provide sufficient energy to break the intermolecular bonds, most solids will undergo a phase change and become liquids. The temperature at which this type of phase change occurs is the *melting point*. The temperature at which a liquid becomes a solid is the *freezing point*. In general, these temperatures are the same for any given substance, but they can differ slightly.

In the liquid phase, molecules of a substance are relatively free to move, and the liquid assumes the shape of its container (Figure 11.3b). Adding heat increases the energy of the molecules of a liquid. When they have enough energy to become separated, the liquid changes to its gaseous (vapor) phase. This change may occur slowly by **evaporation**, or rapidly, at a temperature called the *boiling point*. The reverse of this phase change is called **condensation**. The temperature at which a gas condenses into a liquid is the *condensation point*.

Some solids, such as dry ice (solid carbon dioxide), mothballs, and certain air fresheners, can change directly from the solid to the gaseous phase. This process is called **sublimation**. Like evaporation, the rate of sublimation increases with temperature. The reverse phase change, from gas to solid, is **deposition**. Frost, for example, is solidified water vapor (gas) deposited on grass, car windows, and other objects. Frost is *not* frozen dew (liquid water), as is sometimes mistakenly assumed.

11.3.1 Heat of Transformation

Usually when heat is transferred to a substance its temperature increases. However, while heat is added (or removed) during a phase change, the temperature of the substance does *not* change. For example, if heat is added to a quantity of ice at −10 °C, the temperature of the ice increases until it reaches its melting point of 0 °C. At this point, the addition of more heat does *not* increase the ice's temperature, but causes it to melt, or change phase. (The heat must be added slowly so that the ice and melted water remain in thermal equilibrium; otherwise, the cold melted water *can* warm above 0 °C even though the ice remains at 0 °C.) Only after the ice is completely melted does adding more heat cause the temperature of the water to rise.

A similar situation occurs during the liquid–gas phase change at the boiling point. Adding more heat to boiling water only causes more vaporization. A temperature increase occurs only *after* the water is completely boiled, resulting in *superheated steam*. Keep in mind that ice can be colder than 0 °C and steam can be hotter than 100 °C.

The heat required for a phase change is called the **heat of transformation*** (L), which is defined as the magnitude of the heat to cause the change per unit mass:

$$L = \frac{|Q|}{m} \quad \text{(latent heat)} \tag{11.3}$$

(SI unit of latent heat is J/kg.)

---

* Originally heat of transformation was called *latent heat* (Latin *latent* for hidden). Early investigators knew heat exchanges were occurring but did not observe temperature changes, thus the heat was "hidden" to them.

The heat of transformation for a solid–liquid phase change is called the **heat of fusion** (L_f), and for a liquid–gas phase change it is called the **heat of vaporization** (L_v). The subscript denotes the phase change of interest. (As noted before, these are sometimes called *latent heats*, as in latent heat of fusion, for example, hence the use of the symbol L.) The heats of transformation of some substances, along with their melting and boiling points, are in ▼ **Table 11.2**. (*Heats of sublimation* are symbolized as L_s and are not shown.) When the phase change results in a more ordered phase, such as making solid ice from liquid water, the heat of transformation represents the amount of energy (per kg) that must be *removed*. The same wording, heat of fusion, is used to describe the reverse process in which heat is *added*, such as during the creation of liquid water from solid ice. However, when liquid forms from water vapor, the energy *removed* is referred to as the **heat of condensation**.

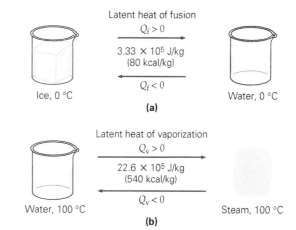

▲ **FIGURE 11.4 Phase changes and latent heats (a)** At 0 °C, 3.33×10^5 J must be added to 1 kg of ice or removed from 1 kg of liquid water to change its phase. **(b)** At 100 °C, 22.6×10^5 J must be added to 1 kg of liquid water or removed from 1 kg of steam to change its phase.

TABLE 11.2 Temperatures of Phase Changes and Latent Heats for Various Substances (at 1 atm)

Substance	Melting Point	L_f (J/kg)	Boiling Point	L_v (J/kg)
Alcohol, ethyl	−114 °C	1.0×10^5	78 °C	8.5×10^5
Gold	1063 °C	0.645×10^5	2660 °C	15.8×10^5
Helium[a]	—	—	−269 °C	0.21×10^5
Lead	328 °C	0.25×10^5	1744 °C	8.67×10^5
Mercury	−39 °C	0.12×10^5	357 °C	2.7×10^5
Nitrogen	−210 °C	0.26×10^5	−196 °C	2.0×10^5
Oxygen	−219 °C	0.14×10^5	−183 °C	2.1×10^5
Tungsten	3410 °C	1.8×10^5	5900 °C	48.2×10^5
Water	0 °C	3.33×10^5	100 °C	22.6×10^5

[a] Not a solid at a pressure of 1 atm; melting point is −272 °C at 26 atm.

When the heat for a phase change is needed, a useful form of Equation 11.3 is obtained by solving for Q (including a ± sign for the two possible directions of heat flow):

$$Q = \pm mL \quad \text{(signs with latent heat)} \quad (11.4)$$

When solving calorimetry problems involving phase changes, one must be careful to *choose* the correct sign for those terms, in agreement with our sign conventions (▶ **Figure 11.4**). For example, if water is condensing from steam into liquid droplets, *removal* of heat is involved, necessitating the choice of the *negative* sign.

11.3.2 Problem-Solving Hint

Recall in Section 11.2 we did not include phase changes. There the expression for heat ($Q = cm\Delta T$) *automatically* gave the

correct sign for Q from the sign of ΔT. But there is no ΔT during a phase change. *Choosing the correct sign is up to you.*

▶ **Figure 11.5** explicitly shows the two types of heat terms (specific heat and heat of transformation) that must be employed in the general situation when temperature changes *and* phase changes are involved. The numbers in the figure are associated with Example 11.5.

EXAMPLE 11.5: FROM COLD ICE TO HOT STEAM

Heat is added to 1.00 kg of cold ice at −10 °C. How much heat is required to change the cold ice to hot steam at 110 °C?

THINKING IT THROUGH. Five steps are involved: (1) heating ice to its melting point (temperature change), (2) melting ice to water at 0 °C (phase change), (3) heating water to its boiling point (temperature change), (4) vaporizing water to steam (water vapor) at 100 °C (phase change), and (5) heating steam (temperature change). (Refer to Figure 11.5.)

SOLUTION

Given:

$$m = 1.00 \text{ kg}$$
$$T_i = -10 \text{ °C}$$
$$T_f = 110 \text{ °C}$$
$$L_f = 3.33 \times 10^5 \text{ J/kg (from Table 11.2)}$$
$$L_v = 22.6 \times 10^5 \text{ J/kg (from Table 11.2)}$$
$$c_{ice} = 2100 \text{ J/(kg·°C) (from Table 11.1)}$$
$$c_{water} = 4186 \text{ J/(kg·°C) (from Table 11.1)}$$
$$c_{steam} = 2000 \text{ J/(kg·°C) (from Table 11.1)}$$

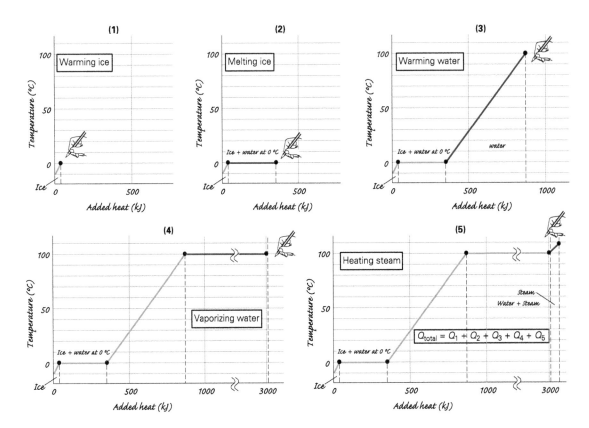

▲ FIGURE 11.5　**From cold ice to superheated steam** See Example 11.5 for details.

1. $Q_1 = c_{ice} m \Delta T_1 = [2100 \text{ J/(kg·°C)}](1.00 \text{ kg})$
$\times [0°C - (-10°C)] = +2.10 \times 10^4 \text{ J}$　(heating ice)

2. $Q_2 = +mL_v = (1.00 \text{ kg})(3.33 \times 10^5 \text{ J/kg})$
$= +3.33 \times 10^5 \text{ J}$　(melting ice)

3. $Q_3 = c_{water} m \Delta T_2 = [4186 \text{ J/(kg·°C)}](1.00 \text{ kg})$
$\times (100°C - 0°C) = +4.19 \times 10^5 \text{ J}$　(heating water)

4. $Q_4 = +mL_v = (1.00 \text{ kg})(22.6 \times 10^5 \text{ J/kg})$
$= +2.26 \times 10^6 \text{ J}$　(vaporizing water)

5. $Q_5 = c_{steam} m \Delta T_3 = [2000 \text{ J/(kg·°C)}](1.00 \text{ kg})$
$\times (110°C - 100°C) = +2.00 \times 10^4 \text{ J}$　(heating steam)

Find:　Q_{total} (total heat required)

The total heat required is thus

$$Q_{total} = \sum_i Q_i = 2.10 \times 10^4 \text{ J} + 3.33 \times 10^5 \text{ J} + 4.19 \times 10^5 \text{ J}$$
$$+ 2.26 \times 10^6 \text{ J} + 2.00 \times 10^4 \text{ J}$$
$$= 3.05 \times 10^6 \text{ J}$$

The heat of vaporization is, by far, the largest. It is actually greater than the sum of the other four terms.

FOLLOW-UP EXERCISE. How much heat must a freezer remove from liquid water (initially at 20 °C) to create 0.250 kg of ice at −10 °C?

11.3.3　Problem-Solving Hint

Note that the heat must be computed at each phase change. It is a common error to use the specific heat with a temperature interval *that includes* a phase change. Also, a complete phase change cannot be assumed until you have checked for it numerically. See Example 11.6.

**EXAMPLE 11.6: PRACTICAL CALORIMETRY –
USING PHASE CHANGES TO SAVE A LIFE**

Organ transplants are becoming commonplace. Many times, the procedure involves removing a healthy organ from a deceased person and flying it to a recipient. To prevent deterioration, the organ is typically packed in ice in an insulated container. Assume a human liver has a mass of 0.500 kg, is initially at 29 °C, and has a specific heat value of 3500 J/(kg·°C). If the liver is initially packed in 2.00 kg of ice at −10 °C, determine the equilibrium temperature.

THINKING IT THROUGH. The liver will cool, and the ice will warm. However, it is not clear what temperature the ice will reach. If it gets to the freezing point, it will begin to melt, and a phase change must be considered. If all of it melts, then additional heat required to warm that water to a temperature above 0 °C must be considered. Thus, care must be taken, since it cannot be assumed that all the ice melts, or even that the ice reaches its melting point. Hence, the calorimetry equation cannot be written down until the terms in it are determined. An initial review of the *possible* heat transfers is needed before the equilibrium temperature can be found.

SOLUTION

Listing the data, the information from the tables, and using the subscript L for liver:

Given:

$m_L = 0.500$ kg
$m_{ice} = 2.00$ kg
$c_L = 3500$ J/(kg·°C)
$c_{ice} = 2100$ J/(kg·°C) (from Table 11.1)
$L_f = 3.33 \times 10^5$ J/kg (from Table 11.2)

Find: T_f (equilibrium temperature)

The heat required to bring the ice from −10 °C to 0 °C is

$$Q_{ice} = c_{ice}m_{ice}\Delta T_{ice} = [2100 \text{ J/(kg·°C)}](2.00 \text{ kg})[0°C - (-10°C)]$$
$$= +4.20 \times 10^4 \text{ J}$$

Since this heat comes from the liver, the *maximum* heat available from the liver needs to be calculated. This would occur if its temperature dropped all the way from 29 °C to 0 °C:

$$Q_{L,max} = c_L m_{L1}\Delta T_{L,max} = [3500 \text{ J/(kg·°C)}](0.500 \text{ kg})[0°C - 29°C]$$
$$= -5.08 \times 10^4 \text{ J}$$

Comparing the two answers, it is seen that this *is* enough heat to bring the ice to 0 °C. But how much ice will melt? Will all of it melt, resulting in an equilibrium temperature above 0 °C, or will not all of it melt, giving us a final temperature of 0 °C?

To check this, let's find the magnitude of the *extra* heat, $|Q'|$, that would be transferred from the liver if its temperature were to drop to 0 °C:

$$|Q'| = |Q_{l,max}| - 4.20 \times 10^4 \text{ J}$$
$$= 5.08 \times 10^4 \text{ J} - 4.20 \times 10^4 \text{ J} = 8.8 \times 10^3 \text{ J}$$

Compare this with the magnitude of the heat needed to melt the ice completely ($|Q_{melt}|$) to decide whether this can, in fact, happen. The heat required to melt *all* the ice is

$$Q_{melt} = +m_{ice}L_{ice} = +(2.00 \text{ kg})(3.33 \times 10^5 \text{ J/kg}) = +6.66 \times 10^5 \text{ J}$$

Since this value is larger than the amount available from the liver, only part of the ice will melt. At equilibrium, the temperature of the liver, the melted water and the remaining ice are all 0 °C, and the remainder of the ice is at 0 °C. Thus heat exchanges stop. The final result is that the liver is in a container with ice and some liquid water, all at 0 °C. Since the container is a good insulator, it should prevent any inward heat flow that might raise the liver's temperature. The liver should arrive in good shape.

FOLLOW-UP EXERCISE. (a) In this Example, how much ice melts? (b) If the ice originally was at its melting point, what would the equilibrium temperature have been?

11.3.4 Problem-Solving Hint

In Example 11.6 the procedure did not jump directly to $\Sigma_i Q_i = 0$. This would be equivalent to assuming all the ice melts. In fact, if this step had been done, we would have been on the wrong track. For calorimetry problems *involving phase changes,* a careful step-by-step numerical "accounting" procedure should be followed.

11.3.5 Evaporation

The **evaporation** of water from an open container becomes evident only after a relatively long period of time. This phenomenon can be explained in terms of the kinetic theory (Section 10.5). The molecules in a liquid are in motion at different speeds. A faster-moving molecule near the surface may momentarily leave the liquid. If its speed is not too large, the molecule will return to the liquid, because of the attractive forces exerted by the other molecules. Occasionally, however, a molecule has a large enough speed to leave the liquid entirely. The higher the temperature of the liquid, the more likely this phenomenon is to occur.

The escaping molecules take their energy with them. Since those molecules with greater-than-average energy are the ones most likely to escape, the average molecular energy, and thus the temperature of the remaining liquid, will be reduced. That is, *evaporation is a cooling process* for the object from which the molecules escape. You have probably noticed this phenomenon when drying off after a bath or shower.

11.4 Heat Transfer

11.4.1 Conduction in Solids

You keep a pot of coffee hot on an electric stove because heat is conducted through the bottom of the pot from the hot metal burner. The process of conduction results from molecular interactions. Molecules in the higher-temperature region of the system (here the burner) are moving, on average, more rapidly. They collide with, and transfer some of their energy to, the less energetic molecules in a nearby cooler part (here the pot bottom). In this way, internal energy is *conductively transferred* on a molecular level from higher-temperature regions to lower-temperature regions.

Our study of thermal conduction will be restricted to solids that can be divided into two general thermal categories: metals and nonmetals. Metals are generally good conductors of heat, or good **thermal conductors**. Materials such as wood and cloth are poor heat conductors sometimes referred to as *thermal insulators*.

The ability of a substance to conduct heat depends on its phase. Gases are poor thermal conductors; their molecules are relatively far apart, and collisions are relatively infrequent. Liquids and solids are better thermal conductors than gases,

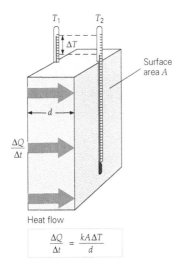

$$\frac{\Delta Q}{\Delta t} = \frac{kA\Delta T}{d}$$

▲ FIGURE 11.6 **Thermal conduction** Heat conduction through a solid slab is measured by the time rate of heat flow ($\Delta Q/\Delta t$ in J/s or watts) that depends directly on the temperature difference across it (ΔT), its cross-sectional area (A) and the material's thermal conductivity (k). The rate is also *inversely* proportional to the slab thickness (d).

because their molecules are closer together and interact more readily.

Heat conduction through a solid material is described using the *time rate of heat flow* ($Q/\Delta t$ in J/s or W) as illustrated in ▲ **Figure 11.6**. Experiment observations show that this rate depends on the temperature difference across the material as well as its size and composition.

Experimentally, the heat flow rate through a slab of material is directly proportional to the material's surface area (A) and the temperature difference across its ends (ΔT), and inversely proportional to its thickness (d). That is,

$$\frac{Q}{\Delta t} \propto \frac{A\Delta T}{d}$$

Using a constant of proportionality k allows us to write this as an equation:

$$\frac{Q}{\Delta t} = \frac{kA\Delta T}{d} \quad \text{(thermal conduction)} \quad (11.5)$$

The constant k, called the **thermal conductivity**, characterizes the heat-conducting ability of a material and depends only on the type of material. The greater the value of k for a material, the better it will conduct heat, all other factors being equal. The SI units of k are J/(m·s·°C) or W/(m·°C). The thermal conductivities of various substances are listed in ▶ **Table 11.3**. Compare the relatively large values of the thermal conductivity of the good thermal conductors (metals), to the relatively small values for the good thermal insulators, such as Styrofoam and wood.

TABLE 11.3 Thermal Conductivities of Some Substances

Substance	Thermal Conductivity (k) J/(m·s·°C) or W/(m·°C)
Metals	
Aluminum	240
Copper	390
Iron	80
Stainless steel	16
Silver	420
Liquids	
Transformer oil	0.18
Water	0.57
Gases	
Air	0.024
Hydrogen	0.17
Oxygen	0.024
Other Materials	
Brick	0.71
Concrete	1.3
Cotton	0.075
Fiberboard	0.059
Floor tile	0.67
Glass (typical)	0.84
Glass wool	0.042
Goose down	0.025
Human tissue (average)	0.20
Ice	2.2
Styrofoam	0.042
Wood, oak	0.15
Wood, pine	0.12
Vacuum	0

As an example of the use of a good conductor, consider that many stainless steel pots are copper clad on the bottom. (▼ **Figure 11.7**). Being a better thermal conductor than steel, copper conducts heat faster to the food as well as promoting the distribution of heat for even cooking.

▲ FIGURE 11.7 **Copper-clad cookware** Copper is used with some stainless steel pots and saucepans. The high thermal conductivity of copper ensures the rapid and even spread of heat from the burner.

EXAMPLE 11.7: THE COST OF HEAT LOSS THROUGH UNINSULATED CEILINGS

A room has a pine ceiling that measures 3.0 m by 5.0 m and is 2.0 cm thick. On a cold evening, the temperature inside the room at ceiling height is 20 °C, and the temperature in the attic is 8.0 °C. Assuming that the temperatures are constant and heat loss is due to conduction only – no insulation in the attic, (a) how much energy (in joules) is conducted through the ceiling overnight (10 h)? (b) Residential electrical bills are typically expressed in units of power multiplied by time $(E = P \times \Delta t)$ called the kilowatt-hour (kwh) where $1 \text{ kwh} = (1.0 \times 10^3 \text{ w})(3.6 \times 10^3 \text{ s}) = 3.6 \times 10^6 \text{ J}$. Assuming this room uses electrical energy for heating, how much does this cost if electrical energy is billed at 15 cents/kwh in this location?

THINKING IT THROUGH. (a) Equation 11.5 is directly applicable here. After the rate is determined, the total heat flow in 10 h can be found. (b) Converting joules into kwh will enable determination of the cost.

SOLUTION

Listing the data, computing some of the quantities in Equation 11.4, and making conversions:

Given:

$A = 3.0 \text{ m} \times 5.0 \text{ m} = 15 \text{ m}^2$
$d = 2.0 \text{ cm} = 0.020 \text{ m}$
$\Delta T = 20 \text{ °C} - 8.0 \text{ °C} = 12 \text{ °C}$
$\Delta t = 10 \text{ h} = 3.6 \times 10^4 \text{ s}$
$k = 0.12 \text{ J/(m·s·°C)}$ (wood, pine from Table 11.3)

Find:

(a) Heat loss in 10 h
(b) Electrical cost

(a) First the heat flow rate is determined (to two significant figures):

$$\frac{Q}{\Delta t} = \frac{kA\Delta T}{d} = \frac{[0.12 \text{ J/(m·s·°C)}](15 \text{ m}^2)(12 \text{ °C})}{0.020 \text{ m}}$$
$$= 1.1 \times 10^3 \text{ J/s}$$

Thus the heat conducted in 10 h is:

$$Q = (1.1 \times 10^3 \text{ J/s})(3.6 \times 10^4 \text{ s}) = 4.0 \times 10^7 \text{ J}$$

(b) Expressed in kwh: $Q = \dfrac{4.0 \times 10^7 \text{ J}}{3.6 \times 10^6 \text{ J/kwh}} = 11 \text{ kwh}$. At 15 cents per kwh, this is a cost of $1.65.

FOLLOW-UP EXERCISE. In this Example, to cut the cost by 25%, at what temperature should the air at the ceiling be kept, assuming the same constant attic temperature?

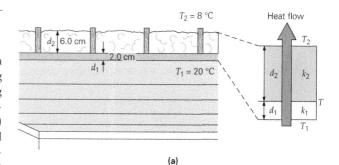

(a)

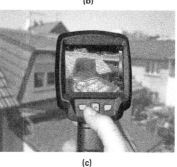

(b)

(c)

▲ **FIGURE 11.8 Insulation and thermal conductivity (a,b)** Attics should be insulated to prevent loss of heat by conduction. See Example 11.8. **(c)** This thermogram allows visualization of a house's heat loss. Here its display shows roof areas with different heat loss rates in various coded colors for easy identification.

EXAMPLE 11.8: THERMAL INSULATION – HELPING PREVENT HEAT LOSS

The homeowner in Example 11.7 decides to cut heating costs by adding a 6.0 cm layer of glass wool insulation above the ceiling (▲ **Figure 11.8**). Assuming the same temperatures and that heat loss is due to conduction only, how much (a) energy, and (b) money does the added layer save in 10 h?

THINKING IT THROUGH. There are two materials, so the heat flow rate for the combination must be found. But at a steady rate, *the heat flow rate must be the same through both.* If this were not true, then thermal energy would pile up at the interface, which is known to not happen. To find the energy saved, the heat loss with the insulation should be compared to that of no insulation (Example 11.7).

SOLUTION

Listing the data, computing some of the quantities in Equation 11.5, and making conversions:

Given:

$A = 3.0 \text{ m} \times 5.0 \text{ m} = 15 \text{ m}^2$
$d_1 = 2.0 \text{ cm} = 0.020 \text{ m}$
$d_2 = 6.0 \text{ cm} = 0.060 \text{ m}$
$\Delta t = 10 \text{ h} = 3.6 \times 10^4 \text{ s}$
$k_1 = 0.12 \text{ J/(m·s·°C)}$ (wood, pine) $\Big\}$ (from Table 11.3)
$k_2 = 0.042 \text{ J/(m·s·°C)}$ (glass wool)

Find:

(a) Energy saved with insulation in 10 h
(b) Money saved with insulation in 10 h

(a) The heat rate through the wood *and* the insulation *combination* must be found. Let T be the temperature at the interface of the materials and T_1 and T_2 be the room and attic temperatures, respectively ($T_1 > T > T_2$ in Figure 11.8a). Writing both temperature differences as positive, the flow rates are:

$$\left(\frac{Q}{\Delta t}\right)_1 = \frac{k_1 A(T_1 - T)}{d_1} \quad \text{and} \quad \left(\frac{Q}{\Delta t}\right)_2 = \frac{k_2 A(T - T_2)}{d_2}$$

T is not known but can be found because the flow rates are the same through both materials:

$$\frac{k_1 A(T_1 - T)}{d_1} = \frac{k_2 A(T - T_2)}{d_2}$$

The As cancel, and solving for T gives

$$T = \frac{k_1 d_2 T_1 + k_2 d_1 T_2}{k_1 d_2 + k_2 d_1}$$
$$= \frac{\left[0.12 \text{ J/(m·s·°C)}\right](0.060 \text{ m})(20°C) + \left[0.042 \text{ J/(m·s·°C)}\right](0.020 \text{ m})(8.0°C)}{\left[0.12 \text{ J/(m·s·°C)}\right](0.060 \text{ m}) + \left[0.042 \text{ J/(m·s·°C)}\right](0.020 \text{ m})}$$
$$= 18.7°C$$

Since the flow rate of both layers is the same, the expression for either can be used to find it. Choosing the wood ceiling, care must be taken to use the correct ΔT. The temperature at the wood-insulation interface is 18.7 °C; thus,

$$\Delta T_{\text{wood}} = |T_1 - T| = |20°C - 18.7°C| = 1.3°C$$

Therefore, the heat flow rate is

$$\left(\frac{Q}{\Delta t}\right)_1 = \frac{k_1 A |\Delta T_{\text{wood}}|}{d_1} = \frac{\left[0.12 \text{ J/(m·s·°C)}\right](15 \text{ m}^2)(1.3°C)}{0.020 \text{ m}}$$
$$= 1.2 \times 10^2 \text{ J/s (or W)}$$

In 10 h, the heat loss would be

$$Q = \left(\frac{Q}{\Delta t}\right)_1 \times \Delta t = (1.2 \times 10^2 \text{ J/s})(3.6 \times 10^4 \text{ s}) = 4.3 \times 10^6 \text{ J}$$

Compared to the uninsulated ceiling in Example 11.7, this represents a decreased heat loss of

$$Q_{\text{saved}} = 4.0 \times 10^7 \text{ J} - 4.3 \times 10^6 \text{ J} = 3.6 \times 10^7 \text{ J}$$

Expressed as a percentage, the insulated heat loss is $\frac{4.3 \times 10^6 \text{ J}}{4.0 \times 10^7 \text{ J}} \times (100\%) = 11\%$ of the uninsulated loss. Clearly this represents a substantial savings to the homeowner.

(b) The difference in heat losses between the two situations, expressed in kwh, is:

$$Q_{\text{saved}} = \frac{3.6 \times 10^7 \text{ J}}{3.6 \times 10^6 \text{ J/kwh}} = 10 \text{ kwh or a savings of } \$1.50$$

FOLLOW-UP EXERCISE. Verify that the heat flow rate through the insulation is the same as that through the wood in this Example.

11.4.2 Convection

In general, compared with solids, liquids and gases are not good thermal conductors. However, the mobility of molecules in fluids permits heat transfer by another process – convection. (The term "fluid" refers to any substance that can flow, and hence includes liquids *and* gases.) **Convection** is heat transfer as a result of mass molecular movement and can be natural or forced.

Natural convection occurs in liquids and gases. For example, when cold water is in contact with a hot object, such as the bottom of a pot on a stove, heat is transferred to the water adjacent to the pot by conduction. Since the water at the bottom is warmer than at the surface, its density is lower, causing it to rise. The top water, being cooler, has a higher density, and thus sinks. This sets up a natural convection flow in which thermal energy is moved.

Such convections can be important in atmospheric processes, as illustrated in ▶ **Figure 11.9**. During the day, the ground heats up more quickly than do large bodies of water, as you may have noticed if you have been to the beach and stepped on burning hot sand. This phenomenon occurs because the water has a higher specific heat than land (recall $\Delta T \propto 1/c$). The air in contact with the warm ground is heated and expands, becoming less dense.

As a result, the warm air rises (vertical air currents or *thermals*) and, to fill the space, other air moves horizontally (onshore winds) – creating a sea breeze near a large body of water. Cooler air descends, and a thermal *convection cycle* is set up, transferring heat away from the land. At night, the

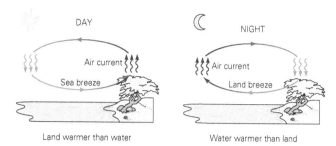

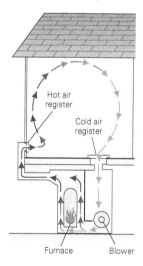

▲ FIGURE 11.9 **Convection cycles** During the day, natural convections give rise to sea breezes near large bodies of water. At night, the pattern of circulation is reversed, and the land breezes blow. See text for detailed explanation.

ground's temperature drops more quickly than the water, and the water surface becomes warmer than the land. As a result, the air current is reversed and an offshore breeze ensues.

In *forced convection*, the fluid is moved mechanically. Common examples of forced convection systems are forced-air heating systems in homes (▼ **Figure 11.10**), the human circulatory system, and the cooling system of an automobile engine. The human body loses a great deal of heat when the surroundings are colder than the body. The body's internal energy is transferred close to the surface of the skin by blood circulation. From there, the heat is transferred to the environment. This circulatory system is highly adjustable and blood flow can be increased or decreased to specific areas depending on the need.

▲ FIGURE 11.10 **Forced convection** Houses are commonly heated by forced convection. Registers or gratings in the floors or walls allow heated air to enter and cooler air to return to the heat source.

In liquid-cooled engines, coolant is circulated (pumped) through the cooling system. The coolant carries internal energy from the engine to the radiator (this is a form of *heat exchanger*), where forced airflow over the radiator, produced by the fan and/or car movement, carries it away. The *radiator* of an automobile is actually misnamed – most of the heat transferred to the environment is by forced convection – not by radiation (discussed in the next section).

CONCEPTUAL EXAMPLE 11.9: FOAM INSULATION – BETTER THAN AIR?

To save energy (and money), insulation is sometimes blown into the space between the inner and outer walls of a house. Since air is a better thermal insulator than foam (Table 11.3), why do you think that foam insulation is used? Is it to prevent heat loss by (a) conduction, (b) convection, or (c) fireproofing?

REASONING AND ANSWER. Foams will generally burn, so (c) isn't likely to be the answer. Air is a poor thermal conductor (alternatively a better thermal insulator), even poorer than foam, so the answer can't be (a). However, as a gas, air is subject to convection *within the wall space*. In the winter, the air near the warm inner wall is heated and rises, thus setting up a convection cycle in the space, and transferring heat to the cold outer wall where it is lost to the outside via conduction. In the summer, with air conditioning on and the house interior kept cool, this cycle is reversed. The use of foam then blocks the movement of air and stops these convection cycles. Hence, the answer is (b).

FOLLOW-UP EXERCISE. Thermal underwear and thermal blankets are loosely knit with lots of small holes. Wouldn't they be more effective if the material were closely knit?

11.4.3 Radiation

Conduction and convection require material as a transport medium. The third mechanism of heat transfer needs no medium; it is called **radiation**, which refers to energy transfer by electromagnetic waves (Section 20.4). Heat is transferred to the Earth from the Sun through empty space by radiation. Visible light and other forms of electromagnetic radiation are commonly referred to as *radiant energy.*

You have experienced heat transfer by radiation if you've ever stood near an open fire (▼ **Figure 11.11**). You can feel energy falling onto your exposed hands and face. This transfer is not due to convection or conduction, since heated air rises and air is a poor conductor. Visible radiation is emitted from the burning material, but most of the heating effect comes from the absorption

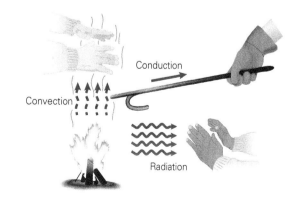

▲ FIGURE 11.11 **Heating by conduction, convection, and radiation** The hands on top of the flame are warmed by the convection of rising hot air (and some radiation). The gloved hand is warmed by conduction. The hands to the right of the flame are warmed by radiation.

of invisible *infrared radiation*. (*Infrared* refers to the fact that its wavelength is longer than that of red light and therefore outside the visible spectrum.)

Infrared radiation is sometimes referred to as *heat* or *thermal radiation*. You may have noticed the reddish infrared lamps used to keep food warm in buffet lines. Heat transfer by infrared radiation is also important in maintaining our planet's warmth by a mechanism known as the *greenhouse effect.*

The greenhouse effect helps regulate the Earth's long-term average temperature, which has been fairly constant for centuries. When a portion of the solar radiation (mostly visible light) reaches and warms the Earth's surface, the Earth, in turn, reradiates energy in the form of infrared radiation (IR). As the reradiated IR radiation passes back through the atmosphere, some of its energy is absorbed by so-called *greenhouse* gases – primarily water vapor, carbon dioxide, and methane – thus warming the atmosphere. These gases are selective absorbers; that is, they absorb radiation at certain IR wavelengths but not at others (▼ **Figure 11.12a**). Without this absorption, the IR radiation would go back into space and life on the Earth would probably not exist. These conditions would result in an average surface temperature of approximately −18 °C, rather than the present value of 15 °C.

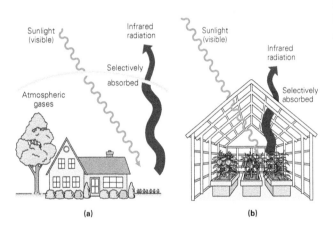

▲ FIGURE 11.12 The greenhouse effect (a) The so-called greenhouse gases in our atmosphere, particularly water vapor, methane, and carbon dioxide, are selective absorbers that have absorption properties similar to those of glass. Visible (solar) light is transmitted and heats the Earth's surface, which emits IR radiation. Much of this re-emitted radiation is absorbed/reflected by the gases and thus its energy is trapped in the Earth's atmosphere. Without these gases, the Earth's surface and atmosphere would be much colder. **(b)** A greenhouse operates in a similar way, with the glass "roof" playing the part of the greenhouse gases.

This atmospheric phenomenon is called the greenhouse effect because the atmosphere serves to function somewhat like the glass "roof" atop a greenhouse. In this case, the visible (solar) radiation is transmitted through the glass and warms the interior materials that in turn reradiate IR radiation. This reradiated radiation is reflected/absorbed by the covering glass, thus trapping the energy resulting in a warm interior even if the outside might be cold (Figure 11.12b). The glass enclosure serves to keeps the warmed air from escaping by eliminating heat losses

by convection. As you probably know, it can be quite warm in a greenhouse on a sunny day, even during a cold winter day. This effect can also make a closed car extremely hot on a sunny day.

The problem on Earth is that human activities (since the beginning of the industrial age) have been adding to the naturally occurring greenhouse gases. With the combustion of hydrocarbon fuels (gas, oil, coal, etc.), vast amounts of carbon dioxide are vented into the atmosphere, where they trap more and more energy. There is strong evidence and wide-ranging scientific agreement that this is leading to an increase in the Earth's average surface temperature – which could lead to large-scale climate change. Such changes could dramatically affect agricultural production and world food supplies, as well as cause melting of the polar ice caps. Climate change could lead to more devastating storms and more severe weather as well as rising sea levels – potentially flooding low-lying regions, thus endangering coastal ports and cities.

As a practical use of radiation, cameras that employ special IR sensors can take pictures or scans, called *thermograms*. These pictures are processed into contrasting colored areas corresponding to different temperatures, such as the skin on the human hand as in ▼ **Figure 11.13**. Industry uses this technique, called *thermography*, to detect product flaws. Modern medicine applies thermography to detect tumors, which are detectable because they are warmer than surrounding healthy flesh and emit slightly different IR wavelengths.

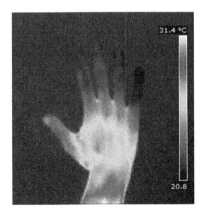

▲ FIGURE 11.13 Applied thermography Thermograms can be used to detect temperature differences around the human body. Here a thermogram of a hand graphically shows skin temperatures using a color-code (colors on bar to right shown as shading here). Cancer can be detected this way because tumor regions are at higher temperatures than normal tissue. See text for more details.

The rate at which an object radiates energy is proportional to the *fourth power* of that object's absolute temperature. This relationship is expressed as **Stefan's law**, developed by the Austrian physicist Josef Stefan (1835–1893):

$$\frac{Q}{\Delta t} = \sigma A e T^4 \quad \text{(radiated rate)} \tag{11.6}$$

[σ (Greek sigma) is the *Stefan-Boltzmann constant*:
$$\sigma = 5.67 \times 10^{-8} \text{ W/(m}^2\cdot\text{K}^4).]$$

The radiated rate is in J/s or in watts. A is the object's surface area in m^2 and T is its absolute temperature in kelvins. The **emissivity** (e) is a unitless number between 0 and 1 that is a characteristic of the material. Dark surfaces have emissivities close to 1, and shiny surfaces have values close to zero. For comparison, the emissivity of warm human skin is almost 1, indicating that it is a very effective radiator of energy. An *ideal*, or *perfect*, absorber (and emitter) is often referred to as a **blackbody** ($e = 1.0$).

Note that dark surfaces are not only better emitters of radiation but are *also* good absorbers. This must be the case, since to maintain constant temperature, the incident energy must equal the emitted energy. Shiny surfaces are poor absorbers, since most of the incident radiation is reflected. This is why it is better to wear light-colored clothes in the summer and dark-colored clothes in the winter.

An object at thermal equilibrium with its surroundings remains at a constant temperature, emitting and absorbing radiation at the same rate. However, if the temperatures of the object and the surroundings (environment) are different, there will be a net flow of radiant energy. If the object's temperature is T and the environment's is T_{env}, then the *net* rate of radiant energy (loss or gain) by the object is the *difference* between these rates:

$$\frac{Q_{net}}{\Delta t} = \sigma A e \left(T_{env}^4 - T^4 \right) \qquad (11.7)$$

This equation is consistent with our heat flow sign conventions. For example, if T_{env} is less than T, then the net rate is *negative*, indicating a net energy *loss*. Last, if $T_{env} = T$ (i.e., no temperature difference), there is an exchange of radiant energy at *equal* rates, but it results in no *net* change to the object's thermal energy.

EXAMPLE 11.10: BODY HEAT – RADIANT HEAT TRANSFER

Assume your skin has an emissivity of 0.90 at 34 °C, and a total area of 1.5 m^2. What is the (net) energy loss per second from your skin if the room temperature is 20 °C?

THINKING IT THROUGH. $Q_{net}/\Delta t$ is found from Equation 11.7. The heat transfers are between the skin and the surroundings. The temperatures *must* be in kelvins.

SOLUTION

Given:

$$T_{env} = (20 + 273)\,K = 293\,K$$
$$T = (34 + 273)\,K = 307\,K$$
$$e = 0.90$$
$$A = 1.5\,m^2$$
$$\sigma = 5.67 \times 10^{-8}\,W/(m^2 \cdot K^4)$$

Find: $Q_{net}/\Delta t$ (net energy loss rate)

From Equation 11.7 directly,

$$\begin{aligned}
\frac{Q_{net}}{\Delta t} &= \sigma A e \left(T_{env}^4 - T^4 \right) \\
&= \left[5.67 \times 10^{-8}\,W/(m^2 \cdot K^4) \right] (1.5\,m^2)(0.90) \times \left[(293\,K)^4 - (307\,K)^4 \right] \\
&= -116\,W \ (or\ -116\,J/s)
\end{aligned}$$

Energy is lost as indicated by the minus sign. That is, the human body loses heat at a rate that is on the order of a 100-W light bulb. No wonder a closed room full of people can get very warm.

FOLLOW-UP EXERCISE. (a) In this Example, suppose the skin had been exposed to an ambient room temperature of only 10 °C. What would the rate of energy loss be? (b) Elephants have huge body masses and large caloric food intakes. Can you explain how their big ear flaps might help stabilize their body temperature? [*Hint:* Think area.]

11.4.4 Problem-Solving Hint

Note that in Example 11.10, the fourth powers of the temperatures were found first, and then their difference found. It is *not* correct to find the temperature difference and then raise it to the fourth power. In other words, $T_{env}^4 - T^4 \neq (T_{env} - T)^4$.

11.4.5 Heat Transfer Mechanisms in Action

There are many real-life examples involving heat transfer using the three mechanisms discussed previously. For example, in the spring, a late frost could kill the buds on fruit trees. To save the buds, some growers spray water on the trees to form ice before a hard (well below 0 °C) frost occurs. Using ice to save buds? Ice is a relatively poor (and inexpensive) thermal conductor, so it has an insulating effect. It will maintain the buds' temperature at 0 °C, not going below that value, and therefore protect the buds, which for the most part do not suffer significant damage until they reach a temperature well below 0 °C.

To protect orchards from freezing on a cold clear (no clouds) night, growers sometimes use smudge pots – containers in which material is burned to create a dense cloud of smoke. At night, when the solar-warmed ground cools off by radiation and there are no atmospheric clouds to trap this energy, the smoke cloud helps by absorbing and reradiating the heat back to the ground, as would happen if there were normal cloud cover. This takes the ground and orchard longer to cool, hopefully without reaching freezing temperatures before the Sun comes up.

A Thermos bottle (▶ **Figure 11.14**) keeps cold beverages cold and hot ones hot. It consists of a double-walled, partially evacuated container with silvered walls (mirrored interior). The bottle is constructed to minimize all three mechanisms of heat transfer. The double-walled and partially evacuated container counteracts conduction and convection because

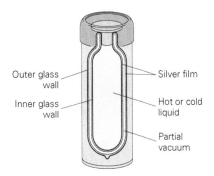

▲ **FIGURE 11.14** **Thermal insulation** The Thermos bottle minimizes all three mechanisms of heat transfer. See text for description.

both processes depend on a medium to transfer the energy (the double walls are more for holding the partially evacuated region than for reducing conduction and convection). The mirrored interior minimizes loss by radiation. The stopper on top of the thermos stops convection off the top of the liquid as well.

Look at ▼ **Figure 11.15**. Why would anyone wear a dark robe in the desert? Dark objects absorb radiation very well. Wouldn't a white robe be better? A dark robe definitely absorbs more radiant energy and warms the air inside near the body. But note that the robe is open at the bottom. The warm air rises and exits at the neck area, and outside cooler air enters the robe at the bottom – providing natural convection air circulation.

▲ FIGURE 11.15   **A dark robe in the desert?** Dark objects absorb more radiation than do lighter ones, and they become hotter. What's going on here? See the text for an explanation.

Finally, consider some of the thermal factors involved in "passive" solar house design used as far back as in ancient China (▶ **Figure 11.16**). The term *passive* means that the design elements require no active use of energy. In Beijing, China, for example, the angles of the sunlight are 76 °, 50 °, and 27 ° above the horizon at the summer solstice, the spring and fall equinoxes, and the winter solstice, respectively. With a proper combination of column height and roof overhang, a maximum amount of

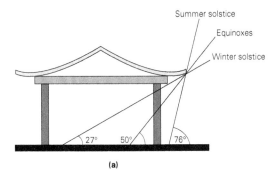

(a)

(b)

▲ **FIGURE 11.16**   **Aspects of passive solar design in ancient China** **(a)** In summer, with the sun angle high, the overhangs provide shade to the building. The brick and mud walls are thick to reduce conductive heat flow to the interior. In winter, the sun angle is low, so the sunlight streams into the building, especially with the help of the upward curved overhangs. **(b)** A photo of such a building in China.

sunlight is allowed into the building in winter, but most of the sunlight will not reach the inside in the summer. The overhangs of the roofs are also curved upward, not just for good looks, but also for letting the maximum amount of light into the building in the winter.

Chapter 11 Review

- **Heat (Q)** is internal energy exchanged between objects at different temperatures.
- **Specific heat (c)** is

$$c = \frac{Q}{m\Delta T} \qquad (11.1)$$

- **Calorimetry** is a technique based on conservation of energy. Assuming a thermally isolated container, the sum of all the heat exchanges must equal zero.
- **Heat of transformation (L)** is the heat required to change the phase of an object *per kilogram*. During a phase change, the temperature of the system does not change.

$$L = \frac{|Q|}{|m|} \quad \text{or} \quad Q = \pm mL \qquad (11.3, 11.4)$$

- Heat transfer due to direct contact of objects at different temperatures is **conduction**. The rate of heat flow by conduction through a slab of material is

$$\frac{Q}{\Delta t} = \frac{kA\Delta T}{d} \tag{11.5}$$

- **Convection** refers to heat transfer due to mass movement of gas or liquid molecules. *Natural convection* is usually driven by density differences caused by temperature differences. In *forced convection*, the movement is driven by mechanical means such as fans.

- **Radiation** refers to heat transferred by electromagnetic radiation between objects at different temperatures such as an object and its environment. The net rate of transfer is

$$\frac{Q_{\text{net}}}{\Delta t} = \sigma Ae\left(T_{\text{env}}^4 - T^4\right) \tag{11.7}$$

where $\sigma = 5.67 \times 10^{-8}$ W/(m²·K⁴).

End of Chapter Questions and Exercises

Multiple Choice Questions

11.1 Definition and Units of Heat
1. The SI unit of heat energy is the (a) calorie, (b) kilocalorie, (c) Btu, (d) joule.
2. Which of the following is the largest unit of heat energy: (a) 1 calorie, (b) 1 Btu, (c) 1 J, or (d) 1 kJ?
3. The mechanical equivalent of heat is (a) 1 kcal = 4.186 J, (b) 1 J = 4.186 cal, (c) 1 cal = 4.186 J, (d) 1 Cal = 4.186 J.

11.2 Specific Heat and Calorimetry
4. The amount of heat necessary to change the temperature of 1 kg of a substance by 1 °C is called its (a) specific heat, (b) heat of combustion, (c) mechanical equivalent of heat.
5. The same amount of heat is added to two objects of the same mass. If object 1 experienced a greater temperature change than object 2, then how do their specific heats compare? (a) $c_1 > c_2$, (b) $c_2 > c_1$, (c) $c_1 = c_2$, (d) can't tell from the data given.
6. The fundamental physical principle used in the method of calorimetry is (a) Newton's second law, (b) conservation of momentum, (c) conservation of energy, (d) equilibrium.
7. For gases, how does their specific heat at constant pressure, c_p, compare to that at constant volume, c_v? (a) $c_p > c_v$, (b) $c_p = c_v$, (c) $c_p < c_v$, or (d) none of the preceding.

11.3 Phase Changes and Heat of Transformation
8. The units of heat of transformation are (a) 1/°C, (b) J/(kg·°C), (c) J/°C, (d) J/kg.
9. Heat of transformation is always (a) part of the specific heat, (b) related to the specific heat, (c) the same as the mechanical equivalent of heat, (d) none of the preceding.

10. Two blocks of ice each have enough heat added to them to completely melt at the melting point. Block A has twice the mass as B. Which has the larger heat of transformation (fusion)? (a) A, (b) B, (c) they are the same, (d) none of the preceding.

11.4 Heat Transfer
11. For the best effect, house insulation materials should have (a) high thermal conductivity, (b) low thermal conductivity, (c) their conductivity values don't matter.
12. Which of the following is the dominant heat transfer mechanism by which the Earth receives energy from the Sun: (a) conduction, (b) convection, (c) radiation, or (d) all of the preceding?
13. Water is a poor heat conductor, but a pot of water can be heated more quickly than you might think. This faster-than-expected heating time is probably due to what additional mechanism? (a) convection, (b) radiation, (c) vaporization.

Conceptual Questions

11.1 Definition and Units of Heat
1. Why would the Btu not be a good unit to use on the Moon? [*Hint:* Look at its definition and gravity.]
2. Discuss the difference between the concepts of internal energy and heat.
3. If someone says that a hot object contains more heat than a cold one, how would you answer? Why?
4. It is possible to increase a gas's temperature by rapidly compressing it (doing work on it). Observing you doing this, a friend says that you "heated up" the gas. Evaluate this statement for correctness.
5. Is it possible for internal energy to be transferred from a cold object to a hotter one? Explain.

11.2 Specific Heat and Calorimetry
6. At a lake, does a kilogram of lake water or a kilogram of lake beach sand experience a larger temperature rise during a hot summer day? Which experiences a bigger temperature drop a cold winter night? Explain.
7. Equal amounts of heat are added to two different objects at the same initial temperature. What factor(s) can cause the final temperature of the two objects to be different?
8. Many people have performed fire walking, in which a bed of red-hot coals is walked on with bare feet. (You should not try this at home!) How is this possible? [*Hint:* Human tissues largely consist of water.]
9. A hot steel ball is dropped into a cold aluminum cup containing some water. (Assume the system is isolated.) If the ball loses 400 J of heat, what can be said about the amount of heat gained by the cup compared to this 400 J? What about the water?

11.3 Phase Changes and Latent Heat

10. You are monitoring the temperature of cold ice cubes (−5.0 °C) in a cup as the ice and cup are heated. Initially, the temperature rises, but it stops at 0 °C. After a while, it begins rising again. Is anything wrong with the thermometer? Explain.

11. Discuss the energy conversion in the process of adding heat to an object that is undergoing a phase change.

12. In general, you would get a more severe burn from coming into contact with a given mass of steam at 100 °C than from the same mass of hot water at 100 °C. Why?

13. When you breathe out in the winter, it is said you can see your breath. Explain.

11.4 Heat Transfer

14. A plastic ice cube tray and a metal ice cube tray are removed from the same freezer, at the same initial temperature. However, the metal one feels cooler to the touch. Why?

15. The approaches to bridges in cold climates are typically preceded by warning signs indicating that the road surface *on the bridge* can ice up before the road surface approaching the bridge. Explain. [*Hint*: A bridge has its underside exposed; the roadbed is in contact with the ground.]

16. Polar bears have an excellent heat insulation system. (Sometimes even infrared cameras cannot detect them.) Polar bear hairs are actually hollow inside. Explain how this helps the bears maintain their body temperature in the cold winter.

17. Explain in your own words how the Thermos bottle (Figure 11.14) uses all the mechanisms of heat transfer to keep coffee hot and also keep a cold drink cold.

Exercises*

Integrated Exercises (IEs) *are two-part exercises. The first part typically requires a conceptual answer choice based on physical thinking and basic principles. The following part is quantitative calculations associated with the conceptual choice made in the first part of the exercise.*

11.1 Definition and Units of Heat

1. • A window air conditioner has a rating of 20 000 Btu/h. What is this expressed in watts?

2. • A person goes on a 1500-Cal/day diet to lose weight. What is his daily energy allowance expressed in joules?

3. • During a game, a typical NBA basketball player will do about 1.0×10^6 J of work in one hour. Express this in Calories.

4. •• A typical adult *metabolic rate* (the rate at which the body "burns" – that is, converts – [chemical] food energy into thermal energy) is about 4×10^5 J/h, and the average food energy in a Big Mac™ is 600 calories. If a person lived on nothing but Big Macs, how many per day would they have to eat to maintain a constant body mass?

5. •• A student ate a Thanksgiving dinner that totaled 2800 Cal. He wants to use up all that energy by lifting a 20-kg mass a distance of 1.0 m. Assume that no work is required in lowering the mass. (a) How many times must he lift the mass? (b) If he can lift and lower the mass once every 0.5 s, how long would this take?

11.2 Specific Heat and Calorimetry

6. • It takes 2.0×10^6 J of heat to bring a pot of water from 20 °C to a boil. What is the water's mass?

7. IE • The temperatures of a lead block and a copper block, both with a mass of 1.0 kg and at 20 °C, are to be raised to 100 °C. (a) The copper will require (1) more heat, (2) the same heat, (3) less heat than the lead. Why? (b) Calculate the difference between the heat required for the two blocks to prove your answer to part (a).

8. • A 5.00-g pellet of aluminum reaches a final temperature of 63 °C when gaining 200 J of heat. What was its initial temperature?

9. • Blood can carry excess internal energy from the interior to the surface of the body, where it is transferred to the outside environment. If 0.250 kg of blood at a temperature of 37.0 °C flows to the surface and loses 1500 J of heat, what is the temperature of the blood when it flows back into the interior? Assume blood has the same specific heat as water.

10. IE •• Equal amounts of heat are added to an aluminum block and a copper block of different masses to achieve the same temperature increase. (a) The mass of the aluminum block is (1) more, (2) the same, (3) less than the mass of the copper block. Why? (b) If the mass of the copper block is 3.00 kg, what is the mass of the aluminum block?

11. •• An engine of alloy construction consists of 25 kg of aluminum and 80 kg of iron. How much heat does the engine absorb as its temperature increases from 20 °C to 100 °C?

12. IE •• Equal amounts of heat are added to different masses of copper and lead. The copper's temperature increases by 5.0 °C and the lead's by 10 °C. (a) The lead has (1) a greater mass than the copper, (2) the same amount of mass as the copper, (3) less mass than the copper. (b) Calculate the mass ratio of lead to copper to prove your answer to part (a).

13. IE •• Initially at 20 °C, 0.50 kg of aluminum and 0.50 kg of iron are heated to 100 °C. (a) The aluminum gains (1) more heat than the iron, (2) the same amount

of heat as the iron, (3) less heat than the iron. Why? (b) Calculate the difference in heat required to prove your answer to part (a).

14. •• A 0.20-kg glass cup at 20 °C is filled with 0.40 kg of hot water at 90 °C. Neglecting any heat losses to the environment, what is the equilibrium temperature of the water?

15. •• A 0.250-kg coffee cup at 20 °C is filled with 0.250 kg of brewed coffee at 100 °C. The cup and the coffee come to thermal equilibrium at 80 °C. If no heat is lost to the environment, what is the specific heat of the cup material? [*Hint:* Treat the coffee as water thermally.]

16. •• An aluminum spoon at 100 °C is placed in a Styrofoam cup containing 0.200 kg of water at 20 °C. If the final equilibrium temperature is 30 °C and no heat is lost to the cup or the environment, what is the mass of the spoon?

17. •• A student doing an experiment pours 0.150 kg of heated copper shot into a 0.375-kg aluminum calorimeter cup containing 0.200 kg of water. The cup and water are both initially at 25 °C. The system comes to thermal equilibrium at 28 °C. What was the initial temperature of the shot?

18. •• At what average rate would heat have to be removed from 1.5 L of (a) water and (b) mercury to reduce each liquid's temperature from 20 °C to its freezing point in 3.0 min?

19. •• While resting, a person gives off heat at a rate of about 100 W. If the person is submerged in a tub containing 150 kg of water at 27 °C and assuming that all the heat from the person goes only into the water, how much time will it take for the water temperature to rise to 28 °C?

20. •• To determine the specific heat of a new metal alloy, a 0.150 kg piece of it is heated to 400 °C and then placed in a 0.200 kg aluminum calorimeter cup containing 0.400 kg of water at 10.0 °C. If the final temperature of the system is 30.5 °C, what is the specific heat of the alloy? (Ignore the calorimeter stirrer and thermometer.)

21. IE •• In a calorimetry experiment, 0.50 kg of a metal at 100 °C is added to 0.50 kg of water at 20 °C in an aluminum calorimeter cup. The cup has a mass of 0.250 kg. (a) If some water splashed out of the cup when the metal was added, the measured specific heat will appear to be (1) higher, (2) the same, (3) lower than the value calculated for the case in which the water does not splash out. Why? (b) If the final temperature of the mixture is 25 °C, and no water splashed out, what is the specific heat of the metal?

22. ••• Lead pellets of total mass 0.60 kg are heated to 100 °C and then placed in a well-insulated aluminum cup of mass 0.20 kg that contains 0.50 kg of water initially at 17.3 °C. What is the equilibrium temperature of the mixture?

23. ••• A student mixes 1.0 L of water at 40 °C with 1.0 L of ethyl alcohol at 20 °C. Assuming no heat exchanges with the container and surroundings, what is the final temperature of the mixture?

24. ••• A room full of people tends to be warmer than when that room is empty. Suppose that ten people are in a 4.0 m × 6.0 m × 3.0 m room initially at 20 °C. If each gives off heat at a rate of 100 W and, assuming no heat exchanges to the furniture, walls, or outside, what would the temperature of the room be after 10 min? Assume that at 20 °C, the air density is 1.2 kg/m³ and specific heat (at constant atmospheric pressure) is 1.0×10^3 J/(kg·°C).

11.3 Phase Changes and Heat of Transformation

25. • How much heat is required to melt a 2.5-kg block of ice at 0 °C?

26. • How much heat is required to boil away 1.50 kg of water initially at 100 °C?

27. IE • (a) Converting 1.0 kg of water at 100 °C to steam at 100 °C requires (1) more heat, (2) the same amount of heat, (3) less heat than converting 1.0 kg of ice at 0 °C to water at 0 °C. Explain. (b) Calculate the difference in heat required to prove your answer to part (a).

28. • Water is boiled to add moisture to the air in a room to help a congested person breathe better. Find the heat required to boil away 1.0 L of water initially at 50 °C.

29. • An artist wants to melt some lead to make a statue. How much heat must be added to 0.75 kg of lead at 20 °C to cause it to melt completely?

30. • First calculate the heat that needs to be removed to convert 1.0 kg of steam at 100 °C to water at 40 °C and then compute the heat that needs to be removed to lower the temperature of water at 100 °C to water at 40 °C. Compare the two results. Are you surprised?

31. • How much heat is required to completely boil away 0.50 L of liquid nitrogen at −196 °C? (The density of liquid nitrogen is 8.0×10^2 kg/m³.)

32. IE •• An alcohol rub can rapidly decrease body (skin) temperature. (a) This is because of (1) the cooler temperature of the alcohol, (2) the evaporation of alcohol, (3) the high specific heat of the human body. (b) To decrease the body temperature of a 65-kg person by 1.0 °C, what mass of alcohol must be evaporated from their skin (consider evaporation only, ignore the heat to raise the alcohol to its boiling point, and treat the human body as, on average, water)?

33. IE •• Internal energy must be removed to condense mercury vapor at 630 K into liquid mercury at that temperature. (a) This heat transfer involves (1) only specific heat, (2) only latent heat, or (3) both specific and latent heats. Explain. (b) If the mass of the mercury vapor is 15 g, how much heat must be removed?

34. •• If 0.050 kg of ice at 0 °C is added to 0.300 kg of water at 25 °C in a 0.100-kg aluminum calorimeter cup, what is the final temperature of the water?

35. •• How much ice (at 0 °C) must be added to 0.500 kg of water at 100 °C in a 0.200-kg aluminum calorimeter cup to end up with all liquid at 20 °C?

36. •• Ice (initially at 0 °C) is added to 0.75 L of tea at 20 °C to make the coldest possible iced tea. If enough ice is added so the final result is *all* liquid, how much liquid is in the final mixture?

37. •• To cool a hot piece of 4.00-kg steel at 900 °C, it is put into a 5.00-kg water bath at 20 °C. What is the final temperature of the steel-water mixture?

38. •• Steam at 100 °C is bubbled into 0.250 kg of water at 20 °C in a calorimeter cup, where it condenses into liquid form. How much steam will have been added when the water in the cup reaches 60 °C? (Ignore the effect of the cup.)

39. IE •• Evaporation of water from our skin is an important mechanism for controlling body temperature. (a) This is because (1) water has a high specific heat, (2) water has a high heat of vaporization, (3) water contains more heat when hot, (4) water is a good thermal conductor. (b) In a 3.5-h intense cycling race, a cyclist can lose 7.0 kg of water through perspiration. Estimate how much internal energy the cyclist loses in the process.

40. IE ••• A 0.400-kg piece of ice at −10 °C is placed into an equal mass of water at 30 °C. (a) When thermal equilibrium is reached between the two, (1) all the ice will melt, (2) some of the ice will melt, (3) none of the ice will melt. (b) How much ice melts?

41. ••• As heat is added to one kilogram of a substance, the data results in the *T*-versus-*Q* graph of ▼ **Figure 11.17**. (a) What are its melting and boiling points? In SI units, what are (b) the specific heats of it *during* the various phases and (c) the heats of transformation of it as the phase *changes*?

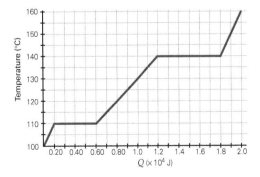

▲ FIGURE 11.17 **Temperature versus heat input** See Exercise 41.

42. ••• In an experiment, a 0.150-kg piece of a ceramic material at 20 °C is placed in liquid nitrogen at its boiling point to cool in a perfectly insulated flask, which allows the gaseous N_2 to immediately escape. How many liters of liquid nitrogen will be boiled away during this operation? (Take the specific heat of the material to be that of glass and the liquid nitrogen density as 8.0×10^2 kg/m^3.)

11.4 Heat Transfer

43. • The single glass pane in a window has dimensions of 2.00 m by 1.50 m and is 4.00 mm thick. How much heat will flow through the glass in 1.00 h if there is a temperature difference of 2.00 °C between the inner and outer surfaces? (Consider conduction only.)

44. IE • Assume that a tile floor and an oak floor are at the same temperature and are equally thick. Both are cemented to the same subfloor (i.e., their undersides are at the same temperature also). (a) Compared with the oak floor, the tile floor will conduct heat away from your bare feet (1) faster, (2) at the same rate, (3) slower. Why? (b) Calculate the ratio of the heat flow rate of the tile floor to that of the oak.

45. IE • A house designer can choose between a brick wall or a concrete wall with the same thickness. (a) During a cold winter day, compared with a concrete wall, a brick wall will conduct energy out of the house (1) faster, (2) at the same rate, (3) slower. Why? (b) Calculate the ratio of heat flow rate through the brick to that of the concrete.

46. • Assume a goose has a 2.0-cm-thick layer of feather down (on average) and a body surface area of 0.15 m^2. What is the rate of heat loss *by conduction* if the goose, with a body temperature of 41 °C, is outside on a day when the air is at 11 °C?

47. • Assume your skin has an emissivity of 0.90, a temperature of 34 °C, and an exposed area of 0.25 m^2. What is the net energy lost per second due to radiation if the outside temperature is 22 °C?

48. • The United States five-cent coin, the nickel, has a mass of 5.1 g, a volume of 0.719 cm^3, and a total surface area of 8.54 cm^2. Treating the coin as an ideal radiator, how much radiant energy per second does it emit if its temperature is 20 °C?

49. IE •• An aluminum bar and a copper bar of identical cross-sectional area have the same temperature difference between their ends and conduct heat at the same rate. (a) The copper bar is (1) longer, (2) of the same length, (3) shorter than the aluminum bar. Why? (b) Calculate the ratio of the length of the copper bar to that of the aluminum bar.

50. •• A copper teakettle has a circular bottom 30.0 cm in diameter that has a uniform thickness of 2.50 mm. It sits on an electric burner whose temperature is 150 °C. (a) If the kettle contains water at its boiling point, what

is the rate of heat conduction through its bottom? (b) Assuming that the heat from the burner is the only heat input, how much water is boiled away in 5.0 min? Is your answer unreasonably large? If yes, explain why.

51. •• Assuming the human body has a 1.0-cm-thick layer of skin tissue and a surface area of 1.5 m², estimate the rate at which internal energy is conducted from inside the body to the surface if the skin temperature is 34 °C. (Assume a normal body temperature of 37 °C for the interior temperature.)

52. IE •• The emissivity of an object is 0.50. (a) Compared with a blackbody at the same temperature, this object would radiate (1) more power, (2) the same amount of power, (3) less power. Why? (b) Calculate the ratio of the power radiated by a blackbody to that radiated by the object.

53. •• A lamp filament has a *net* radiation rate of 100 W when the temperature of the surroundings is 20 °C, and only 99.5 W when the surroundings are at 30 °C. If the temperature of the filament is the same in each case, what is its temperature in Celsius?

54. IE •• (a) If the Kelvin temperature of an object is doubled, its emitted radiation rate increases by (1) 2, (2) 4, (3) 8, (4) 16 times. Explain. (b) If its temperature is increased from 20 °C to 40 °C, by how much does the radiated power change?

55. •• An object with a surface temperature of 100 °C is radiating heat at a rate of 200 J/s. To double the object's rate of radiation energy, what should be its surface temperature, expressed in Celsius?

56. •• Solar heating takes advantage of solar collectors such as in ▼ **Figure 11.18**. During daylight hours, the average intensity of solar radiation at the top of the atmosphere is about 1400 W/m². On average, about 50% reaches the surface. (The rest is reflected, scattered, and absorbed by the atmosphere.) Assuming a constant 700 W/m², how much energy would be received, on average, by the cylindrical collector shown in the figure during 10 h of daylight?

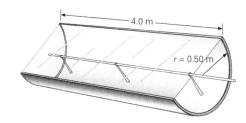

▲ FIGURE 11.18 **Solar collector and solar heating** See Exercise 56.

57. ••• A steel cylinder of radius 5.0 cm and length 4.0 cm is placed in end-to-end thermal contact with a copper cylinder of the same dimensions. The free ends of the two cylinders are maintained at constant temperatures of 95 °C (steel) and 15 °C (copper) and a steady heat flow rate is attained. Under these conditions, how much energy will flow through the cylinders in 20 min?

12

Thermodynamics*

The efficiency of this modern gasoline engine is determined by the laws of thermodynamics.

As the word implies, thermodynamics deals with heat transfers. The development of thermodynamics started about 200 years ago out of efforts to develop heat engines. The steam engine was one of the first such devices, designed to convert thermal energy into mechanical work. Steam engines in factories and locomotives powered the Industrial Revolution, changing the world forever.

Gasoline-powered automobile engines, such as the one in the chapter-opening photo, convert chemical energy into work that results in the kinetic energy of the auto. With decreasing natural resources and concern about greenhouse gas emissions, gasoline engines have, over recent years, improved dramatically in their thermal efficiency – a concept central to understanding some of the practical aspects of thermodynamics.

In this chapter, you'll learn how heat can best be used to perform useful work. The laws governing these conversions include some of the most general and far-reaching laws in all of physics. Although our study is primarily concerned with heat and work, thermodynamics is a broad and comprehensive science that includes a great deal more than heat engine theory.

---

* The mathematics needed in this chapter involves natural logarithms (ln) and common logarithms (log). You may want to review these in Appendix I.

12.1 The First Law of Thermodynamics

Recall from Section 5.1 that mechanical work done on a system represented a way of changing its total mechanical energy (E_m). For example, if you push on a chair and set it into motion, the work (force through a distance) done on the chair goes into increasing its kinetic energy. In Chapter 5, the focus was on mechanics, thus heat exchanges and internal energies (non-mechanical energies) were not considered. There the emphasis was on mechanical work done on a system by the environment, W_{env}, and the important relationship is the work-(mechanical) energy theorem (here only changes in gravitational and elastic potential energies, ΔU_g and ΔU_{sp} respectively, are shown):

$$W_{env} = \Delta E_m = \Delta K + \Delta U_g + \Delta U_{sp} \qquad (12.1)$$

However, there is a second way to change the total energy of a system (total includes all forms of energy, not just mechanical) – that is by adding or removing heat from it. Although the details of this "heat flow" are not seen on a macroscopic level, the heat transfer involves work done on an atomic level. During a conduction process, for example, energy is transferred from a hot object to a cold object, as the more energetic atoms of the hot object do work on the slower atoms of the cold object (▼ Figure 12.1). Thus energy is transferred from atom to atom into the cold object. This ongoing process appears as the "flow" of internal energy from hot to cold that we observe as heat on a macroscopic level.

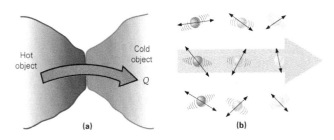

▲ **FIGURE 12.1** **Heat flow (via conduction) on the atomic scale** **(a)** Macroscopically, heat is transferred by conduction from the hot object to the cold one. **(b)** On the atomic scale, heat conduction is explained as the energy transfer from the more energetic atoms (in the hot object) to the less energetic atoms (in the cold object). This transfer of energy from an atom to its neighbor results in the heat transfer observed in part (a).

Thus Equation 12.1 must be generalized to include work (W_{env}) and heat (Q) as the two ways of changing the total energy of the system which is now generalized to include **internal energy U_{int}**:

$$Q + W_{env} = \Delta E_m + \Delta U_{int} \qquad (12.2)$$

For our study of thermodynamics, the system's energy will include only its internal energy - ignoring mechanical

forms – which leads to a specialized form of Equation 12.2, the **first law of thermodynamics** (first form).

$$Q + W_{env} = \Delta U_{int} \qquad (12.3)$$

Thus heat as well as work can change the internal energy of the system. It is important to understand what the symbols mean and their sign conventions (▶ **Figure 12.2**). Q represents the added to (+) or removed (−) from the system, ΔU_{int} is the change in internal energy of the system, and W_{env} is the work done by the environment on the system (in Figure 12.2 the environment is the piston and the system is a gas). For example, if the system is a sample of gas that absorbs 1000 J of heat and simultaneously is compressed by an external force that does 400 J of work on it, then its internal energy would increase by 1400 J since $\Delta U_{int} = (+1000 \text{ J}) + (+400 \text{ J}) = +1400 \text{ J}$. Consider some other possibilities involving a gas sample as the system in the following integrated example.

INTEGRATED EXAMPLE 12.1: BALANCING HEAT AND WORK

A sample of gas is moved into a cold room and loses heat to the air. At the same time, it is compressed. (a) What can you say about the gas's internal energy? It must (1) increase, (2) decrease, (3) stay the same, or (4) it could do any of these. Explain. (b) To prove your answer, determine the work done by the gas for three situations: (i) it loses 400 J of heat and has 500 J of work done on it, (ii) it loses 400 J of heat and has 300 J of work done on it, and (iii) it loses 400 J of heat and has 400 J of work done on it.

(a) **CONCEPTUAL REASONING.** Whether the internal energy of a gas changes (and how it changes) according to Equation 12.3 involves comparing the heat flow and work done on it. Since these two quantities could have any value, then it is possible for the heat flow (negative) to be larger (in magnitude) than the (positive) work resulting in a decrease in internal energy. But the reverse is also possible, resulting in an increase in internal energy. It is also true that they could be equal but opposite in sign, resulting in no change in internal energy. So the correct answer is (4) – because you cannot tell without specific numerical values for the heat and work.

(b) **QUANTITATIVE REASONING AND SOLUTION.** Equation 12.3 will be used to calculate the change in internal energy with the three sets of numbers. Listing the data with the correct signs,

Given:

   i. $Q_1 = -400$ J and $W_{env,1} = +500$ J
  ii. $Q_2 = -400$ J and $W_{env,2} = +300$ J
 iii. $Q_3 = -400$ J and $W_{env,3} = +400$ J

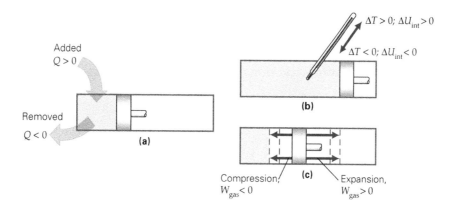

▲ FIGURE 12.2   **Sign conventions for Q, W, and ΔU_{int}** **(a)** A gas sample is useful for understanding the sign convention for heat flow. **(b)** The experimental way to tell if the internal energy of a gas changes is to take its temperature. Since its internal energy is directly related to its temperature, a change in temperature implies a change in internal energy of the same sign. **(c)** Gas expansional work done by the gas on the environment W_{gas} (here the piston) is positive. If the gas is compressed it does negative work on the piston. The work done by the gas on the environment is equal to, but opposite in sign to, the work done by the environment on the gas: $W_{gas} = -W_{env}$. See text for details.

Find: ΔU_{int} (change in internal energy for all three scenarios)

Applying the first law of thermodynamics three times:

   i. $\Delta U_{int} = Q + W_{env} = (-400\ J) + (500\ J) = +100\ J$
  ii. $\Delta U_{int} = Q + W_{env} = (-400\ J) + (300\ J) = -100\ J$
 iii. $\Delta U_{int} = Q + W_{env} = (-400\ J) + (400\ J) = 0\ J$

FOLLOW-UP EXERCISE. Recall that for an ideal gas (Chapter 10) its internal energy is directly related to its temperature. In this Example, which situation would result in the largest gas temperature change?

Our focus will be on heat engines – systems that convert thermal energy into useful mechanical work. For example, the expansion of a gas might cause it to move a piston, as in a car engine. Thus our interest is on W_{gas}, not W_{env}. As shown in Figure 12.2, the gas's force and the external (environment) force are equal and opposite. Hence they perform the same magnitude of work but opposite in sign: $W_{gas} = -W_{env}$. To create a more practical form of Equation 12.3 (one that emphasizes work by the gas) substitute $-W_{gas}$ for W_{env} and rearrange. This leads to an equivalent expression suitable for thermal engine discussions, **first law of thermodynamics** (second form):

$$Q = \Delta U_{int} + W_{gas} \tag{12.4}$$

Equation 12.4 simply provides a different emphasis, one in which heat exchanges can cause both internal energy changes and/or useful work done by the gas on the environment. This latter quantity is what heat engines are all about. This form was used in building the first heat engines (Sections 12.4 and 12.5). For the remainder of this chapter, the focus is on W_{gas} and our system will be an ideal gas. Thus Equation 12.4 will be the key.

The first law is general and is not restricted to just gases. For example, consider its application to exercise and weight loss in the next example.

EXAMPLE 12.2: ENERGY BALANCING – EXERCISING USING PHYSICS

A 65-kg worker shovels coal for 3.0 h. During the shoveling, the worker did work at an average rate of 20 W and lost heat to the environment at an average rate of 480 W. Ignoring the loss of water by the evaporation of perspiration from his skin, how much mass (assumed to be fat) will the worker lose? The energy value of fat (E_{fat}) is 9.3 kcal/g.

THINKING IT THROUGH. Since the time duration of the shoveling, the rate of work done (power) and the rate of heat loss are known, thus totals of these over the three hours can be found. Then the change in the internal energy of the worker can be determined from Equation 12.4. This loss of thermal energy results from "burning" (loss) of fat.

SOLUTION

Here the worker is the system, doing W, shoveling coal. Listing the given values, and converting power to work and heat:

Given:

$\quad W = P\Delta t = (20\ J/s)(3.0\ h)(3600\ s/h) = +2.16 \times 10^5\ J$
$\quad$ (positive – done *by* the worker)
$\quad Q = -(480\ J/s)(3.0\ h)(3600\ s/h) = -5.18 \times 10^6\ J$
$\quad$ (negative – heat is lost)
$\quad E_{fat} = 9.3\ kcal/g = 9.3 \times 10^3\ kcal/kg \times (4186\ J/kcal) = 3.89 \times 10^7\ J/kg$

Find: m_{fat} (mass of fat burned)

Rearranging Equation 12.4 to solve for the worker's internal energy loss (notice the careful use of signs):

$$\Delta U_{int} = Q - W = -5.18 \times 10^6\,\text{J} - (+2.16 \times 10^5\,\text{J}) = -5.40 \times 10^6\,\text{J}$$

Using the energy generated by metabolizing 1 kg of fat:

$$m_{fat} = \frac{|\Delta U_{int}|}{E_{fat}} = \frac{5.40 \times 10^6\,\text{J}}{3.89 \times 10^7\,\text{J/kg}} = 0.140\,\text{kg} = 140\,\text{g}$$

This amounts to weight loss of about 5 ounces, or a third of a pound.

FOLLOW-UP EXERCISE. How much fat would be lost if a person played basketball for 3.0 h, doing work at a rate of 120 W and causing an internal energy loss at a rate of 600 W?

As seen in this example, when applying the laws of thermodynamics, the proper use of signs cannot be overemphasized.

12.2 Thermodynamic Processes for an Ideal Gas

12.2.1 The State of a Gas and Thermodynamics Processes

Just as there are kinematic equations to describe the motion of an object, there are equations of state to describe the conditions, or states, of thermodynamic systems. These equations express a relationship between the thermodynamic variables that describe the system. The ideal gas law, $pV = nRT$ (Section 10.3), is an example of just such an equation since it establishes a relationship between the pressure (p), volume (V), temperature (T), and number of moles (n, or equivalently, N, the number of molecules) of a gas. Because these variables describe the state of the gas completely, these are examples of *state variables*. Thus if a system changes state, it will, in general, have a different set of state variables.

In this chapter the system we will mostly be concerned with is an ideal gas sample. It is useful to create a graphical display of a gas's state by plotting its thermodynamic variables (p, V, T), as locations on a set of axes, similar to using Cartesian coordinates (x, y, z) to display locations. A general two-dimensional illustration of such a plot is shown in ▶ **Figure 12.3**. Just as the coordinates (x, y) specify a specific location on Cartesian axes, for an ideal gas, the coordinates (V, p) specify completely its state on the *p-V* diagram. This is because the ideal gas law can be solved for the gas's temperature if its pressure, volume, and number of moles are known. In other words, on a *p-V* diagram, each point tells the pressure and volume directly, and thus the gas's temperature indirectly. In summary, to describe the state of a known quantity of gas, only its volume and pressure are

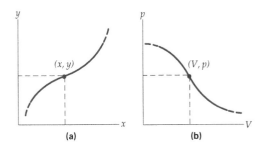

▲ **FIGURE 12.3** **Graphing states (a)** On a Cartesian graph, the coordinates (x, y) represent a individual location. **(b)** Similarly, on a *p-V* graph or diagram, the coordinates (V, p) represent a particular state of a system. (It is common to say *p-V*, rather than *V-p*, because the plot is a *p* vs. *V* graph.)

needed – that is, a specific location on a *p-V* plot. A gas or system that has a well-defined set of state variables such as p, V, T is said to be in *thermal equilibrium*.

A **thermodynamic process** involves changes in the state, or the thermodynamic "coordinates," of a system. For example, a gas initially in state 1, described by state variables (p_1, V_1, T_1), might change to a state 2 described by a different set (p_2, V_2, T_2). Processes are classified as either *reversible* or *irreversible*. Suppose that a system of gas in equilibrium (state variables p, V, T) is allowed to expand quickly as its pressure is reduced. The state of the system will change rapidly and unpredictably, but eventually the system will reach a different state. On a *p-V* diagram (▼ **Figure 12.4**), the initial and final states (labeled 1 and 2, respectively) are known, but what happened in between is not. This type of process is an **irreversible process** – which means that the process path can't be retraced, because of the unknown non-equilibrium conditions between states 1 and 2. An explosion is an example of an irreversible process.

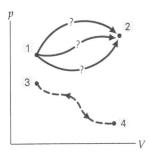

▲ **FIGURE 12.4** **Reversible and irreversible processes** If a gas quickly goes from state 1 to state 2, the process is irreversible, since the "path" details are not known. If, however, the gas is taken through many closely spaced and well-defined equilibrium states (here 3 to 4), the process is reversible, in principle.

If, however, the gas changes state *very slowly*, passing from one equilibrium state to a neighboring one, eventually arriving at its final state (Figure 12.4, states 3 and 4), then the process path *is* known. In such a situation, the system could be brought back to its initial conditions by "traveling" the process path in the reverse direction, recreating the intermediate states along the way. Such a process is a **reversible process**. In practice, all real thermodynamic processes are irreversible to some degree. However, the concept of the ideal reversible process is useful for studying the basic laws of thermal physics and the assumption will be for our purposes that all processes are reversible.

12.2.2 Work Done by a Gas

Because heat engines are our focus, the conversion of heat into mechanical work is clearly important. The work done by a gas, for example, depends on the type of (reversible) process it undergoes during its expansion/contraction. But for a general process, how is this work actually computed? To answer this, consider a cylindrical piston with end area A, containing a sample of gas (▼ **Figure 12.5**). Imagine that the gas is allowed to push the piston (and thus expand) a small distance Δx. If the volume and temperature of the gas does not change appreciably, the pressure remains almost constant. Thus, in pushing the piston slowly and steadily outward, the gas does positive work on the piston because it is pushing in the direction the piston moves. From the definition of work (Chapter 5), the result is $W_{gas} = F\Delta x \cos \theta = F\Delta x \cos 0° = F\Delta x$. For gases, this can be converted into "pressure language" since $F_{gas} = pA$. Thus $W_{gas} = pA\Delta x$. But $A\Delta x$ is the volume of a cylinder with end area A and height Δx. In this case, that volume represents the change in volume of the gas, or $\Delta V = A\Delta x$, and the final result for the work done is $W_{gas} = p\Delta V$. Notice in Figure 12.5 this work is positive because ΔV is positive (volume increase). If the gas should contract, then the gas does negative work (volume change is negative $\Delta V < 0$) because the force exerted by the gas is in the opposite direction of the piston movement.

Of course, gases don't always change their volumes by small amounts and aren't usually subject to constant pressure. How is the calculation of work handled under these more general circumstances? The answer is seen in ▼ **Figure 12.6**, where a reversible path on a p-V diagram is shown. During each small step, the pressure remains approximately constant. For each step, the work done can then be approximated by $W_{gas} = p\Delta V$. Graphically, this is the area of a narrow rectangle, extending from the process curve to the V-axis. To get an exact value, think of the area as made up of a large number of thin rectangles. As the number of rectangles becomes infinitely large, each rectangle's thickness approaches zero. (This process involves calculus and is beyond the scope of this book.) The general result is

The work done by a gas is equal to the area under the process curve on a p-V diagram.

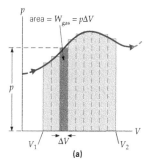

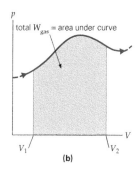

▲ FIGURE 12.6 **Thermodynamic work done by a gas is the area under the process curve (a)** If a gas expands by a significant amount, its work can be approximated by treating the expansion in small steps, each one representing a small amount of work done at constant pressure. The total work is found by adding up the many rectangular strips. **(b)** If the number of rectangular strips becomes very large, and each becomes extremely thin, the calculation of the area becomes exact. Thus the work done by the gas is equal to the area between the process curve and the V-axis.

From this area-under-the-curve idea, it should be clear that the work done by the gas depends on the process path (▶ **Figure 12.7**). Notice that more work is done if the process takes place at higher pressures as represented by a larger area. Fundamentally more work is done by the gas at higher pressures because it exerts a larger force (on average) over the same distance.

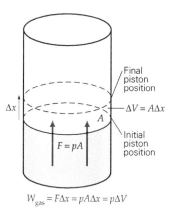

▲ FIGURE 12.5 **Work in thermodynamic terms** If a gas expands slowly by a small amount and does work, its pressure can be taken as constant. During this expansion, a small amount of work is done by the gas ($p\Delta V$).

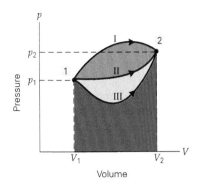

▲ FIGURE 12.7 **Thermodynamic work done by a gas depends on its process path** Here the gas expands by the same amount, but by three different processes. The work done during process I is larger than the work during process II, which in turn is larger than that done during process III. Fundamentally, applying a larger force (pressure) through the same distance (volume change) means more work. Process I includes the medium red, pink, and dark red areas; process II includes just the pink and dark red areas; and process III includes just the dark red area.

12.2.3 Isothermal Processes

An **isothermal process** is a constant-temperature process (*iso* for equal, *thermal* for temperature). In this case, the process path is an *isotherm*, and every point on it is at the same temperature (▼ **Figure 12.8**). For a given gas quantity and constant temperature, the quantity nRT is constant. Therefore, $pV = nRT =$ a constant or $p \propto 1/V$. On a p-V diagram this displays as a hyperbola. (Recall a hyperbola on x-y Cartesian coordinates is written as $y \propto 1/x$, and plots as a downward curve.)

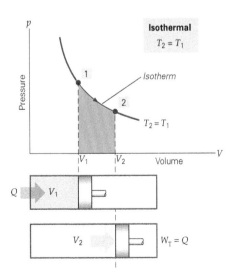

▲ FIGURE 12.8 **Isothermal (constant temperature) process** Under this condition, *all* the heat added to the gas goes into work (the expanding gas moves the piston). Because $\Delta T = 0$, and thus $\Delta E_{th} = 0$, from the first law of thermodynamics: $Q = W_{gas}$. The work is equal to the area (shaded) under the isotherm on the p-V diagram.

In the expansion from state 1 (initial) to state 2 (final) in Figure 12.8, heat is added to the gas, while the pressure and volume vary in a way that the temperature remains constant. It is important to recall that for a given amount (moles) of an ideal gas, its internal energy is all in the form of thermal energy and depends only on its temperature (see Sections 10.5 and 10.6). From now on, our system will assumed to be an ideal gas unless otherwise noted and we will replace ΔU_{int} with ΔE_{th} (expressions for E_{th} are in Chapter 10).

For any isothermal process, $\Delta T = 0$ and thus $\Delta E_{th} = 0$. Applying these conditions with the first law of thermodynamics, the result is $Q = \Delta E_{th} + W_{gas} = 0 + W_{gas}$. The work* done by the gas is equal to the area under the curve (because the pressure varies, this requires calculus to compute). The result is

$$W_T = nRT \ln\left(\frac{V_2}{V_1}\right) \quad \text{(ideal gas isothermal process)} \quad (12.5)$$

Since the product nRT is a constant here, the work done depends on the ratio of the end point volumes.

12.2.4 Problem-Solving Hint

In Equation 12.5, the function "ln" stands for *natural logarithm*. You might be more familiar with *common logarithms* ("log") that are referenced to the base 10 (see Appendix I). For example, since $100 = 10^2$, the logarithm of 100 is 2, or log 100 = 2. The natural logarithm is similar, except it uses a different base, e (≈ 2.7183). That is, if $y = e^x$, then x is called the natural logarithm of y. As a check, before working problems, you should try to find the natural logarithm of 100 on your calculator. (The answer is ln 100 = 4.605.)

12.2.5 Isobaric Process

A constant-pressure process is called an **isobaric process** (*iso* for equal, and *bar* for pressure). An isobaric process for an ideal gas is illustrated in ▶ **Figure 12.9**. On a p-V diagram, an isobaric process is represented by a horizontal line called an *isobar*. As the heat is added and the gas expands, it does work on the piston and its temperature increases. Figure 12.9 illustrates this as the gas moves to a higher temperature isotherm. This temperature increase means that the thermal energy of the gas increases, since $\Delta E_{th} \propto \Delta T$. Here the area representing the work W_p is rectangular in shape and is computed from "length times width":

$$W_p = p(V_2 - V_1) = p\Delta V \quad \text{(ideal gas isobaric process)} \quad (12.6)$$

Note that when heat is added to or removed from a gas under isobaric conditions, the gas's thermal energy must change and

---

* The convention for designating work by the gas will use a subscript indicating what variable is constant. Thus, W_T is isothermal work and later on W_p will indicate work done at constant pressure.

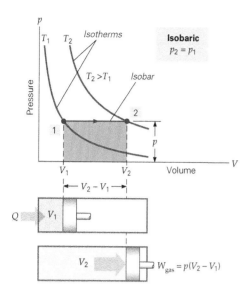

▲ FIGURE 12.9   Isobaric (constant pressure) process Here heat added to the gas becomes work done by the gas *and* a change in its thermal energy. The work is the area under the isobar (from state 1 to 2) which is the shaded rectangle on the p-V diagram. Note the two isotherms. They are not part of the isobaric process, but they show us that the temperature rises during the isobaric expansion.

the gas *must* do work (either positive or negative). This relationship can be written explicitly by combining the first law of thermodynamics with the work expression (Equation 12.6):

$$Q = \Delta E_{th} + p\Delta V \quad \text{(ideal gas isobaric process)} \quad (12.7)$$

For a detailed comparison of an isobaric process and an isothermal process, consider the following Integrated Example (take careful note of the signs).

INTEGRATED EXAMPLE 12.3: ISOTHERMS VERSUS ISOBARS – WHICH AREA?

Two moles of a monatomic ideal gas, initially at 0 °C and 1.00 atm, are expanded to twice their original volume, using two different processes. They are expanded first isothermally and then isobarically, both from the same initial state. (a) Does the gas do more work (1) during the isothermal process, (2) during the isobaric process, or (3) does it do the same work during both processes? Explain. (b) To prove your answer, determine the work done by the gas in each process.

(a) **CONCEPTUAL REASONING**. As shown in ▶ **Figure 12.10**, both processes involve expansion. The isobaric process is at constant pressure while the isothermal process represents a pressure decrease with expansion. Thus, the gas does more work during the isobaric expansion (more area under the curve). Fundamentally, this is because the isobaric process is done at higher (constant) pressure than the isothermal process (lower average pressure as it drops with the expansion). Thus, the correct answer to part (a) is (2), the gas does more work during the isobaric process.

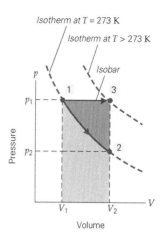

▲ FIGURE 12.10   Comparing work In Integrated Example 12.2, the gas does positive work while expanding. It does more work under isobaric conditions (from state 1 to 3) than under isothermal conditions (from state 1 to 2) because the pressure remains constant on the isobar but decreases along the isotherm. A higher average pressure and thus force means more work. (Graphically just compare the areas under the curves.)

(b) **QUANTITATIVE REASONING AND SOLUTION**. If the volumes are known, Equations 12.5 and 12.6 can be used. These quantities must first be determined from the ideal gas law.

Listing the data,

Given:

$$p_1 = 1.00 \text{ atm} = 1.01 \times 10^5 \text{ N/m}^2$$
$$T_1 = 0 \text{ °C} = 273 \text{ K}$$
$$n = 2.00 \text{ mol}$$
$$V_2 = 2V_1$$

Find: W_T and W_p

From Equation 12.5:

$$W_T = nRT \ln\left(\frac{V_2}{V_1}\right) = (2.00 \text{ mol})[8.31 \text{ J}/(\text{mol·K})](273 \text{ K})(\ln 2)$$
$$= +3.14 \times 10^3 \text{ J}$$

For the isobar, the volumes are determined from the ideal gas law. Using the ideal gas law,

$$V_1 = \frac{nRT_1}{p_1} = \frac{(2.00 \text{ mol})[8.31 \text{ J}(\text{mol·K})](273 \text{ K})}{1.01 \times 10^5 \text{ N/m}^2}$$
$$= 4.49 \times 10^{-2} \text{ m}^3$$

$$V_2 = 2V_1 = 8.98 \times 10^{-2} \text{ m}^3$$

From Equation 12.6 the work is

$$W_p = p(V_2 - V_1) = (1.01 \times 10^5 \text{ N/m}^2)(8.98 \times 10^{-2} \text{ m}^3 - 4.49 \times 10^{-2} \text{ m}^3)$$
$$= +4.53 \times 10^3 \text{ J}$$

This is larger than the isothermal work, as expected from part (a).

FOLLOW-UP EXERCISE. In this Example, what is the heat flow in each process?

12.2.6 Isometric Process

An **isometric process** (short for *isovolumetric*, or constant-volume, process), sometimes called an *isochoric process*, is a constant-volume process. As illustrated in ▼ **Figure 12.11**, the process path on a *p-V* diagram is a vertical line, called an *isomet*. No work is done, since the area under such a curve is zero. (There is no displacement of the piston.) Because the gas does no work, if heat is exchanged with the environment it must go completely into changing the gas's thermal energy, and therefore its temperature. In terms of the first law of thermodynamics,

$$Q = \Delta E_{\text{th}} \quad \text{(ideal gas isometric process)} \quad (12.8)$$

Consider the following example of an isometric process in action.

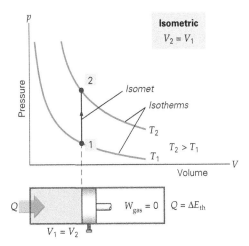

▲ **FIGURE 12.11   Isometric (constant volume) process** Here all of the heat added to the gas goes into increasing the gas's thermal energy, because there is no work done by the gas. (Notice the locking nut on the piston, which prevents any movement.) Again, the isotherms, although not part of the isometric process, enable us to tell for sure that the gas temperature rises.

EXAMPLE 12.4: A PRACTICAL ISOMETRIC EXERCISE – HOW NOT TO RECYCLE A SPRAY CAN

Many "empty" aerosol cans contain remnant propellant gases under approximately 1 atm of pressure (assume 1.00 atm) at 20 °C. They display the warning "Do not dispose of this can in an incinerator or open fire." (a) Explain why it is dangerous to throw such a can into a fire. (b) If there are 0.0100 moles of monatomic gas in the can and its temperature rises to 2000 °F, how much heat was added to the gas? (c) What is the final pressure of the gas?

THINKING IT THROUGH. This is an isovolumetric process; hence, all the heat goes into increasing the gas's thermal energy. A pressure rise is expected, which is where the danger lies. The change in thermal energy can be calculated with Equation 10.16. The final pressure is obtained using the ideal gas law.

SOLUTION

Listing the data and converting temperatures into kelvins:

Given:

$p_1 = 1.00 \text{ atm} = 1.01 \times 10^5 \text{ N/m}^2$
$V_1 = V_2$
$T_1 = 20 \text{ °C} = 293 \text{ K}$
$T_2 = 2000 \text{ °F} = 1.09 \times 10^3 \text{ °C} = 1.37 \times 10^3 \text{ K}$
$n = 0.0100 \text{ mol}$

Find:

(a) Explain the danger in heating the can.
(b) Q (heat added to gas)
(c) p_2 (final pressure of gas)

(a) When the heat is added, it all goes into increasing the gas's thermal energy. Because at constant volume, pressure is proportional to temperature, the final pressure will be greater than 1.00 atm. The danger is that the container could explode into metallic fragments if the maximum design pressure of the container is exceeded.

(b) The work in an isometric process is zero. From Equation 10.16, $E_{\text{th}} = (3/2)nRT$, thus the heat is

$$Q = \Delta E_{\text{th}} = \frac{3}{2}nR\Delta T$$
$$= \frac{3}{2}(0.0100 \text{ mol})[8.31 \text{ J/(mol·K)}](1.37 \times 10^3 \text{ K} - 293 \text{ K})$$
$$= +134 \text{ J}$$

(c) The final pressure of the gas is determined from the ideal gas law:

$$p_2 = p_1\left(\frac{V_1}{V_2}\right)\left(\frac{T_2}{T_1}\right) = (1.00 \text{ atm})\left(\frac{V_1}{V_1}\right)\left(\frac{1.37 \times 10^3 \text{ K}}{293 \text{ K}}\right) = 4.68 \text{ atm}$$

FOLLOW-UP EXERCISE. Suppose the can in this Example was designed to withstand pressures up to 3.50 atm. What would be the highest Celsius temperature it could reach without exploding?

12.2.7 Adiabatic Process

In an **adiabatic process**, no heat is transferred into or out of the system. That is, $Q = 0$ (▶ **Figure 12.12**). (The Greek word *adiabatos* means "impassable.") This condition applies to a system that is completely thermally isolated as if surrounded by "perfect" insulation. Under real-life conditions, adiabatic processes can only be approximated. For example, nearly adiabatic processes can take place if the expansions/contractions occur rapidly enough so there isn't time for significant heat to flow into or out of the system. In other words, *quick* processes can approximate adiabatic conditions.

The curve for this process is called an *adiabat*. During an adiabatic process, all three thermodynamic coordinates (p, V, T) change. For example, if the pressure on a gas is reduced, the gas expands. However, no heat flows into the gas. Without a compensating input of heat, work is done at the expense of the gas's (internal) thermal

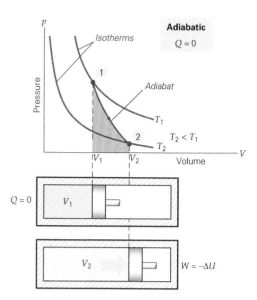

▲ FIGURE 12.12 **Adiabatic ($Q = 0$) process** In an adiabatic process (e.g., a gas in a cylinder – note the heavy insulation), no heat is added to or removed from the system. During expansion (shown), positive work is done by the gas at the expense of its thermal energy: $W_{gas} = -\Delta E_{th}$. Pressure, volume, and temperature all change during an adiabatic process. As usual, the work done by the gas is represented by shaded area between the adiabat and the V-axis.

energy. Therefore, ΔE_{th} is negative. Since the thermal energy of the gas, and therefore its temperature, both decrease, such an expansion is a cooling process. Similarly, an adiabatic compression is a warming process (temperature increase). Mathematically, applying the first law of thermodynamics to an adiabatic process means $Q = 0 = \Delta E_{th} + W_{gas}$ or, upon rearrangement

$$\Delta E_{th} = -W_{gas} \quad \text{(adiabatic process)} \quad (12.9)$$

For completeness, some other relationships related to adiabatic processes are listed in the following text. An important factor is the ratio of the gas's specific heats, defined by a dimensionless quantity (Greek gamma) $\gamma = c_p/c_v$, where c_p and c_v are the specific heats (expressed per mole) at constant pressure and volume, respectively. For the two common types of gas molecules, monatomic and diatomic, the values of γ are 1.67 and 1.40, respectively. The volume and pressure at any two points on an adiabat are related by

$$p_1 V_1^{\gamma} = p_2 V_2^{\gamma} \quad \text{(ideal gas adiabatic process)} \quad (12.10)$$

and the work done by an ideal gas during an adiabatic process W_{ad} can be shown to be

$$W_{ad} = \frac{p_1 V_1 - p_2 V_2}{\gamma - 1} \quad \text{(ideal gas adiabatic process)} \quad (12.11)$$

To clear up confusion that often occurs between isotherms and adiabats, see Integrated Example 12.5.

INTEGRATED EXAMPLE 12.5: ADIABATS VERSUS ISOTHERMS – TWO PROCESSES OFTEN CONFUSED

A sample of helium expands to triple its initial volume adiabatically first, and then again isothermally. In both cases, assume it starts from the same initial state. The sample contains 2.00 mol of helium ($\gamma = 1.67$), initially at 20 °C and 1.00 atm. (a) Does the gas (1) do more work during the adiabatic process, (2) do more work during the isothermal process, or (3) do the same work during both processes? (b) Calculate the work done during each process to verify your reasoning in part (a).

(a) **CONCEPTUAL REASONING.** To determine graphically which process involves more work, note the areas under the process curves (see Figure 12.12). The area under the isothermal process curve is larger; thus, the gas does more work during its isothermal expansion, and the correct answer is (2). Physically, the isothermal expansion involves more work because the pressures are always higher during the isothermal expansion than during the adiabatic one. This larger (average) force acting through the same distance means more work done isothermally.

(b) **QUANTITATIVE REASONING AND SOLUTION.** To determine isothermal work, the ratio of the final to initial volumes is needed and is given. For the adiabatic work, the ratio of specific heats, γ, is important, as are the final pressure and volume. The final pressure can be found using Equation 12.10, and the ideal gas law enables determination of the final volume.

Listing the given values and converting the temperature into kelvins:

Given:

$$p_1 = 1.00 \text{ atm} = 1.01 \times 10^5 \text{ N/m}^2$$
$$n = 2.00 \text{ mol}$$
$$T_1 = (20 + 273) \text{ K} = 293 \text{ K}$$
$$V_2 = 3V_1$$
$$\gamma = 1.67$$

Find: W_T and W_{ad}

The data to calculate the isothermal work from Equation 12.5 are given. The volume ratio is 3 and $\ln 3 = 1.10$, so

$$W_T = nRT \ln\left(\frac{V_2}{V_1}\right)$$
$$= (2.00 \text{ mol})[8.31 \text{ J/(mol·K)}](293 \text{ K})(\ln 3) = +5.35 \times 10^5 \text{ J}$$

For the adiabatic process, the work can be determined from Equation 12.11, but the final pressure and volume are needed first. The final pressure can be determined from a ratio form of Equation 12.10:

$$p_2 = p_1 \left(\frac{V_1}{V_2}\right)^{\gamma} = p_1 \left(\frac{V_1}{3V_1}\right)^{\gamma} = p_1 \left(\frac{1}{3}\right)^{1.67} = 0.160 p_1$$
$$= (0.160)(1.01 \times 10^5 \text{ N/m}^2) = 1.62 \times 10^4 \text{ N/m}^2$$

The initial volume is determined from the ideal gas law:

$$V_1 = \frac{nRT_1}{p_1} = \frac{(2.00\,\text{mol})[8.31\,\text{J/(mol·K)}](293\,\text{K})}{1.01 \times 10^5\,\text{N/m}^2}$$
$$= 4.82 \times 10^{-2}\,\text{m}^3$$

Therefore, $V_2 = 3V_1 = 0.145\,\text{m}^3$. Then, applying Equation 12.11,

$$W_{\text{adiabatic}} = \frac{p_1 V_1 - p_2 V_2}{\gamma - 1}$$
$$= \frac{(1.01 \times 10^5\,\text{N/m}^2)(4.82 \times 10^{-2}\,\text{m}^3) - (1.62 \times 10^4\,\text{N/m}^2)(0.145\,\text{m}^3)}{1.67 - 1}$$
$$= +3.76 \times 10^3\,\text{J}$$

As expected, this result is less than the isothermal work.

FOLLOW-UP EXERCISE. In this Example, (a) calculate the final temperature of the gas in the adiabatic expansion. (b) During the adiabatic expansion, find the change in thermal energy of the gas. Does it equal the negative of the work done by the gas (see the Example)? Explain.

12.2.8 Problem-Solving Hint Using Isotherms

You may have noticed in Figures 12.8 through 12.12 that isotherms were drawn even if the process under study was not isothermal. This was done because superimposing a series of isotherms on the p-V diagram provides a visual way of determining the signs of temperature changes, heat flow, work, and thermal energy changes. The key is to remember that an isotherm is a line of constant temperature, and that the further that line is away from the axes, the higher its constant temperature is. To take advantage of this visualization, follow these steps:

- Sketch a set of isotherms for a series of increasing temperatures on the p-V diagram (▶ **Figure 12.13**).
- Then sketch the process you are analyzing – for example, the isobar shown in Figure 12.13. Remember that on the axes an isochor (constant volume) is a vertical line, an isobar (constant pressure) is a horizontal line, and an adiabat (no heat flow) is a downward sloping curve, steeper than an isotherm.
- Next, use the process line to determine the signs of W_{gas} and ΔE_{th}. Keep in mind that W_{gas} is the area under the p-V curve for the process, and its sign is positive for gas expansion and negative for compression. The sign of ΔT should be clear from the isotherms, since temperature changes as the process "crosses" from one to another. The sign of ΔE_{th} is then determined because it is the same as the sign of ΔT. (Recall $\Delta E_{\text{th}} \propto \Delta T$.)
- Last, the sign of Q can be found by using Equation 12.4, $Q = \Delta E_{\text{th}} + W_{\text{gas}}$. Then the sign of Q tells whether heat was transferred into or out of the system.

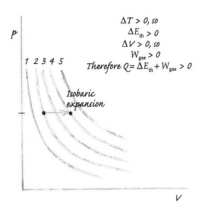

▲ **FIGURE 12.13 Isotherms on a p-V diagram** A useful tool for visualizing the signs of temperature changes, heat flow, work done, and thermal energy changes is to think of the p-V diagram as "overlayed" by "nested" isotherms of increasing temperature (here $T_5 > T_4 > T_3 > T_2 > T_1$). An example using an isobaric expansion is shown. Can you see how the correct signs are determined this way?

The isobaric process in Figure 12.13 shows the power of this visual approach. Here, the direction of heat flow during an isobaric expansion is to be determined. Expansion implies positive work done by the gas. But what is the direction of the heat flow (or is it zero)? After sketching the isobar, it is clear that it crosses from lower-temperature isotherms to higher-temperature ones. Hence, there is a temperature increase, and therefore ΔE_{th} is positive. Since $Q = \Delta U + W$, it is seen that Q is positive since it is the sum of two positive quantities, and therefore heat enters the gas.

As an exercise in using this approach, try analyzing Examples 12.2 through 12.4. As a final summary of our thermal processes, the characteristics and consequences of these processes are listed in ▼ **Table 12.1**.

TABLE 12.1   Summary of Thermodynamic Processes

Process	Definition	Characteristic	First Law Result
Isothermal	$T = $ constant	$\Delta E_{\text{th}} = 0$	$Q = W_{\text{gas}}$
Isobaric	$p = $ constant	$W_{\text{gas}} = p\Delta V$	$Q = \Delta E_{\text{th}} + p\Delta V$
Isometric	$V = $ constant	$W_{\text{gas}} = 0$	$Q = \Delta E_{\text{th}}$
Adiabatic	$Q = 0$	No heat flow	$\Delta E_{\text{th}} = -W_{\text{gas}}$

12.3 The Second Law of Thermodynamics and Entropy

Suppose that a piece of hot metal is placed in an insulated container of cold water. Heat will be transferred from the metal to the water, and the two will eventually reach thermal equilibrium at some intermediate temperature. For a thermally isolated system, the system's total energy remains constant. Could heat have been transferred from the cold water to the hot metal instead?

This process would not happen naturally. But if it did, the total energy of the system could remain constant, and this "impossible" inverse process would not have violated energy conservation.

There must be another principle that specifies the direction in which a process can take place. This principle is embodied in the **second law of thermodynamics**, which states that certain processes cannot take place even though they may be consistent with the first law.

There are several equivalent statements of the second law. Which one is used depends on the situation to which it is being applied. The one applicable to the aforementioned situation is:

Heat will not flow spontaneously from a cooler body to a warmer body.

An equivalent alternative statement of the second law involves **thermal cycles**. A thermal cycle consists of several separate thermal processes in which the system ends up at its starting conditions. If the system is a gas operating as a **heat engine**, heat enters the system and it does useful mechanical work. The second law, stated in terms of a thermal cycle operating as a heat engine is:

In a thermal cycle, heat energy cannot be completely transformed into mechanical work.

Another equivalent form of the second law applies to perpetual motion engines – ones that can do work without a net heat input. Such machines would be able to do work without losing energy. The second law says:

It is impossible to construct an operational perpetual motion machine.

Attempts have been made to construct such perpetual machines, with no success.*

12.3.1 Entropy

A quantity that quantitatively indicates the natural direction a process can take was first described by Rudolf Clausius (1822–1888), a German physicist. This quantity is **entropy (S)**. Entropy is a multifaceted concept, with various different (equivalent) physical interpretations:

- Entropy is a measure of a system's ability to do useful work. As a system loses the ability to do work, its entropy increases.
- Entropy determines the direction of time. It is "time's arrow" that points to the "forward" flow of events, distinguishing past events from future ones.
- Entropy is a measure of disorder. All systems naturally move toward greater disorder, or disarray, or increase entropy.
- The total entropy of the universe is increasing.

---

* Although perpetual motion *machines* cannot exist, (very nearly) perpetual motion is known to exist – for example, the planets have been in motion around the Sun for about 5 billion years.

The definition of the change in a system's entropy (ΔS) when an amount of heat (Q) is added or removed during a reversible process *at constant temperature* is

$$\Delta S = \frac{Q}{T} \quad \text{(change in entropy at constant temperature)} \qquad (12.12)$$

SI unit of entropy: joule per kelvin (J/K)

Notice that ΔS is positive if a system absorbs heat ($Q > 0$) and negative if a system loses heat ($Q < 0$). If the temperature changes during the process, calculating the change in entropy requires advanced mathematics and thus our calculations are limited to isothermal processes. Let's look at an example of a how a change in entropy is calculated at a phase change.

EXAMPLE 12.6: CHANGE IN ENTROPY – AN ISOTHERMAL PROCESS

Estimate the change in entropy of a 40-g bead of perspiration (water) that evaporates entirely from an area of skin at a constant temperature 34 °C.

THINKING IT THROUGH. This phase change occurs at constant temperature; hence, Equation 12.12 applies using temperature in kelvins. The heat added can be computed from the heat of transformation.

SOLUTION

Listing the quantities given:

Given:

$m = 40 \text{ g} = 0.040 \text{ kg}$
$T = (34 + 273) \text{ K} = 307 \text{ K}$
$L_v = 2.26 \times 10^6 \text{ J/kg}$

Find: ΔS (change in entropy of water bead)

The heat absorbed by the bead of water to completely evaporate it is:

$$Q = mL_v = (0.040 \text{ kg})(2.26 \times 10^6 \text{ J/kg}) = 9.04 \times 10^4 \text{ J}$$

Then the change in entropy of that amount of water is

$$\Delta S = \frac{Q}{T} = \frac{+9.04 \times 10^4 \text{ J}}{307 \text{ K}} = +2.94 \times 10^2 \text{ J/K}$$

Q is positive and thus the change in entropy is also positive. In other words, the entropy of the water increases. This is reasonable, because a gaseous state is more random (disordered) than a liquid state.

FOLLOW-UP EXERCISE. What is the change in entropy of a 1.00-kg water sample when it freezes to form ice at 0 °C?

Note that the entropy change of the water in Example 12.5 is positive, because the process is a natural one. In general, the direction of any natural process is toward an increase in total system entropy. That is, the entropy of an isolated system never decreases. Another way to state this is that *the entropy of an isolated system increases for every natural process* ($\Delta S > 0$). As an example, an isolated tray of liquid water will never naturally (spontaneously) cool and freeze into ice. This would require a decrease in the entropy of an isolated system.

However, if a system is not isolated, it may undergo a decrease in entropy. For example, if the tray of water (as previously mentioned) is instead put into a freezer, the water will freeze, undergoing a decrease in entropy. But there will be an increase somewhere else in the universe? In this case, the freezer warms the kitchen as it freezes the ice, and the *total entropy* of the system (ice plus kitchen) increases.

Thus, yet another statement of the second law of thermodynamics, this time in terms of entropy is:

The total entropy of the universe increases in every natural process.

Last, consider the foregoing statements rewritten in terms of order and disorder.

All naturally occurring processes move toward a state of greater disorder or disarray.

A working definition of order and disorder may be extracted from everyday observations. For example, suppose you are making a pasta salad and have chopped tomatoes to toss into cooked pasta. Before you mix the pasta and tomatoes, there is a relative amount of order; that is, the ingredients are separate and unmixed. Upon mixing, the separate ingredients become one dish, and there is less order (or more disorder, if you prefer). The pasta salad, once mixed, will never separate into the individual ingredients on its own (i.e., by a natural process – one that happens on its own). However, you could go in and pick out the individual tomato pieces, but that would not be a natural process. As one more entropy illustration, consider the following Example.

EXAMPLE 12.7: A WARM SPOON INTO COOL WATER – ENTROPY INCREASE OR DECREASE?

A metal spoon at 24 °C is immersed in 1.00 kg of water at 18 °C. This system is thermally isolated and comes to equilibrium at 20 °C. (a) Find the approximate change in the system's entropy. (b) Repeat the calculation, assuming that the water temperature drops to 16 °C and the spoon's increases to 28 °C. Comment on the difference in the signs of total entropy changes in (a) and (b).

THINKING IT THROUGH. The system is thermally isolated, hence $Q_s + Q_w = 0$ (subscripts s and w stand for spoon and water, respectively). Q_w is determined from the water mass, specific heat,

and temperature change. The spoon heat exchange is the same but opposite sign. Here the temperature changes are small, so a good approximation for ΔS results by using each object's *average* temperature $\overline{T}$.

SOLUTION

Listing the given values, with subscripts i and f standing for *initial* and *final*, respectively:

Given:

$T_{s,i} = 24$ °C
$T_{w,i} = 18$ °C
$m_w = 1.00$ kg
$c_w = 4186$ J/(kg·°C) (Table 11.1)
(a) $T_f = 20$ °C
(b) $T_{s,f} = 28$ °C; $T_{w,f} = 16$ °C

Find:

(a) ΔS (change in system entropy)
(b) ΔS (change in system entropy)

(a) The heat gained by the water is, from Equation 11.1:

$$Q_w = c_w m_w \Delta T = [4186 \text{ J/(kg·C°)}](1.00 \text{ kg})(2.0 \text{ °C})$$
$$= +8.37 \times 10^3 \text{ J}$$

and therefore, $Q_s = -8.37 \times 10^3$ J.
The average temperatures are

$$\overline{T}_w = 19 \text{ °C} = 292 \text{ K} \quad \text{and} \quad \overline{T}_s = 22 \text{ °C} = 295 \text{ K}$$

Equation 12.10 gives the approximate entropy changes for both the water and the spoon:

$$\Delta S_w \approx \frac{Q_w}{T_w} = \frac{+8.37 \times 10^3 \text{ J}}{292 \text{ K}} = +28.7 \text{ J/K} \quad \text{and}$$
$$\Delta S_s \approx \frac{Q_s}{T_s} = \frac{-8.37 \times 10^3 \text{ J}}{295 \text{ K}} = -28.4 \text{ J/K}$$

The change in the entropy of the system is the sum of these

$$\Delta S \approx \Delta S_w + \Delta S_s \approx +28.7 \text{ J/K} - 28.4 \text{ J/K} = +0.3 \text{ J/K}$$

The spoon's entropy decreased while the water's increased by more. Thus *system* entropy increased.

(b) Although this situation conserves energy, it should violate the second law of thermodynamics. Let's check the system entropy change to see if this is correct. The heat lost by the water is determined by using its 2.0 °C temperature drop and that gained by the spoon is just the opposite, hence:

$$Q_W = c_W m_W \Delta T = [4186 \text{ J/(kg°C)}](1.00 \text{ kg})(-2.0 \text{ °C})$$
$$= -8.37 \times 10^3 \text{ J and therefore}$$
$$Q_s = -8.37 \times 10^3 \text{ J}$$

The average temperatures of $\bar{T}_w = 17\,°C = 290\,K$ and $\bar{T}_s = 26\,°C = 299\,K$, thus approximate entropy changes of each are:

$$\Delta S_w \approx \frac{Q_w}{\bar{T}_w} = \frac{-8.37 \times 10^3 \, J}{290\ K} = -28.9 \ J/K \quad \text{and}$$

$$\Delta S_s \approx \frac{Q_s}{\bar{T}_s} = \frac{+8.37 \times 10^3 \, J}{299\ K} = +28.0 \ J/K$$

Thus the change in the entropy of the *system* is:

$$\Delta S = \Delta S_w + \Delta S_s \approx -28.9 \ J/K + 28.0 \ J/K = -0.9 \ J/K$$

The spoon's entropy increased, but the water's decreased by more. Thus the system entropy decreased. This result shows that process (b) is inconsistent with the second law and therefore cannot happen naturally.

FOLLOW-UP EXERCISE. What should the initial temperatures in this Example be to make the overall system entropy change zero? Explain in terms of heat transfers.

12.4 Heat Engines and Thermal Pumps

For our purposes, a **heat engine** is any device that takes heat from a high-temperature source (a hot, or high-temperature, reservoir), converts some of it to useful work, and expels the rest to its surroundings (a cold, or low-temperature, reservoir). For example, most turbines that generate electricity are heat engines, using heat extracted from various fuel sources (oil, gas, coal, etc.). They might be cooled by river water, for example, thus losing heat to this low-temperature reservoir. A generalized heat engine is represented in ▶ **Figure 12.14a**. (The concern here is not with the mechanical/engineering details, such as pistons and cylinders, but instead will focus on the fundamental physical principles and results.) Also, it will be assumed that the "working substance" (the material that absorbs the heat and does the work) is an ideal gas, subject to the ideal gas laws and results. Using ideal gases makes the mathematics easier while enabling us to concentrate on the basic physics.

Since a *continuous* output is wanted, practical heat engines operate in a **thermal cycle**, or a series of processes that brings the system (the working substance) back to its initial condition. Cyclic heat engines include steam engines as well as internal combustion engines, such as those used in most automobiles.

An idealized, rectangular thermodynamic cycle is shown in Figure 12.14b. It consists of two isobars and two isomets. When these processes occur in the sequence indicated, the system goes through a cycle (1-2-3-4-1), returning to its original condition. When the gas expands (1 to 2), it does (positive) work equal to the area under the isobar. Doing positive work is the desired output

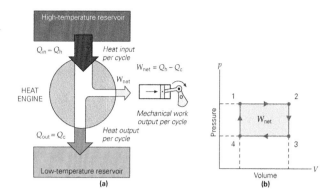

▲ **FIGURE 12.14** **Heat engine (a)** Energy flow for a generalized cyclic heat engine. All Qs are magnitudes only (no signs) in these diagrams. Note that the width of the arrow representing Q_h (heat flow out of hot reservoir) is equal to the combined widths of the arrows representing W_{net} and Q_c (heat flow into cold reservoir), reflecting the conservation of energy: $Q_h = Q_c + W_{net}$. **(b)** This specific cyclic process consists of two isobars and two isomets. The net work output per cycle is the area of the rectangle formed by the process paths. (See Example 12.11 for the analysis of this particular cycle.)

of an engine. (Think of a car engine piston moving the crankshaft.) However, there must be a compression of the gas (3 to 4) to bring it back to its initial conditions. During this phase, the work done by the gas is negative, which is *not* the purpose of an engine. Thus the important quantity in engine design is the net work done by the gas (W_{net}) per cycle. As shown graphically in ▶ **Figure 12.15**, this is just the area *enclosed* by the process curves that make up the cycle. In Figure 12.14b, the area was rectangular but in general it can be any shape. When the paths are not simple, numerical calculations of the areas may be difficult, but the concept remains the same regardless.

12.4.1 Thermal Efficiency

The **thermal efficiency** (ε) of a heat engine is defined as

$$\varepsilon = \frac{\text{net work output}}{\text{heat input}} = \frac{W_{net}}{Q_{in}} \qquad \begin{array}{l} \text{(thermal efficiency} \\ \text{of a heat engine)} \end{array} \qquad (12.13a)$$

This definition tells how much useful work (W_{net}) the engine does in comparison with the input heat it receives (Q_{in}). For example, automobile engines have an efficiency of about 20% to 25% which means that about one-fourth of the heat generated by igniting the air-gasoline mixture is converted into mechanical work, such as turning the car wheels. This also means that about three-fourths of the heat is wasted, being transferred to the environment from the hot exhaust system, radiator system, and metal engine.

W_{net} is determined by applying the first law of thermodynamics (energy conservation) to a cycle. For simplicity, during the discussion of heat engines and pumps, all heat symbols (Qs and Ws) will be magnitude only.

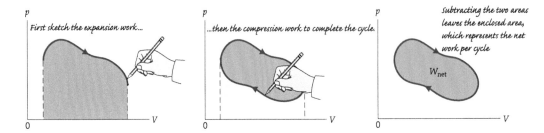

▲ FIGURE 12.15 **Representing the net work done during a thermal cycle** Subtracting the negative work done by the gas during compression from the positive work done during expansion gives the net work done by the gas per cycle.

For an ideal gas in a cycle, the overall change in thermal energy is zero because the overall temperature change is zero. As can be seen from Figure 12.14, $W_{net} = Q_h - Q_c$. Thus, the thermal efficiency (Equation 12.13a) can be rewritten entirely in terms of heat flows (remember all Qs are magnitude only)

$$\varepsilon = \frac{W_{net}}{Q_h} = \frac{Q_h - Q_c}{Q_h} = 1 - \frac{Q_c}{Q_h} \qquad (12.13b)$$

(efficiency of an ideal gas heat engine)

Like mechanical efficiency, thermal efficiency is a dimensionless fraction and is sometimes expressed as a percentage. From Equation 12.13b, to maximize the work output per cycle, the ratio Q_c/Q_h should be minimized. This means that the wasted heat should be as small as possible. But what is this smallest possible value? From Equation 12.13b it seems that a heat engine could have 100% efficiency if there were no exhaust heat during the cycle ($Q_c = 0$). This would require all input heat converted to useful work. However, because the gas returns to its initial state at the end of each cycle, there must always be some compressional

(negative) work done. Thus complete conversion of input heat to useful work by a heat engine cannot happen. This observation led Lord Kelvin (of Kelvin temperature scale fame) to re-state the second law in yet another equivalent manner:

No cyclic heat engine can convert its heat input completely to work.

One of the most common thermal engines is the gasoline (automotive) engine, which employs a *four-stroke cycle.* An approximation of this important cycle is shown in ▼ **Figure 12.16.** This cycle is called the *Otto cycle,* after the German engineer Nikolaus Otto (1832–1891), who built one of the first successful gasoline engines. During the *intake* stroke (1–2), an isobaric expansion, the air-fuel mixture is admitted at atmospheric pressure through the open intake valve as the piston drops. The rotational inertia of the engine's camshaft causes the piston to adiabatically (quickly) compress the air-fuel mixture during the *compression* stroke (2–3). This is followed by fuel ignition (3–4; the spark plug fires, causing an isometric pressure rise). Next, an adiabatic expansion produces the *power* stroke (4–5). Following this, an isometric

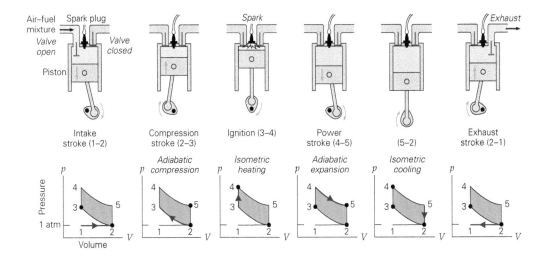

▲ FIGURE 12.16 **The four-stroke cycle of a heat engine** The steps of the four-stroke Otto cycle. The piston moves up and down twice each cycle, for a total of four strokes per cycle. See text for description.

cooling of the system takes place when the piston is at its lowest position (5–2). The final *exhaust* stroke is an isobaric leg (2–1) that expels the burned fuel products. Notice that during a cycle there are *two* up and down piston motions but only *one* power stroke.

EXAMPLE 12.8: THERMAL EFFICIENCY – WHAT YOU GET OUT OF WHAT YOU PUT IN

The small, gasoline-powered engine of a leaf blower absorbs 800 J of heat energy from a high-temperature reservoir (the ignited gas-air mixture) and exhausts 700 J to a low-temperature reservoir (the outside air, through its cooling fins) during one cycle. What is the engine's thermal efficiency?

THINKING IT THROUGH. The definition of thermal efficiency (Equation 12.13a) can be used if W_{net} can be determined. (Remember the Qs are magnitudes.)

SOLUTION

Given:

$$Q_h = 800 \, J$$
$$Q_c = 700 \, J$$

Find: ε (thermal efficiency)

The net work per cycle is

$$W_{net} = Q_h - Q_c = 800 \, J - 700 \, J = 100 \, J$$

Therefore, the thermal efficiency is

$$\varepsilon = \frac{W_{net}}{Q_h} = \frac{100 \, J}{800 \, J} = 0.125 \, (or \, 12.5\%)$$

FOLLOW-UP EXERCISE. (a) What would be the net work per cycle in this Example if the efficiency were raised to 15% and the input heat per cycle raised to 1000 J? (b) How much heat would be exhausted in this case?

The next example illustrates a practical application for a small heat engine.

EXAMPLE 12.9: THERMAL EFFICIENCY – PUMPING WATER

A gasoline-powered water pump can pump 7.6×10^3 kg/h (about 2000 gal) of water from a basement floor up to the ground. This requires 1.0 gal of gasoline (which has an energy content of 1.3×10^6 J/gal). Assume the basement floor is 3.0 m below ground level. (a) What is the thermal efficiency of this pump? (b) How much heat is wasted to the environment in 1.0 h? Assume the kinetic energy of the water is the same at both basement and ground level (at discharge).

THINKING IT THROUGH. The definition of thermal efficiency of a heat engine applies (Equation 12.13a). However, the heat input from the energy content of gasoline and the (net) work output to raise the water (i.e., increase its gravitational potential energy) are needed.

SOLUTION

The heat input is the released chemical energy stored in 1.0 gal of gasoline.

Given:

$$Q_h = (1.0 \, gal)(1.3 \times 10^6 \, J/gal) = 1.3 \times 10^6 \, J$$
$$m = (7.6 \times 10^3 \, kg/h)(1.0 \, h) = 7.6 \times 10^3 \, kg$$
$$\Delta y = 3.0 \, m$$

Find:

(a) ε (thermal efficiency)

(b) Q_c (heat to environment)

(a) The work output is equal to the increase in the water's gravitational potential energy:

$$W_{net} = mg\Delta y = (7.6 \times 10^3 \, kg)(9.80 \, m/s^2)(3.0 \, m) = 2.2 \times 10^5 \, J$$

and thermal efficiency is

$$\varepsilon = \frac{W_{net}}{Q_h} = \frac{2.2 \times 10^5 \, J}{1.3 \times 10^6 \, J} = 0.17 \, (or \, 17\%)$$

(b) The heat exhausted to the environment in 1 h is

$$Q_c = Q_h - W_{net} = 1.3 \times 10^6 \, J - 2.2 \times 10^5 \, J = 1.1 \times 10^6 \, J$$

FOLLOW-UP EXERCISE. If the heat exhausted to the environment were completely absorbed by the pumped water, what would be the temperature change of the water?

The human body is an example of a biological heat engine. The energy source is the energy metabolized from food and fatty tissues. Some of this energy is converted into work, and the rest is expelled to the environment as heat. Thus the efficiency of the human body is

$$\varepsilon = \frac{\text{work output}}{|\text{internal energy loss}|} = \frac{W_{net}}{|\Delta U_{int}|}$$

This efficiency is typically determined by measuring the *time rates* of these quantities during an exercise – that is, the power expended ($W_{net}/\Delta t$), is compared to the energy consumed per unit time (*metabolic rate*), $|\Delta U_{int}|/\Delta t$:

$$\varepsilon = \frac{P}{\left(|\Delta U_{int}| / \Delta t\right)}$$

The power output during a particular activity (e.g., pushing pedals on a bicycle) can be measured by a *dynamometer* (▶ **Figure 12.17**). The metabolic rate is typically measured using breathing devices, also shown in Figure 12.17. This is because the metabolic rate is known to be proportional to the rate of oxygen consumption. In this way, comparing the rate of oxygen consumption to the output power enables efficiencies to be measured.

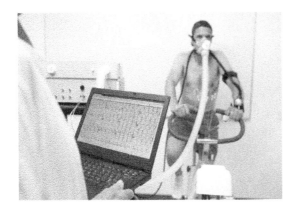

▲ FIGURE 12.17 **Measuring human efficiency during exercise** A cyclist is tested with a breathing device and a dynamometer so that both his power output and metabolic rate can be measured.

The efficiency of the human body depends on the muscles that are used. For activities involving the largest muscles in the body, such as leg muscles, the efficiencies are relatively high. For example, some professional bicycle racers can achieve efficiencies as high as 20%, generating more than 2 hp of power in short bursts. Arm muscles, conversely, are relatively small, so activities such as bench pressing have efficiencies of less than 5%. Like any heat engine, the human body must exhaust waste heat to the environment through processes such as perspiring so as to avoid overheating.

12.4.2 Thermal Pumps: Refrigerators, Air Conditioners, and Heat Pumps

The function performed by a thermal pump is basically the reverse of that of a heat engine. The name **thermal pump** is a generic term for *any* device that transfers heat from a low-temperature reservoir to a high-temperature reservoir (▼ **Figure 12.18a**). Here the term thermal pump includes all types: refrigerators, air conditioners, as well machines called *heat pumps* (see later in this section). For such a transfer to occur, there must be work input because the second law of thermodynamics says that heat will not spontaneously flow from cold to hot. For this to happen, work must be done on the system.

12.4.2.1 Air Conditioners and Refrigerators

A familiar example of a thermal pump is an air conditioner. By using electrical energy to do work on the system, heat is transferred from the inside of the house (low-temperature reservoir) to the outside of the house (high-temperature reservoir), as in Figure 12.18b. A refrigerator (▶ **Figure 12.19**) uses the same principles. In essence, a refrigerator or air conditioner "pumps" heat up *a* temperature gradient, or "temperature hill." Of course, the natural flow of heat would be back into the cool interior, which is the reason for insulated walls. (Think of pumping water out of a basement only to have it want to seep back in from ground level.)

The "performance" of a refrigerators and air conditioners is based on their purpose – to extract heat from the low-temperature reservoir (refrigerator interior, freezer, or cool house interior). Thus their performance is based on the amount of heat removed compared to the work (W_{in}) needed to do so. (As a consumer, more heat for less work is what is wanted.) Thus, because their end use is different, air conditioner/refrigerator system performance is defined differently from heat engine efficiencies.

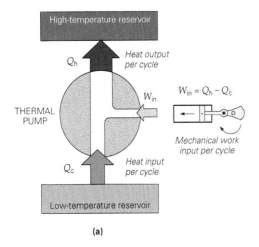

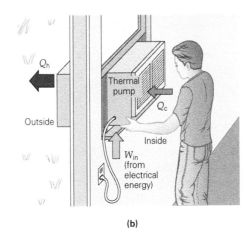

(a)   (b)

▲ FIGURE 12.18 **Thermal pumps (a)** An energy flow diagram for a generalized cyclic thermal pump. Notice the width of the arrow representing Q_h, the heat transferred to the high-temperature reservoir, is equal to the combined widths of the arrows representing W_{in} and Q_c, reflecting conservation of energy: $Q_h = W_{in} + Q_c$. **(b)** An air conditioner is an example of a thermal pump. Using the input work, it transfers heat from a low-temperature reservoir (house interior) to a high-temperature reservoir (outside).

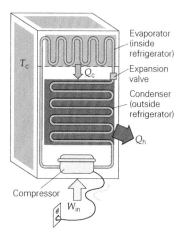

Evaporator (inside refrigerator)

T_c

Q_c

Expansion valve

Condenser (outside refrigerator)

Q_h

Compressor

W_{in}

▲ **FIGURE 12.19 Refrigerator** Heat (Q_c) is taken from the cool interior by the refrigerant as heat of transformation. Q_h is transferred from the condenser to the warmer surroundings.

For these cooling appliances, a measure of their "efficiency" is more correctly expressed as a **coefficient of performance (COP)**. Since the purpose is to extract the most heat compared to the work needed, it is defined as their ratio, or:

$$\text{COP}_{ref} = \frac{Q_c}{W_{in}} \quad \text{(refrigerator/air conditioner)} \quad (12.14a)$$

An alternative expression is based on the fact that a heat pump operates in a cycle. Thus $\Delta E_{cycle} = 0$ and by energy conservation $Q_c + W_{in} = Q_h$ or $W_{in} = Q_h - Q_c$. Substituting this into Equation 12.14a:

$$\text{COP}_{ref} = \frac{Q_c}{Q_h - Q_c} \quad \text{(refrigerator/air conditioner)} \quad (12.14b)$$

The greater the COP, the better the performance – that is, more heat is extracted for each unit of work done. For normal operation, the work input is less than the heat removed, so the COP is greater than 1. The COPs of typical refrigerators and air conditioners range from 3 to 5, depending on operating conditions and design details. This means that the amount of heat removed from the cold reservoir (the refrigerator, freezer, or house interior) is three to five times the amount of work needed to remove it. So for every joule of work done (and paid for), three to five joules of heat are removed – not a bad deal!

12.4.2.2 Heat Pumps

For our purposes, the term *heat pump* applies to commercial devices used to cool homes and offices in summer and to heat them in winter. The summer operation is that of an air conditioner. However, in its winter heating mode, a heat pump heats the interior and cools the outdoors, usually by taking

thermal energy from the cold outside air or ground. For a heat pump in its heating mode, the aim is to keep the house interior warm in spite of the fact that the heat naturally flows out to the cold exterior – again the reason for good insulation. Thus the quantity of interest is Q_h, not Q_c. Hence the COP for heat pump (designated with subscript hp) is the ratio of Q_h to W_{in}:

$$\text{COP}_{hp} = \frac{Q_h}{W_{in}} \quad \text{(heat pump in heating mode)} \quad (12.15a)$$

But from energy conservation, $Q_c + W_{in} = Q_h$ or $W_{in} = Q_h - Q_c$. This can be substituted into Equation 12.15a to create an alternative expression entirely in terms of the heat flows:

$$\text{COP}_{hp} = \frac{Q_h}{Q_h - Q_c} \quad \text{(heat pump in heating mode)} \quad (12.15b)$$

Typical COP_{hp} values range from 2 and 4, again depending on the operating conditions and design. Compared with electrical heating, heat pumps are very efficient. For a joule of electric energy, a heat pump typically delivers from two to four joules as heat, while direct electric heating systems (hot coils) could provide at most 1 joule.

EXAMPLE 12.10: AIR CONDITIONER/HEAT PUMP – THERMAL SWITCH HITTING

A thermal pump operating as an air conditioner in summer extracts 1000 J of heat from the interior of a house for every 400 J of electric energy required to operate it. Determine (a) the air conditioner's COP and (b) its COP if it runs as a heat pump in the winter. Assume it is capable of moving the same amount of heat for the same amount of electric energy, regardless of the direction in which it runs.

THINKING IT THROUGH. The input work and input heat in (a) is known, so the COP for a refrigerator (Equation 12.14a) can be applied. For the reverse operation, the definition of COP for a heat pump (Equations 12.15a or 12.15b) is appropriate.

SOLUTION
Given:

$$Q_c = 1000 \text{ J}$$
$$W_{in} = 400 \text{ J}$$

Find:

(a) COP_{ref}
(b) COP_{hp}

(a) From Equation 12.14a, the COP as an air conditioner is

$$\text{COP}_{ref} = \frac{Q_c}{W_{in}} = \frac{1000 \text{ J}}{400 \text{ J}} = 2.5$$

(b) Here the relevant quantity is the output heat

$$Q_h = Q_c + W_{in} = 1000\,J + 400\,J = 1400\,J$$

Thus, acting as a heat pump, the COP is

$$COP_{hp} = \frac{Q_h}{W_{in}} = \frac{1400\,J}{400\,J} = 3.5$$

FOLLOW-UP EXERCISE. (a) Suppose you redesigned the thermal pump in this Example to perform the same operation, but with 25% less work input. What would be the two COPs then? (b) Which COP would have the larger percentage increase?

12.5 The Carnot Cycle and Ideal Heat Engines

From the second law of thermodynamics, all heat engines must exhaust some heat. But how much must be lost in the process? What is their *maximum* efficiency? It must be less than 100%, but how much less? The answers came from Sadi Carnot (1796–1832), a French engineer. He found the thermodynamic cycle that an ideal, or most efficient, heat engine would use.

Specifically, he found that the ideal engine absorbs heat from a constant high-temperature reservoir (T_h) and exhausts it to a constant low-temperature reservoir (T_c). These processes were therefore isothermal. To complete the ideal cycle, Carnot showed the remaining processes were adiabatic. He concluded that any real engine operating between two constant temperature reservoirs could *not* have an efficiency greater than that of his ideal heat engine, operating between the same temperatures. This ideal cycle is called the **Carnot cycle** and is shown in ▼ **Figure 12.20**.

Carnot showed that the maximum efficiency, called the **Carnot efficiency** (ε_C), depends only upon the two temperature extremes (temperatures in kelvins, as usual):

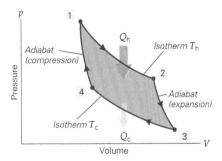

▲ **FIGURE 12.20 The Carnot (ideal) cycle** The Carnot, or ideal, cycle consists of two isotherms and two adiabats. Heat is absorbed during the isothermal expansion and exhausted during the isothermal compression. An engine following this cycle has the maximum possible efficiency of any engine operating between the same two temperatures. (See text for details.)

$$\varepsilon_C = 1 - \frac{T_c}{T_h} \quad \text{(Carnot efficiency, ideal heat engine)} \quad (12.16)$$

As usual, this efficiency is often expressed as a percentage rather than a fraction. Note that the Carnot efficiency is the theoretical upper limit on the efficiency of a cyclic heat engine operating between two known temperatures. In practice, this limit cannot be achieved because the reversible processes can only be approximated. However, the Carnot efficiency does illustrate a general idea of how to achieve greater efficiency: The greater the difference in the temperatures of the heat reservoirs, the greater the Carnot efficiency. For example, if T_h is twice T_c or $T_c/T_h = 0.50$, the Carnot efficiency is

$$\varepsilon_C = 1 - \frac{T_c}{T_h} = 1 - 0.50 = 0 - 5.0 (\times 100\%) = 50\%$$

However, if T_h is four times T_c, the Carnot efficiency would be 75% (you should be able to show this).

Since a heat engine can never attain 100% thermal efficiency, it is useful to compare its actual efficiency ε with its maximum efficiency, ε_C. To see this, study the next Example carefully.

EXAMPLE 12.11: CARNOT EFFICIENCY – THE YARDSTICK FOR REAL ENGINES

An engineer is designing a cyclic heat engine to operate between the temperatures of 150 °C and 27 °C. (a) What is the maximum efficiency that can be achieved? (b) Suppose the engine, when built, does 1000 J of work per cycle for every 5000 J of input heat per cycle. What is its efficiency, and how close is it to the Carnot efficiency?

THINKING IT THROUGH. The Carnot efficiency is given by Equation 12.16. Temperatures must first be converted to kelvins. In (b), the actual efficiency should be calculated, and it should be less than the result in (a).

SOLUTION

Given:

$$T_h = (150 + 273)\,K = 423\,K$$
$$T_c = (27 + 273)\,K = 300\,K$$
$$W_{net} = 1000\,J$$
$$Q_h = 5000\,J$$

Find:

(a) ε_C (Carnot efficiency)
(b) ε (actual efficiency) compare it to ε_c

(a) Using Equation 12.16 to find the maximum efficiency (both as a fraction and a percentage)

$$\varepsilon_C = 1 - \frac{T_c}{T_h} = 1 - \frac{300\,K}{423\,K} = 0.291(\times 100\%) = 29.1\%$$

(b) The actual efficiency is, from Equation 12.13a,

$$\varepsilon = \frac{W_{net}}{Q_h} = \frac{1000\,J}{5000\,J} = 0.200\,(\text{or } 20.0\%)$$

Thus expressing their comparison as a ratio and percentage

$$\frac{\varepsilon}{\varepsilon_C} = \frac{0.200}{0.291} = 0.687\,(\text{or } 68.7\%)$$

In other words, the heat engine is operating at 68.7% of its theoretical maximum. That's pretty good. It should not be compared to 100% – an impossible goal for any engine.

FOLLOW-UP EXERCISE. If the operating high temperature of the engine in this Example were increased to 200 °C, what would be the change in the theoretical efficiency?

Chapter 12 Review

- A version of the **first law of thermodynamics** useful in the study of heat engines and refrigerators is

$$Q = \Delta U_{int} + W_{gas} \tag{12.4}$$

- Important **thermodynamic processes** (for gases) are:

  isothermal: a process at constant temperature
  isobaric: a process at constant pressure
  isometric: a process at constant volume
  adiabatic: a process involving no heat flow ($Q = 0$)

- The expressions for **thermodynamic work** done by an ideal gas during various processes are

$$W_T = nRT \ln\left(\frac{V_2}{V_1}\right) \quad \text{(ideal gas isothermal process)} \tag{12.5}$$

$$W_p = p(V_2 - V_1) = p\Delta V \quad \text{(ideal gas isobaric process)} \tag{12.6}$$

$$W_{ad} = \frac{p_1 V_1 - p_2 V_2}{\gamma - 1} \quad \text{(ideal gas adiabatic process)} \tag{12.11}$$

($\gamma = c_p/c_v$ is the ratio of specific heats at constant pressure and volume, respectively.)

- The **second law of thermodynamics** determines whether a process can take place naturally or alternatively, and it specifies the direction a process can take.

- **Entropy (*S*)** is a measure of the disorder of a system. The **change in entropy** of a system at constant temperature is

$$\Delta S = \frac{Q}{T} \quad \begin{array}{l}\text{(change in entropy}\\ \text{at constant temperature)}\end{array} \tag{12.12}$$

- A **heat engine** is a device that converts heat into work. The **thermal efficiency** ε of an engine is

$$\varepsilon = \frac{W_{net}}{Q_h} = \frac{Q_h - Q_c}{Q_h} = 1 - \frac{Q_c}{Q_h} \tag{12.13b}$$

- A **thermal pump** is a device that transfers heat energy from a low-temperature reservoir to a high-temperature reservoir. The coefficient of performance (COP) is the ratio of heat transferred to the input work. The expression for COP depends on whether the thermal pump is used as a heat pump or as an air conditioner/refrigerator.

$$COP_{ref} = \frac{Q_c}{W_{in}} \quad \text{(refrigerator/air conditioner)} \tag{12.14a}$$

$$COP_{hp} = \frac{Q_h}{W_{in}} \quad \text{(heat pump in heating mode)} \tag{12.15a}$$

- A **Carnot cycle** is a theoretical heat engine cycle consisting of two isotherms and two adiabats. Any real heat engine has a lower efficiency that the Carnot efficiency, which is given by

$$\varepsilon_C = 1 - \frac{T_c}{T_h} \quad \begin{array}{l}\text{(Carnot efficiency, ideal}\\ \text{heat engine)}\end{array} \tag{12.16}$$

End of Chapter Questions and Exercises

Multiple Choice Questions

12.1 The First Law of Thermodynamics
and
12.2 Thermodynamic Processes for an Ideal Gas

1. During an isobaric expansion of an ideal gas, which of these do not change: the gas's (a) thermal energy, (b) temperature, (c) volume, (d) pressure.

2. There is no heat flow into or out of the system in an (a) isothermal process, (b) adiabatic process, (c) isobaric process, (d) isometric process.

3. If the work done by a system is equal to zero, the process must be (a) isothermal, (b) adiabatic, (c) isobaric, (d) isometric.

4. According to the first law of thermodynamics, if compressional work is done on an ideal gas, then (a) its thermal energy must change, (b) heat must be transferred from it, (c) its thermal energy may change and/or heat may be transferred from it, (d) heat must be transferred to it.

5. When heat is added to an ideal gas during an isothermal expansion, (a) work is done by the ideal gas, (b) the ideal gas's thermal energy increases, (c) its thermal energy decreases.

6. When an ideal gas expands adiabatically, (a) work is done by it, (b) its thermal energy does not change, (c) its temperature increases.

7. When an ideal gas is compressed isobarically, (a) its thermal energy increases, (b) it does negative work on the environment, (c) its pressure increases.

12.3 The Second Law of Thermodynamics and Entropy

8. During any natural process, the overall change in the entropy of the system involved could not be (a) negative, (b) zero, (c) positive.

9. For which type of thermodynamic process is the change in entropy of an ideal gas zero: (a) isothermal, (b) isobaric, (c) isometric, or (d) none of the preceding?

10. Which one of the following statements is a violation of the second law of thermodynamics: (a) heat flows naturally from hot to cold, (b) heat can be completely converted to mechanical work, (c) the entropy of the universe can never decrease?

11. An ideal gas is compressed isothermally. The change in entropy of the gas for this process is (a) positive, (b) negative, (c) zero, (d) none of the preceding.

12.4 Heat Engines and Thermal Pumps

12. When the first law of thermodynamics is applied to a heat engine, the result is: (a) $W_{net} = Q_h + Q_c$, (b) $W_{net} = Q_h - Q_c$, (c) $W_{net} = Q_c - Q_h$, (d) $Q_c = 0$.

13. For a cyclic heat engine: (a) $\varepsilon = 1$, (b) $Q_h = W_{net}$, (c) $\Delta E_{th} = W_{net}$, (d) $Q_h > Q_c$.

14. A thermal pump (a) is rated by thermal efficiency, (b) requires work input, (c) has $Q_h = Q_c$, (d) must have a COP < 1.

15. Which of the following is the determining factor in the thermal efficiency of a heat engine?
(a) $Q_c \times Q_h$, (b) Q_c / Q_h, (c) $Q_h - Q_c$, or (d) $Q_h + Q_c$?

12.5 The Carnot Cycle and Ideal Heat Engines

16. The Carnot cycle consists of (a) two isobaric and two isothermal processes, (b) two isometric and two adiabatic processes, (c) two adiabatic and two isothermal processes, or (d) any four processes that return the system to its initial state.

17. Which of the following temperature reservoir relationships would yield the lowest efficiency for a Carnot engine? (a) $T_c = 0.15T_h$, (b) $T_c = 0.25T_h$, (c) $T_c = 0.50T_h$, or (d) $T_c = 0.90T_h$?

18. For a heat engine that operates between two reservoirs with temperatures T_c and T_h, the Carnot efficiency is the (a) highest possible, (b) lowest possible, (c) an average value, or (d) none of the preceding.

19. For a given amount of heat removed from its interior Q_c, which refrigerator has the highest COP?
(a) $Q_h = 1.1\,Q_c$, (b) $Q_h = 1.5\,Q_c$, (c) $Q_h = 2.5\,Q_c$, or (d) $Q_h = 5.5\,Q_c$?

Conceptual Questions

12.1 The First Law of Thermodynamics and
12.2 Thermodynamic Processes for an Ideal Gas

1. On a p-V diagram, sketch an isobaric process for an ideal gas that results in a temperature drop.

2. On a p-V diagram, sketch a cyclic process that consists of an isothermal expansion followed by an isobaric compression and ending with an isometric process.

3. In ▼ **Figure 12.21**, the plunger of a syringe is pushed in quickly, and small pieces of paper at the bottom of the syringe get hot enough to catch fire. Explain this using the first law of thermodynamics.

▲ FIGURE 12.21 **Syringe fire** See Conceptual Question 3 for details.

4. Discuss heat, work done by you, and the change in internal energy of your body when you shovel snow.

5. In an adiabatic process, there is no heat exchange between the system and the environment, but the temperature of the ideal gas changes. How can this be? Explain.

6. In an isobaric expansion, an ideal gas sample can do work on the environment, but its temperature also increases. How can this be? Explain.

7. An ideal gas initially at temperature T_o, pressure p_o, and volume V_o is compressed to one-half its initial volume. As shown in ▼ **Figure 12.22**, it undergoes three processes always restarting at the same initial conditions:

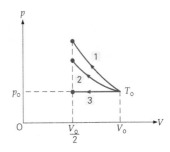

▲ FIGURE 12.22 **Thermodynamic processes** See Conceptual Question 7 for details.

1 is adiabatic, 2 is isothermal, and 3 is isobaric. Rank the work done on the gas and the final temperatures of the gas, from highest to lowest, for all three processes, and explain your reasoning for the rankings.

8. If ideal gas sample A receives more heat than ideal gas sample B, will A necessarily experience a higher increase in thermal energy? Explain.

9. Using the kinetic theory (microscopic) picture of an ideal gas, explain the increase in temperature during an adiabatic compression.

10. Using the kinetic theory (microscopic) picture of an ideal gas, explain the decrease in thermal energy that occurs during an adiabatic expansion.

12.3 The Second Law of Thermodynamics and Entropy

11. Heat is converted to mechanical energy in many applications, such as car engines. Is this a violation of the second law of thermodynamics? Explain.

12. Does the entropy of each of the following objects increase or decrease? (a) *Ice* as it melts; (b) *water vapor* as it condenses; (c) *water* as it is heated on a stove; (d) *food* as it is cooled in a refrigerator.

13. When a quantity of hot water is mixed with a quantity of cold water, the combined system comes to thermal equilibrium at some intermediate temperature. How does the entropy of each liquid change? The system?

14. A student challenges the second law of thermodynamics by saying that entropy does not have to increase in all situations, such as when water freezes to ice. Is this challenge valid? Why or why not?

15. The Sun formed into a more ordered system by gravitational collapse of a huge gas cloud about five billion years ago. How could this apparent entropy decrease happen if natural processes are supposed to result in an entropy increase? Explain.

16. A student tries to cool his dormitory room by opening the refrigerator door. Will that work? Explain.

12.4 Heat Engines and Thermal Pumps

17. What is the net change in pressure and thermal energy of a cyclic ideal gas heat engine at the end a complete cycle?

18. Lord Kelvin's statement of the second law of thermodynamics as applied to heat engines ("No heat engine operating in a cycle can convert its heat input completely to work") refers to their operation *in a cycle*. Why is the phrase "in a cycle" included?

19. If heat engine A absorbs more heat than heat engine B from a hot reservoir, will engine A necessarily do more net work than engine B? Explain your reasoning.

20. The heat output of a thermal pump is greater than the energy used to operate the pump. Does this device violate the first law of thermodynamics? Explain.

21. The maximum efficiency of a heat engine is less than 1 (<100%). Then how can the COP of a thermal pump be greater than 1? Explain.

12.5 The Carnot Cycle and Ideal Heat Engines

22. Diesel engines are more efficient than gasoline engines. Which type of engine would you expect to run hotter? Why?

23. If you have the choice of running your heat engine between either of the following two sets of temperatures for the cold and hot reservoirs, which would you choose, and why: between 100 °C and 300 °C or between 50 °C and 250 °C? Explain.

24. Carnot engine A operates at a higher hot reservoir temperature than Carnot engine B. Will engine A necessarily have a higher Carnot efficiency? Explain.

Exercises*

Integrated Exercises (IEs) are two-part exercises. The first part typically requires a conceptual answer choice based on physical thinking and basic principles. The following part is quantitative calculations associated with the conceptual choice made in the first part of the exercise.

12.1 The First Law of Thermodynamics
and
12.2 Thermodynamic Processes for an Ideal Gas

1. • When playing in a tennis match, you lose 6.5×10^5 J of heat, and your internal energy decreases by 1.2×10^6 J. How much work did you do in the match?

2. IE • A rigid container contains 1.0 mol of an ideal gas that slowly receives 2.0×10^4 J of heat. (a) The work done by the gas is (1) positive, (2) zero, (3) negative. Why? (b) What is the change in the thermal energy of the gas?

3. IE • A quantity of ideal gas goes through an isothermal process and does 400 J of net work. (a) The thermal energy of the gas is (1) higher than, (2) the same as, (3) less than when it started. Why? (b) Is heat added to or removed from the system, and how much is involved?

4. • An ideal gas goes through a thermodynamic process in which 500 J of work is done on the gas and the gas loses 300 J of heat. What is the change in thermal energy of the gas?

5. IE • While doing 500 J of work, an ideal gas expands adiabatically to 1.5 times its initial volume. (a) The temperature of the gas (1) increases, (2) remains the same, (3) decreases. Why? (b) What is the change in the thermal energy of the gas?

6. IE • An ideal gas expands from 1.0 m³ to 3.0 m³ at atmospheric pressure while absorbing 5.0×10^5 J of heat. (a) The temperature of the gas (1) increases, (2) stays the same, (3) decreases. Explain. (b) What is the change in thermal energy of the gas?

---

* The bullets denote the degree of difficulty of the exercises: •, simple; ••, medium; and •••, more difficult.

7. •• An ideal gas is under an initial pressure of 2.45×10^4 Pa and occupies a volume of 0.20 m³. The slow addition of 8.4×10^3 J of heat to this gas causes it to expand isobarically to a volume of 0.40 m³. (a) How much work is done by the gas? (b) Does the thermal energy of the gas change? If so, by how much?

8. •• An Olympic weight lifter lifts 145 kg a vertical distance of 2.1 m. When he does so, 6.0×10^4 J of heat is transferred to air through perspiration. Does he gain or lose internal energy, and how much?

9. IE •• An ideal gas is taken through the processes shown in ▼ **Figure 12.23**. (a) Is the overall change in the thermal energy of the gas (1) positive, (2) zero, or (3) negative? Explain. (b) In terms of p and V, how much work is done by or on the gas, and (c) what is the net heat transfer in the overall process?

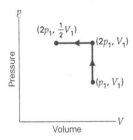

▲**FIGURE 12.23** **A p-V diagram for an ideal gas** See Exercise 9 for details.

10. •• A fixed quantity of ideal gas undergoes a series of processes shown in the p-V diagram in ▼ **Figure 12.24**. How much work is done in each process?

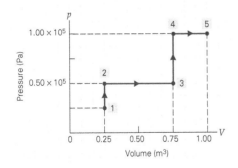

▲**FIGURE 12.24** **A p-V diagram and work** See Exercises 10 and 11 for details.

11. •• Suppose that after the final process in Figure 12.24, the pressure of the gas is first decreased isometrically from 1.0×10^5 Pa to 0.70×10^5 Pa, and then compressed isobarically from 1.0 m³ to 0.80 m³. What is the overall net work done now, from start to the new finish?

12. IE •• An ideal gas is enclosed in a cylindrical piston with a 12.0-cm radius. Heat is slowly added to the gas while the pressure is maintained at 1.00 atm. During the process, the piston moves 6.00 cm. (a) This is an (1) isothermal, (2) isobaric, (3) adiabatic process. Explain. (b) If the heat transferred to the gas during the expansion is 420 J, what is the change in the thermal energy of the gas?

13. IE •• 2.0 mol of an ideal gas expands isothermally from a volume of 20 L to 40 L at 20 °C. (a) The work done by the gas is (1) positive, (2) negative, (3) zero. Explain. (b) What is the magnitude of the work?

14. •• A monatomic ideal gas ($\gamma = 1.67$) is compressed adiabatically from a pressure of 1.00×10^5 Pa and volume of 240 L to a volume of 40.0 L. (a) What is the final pressure of the gas? (b) How much work is done on the gas?

15. •• An ideal gas sample expands isothermally by tripling its volume while doing 5.0×10^4 J of work at 40 °C. (a) How many moles of gas are in the sample? (b) Was heat added to or removed from the sample, and how much?

16. IE ••• The temperature of 2.0 mol of ideal gas is increased from 150 °C to 250 °C by two different processes. In process A, 2500 J of heat is added to the gas; in B, 3000 J of heat is added. (a) In which case is more work done: (1) A, (2) B, or (3) the same amount of work is done during both? Explain. (b) Calculate the change in thermal energy and work done for each process.

17. IE ••• 100 mol of an ideal monatomic gas is compressed as shown in ▼ **Figure 12.25**. (a) Is the work done by the gas (1) positive, (2) zero, or (3) negative? Explain. (b) How much work is done by the gas? (c) What is the change in the gas's temperature? (d) What is the change in the gas's thermal energy? (e) How much heat exchange involved?

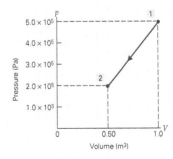

▲**FIGURE 12.25** **A variable p-V process and work** See Exercise 17 for details.

18. ••• One mole of an ideal gas is taken through the cyclic process in ▶ **Figure 12.26**. (a) Compute the work done during each of the processes. (b) Find ΔE_{th}, W_{gas}, and Q for the complete cycle. (c) What is T_3?

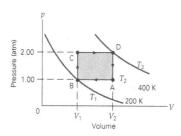

▲ FIGURE 12.26 **A cyclic process** See Exercise 18 for details.

12.3 The Second Law of Thermodynamics and Entropy

19. • What is the change in entropy of mercury vapor ($L_v = 2.7 \times 10^5$ J/kg) when 0.50 kg of it condenses to a liquid at its boiling point of 357 °C?

20. IE • 2.0 kg of ice melts completely into liquid water at 0 °C. (a) The change in entropy of the ice (water) in this process is (1) positive, (2) zero, (3) negative. Explain. (b) Compute the change in entropy of the ice (water).

21. IE • A process involves 1.0 kg of steam condensing to water at 100 °C. (a) The change in entropy of the steam (water) is (1) positive, (2) zero, (3) negative. Why? (b) Compute the change in entropy of the steam (water)?

22. • During a liquid-to-solid phase change of a substance, its change in entropy is -4.19×10^3 J/K. If 1.67×10^6 J of heat is removed in the process, what is the freezing point of this substance in degrees Celsius?

23. •• In an isothermal expansion at 27 °C, an ideal gas does 60 J of work. What is the change in entropy of the gas?

24. IE •• One mole of an ideal gas undergoes an isothermal compression at 0 °C, and 7.5×10^3 J of work is done in compressing the gas. (a) Will the entropy of the gas (1) increase, (2) remain the same, or (3) decrease? Why? (b) What is the change in entropy of the gas?

25. IE •• A quantity of an ideal gas undergoes an isothermal expansion at 20 °C and does 3.0×10^3 J of work on its surroundings in the process. (a) Will the entropy of the gas (1) increase, (2) remain the same, or (3) decrease? Explain. (b) What is the change in the entropy of the gas?

26. •• In the winter, heat from a house with an inside temperature of 18 °C leaks out at a rate of 2.0×10^4 J/s. The outside temperature is 0 °C. (a) What is the change in entropy per second of the house? (b) What is the total change in entropy per second of the house-outside system?

27. IE •• A perfectly isolated system consists of two large thermal reservoirs at constant temperatures of 100 °C and 0 °C. Assume the reservoirs make contact and 1000 J of heat flowed from the cold reservoir to the hot reservoir spontaneously with no appreciable reservoir temperature changes. (a) The total change in entropy of this isolated system (both reservoirs) would be (1) positive, (2) zero, (3) negative. Explain. (b) Calculate the total change in entropy of this isolated system.

28. IE •• Two large heat reservoirs at temperatures 200 °C and 60 °C, respectively, are brought into thermal contact, and 1.5×10^3 J of heat spontaneously flows from one to the other with no significant temperature change. (a) The change in the entropy of the two-reservoir system is (1) positive, (2) zero, (3) negative. Explain. (b) Calculate the change in the entropy of the two-reservoir system.

12.4 Heat Engines and Thermal Pumps

29. • If an engine does 400 J of net work each cycle while absorbing 2000 J of heat each cycle, what is its efficiency?

30. • If an engine does 200 J of net work and exhausts 800 J of heat per cycle, what is its efficiency?

31. • A gasoline engine has a thermal efficiency of 28%. If the engine absorbs 2000 J of heat per cycle, (a) what is the net work output per cycle? (b) How much heat is exhausted per cycle?

32. • A heat engine with a thermal efficiency of 20% does 500 J of net work each cycle. How much heat per cycle is lost to the low-temperature reservoir?

33. • A heat engine with a thermal efficiency of 15.0% absorbs 1.75×10^5 J of heat from the hot reservoir. How much heat is lost per each cycle?

34. IE • The heat output of an engine is 1.75×10^3 J per cycle, and the net work out is 4.0×10^3 J per cycle. (a) The heat input is (1) less than 44.0×10^3 J, (2) between 4.0×10^3 J and 7.5×10^3 J, (3) greater than 7.5×10^3 J. Explain. (b) What is the heat input and thermal efficiency of the engine?

35. •• A gasoline engine burns fuel that releases 3.3×10^8 J of heat per hour. (a) What is the energy input during a 2.0 h period? (b) If the engine delivers 25 kW of power during this time, what is its thermal efficiency?

36. IE •• A steam engine is to have its thermal efficiency improved from 8.00% to 10.0% while continuing to produce 4500 J of useful work per cycle. (a) Must the ratio of the heat output to heat input (1) increase, (2) remain the same, or (3) decrease? Why? (b) What is the change in Q_c/Q_h if this improvement happens?

37. IE •• An engineer redesigns a heat engine and improves its thermal efficiency from 20% to 25%. (a) Must the ratio of the heat input to heat output (1) increase, (2) remain the same, or (3) decrease? Explain. (b) What is the engine's change in Q_h/Q_c after the redesign is implemented?

38. •• When running, a refrigerator exhausts heat to the kitchen at a rate of 10 kW when the required input work is at a rate of 3.0 kW. (a) At what rate is heat removed from its cold interior? (b) What is the COP of the refrigerator?

39. •• A refrigerator with a COP of 2.2 removes 4.2×10^5 J of heat from its interior each cycle. (a) How much heat is exhausted each cycle? (b) What is the total work input in joules for 10 cycles?

40. •• An air conditioner has a COP of 2.75. What is the power rating of the unit if it is to remove 1.00×10^7 J of heat from a house interior in 20 min?

41. •• A heat pump removes 2.2×10^3 J of heat from the outdoors and delivers 4.3×10^3 J of heat to the inside of a house each cycle. (a) How much work is required per cycle? (b) What is the COP of this pump?

42. •• A steam engine has a thermal efficiency of 15.0%. If its heat input for each cycle is supplied by the condensation of 8.00 kg of steam at 100 °C (a) what is the net work output per cycle, and (b) how much heat is lost to the surroundings in each cycle?

43. ••• A coal-fired power plant produces 900 MW of electric power and operates at a thermal efficiency of 25% (a) What is the input heat rate from the burning coal? (b) What is the rate of heat discharge from the plant? (c) Water at 15 °C from a nearby river is used to cool the discharged heat. If the cooling water is not to exceed a temperature of 40 °C, how many gallons per minute of the cooling water are required?

12.5 The Carnot Cycle and Ideal Heat Engines

44. • A Carnot engine has an efficiency of 35% and takes in heat from a high-temperature reservoir at 178 °C. What is the Celsius temperature of the engine's low-temperature reservoir?

45. • A steam engine operates between 100 °C and 20 °C. What is the Carnot efficiency of the ideal engine that operates between these temperatures?

46. • It has been proposed that temperature differences in the ocean could be used to run a heat engine to generate electricity. In tropical regions, the water temperature is about 25 °C at the surface and about 5 °C at very deep depths. (a) What would be the maximum theoretical efficiency of such an engine? (b) Would a heat engine with such a low efficiency be practical? Explain.

47. • What is the Celsius temperature of the hot reservoir of a Carnot engine that is 32% efficient and has a 20 °C cold reservoir?

48. • An engineer wants to run a heat engine with an efficiency of 40% between a high-temperature reservoir at 300 °C and a low-temperature reservoir. What is the maximum Celsius temperature of the low-temperature reservoir?

49. •• A Carnot engine with an efficiency of 40% operates with a low-temperature reservoir at 40 °C and exhausts 1200 J of heat each cycle. What are (a) the heat input per cycle and (b) the Celsius temperature of the high-temperature reservoir?

50. •• A Carnot engine takes 2.7×10^4 J of heat per cycle from a high-temperature reservoir at 320 °C and exhausts some of it to a low-temperature reservoir at 120 °C. How much net work is done by the engine per cycle?

51. IE •• A Carnot engine takes in heat from a reservoir at 350 °C and has an efficiency of 35%. The exhaust temperature has not changed, and the efficiency is increased to 40%. (a) The temperature of the hot reservoir is (1) lower than, (2) equal to, (3) higher than 350 °C. Explain. (b) What is the new Celsius temperature of the hot reservoir?

52. •• An inventor claims to have created a heat engine that produces 10.0 kW of power for a 15.0-kW heat input while operating between reservoirs at 27 °C and 427 °C. (a) Is this claim valid? (b) To produce 10.0 kW of power, what is the minimum heat input required?

53. •• An inventor claims to have developed a heat engine that, each cycle, takes in 5.0×10^5 J of heat from a high-temperature reservoir at 400 °C and exhausts 2.0×10^5 J to the surroundings at 125 °C. Would you invest your money in the production of this engine? Explain.

54. •• A heat engine operates at a thermal efficiency that is 45% of the Carnot efficiency. If the temperatures of the high-temperature and low-temperature reservoirs are 400 °C and 50 °C, respectively, what are the Carnot efficiency and the thermal efficiency of the engine?

55. IE •• A Carnot engine operating between reservoirs at 27 °C and 227 °C does 1500 J of work in each cycle. (a) The change in entropy for the engine for each cycle is (1) negative, (2) zero, (3) positive. Why? (b) What is the heat input of the engine?

56. •• The *autoignition temperature* of a fuel is defined as the temperature at which a fuel-air mixture would self-explode and ignite. Thus, it sets an upper limit on the temperature of the hot reservoir in an automobile engine. The autoignition temperatures for commonly available gasoline and diesel fuel are about 495 °F and 600 °F, respectively. What are the maximum Carnot efficiencies (so as *not* to autoignite) of a gasoline engine and a diesel engine if the cold reservoir temperature is taken to be 40 °C?

57. •• Because of limitations on materials, the maximum temperature of the superheated steam used in a turbine for the generation of electricity is about 540 °C. (a) If the steam condenser operates at 20 °C, what is the maximum Carnot efficiency of a steam turbine generator? (b) The actual efficiency of such generators is about 35% to 40%. What does this range tell you?

58. •• The working substance of a cyclic heat engine is 0.75 kg of an ideal gas. The cycle consists of two isobaric processes and two isometric processes, as shown in ▼ **Figure 12.27**. What would be the efficiency of a Carnot engine operating with the same high-temperature and low-temperature reservoirs?

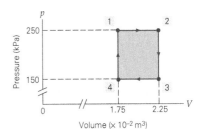

▲ **FIGURE 12.27 Thermal efficiency** See Exercise 58 for details.

13

Vibrations and Waves

The energy transported by waves across the ocean becomes the surfer's kinetic energy on the way to the beach.

The chapter-opening photograph depicts what a lot of people probably first think of when hearing the word *wave*. We're all familiar with ocean waves and their smaller relatives, the ripples that form on a lake or pond when something disturbs the surface. However, many of the waves that are important to us do not exhibit their wave nature as obviously as do these water waves. In fact, sound and visible light are both wave phenomena. Actually all electromagnetic radiations, not just visible light, behave as waves – radio, microwaves, X-rays, and so on. Chapter 28 discusses how even particles can exhibit wavelike properties.

Waves share many similarities to vibrations or oscillations – that is, back-and-forth motion – such as a swinging pendulum. In fact, waves can be thought of as traveling oscillations. But before looking at waves, the investigation of oscillation of systems of a single mass is in order.

13.1 Simple Harmonic Motion

The motion of an oscillating object depends on a *restoring force* to keep that object in repeating motion about a central point – called the *equilibrium point* – where the force is zero. Our study of such motion begins by considering a simple type of restoring force (here acting along the *x*-axis): a force who magnitude is proportional to the object's displacement from equilibrium. A common example is the (ideal) spring force, described by Hooke's law (Section 5.2),

$$F_s = -kx \quad \text{(Hooke's law)} \tag{13.1}$$

where k is the spring constant which is a measure of the spring's stiffness. The negative sign indicates that the direction of the spring force is opposite to its displacement. The sign is crucial so as to denote that the force acts to restore the object to its equilibrium position.

Suppose that an object on a horizontal frictionless surface is connected to a Hooke's law spring as in ▶ **Figure 13.1**. When the object is displaced from equilibrium and released, it will move back and forth – that is, it will vibrate, or oscillate, about the equilibrium position. An oscillation or a vibration is an example of a *periodic motion* – one that repeats itself along the same path. For linear oscillations, like an object attached to a spring, this path is along a straight axis that may be vertical, horizontal, or at an angle. For the angular oscillation of a pendulum, the repeating path is a circular arc. The motion under the influence of a linear restoring force (such as a Hooke's law spring) is called **simple harmonic motion (SHM)**. The name arises because for a linear restoring force the motion can be described by *harmonic functions* (sine and cosines), as will be seen later.

The distance of an object in SHM from its equilibrium position is its displacement and is a vector quantity (Section 2.2). Often, the equilibrium position is chosen to be the origin of the axis ($x_o = 0$). With this convention, the displacement can be replaced by x ($\Delta x = x - x_o = x$). As shown in Figure 13.1 the vector displacement can be either positive or negative, indicating its direction. The magnitude of the object's maximum displacement from equilibrium is the **amplitude (A)**, a *scalar* quantity (no direction). To account for displacement direction using our signs, the maximum displacements are written as $+A$ and $-A$ (Figure 13.1b, d).

Besides amplitude, two other important quantities used in describing an oscillation are period and frequency. The **period (T)** is the time it takes to complete one cycle, that is a *complete* round trip or oscillation, and is usually measured in seconds (per cycle). For example, if an object starts at $x = +A$ (Figure 13.1b), when it returns to $x = +A$ (Figure 13.1f), it will have completed one cycle and taken a time of one period. Note that if an object was initially at equilibrium, then its *second* return to this point would mark a complete cycle, not its first return. (Why its *second* return?) In either case, the object would travel a total distance of $4A$ during one cycle or period.

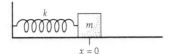

(a) Equilibrium

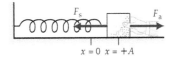

(b) $t = 0$ Just before release

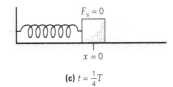

(c) $t = \frac{1}{4}T$

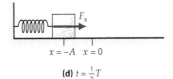

(d) $t = \frac{1}{2}T$

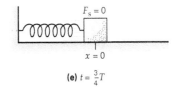

(e) $t = \frac{3}{4}T$

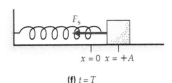

(f) $t = T$

▲ **FIGURE 13.1 Simple harmonic motion (SHM)** When an object on a spring **(a)** is at its equilibrium position ($x = 0$) and then **(b)** is displaced and released, the object undergoes SHM (assuming no frictional losses). The time it takes to complete one cycle is the period of oscillation (T). (Here, F_s is the spring force and F_a is the applied force.) **(c)** At $t = T/4$, the object is back at its equilibrium position; **(d)** at $t = T/2$, it is at $x = -A$. **(e)** During the next half-cycle, the motion is to the right; **(f)** at $t = T$, the object is back at its initial ($t = 0$) starting position as in (b).

The **frequency (f)** of an oscillation is the number of cycles that occur per second. The frequency and the period are inversely proportional, that is,

$$f = \frac{1}{T} \quad \text{(frequency and period)} \tag{13.2}$$

The SI unit of frequency is the hertz (Hz), or cycles per second (cycles/s or 1/s or s^{-1}). The hertz is named for Heinrich Hertz (1857–1894), a German physicist and early investigator

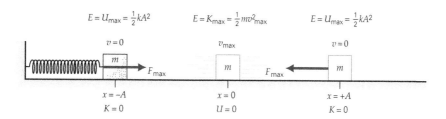

$$E = U_{max} = \tfrac{1}{2}kA^2 \qquad E = K_{max} = \tfrac{1}{2}mv^2_{max} \qquad E = U_{max} = \tfrac{1}{2}kA^2$$

▲ **FIGURE 13.2   Oscillations and energy trade-offs** For a mass oscillating in SHM on a spring (frictionless surface), the total energy at the end points ($x = \pm A$) is all potential, U_{max}, thus $E = \tfrac{1}{2}kA^2$. At equilibrium ($x = 0$), the total energy is all kinetic, K_{max}, thus $E = \tfrac{1}{2}mv^2_{max}$.

of electromagnetic waves. The inverse relationship is reflected in the units. Period is measured in seconds per cycle, and frequency in cycles per second. For example, if $T = 1/2$ s/cycle, then it would complete 2 cycles each second, or $f = 2$ cycles/s $= 2$ Hz. Note that although a cycle is not a unit, it is convenient to express frequency in cycles per second to help with unit analysis. The terms used to describe SHM are summarized in ▼ **Table 13.1**.

TABLE 13.1   Terms Used to Describe Simple Harmonic Motion

displacement—the change in position of an object measured from its equilibrium position ($x - x_o = x$ with $x_o = 0$).

amplitude (A)—the magnitude of the maximum displacement, or the maximum distance, of an object from its equilibrium position.

period (T)—the time for one complete cycle of motion.

frequency (f)—the number of cycles per second (in hertz or inverse seconds, where $f = 1/T$).

13.1.1 Energy and Speed of a Spring Mass: Spring System in SHM

Recall from Section 5.4 that the potential energy stored in a spring that is stretched or compressed a distance x from equilibrium is given by

$$U = \frac{1}{2}kx^2 \quad \text{(potential energy stored in a spring)} \quad (13.3)$$

An object of mass m oscillating on a spring also has kinetic energy, thus the total mechanical energy E of the system is:*

$$E = K + U = \frac{1}{2}mv^2 + \frac{1}{2}kx^2 \quad (13.4)$$

$$\text{(spring-mass mechanical energy)}$$

When the object is at maximum displacement, either at $x = +A$ or $-A$ (called *end points*) it is at rest, $v = 0$ (▲ **Figure 13.2**). At the end points, the total energy is in the form of potential energy, thus an expression for the total energy is $E = \tfrac{1}{2}m(0)^2 + \tfrac{1}{2}k(\pm A)^2 = \tfrac{1}{2}kA^2$ or

$$E = \frac{1}{2}kA^2 \quad \text{(total energy in spring SHM)} \quad (13.5)$$

It turns out that in general, the total energy of an object in simple harmonic motion is directly proportional to the square of its amplitude.

Since E is constant, Equations 13.4 and 13.5 can be combined to allow the object's velocity to be determined as a function of its displacement. First:

$$E = \frac{1}{2}mv^2 + \frac{1}{2}kx^2 = \frac{1}{2}kA^2 \quad (13.6)$$

and then solving for v^2 and taking the square root:

$$v = \pm\sqrt{\frac{k}{m}(A^2 - x^2)} \quad \text{(velocity of an object in SHM)} \quad (13.7)$$

Here the positive and negative signs indicate velocity direction. As can be seen, when the object is at its end points and $x = \pm A$, this equation gives the correct value of zero for velocity.

When the oscillating object passes through equilibrium (unstretched spring, or $x = 0$), at that instant, the potential energy is zero. Thus the energy is all in the form of kinetic energy, and the object is traveling at its maximum speed v_{max}. The expression for the energy in this case is $E = \tfrac{1}{2}kA^2 = K_{max} = \tfrac{1}{2}mv^2_{max}$, which can be solved for the maximum speed:

$$v_{max} = \sqrt{\frac{k}{m}}\,A \quad \text{(maximum speed of mass on spring)} \quad (13.8)$$

The next Example illustrates the continuous trade-off between kinetic and potential energy.

EXAMPLE 13.1: A BLOCK AND A SPRING – USING ENERGY METHODS TO DESCRIBE SHM

A block with a mass of 0.50 kg sitting on a frictionless surface is connected to a light spring that has a spring constant of 180 N/m (Figures 13.1 and 13.2). If the block is displaced 15 cm from equilibrium and released, what are (a) the total energy of the system, (b) the speed of the block when it is 10 cm from equilibrium, and (c) the maximum speed of the block?

---

* In our spring-mass oscillation studies, the mass of the spring will be neglected compared to that of the attached object. This is equivalent to neglecting the kinetic energy of the spring. Corrections for spring mass to equations like 13.4 can be made but are beyond the scope of this text.

THINKING IT THROUGH. The total energy depends on the spring constant (k) and the amplitude (A), which are given. At $x = 10$ cm, the speed should be less than the maximum speed. (Why?)

SOLUTION

First the given data and what is to be found are listed, as usual. Here the initial displacement is the amplitude. (This is not the general case; why not?) The speeds can be calculated using energy methods. In (b) since the question asks for speed not velocity, it does not include direction, thus the answer is the same regardless of the sign of x; that is, regardless of on which side of the equilibrium position the block is.

Given:

(a) $m = 0.50$ kg
$k = 180$ N/m
$A = 15$ cm
(b) $x = \pm 10$ cm

Find:

(a) E (total energy)
(b) v (speed)
(c) v_{max}

(a) The total energy is given by Equation 13.6 (after converting A into meters):

$$E = \frac{1}{2}kA^2 = \frac{1}{2}(180\,\text{N/m})(0.15\,\text{m})^2 = 2.0\,\text{J}$$

(b) The speed at a displacement 10 cm on either side of equilibrium (i.e., for $x = \pm 10$ cm $= \pm 0.10$ m) can be found from Equation 13.7:

$$v = \sqrt{\frac{k}{m}(A^2 - x^2)} = \sqrt{\frac{180\,\text{N/m}}{0.50\,\text{kg}}[(0.15\,\text{m})^2 - (\pm 0.10\,\text{m})^2]}$$
$$= \sqrt{4.5\ \text{m}^2/\text{s}^2} = 2.1\,\text{m/s}$$

As an alternative, basic energy conservation principles can be applied. At $x = \pm 10$ cm $= \pm 0.10$ m, the potential energy is

$$U = \frac{1}{2}kx^2 = \frac{1}{2}(180\,\text{N/m})(0.10\,\text{m})^2 = 0.90\,\text{J}$$

The kinetic energy is the difference between potential energy and total energy or 1.10 J. Using the definition of kinetic energy, the speed can be found (and agrees):

$$K = \frac{1}{2}mv^2 \quad \text{or} \quad v = \sqrt{\frac{2K}{m}} = \sqrt{\frac{2(1.10\,\text{J})}{0.50\,\text{kg}}} = 2.1\,\text{m/s}$$

(c) The maximum speed occurs at $x = 0$, when the kinetic energy is the total energy 2.0 J. Thus

$$v_{max} = \sqrt{\frac{2K_{max}}{m}} = \sqrt{\frac{2(2.0\,\text{J})}{0.50\,\text{kg}}} = 2.8\,\text{m/s}$$

Alternatively, Equation 13.7 could be applied

$$v_{max} = \sqrt{\frac{180\,\text{N/m}}{0.50\,\text{kg}}[(0.15\,\text{m})^2 - (0)^2]} = 2.8\,\text{m/s}$$

FOLLOW-UP EXERCISE. What is the magnitude of the acceleration when $x = \pm 0.10$ m? What is the maximum acceleration?

13.2 Equations of Motion

The equation that gives an object's position as a function of time is referred to as the *equation of motion*. For example, the equation of motion for constant linear acceleration (see Chapter 2) is $x = x_0 + v_0t + (1/2)at^2$, which does *not* apply here because in SHM acceleration is *not* constant.

In general, determination of equations of motion requires the use of Newton's laws. However, with non-constant acceleration, calculus is required. A more visual approach to obtaining the equation of motion for SHM can come from studying the relationship between simple harmonic and uniform circular motions. For now, let's concentrate on vertical SHM – for example, a mass hung from a spring. This vertical SHM can be viewed as a component of uniform circular motion, as illustrated in ▶ **Figure 13.3**. As the object orbits in uniform circular motion (constant angular speed ω) in a vertical plane, its shadow moves back and forth vertically, following the same path as an object on a spring would; that is, simple harmonic motion. Thus, the equation of motion of the shadow is the same as the equation of motion for an object oscillating on a spring.

At this point you may rightly ask how might horizontal SHM – such as a spring-mass system on a frictionless surface – be described. Although the following development is discussed assuming a vertical SHM, the same ideas and equations apply to horizontal SHM. To use them for horizontal SHM, simply replace the position variable y with x. It turns out that the expressions for frequency (and period) are the same for both SHMs. The only difference is that the vertical motion equilibrium location is changed by the constant force of gravity, whereas in horizontal SHM equilibrium is at the end of the relaxed spring. Thus it is important to remember that y and x refer to the displacement relative to equilibrium, stretched or not.

Getting back to vertical motion, from the reference circle in Figure 13.3b, the y-coordinate (position) of the orbiting object is given by $y = A\sin\theta$. But this object moves at a constant angular speed ω. Assuming $\theta = 0$ at $t = 0$, the angle varies linearly with time as $\theta = \omega t$, giving rise to a specialized SHM equation of motion to be used only if the object starts at equilibrium:

$$y = A\sin(\omega t) \quad \text{(SHM if } y_0 = 0, \text{initial velocity upward)} \quad (13.9)$$

Note as t increases from zero, y increases in the positive direction, so this equation describes initial upward motion only. But what if the mass were initially at its maximum

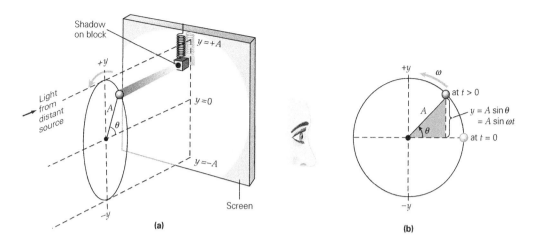

▲ **FIGURE 13.3** **Reference circle for vertical motion (a)** The shadow of an object in uniform circular motion has the same motion as an object oscillating on a spring in simple harmonic motion. **(b)** The shadow location on the vertical axis is $y = A \sin \theta$. However, for uniform circular motion $\theta = \omega t$. Thus the location as a function of time is $y = A \sin \omega t$ (assuming $y = 0$ at $t = 0$).

displacement; that is, the amplitude position $y_0 = +A$? In this case, the sine function is incorrect, because Equation 13.9 gives $y_0 = 0$. For this case, the equation that correctly matches the *initial conditions* involves the cosine and is $y = A \cos(\omega t)$. To check, note at $t = 0$, it gives the correct initial location: $y_0 = A \cos \omega(0) = +A$. Thus

$$y = A \cos(\omega t) \quad \text{(SHM if } y_0 = +A, \text{ zero initial velocity)} \quad (13.10)$$

Here, the initial motion from rest will be downward, because after the start, y begins to decrease.

In general, the equation of motion for an oscillating object may involve either a sine or a cosine function. Both of these functions are termed *sinusoidal*. For a visual description of this, see ▼ **Figure 13.4**.

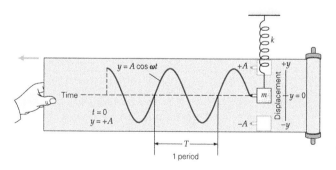

▲ **FIGURE 13.4** **Sinusoidal equation of motion** As time passes, the oscillating object traces out a sinusoidal curve on the moving paper. In this case, $y = A \cos(\omega t)$, because the object's initial location is $y_0 = +A$.

The angular speed ω (in rad/s) of the reference circle object (Figure 13.3), when used to describe the SHM motion (the "shadow"), is renamed the angular *frequency* of the oscillating

object, because the shadow object is not rotating. Recall that $\omega = 2\pi f$ where f is the frequency of revolution of the object (Section 7.2). From Figure 13.3, the frequency of the "orbiting" object is the same as the frequency of the "oscillating" object. Thus, using $f = 1/T$, Equations 13.9 and 13.10 may be rewritten as

$$y = A \sin(2\pi ft) = A \sin\left(\frac{2\pi t}{T}\right) \quad (13.11)$$
$$\text{(SHM if } y_0 = 0 \text{ and initial velocity upward)}$$

and

$$y = A \cos(2\pi f) = A \cos\left(\frac{2\pi t}{T}\right) \quad (13.12)$$
$$\text{(SHM if } y_0 = +A, \text{ zero initial velocity)}$$

The most general SHM equations of motion are actually a combination of sines and cosines. The general case is beyond the scope of this text. Thus our study of SHM is restricted to the four special cases (i.e., specific initial conditions) shown in ▶ **Figure 13.5** and summarized as follows.

- (Figure 13.5a) An object in vertical SHM initially at equilibrium moving upward.
- (Figure 13.5b) An object in vertical SHM released from rest from its positive amplitude location (+A).
- (Figure 13.5c) An object in vertical SHM initially at equilibrium moving downward.
- (Figure 13.5d) An object in vertical SHM released from rest from its negative amplitude location ($y = -A$).

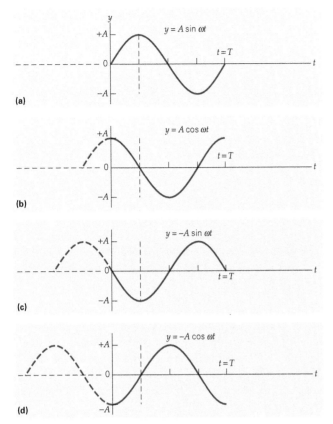

(a) $y = A \sin \omega t$

(b) $y = A \cos \omega t$

(c) $y = -A \sin \omega t$

(d) $y = -A \cos \omega t$

▲ **FIGURE 13.5 Initial conditions and equations of motion** The initial conditions (location and velocity) determine the form of the equation of motion – for the cases shown here, either a sine or a cosine. At $t = 0$, the initial conditions are **(a)** $y_o = 0$, upward velocity; **(b)** $y_o = +A$, starts at rest; **(c)** $y_o = 0$, downward velocity; and **(d)** $y_o = -A$, starts at rest. The equations of motion must match the initial conditions. (See text for a description, equation summaries, as well as a worked example.)

The equations of motion for these four special vertical SHM cases are as follows. (*Note:* To adapt these to horizontal SHM, replace y by x everywhere. In that case, $x = +A$ means maximum spring extension and $x = -A$ means maximum spring compression.):

$$y = \pm A \sin \omega t = \pm A \sin(2\pi f t) = \pm A \sin\left(\frac{2\pi t}{T}\right)$$

(choose $+$ sign for initial motion upward at $y_o = 0$; $\quad$ (13.13)
choose $-$ sign for initial motion downward at $y_o = 0$)

$$y = \pm A \cos \omega t = \pm A \cos(2\pi f t) = \pm A \cos\left(\frac{2\pi t}{T}\right)$$

(choose $+$ sign for initially at rest at $y_o = +A$; $\quad$ (13.14)
choose $-$ sign for initially at rest at $y_o = -A$)

In addition to illustrating the use of these equations, the following example emphasizes that the angles in the trig functions must be expressed in *radians*.

EXAMPLE 13.2: AN OSCILLATING MASS – APPLYING THE EQUATIONS OF MOTION

A mass on a spring oscillates vertically with an amplitude of 15 cm and a frequency of 0.20 Hz. At $t = 0$, it is at the equilibrium location and moving upward. (a) Write the equation of motion for this oscillation. (b) What are its position and direction of motion after 3.1 s have elapsed? (c) How many oscillations (cycles) does the mass make in 12 s?

THINKING IT THROUGH. For part (a), the equation of motion has to be a sine function because the motion starts at equilibrium. (Why not cosine? What is the correct sign?) Since the mass is initially moving upward, the equation of motion should be in the form of Equation 13.13. Part (b) is a lesson in using the equation generated in (a). In part (c), recall that frequency is in cycles per second. Hence, multiplying frequency by the elapsed time will give the number of cycles.

SOLUTION

Given:

(a) $A = 15$ cm $= 0.15$ m
$\quad\ f = 0.20$ Hz
(b) $t = 3.1$ s
(c) $t = 12$ s

Find:

(a) Equation of motion
(b) y (position and direction of motion)
(c) n (number of oscillations or cycles)

(a) Since the motion starts at equilibrium and the mass is moving upward, the appropriate equation is 13.13 with the $+$ sign. Using the known quantities (and making sure the sine function ends up evaluating its angle in radians), the numerical equation is

$$y = (0.15\,\text{m}) \sin[2\pi(0.20\,\text{Hz})t]$$
$$= (0.15\,\text{m}) \sin[(0.4\pi\,\text{rad/s})t]$$

(b) At $t = 3.1$ s (make sure your calculator is set to radians)

$$y = (0.15\,\text{m}) \sin[(0.4\pi\,\text{rad/s})(3.1\,\text{s})] = (0.15\,\text{m}) \sin(3.9\,\text{rad})$$
$$= -0.10\,\text{m} = -10\,\text{cm}$$

Because of the minus sign, at this time, the mass is 10 cm below equilibrium. But what is its direction of motion? For that, let's check the period (T) to see what part of its cycle it is in.

$$T = \frac{1}{f} = \frac{1}{0.20\,\text{Hz}} = 5.0\,\text{s}$$

So after 3.1 s, the mass will have gone through 3.1 s/5.0 s $= 0.62$, or 62%, of a cycle The motion is as follows: first up (first quarter cycle) and then back down to equilibrium (second quarter cycle). Taken together these make up 50% of the cycle. Between there and the next quarter cycle, 75% of the cycle will have been completed. Since it is only at 62%, it must be in the third quarter cycle and moving downward toward $y = -A$.

(c) The number of cycles is given by the product of frequency (cycles/s) and elapsed time (s):

$$n = ft = (0.20 \text{ cycles/s})(12 \text{ s}) = 2.4 \text{ cycles}$$

This same result can be obtained by comparing the 12 s to the period of 5.0 s:

$$n = \frac{t}{T} = \frac{12 \text{ s}}{5.0 \text{ s/cycle}} = 2.4 \text{ cycles}$$

After 12 s, the mass has gone through two cycles and 40% of the third and is on its way back to equilibrium from $y = +A$.

FOLLOW-UP EXERCISE. Find what is asked for in this Example at two other times: (1) $t = 4.5$ s and (2) $t = 7.5$ s.

13.2.1 Problem-Solving Hint

In part (b) of Example 13.2, the argument of the sine is in radians, *not* degrees. Don't forget to set your calculator to radians when evaluating trig functions both in SHM and circular motion.

In the previous Example one might ask whether the period was unique to that amplitude or, more generally, what *does* the period depend upon? To answer, let's find the period of a vertical spring-mass system by comparing it to the reference circle motion (Figure 13.3).

Note that the time for the reference circle object to make a complete "orbit" is the same time it takes for the oscillating object (shadow) to make one complete cycle, which *is* the oscillation period. The object in the reference circle is in uniform circular motion – that is, moving at constant speed. This speed is, in fact, equal to the *maximum* speed of the oscillator, v_{max}, as can again be seen in Figure 13.3.

Since the orbiting object travels one circumference in one period, the period T is simply the distance traveled divided by that speed or $T = $ circumference/v_{max}. But v_{max} is known from Equation 13.8. Putting these together gives $T = \dfrac{d}{v_{max}} = \dfrac{2\pi A}{A\sqrt{k/m}}$ and the amplitude cancels:

$$T = 2\pi\sqrt{\frac{m}{k}} \quad \text{(period of spring-driven SHM)} \quad (13.15)$$

Since $f = 1/T$, an alternative expression to Equation 13.15 is

$$f = \frac{1}{2\pi}\sqrt{\frac{k}{m}} \quad \text{(frequency of spring-driven SHM)} \quad (13.16)$$

Thus *period (and thus frequency) are independent of the amplitude in spring-driven SHMs*. The amplitude cancellation results from the fact that a larger amplitude (a longer travel distance) is associated with a larger average speed. It turns out to be exactly the increase needed to keep the period unchanged.

From Equation 13.15 it can be seen that the greater the mass, the longer the period and the greater the spring constant (stiffer spring), the shorter the period. Thus it is the ratio of mass to stiffness that determines period. For example, an increase in period due to an increase in mass can be offset by using a stiffer spring. Notice also that the stiffer the spring, the higher the frequency, also as expected.

Last, since $\omega = 2\pi f$, an expression for the angular frequency is

$$\omega = \sqrt{\frac{k}{m}} \quad \begin{array}{l}\text{(angular frequency} \\ \text{of spring-driven SHM)}\end{array} \quad (13.17)$$

Another common example of an oscillator is a *simple pendulum* (a small, heavy object on a very light string). It will undergo simple harmonic motion but only if the angle of oscillation is kept small. For large angles, the period does depend on the amplitude and thus the pendulum does not qualify for SHM. If the string's maximum angle from vertical is less than about 10°, the period of a simple pendulum is

$$T = 2\pi\sqrt{\frac{L}{g}} \quad \text{(period of a simple pendulum)} \quad (13.18)$$

where L is the length of the pendulum and g is the acceleration due to gravity. As a practical note, this means that a pendulum-driven clock, as its pendulum swings and decreases in amplitude, still keeps correct time. The period remains unchanged as the amplitude decreases.

Note that unlike the mass-spring system, the period/frequency of a simple pendulum is independent of the mass of the bob. Can you explain why? This may seem strange at first. Think about the force that supplies the restoring force – the gravitational force. Since the force of gravity accelerates all masses at the same rate (see Chapter 2), it is expected that the pendulum parameters should be independent of mass.

Let's take a look at two Examples related to the frequency and period of SHM.

EXAMPLE 13.3: FUN (?) WITH A POTHOLE – FREQUENCY AND SPRING CONSTANT

A typical family automobile body has a mass of 1500 kg. Assume that the car has one spring on each wheel, that the springs are identical, and that the mass is equally distributed over the four springs. (a) What is the spring constant of each spring if the empty car bounces up and down 1.2 times each second when it hits a pothole? (b) What will be the car's oscillation frequency when four 75-kg people are in the car?

THINKING IT THROUGH. (a) The frequency is given by Equation 13.16, and from this the spring constant can be found. Each spring will carry only 1/4 of the total car mass. (b) Once the spring constant is found, Equation 13.16 can be used to find this new frequency, which is expected to be reduced (why?). Note that spring constant is the same with or without the people in the car.

SOLUTION

The data are listed as follows, where m represents the mass on each individual spring:

Given:

(a) $m_1 = \dfrac{1500\,\text{kg}}{4} = 378\,\text{kg}$

      $f_1 = 1.2\,\text{Hz}$

(b) $m_2 = \dfrac{1500\,\text{kg} + 4(75\,\text{kg})}{4} = 450\,\text{kg}$

Find:

(a) k (spring constant)
(b) f (new frequency)

(a)   Using Equation 13.16 to solve for the spring constant, k

$$k = 4\pi^2 f^2 m_1 = 4\pi^2 (1.2\,\text{Hz})^2 (375\,\text{kg}) = 2.13 \times 10^4\,\text{N/m}$$

(b)   The new frequency is found from Equation 13.16

$$f = \frac{1}{2\pi}\sqrt{\frac{k}{m_2}} = \frac{1}{2\pi}\sqrt{\frac{2.13 \times 10^4\,\text{N/m}}{450\,\text{kg}}} = 1.1\,\text{Hz}$$

FOLLOW-UP EXERCISE. Research has shown that the human body feels most comfortable if the oscillation frequency of a car is 1.0 Hz. To achieve this frequency, what spring constant would you use for this half-loaded car (assume two 75-kg people and an even distribution of mass on all four springs)?

EXAMPLE 13.4: FUN WITH A PENDULUM – FREQUENCY AND PERIOD

A helpful older brother takes his sister to play on the swings in the park. He pushes her from behind only on each return to maintain the swing amplitude. Assuming that the swing behaves as a simple pendulum with a length of 2.50 m, (a) what would be the frequency of the oscillations, and (b) what would be the interval between the brother's pushes?

THINKING IT THROUGH. (a) The period is given by Equation 13.18, and the frequency and period are inversely related: $f = 1/T$. (b) Since the brother pushes from one side on each return, he must push once every cycle, so the time between his pushes is equal to the swing's period.

SOLUTION

Given:

   $L = 2.50$ m

Find:

(a) f (frequency)
(b) T (period)

(a)   Taking the reciprocal of Equation 13.18 gives the frequency:

$$f = \frac{1}{T} = \frac{1}{2\pi}\sqrt{\frac{g}{L}} = \frac{1}{2\pi}\sqrt{\frac{9.80\,\text{m/s}^2}{2.50\,\text{m}}} = 0.315\,\text{Hz}$$

(b)   The period is then found from the frequency:

$$T = \frac{1}{f} = \frac{1}{0.315\,\text{Hz}} = 3.17\,\text{s}$$

The brother must push every 3.17 s to maintain a steady swing (and to keep his sister from complaining).

FOLLOW-UP EXERCISE. In this Example, the brother, a physics buff, carefully measures the period of the swing to be 3.18 s, not 3.17 s. If the length of 2.50 m is accurate, what is the acceleration due to gravity at the location of the park? Assume your answer is accurate to three significant figures. Considering this value as exact, do you think the park is at sea level?

13.2.2 Velocity and Acceleration in SHM

Expressions for the velocity and acceleration of an object in SHM can be obtained from the equations of motion. The methods to derive these are beyond the scope of this book, so the results are simply quoted here. First, let's do instantaneous velocity in SHM. (Note again – these are only for use with the special initial conditions as stated.) As before, if these are to be used to describe horizontal SHM, then replace y by x everywhere and remember that the $+$ sign will then stand for spring extension and the $-$ sign for compression.

$$v_y = \pm \omega A \cos \omega t = \pm \omega A \cos(2\pi ft) = \pm \omega A \cos\left(\frac{2\pi t}{T}\right)$$

(choose $+$ sign for initial motion upward at $y_o = 0$;   (13.19)
choose $-$ sign for initial motion downward at $y_o = 0$)

$$v_y = \mp \omega A \sin \omega t = \mp \omega A \sin(2\pi ft) = \mp \omega A \sin\left(\frac{2\pi t}{T}\right)$$

(choose $-$ sign for initially at rest at $y_o = +A$;   (13.20)
choose $+$ sign for initially at rest at $y_o = -A$)

Last, there are the expressions for instantaneous acceleration in SHM with the same warnings about signs and usage for horizontal oscillations as for the velocity expressions previously mentioned.

$$a_y = \mp \omega^2 A \sin \omega t = \mp \omega^2 A \sin(2\pi ft) = \mp \omega^2 A \sin\left(\frac{2\pi t}{T}\right)$$

(choose $-$ sign for initial motion upward at $y_o = 0$;   (13.21)
choose $+$ sign for initial motion downward at $y_o = 0$)

$$a_y = \mp \omega^2 A \cos \omega t = \mp \omega^2 A \cos(2\pi f t) = \mp \omega^2 A \cos\left(\frac{2\pi t}{T}\right)$$

(choose − sign for initially at rest at $y_o = +A$; (13.22)

choose + sign for initially at rest at $y_o = -A$)

As an illustration of the use of these equations, let's study further the oscillator in Example 13.2.

EXAMPLE 13.5: AN OSCILLATING MASS REVISITED

For the oscillator and conditions in Example 13.2, determine the following. (a) The equations for the velocity and acceleration as a function of time. Then use these results to find its (b) velocity and (c) acceleration after 3.1 s have elapsed. Comment on the signs of the answers based on the location of the oscillator at this time as determined in Example 13.2.

THINKING IT THROUGH. For part (a), the correct equations are determined by choosing those that match the initial conditions. Parts (b) and (c) involve evaluation of the trig functions – remembering their arguments are in radians.

SOLUTION

Given:

(a) $A = 15\,\text{cm} = 0.15\,\text{m}$
(b) $f = 0.20\,\text{Hz}$
(c) $t = 3.1\,\text{s}$

Find:

(a) Equations for velocity and acceleration as functions of time
(b) Velocity $(\vec{\mathbf{v}})$ after 3.1 s
(c) Acceleration $(\vec{\mathbf{a}})$ after 3.1 s

(a) The motion starts at equilibrium with the mass moving upward, thus the appropriate equation for velocity is Equation 13.19 with the + sign. Since $\omega = 2\pi f = 1.26\,\text{rad/s}$ and $A = 15\,\text{cm}$, the result for velocity is

$$v_y = +\omega A \cos \omega t = +(1.26\,\text{rad/s})(15\,\text{cm})\cos(1.26t)$$
$$= +18.9 \cos(1.26t)\,\text{cm/s}$$

The motion starts at equilibrium with the mass that is moving upward, thus appropriate equation for acceleration is Equation 13.21 with the − sign.

$$a_y = -\omega^2 A \sin \omega t = -(1.26\,\text{rad/s})^2(15\,\text{cm})\sin(1.26t)$$
$$= -23.8 \cos(1.26t)\,\text{cm/s}^2$$

(b) At $t = 3.1\,\text{s}$ (make sure your calculator is set to radians!) the velocity is

$$v_y = +18.9 \cos[(1.26\,\text{rad/s})(3.1\,\text{s})]\,\text{cm/s} = -13.6\,\text{cm/s}$$

and the acceleration is

$$a_y = -23.8 \sin[(1.26\,\text{rad/s})(3.1\,\text{s})]\,\text{cm/s}^2$$
$$= +16.5\,\text{cm/s}^2$$

In Example 13.2 it was found that at this time the oscillator was 10 cm below equilibrium. Thus its y value was negative. It was also in the third quarter of the cycle, meaning it was traveling downward, and a negative velocity sign is expected and our calculation produces it. Similarly, at that location it is slowing, thus a positive value for acceleration is expected (moving down but slowing up), and that is our result. Just as in the location calculation in Example 13.2, here it was crucial to use radians in the trig functions to get the correct sign, and thus directions, for velocity and location.

FOLLOW-UP EXERCISE. Use the equations generated in this Example to show that after one period has elapsed, the oscillator is back to its initial conditions; that is, at equilibrium and moving upward.

13.2.3 Damped Harmonic Motion

Simple harmonic motion with constant amplitude implies no loss of energy, but in practical applications there are always frictional losses. Therefore, to maintain a constant amplitude motion, energy must be added to a system by some external driving force, such as someone pushing a swing. When the external driving force stops or is removed, the amplitude and the energy of an oscillator decrease with time, giving rise to **damped harmonic motion** (▼ Figure 13.6a). The time required for the oscillations to cease, or damp out, depends on the magnitude and type of the damping force (such as air resistance).*

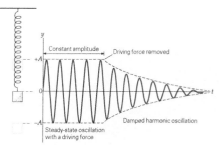

▲ FIGURE 13.6 **Damped harmonic motion (a)** When a driving force adds energy to a system in an amount equal to the energy losses of the system, the oscillation becomes steady with a constant amplitude. When the driving force is removed, the oscillations decay (i.e., they are damped), and the amplitude decreases (nonlinearly) with time. **(b)** In some applications, damping is desirable and even promoted, as with shock absorbers in automobile suspension systems. Otherwise, the passengers would be in for a bouncy ride.

---

* The mathematical analysis of both driven and damped oscillatory systems is beyond the scope of this text.

In many applications involving continuous periodic motion, damping is unwanted and necessitates an energy input (work done by an external force) to keep the oscillations going – such as pushing a child on a swing. However, in many instances, stopping the oscillations by damping is desirable. For example, shock absorbers provide damping in suspension systems of autos (Figure 13.6b). Without these systems to dissipate energy, say after hitting a bump, the ride would be bouncy for a long time. In California and around the world in earthquake-prone areas, many new buildings incorporate damping mechanisms (giant shock absorbers) to dampen their oscillatory motion quickly after they are set in motion by earthquake waves.

13.3 Wave Motion

Waves come in various types, such as water waves, sound waves, earthquake waves, and light waves. All waves result from a disturbance, the source of the wave. In this chapter, the focus is on *mechanical waves* – those that are propagated in, or travel through, a material medium. (Light waves do *not* have this requirement and will be considered separately in later chapters.)

When a medium is disturbed, work is done on it and thus energy imparted to it. The addition of the energy sets some of the particles in the medium vibrating. Because they are linked by inter-molecular forces, the oscillation of each affects its neighbors. Thus the energy propagates, or spreads, by means of interactions between the particles comprising the medium. An analogy to this process is in ▼ **Figure 13.7**, where the "particles" are dominoes.

▲ **FIGURE 13.7** **Energy transfer – a wave pulse** The propagation of a disturbance, or a transfer of energy by wave action, is seen in a row of falling dominoes. Here the medium is the row of dominos. The disturbance is the external push on the first one and this type of wave is called a wave pulse.

As each one falls, it topples the next. Thus, energy is transferred from domino to domino, and the disturbance propagates through the medium (here the row of dominos) – note that the energy travels, *not* the medium. The wave action shown in the domino photo is actually a *wave pulse*, which results from a one-time disturbance, here a push of the finger. There is no restoring force between the dominoes, so they do not oscillate. Therefore, the "domino pulse" travels but does *not* repeat itself at any location.

Similarly, if the end of a stretched rope is given a quick shake, the disturbance transfers energy from the hand to the rope, as illustrated in ▼ **Figure 13.8**. The forces acting between the "particles" in the rope cause them to move in response to the motion of the hand, and a *string wave pulse* travels down the rope. Each piece of the rope goes up and then down as the pulse passes. This motion of individual rope pieces and the propagation of the wave pulse can be observed by tying pieces of ribbon onto the rope (at x_1 and x_2 in the figure). As the disturbance passes point x_1, the ribbon rises and falls, as do the rope's "particles." Later, the same thing happens at x_2, indicating that the energy (disturbance) is propagating, or traveling, on the rope.

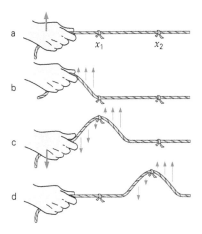

▲ **FIGURE 13.8** **String wave pulse** The hand disturbs the stretched rope in a quick up-and-down motion, and a string wave pulse propagates along the rope. (The red arrows represent the velocities of hand and rope pieces at different times and locations.) Pieces of the rope move up and down as the pulse passes. The energy in the pulse is both kinetic (motion) and potential (elastic).

In a continuous material medium, particles interact with their neighbors, and restoring forces cause them to oscillate when disturbed. Thus, a disturbance may be repeated over and over in time at each position. Such a regular, rhythmic, and continuous disturbance is a *periodic wave*, or just **wave motion**. In this case we say that energy is carried through the medium by the wave motion.

A periodic wave requires a disturbance produced by an oscillating source (▶ **Figure 13.9**) so the particles will move up and down continuously. If the driving source oscillates in simple harmonic motion, the resulting particle motion of each rope piece is also simple harmonic.

Such periodic wave motion will have sinusoidal forms (sine or cosine) in both time and space. Being *sinusoidal in space* means that a photograph of the wave at any instant ("freezing" it in time) would show a sinusoidal waveform (the curves in Figure 13.9). However, if you observe at a given location in space as a wave passes, a particle of the medium would oscillate up and down *sinusoidally with time* (see SHM in Section 13.2).

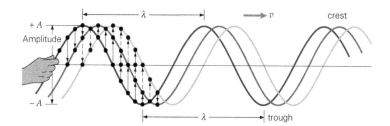

▲ **FIGURE 13.9** **Periodic traveling wave** A continuous harmonic disturbance can set up a sinusoidal wave in a stretched rope, for example, and the wave travels down the rope with wave speed v. Each piece of the rope oscillates vertically in simple harmonic motion. The distance between two successive points that are in phase (e.g., at two successive crests) on the waveform is the wavelength λ of the wave. Can you tell how much time has elapsed, as a fraction of the period T, between the darkest red and lightest red waves?

13.3.1 Wave Characteristics

Specific physical quantities are used to describe sinusoidal waves. As with SHM, the **amplitude (A)** of a wave is the maximum distance from the particle's equilibrium position (Figure 13.9). For a water wave, for example, this corresponds to the height of a wave crest or the depth of a trough. Recall that in SHM, the total energy of the oscillator is proportional to the square of the amplitude. Similarly, the energy *transported* by a wave is proportional to the square of its amplitude ($E \propto A^2$). Note the difference, though: A wave *transmits* energy through space, whereas an oscillator's energy is localized.

For a periodic or sinusoidal wave, the distance between successive crests (or troughs) is the **wavelength (λ)** (Figure 13.9). Actually, it is the distance between *any* two successive parts of the wave that are in phase (successive identical points on the waveform). The crest and trough notation is used for convenience.

The **frequency (f)** of a periodic wave is the number of waves per second – that is, the number of complete waveforms, or wavelengths – that pass a given point per second. The frequency of the wave is the same as the frequency of the SHM source that created it.

A periodic wave can be described by its **period (T)**. The period $T = 1/f$ is the time for one complete waveform (a full wavelength) to pass a point. Since a wave (energy) travels, it has a **wave speed (v)**. Any particular point on the wave (such as a crest) travels a distance of one wavelength λ in one period T. Thus, since $v = d/t$ and $f = 1/T$, the general relationship is

$$v = \frac{\lambda}{T} = \lambda f \quad \text{(wave speed)} \tag{13.23}$$

The SI units for v are, as expected, m/s. In general, the wave speed depends on the properties of the medium as well as the source frequency f.

EXAMPLE 13.6: DOCK OF THE BAY – FINDING WAVE SPEED

A person on a pier observes a set of incoming sinusoidal water waves with a distance of 5.6 m between the crests. If a wave laps against the pier every 2.0 s, what are (a) the frequency and (b) the speed of the waves?

THINKING IT THROUGH. Since the period and wavelength are known, the definition of frequency and Equation 13.23 for wave speed can be used.

SOLUTION

The distance between crests is the wavelength.

Given:

$$\lambda = 5.6 \text{ m}$$
$$T = 2.0 \text{ s}$$

Find:

(a) f (frequency)
(b) v (wave speed)

(a) The lapping indicates the arrival of a wave crest; hence, 2.0 s is the wave period – the time it takes to travel one wavelength (the crest-to-crest distance). Thus

$$f = \frac{1}{T} = \frac{1}{2.0 \text{ s}} = 0.50 \text{ Hz}$$

(b) Then Equation 13.23 allows the wave speed to be found:

$$v = \lambda f = (5.6 \text{ m})(0.50 \text{ s}^{-1}) = 2.8 \text{ m/s}$$

Alternatively,

$$v = \frac{\lambda}{T} = \frac{5.6 \text{ m}}{2.0 \text{ m}} = 2.8 \text{ m/s}$$

FOLLOW-UP EXERCISE. On another day, the person measures the speed of sinusoidal water waves at 2.5 m/s. (a) How far does a wave crest travel in 4.0 s? (b) If the distance between successive crests is 6.3 m, what is the frequency of these waves?

13.3.2 Types of Waves

In general, waves may be divided into two types, based on the direction of the particles' oscillations relative to that of the wave velocity (direction). In a **transverse wave**, the particle motion is *perpendicular* to the wave velocity. The wave traveling on a stretched string (Figure 13.9) is an example of a transverse wave, as is the wave shown in ▼ **Figure 13.10a**. A transverse wave is sometimes called a *shear wave*, because the disturbance supplies a force that tends to shear the medium – that is, to separate layers of the material at right angles to the wave velocity. Shear waves can propagate only in solids. A liquid or a gas does not have sufficient restoring forces between particles to propagate a transverse wave.

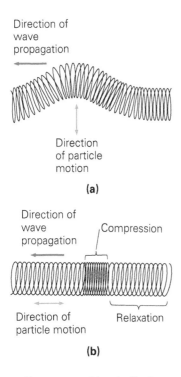

(a)

(b)

▲ FIGURE 13.10 **Transverse and longitudinal waves (a)** In a transverse wave, the motion of the particles is perpendicular to the direction of the wave velocity, as shown here in a spring for a wave moving to the left. **(b)** In a longitudinal wave, the particle motion is parallel to (or *along*) the direction of the wave velocity. Here, a wave pulse also moves to the left. Can you explain the motion of the wave *source* for both types of waves?

In a **longitudinal wave**, the particle oscillation is *parallel* to the wave velocity. A longitudinal wave can be produced, for example, in a stretched spring by a source that moves one end of the coils back and forth along the spring axis (Figure 13.10b). Alternating pulses of compression and relaxation then travel along the spring coils. A longitudinal wave is sometimes called a *compressional wave*, because the force tends to alternately compress and stretch the medium. Longitudinal waves can propagate in solids, liquids, and gases, because all of these can be compressed to some extent.

Sound waves are an everyday example of a longitudinal wave and are produced by a periodic disturbance, typically a speaker. A speaker alternately produces compressions and *rarefactions* (regions where the air density is reduced, or rarefied) that travel through the air as what we call sound (waves). These will be discussed in greater detail in Chapter 14.

13.4 Wave Properties

Among the properties exhibited by all waves are superposition, interference, reflection, refraction, dispersion, and diffraction.

13.4.1 Superposition and Interference

When two or more waves meet or pass through the same region of a medium, they pass through each other and each wave proceeds without being altered. While they are in the same region, the waves are said to be interfering. What happens during interference? That is, what does the combined waveform look like? The answer is given by the **principle of superposition**:

> At any time, the combined waveform of two or more interfering waves is given by the sum of the displacements of the individual waves at each point in the medium.

Using the principle of superposition, interference is illustrated in ▼ **Figure 13.11**. The displacement of the combined waveform at any point is $y = y_1 + y_2$, where y_1 and y_2 are the displacements of the individual pulses at that point. (Directions are indicated by plus and minus signs.) In combining waves, one must take into account the possibility that they can produce disturbances in opposite directions; that is, the wave disturbances must be treated like vector addition.

In Figure 13.11, the vertical displacements of the pulses are in the same direction, and the amplitude of the combined waveform is greater than that of either pulse. This is called **constructive interference**. Conversely, if one pulse has a negative displacement, the two pulses tend to cancel each other when they overlap, and the amplitude of the combined waveform is smaller than that of either pulse. This is called **destructive interference**.

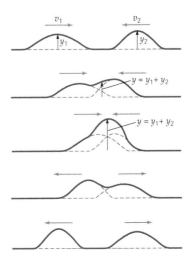

▲ FIGURE 13.11 **Principle of superposition** When two waves meet, they interfere according to the principle of superposition. The beige tint marks the area where the two waves, moving in opposite directions, overlap and thus combine. The displacement at any point on the combined wave is equal to the sum of the (signed) displacements on the individual waves: $y = y_1 + y_2$.

The two special cases of *total* constructive and *total* destructive interference for traveling wave pulses (same amplitude) are illustrated in ▼ **Figure 13.12**. In Figure 13.12a, the instant the pulses overlap, the amplitude of the combined pulse is twice that of either. This represents **total constructive interference**. When the interfering pulses have opposite displacements (Figure 13.12b) when they overlap the waveform will momentarily disappear. This represents **total destructive interference**.

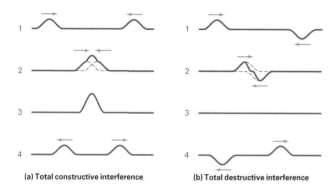

(a) Total constructive interference   **(b) Total destructive interference**

▲ FIGURE 13.12   **Interference (a)** This shows two wave pulses meeting and interfering constructively. When the pulses are exactly superimposed (at 3), total constructive interference occurs. **(b)** When the interfering pulses are out of phase and exactly superimposed (at 3), total destructive interference occurs.

The word *destructive* unfortunately tends to imply that wave energy, as well as the waveform, is momentarily zero. This is *not* the case. At the point of total destructive interference, when the net wave shape drops to zero, the wave energy exists in the medium as kinetic energy. That is, pieces of the straight-looking string are moving.

There are practical applications that utilize destructive interference. For example, it is used for sound waves in automobile mufflers. Inside the muffler there are pipes and chambers arranged so the sound waves (i.e., pressure waves) of the exhaust gases are reflected back and forth in such a way as to give rise to destructive interference, thus greatly reducing the noise of the exhaust system itself.

Another such application is "active noise cancellation," which involves sound wave cancellation by electro-acoustical means. An important application is for pilots who need to hear what's going on around them over the low-frequency engine noise (▼ **Figure 13.13**). Special headphones include a microphone that picks up this engine noise. Electronic circuitry in the headphone creates a wave that is the inverse of the engine noise. When this played through the headphones, the resulting destructive interference with the engine noise produces a quieter background. This enables pilots to better hear mid- and high-frequency sounds, such as conversation and instrument warning sounds.

13.4.2 Reflection, Refraction, Dispersion, and Diffraction

Besides meeting other waves, waves can (and do) meet objects or a boundary with another medium. In such cases, several things may occur. One of these is **reflection**, which occurs when a wave strikes an object or comes to a boundary with another medium and is at least partly diverted back into the original medium. An echo is the reflection of sound waves, and mirrors reflect light waves.

Two cases of reflection for a string wave are illustrated in ▶ **Figure 13.14**. If the end of the string is fixed, the reflected pulse is inverted (Figure 13.14a). The incoming pulse causes the string to exert an upward force on the wall, and the wall exerts an equal and opposite downward force on the string (by Newton's third law). The downward force creates the downward, or inverted, reflected pulse. If the end of the string is free to move, then the reflected pulse is not inverted, as illustrated in Figure 13.14b. In this sketch the string is attached to a ring that can move freely on a smooth pole. The ring is forced upward by the incoming pulse and then comes back down, creating a non-inverted reflected pulse.

More generally, when a wave strikes a boundary, the wave is not completely reflected. Instead, some of the wave's energy is reflected and some is transmitted. When a wave crosses a boundary into another medium, its speed generally changes because the new material has different characteristics. When entering the medium at an angle, the transmitted wave moves in a direction different from that of the incident wave. This phenomenon is called **refraction** (▶ **Figure 13.15**).

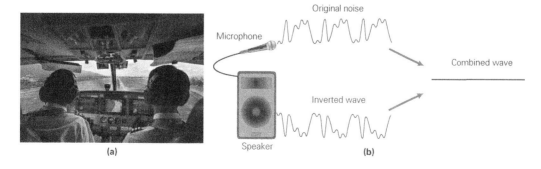

(a)   Microphone   Original noise   Combined wave   Inverted wave   Speaker   **(b)**

▲ FIGURE 13.13   **Destructive interference in action (a)** Pilots use headphones mounted with a microphone that picks up low-frequency engine noise. **(b)** A wave is generated that is inverse that of the engine noise. When played back through the headphones, destructive interference produces less engine noise. This process is called "active noise cancellation."

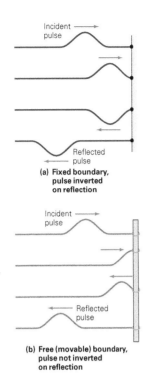

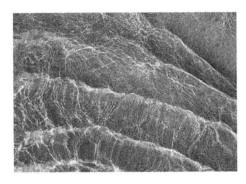

▲ FIGURE 13.14 Reflection (a) When a wave (pulse) on a string is reflected from a fixed boundary, the reflected wave is inverted. **(b)** If the string is free to move at the boundary, the reflected wave is not inverted.

▲ FIGURE 13.15 Refraction Waves approaching a beach are shown from overhead. Water depth on the right is less than on the left. Since water wave speed decreases with water depth, the right side of each crest slows and thus changes direction (refracts) while the left side does not. (Note that the wavelength changes also.)

Since refraction depends on changes in the wave speed, you might be wondering what physical parameters determine wave speed. Generally, there are two types of waves. The simplest kinds are those whose speed does *not* depend on wavelength (or frequency). Thus all such waves travel at the same speed, determined solely by the properties of the medium. These waves are called **non-dispersive waves**, because they do not disperse, or spread apart from one another as they travel. An example of a non-dispersive transverse wave is a wave on a string, whose speed, as will be seen, is determined only by the tension and mass density of the string (Section 13.5). Sound is an example of

a non-dispersive longitudinal wave. For example, the speed of sound in air is determined only by the temperature, compressibility, and density of the air. Indeed, if the speed of sound did depend on the frequency, at the back of the symphony hall you might hear the violins well before the cellos, even though the two sound waves were in perfect synchronization when they left the orchestra.

If the wave speed *does* depend on wavelength (or frequency), the waves are said to be **dispersive waves** and exhibit **dispersion**. For these waves, different frequencies (or wavelengths) have different speeds and thus spread apart. For example, water waves created by a storm travel over the ocean surface travel at different speeds, and thus separate out after a while. Similarly, light waves of different wavelength, when traveling through transparent materials, are dispersive. This is the basis for prisms separating sunlight into a color spectrum and for the formation of a rainbow, as will be seen in Section 22.5.

Diffraction[*] is the term applied to waves when they bend or "wrap" around an edge of an object. For example, if you are outdoors and near an open door of a room, you can often hear people in that room talking even if you cannot see them. This would not be possible if sound waves traveled in a straight line. As they pass through the door opening, instead of being sharply cut off, they "wrap around" the edge, and you can hear the sound.

In general, the effects of diffraction are evident only when the size of the diffracting object or opening is about the same as or smaller than the wavelength of the waves. The dependence of diffraction on the wavelength and size of the object or opening is illustrated in **▶ Figure 13.16**. For many waves, diffraction is negligible. For instance, visible light has wavelengths on the order of 10^{-6} m. Such wavelengths are much too small to exhibit diffraction when they pass through common-sized openings, such as an eyeglass lens or a door opening.

Reflection, refraction, dispersion, and diffraction will be considered in more detail when light waves and the study of optics are considered in Chapters 22 through 25.

13.5 Standing String Waves and Resonance

If you shake one end of a stretched rope (or wire or string) that is fixed at the other end, waves travel on it to the fixed end and are reflected back. The waves going to and from the fixed end interfere with each other. In most cases, the combined waveforms have a changing, jumbled appearance. But if the rope is shaken at just the right frequency, a steady waveform, or series of uniform loops, appears to stand in place along the rope. Appropriately, this phenomenon is called a **standing wave** (**▶ Figure 13.17**). It arises because of interference between the opposite-traveling waves of the same wavelength, frequency, and speed. Neglecting any decrease in amplitude (i.e., energy

---

[*] Although diffraction involves a directional change, it is not directly related to refraction.

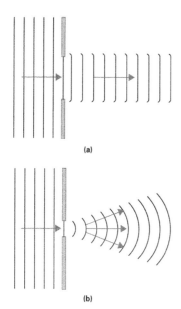

(a)

(b)

▲ **FIGURE 13.16 Diffraction** Diffraction effects are greatest when the opening (or object) is about the same size as or smaller than the wavelength of the waves. **(a)** Here, with an opening much larger than the wavelength of the waves, diffraction is barely noticeable and even then, only near the edges. **(b)** With an opening about the same size as the wavelength, diffraction is obvious and produces nearly semicircular waves.

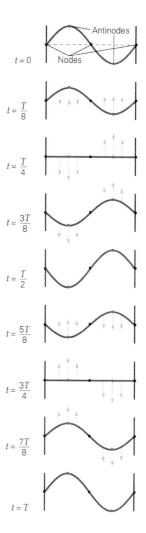

▲ **FIGURE 13.17 Standing waves** Standing waves can be formed by identical interfering waves that are traveling in opposite directions. The standing wave condition of alternating destructive and constructive interference recurs periodically. Here the velocities of selected rope locations are shown by arrows. This resulting motion gives rise to standing waves with stationary nodes (zero amplitude) and antinodes (maximum amplitude).

losses at reflection, etc.) the resulting, or net, energy flow down the rope is thus zero. In effect then, the energy is "standing" in the loops.

Some locations on the rope remain stationary and are called **nodes**. At these points, the displacements of the interfering waves are *always* equal and opposite. Thus, by the principle of superposition, the interfering waves cancel each other completely at these points, and the rope does not undergo a net displacement. At all other points, the rope oscillates back and forth at the same frequency as the underlying waves. The points of maximum amplitude, where constructive interference is greatest, are called **antinodes**. As you can see in Figure 13.17, adjacent antinodes and nodes are separated by a half-wavelength ($\lambda/2$ – "one loop").

Standing waves can be generated in a rope by more than one frequency; the higher the frequency, the shorter the wavelength and thus the more half-wavelength "loops" are on the rope. The requirement for a string standing waves is that an integer number of half-wavelengths "fit" into the length of the rope. The frequencies at which these standing waves are produced are called **natural frequencies**, or **resonant frequencies**. The resulting standing wave patterns are called *normal*, or *resonant*, modes of vibration. Unlike a pendulum or spring-mass system, most systems that oscillate have several natural frequencies. As might be expected, these depend on such factors as mass, elasticity or restoring force, and geometry (boundary conditions).

The natural frequencies of a stretched string or rope can be determined as follows. First, note that the boundary conditions are that both ends remain fixed (i.e., each end must be a node).

Then notice that the number of half-wavelength segments (along the string) that fit between the end nodes is always an integer number n of half-wavelengths (▶ **Figure 13.18** – where $L = 1(\lambda_1/2)$, $L = 2(\lambda_2/2)$, $L = 3(\lambda_3/2)$, etc.). Generalizing these leads to

$$L = n\left(\frac{\lambda_n}{2}\right) \quad \text{or} \quad \lambda_n = \frac{2L}{n} \quad \text{for } n = 1, 2, 3\ldots$$

The natural frequencies of oscillation are therefore

$$f_n = \frac{v}{\lambda_n} = n\left(\frac{v}{2L}\right) = nf_1 \quad \text{for } n = 1, 2, 3\ldots \tag{13.24}$$

(natural frequencies for a stretched string)

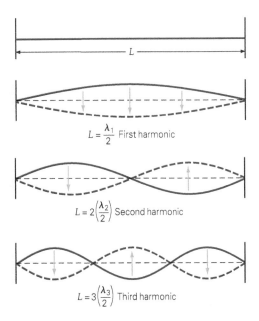

$L = \dfrac{\lambda_1}{2}$ First harmonic

$L = 2\left(\dfrac{\lambda_2}{2}\right)$ Second harmonic

$L = 3\left(\dfrac{\lambda_3}{2}\right)$ Third harmonic

▲ FIGURE 13.18 **Natural frequencies** A stretched string can have standing waves only at certain frequencies. These correspond to the number of half-wavelength loops that fit on the length of string between the nodes at the fixed ends.

where v is the speed of the string waves. The lowest natural frequency ($f_1 = v/2L$ for $n = 1$) is called the **fundamental frequency**. From this result it can be seen that the other (higher) natural frequencies are integer multiples of the fundamental frequency f_1 that is $f_n = nf_1$ (for $n = 2, 3, …$) The set of these frequencies f_1, $f_2 = 2f_1, f_3 = 3f_1, …$ is called a **harmonic series**: f_1 (the fundamental frequency) is the *first harmonic*, f_2 the *second harmonic*, and so on.

Strings that are fixed at each end are found in musical instruments such as violins, pianos, and guitars. When such a string is disturbed – that is, plucked, struck, or bowed – the resulting vibration is complicated because it generally includes higher harmonics in addition to the first harmonic. The number of harmonics depends on how and where the string is disturbed. It is the combination of harmonic frequencies that gives a particular instrument its characteristic sound quality (see Section 14.6). As Equation 13.24 shows, all the harmonic frequencies depend on the length of the string. Hence different frequencies can be obtained on a violin string by pressing the string at a particular location so as to change its vibrating length (▶ **Figure 13.19**).

As is seen in Equation 13.24, the natural frequencies of a stretched string depend on the string wave speed. This is determined by other physical parameters associated with the string (but *not* the frequency – recall string waves are non-dispersive). For a string, the wave speed (v) is

$$v = \sqrt{\dfrac{F_{\mathrm{T}}}{\mu}} \quad \text{(wave speed on a stretched string)}$$

(F_{T} is string tension and μ is its linear mass density $m_{\mathrm{s}}/L_{\mathrm{s}}$)

$$(13.25)$$

▲ FIGURE 13.19 **Fundamental frequencies** Performers using stringed instruments such as the violin employ their fingers to play notes. By pressing a string against the fingerboard, this violinist changes the length that is free to vibrate. This change results in changes to the string harmonic frequencies and the exact tone the instrument produces.

Here we use the symbol F_{T} rather than T for tension so as not to confuse it with period. Equation 13.24 can thus be written as

$$f_n = n\left(\dfrac{v}{2L}\right) = \dfrac{n}{2L}\sqrt{\dfrac{F_{\mathrm{T}}}{\mu}} \quad n = 1, 2, 3, …$$

(natural frequencies of a stretched string)

$$(13.26)$$

Note the greater the linear mass density of a string, the lower its natural frequencies. Thus the low-frequency strings on a violin or guitar are thicker, and thus denser (linearly), than the high-frequency strings. By tightening a string, all the frequencies of that string are increased. Changing the tension in the string is how violinists, for example, tune their instruments before a performance.

EXAMPLE 13.7: A PIANO STRING – FUNDAMENTAL FREQUENCY AND HARMONICS

A piano string with a length of 1.15 m and a mass of 20.0 g is under a tension of 6.30×10^3 N. (a) What is the fundamental frequency of this string? (b) What are the frequencies of the next two harmonics?

THINKING IT THROUGH. The linear mass density can be determined from the data. This, with the given string tension, can be used to find the fundamental frequency f_1. From this, the next two harmonic frequencies (f_2 and f_3) can be found.

SOLUTION

Given:

$L_{\mathrm{s}} = 1.15$ m
$m_{\mathrm{s}} = 20.0$ g $= 0.0200$ kg
$F_{\mathrm{T}} = 6.30 \times 10^3$ N

Find:

(a) f_1 (fundamental)
(b) f_2 and f_3

(a) The linear mass density of the string is

$$\mu = \frac{m_s}{L_s} = \frac{0.0200\,\text{kg}}{1.15\,\text{m}} = 0.0174\,\text{kg/m}$$

Then, using Equation 13.26,

$$f_1 = \frac{1}{2L}\sqrt{\frac{F_T}{\mu}} = \frac{1}{2(1.15\,\text{m})}\sqrt{\frac{6.30\times10^3\,\text{N}}{0.0174\,\text{kg/m}}} = 262\,\text{Hz}$$

(b) Since $f_2 = 2f_1$ and $f_3 = 3f_1$, then

$$f_2 = 2f_1 = 2(262\,\text{Hz}) = 524\,\text{Hz}$$

and

$$f_3 = 3f_1 = 3(262\,\text{Hz}) = 786\,\text{Hz}$$

FOLLOW-UP EXERCISE. A musical note is referenced to the fundamental frequency, or first harmonic. In musical terms, the second harmonic is sometimes referred to as the first *overtone*, the third harmonic is the second overtone, and so on. If an instrument has a third overtone with a frequency of 880 Hz, what is the frequency of the first overtone?

13.5.1 Resonance

When an oscillating system is driven or shaken at one of its natural, or resonant, frequencies, the maximum amount of energy is transferred to the system. The natural frequencies of a system are the frequencies at which the system "prefers" to vibrate, so to speak. When a system is being driven at one of its natural frequencies, this is called **resonance**.

A common example of a mechanical system in resonance is the act of pushing a swing. Basically, a swing is a simple pendulum and has only one natural frequency for a given length (recall $f = \left(\frac{1}{2\pi}\right)\sqrt{\frac{g}{L}}$). If you push the swing at this frequency and in time with its motion, its amplitude and energy increase (▼ **Figure 13.20**) and the person on the swing can enjoy an exciting, large amplitude ride.

▲ **FIGURE 13.20 Resonance in the playground** The swing behaves like a pendulum in SHM. To transfer energy efficiently, and thus maintain a large amplitude, the man must time his pushes to the natural frequency of the swing.

As has been shown, a stretched string has many natural frequencies. Any driving frequency will cause a disturbance in the string. However, if the frequency of the driving force is not equal to one of the natural frequencies, the resulting pattern will be relatively small and jumbled. In contrast, when the driving force frequency matches one of the natural frequencies, a maximum amount of energy is transferred to the string and a steady standing wave pattern results. The antinode amplitude becomes relatively large while the stationary node locations become obvious.

While resonance can be a positive goal, there are situations when it might be unwanted. For example, if a large number of soldiers march over a small bridge, they are generally ordered to "break step," so as not to make their stepping force occur in unison. This is to avoid having the marching (stepping) frequency correspond to one of the bridge's natural frequencies. If this occurred, it might be set the bridge into a large resonant vibration, and possibly cause a collapse. This actually occurred on a suspension bridge in England (in 1831) that collapsed as a direct result of the resonance vibrations induced by the marching soldiers.

Another incident of a bridge vibrating into collapse is that of the famed "Galloping Gertie," the Tacoma Narrows Bridge (Washington State). After being open for only 4 months, on November 7, 1940, sustained winds with speeds of 65–72 km/h (40–45 mi/h) started the main span vibrating. The main span vibrated in a transverse mode of frequency 0.6 Hz and a torsional (twisting) mode at 0.2 Hz. Hours after the start, the main span finally collapsed (▼ **Figure 13.21**). The driving force of the collapse is quite complicated, but the energy provided by the wind was a major factor.

▲ **FIGURE 13.21 Galloping Gertie** The collapse of the Tacoma Narrows Bridge on November 7, 1940, is captured in this frame from a movie camera.

Chapter 13 Review

- **Simple harmonic motion (SHM)** requires a restoring force directly proportional to the displacement from equilibrium, such as an ideal (Hooke's law) spring:

$$F_s = -kx \quad \text{(Hooke's law)} \tag{13.1}$$

- **Frequency (f)** and **period (T)** are inversely related:

$$f = \frac{1}{T} \quad \text{(frequency and period)} \quad (13.2)$$

- **Total energy of a mass-spring system in SHM:**

$$E = \frac{1}{2}mv^2 + \frac{1}{2}kx^2 = \frac{1}{2}kA^2 \quad (13.6)$$

- The **equations of motion for spring-driven SHM** depend on initial conditions. (All SHM equations – location, velocity and acceleration – can be adapted to describe horizontal SHM by replacing y by x everywhere and using the + sign to indicate spring extension and the – sign for compression):

$$y = \pm A \sin \omega t = \pm A \sin(2\pi f t) = \pm A \sin\left(\frac{2\pi t}{T}\right)$$

(+ sign for initial motion upward at $y_0 = 0$; (13.13)
 – sign for initial motion downward at $y_0 = 0$)

$$y = \pm A \cos \omega t = \pm A \cos(2\pi f t) = \pm A \cos\left(\frac{2\pi t}{T}\right)$$

(+ sign for initially at rest at $y_0 = +A$; (13.14)
 – sign for initially at rest at $y_0 = -A$)

- **Period of a spring-mass system:**

$$T = 2\pi\sqrt{\frac{m}{k}} \quad \text{(period of spring-driven SHM)} \quad (13.15)$$

- **Angular frequency of a spring-mass system:**

$$\omega = \sqrt{\frac{k}{m}} \quad \text{(angular frequency of spring-driven SHM)} \quad (13.17)$$

- *Period of a simple pendulum (small-angle approximation):*

$$T = 2\pi\sqrt{\frac{L}{g}} \quad \text{(period of a simple pendulum)} \quad (13.18)$$

- **Velocity in SHM:**

$$v_y = \pm \omega A \cos \omega t = \pm \omega A \cos(2\pi f t) = \pm \omega A \cos\left(\frac{2\pi t}{T}\right)$$

(choose + sign for initial motion upward at $y_0 = 0$; (13.19)
 choose – sign for initial motion downward at $y_0 = 0$)

$$v_y = \mp \omega A \sin \omega t = \mp \omega A \sin(2\pi f t) = \mp \omega A \sin\left(\frac{2\pi t}{T}\right)$$

(choose – sign for initially at rest at $y_0 = +A$; (13.20)
 choose + sign for initially at rest at $y_0 = -A$)

- **Acceleration in SHM:**

$$a_y = \mp \omega^2 A \sin \omega t = \mp \omega^2 A \sin(2\pi f t) = \mp \omega^2 A \sin\left(\frac{2\pi t}{T}\right)$$

(choose – sign for initial motion upward at $y_0 = 0$; (13.21)
 choose + sign for initial motion downward at $y_0 = 0$)

$$a_y = \mp \omega^2 A \cos \omega t = \mp \omega^2 A \cos(2\pi f t) = \mp \omega^2 A \cos\left(\frac{2\pi t}{T}\right)$$

(choose – sign for initially at rest at $y_0 = +A$; (13.22)
 choose + sign for initially at rest at $y_0 = -A$)

- A **wave** is a disturbance in time and space; energy is propagated by wave motion.

$$v = \frac{\lambda}{T} = \lambda f \quad \text{(wave speed)} \quad (13.23)$$

- **Superpostion principle:** At any time, the combined waveform of two or more interfering waves is given by the sum of the displacements of the individual waves at each point in the medium.
- The natural frequencies of **standing waves** on a string:

$$f_n = n\left(\frac{v}{2L}\right) = \frac{n}{2L}\sqrt{\frac{F_T}{\mu}} \quad n = 1, 2, 3, \dots \quad (13.26)$$

End of Chapter Questions and Exercises

Multiple Choice Questions

13.1 Simple Harmonic Motion

1. For an object in SHM, the net force on it and its displacement from equilibrium are (a) in the same direction, (b) opposite in direction, (c) perpendicular, (d) none of the preceding.
2. The maximum kinetic energy of a mass-spring system in SHM is equal to (a) A, (b) A^2, (c) kA, (d) $kA^2/2$.
3. If the frequency of a system in SHM is doubled, its period becomes (a) doubled, (b) halved, (c) four times larger, (d) one-quarter as large.
4. When an object in horizontal SHM is at its equilibrium position, the kinetic energy of the system is (a) zero, (b) maximum, (c) half its maximum, (d) none of the preceding.

13.2 Equations of Motion

5. The equation of motion for an object in SHM (a) is a sine or cosine function, (b) is a tangent or cotangent function, (c) could be any mathematical function, (d) none of these.
6. Given the SHM equation of motion: $y = A\sin[200\,\pi t]$. The frequency of this oscillation is (a) 50 Hz, (b) 100 Hz, (c) 200 Hz, (d) 200π Hz.

7. This equation describes the SHM of an object: $y = A\sin(2\pi t/T)$. Its position (y) three-quarters of the period after the motion starts is (a) $+A$, (b) $-A$, (c) $A/2$, (d) 0.

13.3 Wave Motion

8. Wave motion in a medium involves (a) the propagation of a disturbance, (b) inter-particle interactions, (c) the transfer of energy, (d) all of the preceding.

9. For a longitudinal wave, the direction between the wave velocity and medium particle oscillation is (a) 90°, (b) 0°, (c) 45°, (d) none of these.

10. A string wave is an example of what type of wave? (a) transverse, (b) longitudinal, (c) compressional, (d) none of these.

13.4 Wave Properties

11. When two waves overlap, the resultant waveform is determined by (a) reflection, (b) refraction, (c) diffraction, (d) superposition.

12. When two identical waves of the same wavelength and amplitude A overlap in phase, the amplitude of the resulting wave is (a) A, (b) $2A$, (c) $3A$, (d) $4A$.

13. You can often hear people talking from around a corner of a building. This is due primarily to (a) reflection, (b) refraction, (c) interference, (d) diffraction.

13.5 Standing Waves and Resonance

14. For two traveling waves to form a standing wave, the waves must have the same (a) wavelength, (b) amplitude, (c) speed, (d) all of the preceding.

15. The points of zero amplitude on string standing wave are (a) nodes, (b) antinodes, (c) fundamentals, (d) resonance points.

16. For a standing wave on a rope, the distance between adjacent antinodes is (a) 1/4 wavelength, (b) 1/2 wavelength, (c) one wavelength, (d) two wavelengths.

17. When a stretched violin string oscillates in its second harmonic mode, the standing wave pattern on the string exhibits a total of (a) 1/4 wavelength, (b) 1/2 wavelength, (c) one wavelength, (d) two wavelengths.

Conceptual Questions

13.1 Simple Harmonic Motion

1. If the amplitude of an object in SHM is doubled, how are (a) the system total energy and (b) the object's maximum speed affected?

2. How does the speed of a mass in SHM change as it leaves equilibrium? Explain.

3. A mass-spring system in SHM has an amplitude A and period T. How long, in units of T, does the mass take to travel a distance A? How about $2A$?

4. A very elastic ball rattles back and forth horizontally between the walls of a box. Is this a simple harmonic motion? Explain.

13.2 Equations of Motion

5. If a mass-spring system were taken to the Moon, would its period change? How about the period of a simple pendulum if taken to the Moon? Explain both.

6. If you want to increase the frequency of the SHM of a mass-spring system, would you increase or decrease the mass? Explain.

7. If the length of a simple pendulum is doubled, what is the ratio of the new period to the old one?

8. Would the period of a pendulum in an upward-accelerating elevator be increased or decreased compared with its period in a non-accelerating elevator? Explain.

9. One simple harmonic motion is described by a sine function, $y = A\sin(\omega t)$ and another is described by a cosine function, $y = A\cos(\omega t)$. How do their initial positions, speeds, and accelerations compare?

13.3 Wave Motion

10. When a wave pulse travels along a rope, what physical quantity actually travels?

11. ▼ **Figure 13.22** shows a picture of a mechanical waves traveling along a stretched coil (spring). Is it transverse or longitudinal? Explain your reasoning.

12. What type(s) of wave(s), transverse or longitudinal, are able to propagate through (a) solids, (b) liquids, and (c) gases?

13. Sometimes wheat in a field can be seen swaying when a breeze blows. What type of motion are these stalks undergoing, transverse or longitudinal? Explain.

▲ FIGURE 13.22 **Transverse or longitudinal?** See Conceptual Question 11.

13.4 Wave Properties

14. What physical quantity(ies) is (are) cancelled when destructive interference occurs? What happens to the wave energy in such a situation? Explain.

15. Dolphins and bats can determine the location of prey by emitting ultrasonic sound waves. Which wave phenomenon is involved?

16. If sound waves were dispersive, what would be the consequences of someone listening to an orchestra in a concert hall?

13.5 Standing Waves and Resonance

17. Is it possible to generate harmonic sound of any frequency using a violin string with a fixed tension and length? Explain.

18. If they have the same tension, mass, and length, will a thicker or a thinner guitar string sound higher in frequency? Why?

19. A child's swing (treated as a pendulum) has only one natural frequency, f_1, yet it can be driven so as to maintain or increase its energy at frequencies of $f_1/2$, $f_1/3$, and $2f_1$. How is this possible?

20. By rubbing the circular lip of a wide, thin wine glass with a moist finger, you can make the glass "sing." (Try it.) What causes this?

Exercises*

*Integrated Exercises (**IEs**) are two-part exercises. The first part typically requires a conceptual answer choice based on physical thinking and basic principles. The following part is quantitative calculations associated with the conceptual choice made in the first part of the exercise.*

13.1 Simple Harmonic Motion

1. • A particle oscillates in SHM with an amplitude A. What is the total *distance* (in terms of A) the particle travels in three periods?

2. • If it takes a particle in SHM 0.50 s to travel from equilibrium to maximum displacement from equilibrium, what is the period of this oscillation?

3. • A 0.75-kg object on a spring completes a cycle every 0.50 s. What is the oscillation frequency?

4. • A particle in SHM oscillates at frequency of 40 Hz. What is its period?

5. • An oscillator's frequency changes from 0.25 Hz to 0.50 Hz. What is the period change?

6. • An object (mass 0.50 kg) is attached to a spring with spring constant of 10 N/m. If the object is pulled 5.0 cm from equilibrium and released, what is its maximum speed?

7. • An object (mass 1.0 kg) is attached to a spring with spring constant 15 N/m. If the object has a maximum speed of 50 cm/s, what is the amplitude of its oscillation?

8. **IE** •• (a) At what position is the magnitude of the force on a mass in a horizontal mass-spring system minimum: (1) $x = 0$, (2) $x = -A$, or (3) $x = +A$? Why? (b) If $m = 0.500$ kg, $k = 150$ N/m, and $A = 0.150$ m, what are the magnitude of the force on the mass and its acceleration at $x = 0$, 0.050 and 0.150 m?

9. **IE** •• (a) At what position is the speed of a mass in a horizontal mass-spring system maximum: (1) $x = 0$, (2) $x = -A$, or (3) $x = +A$? Why? (b) If $m = 0.250$ kg, $k = 100$ N/m, and $A = 0.10$ m for such a system, what is the mass's maximum speed?

10. •• A mass-spring system is in horizontal SHM. If the mass is 0.25 kg, the spring constant is 12 N/m, and the amplitude is 15 cm, (a) what is the maximum speed of the mass, and (b) where does this occur? (c) What is the speed at a half-amplitude position?

11. •• A horizontal spring on a frictionless and level air track has a 0.150-kg object attached to it and then it is stretched 6.50 cm. Then the object is given an outward velocity of 2.20 m/s. If the spring constant is 35.2 N/m, determine how much farther the spring stretches.

12. •• A 0.25-kg object is attached to a light spring of spring constant 49 N/m and gently eased to its equilibrium position where it is at rest. (a) How far does the mass stretch the string? (b) By how much does the gravitational potential energy of the spring-mass system change? (c) By how much does the elastic potential energy of the spring-mas system change?

13. •• A 0.25-kg object is attached to a vertical spring of spring constant 49 N/m and the system is brought to rest at its equilibrium position. The object is then pulled down 0.10 m from equilibrium and released. What is the object's speed as it passes equilibrium?

14. •• A 0.350-kg block moving vertically upward collides with a light vertical spring and compresses it 4.50 cm before coming to rest. If the spring constant is 50.0 N/m, what was the initial speed of the block? (Ignore energy losses during the collision.)

15. ••• A 0.250-kg ball is dropped from a height of 10.0 cm onto a spring, as illustrated in ▼ **Figure 13.23**. If the spring has a spring constant of 60.0 N/m, (a) what distance will the spring be compressed? (Neglect energy loss during collision.) (b) On recoiling upward, how high will the ball go?

▲ FIGURE 13.23 **How far down?** See Exercise 15.

13.2 Equations of Motion

16. • A 0.50-kg mass oscillates in SHM on a spring with a spring constant of 200 N/m. What are (a) the period and (b) the frequency of the oscillation?

17. • The simple pendulum in a tall clock is 0.75 m long. What are (a) the period and (b) the frequency of this pendulum?

18. • How much mass should be on the end of a spring ($k = 100$ N/m) so the period is 2.0 s?

---

* The bullets denote the degree of difficulty of the exercises: •, simple; ••, medium; and •••, more difficult.

19. • If the frequency of a mass-spring system is 1.50 Hz and the mass on the spring is 5.00 kg, what is the spring constant?

20. • A breeze sets a suspended lamp swinging. If the period is 1.0 s, what is the distance from ceiling to lamp at the low point? Assume the lamp acts as a simple pendulum.

21. • In each case, choose sine or cosine (with the proper sign) for the equation of motion that best describes the location (x) of a mass attached to a spring on a horizontal frictionless surface. (a) The spring is stretched and released, (b) the spring is compressed and released, and (c) the mass is shoved outward (spring extension) from equilibrium.

22. • The equation of motion for an oscillator in vertical SHM is $y = (0.10 \text{ m})\sin(100t)$. What are the (a) amplitude, (b) frequency, and (c) period of this motion?

23. • The location of an object in vertical SHM is $y = (5.0 \text{ cm})\cos(20\pi t)$. What are the system's (a) amplitude, (b) frequency, and (c) period?

24. • The location of an object in vertical SHM is given by $y = (0.25 \text{ m})\cos(314t)$ where y is in meters and t is in seconds. What is the location of the oscillator at (a) $t = 0$, (b) $t = 5.0$ s, and (c) $t = 15$ s?

25. •• The equation of motion of a horizontal SHM oscillator is $x = (0.50 \text{ m})\sin(2\pi ft)$ where x is in meters and t is in seconds. If the position of the oscillator is $x = 0.25$ m at $t = 0.25$ s, what is the frequency of the oscillator?

26. IE •• The oscillations of two oscillating mass-spring systems are graphed in ▼ **Figure 13.24**. The mass in System A is four times that in System B. (a) Compared with System B, System A has (1) more, (2) the same, or (3) less energy. Why? (b) Calculate the ratio of energy between System B and System A.

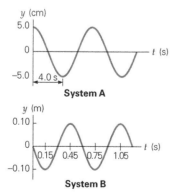

▲ FIGURE 13.24 **Simple harmonic oscillations** See Exercises 26, 34, and 35.

27. •• The velocity of an object in a vertically oscillating mass-spring system is $v = (0.750 \text{ m/s})\sin(4t)$. Find the object's (a) amplitude and (b) maximum acceleration.

28. IE •• (a) If the mass in an oscillating mass-spring system is halved, the new period is (1) 2, (2) $\sqrt{2}$, (3) $1/\sqrt{2}$, (4) 1/2 times the old period. Explain. (b) If its initial period is 3.0 s and the mass is reduced to 1/3 of its initial value, what is the new period?

29. IE •• (a) If the spring constant in an oscillating mass-spring system is halved, the new period is (1) 2, (2) $\sqrt{2}$, (3) $1/\sqrt{2}$, (4) 1/2 times the old period. Explain. (b) If the initial period is 2.0 s and the spring constant is reduced to 1/3 of its initial value, what is the new period?

30. •• What is the maximum spring potential energy of a horizontal mass-spring oscillator whose equation of motion is $x = (0.350 \text{ m})\sin(7t)$ if mass on the spring is 0.900 kg?

31. •• Two masses oscillate on light springs. The second mass is half of the first and its spring constant is twice that of the first. Which system will have the greater frequency, and what is the ratio of the frequency of the second mass to that of the first mass?

32. •• During an earthquake, the floor of an apartment building is observed to oscillate in approximately SJM with a period of 1.95 s and an amplitude of 8.65 cm. Determine the maximum speed and acceleration of the floor during this motion.

33. IE •• (a) If a pendulum clock were taken to the Moon, where the acceleration due to gravity is only one-sixth (take this to be exact) that on the Earth, will the period of oscillation (1) increase, (2) remain the same, or (3) decrease? Why? (b) If the period on the Earth is 2.0 s, what is the period on the Moon?

34. •• Assume that the motion of an object oscillating on a light spring is described by the graph for System A in Figure 13.24. (a) Write the equation of motion in terms of a sine or cosine function. (b) If the spring constant is 20 N/m, what is the mass of the object?

35. •• Assume that the motion of a 0.25-kg mass oscillating on a light spring is described by the graph for System B in Figure 13.24. (a) Write the equation for the displacement of the mass as a function of time. (b) What is the spring constant of the spring?

36. ••• A grandfather clock behaves like a simple pendulum 75 cm long. The clock is accidentally broken, and when it is repaired, the length of the pendulum is shortened by 2.0 mm. (a) Will the repaired clock gain or lose time? (b) By how much will the time on the repaired clock differ from the correct time (the time determined by the original pendulum) in 24 h? (c) If the pendulum rod were metal, would the surrounding temperature make a difference in the timekeeping of the clock? Explain.

37. ••• The velocity of a vertically oscillating 5.00-kg mass on a spring is given by $v = (-0.600 \text{ m/s})\sin(6t)$ (a) Determine an expression for its location, y, as a function of time. (b) Where does the motion start and in what direction

does the object move initially and with what speed? (c) Determine the period of the motion. (d) Determine the maximum force on the mass.

13.3 Wave Motion

38. • A sound wave travels at 340 m/s in air. If this wave has a frequency of 1000 Hz, what is its wavelength?

39. • A wave on a very long rope takes 2.0 s to travel 10 m of the rope. If the wavelength of the wave is 2.5 m, what is the frequency of oscillation of any piece of the rope?

40. • A student reading his physics book on a dock notices that the distance between two successive incoming wave crests is 0.75 m, and the difference in their times of arrival is 1.6 s. What is the speed of the water waves?

41. • Bats can determine the location of prey using *echolocation* (measuring the travel time for a sound echo to return). If it takes 15 ms for a bat to receive an *ultrasonic* (i.e., above the frequency of human hearing) sound wave reflected off a mosquito, how far is the mosquito from the bat? Take the speed of sound as 345 m/s.

42. • Light waves travel in a vacuum at a speed of 3.00×10^8 m/s. The frequency of blue light is about 6.0×10^{14} Hz. What is the approximate wavelength of this blue light?

43. •• A submarine's sonar generator produces ultrasonic sound waves at 2.50 MHz. The wavelength of the waves in seawater is 4.80×10^{-4} m. When the generator is directed downward, an echo reflected from the ocean floor is received 10.0 s later. How deep is the ocean at that point, assuming the sub to be on the ocean surface?

44. •• ▼ **Figure 13.25a** shows a snapshot of a wave on a rope, and Figure 13.25b describes the position as a function of time of a fixed point on that rope. (a) What is the amplitude of the wave? (b) What is its wavelength? (c) What is its period? (d) What is the wave speed?

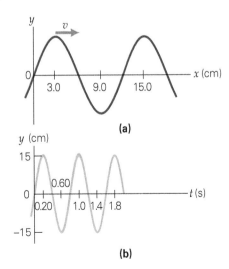

▲ **FIGURE 13.25 How high and how fast?** See Exercise 44.

13.5 Standing Waves and Resonance

45. • If the frequency of the third harmonic on a tight string is 600 Hz, what is the frequency of the first harmonic?

46. • The fundamental frequency of a stretched string is 150 Hz. What are the frequencies of (a) the second harmonic and (b) the third harmonic?

47. • If the frequency of the fifth harmonic of a vibrating string is 425 Hz, what is the frequency of the second harmonic?

48. • A standing wave is formed in a stretched string that is 3.0 m long. What are the wavelengths of (a) the first harmonic and (b) the second harmonic?

49. • If the wavelength of the third harmonic on a string is 5.0 m, what is the string length?

50. IE •• A piece of thin steel cable is under tension. (a) If the tension doubles, the wave speed on the cable (1) doubles, (2) halves, (3) increases by $\sqrt{2}$, (4) decreases by $\sqrt{2}$. Why? (b) If the linear mass density of the cable is 0.125 kg/m and it is under a tension of 9.00 N, what is the wave speed on it? (c) If the cable is 10.0 m long, what are its lowest three standing wave frequencies?

51. •• On a violin, a correctly tuned "A" string has a frequency of 440 Hz. If an "A" string produces sound at 450 Hz under a tension of 500 N, what should the tension be to produce the correct frequency, assuming the string length remains unchanged?

52. •• Will a standing wave be formed in a 4.0-m length of stretched string that has a wave speed of 12 m/s if it is driven at a frequency of (a) 15 Hz or (b) 20 Hz?

53. •• Two waves of equal amplitude and frequency of 250 Hz travel in opposite directions at a speed of 150 m/s in a string. If the string is 0.90 m long, for which harmonic mode is the standing wave set up in the string?

54. •• A physics professor buys 100 m of string with a total mass of 0.150 kg. This string is used to set up a standing wave lecture demonstration between two posts 3.0 m apart. If the desired second harmonic frequency is 35 Hz, what should be the string tension?

55. IE •• String A has twice the tension but half the linear mass density of string B, and both have the same length. (a) The frequency of the first harmonic on string A is (1) four times, (2) twice, (3) half, (4) 1/4 times that of string B. Explain. (b) If the length of the strings is 2.5 m and the wave speed on A is 500 m/s, what are the frequencies of the first harmonic on both strings?

56. •• You want to set up two standing string waves using a length of uniform piano wire that is 3.0 m long with a mass of 0.150 kg. You cut it into two segments, one with a length of 1.0 m and the other 2.0 m, and then place each under tension. What should be the ratio of tensions (expressed as short/long) so their fundamental frequencies are the same?

57. IE •• A violin string is tuned to its fundamental, or first harmonic, frequency. (a) If a violinist wants a higher frequency by changing only its length, should the string be (1) lengthened, (2) kept the same, or (3) shortened? Why? (b) If the string is tuned to 520 Hz and the violinist puts a finger down on the string one-eighth of the string length from the neck end, what is the frequency of the string when the instrument is played this way?

58. ••• A uniform string with a length of 1.80 m is tied tightly at both ends under a tension of 100 N. When it vibrates in its third harmonic (draw a sketch?), the sound given off has a frequency of 75.0 Hz. What is the mass of the string?

14

Sound

Although the different instruments in this photo emit sounds of various frequencies, they all travel at the same speed.

The band shown in this chapter-opening photo is clearly giving good vibrations! We owe a lot to sound waves. Not only do they provide us with a source of enjoyment in the form of music, but they also bring a wealth of vital information about our environment, from the chime of a doorbell to the shrill of a police siren to the song of a bird. Indeed, sound waves are the basis for our major form of communication – speech. These waves can also constitute irritating distractions (noise). But sound waves become music, speech, or noise only when our ears perceive them. Physically, sound is simply wave energy that propagates in solids, liquids, and gases. Without a medium, there can be no sound; in a vacuum, as in outer space, there is utter silence.

This distinction between the sensory and physical meanings of sound provides an answer to the old philosophical question: If a tree falls in the forest where there is no one to hear it, is there sound? The answer depends on how sound is defined – the answer is no if thinking in terms of sensory hearing, but yes if considering physical waves.

Since sound waves are all around us most of the time, we are exposed to many interesting sound phenomena. Some of the most important of these will be considered in this chapter.

14.1 Sound Waves

For sound to exist there must be a disturbance to a medium, giving rise to the traveling sound waves. This disturbance may be the clapping of hands or the skidding of car tires. Under water, it might be the click of rocks against one another. **Sound waves** in gases and liquids (both are fluids; see Chapter 9) are primarily longitudinal waves. However, sound disturbances moving through solids can have both longitudinal and transverse components. The intermolecular interactions in solids are much stronger than in fluids and allow transverse components to propagate.

The characteristics of sound waves can be visualized by considering those produced by a tuning fork, which is essentially a metal bar bent into a U shape (▼ **Figure 14.1**). The prongs, or tines, vibrate when struck. The fork vibrates at its fundamental frequency, so a single tone is heard. (A *tone* is sound of a definite frequency.) The vibrations disturb the air, producing alternating high-pressure regions called *condensations* and low-pressure regions called *rarefactions*. As the fork continues to vibrate, the disturbances propagate outward, and a series of them can be represented by a sinusoidal wave traveling through space.

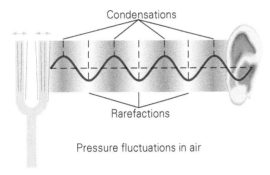

▲ **FIGURE 14.1 Vibrations make waves** A vibrating tuning fork disturbs the air, producing alternating high-pressure regions (condensations) and low-pressure regions (rarefactions), which form sound waves.

When the pressure disturbances (i.e., the sound) reach the ear, the eardrum (a thin membrane) is set into vibration by the alternating high- and low-pressure regions impacting it (the normal pressure on the inside of the eardrum is atmospheric and constant). On the interior of the eardrum, tiny bones carry the vibrations to the inner ear, where they are picked up by the auditory nerve and relayed to the brain in the form of electrical waves where the brain interprets the vibrations as sound.

Physical characteristics of the ear limit the perception of sound. Only sound waves with frequencies between about 20 Hz and 20 kHz (kilohertz) can initiate nerve impulses that are interpreted by the brain as sound. This frequency range is called the **audible region** of the **sound frequency spectrum**. Hearing is most acute in the 1000 Hz–10 000 Hz range, with speech mainly in the frequency range between 300 Hz and 3400 Hz.

14.1.1 Infrasound

Sound wave frequencies lower than 20 Hz are in the **infrasonic region** (*infrasound*). Waves in this region, which humans are unable to hear, are found in nature. Longitudinal waves generated by earthquakes have infrasonic frequencies, and these waves are used to study the Earth's interior. Infrasonic waves are also generated by wind and weather patterns. Elephants and cattle have hearing responses in the infrasonic region and may give early warnings of earthquakes and weather disturbances, such as tornadoes. It has been found that the vortex of a tornado produces infrasound and can be detected miles away from a tornado, thus providing a method for gaining increased warning times of tornado approaches. Nuclear explosions produce infrasound, and after the Nuclear Test Ban Treaty of 1963, infrasound listening stations were set up to detect possible violations. Now these stations are also used to detect other sources such as earthquakes and tornadoes.

14.1.2 Ultrasound

Above 20 kHz in the sound frequency spectrum is the **ultrasonic region** (*ultrasound*). Ultrasonic waves can be generated by high-frequency vibrations in crystals. Ultrasonic waves cannot be detected by humans, but can be by other animals. The audible region for dogs extends to about 40 kHz, which is how ultrasonic or "silent" whistles can be used to call dogs without disturbing people.

There are many practical uses for ultrasound. Because ultrasound can travel for kilometers in water, it is used in sonar to detect underwater objects and their ranges (distances), much like radar uses radio waves. Sound pulses generated by the sonar apparatus are reflected by underwater objects, and the resulting echoes are picked up by a detector. The time for a pulse to make a round trip, together with the speed of sound in water, gives the distance to the object. Sonar is widely used by fishermen to detect schools of fish, and in a similar manner can provide oceanographers with accurate ocean depth data.

Probably the best-known applications of ultrasound are in medicine. For instance, ultrasound is commonly used to obtain an image of a fetus, avoiding potentially dangerous X-rays. Ultrasonic generators produce high-frequency sound wave pulses that scan a designated region of the body. When the waves encounter a boundary between tissues of differing densities, they are partially reflected. These reflections are monitored by a receiving transducer and a computer constructs an image from the reflected signals.

Ultrasonic baths are routinely used to clean metal machine parts, dentures, and jewelry. The high-frequency vibrations loosen particles in otherwise inaccessible places. Ultrasound waves can also be used to locate and diagnose gallstones and kidney stones. In many cases they can be used to break the stone up through a technique called lithotripsy (Greek, "stone breaking").

There are many applications of ultrasonic sonar in nature. Sonar appeared in the animal kingdom long before it was developed by humans. On nocturnal hunting flights, bats use a kind of natural sonar to navigate in and out of their caves and to locate and catch flying insects (▼ **Figure 14.2**). The bats emit pulses of ultrasound and track their prey by means of the reflected echoes. The technique is known as *echolocation*.

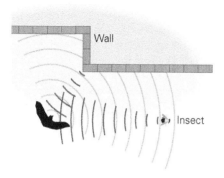

▲ **FIGURE 14.2   Echolocation** With the aid of their natural sonar systems, bats can hunt flying prey. The bats emit pulses of ultrasonic waves, which are in their audible region, and use the echoes reflected to judge distance and guide their attack.

14.2 The Speed of Sound

In general, the speed at which a disturbance moves through a medium depends on its physical quantities. For example, as learned in Section 13.5, the wave speed on a stretched string is $v = \sqrt{F_\text{T}/\mu}$, where F_T is the string tension and μ is its linear mass density. Another example is the speed of sound in a solid, given by $v = \sqrt{Y/\rho}$, where Y is its Young's modulus and ρ is its density. For liquids, the analogous expression is $v = \sqrt{B/\rho}$, where B is the bulk modulus of the liquid and ρ is its density. Solids are generally more elastic than liquids, which in turn are more elastic than gases. In elastic solid materials, the relatively large restoring forces between the atoms or molecules cause a sound wave to propagate through them faster than through liquids and gases. The speed of sound is generally about two to four times as fast in solids as in liquids and about ten to fifteen times as fast in solids as in gases (▶ **Table 14.1**).

The speed of sound in a gas generally depends on the temperature of that gas. In dry air, for example, the speed of sound is 331 m/s (about 740 mi/h) at 0 °C. As the temperature increases, so does the speed of sound. For everyday temperature ranges, the speed of sound in air increases by about 0.6 m/s for each degree Celsius above 0 °C. A good approximation of the speed of sound in air for a particular temperature is

$$v = (331 + 0.6T_\text{C}) \text{ m/s} \quad \text{(speed of sound in air)} \quad (14.1)$$

where T_C is the air temperature in degrees Celsius.

Let's use the first Example to take a comparative look at the speed of sound in different media.

TABLE 14.1   Speed of Sound in Various Media (Typical Values)

Medium	Speed (m/s)
Solids	
Aluminum	5100
Copper	3500
Glass	5200
Iron	4500
Polystyrene	1850
Zinc	3200
Liquids	
Alcohol, ethyl	1125
Mercury	1400
Water	1500
Gases	
Air (0 °C)	331
Air (100 °C)	387
Helium (0 °C)	965
Hydrogen (0 °C)	1284
Oxygen (0 °C)	316

EXAMPLE 14.1: SOLID, LIQUID, GAS – SPEED OF SOUND IN DIFFERENT MEDIA

From their material properties, find the speed of sound in (a) a solid copper rod, (b) liquid water, and (c) air at room temperature (20 °C).

THINKING IT THROUGH. The speed of sound in a solid or a liquid depends on the elastic modulus and the density of the solid or liquid. These values are available in Tables 9.1 and 9.2. The speed of sound in air is given by Equation 14.1.

SOLUTION

Given:

(a) $Y_\text{Cu} = 11 \times 10^{10}\,\text{N/m}^2$
      $\rho_\text{Cu} = 8.9 \times 10^3\,\text{kg/m}^3$
(b) $B_{\text{H}_2\text{O}} = 2.2 \times 10^9\,\text{N/m}^2$
      $\rho_{\text{H}_2\text{O}} = 1.0 \times 10^3\,\text{kg/m}^3$
      (from Tables 9.1 and 9.2 for moduli and density values)
      $T_\text{C} = 20\,°\text{C}$ (for air)

Find:

(a) v_Cu (sound in copper)
(b) $v_{\text{H}_2\text{O}}$ (sound in water)
(c) v_air (sound in air)

(a)   For a solid, $v = \sqrt{Y/\rho}$:

$$v_\text{Cu} = \sqrt{\frac{Y}{\rho}} = \sqrt{\frac{11 \times 10^{10}\,\text{N/m}^2}{8.9 \times 10^3\,\text{kg/m}^3}} = 3.5 \times 10^3\,\text{m/s}$$

(b) For a liquid, $v = \sqrt{B/\rho}$:

$$v_{H_2O} = \sqrt{\frac{B}{\rho}} = \sqrt{\frac{2.2 \times 10^9 \, N/m^2}{1.0 \times 10^3 \, kg/m^3}} = 1.5 \times 10^3 \, m/s$$

(c) For air at 20°C, using Equation 14.1, the result is

$$v_{air} = \left(331 + 0.6T_C\right)m/s = \left[331 + 0.6(20)\right]m/s = 343 \, m/s$$

FOLLOW-UP EXERCISE. At what temperatures would the speed of sound in air be (a) 1% faster, (b) 2% slower than the answer to part (c) of this example? *(Answers to all Follow-Up Exercises are given in Appendix V at the back of the book.)*

The answer for part (c) of the previous example may seem fast to you, but compared to other wave speeds it can be slow. For example, when sitting in the outfield bleachers at a baseball game, you may notice that the sound from a batted ball reaches you well after you actually see the ball hit. This difference is due to the fact that the speed of sound is very much less than that of light. This effect can be put to practical use as a quick way to estimate the distance to a lightning strike during a thunderstorm, as shown in the next example.

EXAMPLE 14.2: TRACKING THUNDERSTORMS?

(a) Show that $\frac{1}{3}$ km/s and $\frac{1}{5}$ mi/s are reasonable approximations for the speed of sound in air at room temperature (20°C) by computing their percent difference from the correct value. (b) Suppose you see a lightning flash. To estimate the elapsed time in seconds, you immediately start counting "one thousand one, one thousand two, ..." stopping only when you hear the thunder. If you get to the count of ten, use the answers in (a) to estimate the distance to the lightning strike in miles.

THINKING IT THROUGH. The actual speed of sound was found in part (c) of Example 14.1. By converting $\frac{1}{3}$ km/s and $\frac{1}{5}$ mi/s to m/s, comparisons can be made.

SOLUTION

Listing what is given, and making use of the answer to (c) of Example 14.1:

Given:

$T_C = 20°C$
$v_{air} = 343$ m/s (exact)
$v_{km} = \frac{1}{3}$ km/s (approx.)
$v_{mi} = \frac{1}{5}$ mi/s (approx.)

Find:

(a) How approximations compare to actual value
(b) Distance to lightning strike

(a) Converting both approximations to SI units:

$$v_{km} = \frac{1}{3} \, km/s\left(10^3 \, m/km\right) = 333 \, m/s$$

$$v_{mi} = \frac{1}{5} \, mi/s\left(1609 \, m/mi\right) = 322 \, m/s$$

The percentage differences* are

for $v_{km} = \frac{1}{3}$ km/s

$$\% \, difference = \frac{(333 - 343)}{343} \times 100\% = -2.9\% \quad (low)$$

for $v_{mi} = \frac{1}{5}$ mi/s

$$\% \, difference = \frac{(322 - 343)}{343} \times 100\% = -6.1\% \quad (even \, lower)$$

Both approximations are low but close, however the one expressed in km/s is considerably better.

(b) As can easily be seen from the speed in mi/s, counting for 10 seconds means a distance of about 2 miles. In general, every count of 5 seconds means a distance of about one mile.

FOLLOW-UP EXERCISE. (a) Show that the speed of sound in air at 20°C is approximately 1100 ft/s. (b) What is the time delay between seeing a batted ball being hit and the sound that it makes if you are seated in the outfield bleachers 420 feet away?

14.3 Sound Intensity and Sound Intensity Level

Wave motion is one way of propagating energy through space. The rate of energy transfer is expressed in terms of **intensity (I)**, which is defined as energy transported per unit time across a unit area. Since power is energy per second, intensity can be thought of as power per unit area or:

$$intensity = \frac{energy/time}{area} = \frac{power}{area}$$

or in symbolic terms

$$I = \frac{E/T}{A} = \frac{P}{A}$$

The SI units of intensity are therefore watts per square meter (W/m²).

A very small (point) source of sound energy will emit spherical sound waves, as shown in ▶ **Figure 14.3**. If there are no losses in the medium, the intensity at a distance R from the source is

$$I = \frac{P}{A} = \frac{P}{4\pi R^2} \quad (point \, source) \qquad (14.2)$$

where P is the power of the source and $4\pi R^2$ is the area of a sphere of radius R, through which the sound energy passes. The intensity of a point source of sound is therefore *inversely proportional to the square of the distance from the source* (an inverse-square

---

* The percentage difference of a quantity x from its accepted (or true) value is typically defined as: % difference of x from $x_{true} \equiv ((x - x_{true})/x_{true}) \times 100$.

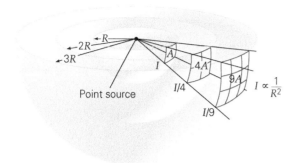

▲ FIGURE 14.3 **Intensity of a point source** The energy emitted from a point source spreads out equally in all directions. Since intensity is power divided by area, $I = P/A = P/(4\pi R^2)$, where the area is that of a spherical surface. The intensity then decreases with the distance from the source as $1/R^2$.

relationship). Intensities at different distances from a point source (same power) can be compared as a ratio:

$$\frac{I_2}{I_1} = \frac{P/\left(4\pi R_2^2\right)}{P/\left(4\pi R_1^2\right)} = \frac{R_1^2}{R_2^2} \quad \text{(point sources)} \qquad (14.3)$$

For example, to determine the intensity ratio at twice the distance from a point source, we have $R_2 = 2R_1$. Then

$$\frac{I_2}{I_1} = \left(\frac{R_1}{R_2}\right)^2 = \left(\frac{1}{2}\right)^2 = \frac{1}{4} \quad \text{or} \quad I_2 = \frac{I_1}{4}$$

Since intensity decreases as $1/R^2$, doubling the distance decreases the intensity to one-fourth its original value. A visual way to understand this inverse-square relationship for a point source of sound energy is to look at the geometry. As seen in Figure 14.3, the greater the distance from the source, the larger the area over which the sound energy is spread, resulting in reduced intensity. Since this area increases as the square of R, the intensity decreases accordingly – that is, as $1/R^2$.

Sound intensity as perceived by the ear is called **loudness**. On the average, the human ear can detect sound waves (at 1 kHz) of intensity as low as 10^{-12} W/m². This intensity (I_o) is referred to as the *threshold of hearing*. Thus, to hear a sound, it must not only have a frequency in the audible range but must also be of sufficient intensity.

As intensity is increased, the sound becomes louder. At an intensity of 1.0 W/m², the sound is uncomfortably loud and may be painful to the ear. This intensity (I_p) is called the *threshold of pain*. Note that the thresholds of pain and hearing differ by a factor of 10^{12} since $\dfrac{I_p}{I_o} = \dfrac{1.0\,\text{W/m}^2}{10^{-12}\,\text{W/m}^2} = 10^{12}$. That is, the intensity at the threshold of pain is a *trillion* times that at the threshold of hearing. Because of this huge range, the perceived loudness is *not* proportional to intensity. That is, if intensity is doubled, the perceived loudness does *not* double. In fact, a doubling of perceived loudness corresponds approximately to a tenfold increase

in intensity. Thus a sound of intensity 10^{-5} W/m² is perceived to be twice as loud as one of 10^{-6} W/m².

14.3.1 Sound Intensity Level: The Bel and the Decibel

It is convenient to compress the large range of sound intensities by using a logarithmic scale (base 10) to express **intensity levels** (*not* to be confused with sound intensity). The intensity level of a sound is referenced to a standard intensity, the threshold of hearing, $I_o = 10^{-12}$ W/m². For a sound intensity of I, its intensity level is defined as the logarithm (abbreviated log) of the ratio of I to I_o, that is, $\log(I/I_o)$. For example, if at a certain distance a sound has an intensity of $I = 10^{-6}$ W/m², then its intensity *level* is 6 since $\log\left(\dfrac{I}{I_o}\right) = \log\left(\dfrac{10^{-6}\,\text{W/m}^2}{10^{-12}\,\text{W/m}^2}\right) = \log 10^6 = 6$ (recall in general, $\log 10^n = n$).

Although intensity level has no units or dimensions, it is common to describe it using the **bel (B)** in honor of Alexander Graham Bell, the first practical user of the telephone. Thus, a sound with an intensity of 10^{-6} W/m² has an intensity level of 6 bel or 6 B. With this definition, the huge intensity range from 10^{-12} W/m² to 1.0 W/m² is compressed into an intensity level scale from 0 B to 12 B.

A finer intensity level scale uses the **decibel (dB)**, which is a tenth of a bel. Thus the range from 0 to 12 B corresponds to 0 to 120 dB. In this case, the equation for the relative **sound intensity level**, or **decibel level (β)**, is

$$\beta = 10 \log\left(\frac{I}{I_o}\right) \quad \text{(sound intensity level in dB)} \qquad (14.4)$$

where $I_o = 10^{-12}$ W/m². The decibel intensity scale and familiar sounds at representative intensity levels are in ▼ **Figure 14.4**.

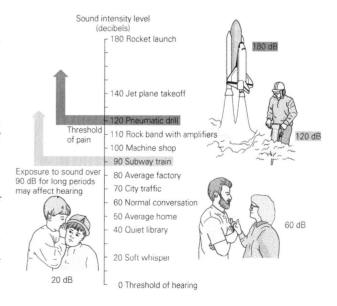

Sound intensity level (decibels)

- 180 Rocket launch
- 140 Jet plane takeoff
- 120 Pneumatic drill
- 110 Rock band with amplifiers
- 100 Machine shop
- 90 Subway train
- 80 Average factory
- 70 City traffic
- 60 Normal conversation
- 50 Average home
- 40 Quiet library
- 20 Soft whisper
- 0 Threshold of hearing

Threshold of pain

Exposure to sound over 90 dB for long periods may affect hearing

180 dB · 120 dB · 60 dB · 20 dB

▲ FIGURE 14.4 **Sound intensity levels and the decibel scale** The intensity levels of some common sounds on the decibel (dB) scale.

EXAMPLE 14.3: COMPUTING SOUND INTENSITY LEVELS – WORKING WITH LOGARITHMS

What is the intensity level of a sound with an intensity of $5.0 \times 10^{-6} \, \text{W/m}^2$?

THINKING IT THROUGH. The sound intensity levels are found using Equation 14.4.

SOLUTION

Given:

$I = 5.0 \times 10^{-6} \, \text{W/m}^2$

$I_\text{o} = 10^{-12} \, \text{W/m}^2$

Find: β (sound intensity level)

From Equation 14.4 (recall that the logarithm of a product is the sum of the logarithms):

$$\beta = 10 \log\left(\frac{I}{I_\text{o}}\right) = 10 \log\left(\frac{5.0 \times 10^{-6} \, \text{W/m}^2}{10^{-12} \, \text{W/m}^2}\right)$$

$$= 10 \log(5.0 \times 10^{6}) = 10(\log 5.0 + \log 10^{6})$$

$$= 10(0.70 + 6) = 67 \, \text{dB}$$

FOLLOW-UP EXERCISE. Note in this Example that an intensity of $5.0 \times 10^{-6} \, \text{W/m}^2$ is halfway between 10^{-6} and 10^{-5} (or 60 and 70 dB), yet its intensity level is *not* the midway value of 65 dB. (a) Why? (b) What intensity *does* correspond to 65 dB? (Compute it to three significant figures.)

EXAMPLE 14.4: COMBINING SOUNDS – ADDING INTENSITIES BUT NOT INTENSITY LEVELS

Sitting at a sidewalk restaurant table in Paris, a friend talks to you at a normal conversation intensity level of 60 dB. At the same time, the intensity level from street traffic at your location is known to be 60 dB. What is the intensity level of the combined sounds?

THINKING IT THROUGH. It is tempting simply to add the two sound intensity levels to a total of 120 dB. But intensity levels in decibels are *logarithmic*, so you can't add them in the normal way. However, energy and power combine additively, so intensities (I) can be added arithmetically. Thus the correct method to find the combined intensity level is to use the sum of the intensities.

SOLUTION

Listing the data and what is to be found:

Given:

$\beta_1 = 60 \, \text{dB}$

$\beta_2 = 60 \, \text{dB}$

Find: total β

To find the intensity of a sound of intensity level of 60 dB (i.e., both sounds here) involves reversing the logarithms as follows:

$$\beta_1 = \beta_2 = 60 \, \text{dB} = 10 \log\left(\frac{I_1}{I_\text{o}}\right) = 10 \log\left(\frac{I_1}{10^{-12} \, \text{W/m}^2}\right)$$

$$\therefore \log\left(\frac{I_1}{10^{-12} \, \text{W/m}^2}\right) = 6 \quad \text{or} \quad \frac{I_1}{10^{-12} \, \text{W/m}^2} = 10^{6}$$

Thus

$$I_1 = I_2 = 10^{-6} \, \text{W/m}^2$$

The total intensity is

$$I_\text{total} = I_1 + I_2 = 1.0 \times 10^{-6} \, \text{W/m}^2 + 1.0 \times 10^{-6} \, \text{W/m}^2$$

$$= 2.0 \times 10^{-6} \, \text{W/m}^2$$

Then, the intensity level of the total is

$$\beta = 10 \log\left(\frac{I_\text{total}}{I_\text{o}}\right) = 10 \log\left(\frac{2.0 \times 10^{-6} \, \text{W/m}^2}{10^{-12} \, \text{W/m}^2}\right) = 10 \log(2.0 \times 10^{6})$$

$$= 10(\log 2.0 + \log 10^{6}) = 10(0.30 + 6.0) = 63 \, \text{dB}$$

This value is a long way from 120 dB. Notice that the combined intensities doubled the intensity value, but the intensity level increased by only 3 dB because of the logarithmic definition of intensity level.

FOLLOW-UP EXERCISE. In this Example, suppose the added noise gave a total that *tripled* the sound intensity level of the conversation. What would be the total combined intensity level in this case?

Hearing may be damaged by excessive noise. Hearing damage depends on the sound intensity level (decibel level) and exposure time. The exact combinations vary for different people, but a general guide to noise levels is given in ▶ **Table 14.2**.

14.4 Sound Phenomena

14.4.1 Reflection, Refraction, and Diffraction of Sound

An echo is a familiar example of the *reflection* of sound – sound "bouncing" off a surface. Sound *refraction* is usually a less familiar phenomenon. Refraction refers to a change in wave direction due to a change in wave speed. You may have experienced it on a calm summer evening, when it is possible to hear distant voices or other sounds that you might think would be inaudible. For the sound waves to be refracted downward (toward you) there must be a layer of cooler air near the ground or water and a layer of warmer air above it. Together these provide a wave speed change

TABLE 14.2   Sound Intensity Levels and Ear Damage Exposure Times

Sound	Decibels (dB)	Examples	Time of Nonstop Exposure That Can Cause Damage
Faint	30	Quiet library, whispering	
Moderate	60	Normal conversation, sewing machine	
Very loud	80	Heavy traffic, noisy restaurant, screaming child	10 h
	90	Lawnmower, motorcycle, loud party	Less than 8 h
	100	Chainsaw, subway train, snowmobile	Less than 2 h
Extremely loud	110	Stereo headset at full blast, rock concert	30 min
	120	Dance clubs, car stereos, action movies, some musical toys	15 min
	130	Jackhammer, loud computer games, loud sporting events	Less than 15 min
Painful	140	Boom stereos, gunshot blast, firecrackers	Only seconds (e.g., hearing loss can occur from a few shots of a high-powered gun if protection is not worn)

and resulting refraction. These conditions occur frequently over bodies of water, which cool after sunset. As a result of the cooler water, a layer of cool air forms just above the water's surface. The sound waves are then refracted in an arc that may allow a distant person to receive an increased sound intensity (see ▼ **Figure 14.5**).

Another phenomenon is *diffraction*, described in Section 13.4. Sound may be diffracted, or spread out, when traveling around corners, objects or through openings. For example, a person speaking in a room can be heard outside as the sound diffracts (spreads out) as it passes through the doorway, even though that person cannot be seen. Recall that conversational sound wavelengths are on the order of a meter, about the same order of magnitude as the door width, leading to significant sound diffraction. However, the typical wavelength of visible light is much smaller than the opening, thus the light travels essentially in straight lines with no observable diffraction.

Reflection, refraction, and diffraction are described in a general sense here for sound. These phenomena are important considerations for light waves as well, and will be discussed more fully in Chapters 22 and 24 (Volume Two).

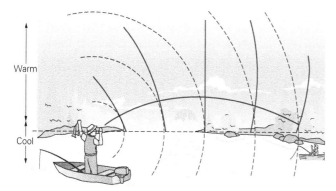

▲**FIGURE 14.5 Sound refraction** Sound travels more slowly in the cool air near the water surface than in the upper, warmer air. As a result, the waves are refracted downward (red curved line). Here the dotted lines refer to the boater's voice with no refraction and the red wave fronts depict how the speed change results in refraction which, in turn, increases sound intensity on the far shore over what might be expected.

14.4.2 Interference

Like waves of any kind, sound waves *interfere* when they meet. But what is the resultant waveform as they overlap or interfere? This is where the *principle of superposition* comes into play. (See the superposition principle in Section 13.4.) Suppose that two small loudspeakers are separated in space and emit sound waves in phase with the same frequency. If the speakers are treated as point sources, then the waves will spread out spherically and interfere (▶ **Figure 14.6a**). The lines from a particular speaker represent wave crests (air compression), and the troughs (air rarefaction) lie in the intervening white areas.

At various locations, of course, there will be constructive or destructive interference. But, if two waves meet at a location where they are exactly in phase (two crests or two troughs coincide), there will be **total constructive interference** (Figure 14.6b). Notice that the waves are in phase at point C. If, instead, the waves meet such that the crest of one coincides with the trough of the other (point D), the waves will cancel out (Figure 14.6c). The result will be **total destructive interference**.

It is convenient to describe the path lengths traveled by the waves in terms of wavelength (λ) to determine whether or not they arrive in phase. Consider the waves arriving at point C in Figure 14.6b. The two path lengths in this case differ by one wavelength since $L_{AC} = 4\lambda$ and $L_{BC} = 3\lambda$. In general a **phase difference** ($\Delta\theta$) is caused by this *path length difference* (ΔL). They are related as follows

$$\Delta\theta = \frac{2\pi}{\lambda}(\Delta L) \quad \begin{array}{l}\text{(phase difference due} \\ \text{to path length difference)}\end{array} \quad (14.5)$$

Since 2π rad is equivalent, in angular terms, to a full wave cycle or wavelength, multiplying the path length difference by $2\pi/\lambda$ gives the phase difference in radians. For the example illustrated in Figure 14.6b,

$$\Delta\theta = \frac{2\pi}{\lambda}(L_{AC} - L_{BC}) = \frac{2\pi}{\lambda}(4\lambda - 3\lambda) = 2\pi \, \text{rad}$$

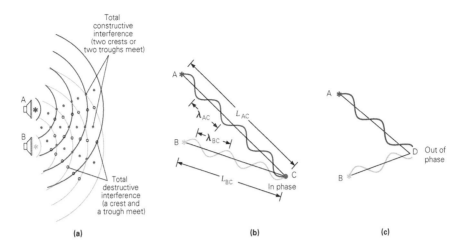

▲ **FIGURE 14.6** **Interference (a)** Sound waves from two-point sources spread out and interfere. **(b)** At points where the waves arrive in phase (zero phase difference), such as point C, constructive interference occurs. **(c)** At points where the waves arrive completely out of phase (phase difference of 180°), such as point D, destructive interference occurs. The phase difference at a particular point depends on the path lengths the waves travel to reach that point.

Because $\Delta\theta = 2\pi$ rad $= 360°$, the waves arrive at point C shifted from one another by one wavelength, which means they are back in phase. Thus, point C is a location of total constructive interference resulting in an increase in intensity, or sound loudness. In general, at a given location, two sound waves are in phase when the path length difference is zero or an integral multiple of the wavelength.

$$\Delta L = n\lambda \text{ (where } n = 0,1,2,3,\ldots\text{)} \quad \text{(constructive interference) (14.6)}$$

A similar analysis at point D in Figure 14.6c ($L_{AD} = 2\frac{1}{4}\lambda$ and $L_{BD} = 1\frac{3}{4}\lambda$) gives

$$\Delta\theta = \frac{2\pi}{\lambda}\left[2\left(\frac{1}{4}\right)\lambda - 1\left(\frac{3}{4}\right)\lambda\right] = \pi \text{ rad}$$

Because $\Delta\theta = 180°$, the waves arrive at point D completely out of phase, and destructive interference results there.

In general, at a given location, two sound waves will be out of phase when the path length difference is an odd number of half-wavelengths ($\lambda/2$).

$$\Delta L = m\left(\frac{\lambda}{2}\right) \quad \text{(where } m = 1,3,5,\ldots\text{)} \quad \text{(destructive interference)} \quad (14.7)$$

At these points, a softer, or less intense, sound will be heard or detected. If the amplitudes of the waves are exactly equal, the destructive interference is total and no sound is heard.

Destructive interference of sound waves provides a way to reduce loud noises, which can be distracting and cause hearing discomfort. The procedure is to have a reflected wave or an introduced wave with a phase difference that cancels out the original sound as much as possible. As an example, the use of this idea in the design of automobile mufflers was discussed in Section 13.4.

EXAMPLE 14.5: PUMP UP THE VOLUME – SOUND INTERFERENCE

At an open-air concert on a hot day (air temperature 25 °C), you sit 7.00 and 9.10 m, respectively, from a pair of speakers, located at either side of the stage. A musician, warming up, plays a single 494-Hz tone through both the speakers simultaneously and in phase. Determine the type of interference that you hear. (Consider the speakers to be point sources and ignore any energy losses during the travel.)

THINKING IT THROUGH. The sounds from the speakers will interfere – but is it constructive, destructive, or something in between? This depends on the path length difference, which can be computed from the given distances.

SOLUTION

Given:

$$d_1 = 7.00 \text{ m and } d_2 = 9.10 \text{ m}$$
$$f = 494 \text{ Hz}$$
$$T = 25 \text{ °C}$$

Find: Type of interference heard

The path length difference can be expressed in terms of the wavelength. This is found from the wave relationship $\lambda = v/f$. First the speed of sound, v, must be determined using Equation 14.1:

$$v = (331 + 0.6T_C)\text{ m/s} = [331 + 0.6(25)]\text{ m/s} = 346 \text{ m/s}$$

Therefore the wavelength is

$$\lambda = \frac{v}{f} = \frac{346\,\text{m/s}}{494\,\text{Hz}} = 0.700\,\text{m}$$

and the distances in terms of wavelength are

$$d_1 = (7.00\,\text{m})\left(\frac{\lambda}{0.700\,\text{m}}\right) = 10.0\lambda \quad \text{and}$$

$$d_2 = (9.10\,\text{m})\left(\frac{\lambda}{0.700\,\text{m}}\right) = 13.0\lambda$$

Finally, the path length difference (in terms of wavelengths) is

$$\Delta L = d_2 - d_1 = 13.0\lambda - 10.0\lambda = 3.0\lambda$$

which is an integral number of wavelengths ($n = 3$); thus total constructive interference occurs at your location. The waves reinforce each other, and an extra loud tone at 494 Hz would be heard.

FOLLOW-UP EXERCISE. Suppose that in this Example the tone traveled to a person sitting 7.00 and 8.75 m, respectively, from the two speakers. What would be the situation in that case?

14.4.3 Beats

Another interesting interference effect occurs when two tones of nearly the same frequency ($f_1 \approx f_2$) are sounded simultaneously. The ear senses "pulsations" in loudness, known as **beats**. The human ear can detect up to about seven beats or pulses per second. A greater number of these per second sounds "smooth" (or continuous, without any obvious pulsations).

Suppose two sinusoidal waves of the same amplitude, but slightly different frequencies, interfere by arriving together at a given location (▼ **Figure 14.7a**). Figure 14.7b represents the resulting sound wave. The amplitude of the combined wave varies sinusoidally, as shown by the black curves (known as *envelopes*) that outline the wave.

What does this variation in amplitude mean in terms of what the listener perceives? A listener will hear a pulsating sound (beats), determined by the envelope. As the combined

wave travels toward the listener, loud sound will be heard when the maximum combined amplitude reaches her (where waves interfere constructively) and a minimum of sound will be heard when the regions of destructive interference (cancellation) arrive. Detailed mathematics shows that a listener will hear the beats at a frequency called the **beat frequency** (f_b), given by

$$f_b = |f_1 - f_2| \tag{14.8}$$

The absolute value is taken because the frequency f_b cannot be negative, even if $f_2 > f_1$. A negative beat frequency would be meaningless. It should be noted that the underlying tone has a frequency of the average of the two frequencies. It is said that the underlying tone is "modulated" at the beat frequency; that is, it oscillates in loudness at the beat frequency. For example, consider two vibrating tuning forks with individual frequencies of 517 and 513 Hz. A nearby listener would hear a tone of 515 Hz (the average of the frequencies) but with a loudness that would vary at the beat frequency of $f_b = 517 - 513\,\text{Hz} = 4\,\text{Hz}$. Musicians use this phenomenon to tune stringed instruments. They listen simultaneously to both the instrument and a calibrated sound source (of known frequency), adjusting the tension in the strings until the beats disappear ($f_1 = f_2$), and the frequencies match.

14.5 The Doppler Effect

The **pitch** (or *perceived frequency*) of the sound emitted by the siren of an ambulance, for example, is increased as it approaches you and reduced as it recedes. Similarly, variations in the frequency of motor noise can be heard as a race car passes by. This change in perceived sound frequency due to source motion is an example of the **Doppler effect**. (The Austrian physicist Christian Doppler [1803–1853] first described this effect.)

As ▶ **Figure 14.8** shows, the sound waves emitted by a moving source tend to bunch up in front of the source and spread out in back. An expression for the shift in frequency due to a moving source in still air can be found by studying the situation depicted in ▶ **Figure 14.9**. Let speed of sound in air be v, the speed of the moving source be v_s, and the frequency of the sound produced by the source be f_s. In one period, $T = 1/f_s$, a wave crest moves

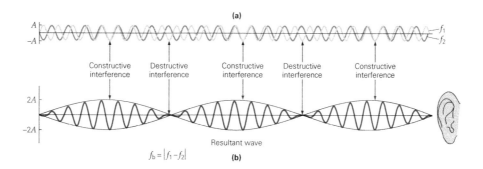

▲ FIGURE 14.7 **Beats** Two traveling waves of equal amplitude and slightly different frequencies interfere and give rise to a pulsating loudness called beats. The beat frequency, or the rate at which the loudness pulsates, is given by $f_b = |f_1 - f_2|$.

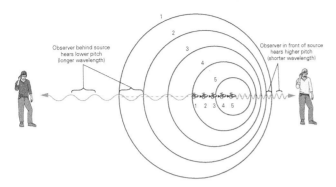

▲ FIGURE 14.8 **The Doppler effect for a moving source** The sound waves bunch up in front of a moving source – the whistle – giving a higher frequency there. The waves trail out behind the source, giving a lower frequency there.

a distance $d = vT = \lambda$, which is one wavelength in the still air, regardless of whether the source is moving. However, during the course of one period, the source travels a distance $d_s = v_sT$ before emitting another wave crest. Thus distance between the successive crests is *shortened*, creating a new wavelength λ':

$$\lambda' = d - d_s = vT - v_sT = (v - v_s)T = \frac{v - v_s}{f_s}$$

The frequency heard by the listener (or observer – hence subscript o) f_o is related to the shortened wavelength by $f_o = v/\lambda'$, and substituting λ' gives

$$f_o = \frac{v}{\lambda'} = \left(\frac{v}{v - v_s}\right)f_s = \left(\frac{1}{1 - (v_s/v)}\right)f_s$$

The last expression shows that the important factor in determining the frequency is not the values of the two speeds, but their *ratio* (in the denominator). Analysis of the situation when the source is moving away from a listener gives a similar result, but with a + sign in the denominator. Thus in general

$$f_o = \left(\frac{v}{v \pm v_s}\right)f_s = \left(\frac{1}{1 \pm (v_s/v)}\right)f_s$$

$$\begin{cases} - & \text{for source moving toward a stationary observer} \\ + & \text{for source moving away from a stationary observer} \end{cases}$$

(14.9)

If the source is approaching, then the denominator $1 - (v_s/v)$ is less than 1, ensuring that f_o is larger than f_s, as expected. So if the approaching source speed is one-tenth the speed of sound, then $v_s/v = 1/10$ and from Equation 14.9,

$$f_o = \left(\frac{1}{1 - (1/10)}\right)f_s = \frac{10}{9}f_s.$$

A Doppler frequency shift also occurs with a moving observer and a stationary source, although this situation is a bit different. As the observer moves toward the source, the distance between successive wave crests is unchanged and stays at the normal wavelength. However the wave speed *as measured by the moving observer* (relative to the still air) is different from v_s. According to the approaching observer, the sound from the source passes her at a speed of $v' = v + v_o$, where v_o is the speed of the observer and v is the speed of sound in still air. Thus the observer, by moving toward the source, meets more wave crests in a given time, resulting in a higher perceived frequency (or pitch).

For the approaching listener, the observed frequency is $f_o = (v'/\lambda) = [(v + v_o)/v]f_s$. For a receding listener, the only difference is a minus sign in the numerator (resulting in a lowering of the perceived frequency). Combining the two situations and writing the speeds again as a ratio

$$f_o = \left(\frac{v \pm v_o}{v}\right)f_s = \left(1 \pm \frac{v_o}{v}\right)$$

$$\begin{cases} - & \text{for an observer moving away from a stationary source} \\ + & \text{for an observer moving toward a stationary source} \end{cases}$$

(14.10)

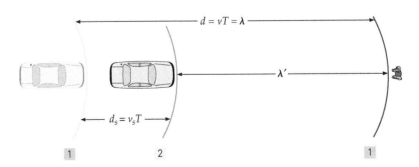

▲ FIGURE 14.9 **The Doppler effect and wavelength** Sound from a moving car's horn travels a distance d (one wavelength) during a time T (the horn vibration period). During this time, the car (the source) travels a distance d_s before putting out a second pulse, thereby shortening the observed wavelength of the sound in the approaching direction.

14.5.1 Problem-Solving Hint

You may find it difficult to remember whether a plus or minus sign is used in the Doppler effect equations. Let your experience help you. For the stationary observer, the frequency of the sound increases when the source approaches, thus the denominator in Equation 14.9 must be *smaller* than the numerator and you logically choose to use the minus sign. With a receding source, the frequency is lower thus the denominator in Equation 14.9 must be *larger* than the numerator, and the correct choice is clearly the plus sign. Similar reasoning can help you choose a plus or minus sign for the numerator in Equation 14.10.

EXAMPLE 14.6: ON THE ROAD AGAIN – THE DOPPLER EFFECT

As a truck traveling at 96 km/h approaches and passes a person standing along the highway, the driver sounds the horn. If the horn has a frequency of 400 Hz, what are the frequencies of the sound waves heard by the person (a) as the truck approaches and (b) after it has passed? (Assume that the speed of sound is 346 m/s.)

THINKING IT THROUGH. This situation is an application of Equation 14.9, with a moving source and stationary observer. In such problems, it is important to identify the data correctly. The frequency in (a) is expected to be greater than 400 Hz and less than 400 Hz in (b).

SOLUTION

Both speeds need to be in the same units, here m/s.

Given:

$$v_s = 96 \text{ km/h} = 27 \text{ m/s}$$
$$f_s = 400 \text{ Hz}$$
$$v = 346 \text{ m/s}$$

Find:

(a) f_o (truck approaching)
(b) f_o (truck receding)

(a) From Equation 14.9 and choosing a minus sign (to insure a higher frequency)

$$f_o = \left(\frac{1}{1-(v_s/v)}\right)f_s$$

$$= \left(\frac{1}{1-(27 \text{ m/s}/346 \text{ m/s})}\right)(400 \text{ Hz}) = 434 \text{ Hz}$$

(b) A plus sign is used in Equation 14.11 when the source is moving away:

$$f_o = \left(\frac{1}{1-(v_s/v)}\right)f_s = \left(\frac{1}{1+(27 \text{ m/s}/346 \text{ m/s})}\right)(400 \text{ Hz}) = 371 \text{ Hz}$$

FOLLOW-UP EXERCISE. Instead, suppose that the observer in this Example were initially moving toward and then past a stationary 400-Hz source at a speed of 96 km/h. What would be the observed frequencies? (Would they differ from those for the moving source?)

A more general case for the Doppler effect occurs when *both* the source and the observer are moving. Let's consider this conceptually in the next Example.*

CONCEPTUAL EXAMPLE 14.7: IT'S ALL RELATIVE – MOVING SOURCE AND MOVING OBSERVER

Suppose a sound source and an observer are moving away from each other in opposite directions, each at half the speed of sound in air. In this case, the observer would (a) receive sound with a frequency higher than the source frequency, (b) receive sound with a frequency lower than the source frequency, (c) receive sound with the same frequency as the source frequency, or (d) receive no sound from the source.

REASONING AND ANSWER. When a source moves away from a stationary observer, the observed frequency is lower (Equation 14.10). Similarly, when an observer moves away from a stationary source, the observed frequency is also lower (Equation 14.13). With both source and observer moving away from each other in opposite directions, the combined effect would make the observed frequency even less, so neither (a) nor (c) is the answer.

It would appear that (b) is the correct answer, but (d) must logically eliminated for completeness. Remember that the speed of sound relative to the air is constant. Therefore, (d) would be correct *only if the observer is moving faster than the speed of sound* relative to the air. Since the observer is moving at only half the speed of sound, (b) is the correct answer.

FOLLOW-UP EXERCISE. In this Example, what would be the result if both the source and the observer were traveling in the same direction with the same speed (both less than the speed of sound)?

The Doppler effect also applies to light waves, although the equations describing the effect are different from those for sound. If a distant light source such as a star is receding from us, the frequency of the light we receive from it is lowered. That is, the light is shifted toward the red (long-wavelength) end of the spectrum, an effect known as a *Doppler red shift*. Similarly, the frequency of light from an object approaching is increased – the

---

* In the case of both observer and source moving through still air,

$$f_o = \left(\frac{v \pm v_o}{v \mp v_s}\right)f_s$$

The upper signs in the numerator and denominator apply if the observer and source move toward each other, and the lower signs apply if they are moving away.

light is shifted toward the blue (short-wavelength) end of the spectrum, producing a *Doppler blue shift*. The magnitude of the shift is related to the speed of the source. The Doppler shift of light from astronomical objects is thus a vital tool used by astronomers for measuring speeds (and direction of motion) of various astronomical objects.

You may have been subjected to a practical application of the Doppler effect if you have ever been caught speeding in your car by police radar, which uses reflected radio waves (which are long-wavelength light, not sound). If radio waves are reflected from a parked car, the reflected waves return at the same frequency. But for a car that is moving toward a patrol car, the reflected waves arrive at a higher frequency than the emitted waves. Interestingly enough in this case, there is actually a *double* Doppler shift: As it intercepts and receives the wave, the moving car acts like a moving observer (the first Doppler shift), but then, in reflecting the wave back, the car acts like a moving source (the second Doppler shift). The magnitudes of these combined shifts depend on the speed of the car. Comparing the received frequency to the frequency of the source, a computer can quickly calculate and display this speed for the officer.

14.5.2 Sonic Booms

Consider a jet plane that can travel at supersonic speeds. As the speed of a moving source of sound approaches the speed of sound, the waves ahead of the source come close together (▶ Figure 14.10a). When a plane is traveling at the speed of sound, the waves can't outrun it, and they pile up in front. At supersonic speeds, the waves overlap. This overlapping of a large number of waves forms a large pressure ridge, or *shock wave*. This kind of wave is sometimes called a *bow wave* because it is analogous to the wave produced by the bow of a boat moving through water at a speed greater than the speed of the water waves. As another example, Figure 14.10b shows the shock wave trailing a high-speed bullet.

For aircraft traveling at supersonic speed, the shock wave trails out to the sides and downward in a conical shape whose angle depends on the aircraft's speed (see Figure 14.10a). When this pressure ridge passes over an observer on the ground, the large concentration of energy produces what is known as a **sonic boom**. There is really a double boom, because shock waves are formed at both ends of the aircraft. Under certain conditions, the shock waves can break windows and cause other damage to structures on the ground.

The ratio of the speeds, that of the source to that of sound, v_s/v, is the **Mach number (M)**, named after Ernst Mach (1838–1916), an Austrian physicist who used it in studying supersonics. From the geometry in ▶ Figure 14.11 it can be seen that $\sin\theta = (vt/v_s t) = v/v_s = 1/M$, thus

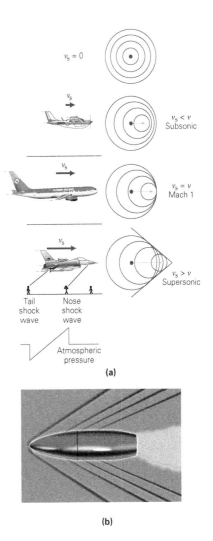

$$M = \frac{v_s}{v} = \frac{1}{\sin\theta} \tag{14.11}$$

▲ FIGURE 14.10   **Bow waves and sonic booms (a)** When an aircraft exceeds the speed of sound in air, v_s, the sound waves form a pressure ridge, or shock wave. As the trailing shock wave passes over the ground, observers hear a sonic boom (actually, two booms, because shock waves are formed at the front and tail of the plane). **(b)** A bullet traveling greater than the speed of sound in air. Note the shock waves (and the turbulence behind the bullet).

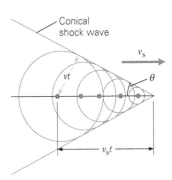

▲ FIGURE 14.11   **Shock wave cone and Mach number** When the speed of the source (v_s) is greater than the speed of sound in air (v), the interfering spherical sound waves form a conical shock wave that appears as a V-shaped pressure ridge when viewed in two dimensions. The angle θ is given by $\sin\theta = v/v_s$, and the inverse ratio v_s/v is called the Mach number.

Notice that if the plane is flying just at the speed of sound then $v_s = v$ and M is 1. A value for M less than 1 indicates a *subsonic* speed (less than that of sound), and a value greater than 1 indicates a *supersonic* speed (greater than that of sound). In the latter case, M tells the speed of the aircraft in terms of a multiple of the speed of sound. A Mach number of 2, for instance, indicates a speed twice the speed of sound. Since we know that $\sin \theta \leq 1$, no shock wave can exist unless $M \geq 1$.

On a smaller scale, you have probably heard a "mini" sonic boom – the "crack" of a whip. This is because the whip's tip is moving at supersonic speed! The "crack" is actually the sound made by air rushing back into the region of reduced pressure created by the final flip of the whip's tip.

14.6 Musical Instruments and Sound Characteristics

Musical instruments provide good examples of standing waves and boundary conditions. As was learned in Section 13.5, there are certain natural or standing wave frequencies that can exist on a stretched string (as the strings on a guitar, for example). Initially adjusting the tension in a string tunes it to a particular (fundamental) frequency. Then the effective length of the string is varied by finger location and pressure.

Standing waves can also exist in air columns. You have probably done this in blowing across the open top of a soda bottle, producing an audible tone.* The blowing across the bottle excites the fundamental mode of the column of air in the bottle. The frequency of the tone depends on the length of the air column. With more liquid in the bottle, the air column is shorter and the frequency higher. There will be an antinode (maximum air molecule motion) at the open end of the bottle and a node (no molecular motion) at the liquid surface or the bottom of an empty bottle.

Standing waves are the basis of wind musical instruments. For example, consider a pipe organ with fixed lengths of pipe, which may be open-open or open-closed (▶**Figure 14.12a**). At the open ends, the boundary conditions for a standing sound wave require the molecules be free to move and thus be a motion antinode. Similarly, any closed end cannot allow molecular movement and thus must be a motion node for a standing wave. It is important to realize that although these diagrams may look like string waves which are transverse, they actually depict sound waves which are longitudinal. Hence the molecular motion is actually parallel to the axis of the pipe, not perpendicular to it like a piece of a vibrating string would be.

Analysis similar to that done in Section 13.5 for a stretched string but with the proper boundary conditions for sound waves

---

* A more complicated phenomenon, called Helmholtz resonance, occurs here, but for simplicity and standing waves, we assume the bottle to be a circular cylinder.

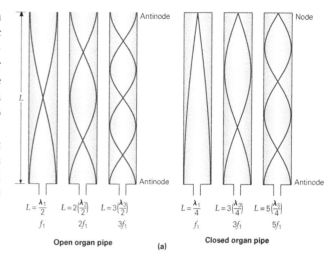

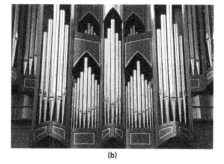

▲**FIGURE 14.12 Standing waves** **(a)** Longitudinal standing waves (illustrated here as sinusoidal curves) are formed in vibrating air columns in pipes. An open pipe has motion antinodes at both open ends. A closed pipe has a motion node at the closed end and a motion antinode at the open end. **(b)** A modern pipe organ. The pipes can be open or closed.

leads to the following expressions for the natural frequencies of the two types of pipes (length L)

$$f_n = \frac{v}{\lambda_n} = n\left(\frac{v}{2L}\right) = nf_1 \quad n = 1, 2, 3, \ldots \qquad (14.12)$$

(open-open pipe)

and

$$f_m = \frac{v}{\lambda_m} = m\left(\frac{v}{4L}\right) = mf_1 \quad m = 1, 3, 5, \ldots \qquad (14.13)$$

(open-closed pipe)

where v is the speed of sound in air. Note that the natural frequencies depend on the length of the pipe. This is an important consideration in a pipe organ (Figure 14.12b), particularly in selecting the dominant or fundamental frequency. (The diameter of the pipe is also a factor but is not considered in our simple analysis.)

The same physical principles apply to wind and brass instruments. In all of these, human breath is used to create standing waves in a tube. Most such instruments allow the player to vary the effective length of the tube and thus the frequency or pitch produced – either with the help of slides or valves that vary the actual length of tubing in which the air can resonate, as in most brasses, or by opening and closing holes in the tube, as in woodwinds (▼ **Figure 14.13**).

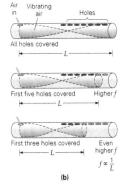

(a)   (b)

▲ **FIGURE 14.13   Wind instruments (a)** Wind instruments, such as clarinets, are essentially open tubes. **(b)** The effective length of the air column, and hence the pitch of the sound, is varied by opening and closing holes along the tube. The frequency f is inversely proportional to the effective length L of the air column.

Recall from Section 13.5 that a musical note or tone is referenced to the fundamental vibrational frequency of an instrument. In musical terms, the first overtone is the second harmonic, the second overtone is the third harmonic, and so on. Note that for a closed organ pipe (Equation 14.13), the even harmonics are missing.

EXAMPLE 14.8: PIPE DREAMS – FUNDAMENTAL FREQUENCY

A particular open organ pipe has a length of 0.653 m. Taking the speed of sound in air to be 345 m/s, what is the fundamental frequency of this pipe?

THINKING IT THROUGH. The fundamental frequency ($n = 1$) of an open-open pipe can be gotten directly by Equation 14.12. Or realize that in the fundamental mode, there is a half-wavelength so $\lambda = 2L = 1.31$ m.

SOLUTION

Let's try both methods to find the fundamental frequency:

Given:

$L = 0.653$ m
$v = 345$ m/s (speed of sound)

Find: f_1 (fundamental frequency)

Using $n = 1$ in Equation 14.12,

$$f_1 = \frac{v}{2L} = \frac{345 \text{ m/s}}{2(0.653 \text{ m})} = 264 \text{ Hz}$$

and because the wavelength is known, alternatively

$$f_1 = \frac{v}{\lambda_1} = \frac{345 \text{ m/s}}{1.31 \text{ m}} = 264 \text{ Hz}$$

FOLLOW-UP EXERCISE. A open-closed organ pipe has a fundamental frequency of 256 Hz. What would be the frequency of its first overtone? Is this frequency audible?

Perceived sounds are described by terms whose meanings are similar to those used to describe the physical properties of sound waves. Physically, a wave is generally characterized by *intensity*, *frequency*, and *waveform* (harmonics). The corresponding terms used to describe the sensations of the ear are *loudness*, *pitch*, and *quality* (or timbre). These general correlations are shown in ▼ **Table 14.3**. However, the correspondence is not perfect. The physical properties are objective and can be measured directly. The sensory effects are subjective and vary from person to person. (Think of temperature as measured by a thermometer and by the sense of touch.)

Sound intensity and its measurement on the decibel scale were covered in Section 14.3. Loudness is related to intensity, but the human ear responds differently to sounds of different frequen-

TABLE 14.3   General Correlation between Perceptual and Physical Characteristics of Sound

Sensory Effect	Physical Wave Property
Loudness	Intensity
Pitch	Frequency
Quality (timbre)	Waveform (harmonics)

cies. For example, two tones with the same intensity (in watts per square meter) but different frequencies might be judged by the ear to be different in loudness.

Frequency and *pitch* are often used synonymously, but again there is an objective-subjective difference: If the same low-frequency tone is sounded at two intensity levels, most people say that the more intense sound has a lower pitch, or perceived frequency.

The curves in the graph of intensity level versus frequency shown in ▶ **Figure 14.14** are called *equal-loudness contours* (or Fletcher-Munson curves, after the researchers who generated them). These contours join points representing intensity-frequency combinations that a person with average hearing judges to be equally loud. The top curve shows that the decibel level of the threshold of pain (120 dB) does not vary a great deal over the normal hearing range, regardless of the frequency of the sound. In contrast, the threshold of hearing, represented by the lowest contour, varies widely with frequency. For a tone with a frequency of 2000 Hz, the threshold of hearing is 0 dB, but a 20-Hz tone would have to have an intensity level of over 70 dB just to be heard (the extrapolated *y*-intercept of the lowest curve).

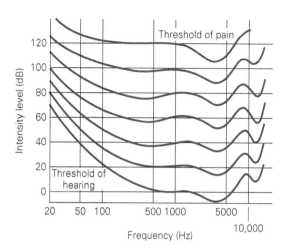

▲ **FIGURE 14.14 Equal-loudness contours** The curves indicate tones that are judged to be equally loud, although they have different frequencies and intensity levels. For example, on the lowest contour, a 1000-Hz tone at 0 dB sounds as loud as a 50-Hz tone at 40 dB. Note that the frequency axis is logarithmic to compress the large frequency range.

It is interesting to note that there are minima in the curves. The hearing curves show a significant dip in the 2000–5000 Hz range, the ear being most sensitive around 4000 Hz. A tone with a frequency of 4000 Hz can be heard at intensity levels *below* 0 dB. Another dip in the curves, or region of extra sensitivity, occurs at about 12 000 Hz.

The minima occur as a result of resonance in an open-closed cavity in the auditory canal (similar to an open-closed pipe). The length of the cavity is such that it has a fundamental resonance frequency of about 4000 Hz, resulting in extra sensitivity. Since it behaves as an open-closed pipe, the next natural frequency is the third harmonic (see Equation 14.13), which is three times the fundamental frequency, or about 12 000 Hz.

EXAMPLE 14.9: THE HUMAN EAR CANAL – STANDING WAVES

Consider the human ear canal to be a cylindrical tube of depth 2.54 cm (1.0 in.). What would be the lowest standing sound wave frequency? For realism, assume the air in the canal is a bit below normal body temperature or 30 °C.

THINKING IT THROUGH. The auditory ear canal as described is essentially a closed pipe – open at one end (outer ear canal) and closed at the other (eardrum). The lowest-frequency resonance standing wave that will fit in the pipe is $L = \lambda/4$ (Figure 14.12), and thus $\lambda = 4L$. Then, $f_1 = v/\lambda_1 = v/4L$, where v is the speed of sound in air.

SOLUTION

Given:

$L = 2.54 \text{ cm} = 0.0254 \text{ m}$
$T = 30 \text{°C}$

Find: f_1 (lowest ear canal resonance frequency)

First, the speed of sound at 37 °C is

$$v = (331 + 0.6T_C) \text{ m/s} = [331 + 0.6(30)] \text{ m/s} = 349 \text{ m/s}$$

and therefore

$$f_1 = \frac{v}{4L} = \frac{349 \text{ m/s}}{4(0.0254 \text{ m})} = 3.44 \times 10^3 \text{ Hz} = 3.44 \text{ kHz}$$

This is in good agreement with the curves in Figure 14.14, that show a dip (maximum sensitivity) at about this frequency. How about the other dip just above 10 kHz? Check out the next natural frequency for the ear canal, f_3.

FOLLOW-UP EXERCISE. Children have smaller ear canals than adults, on the order of 1.30 cm in length. What is the lowest fundamental frequency for a child's ear canal? Use the same air temperature as in the Example.

The **quality** of a tone is the characteristic that enables it to be distinguished from another tone of basically the same intensity and frequency. Tone quality depends on the waveform – specifically, the number of harmonics (overtones) present and their relative intensities (▼ **Figure 14.15**). The tone of a voice depends in large part on the vocal resonance cavities. One person can sing a tone with the same basic frequency and intensity as another, but different combinations of overtones give the two voices different qualities.

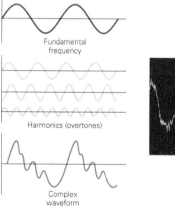
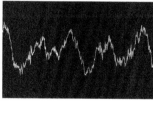

▲ **FIGURE 14.15 Waveform and quality (a)** The superposition of sounds of different frequencies and amplitudes gives a complex waveform. The harmonics, or overtones, determine the quality of the sound. **(b)** A stringed instrument waveform displayed on an oscilloscope. Notice the fundamental frequency (big peaks) and superimposed on that several higher frequencies (the small, closer together peaks).

The notes of a musical scale correspond to certain frequencies; for instance, the frequency of 264 Hz corresponds closely to a note called middle C (or C4). When a note is played on an instrument, its assigned frequency is that of the first harmonic, which is the fundamental frequency. Thus for each note, there are several overtones or harmonics. The fundamental frequency is dominant over the accompanying overtones that determine the sound quality of the instrument. Recall from Section 13.5 that the overtones that are produced depend on how an instrument is played. Whether a violin string is plucked or bowed, for example, can be discerned from the quality of idential notes.

Chapter 14 Review

- The **sound frequency spectrum** is divided into infrasonic ($f < 20$ Hz), audible (20 Hz $< f <$ 20 kHz), and ultrasonic ($f > 20$ kHz) frequency regions.
- The speed of sound in a medium depends on the elasticity of the medium and its density. In general, $v_{\text{solids}} > v_{\text{liquids}} > v_{\text{gases}}$.

Speed of sound in dry air:

$$v = (331 + 0.6T_{\text{C}})\,\text{m/s} \qquad (14.1)$$

- The intensity of a point source of sound energy is inversely proportional to the square of the distance from the source.

Intensity of a point source:

$$I = \frac{P}{4\pi R^2} \quad \text{and} \quad \frac{I_2}{I_1} = \left(\frac{R_1}{R_2}\right)^2 \qquad (14.2, 14.3)$$

- The sound intensity level is a logarithmic function of the sound intensity and is expressed in decibels (dB).

Intensity level (in decibels, dB):

$$\beta = 10\log\left(\frac{I}{I_{\text{o}}}\right) \quad \text{where} \quad I_{\text{o}} = 10^{-12}\,\text{W/m}^2 \qquad (14.4)$$

- Sound wave interference at a location due to two-point sources depends on phase difference that is related to path length differences between the sources and the location. Sound waves that arrive at a location in phase reinforce each other (constructive interference); sound waves that arrive at a location out of phase cancel each other (destructive interference).

Phase difference (where ΔL is the path length difference):

$$\Delta\theta = \frac{2\pi}{\lambda}(\Delta L) \qquad (14.5)$$

Condition for constructive interference:

$$\Delta L = n\lambda \quad (n = 0, 1, 2, 3, \ldots) \qquad (14.6)$$

Condition for destructive interference:

$$\Delta L = m\left(\frac{\lambda}{2}\right) \quad (m = 1, 3, 5, \ldots) \qquad (14.7)$$

- **Beat frequency:**

$$f_{\text{b}} = |f_1 - f_2| \qquad (14.8)$$

- **Doppler effect:**
  The Doppler effect depends on the velocities of the sound source and observer relative to still air. When the relative motion of the source and observer is toward each other, the observed pitch/frequency increases; when the relative motion of the source and observer is away from each other, the observed pitch/frequency decreases.

  Moving source, stationary observer

$$f_{\text{o}} = \left(\frac{v}{v \pm v_{\text{s}}}\right)f_{\text{s}} = \left(\frac{1}{1 \pm (v_{\text{s}}/v)}\right)f_{\text{s}}$$

$$\begin{cases} - & \text{for source moving toward a stationary observer} \\ + & \text{for source moving away from a stationary observer} \end{cases}$$

$$(14.9)$$

  Moving observer, stationary source

$$f_{\text{o}} = \left(\frac{v \pm v_{\text{o}}}{v}\right)f_{\text{s}} = \left(1 \pm \frac{v_{\text{o}}}{v}\right)f_{\text{s}}$$

$$\begin{cases} - & \text{for an observer moving away from a stationary source} \\ + & \text{for an observer moving toward a stationary source} \end{cases}$$

$$(14.10)$$

- **Mach number related to angle of shock wave:**

$$M = \frac{v_{\text{s}}}{v} = \frac{1}{\sin\theta} \qquad (14.11)$$

- **Natural frequencies of an open-open organ pipe:**

$$f_n = \frac{v}{\lambda_n} = n\left(\frac{v}{2L}\right) = nf_1 \quad (n = 1, 2, 3, \ldots) \qquad (14.12)$$

- **Natural frequencies of a open-closed organ pipe:**

$$f_m = \frac{v}{\lambda_m} = m\left(\frac{v}{4L}\right) = mf_1 \quad (m = 1, 3, 5, \ldots) \qquad (14.13)$$

End of Chapter Questions and Exercises

Multiple Choice Questions

14.1 Sound Waves
and
14.2 The Speed of Sound

1. A sound wave with a frequency of 15 Hz is in what region of the sound spectrum: (a) audible, (b) infrasonic, (c) ultrasonic, or (d) supersonic?

2. A sound wave in air (a) is longitudinal, (b) is transverse, (c) has longitudinal and transverse components, (d) travels faster than a sound wave through a liquid.

3. The speed of sound is generally greatest in (a) solids, (b) liquids, (c) gases, (d) a vacuum.

4. The speed of sound in air (a) is about 1/3 km/s, (b) is about 1/5 mi/s, (c) depends on temperature, (d) all of the preceding.

5. The speed of sound in water is about 4 times that in air. A single-frequency sound source in air is f_o. When its sound goes from air into water, the frequency of the sound in the water will be (a) $4f_o$, (b) $f_o/4$, (c) f_o.

14.3 Sound Intensity and Sound Intensity Level

6. If the air temperature increases, would the sound intensity from a constant-output point source (a) increase, (b) decrease, or (c) remain unchanged?

7. The decibel scale is referenced to a standard intensity of (a) 1.0 W/m², (b) 10^{-12} W/m², (c) normal conversation, (d) the threshold of pain.

8. If the intensity level of a sound at 20 dB is increased to 40 dB, the intensity would increase by a factor of (a) 10, (b) 20, (c) 40, (d) 100.

9. The intensity of a sound wave is directly proportional to the (a) amplitude, (b) frequency, (c) square of the amplitude, (d) square of the frequency.

10. A sound with an intensity level of 30 dB is how many times more intense than the threshold of hearing: (a) 10, (b) 100, (c) 1000, or (d) 3000?

14.4 Sound Phenomena
and
14.5 The Doppler Effect

11. Constructive and destructive interference of sound waves is based on (a) the speed of sound, (b) diffraction, (c) phase difference, (d) the principle of superposition.

12. Beats are the direct result of (a) the principle of superposition, (b) refraction, (c) diffraction, (d) the Doppler effect.

13. Police radar equipment measure speeds based on (a) refraction, (b) the Doppler effect, (c) diffraction, (d) sonic boom.

14.6 Musical Instruments and Sound Characteristics

14. Given open-open and open-closed pipes of the same length, which would have the lowest natural frequency: (a) the open-open pipe, (b) the open-closed pipe, or (c) they both would have the same lowest frequency?

15. The human ear can hear tones best at about what frequency? (a) 1000 Hz, (b) 4000 Hz, (c) 6000 Hz, (d) all frequencies.

Conceptual Questions

14.1 Sound Waves
and
14.2 The Speed of Sound

1. The speed of sound in air does not depend on frequency. Explain how this enables concert goers in all seats to hear the same music in the right sequence/combination.

2. As a sound wave travels from warm air into cold air, explain how its frequency, wavelength, and speed change.

3. Explain why the speed of sound in helium is a lot faster than in air. [*Hint:* Check out the molecular masses.]

4. What role, if any, do you think humidity should play on the speed of sound in air?

14.3 Sound Intensity and Sound Intensity Level

5. Where is the intensity greater and by what factor: (1) at a distance R from a small sound source of power P, or (2) at a distance $2R$ from a small sound source of power $2P$? Explain.

6. The Richter scale, used to measure the intensity level of earthquakes, is a logarithmic scale, as is the decibel scale. Why are such logarithmic scales used?

7. Can there be negative decibel levels, such as −10 dB? If so, what would these mean?

14.4 Sound Phenomena
and
14.5 The Doppler Effect

8. Do interference beats have anything to do with the "beat" of music? Explain.

9. (a) Is there a Doppler effect if a sound source and an observer are moving through still air at the same velocity? (b) What would be the effect on the frequency of sound heard by a stationary observer if the source were accelerating toward that observer?

10. As a person walks between a pair of in-phase loudspeakers that produce tones of the same amplitude and frequency, she hears a varying sound intensity. Explain.

11. Explain how Doppler radar instruments are used to monitor both location and internal motions of a storm.

12. A stationary sound source and a stationary observer are a fixed distance apart. However, the air between them is moving toward the observer with a constant speed. How do you think the frequency received by the observer might be affected? Explain your reasoning.

14.6 Musical Instruments and Sound Characteristics

13. The frets on a guitar fingerboard are spaced closer together the farther they are from the neck. Why is this? What would be the result if they were evenly spaced?

14. Is it possible for an open-open organ pipe and an open-closed organ pipe of the same length to produce notes of the same frequency? Justify your answer.

15. How would an increase in air temperature affect the frequencies of an organ pipe?

16. Why are there no even harmonics in a closed organ pipe?

Exercises*

Integrated Exercises (IEs) are two-part exercises. The first part typically requires a conceptual answer choice based on physical thinking and basic principles. The following part is quantitative calculations associated with the conceptual choice made in the first part of the exercise.

14.1 Sound Waves
and
14.2 The Speed of Sound

1. • What is the speed of sound in air at (a) 10 °C and (b) 20 °C?

2. • The speed of sound in air on a summer day is measured as 350 m/s. What is the air temperature?

3. • Sonar is used to map the ocean floor. If an ultrasonic signal is received 2.0 s after it is emitted, how deep is the ocean floor at that location?

4. • What temperature change from 0 °C would increase the speed of sound by 1.0%?

5. • The expression for the speed of a sound wave in a liquid is $v = \sqrt{Y/\rho}$. Show that this equation is dimensionally correct. Repeat this using $v = \sqrt{Y/\rho}$ for speed of sound in a solid.

6. •• Particles approximately 3.0×10^{-2} cm in diameter are to be scrubbed loose from machine parts immersed in an aqueous (watery) ultrasonic cleaning bath. Above what frequency should the bath be operated to produce wavelengths of this size and smaller?

7. •• Medical ultrasound uses a frequency of around 20 MHz to diagnose human conditions and ailments by reflection. Objects and structures down to the size of the wavelength can be detected using these waves. (a) If the speed of sound in tissue is 1500 m/s, estimate the smallest detectable object. (b) If the penetration depth in tissue is about 200 wavelengths, how deep can these waves penetrate into tissue?

8. •• A person holds a rifle horizontally and fires at a target. The bullet has a muzzle speed of 200 m/s, and the person hears the bullet strike the target 1.00 s after firing it. The air temperature is 72 °F. What is the distance to the target?

9. •• A freshwater dolphin sends an ultrasonic sound to locate a prey. If the echo off the prey is received by the dolphin 0.12 s after being sent, how far is the prey from the dolphin?

10. •• A submarine on the ocean surface receives a sonar echo indicating an underwater object. The echo comes back at an angle of 20° above the horizontal and the echo took 2.32 s to get back to the submarine. What is the object's depth?

11. •• The speed of sound in human tissue is on the order of 1500 m/s. A 3.50-MHz probe is used for an ultrasonic procedure. (a) If the effective depth of the ultrasound is 250 wavelengths, what is this depth in meters? (b) What is the time lapse for the ultrasound to make a round trip if reflected from an object at the effective depth? (c) The smallest detail capable of being detected is on the order of one wavelength of the ultrasound. What would this be?

12. •• The size of your eardrum partially determines the upper frequency limit of your audible region, usually between 16 000 and 20 000 Hz. If the wavelength at the upper limit is on the order of twice the diameter of the eardrum and the air temperature is 20 °C, how wide is your eardrum? Is your answer reasonable?

13. IE ••• On hiking up a mountain that has several overhanging cliffs, a climber drops a stone off the first cliff to determine its height by measuring the time it takes to hear the stone hit the ground. (a) At a second cliff that is twice the height of the first, the measured time of the sound from the dropped stone is (1) less than double, (2) double, or (3) more than double that of the first. Why? (b) If the measured time is 4.8 s for the stone dropping from the first cliff, and the air temperature is 20 °C, how high is the cliff? (c) If the height of a third cliff is three times that of the first one, what would be the measured time for a stone dropped from that cliff to reach the ground?

14. ••• A bat moving at 15.0 m/s emits a high-frequency sound as it approaches a wall that is 25.0 m away. Assuming that the bat continues straight toward the wall, how far away is it when it receives the echo? (Assume the air temperature in the cave to be 0 °C.)

15. ••• Sound propagating through air at 30 °C passes through a vertical cold front into air that is 4.0 °C. If the sound has a frequency of 2500 Hz, by what percentage does its wavelength change in crossing the boundary?

14.3 Sound Intensity and Sound Intensity Level

16. • Calculate the intensity generated by a 1.0-W point source of sound at a location (a) 3.0 m and (b) 6.0 m from it.

17. IE • (a) If the distance from a point sound source triples, the sound intensity will be (1) 3, (2) 1/3, (3) 9, (4) 1/9 times the original value. Why? (b) By how much must the distance from a point source be increased to reduce the sound intensity by half?

---

* The bullets denote the degree of difficulty of the exercises: •, simple; ••, medium; and •••, more difficult.

18. • Assuming that the diameter of your eardrum is 1 cm (see Exercise 12), what is the sound power received by the eardrum at the threshold of (a) hearing and (b) pain?

19. • Find the intensity levels in decibels for sounds with intensities of (a) 10^{-2} W/m², (b) 10^{-6} W/m², and (c) 10^{-15} W/m².

20. •• During a weapons test, a compact bomb is exploded and it produces an intensity level of 160 dB at a distance of 10 m. What would be the intensity level at 100 m away? (Assume no energy is lost due to reflections, etc.)

21. IE •• (a) If the power of a sound source doubles, the intensity level at a certain distance from the source (1) increases, (2) exactly doubles, or (3) decreases. Why? (b) What are the intensity levels at a distance of 10 m from a 5.0-W and a 10-W source, respectively?

22. •• The intensity levels of two people holding a conversation are 60 and 70 dB, respectively. What is the intensity of the combined sounds?

23. •• A point source emits radiation in all directions at a rate of 7.5 kW. What is the intensity of the radiation 5.0 m from the source?

24. •• Two sound sources have intensities of 10^{-9} and 10^6 W/m², respectively. Which source is more intense and by how many times more?

25. •• Assuming that 20 people, each capable of individually producing an intensity level of 60 dB at a given location, all speak simultaneously. What is the total sound intensity at that location?

26. •• A rock band using one speaker produces sound that creates an intensity level of 110 dB at a distance of 15 m from the speaker. Assuming the sound is radiated equally over a hemisphere in front of the band, what is the total power output?

27. •• A person has a hearing loss of 30 dB for a particular frequency. What is the sound intensity that is heard at this frequency that has an intensity of the threshold of pain?

28. •• A compact speaker puts out 100 W of sound power. (a) Neglecting losses to the air, at what distance would the sound intensity be at the pain threshold? (b) Neglecting losses to the air, at what distance would the sound intensity be that of normal speech? Does your answer seem reasonable? Explain.

29. •• What is the intensity level of a 23-dB sound after being amplified (a) ten thousand times, (b) a million times, (c) a billion times?

30. •• In a neighborhood challenge to see who can climb a tree the fastest, you are ready to climb. Your friends have surrounded you in a circle as a cheering section; each individual alone would cause a sound intensity level of 80 dB at your location. If the actual sound level at your location is 87 dB, how many people are rooting for you?

31. IE •• A dog's bark has a sound intensity level of 40 dB. (a) If two of the same dogs were barking, the intensity level is (1) less than 40 dB, (2) between 40 dB and 80 dB, (3) 80 dB. (b) What would be the intensity level?

32. •• At a rock concert, the average sound intensity level for a person in a front-row seat is 110 dB for a single band. If all the bands scheduled to play produce sound of that same intensity, how many of them would have to play simultaneously for the sound level to be at or above the threshold of pain?

33. •• At a distance of 12.0 m from a point source, the intensity level is measured to be 70 dB. At what distance from the source will the intensity level be 40 dB?

34. •• At a Fourth of July celebration, a firecracker explodes (▼ **Figure 14.16**). Considering the firecracker to be a point source, what are the intensities heard by observers at points B, C, and D, relative to that heard by the observer at A?

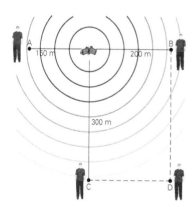

▲ FIGURE 14.16 **A big bang** See Exercise 34.

35. •• An office in an e-commerce company has fifty computers, which generate a sound intensity level of 40 dB. The office manager tries to cut the noise to half as loud by removing twenty-five computers. Does he achieve his goal? What is the intensity level generated by twenty-five computers?

36. ••• A bee produces a buzzing sound that is barely audible to a person 3.0 m away. How many bees would have to be buzzing at that distance to produce a sound with an intensity level of 50 dB?

14.4 Sound Phenomena
and
14.5 The Doppler Effect

37. • What is the frequency heard by a person driving at 60 km/h directly toward a factory whistle emitting sound of frequency 800 Hz, if the air temperature is 0 °C?

38. **IE** • On a day with a temperature of 20 °C and no wind blowing, the frequency heard by a moving person from a 500-Hz stationary siren is 520 Hz. (a) The person is (1) moving toward, (2) moving away from, or (3) stationary relative to the siren. Explain. (b) What is the person's speed? Assume any movement is along the line joining the siren and the person.

39. •• While standing near a railroad crossing, you hear a train horn. The frequency emitted by the horn is 400 Hz. If the train is traveling at 90.0 km/h and the air temperature is 25 °C, what is the frequency you hear (a) when the train is approaching and (b) after it has passed?

40. •• Two identical strings on different cellos are tuned to the 440-Hz A note. The peg holding one of the strings slips, so its tension is decreased by 1.5%. What is the beat frequency heard when the strings are then played together?

41. •• The frequency of an ambulance siren is 700 Hz. What are the frequencies heard by a stationary pedestrian as the ambulance approaches and then later moves away from her at a constant speed of 90.0 km/h? (Assume that the air temperature is 20 °C.)

42. •• A jet flies at a speed of Mach 2.0. What is the half-angle θ of the conical shock wave formed by the aircraft? Can you tell the speed of the shock wave?

43. **IE** •• A fighter jet flies at a speed of Mach 1.5. (a) If the jet were to fly faster than Mach 1.5, the half-angle θ of the conical shock wave would (1) increase, (2) remain the same, (3) decrease. Why? (b) What is the half-angle of the conical shock wave formed by the jet plane at Mach 1.5?

44. •• The half-angle θ of the conical shock wave formed by a supersonic jet is 30°. What are (a) the Mach number of the aircraft and (b) the actual speed of the aircraft if the air temperature is −20 °C?

45. •• An observer is traveling between two identical sources of sound (frequency 100 Hz). His speed is 10.0 m/s as he approaches one and recedes from the other. (a) What frequency does he hear from each source? (b) How many beats per second does he hear? Assume normal room temperature.

46. ••• A bystander hears a siren vary in frequency from 476 Hz to 404 Hz as a fire truck approaches, passes, and then moves away on a straight street (▼ **Figure 14.17**). What is the speed of the truck? (Take the speed of sound in air to be 343 m/s.)

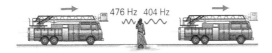

▲ **FIGURE 14.17**   **The siren's wail** See Exercise 46.

47. ••• Bats emit sounds of frequencies around 35.0 kHz and use echolocation to find their prey. If a bat is moving with a speed of 12.0 m/s toward a hovering, stationary insect, (a) what is the frequency received by the insect if the air temperature is 20 °C? (b) What frequency of the reflected sound is heard by the bat? (c) If the insect were initially moving directly away from the bat, would this affect the frequencies? Explain.

48. ••• A supersonic jet flies directly overhead relative to an observer, at an altitude of 2.0 km (▼ **Figure 14.18**). When the observer finally hears the first sonic boom, the plane has flown a horizontal distance of 2.5 km beyond him at a constant speed. (a) What is the angle of the shock wave cone? (b) At what Mach number is the plane flying? (Assume a value for the speed of sound based on an average constant temperature of 15 °C.)

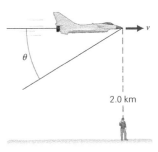

▲ **FIGURE 14.18**   **Faster than a speeding bullet** See Exercise 48.

14.6 Musical Instruments and Sound Characteristics

49. • The first three natural frequencies of an organ pipe are 126, 378, and 630 Hz. (a) Is the pipe an open-open or open-closed? (b) Taking the speed of sound in air to be 340 m/s, find the length of the pipe.

50. • An open-closed organ pipe has a fundamental frequency of 528 Hz (a C note) at 20 °C. What is the fundamental frequency of the pipe when the temperature is 0 °C?

51. • The human ear canal is about 2.5 cm long. It is open at one end and closed at the other. (a) What is the fundamental frequency of the ear canal at 20 °C? (b) To what frequency is the ear most sensitive? (c) If a person's ear canal is longer than 2.5 cm, is the fundamental frequency higher or lower than that in part (a)? Explain.

52. •• An organ pipe that is closed at one end has a length of 0.80 m. At 20 °C, what is the distance between a node and an adjacent antinode for (a) the second harmonic and (b) the third harmonic?

53. •• An open-open organ pipe and one that is closed at one end both have lengths of 0.52 m. Assuming a temperature of 20 °C, what is the fundamental frequency of each pipe?

54. •• An open organ pipe is 0.50 m long. If the speed of sound is 340 m/s, what are the pipe's fundamental frequency and the frequencies of the first two overtones?

55. •• An organ pipe that is closed at one end is 1.10 m long. It is oriented vertically and filled with carbon dioxide gas (which is denser than air and thus will stay in the pipe). A tuning fork with a frequency of 60.0 Hz is used to set up a standing wave in the fundamental mode. What is the speed of sound in carbon dioxide?

56. •• An open-open organ pipe 0.750 m long has its first overtone at a frequency of 441 Hz. What is the temperature of the air in the pipe?

57. IE •• When all of its holes are closed, a flute is essentially a tube that is open at both ends, with the length measured from the mouthpiece to the far end. If a hole is opened, then the length of the tube is effectively measured from the mouthpiece to that hole. (a) Is the position at the mouthpiece (1) a node, (2) an antinode, or (3) neither a node nor an antinode? Why? (b) If the lowest fundamental frequency on a flute is 262 Hz, what is the minimum length of the flute at 20 °C? (c) If a note of frequency 440 Hz is to be played by opening a hole somewhere along the length of the flute, what should be the distance between that opened hole and the mouthpiece?

58. ••• An organ pipe that is closed at one end is filled with helium. The pipe has a fundamental frequency of 660 Hz in air at 20 °C. What is the pipe's fundamental frequency with the helium in it?

59. ••• An open-open organ pipe, in its fundamental mode, has a length of 50.0 cm. A second pipe, closed at one end, is also in its fundamental mode. A beat frequency of 2.00 Hz is heard when both are making sound. Determine the possible lengths of the closed pipe. Assume normal room temperature.

60. ••• Bats typically give off an ultrahigh-frequency sound at about 50 000 Hz. If a bat is approaching a stationary object at 18.0 m/s, what will be the reflected frequency it detects? Assume the air in the cave is at 5 °C. [*Hint:* You will need to apply the Doppler equations twice. Why?]

Appendices

Appendix I*   Mathematical Review for College Physics

A   Symbols, Arithmetic Operations, Exponents, and Scientific Notation

Commonly Used Symbols in Relationships

$=$ means two quantities are equal, such as $2x = y$.

$\equiv$ means "defined as," such as the definition of pi:

$$\pi \equiv \frac{\text{circumstance of a circle}}{\text{the diameter of that circle}}$$

$\approx$ means approximately equal, as in 30 m/s $\approx$ 60 m/h.

$\neq$ means inequality, such as $\pi \neq 22/7$.

$\geq$ means that one quantity is greater than or equal to another. For example, if the age of the universe ≥ 10 billion years, its minimum age is 10 billion years.

$\leq$ means that one quantity is less than or equal to another. For example, if a lecture room holds ≤ 45 students, the maximum number of students is 45.

$>$ means that one quantity is greater than another, such as 14 eggs > 1 dozen eggs.

$\gg$ means that one quantity is *much* greater than another. For example, the number of people on Earth $\gg$ 1 million.

$<$ means that one quantity is less than another, such as $3 \times 10^{22} < 10^{24}$.

$\ll$ means that one quantity is much less than another, such as $10 \ll 10^{11}$.

$\propto$ means proportional to. That is, if $y = 2x$ then $y \propto x$.

This means that if x is increased by a certain multiplicative factor, y is also increased the same way. For example, if $y = 3x$, if x is changed by a factor of n (i.e., if x becomes nx), then so is y, because $y' = 3x' = (3nx) = n(3x) = ny$.

ΔQ means "change in the quantity Q." This means "final minus initial." For example, if the value V of an investor's stock portfolio in the morning is $V_i = \$10\ 100$ and at the close of trading it is $V_f = \$10\ 050$, then $\Delta V = V_f - V_i = \$10\ 050 - \$10\ 100 = -\$50$.

The Greek letter capital sigma (Σ) indicates the sum of a series of values for the quantity Q_i where $i = 1, 2, 3, ..., N$, that is,

$$\sum_{i=1}^{N} Q_i = Q_1 + Q_2 + Q_3 + \cdots Q_N$$

---

* This appendix does not include a discussion of significant figures, since a thorough discussion is presented in Section 1.6.

$|Q|$ denotes the absolute value of a quantity Q without a sign. If Q is positive, then $|Q| = Q$; if Q is negative then $|-Q| = Q$. Thus $|-3| = 3$.

Arithmetic Operations and Their Order of Usage

Basic arithmetic operations are addition (+) subtraction (–), multiplication (× or ·), and division (÷ or /). Another common operation, exponentiation (x^n), involves raising a quantity (x) to a power (n). If several of these operations are included in one equation, they are performed in this order: (a) parentheses, (b) exponentiation, (c) multiplication, (d) division, (e) addition, and (f) subtraction.

A handy mnemonic used to remember this order is: "Please Excuse My Dear Aunt Sally," where the capital letters stand for the various operations in order: Parentheses, Exponents, Multiplication, Division, Addition, Subtraction. Note that operations within parentheses are always first, so to be on the safe side, appropriate use of parentheses is encouraged. For example, $24^2/8{\cdot}4 + 12$ could be evaluated several ways. However, according to the agreed-on order, it has a unique value: $24^2/8{\cdot}4 + 12 = 576/8{\cdot}4 + 12 = 576/32 + 12 = 18 + 12 = 30$. To avoid possible confusion, the quantity could be written using two sets of parentheses as follows: $[24^2/(8{\cdot}4)] + 12 = [576/(32)] + 12 = 18 + 12 = 30$.

Exponents and Exponential Notation

Exponents and exponential notation are very important when employing scientific notation (see the next section). You should be familiar with power and exponential notation (both positive and negative, fractional and integral) such as the following:

$$x^0 = 1$$

$$x^1 = x \qquad x^{-1} = \frac{1}{x}$$

$$x^2 = x \cdot x \qquad x^{-2} = \frac{1}{x_2} \qquad x^{1/2} = \sqrt{x}$$

$$x^3 = x \cdot x \cdot x \quad x^{-3} = \frac{1}{x^3} \quad x^{1/3} = \sqrt[3]{x} \quad \text{etc.}$$

Exponents combine according to the following rules:

$$x^a \cdot x^b = x^{(a+b)} \quad x^a/x^b = x^{(a-b)} \quad (x^a)^b = x^{ab}$$

Scientific Notation (Powers-of-10 Notation)

In physics, many quantities have values that are very large or very small. To express them, **scientific notation** is frequently used. This notation is sometimes referred to as *powers-of-10* notation for obvious reasons. (See the previous section for a discussion of exponents.) When the number 10 is squared or cubed, we can write it as $10^2 = 10 \times 10 = 100$ or $10^3 = 10 \times 10 \times 10 = 1000$. You can see that the number of zeros is equal to the power of 10. Thus 10^{23} is a compact way of expressing the number 1 followed by 23 zeros.

A number can be represented in many different ways. For example, the distance from the Earth to the Sun is 93 million miles. This value can be written as 93 000 000 miles. Expressed in a more compact scientific notation, there are many correct

forms, such as 93×10^6 miles, 9.3×10^7 miles, or 0.93×10^8 miles. Any of these is correct, although the second is preferred, because when using powers of 10 notation, it is customary to leave only one digit to the left of the decimal point, in this case 9.3. (This is called customary or standard form.) Note that the exponent, or power of 10, changes when the decimal point of the prefix number is shifted.

Negative powers of 10 also can be used. For example, $10^{-2} = (1/10^2) = (1/100) = 0.01$. So, if a power of 10 has a negative exponent, the decimal point may be shifted to the left once for each power of 10. For example, 5.0×10^{-2} is equal to 0.050 (two shifts to the left).

The decimal point of a quantity expressed in powers-of-10 notation may be shifted to the right or left irrespective of whether the power of 10 is positive or negative. General rules for shifting the decimal point are as follows:

1. The exponent, or power of 10, is *increased* by 1 for every place the decimal point is shifted to the left.
2. The exponent, or power of 10, is *decreased* by 1 for every place the decimal point is shifted to the right.

This is simply a way of saying that as the coefficient (prefix number) gets smaller, the exponent gets correspondingly larger, and vice versa. Overall, the number is the same.

When multiplying using this notation, the exponents are added. Thus, $10^2 \times 10^4 = (100)(10\,000) = 1\,000\,000 = 10^6 = 10^{2+4}$. Division follows similar rules using negative exponents, for example, $(10^5/10^2) = (100\,000/100) = 1000 = 10^3 = 10^{5+(-2)}$.

Care should be taken when adding and subtracting numbers written in scientific notation. Before doing so, all numbers must be converted to the same power of 10. For example,

$$1.75 \times 10^3 - 5.0 \times 10^2 = 1.75 \times 10^3 - 0.50 \times 10^3$$
$$= (1.75 - 0.50) \times 10^3 = 1.25 \times 10^3$$

B   Algebra and Common Algebraic Relationships

General

The basic rule of algebra, used for solving equations, is that if you perform any legitimate operation on both sides of an equation, it remains an equation, or equality. (An example of an illegal operation is dividing by zero; why?) Thus such operations as adding a number to both sides, cubing both sides, and dividing both sides by the same number all maintain the equality.

For example, suppose you want to solve $\dfrac{x^2 + 6}{2} = 11$ for x.

To do this, first multiply both sides by 2, giving $\left(\dfrac{x^2 + 6}{2}\right) \times 2 = 11 \times 2 = 22$ or $x^2 + 6 = 22$. Then subtract 6 from both sides to obtain $x^2 + 6 - 6 = 22 - 6 = 16$ or $x^2 = 16$. Finally, taking the square root of both sides, the solutions are $x = \pm 4$.

Some Useful Results

Many times, *the square of the sum or difference of two numbers* is required. For any numbers a and b:

$$(a \pm b)^2 = a^2 \pm 2ab + b^2$$

Similarly, *the difference of two squares* can be factored:

$$(a^2 - b^2) = (a + b)(a - b)$$

A quadratic equation is one that can be expressed in the form $ax^2 + bx + c = 0$. In this form it can always be solved (usually for two different roots) using the *quadratic formula*: $x = \dfrac{-b \pm \sqrt{b^2 - 4ac}}{2a}$. In kinematics, this result can be especially useful as it is common to have equations of this form to solve; for example: $4.9t^2 - 10t - 20 = 0$. Just insert the coefficients (making sure to include the sign) and solve for t (here t represents the time for a ball to reach the ground when thrown straight upward from the edge of a cliff; see Chapter 2). The result is

$$t = \frac{10 \pm \sqrt{10^2 - 4(4.9)(-20)}}{2(4.9)} = \frac{10 \pm 22.2}{9.8}$$

$$= +3.3\,\text{s} \quad \text{or} \quad -1.2\,\text{s}$$

In all such problems, time is "stopwatch" time and starts at zero; hence the negative answer can be ignored as physically unreasonable although it is a solution to the equation.

Solving Simultaneous Equations

Occasionally solving a problem might require solving two or more equations simultaneously. In general, if there are N unknowns in a problem, exactly N independent equations will be needed. If there are less than N equations, there are not enough for a complete solution. If there are more than N equations, then some are redundant, and a solution is usually still possible, although more complicated. In general in this textbook, such concerns will usually be with two simultaneous equations, and both will be linear. Linear equations are of the form $y = mx + b$. Recall that when plotted on an x–y Cartesian coordinate system, the result is a straight line with a slope of m ($=\Delta y/\Delta x$) and a y-intercept of b, as shown for the red line here.

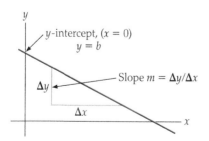

To solve two linear equations simultaneously graphically, simply plot them on the axes and evaluate the coordinates at their intersection point. While this can always be done in principle, it is only an approximate answer and usually takes quite a bit of time.

The most common (and exact) method of solving simultaneous equations involves the use of algebra. Essentially, you solve one equation for an unknown and substitute the result into the other equation, ending up with one equation and one unknown. Suppose you have two equations and two unknown quantities (x and y), but in general, any two unknown quantities:

$$3y + 4x = 4 \quad \text{and} \quad 2x - y = 2$$

Solving the second equation for y yields $y = 2x - 2$. Substituting this value for y into the first equation, $3(2x - 2) + 4x = 4$. Thus, $10x = 10$ and $x = 1$. Putting this value into the second of the original two equations, the result is $2(1) - y = 2$ and therefore $y = 0$. (Of course, at this point a good double-check is to substitute the answers and see if they solve both equations.)

C Geometric Relationships

In physics and many other areas of science, it is important to know how to find circumferences, areas, and volumes of some common shapes. Here are some equations for such shapes.

Circumference (c), Area (A), and Volume (V)

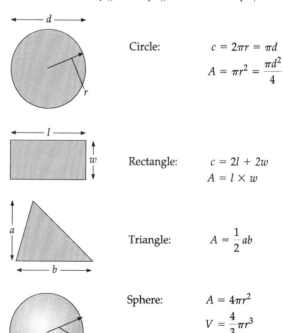

Circle: $\quad c = 2\pi r = \pi d$

$$A = \pi r^2 = \frac{\pi d^2}{4}$$

Rectangle: $\quad c = 2l + 2w$

$$A = l \times w$$

Triangle: $\quad A = \frac{1}{2}ab$

Sphere: $\quad A = 4\pi r^2$

$$V = \frac{4}{3}\pi r^3$$

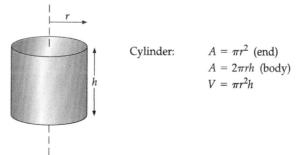

Cylinder: $\quad A = \pi r^2$ (end)

$A = 2\pi rh$ (body)

$V = \pi r^2 h$

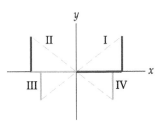

D   Trigonometric Relationships

Understanding elementary trigonometry is crucial in physics, especially since many of the quantities are vectors. Here is a summary of definitions of the three most common trig functions, which you should commit to memory.

Definitions of Trigonometric Functions

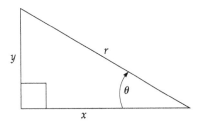

$$\sin\theta = \frac{y}{r} \quad \cos\theta = \frac{x}{r} \quad \tan\theta = \frac{\sin\theta}{\cos\theta} = \frac{y}{x}$$

$\theta°$ (rad)	$\sin\theta$	$\cos\theta$	$\tan\theta$
0° (0)		1	0
30° ($\pi/6$)	0.500	0.866	0.577
45° ($\pi/4$)	0.707	0.707	1.00
60° ($\pi/3$)	0.866	0.500	1.73
90° ($\pi/2$)	1	0	$\to \infty$

For very small angles,

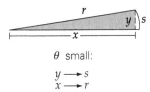

θ small:

$$y \longrightarrow s$$
$$x \longrightarrow r$$

$$\theta \text{ (in rad)} = \frac{s}{r} \approx \frac{y}{r} \approx \frac{y}{x}$$
$$\theta \text{ (in rad)} \approx \sin\theta \approx \tan\theta$$
$$\cos\theta \approx 1 \quad \sin\theta \approx \theta \quad \text{(radians)}$$
$$\tan\theta = \frac{\sin\theta}{\cos\theta} \approx \theta \quad \text{(radians)}$$

The sign of a trigonometric function depends on the quadrant, or the signs of x and y. For example, in the second quadrant x is negative and y is positive; therefore, $\cos\theta = x/r$ is negative and $\sin\theta = y/r$ is positive. [Note that r (shown as the dashed lines) is always taken as positive.] In the figure, the dark red lines are positive and the light red lines negative.

Some Useful Trigonometric Identities

$$1 = \sin^2\theta + \cos^2\theta$$

$$\sin 2\theta = 2\sin\theta\cos\theta$$

$$\cos 2\theta = \cos^2\theta - \sin^2\theta = 2\cos^2\theta - 1 = 1 - 2\sin^2\theta$$

$$\sin^2\theta = \frac{1}{2}(1 - \cos 2\theta)$$

$$\cos^2\theta = \frac{1}{2}(1 + \cos 2\theta)$$

For half-angle ($\theta/2$) identities, simply replace θ with $\theta/2$; for example,

$$\sin^2(\theta/2) = \frac{1}{2}(1 - \cos\theta)$$

$$\cos^2(\theta/2) = \frac{1}{2}(1 + \cos\theta)$$

Trigonometric values of sums and differences of angles are sometimes of interest. Here are several basic relationships.

$$\sin(\alpha \pm \beta) = \sin\alpha\cos\beta \pm \cos\alpha\sin\beta$$

$$\cos(\alpha \pm \beta) = \cos\alpha\cos\beta \mp \sin\alpha\sin\beta$$

$$\tan(\alpha \pm \beta) = \frac{\tan\alpha \pm \tan\beta}{1 \mp \tan\alpha\tan\beta}$$

Law of Cosines

For a triangle with angles A, B, and C with opposite sides a, b, and c, respectively:

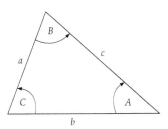

$a^2 = b^2 + c^2 - 2bc\cos A$ (with similar results for $b^2 = \ldots$ and $c^2 = \ldots$)

If $A = 90°$ then $\cos A = 0$ and this reduces to the Pythagorean theorem as it should: $a^2 = b^2 + c^2$.

Law of Sines

For a triangle with angles A, B, and C with opposite sides a, b, and c, respectively:

$$\frac{a}{\sin A} = \frac{b}{\sin B} = \frac{c}{\sin C}$$

E   Logarithms

Presented here are some of the fundamental definitions and relationships for logarithms. Logarithms are commonly used in science, so it is important that you know what they are and how to use them. Logarithms are useful because, among other things, they allow you to more easily multiply and divide very large and very small numbers.

Definition of a Logarithm

If a number x is written as another number a to some power n, as $x = a^n$, then n is defined to be the *logarithm of the number x to the base a*. This is written compactly as

$$n \equiv \log_a x$$

Common Logarithms

If the base a is 10, the logarithms are called *common logarithms*. When the abbreviation *log* is used, without a base specified, base 10 is assumed. If another base is being used, it will be specifically shown. For example, $1000 = 10^3$; therefore, $3 = \log_{10} 1000$, or simply $3 = \log 1000$. This is read "3 is the log of 1000."

Identities for Common Logarithms

For any two numbers x and y:

$$\log(10^x) = x$$
$$\log(xy) = \log x + \log y$$
$$\log\left(\frac{x}{y}\right) = \log x - \log y$$
$$\log(x^y) = y \log x$$

Natural Logarithms

The natural logarithm uses as its base the irrational number e. To six significant figures, its value is $e \approx 2.71828...$ Fortunately, most calculators have this number (along with other irrational numbers, such as pi) in their memories. (You should be able to find both e and π on yours.) The natural logarithm received its name because it occurs naturally when describing a quantity that grows or decays at a constant percentage (rate). The natural logarithm is abbreviated *ln* to distinguish it from the common logarithm, *log*. That is, $\log_e x \equiv \ln x$, and if $n = \ln x$, then $x = e^n$. Similarly to the common logarithm, we have the following relationships for any two numbers x and y:

$$\ln(e^x) = x$$
$$\ln(xy) = \ln x + \ln y$$
$$\ln\left(\frac{x}{y}\right) = \ln x - \ln y$$
$$\ln(x^y) = y \ln x$$

Occasionally you must convert between the two types of logarithms. For that, the following relationships can be handy:

$$\log x = 0.43429 \ln x$$
$$\ln x = 2.3026 \log x$$

Appendix II   Planetary Data

Name	Equatorial Radius (km)	Mass (compared with Earth's)[a]	Mean Density ($\times 10^3$ kg/m³)	Surface Gravity (compared with Earth's)	Semimajor Axis $\times 10^6$ km	Semimajor Axis AU[b]	Orbital Period Years	Orbital Period Days	Eccentricity	Inclination to Ecliptic
Mercury	2439	0.0553	5.43	0.378	57.9	0.3871	0.24084	87.96	0.2056	7°00′26″
Venus	6052	0.8150	5.24	0.894	108.2	0.7233	0.615 15	224.68	0.0068	3°23′40″
Earth	6378.140	1	5.515	1	149.6	1	1.000 04	365.25	0.0167	0°00′14″
Mars	3397.2	0.1074	3.93	0.379	227.9	1.5237	1.8808	686.95	0.0934	1°51′09″
Jupiter	71 398	317.89	1.36	2.54	778.3	5.2028	11.862	4337	0.0483	1°18′29″
Saturn	60 000	95.17	0.71	1.07	1427.0	9.5388	29.456	10 760	0.0560	2°29′17″
Uranus	26 145	14.56	1.30	0.8	2871.0	19.1914	84.07	30 700	0.0461	0°48′26″
Neptune	24 300	17.24	1.8	1.2	4497.1	30.0611	164.81	60 200	0.0100	1°46′27″
Pluto[c]	1500–1800	0.02	0.5–0.8	~0.03	5913.5	39.5294	248.53	90 780	0.2484	17°09′03″

[a] Planet's mass/Earth's mass, where $M_E = 6.0 \times 10^{24}$ kg.
[b] Astronomical unit: 1 AU $= 1.5 \times 10^8$ km, the average distance between the Earth and the Sun.
[c] Pluto is now classified as a "dwarf" planet.

Appendix III   Alphabetical Listing of the Chemical Elements through Atomic Number 109
(The periodic table is provided inside the back cover.)

Element	Symbol	Atomic Number (Proton Number)	Atomic Mass
Actinium	Ac	89	227.0278
Aluminum	Al	13	26.981 54
Americium	Am	95	(243)
Antimony	Sb	51	121.757
Argon	Ar	18	39.948
Arsenic	As	33	74.9216
Astatine	At	85	(210)
Barium	Ba	56	137.33
Berkelium	Bk	97	(247)
Beryllium	Be	4	9.01218
Bismuth	Bi	83	208.9804
Bohrium	Bh	107	(264)
Boron	B	5	10.81
Bromine	Br	35	79.904
Cadmium	Cd	48	112.41
Calcium	Ca	20	40.078
Californium	Cf	98	(251)
Carbon	C	6	12.011
Cerium	Ce	58	140.12
Cesium	Cs	55	132.9054
Chlorine	Cl	17	35.453
Chromium	Cr	24	51.996
Cobalt	Co	27	58.9332
Copper	Cu	29	63.546
Curium	Cm	96	(247)
Dubnium	Db	105	(262)
Dysprosium	Dy	66	162.50
Einsteinium	Es	99	(252)
Erbium	Er	68	167.26
Europium	Eu	63	151.96
Fermium	Fm	100	(257)
Fluorine	F	9	18.998 403
Francium	Fr	87	(223)
Gadolinium	Gd	64	157.25
Gallium	Ga	31	69.72
Germanium	Ge	32	72.561
Gold	Au	79	196.9665

Element	Symbol	Atomic Number (Proton Number)	Atomic Mass
Hafnium	Hf	72	178.49
Hahnium	Ha	105	(262)
Hassium	Hs	108	(265)
Helium	He	2	4.002 60
Holmium	Ho	67	164.9304
Hydrogen	H	1	1.007 94
Indium	In	49	114.82
Iodine	I	53	126.9045
Iridium	Ir	77	192.22
Iron	Fe	26	55.847
Krypton	Kr	36	83.80
Lanthanum	La	57	138.9055
Lawrencium	Lr	103	(260)
Lead	Pb	82	207.2
Lithium	Li	3	6.941
Lutetium	Lu	71	174.967
Magnesium	Mg	12	24.305
Manganese	Mn	25	54.9380
Meitnerium	Mt	109	(268)
Mendelevium	Md	101	(258)
Mercury	Hg	80	200.59
Molybdenum	Mo	42	95.94
Neodymium	Nd	60	144.24
Neon	Ne	10	20.1797
Neptunium	Np	93	237.048
Nickel	Ni	28	58.69
Niobium	Nb	41	92.9064
Nitrogen	N	7	14.0067
Nobelium	No	102	(259)
Osmium	Os	76	190.2
Oxygen	O	8	15.9994
Palladium	Pd	46	106.42
Phosphorus	P	15	30.973 76
Platinum	Pt	78	195.08
Plutonium	Pu	94	(244)
Polonium	Po	84	(209)
Potassium	K	19	39.0983

Element	Symbol	Atomic Number (Proton Number)	Atomic Mass
Praseodymium	Pr	59	140.9077
Promethium	Pm	61	(145)
Protactinium	Pa	91	231.0359
Radium	Ra	88	226.0254
Radon	Rn	86	(222)
Rhenium	Re	75	186.207
Rhodium	Rh	45	102.9055
Rubidium	Rb	37	85.4678
Ruthenium	Ru	44	101.07
Rutherfordium	Rf	104	(261)
Samarium	Sm	62	150.36
Scandium	Sc	21	44.9559
Seaborgium	Sg	106	(263)
Selenium	Se	34	78.96
Silicon	Si	14	28.0855
Silver	Ag	47	107.8682
Sodium	Na	11	22.989 77
Strontium	Sr	38	87.62
Sulfur	S	16	32.066
Tantalum	Ta	73	180.9479
Technetium	Tc	43	(98)
Tellurium	Te	52	127.60
Terbium	Tb	65	158.9254
Thallium	Tl	81	204.383
Thorium	Th	90	232.0381
Thulium	Tm	69	168.9342
Tin	Sn	50	118.710
Titanium	Ti	22	47.88
Tungsten	W	74	183.85
Uranium	U	92	238.0289
Vanadium	V	23	50.9415
Xenon	Xe	54	131.29
Ytterbium	Yb	70	173.04
Yttrium	Y	39	88.9059
Zinc	Zn	30	65.39
Zirconium	Zr	40	91.22

Appendix IV Properties of Selected Isotopes

Atomic Number (Z)	Element	Symbol	Mass Number (A)	Atomic Mass[a]	Abundance (%) or Decay Mode[b] (If Radioactive)	Half-Life (If Radioactive)
0	(Neutron)	n	1	1.008 665	β^-	10.6 min
1	Hydrogen	H	1	1.007 825	99.985	
	Deuterium	D	2	2.014 102	0.015	
	Tritium	T	3	3.016 049	β^-	12.33 y
2	Helium	He	3	3.016 029	0.00014	
			4	4.002 603	$\approx$100	
3	Lithium	Li	6	6.015 123	7.5	
			7	7.016 005	92.5	
4	Beryllium	Be	7	7.016 930	EC, γ	53.3 d
			8	8.005 305	2α	6.7×10^{-17} s
			9	9.012 183	100	
5	Boron	B	10	10.012 938	19.8	
			11	11.009 305	80.2	
			12	12.014 353	β^-	20.4 ms
6	Carbon	C	11	11.011 433	β^-, EC	20.4 ms
			12	12.000 000	98.89	
			13	13.003 355	1.11	
			14	14.003 242	β^-	5730 y
7	Nitrogen	N	13	13.005 739	β^-	9.96 min
			14	14.003 074	99.63	
			15	15.000 109	0.37	
8	Oxygen	O	15	15.003 065	β^+, EC	122 s
			16	15.994 915	99.76	
			18	17.999 159	0.204	
9	Fluorine	F	19	18.998 403	100	
10	Neon	Ne	20	19.992 439	90.51	
			22	21.991 384	9.22	
11	Sodium	Na	22	21.994 435	β^+, EC, γ	2.602 y
			23	22.989 770	100	
			24	23.990 964	β^-, γ	15.0 h
12	Magnesium	Mg	24	23.985 045	78.99	
13	Aluminum	Al	27	26.981 541	100	
14	Silicon	Si	28	27.976 928	92.23	
			31	30.975 364	β^-, γ	2.62 h
15	Phosphorus	P	31	30.973 763	100	
			32	31.973 908	β^-	14.28 d
16	Sulfur	S	32	31.972 072	95.0	
			35	34.969 033	β^-	87.4 d
17	Chlorine	Cl	35	34.968 853	75.77	
			37	36.965 903	24.23	
18	Argon	Ar	40	39.962 383	99.60	
19	Potassium	K	39	38.963 708	93.26	
			40	39.964 000	β^-, EC, γ, β^+	1.28×10^9 y
20	Calcium	Ca	40	39.962 591	96.94	
24	Chromium	Cr	52	51.940 510	83.79	
25	Manganese	Mn	55	54.938 046	100	
26	Iron	Fe	56	55.934 939	91.8	
27	Cobalt	Co	59	58.933 198	100	
			60	59.933 820	β^-, γ	5.271 y

(Continued)

Atomic Number (Z)	Element	Symbol	Mass Number (A)	Atomic Mass[a]	Abundance (%) or Decay Mode[b] (If Radioactive)	Half-Life (If Radioactive)
28	Nickel	Ni	58	57.935 347	68.3	
			60	59.930 789	26.1	
			64	63.927 968	0.91	
29	Copper	Cu	63	62.929 599	69.2	
			64	63.929 766	β^-, β^+	12.7 h
			65	64.927 792	30.8	
30	Zinc	Zn	64	63.929 145	48.6	
			66	65.926 035	27.9	
33	Arsenic	As	75	74.921 596	100	
35	Bromine	Br	79	78.918 336	50.69	
36	Krypton	Kr	84	83.911 506	57.0	
			89	88.917 563	β^-	3.2 min
38	Strontium	Sr	86	85.909 273	9.8	
			88	87.905 625	82.6	
			90	89.907 746	β^-	28.8 y
39	Yttrium	Y	89	89.905 856	100	
43	Technetium	Tc	98	97.907 210	β^-, γ	4.2×10^6 y
47	Silver	Ag	107	106.905 095	51.83	
			109	108.904 754	48.17	
48	Cadmium	Cd	114	113.903 361	28.7	
49	Indium	In	115	114.903 88	95.7; β^-	5.1×10^{14} y
50	Tin	Sn	120	119.902 199	32.4	
53	Iodine	I	127	126.904 477	100	
			131	130.906 118	β^-, γ	8.04 d
54	Xenon	Xe	132	131.904 15	26.9	
			136	135.907 22	8.9	
55	Cesium	Cs	133	132.905 43	100	
56	Barium	Ba	137	136.905 82	11.2	
			138	137.905 24	71.7	
			144	143.922 73	β^-	11.9 s
61	Promethium	Pm	145	144.912 75	EC, α, γ	17.7 y
74	Tungsten	W	184	183.950 95	30.7	
76	Osmium	Os	191	190.960 94	β^-, γ	15.4 d
			192	191.961 49	41.0	
78	Platinum	Pt	195	194.964 79	33.8	
79	Gold	Au	197	196.966 56	100	
80	Mercury	Hg	202	201.970 63	29.8	
81	Thallium	Tl	205	204.974 41	70.5	
			210	209.990 069	β^-	1.3 min
82	Lead	Pb	204	203.973 044	β^-, 1.48	1.4×10^{17} y
			206	205.974 46	24.1	
			207	206.975 89	22.1	
			208	207.976 64	52.3	
			210	209.984 18	α, β^-, γ	22.3 y
			211	210.988 74	β^-, γ	36.1 min
			212	211.991 88	β^-, γ	10.64 h
			214	213.999 80	β^-, γ	26.8 min
83	Bismuth	Bi	209	208.980 39	100	
			211	210.987 26	α, β^-, γ	2.15 min
84	Polonium	Po	210	209.982 86	α, γ	138.38 d
			214	213.995 19	α, γ	164 ms
86	Radon	Rn	222	222.017 574	α, β	3.8235 d

(Continued)

Atomic Number (Z)	Element	Symbol	Mass Number (A)	Atomic Mass[a]	Abundance (%) or Decay Mode[b] (If Radioactive)	Half-Life (If Radioactive)
87	Francium	Fr	223	223.019 734	α, β^-, γ	21.8 min
88	Radium	Ra	226	226.025 406	α, γ	1.60×10^3 y
			228	228.031 069	β^-	5.76 y
89	Actinium	Ac	227	227.027 751	α, β^-, γ	21.773 y
90	Thorium	Th	228	228.028 73	α, γ	1.9131 y
			232	232.038 054	100; α, γ	1.41×10^{10} y
92	Uranium	U	232	232.037 14	α, γ	72 y
			233	233.039 629	α, γ	1.592×10^5 y
			235	235.043 925	0.72; α, γ	7.038×10^8 y
			236	236.045 563	α, γ	2.342×10^7 y
			238	238.050 786	99.275; α, γ	4.468×10^9 y
			239	239.054 291	β^-, γ	23.5 min
93	Neptunium	Np	239	239.052 932	β^-, γ	2.35 d
94	Plutonium	Pu	239	239.052 158	α, γ	2.41×10^4 y
95	Americium	Am	243	243.061 374	α, γ	7.37×10^3 y
96	Curium	Cm	245	245.065 487	α, γ	8.5×10^3 y
97	Berkelium	Bk	247	247.070 03	α, γ	1.4×10^3 y
98	Californium	Cf	249	249.074 849	α, γ	351 y
99	Einsteinium	Es	252	254.088 02	α, γ, β^-	276 d
100	Fermium	Fm	257	253.085 18	EC, α, γ	3.0 d

[a] The masses given throughout this table are those for the neutral atom, including the Z electrons.
[b] "EC" stands for electron capture.

Appendix V   Answers to Follow-Up Exercises

Chapter 1

1.1 $L = 10$ m

1.2 Yes. $(m/s^2)(m) = m^2/s^2 = (m/s)^2$

1.3 30 days(24 h/day)(60 min/h)(60 s/min) $= 2.6 \times 10^6$ s

1.4 13.3 times

1.5 1 m$^3 = 10^6$ cm^3

1.6 2-L cost: \$1.35/(2 L) = \$0.68/L; (1/2 gal)(3.785 L/gal) = 1.893 L, cost: \$1.32/(1.893 L) = \$0.70/L. Costs are different because of rounding errors, but still there is a 2 cent difference.

1.7 (a) 7.0×10^5 kg^2 (b) 3.02×10^2 (no units)

1.8 (a) 23.70 (b) 22.09

1.9 $V = \pi r^2 h = \pi(0.490$ m$)^2(1.28$ m$) = 0.965$ m^3

1.10 11.6 m

1.11 750 cm$^3 = 7.5 \times 10^{-4}$ m$^3 \approx 10^{-3}$ m^3,
$m = \rho V \approx (10^3$ kg/m$^3)(10^{-3}$ m$^3) = 1$ kg
(by direct calculation, $m = 0.79$ kg)

1.12 $V \approx 10^{-2}$ m^3, cells/vol $\approx 10^4$ cells/mm^3 $(10^9$ mm^3/m$^3) = 10^{13}$ cells/m^3, and (cells/vol)(vol) $\approx 10^{11}$ white cells.

Chapter 2

2.1 $\Delta t = (8 \times 5.0$ s$) + (7 \times 10$ s$) = 110$ s

2.2 (a) $s_1 = 2.00$ m/s (b) $s_2 = 1.52$ m/s (c) $s_3 = 1.72$ m/s $\neq 0$ although the velocity is zero

2.3 No. If the velocity is also in the negative direction, the object will speed up.

2.4 9.0 m/s in the direction of the original motion.

2.5 Yes, 96 m. (A lot quicker, isn't it?)

2.6 No, changes x_o positions, but the separation distance is the same.

2.7 $x = v^2/2a$, $x_B = 48.6$ m, $x_C = 39.6$ m; the Blazer should not tailgate within at least 9.0 m.

2.8 1.16 s longer

2.9 The time for the bill to fall its length is 0.179 s. This time is less than the average reaction time (0.192 s) computed in the Example, so most people cannot catch the bill.

2.10 $y_u = y_d = 5.12$ m, as measured from reference $y = 0$ at the release point.

2.11 $t = \dfrac{v - v_o}{-g_M} = \dfrac{-4.6 \text{ m/s} - (-1.5 \text{ m/s})}{-1.6 \text{ m/s}^2} = 1.9$ s

Chapter 3

3.1 $v_x = -0.40$ m/s, $v_y = +0.30$ m/s, the distance is unchanged

3.2 $x = 9.00$ m, $y = 12.6$ m (same)

3.3 $\vec{v} = (0)\hat{x} + (3.7$ m/s$)\hat{y}$

3.4 $\vec{C} = (-7.7$ m$)\hat{x} + (-4.3$ m$)\hat{y}$

3.5 (a) $y_o = +25$ m and $y = 0$; the equation is the same (b) $\vec{v} = (8.25$ m/s$)\hat{x} + (-22.1$ m/s$)\hat{y}$

3.6 Both increase by six times.

3.7 (a) If not, the stone would hit the side of the block.
(b) Equation 3.11 does not apply; the initial and final heights are not the same. $R = 15$ m, which is far off the 27 m answer.

3.8 The ball thrown at 45°. It would have a greater initial speed.

3.9 At the top of the parabolic arc, the player's vertical motion is zero and is very small on either side of this maximum height. Here, the player's horizontal velocity component dominates, and he moves horizontally, with little obvious motion vertically. This gives the illusion of "hanging" in the air.

3.10 4.15 m from the net.

3.11 $v_{bs}t = (2.33$ m/s$)(225$ m$) = 524$ m

3.12 14.5° W of N

Chapter 4

4.1 6.0 m/s in the direction of the net force

4.2 (a) 11 lb (b) Your weight in pounds divided by ≈ 2.2 lb/kg

4.3 8.3 N

4.4 (a) 50° above the $+x$-axis (b) x- and y-components reversed: $\vec{v} = (9.8$ m/s$)\hat{x} + (4.5$ m/s$)\hat{y}$

4.5 Yes, mutual gravitational attractions between the briefcase and Earth.

4.6 (a) $m_2 > 1.7$ kg (b) $\theta < 17.5°$

4.7 (a) 7.35 N (b) Neglecting air resistance, 7.35 N, downward

4.8 Increase; $\tan\theta = \dfrac{T}{mg} = \dfrac{55\text{N}}{(5.0\text{ kg})(9.8\text{ m/s}^2)} = 1.1$, $\theta = 48°$

4.9 (a) $F_1 = 3.5w$; even greater than F_2 (b) $\Sigma F_y = ma$, and F_1 and F_2 would both increase

4.10 $\mu_s = 1.41 \mu_k$ (for the listings in Table 4.1)

4.11 No. F varies with angle, with the angle for minimum applied force being around 33° in this case. (Greater forces are required for 20° and 50°.) In general, the optimum angle depends on the coefficient of friction.

4.12 Friction is kinetic, and f_k is in the $+x$ direction. Acceleration is in the $-x$ direction.

4.13 Air resistance depends not only on speed, but also on size and shape. If the heavier ball were larger, it would have more exposed area to collide with air molecules, and the retarding force would increase faster. Depending on the size difference, the heavier ball might reach terminal velocity first, and the lighter ball would strike the ground first. Alternatively, they might reach terminal velocity together.

Chapter 5

5.1 -2.0 J

5.2 $d = \dfrac{W}{F\cos\theta} = \dfrac{3.80 \times 10^4 \text{ J}}{(189 \text{ N})(0.866)} = 232$ m

5.3 No, speed would decrease and it would stop moving.

5.4 $W_{x1} = 0.034$ J, $W_x = 0.64$ J (measured from x_o)

5.5 No, $W_2/W_1 = 4$, or four times as much.

5.6 $m_s = m_g/2$ as before. However, $v_s/v_g = (6.0\,\text{m/s})/(4.0\,\text{m/s}) = 3/2$. Using a ratio, $K_s/K_g = 9/8$, the safety still has more kinetic energy than the guard. (Answer could also be obtained from calculations of kinetic energies, but for a comparison, ratio is quicker.)

5.7 $W_3/W_2 = 1.4$, or 40% larger; more work, but a smaller percentage increase.

5.8 $\Delta U = mgh = (60\,\text{kg})(9.8\,\text{m/s}^2)(1000\,\text{m})\ \sin 10° = 10.2 \times 10^4$ J; yes, doubled.

5.9 $\Delta K_{\text{total}} = 0$, $\Delta U_{\text{total}} = 0$

5.10 Without friction, the liquid would oscillate back and forth between the containers.

5.11 9.9 m/s

5.12 No. $E_0 = E$ or $\frac{1}{2}mv_0^2 + mgh = \frac{1}{2}mv^2$. The mass cancels and speed is independent of mass. (Recall in free fall, all objects have with the same vertical acceleration g.)

5.13 0.025 m

5.14 (a) 59% (b) $E_{\text{loss}}/t = mg(y/t) = mgv = (60mg)$ J/s

5.15 Block would stop in rough area.

5.16 52%

5.17 (a) Same work in twice the time (b) Same work in half the time

5.18 (a) No (b) Creation of energy

Chapter 6

6.1 5.0 m/s. Yes, this is 18 km/h or 11 mi/h, a speed at which humans can run.

6.2 (1) Ship has the greatest KE, (2) Bullet has the least KE.

6.3 $(-3.0\,\text{kg·m/s})\hat{\mathbf{x}} + (4.0\,\text{kg·m/s})\hat{\mathbf{y}}$

6.4 It would increase to 60 m/s, and a greater speed means a longer drive, ideally. (There is also a directional consideration.)

6.5 $F_{\text{avg}} = \dfrac{\Delta p}{\Delta t} = \dfrac{-310\,\text{kg·m/s}}{0.600\,\text{s}} = -517\,\text{N}$

6.6 (a) No, for the m_1/m_2 system, there is an external force on block. Yes, for the m_1/m_2 Earth system. But with m_2 attached to the Earth, the mass of this part of the system would be vastly greater than that of m_2, so its change in speed would be negligible. (b) Assuming the ball is tossed in the +direction: for the tosser, $v_t = -0.50$ m/s; for the catcher, $v_c = 0.48$ m/s. For the ball: $p = 0, +25$ kg·m/s, +1.2 kg·m/s.

6.7 No. Energy went into work to break the brick, and some was lost as heat and sound.

6.8 No

6.9 No. All of the kinetic energy cannot be lost to make the dent. The momentum after the collision cannot be zero, since it was not zero initially. Thus, the balls must be moving and have kinetic energy. This can also be seen from Equation 6.11: $K_f/K_i = m_1/(m_1 + m_2)$, and K_f cannot be zero (unless m_1 is zero, which is not possible).

6.10 $x_1 = v_1 t = (-0.80\,\text{m/s})(2.5\,\text{s}) = -2.0$ m,
$x_2 = v_2 t = (1.2\,\text{m/s})(2.5\,\text{s}) = 3.0$ m
$\Delta x = x_2 - x_1 = 3.0\,\text{m} - (-2.0\,\text{m}) = 5.0$ m. The objects are 5.0 m apart.

6.11 (a) $\Delta p_1 = p_{1_f} - p_{1_0} = 32\,\text{kg·m/s} - 40\,\text{kg·m/s}$
$\qquad = -8.0\,\text{kg·m/s}$
$\quad \Delta p_2 = p_{2_f} - p_{2_0} = 13\,\text{kg·m/s} - 5.0\,\text{kg·m/s}$
$\qquad = +8.0\,\text{kg·m/s}$

(b) $\Delta p_1 = p_{1_f} - p_{1_0} = (-20\,\text{kg·m/s}) - (12\,\text{kg·m/s})$
$\qquad = -32\,\text{kg·m/s}$
$\quad \Delta p_2 = p_{2_f} - p_{2_0} = (8.0\,\text{kg·m/s}) - (-24\,\text{kg·m/s})$
$\qquad = +32\,\text{kg·m/s}$

6.12 $p_{1_0} = mv_{1_0}$, $p_{2_0} = -mv_{2_0}$ and $p_1 = mv_1 = -mv_{2_0}$,
$p_2 = mv_2 = mv_{1_0}$, so conserved. $K_i = \dfrac{m}{2}\left(v_{1_0}^2 + v_{2_0}^2\right)$ and
$K_f = \dfrac{m}{2}\left(v_1^2 + v_2^2\right) = \dfrac{m}{2}\left[\left(-v_{2_0}\right)^2 + \left(v_{1_0}\right)^2\right]$, so conserved.

6.13 All of the balls swing out, but to different degrees. With $m_1 > m_2$, the stationary ball (m_1) moves off with a greater speed after collision than the incoming, heavier ball (m_1), and the heavier ball's speed is reduced after collision, in accordance with Equation 6.16 (see Figure 6.14b). Hence, a "shot" of momentum is passed along the row of balls with equal mass (see Figure 6.14a), and the end ball swings out with the same speed as was imparted to m_2. Then, the process is repeated: m_1, *now moving more slowly*, collides again with the initial ball in the row (m_2), and another, but smaller, shot of momentum is passed down the row. The new end ball in the row receives less kinetic energy than the one that swung out just a moment previously and so doesn't swing as high. This process repeats itself instantaneously for each ball, with the observed result that all of the balls swing out to different degrees.

6.14 $X_{\text{CM}} = \dfrac{(\text{same as in example}) + (8.0\,\text{kg})x_4}{(\text{same as in example}) + (8.0\,\text{kg})}$
$\qquad = \dfrac{0 + (8.0\,\text{kg})x_4}{19\,\text{kg}} = +1.0\,\text{m}$
$\quad x_4 = \left(\tfrac{19}{8}\right)\text{m} = 2.4\,\text{m}$

6.15 $(X_{\text{CM}}, Y_{\text{CM}}) = (0.47$ m, 0.10 m); same location as in Example, two-thirds of the length of the bar from m_1. (*Note:* This shows that the location of the CM does not depend on the frame of reference.)

6.16 Yes, the CM does not move.

Chapter 7

7.1 $210°(2\pi\ \text{rad}/360°)(256\ \text{m}) = 938$ m

7.2 0.35% for 10°, 1.2% for 20°

7.3 (a) 4.7 rad/s, 0.38 m/s; 4.7 rad/s, 0.24 m/s. (b) To equalize the running distances, because the curved sections of the track have different radii and thus different lengths.

7.4 120 rpm

7.5 106 rpm

7.6 The string can't be exactly horizontal; it must be at an angle above horizontal so as to supply an upward tension component to balance the (downward) weight force.

7.7 No, it depends on mass: $F_c = \mu_s mg$.

7.8 No. Both masses have the same angular speed ω, and $a_c = r\omega^2$, so actually $a_c \propto r$. Remember, $v = 2\pi r/T$, and note that $v_2 > v_1$, with $a_c = v^2/r$.

7.9 (a) $T = 5.2$ N (b) 1.9 s, $v = r(2\pi/T)$ and $r = L\sin 20°$

7.10 (a) ω and α point away from the observer, perpendicular to the plane of the DVD, (b) ω keeps the same direction, but α points toward the observer because the DVD is slowing.

7.11 (a) -0.031 rad/s^2 (b) 14.2 s

7.12 2.8×10^{-3} m/s^2. The centripetal force of the Earth's gravitational pull and the resulting centripetal acceleration keeps the Moon in orbit.

7.13 $T^2 = \left(\dfrac{4\pi^2}{GM_E}\right)r^3 = \left(\dfrac{4\pi^2}{GM_E}\right)(R_E + h)^3 \approx \left(\dfrac{4\pi^2}{g}\right)R_E \approx 4R_I$

$T = 2\sqrt{R_E} = 2\sqrt{6.4 \times 10^6 \, \text{m}} = 5.1 \times 10^3$ s (Why are the units not consistent?)

7.14 No, they do not vary linearly. $\Delta U = 2.4 \times 10^9$ J, so only a 9.1% increase.

7.15 This is the amount of *negative* work done by an external force or agent when the masses are brought together. To separate the masses to an infinite distance apart, an equal amount of positive work (against gravity) would have to be done.

7.16 $T^2 = \left(\dfrac{4\pi^2}{GM_s}\right)r^3$ and thus

$M_s = \dfrac{4\pi^2 r^3}{GT^2} = \dfrac{4\pi^2(1.50 \times 10^{11} \, \text{m})^3}{(6.67 \times 10^{-11} \, \text{N·m}^2/\text{kg}^2)(3.16 \times 10^7 \, \text{s})^2}$

$= 2.00 \times 10^{30}$ kg

Chapter 8

8.1 $s = r\omega = 5(0.12 \, \text{m})(1.7) = 0.20$ m;
$s = v_{CM}t = (0.10 \, \text{m/s})(2.00 \, \text{s}) = 0.20$ m

8.2 The weights of the balls and the forearm produce torques in the direction opposite that of the applied torque.

8.3 More strain.

8.4 $T \propto 1/\sin\theta$, and as θ gets smaller, so does $\sin\theta$, and T increases. In the limit $\sin\theta \to 0$ and $T \to$ infinity (unrealistic).

8.5 $\sum \tau: Nx - m_1gx_1 - m_2gx_2 - m_3gx_3$
$= (200 \, \text{g})g(50 \, \text{cm}) - (25 \, \text{g})g(0 \, \text{cm})$
$\quad - (75 \, \text{g})g(20 \, \text{cm}) - (100 \, \text{g})g(50 \, \text{cm}) = 0$,
and here $N = Mg$

8.6 No. With f_{s_1}, the reaction force N would not generally be the same (f_{s_2} and N are perpendicular components of the force exerted on the ladder by the wall). In this case, we still have $N = f_{s_1}$, but $Ny - (m_1g)x_1 - (m_mg)x_m - f_{s_2}$, and $x_3 = 0$.

8.7 Five bricks

8.8 (d) No (equal masses), (e) Yes; with larger mass farther from axis of rotation, $I = 360$ kg·m^2.

8.9 $t = 0.63$ s

8.10 $\alpha = \dfrac{2mg - (2\tau_f/R)}{(2m+M)R}; \dfrac{\text{N}}{\text{kg·m}}; \dfrac{\text{N}}{\text{kg·m}} = \dfrac{\text{kg·m/s}^2}{\text{kg·m}} = \dfrac{1}{\text{s}^2}$

8.11 The yo-yo would roll back and forth, oscillating about the critical angle.

8.12 (a) 0.24 m (b) The force of *static* friction, f_s, acts at the point of contact, which is always instantaneously at rest, and so does no work. Some frictional work may be done due to rolling friction, but this is negligible for hard objects and surfaces.

8.13 $v_{CM} = 2.2$ m/s, using a ratio, 1.4 times greater; no rotational energy.

8.14 You already know the answer: 5.6 m/s. (It doesn't depend on the mass of the ball!)

8.15 $M_a = 0.75(75 \, \text{kg}) = 56$ kg; [math not shown];
$L_1 = 13$ kg·m^2/s and $L_2 = (1.3 \, \text{kg·m}^2)\omega$;
$L_2 = L_1$ or $(1.3 \, \text{kg·m}^2)\omega = 13$ kg·m^2/s and $\omega = 10$ rad/s

Chapter 9

9.1 (a) $+0.10\%$ (b) 39 kg

9.2 2.3×10^{-4} L, or 2.3×10^{-7} m^3

9.3 (1) Having enough nails, and (2) having them all of equal height and with not so sharp a point. This could be achieved by filing off the tips of the nails so as to have a "uniform" surface and also increase the effective area.

9.4 3.03×10^4 N (or 6.82×10^3 lb–about 3.4 tons!). This is roughly the force on your back right now. Our bodies don't collapse under atmospheric pressure because cells are filled with incompressible fluids (mostly water), bone, and muscle, which exert with an equal outward force and pressure. It is a pressure *differences* that gives rise to dynamic effects.

9.5 $d_o = \sqrt{\dfrac{F_o}{F_i}}d_i = \sqrt{\dfrac{1}{10}}(8.0 \, \text{cm}) = 2.5 \, \text{cm}$

9.6 Pressure in veins is lower than that in arteries.

9.7 As the balloon rises, the buoyant force decreases as a result of the temperature decrease (less helium pressure, less volume) and the less dense air ($F_b = m_f g = \rho_f gV_f$). When the net force is zero, the balloon's speed will remain constant. The cooling effect continues with altitude and the balloon will eventually start to sink when the net force on it is downward.

9.8 $r \approx 1.0$ m; $F_b = \rho gV = \rho g\left(\dfrac{4}{3}\pi r^3\right)$

$= (1.29 \, \text{kg/m}^3)\left(\dfrac{4g\pi}{3}\right)(1.0 \, \text{m})^3$

$= 53$ N, much more

9.9 (a) The object would sink, so the buoyant force is less than the object's weight. Hence, the scale would have read > 40 N. With a greater density, the object would not be as large, and less water would be displaced. (b) 41.8 N.

9.10 11%

9.11 -18%

9.12 $r = 9.00 \times 10^{-3}$ m, $v = \dfrac{\text{constant}}{A} = \dfrac{8.33 \times 10^{-5} \, \text{m}^3/\text{s}}{\pi(9.00 \times 10^{-3} \, \text{m})^2}$

$= 0.327$ m/s; an increase of 23%

9.13 69%

9.14 As the water falls, speed (v) increases and area (A) must decrease to have the volume flow rate (Av) remain constant.

Chapter 10

10.1 (a) $-40\,°C$ (b) $-40\,°F$. (You should know the answer – this is the temperature at which the Fahrenheit and Celsius temperatures are numerically equal.)

10.2 (a) $T_R = T_F + 460$ (b) $T_R = (9/5)T_C + 492$ (c) $T_R = (9/5)T$

10.3 $96\,°C$

10.4 $n = \dfrac{pV}{RT} = \dfrac{(6.58\times10^6\,\text{Pa})(0.0025\,\text{m}^3)}{[8.31\,\text{J}/(\text{mol·K})](303\,\text{K})} = 6.53\,\text{mol}$, and converting to mass: $m_{O_2} = (6.53\,\text{mol})(32.0\,\text{g/mol}) = 0.209\,\text{g}$

10.5 $70\,°C$ or $158\,°F$

10.6 The student should heat the bearing so its inside diameter will be larger.

10.7 $899\,°C$

10.8 (a) The rotational kinetic energy of the oxygen molecules is the difference between the total energies of oxygen and radon molecules, $2.44\times10^3\,\text{J}$. (b) The oxygen molecule is less massive and has the higher v_{rms}.

Chapter 11

11.1 The work per lift is $W = mgh = 392\,\text{J}$. From the Example, $200\,\text{Cal} = 8.37\times10^5\,\text{J}$. Thus the number of lifts is $(8.37\times10^5\,\text{J})/(392\,\text{J/lift}) = 2.14\times10^3$ lifts. At 1 s per lift, the required time is $2.14\times10^3\,\text{s} \approx 36\,\text{min}$.

11.2 $5.00\,\text{kg}$

11.3 (a) The ratio will be smaller because the specific heat of aluminum is greater than that of copper (b) $Q_w/Q_{\text{pot}} = 15.2$

11.4 The final temperature is expected to be higher because the water has a higher initial temperature. $T_f = 34.4\,°C$.

11.5 $-1.09\times10^5\,\text{J}$ (negative because heat is transferred out, a reduction in internal energy)

11.6 (a) $2.64\times10^{-2}\,\text{kg}$ or $26.4\,\text{g}$ of ice melts (b) The final temperature is still $0\,°C$ because the liver cannot lose enough internal energy to melt all the ice, even if the ice started at $0\,°C$. The final result is an ice/water/liver system at $0\,°C$, but with more water than in the Example.

11.7 To cut cost by 25% means reducing the heat flow rate to 75% of that in the Example. Since this rate is directly proportional to the temperature difference across the wood ceiling and nothing else changes, the temperature difference should be reduced to 75% of $12\,°C$ or $9.0\,°C$. Since the attic is at $8.0\,°C$, the air temperature near the ceiling should be $17\,°C$.

11.8 Use $Q/\Delta t = kA\Delta T/d$ with the appropriate temperature differences, conductivities, and thicknesses. The result is both are $1.1\times10^5\,\text{J/s}$.

11.9 No, the air spaces provide good insulation because air is a poor thermal conductor. The many small pockets of air between the body and the outer garment form an insulating layer that minimizes conduction and so decreases the loss of body heat. (There is little convection in the small spaces.)

11.10 (a) $1.5\times10^2\,\text{W}$ (b) The huge flaps have a large surface area so more internal energy can be radiated out

Chapter 12

12.1 The internal energy depends on the temperature. Thus the maximum temperature change is associated with the maximum internal energy change. In this case both (i) and (ii) have the same size change, with the former being an increase and the latter a decrease.

12.2 $0.20\,\text{kg}$

12.3 In both cases, heat flows into the gas. During the isothermal expansion there is no change in thermal energy thus $Q = W_{\text{gas}} = +3.14\times10^3\,\text{J}$. During the isobaric expansion, $W_{\text{gas}} = +4.53\times10^3\,\text{J}$ and $T_2 = 2T_1 = 546\,\text{K}$, $\Delta E_{\text{th}} = E_{\text{th2}} - E_{\text{th1}} = \frac{3}{2}nR(T_2 - T_1) = +6.80\times10^3\,\text{J}$, therefore, $Q = \Delta E_{\text{th}} + W_{\text{gas}} = +1.13\times10^4\,\text{J}$.

12.4 $753\,°C$

12.5 (a) $142\,\text{K}$ or $-131\,°C$
(b) For monatomic gas, $\Delta E_{\text{th}} = \left(\frac{3}{2}\right)nR\Delta T = -3.76\times10^3\,\text{J}$
This should be the same as $-W_{\text{gas}}$ since, for an adiabatic process, $Q = 0 = \Delta E_{\text{th}} + W_{\text{gas}}$; therefore, $\Delta E_{\text{th}} = -W_{\text{gas}}$.

12.6 $-1.22\times10^3\,\text{J/K}$

12.7 Overall zero entropy change requires $|\Delta S_w| = |\Delta S_s|$ or $|Q_w/T_w| = |Q_s/T_s|$. Because the system is isolated, the magnitudes of the two heat flows *must* be the same, $|Q_w| = |Q_s|$. Thus no overall entropy change requires the water and the spoon to have the same average temperature, $\overline{T}_W = \overline{T}_s$. This is not possible, unless they are initially at the *same* temperature. Thus, this can only happen if there is no net heat flow.

12.8 (a) $150\,\text{J/cycle}$ (b) $850\,\text{J/cycle}$

12.9 $0.035\,°C$

12.10 (a) The new values are $\text{COP}_{\text{ref}} = 3.3$ and $\text{COP}_{\text{hp}} = 4.3$.
(b) The COP of the air conditioner has the largest percentage increase.

12.11 It would show an increase of 7.5%.

Chapter 13

13.1 $36\,\text{m/s}^2$; $54\,\text{m/s}^2$

13.2 (1) $y = -0.088\,\text{m}$, going up; $n = 0.90$ cycle, (2) $y = 0$, going up; $n = 1.5$ cycles.

13.3 $1.6\times10^4\,\text{N/m}$

13.4 $9.75\,\text{m/s}^2$; no. This is less than the sea level values, so the park is above sea level.

13.5 The velocity is $v_y = +18.9\cos(1.26t)$ cm/s which, at $t = 0$, puts the mass at equilibrium because its speed is a maximum at $v_y = +18.9$ cm/s. The period is $1/f = 5.00\,\text{s}$, and plugging this time into the velocity, we get the maximum (upward) velocity once again:
$v_y = +18.9\cos(1.26\times5.00)\text{cm/s} = +18.9\cos(6.28)\text{cm/s}$
$= +18.9\cos(2\pi)\text{cm/s} = +18.9\,\text{cm/s}$

13.6 (a) $10\,\text{m}$ (b) $0.40\,\text{Hz}$

13.7 $440\,\text{Hz}$

Chapter 14

14.1 (a) 25.7 °C (b) 8.7 °C

14.2 (a) (343 m/s) × (3.28 ft/m) = 1125 ft/s ≈ 1100 ft/s

(b) $\Delta t = 420$ ft/1100 ft/s = 0.38 s

14.3 (a) The dB scale is logarithmic, not linear

(b) 3.16×10^{-6} W/m²

14.4 65 dB

14.5 Destructive interference: $\Delta L = 2.5\lambda = 5(\lambda/2)$, and $m = 5$. No sound would be heard if the waves from the speakers had equal amplitudes. Of course, during a concert the sound would not be single-frequency tones but would have a variety of frequencies and amplitudes. Listeners at certain locations might not hear certain parts of the audible spectrum, but this probably wouldn't be noticed.

14.6 Toward, 431 Hz; past, 369 Hz

14.7 With the source and the observer traveling in the same direction at the same speed, their relative velocity would be zero. That is, the observer would consider the source to be stationary. Since the speed of the source and observer is subsonic, the sound from the source would overtake the observer without a shift in frequency. Generally, for motions involved in a Doppler shift, the word *toward* is associated with an *increase* in frequency and *away* with a *decrease* in frequency. Here, the source and observer remain a constant distance apart. (What would be the case if the speeds were supersonic?)

14.8 768 Hz; yes

14.9 $f_1 = \dfrac{v}{4L} = \dfrac{353\,\text{m/s}}{4(0.0130\,\text{m})} = 6790\,\text{Hz}$

Appendix VI   Answers to Odd-Numbered Questions and Exercises

Chapter 1 Multiple Choice

1. (c)
3. (c)
5. (b)
7. (d)
9. (d)
11. (c)
13. (c)
15. (c)
17. (a)
19. (d)

Chapter 1 Conceptual Questions

1. Weight depends on the force of gravity, which can vary with location.
3. One difference is decimal versus duodecimal. Another is basic units: m, kg, s, vs. ft, lb, s.
5. This is by definition: 1 L = 1000 mL and 1 L = 1000 cm³.
7. No, it only tells if it is dimensionally correct. It might have dimensionless numbers such as π.
9. By putting in units and solving for those of the unknown quantity.
11. Yes, whether you multiply or divide must be consistent with unit analysis in the final answer.
13. To provide an estimate of the accuracy of a quantity.
15. For (a) and (b), the result should have the least number of significant figures. For (c) and (d), the result should have the least number of decimal places.
17. No, since an order-of-magnitude calculation is only an estimate.
19. 1 L is close to 1 qt. There are 4 qt/gal, this is about 75 gallons and is *not* reasonable for a car.

Chapter 1 Exercises

1. The decimal system (base 10) has a dime worth 10¢ and a dollar worth 10 dimes, or 100¢. By analogy, a duodecimal system would have a dime worth 12¢ and a dollar worth 12 "dimes," or $1.44 in decimal dollars. Then a penny would be 1/144 of a dollar.
3. (a) 40 MB (b) 5.722×10^{-4} L (c) 268.4 cm (d) 5.5 kilobucks
5. (a) 8.0 L (b) 8.0 kg
7. (d)
9. a is in 1/m; b is dimensionless; c is in m
11. $kg/(m \cdot s^2)$, and no this does not prove that this relationship is physically correct.
13. (a) $(kg)(m/s^2)$ (b) Yes $(kg) \times [(m^2/s^2)/m] = kg \cdot m/s^2$
15. Yes, 1000 ft = 304.9 m $\cong$ 305 m
17. 37 000 000 times
19. (a) 91.5 m by 48.8 m (b) 27.9 cm–28.6 cm
21. 320 m

23. (a) (1) 1 m/s represents the greatest speed (b) 33.6 mi/h
25. (a) 77.3 kg (b) 0.0773 m³ or 77.3 L
27. 6.5×10^3 L/day
29. 1.9×10^{10}/s
31. 15.0 min of arc
33. 5.05 cm = 5.05×10^{-1} dm = 5.05×10^{-2} m
35. (a) 4 (b) 3 (c) 5 (d) 2
37. 6.08×10^{-2} m²
39. (a) (2) cm (b) 0.946 m²
41. (a) 14.7 (b) 11.4 (c) 0.20 m² (d) 0.82
43. (a) 2.0 kg·m/s (b) 2.1 kg·m/s (c) No, the results are not the same due to rounding
45. 100 kg
47. (a) 27% (b) 75 g
49. 5.4×10^3 kg/m³
51. 0.87 m
53. 25 min
55. (a) 1950 hairs (b) 2.0×10^4 hairs
57. (a) (3) less than 190 mi/h (b) 187 mi/h
59. (a) (2) between 5° and 7° (b) 6.2°

Chapter 2 Multiple Choice

1. (a)
3. (c)
5. (c)
7. (d)
9. (d)
11. (c)
13. (d)
15. (d)
17. (a)

Chapter 2 Conceptual Questions

1. Yes, for a round trip. No, distance is always greater than or equal to the magnitude of displacement.
3. The distance traveled is greater than or equal to 300 m. The object could travel a variety of ways as long as it ends up 300 m north. If it travels straight north, then the minimum distance is 300 m.
5. Yes, this is possible. The jogger can jog in the opposite direction during part of the jog (negative instantaneous velocity) as long as the overall jog is in the forward direction (positive average velocity).
7. Not necessarily. The change in velocity is the key. If a fast-moving object does not change its velocity, its acceleration is zero. However, if a slow-moving object changes its velocity, it will have some nonzero acceleration.
9. In part (a), the object accelerates uniformly first, maintains constant velocity (zero acceleration) for a while, and then accelerates uniformly at the same rate as in the first segment. In (b), the object accelerates uniformly.

11. It is zero because the velocity is constant.

13. Yes, if the displacement is negative, meaning the object accelerates to the left.

15. No, since one value of the instantaneous velocity does not tell you if the velocity is changing. It could be zero just for an instant and not zero either before or after that instant, thus it could be changing and the object could be accelerating. You need two values of instantaneous velocity to determine if an object is accelerating.

17. Since the first stone has been accelerating downward for a longer time, it will always have a higher speed. As time goes by it will have fallen farther and thus the gap between them (Δy) will increase.

Chapter 2 Exercises

1. 300 m; zero m. Displacement is the change in position.
3. 9.8 s
5. 75 km/h; zero km/h
7. (a) 1.4 h (b) 0.27 h or 16 min
9. (a) (3) between 40 m and 60 m (b) 45 m 27° west of north
11. (a) 2.7 cm/s (b) 1.9 cm/s
13. (a) 0–2.0 s, $s = 1.0$ m/s; 2.0–3.0 s, $s = 0$ m/s; 3.0–4.5 s, $s = 1.3$ m/s; 4.5–6.5 s, $s = 2.8$ m/s; 6.5–7.5 s, $s = 0$ m/s; 7.5–9.0 s, $s = 1.0$ m/s (b) 0–2.0 s, $\bar{v} = 1.0$ m/s, 2.0–3.0 s, $\bar{v} = 0$ m/s; 3.0–4.5 s, $\bar{v} = 1.3$ m/s; 4.5–6.5 s, $\bar{v} = -2.8$ m/s; 6.5–7.5 s, $\bar{v} = 0$ m/s; 7.5–9.0 s, $\bar{v} = 1.0$ m/s (c) At 1.0 s, $v = 1.0$ m/s; at 2.5 s, $v = 0$ m/s; at 4.5 s, $v = 0$ m/s; at 6.0 s, $v = -2.8$ m/s (d) 4.5–9.0 s, $\bar{v} = -0.89$ m/s
15. (a) 0 (b) 14 m
17. 59.9 mi/h. No, she does not have to exceed the 65 mi/h speed limit.
19. 2.32 m/s²
21. 3.7 s
23. −2.1 m/s² (opposite the velocity)
25. $\bar{a}_{0-4.0} = 2.0$ m/s²; $\bar{a}_{4.0-10.0} = 0$; $a_{10.0-18.0} = -1.0$ m/s²
27. (a)

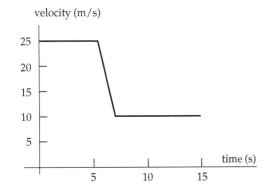

(b) 10 m/s (c) 2.4×10^2 m (d) 17 m/s
29. No, the acceleration must be at least 9.9 m/s².
31. (a) −1.8 m/s² (b) 6.3 s
33. (a) 81.4 km/h (b) 0.794 s

35. 3.09 s and 13.7 s
37. Since $a = 3.33$ m/s², the incline is not frictionless.
39. -2.2×10^5 m/s²
41. 13.3 > 13 m, so no, the car will not stop before hitting the child.
43. (a) The total area equals triangle plus rectangle: $A = v_0 t + \frac{1}{2}at^2 = x - x_0 = \Delta x$ (b) 96 m
45. (a) $v(8.0 \text{ s}) = -12$ m/s; $v(11.0 \text{ s}) = -4.0$ m/s (b) −18 m (c) 50 m
47. (a) 27 m/s (b) 38 m
49. (a) A straight line (linear), slope = $-g$ (b) A parabola
51. 4 times as high
53. 67 m
55. 39.4 m
57. (a) (1) less than 95% (b) 3.61 m
59. (a) 5.00 s (b) 36.5 m/s
61. 1.49 m above the top of the window
63. (a) 8.45 s (b) car: 132 m; motorcycle: 157 m (c) 13 m

Chapter 3 Multiple Choice

1. (a)
3. (c)
5. (c)
7. (d)
9. (c)
11. (b)

Chapter 3 Conceptual Questions

1. The answer is no to both. The component of a vector can never be greater than the magnitude of a vector since the magnitude is the hypotenuse of the triangle representing the vector and its component are the sides of the triangle.
3. (a) Its velocity either increases (it speeds up) or decreases (it slows down) in magnitude only. (b) It follows a parabolic path. (c) It moves along a circular path.
5. (a) The magnitude is $A = \sqrt{A_x^2 + A_y^2}$, which cannot be less than any component. (b) Yes, if all but one of the components are zero.
7. No, a vector quantity cannot be added to a scalar quantity.
9. 45°
11. Vertical and horizontal motion are independent, thus the *horizontal* motion of the ball is identical to that of the car.
13. Motion is relative. Uniform motion is relative, not absolute. You cannot tell if you are moving or if the other bus is moving. The same effect would occur with the buses moving in opposite directions except that you would appear to be moving away from or approaching the other bus.
15. When the player is driving to the basket for a layup, she already has an upward motion. Since the ball is with the player, the ball already has a velocity relative to the ground as the player jumps.

Chapter 3 Exercises

1. (a) 210 km/h (b) 54 km/h
3. (a) (3) between 4.0 m/s² and 7.0 m/s² (b) 5.0 m/s² and 53° above the +x-axis
5. (a) 6.0 m/s (b) 3.6 m/s
7. (a) 70 m (b) 0.57 min and 0.43 min
9. 2.5 m at 53° above the +x-axis
11. (a) (1) $v_y > v_x$ (b) $v_x = 306$ m/s; $v_y = 840$ m/s; $y = 50.4$ km
13. (3.0 m, −0.38 m)
15. (a) $4.0\hat{\mathbf{x}} + 2.0\hat{\mathbf{y}}$ (b) magnitude of 4.5 at 27° above the +x-axis
17. (a) $(-3.4\,\text{cm})\hat{\mathbf{x}} + (-2.9\,\text{cm})\hat{\mathbf{y}}$ (b) magnitude of 4.5 cm at 63° above the −x-axis (c) $(4.0\,\text{cm})\hat{\mathbf{x}} + (-6.9\,\text{cm})\hat{\mathbf{y}}$.
19. (a) $5.0\hat{\mathbf{x}} + 3.0\hat{\mathbf{y}}$ (b) $-3.0\hat{\mathbf{x}} + 7.0\hat{\mathbf{y}}$ (c) $-5.0\hat{\mathbf{x}} - 3.0\hat{\mathbf{y}}$.
21. (a)

(b) $(-5.1)\hat{\mathbf{x}} + (-3.1)\hat{\mathbf{y}}$ or a magnitude of 5.9 at an angle of 31° below the −x-axis.
23. 16 m/s at an angle of 79° above the −x-axis.
25. This happens only when the two vectors are oppositely directed.
27. From the parallelogram in the text, $\vec{\mathbf{A}} + \vec{\mathbf{C}} = \vec{\mathbf{B}}$, which gives $\vec{\mathbf{C}} = \vec{\mathbf{B}} - \vec{\mathbf{A}}$, and also, we get $\vec{\mathbf{C}} - \vec{\mathbf{B}} = -\vec{\mathbf{A}}$. From the same figure, $-\vec{\mathbf{D}} = \vec{\mathbf{B}}$ and $\vec{\mathbf{E}} = \vec{\mathbf{C}}$. Using these, and $\vec{\mathbf{C}} = \vec{\mathbf{B}} - \vec{\mathbf{A}}$, it follows that $\vec{\mathbf{E}} - \vec{\mathbf{D}} + \vec{\mathbf{C}} = 3\vec{\mathbf{B}} - 2\vec{\mathbf{A}}$.
29. (a) Same direction, (b) opposite direction, (c) at right angles.
31. (a) Same direction with magnitude of 35.0 m, (b) opposite direction with magnitude of 5.0 m. (c) When two vectors are in the same direction, the magnitude of the resultant is the sum of their magnitudes. When two vectors are in opposite directions, the magnitude of the resultant is the absolute value of the difference of their magnitudes.
33. (a) (2) north of west (b) 26.7 mi/h at 37.6° north of west
35. 242 N at 48° below the −x-axis.
37. 2.7×10^{-13} m. So no, the designer need not worry about gravity.
39. (a) 13 m horizontally from its original position and 31 m below its original position. The position vector is $\vec{\mathbf{r}} = (13\,\text{m})\hat{\mathbf{x}} + (-31\,\text{m})\hat{\mathbf{y}}$
(b) $\vec{\mathbf{v}} = (5.0\,\text{m/s})\hat{\mathbf{x}} - (25\,\text{m/s})\hat{\mathbf{y}}$.
41. (a) 0.123 s (b) 1.10 m
43. (a) 46° (b) 1.5 km directly above the soldiers
45. 1.3 m past its original position.
47. 3.8×10^2 m
49. (a) 0.43 s (b) 3.5 m/s
51. 63°
53. 40.9 m/s and 11.9° above the horizontal

55. (a) (1) A longer time, (b) Range = 13.3 m and the velocity is $(11.3\,\text{m/s})\hat{\mathbf{x}} + (-7.46\,\text{m/s})\hat{\mathbf{y}}$
57. 6.7 s
59. (a) +55 km/h (b) −35 km/h
61. 240 s or 4 minutes
63. (a) (1) north of east (b) 0.25 m/s at 37° north of east
65. She should orient her umbrella so that its handle points 14° east of due north.
67. Use the following subscripts: b = boat, w = water, and g = ground. For the boat to make the trip straight across, v_{bw} must be the hypotenuse of the right-angle triangle. Hence it must be greater in magnitude than v_{wg}. So if the reverse is true, that is, if $v_{wg} > v_{bw}$, the boat could not make the trip directly across the river.
69. (a) −125 km/h (b) 3.5 m/s² to the north (c) 3.5 m/s² to the north

Chapter 4 Multiple Choice

1. (d)
3. (c)
5. (d)
7. (c)
9. (c)
11. (c)
13. (d)
15. (c)

Chapter 4 Conceptual Questions

1. (a) No, there could be forces but the net force must be zero. (b) No, it could be at constant velocity.
3. No, same mass, same inertia.
5. (a) Balloon moves forward. The air has more inertia and tends to stay at the rear of the car. (b) Balloon moves backward. The air has inertia and tends to keep moving forward.
7. (a) Gradually increase the downward pull of the lower string. For balance, the tension in the upper string must equal the pull plus the object's weight, so it will break first. (b) Pull the lower string with a sudden jerk. By Newton's third law, the object will tend to remain at rest, so the tension in the upper string will not increase as much as the tension in the lower, hence breaking the lower one.
9. The weight is zero in space, because there is no gravity. The mass is still 70 kg.
11. No, both mass and the pull of gravity decrease. This is not a violation of Newton's laws. The thrust needs to overcome (be greater than) the gravitational force to accelerate the rocket. However, as fuel is burned, the mass of the rocket decreases and farther up in space, there is a decreased pull of gravity. So the constant thrust has less gravitational force to overcome and less rocket mass to accelerate. This in turn increases the speed and acceleration of the rocket.

13. The forces act on different objects, one on horse, one on car, and so they cannot cancel.

15.

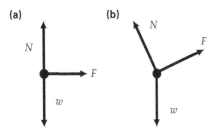

(a)   (b)

The force F is the combination of the pushing by the seat back and the friction force by the seat surface.

17. When the arms are quickly raised, an upward force is required to do this. This force, ultimately, is provided by the normal force of the scale on the person. From Newton's third law, the person will push down on the scale with a force larger than his weight, so the scale reading increases. Conversely, a downward force is required to lower the arms, and the end result is a decrease in the scale reading.

19. Kinetic friction (sliding) is less than static friction.

Chapter 4 Exercises

1. Aluminum mass is 1.4 times the mass of the water.
3. 0.64 m/s^2
5. (a) 20 N on the 2.0 kg and 59 N on the 6 kg (b) Same for both $= g$
7. $(-7.6\text{N})\hat{x} + (0.64\text{N})\hat{y}$
9. (a) (3) The tension is the same in both situations (b) 25 lb
11. (a) (2) the second quadrant (b) 184 N at 12.5° above the $-x$-axis
13. (a) 1.8 m/s^2 (b) 4.5 N
15. (a) (3) 6.0 kg (b) 9.8 N
17. (a) (1) On the Earth (b) 5.4 kg
19. (a) (4) One-fourth as great (b) 4.0 m/s^2
21. 1.23 m/s^2
23. 2.40 m/s^2
25. (a) 30 N (b) -4.6 m/s^2 (The minus sign means that the object is slowing while sliding upward.)
27. (a) backward (b) 0.51 m/s^2
29. (a) (2) Two forces on the book: a gravitational force (weight, w) and a normal force N from the surface. (b) The reaction to w is an upward force on the Earth by the book, and the reaction force to N is a downward force on the horizontal surface by the book.
31. (a) (3) The force the blocks exerted forward on him (b) 3.08 m/s^2
33. (a) 735 N (b) 735 N (c) 885 N (d) 585 N
35. (a) (4) All of the preceding (b) Downward at 1.8 m/s^2
37. (a) 0.96 m/s^2 (b) 2.6×10^2 N
39. 123 N up the incline

41. 64 m
43. (a) (3) both the tree separation and the sag (b) 6.1×10^2 N
45. 2.63 m/s^2
47. 1.5×10^3 N
49. (a) 1.8 m/s^2 (b) 6.4 N
51. (a) 1.2 m/s^2 (m_1 up and m_2 down) (b) 21 N
53. (a) 1.1 m/s^2 (b) 0.
55. 2.7×10^2 N
57. (a) 38 m (b) 53 m
59. 33 m
61. 0.77 m
63. (a) 30° (b) 0.58
65. (a) No, the incline is not frictionless because the person uses a force larger than the minimum (frictionless) of 117 N. (b) 33 N

Chapter 5 Multiple Choice

1. (d)
3. (b)
5. (d)
7. (d)
9. (c)
11. (d)
13. (d)
15. (a)
17. (b)

Chapter 5 Conceptual Questions

1. (a) No – no displacement, no work. (b) Yes, positive work is done by the weightlifter. (c) No, as in (a). (d) Yes, positive work is done by gravity.
3. Positive on the way down, negative on the way up.
5. (d) $\dfrac{W_2}{W_1} = \dfrac{x_2^2 - x_1^2}{x_1^2 - x_o^2} = \dfrac{4.0^2 - 2.0^2}{2.0^2 - 0^2} = 3$ so three times as much.
7. $\dfrac{W'}{W} = \dfrac{v'^2}{v^2} = \dfrac{(v/2)^2}{v^2} = \dfrac{1}{4}$, so $W' = W/4$.
9. They have the same kinetic energy:
   $K_A = (1/2)(4m_B)v_A^2 = 2m_B v_A^2$ and
   $K_B = (1/2)m_B(2v_A)^2 = 2m_B v_A^2$.
11. Both are correct. Potential energy is defined with respect to a reference point. Depending on where the reference point is located, the potential energy of the notebook on a table can be positive (with respect to floor), zero (with respect to table), or even negative (with respect to ceiling).
13. Yes. When the thrown-up ball is at its maximum height, its velocity is zero so it has the same potential energy as the dropped ball. Since the total energy of each ball is conserved, both balls will have the same mechanical energy at half the height of the window. As a matter of fact, both balls will have the same kinetic and potential energy at the half-height position.

15. (a) No, efficiency is only a measure of how much work is done for each unit of energy input. (b) Not necessarily. It also depends on time because work is power multiplied by time.

Chapter 5 Exercises

1. 5.0 N
3. 1.8×10^3 J
5. 3.7 J
7. 2.3×10^3 J
9. -2.2 J
11. (a) (2) Negative work (b) 1.47×10^5 J
13. 80 N/m
15. 1.25×10^5 N/m
17. (a) (1) $\sqrt{2}$ (b) 900 J
19. (a) (1) more than (b) 0.25 J and 0.75 J
21. (a) (3) 75% (b) 7.5 J
23. (a) 45 J (b) 21 m/s
25. 200 m
27. 2.9 J
29. 4.7 J
31. (a) 0.16 J (b) -0.32 J
33. (a) $K_o = 15.0$ J; $U_o = 0$ J; $E_o = K_o + U_o = 15.0$ J (b) $K = 7.65$ J; $U = 7.35$ J; $E = 15.0$ J (c) $K = 0$ J; $U = 15.0$ J; $E = 15.0$ J
35. (a) (3) At the bottom of the swing (b) 5.42 m/s
37. (a) 4.4 m/s (b) 3.7 m/s (c) 0.59 J, converted to heat and sound
39. (a) 11 m/s (b) No, it will not reach the point (c) 7.7 m/s
41. (a) 2.7 m/s (b) 0.38 m (c) 29°
43. (a) 160 N/m (b) 0.33 kg (c) 0.13 J
45. 1.9 m/s
47. 12 m/s
49. 5.7×10^{-5} W
51. 2700 N
53. 1.37×10^2 kg
55. 48.7%

Chapter 6 Multiple Choice

1. (b)
3. (d)
5. (d)
7. (c)
9. (d)
11. (c)
13. (a)
15. (d)

Chapter 6 Conceptual Questions

1. No, mass is also a factor.
3. No, mass is also a factor, because $K = p^2/(2m)$.
5. By stopping abruptly, the contact time is short. From the impulse momentum theorem, a shorter contact time will result in a greater force if all other factors (m, v_o, v) remain the same.
7. With a stiff-legged landing, the floor's force on the jumper acts for a short time, thus requiring a large force to change the jumper's momentum. With bent legs, the force acts for a longer time, and thus a smaller force is needed.
9. No, it is impossible. Before the hit there is initial momentum of the two-object system due to the one moving. According to momentum conservation, the system should also have momentum after the hit. Therefore it is not possible for both to be at rest (zero total system momentum).
11. This is due to the fact that momentum is a vector and kinetic energy is a scalar. For example, two objects of equal mass traveling with the same speed in opposite directions have positive total kinetic energy but zero total momentum. After they collide inelastically, both stop, resulting in zero total kinetic energy and zero total momentum. Therefore, kinetic energy is lost and momentum is conserved.
13. The modern auto bumper crumples upon impact, thus increasing the time of impact. This results in a reduction of the force on the car in a given situation, compared to old rigid bumpers. Force reduction means that all objects in the car require less force to accelerate with the car and thus the passengers and car contents are less likely to be hurt, killed, or damaged.
15. The flamingo's center of mass must be above the foot on the ground to be in equilibrium.

Chapter 6 Exercises

1. (a) 1.5×10^3 kg·m/s (b) zero momentum
3. (a) 85 kg·m/s (b) 3.0×10^4 kg·m/s
5. 5.88 kg·m/s in the opposite direction of the initial velocity.
7. (a) 0.45 (b) 99%
9. $(-3.0\,\text{kg·m/s})\hat{\mathbf{y}}$
11. -6.5×10^3 N
13. (a) (4) to the east of northeast (b) $(823\,\text{kg·m/s})\hat{\mathbf{x}} + (283\,\text{kg·m/s})\hat{\mathbf{y}}$; 870 kg·m/s and 19°
15. 15 m/s
17. 13 m/s
19. (a) (2) the driver putting on the brakes (b) 28.7 m/s
21. $F_{avg1} = 1.1 \times 10^3$ N; $F_{avg2} = 4.7 \times 10^2$ N
23. 15 N upward
25. 370 N in the direction of the 60 m/s velocity.

27. $F_{avg} = -3.0 \times 10^2$ N
29. 0.083 m/s in the opposite direction.
31. 18.1 minutes
33. (a) 1.4 m/s (b) 2.5 cm
35. 4.7 km/h at 62° south of east
37. (a) 11 m/s to the right (b) 22 m/s to the right (c) at rest
39. (a) -66.7 m/s east (b) 3.33×10^4 J
41. (a) (1) right (b) 0.70 m/s to the right
43. (a) No, he does not get back in time (b) 4.5 m/s
45. $v_p = -1.8 \times 10^6$ m/s; $v_a = 1.2 \times 10^6$ m/s.
47. 92%
49. $v_p = 38.2$ m/s; $v_c = 40.2$ m/s
51. 0.94 m
53. (a) 1.0 m/s in the x-direction; 3.3 m/s in the y-direction (b) 73°
55. 0.34 m
57. 50%
59. $(-0.44$ m, 0)
61. CM of the remaining portion is still at the center of the sheet.
63. (a) 65 kg travels 3.3 m and 45 kg travels 4.7 m. (b) Same distances as in (a).

Chapter 7 Multiple Choice

1.   (c)
3.   (a)
5.   (d)
7.   (d)
9.   (b)
11. (a)
13. (d)
15. (c)
17. (d)
19. (c)

Chapter 7 Conceptual Questions

1.   This is by definition. 2π rad $= 360°$ so 1 rad $\approx 57.3°$.
3.   Yes, they all sweep through the same angle. No, they do not have the same tangential speed, because the distances to the center of the wheel are different.
5.   Your tangential speed decreases, using $v = r\omega$, with a decreasing r and constant ω, v decreases.
7.   There is insufficient centripetal force (provided by friction and adhesive forces) on the water drops, so the water drops fly out along a tangent, and the clothes get dry.
9.   (a) At the equator. (b) At either geographic pole. (c) The accelerations are the same at both.
11. Centripetal force is required for a car to maintain its circular path. When a car is on a banked turn, the horizontal component of the normal force on the car is pointing toward the center of the circular path. This component will enable the car to negotiate the turn even when there is no friction.

13. No, this is not possible. As long as the car is traveling on a circular track, it always has centripetal acceleration.
15. It would not appreciably affect its orbit, which is determined primarily by the mass of the Earth. This assumes the Moon's mass is negligible compared to that of the Earth's. In actuality, there would be an effect because the center of mass of the Earth–Moon system would be closer to the Moon.
17. Yes, if you also know the radius of the Earth. The acceleration due to gravity near the surface of the Earth is $a_g = GM_E/R_E^2$. Thus by measuring a_g, you can find M_E.

Chapter 7 Exercises

1.   (4.5 m, 2.8 m)
3.   (a) 30° (b) 75° (c) 135° (d) 180°
5. 1.9×10^2 m
7.   hour hand: 0.065 m, minute hand: 0.94 m, second hand: 66 m
9.   60 rad
11. (a) No, it is not a multiple (b) six such pieces and one 0.28 rad piece
13. 0.087 rad/s
15. (a) 4.80×10^{-3} s (b) 6.32×10^{-3} s
17. 0.42 s
19. 0.634 rad/s; 1.11 m/s
21. 1.3 m/s
23. 64 m
25. 11.3°
27. (a) The gravitational force is supplying the centripetal force (b) 3.1 m/s
29. 29.5 N < 100 N so the string will work.
31. (a) $v = \sqrt{rg}$ (b) (5/2)r
33. 1.1×10^{-3} rad/s^2
35. (a) (3) both angular and centripetal accelerations (b) 53 s (c) $a_c = 8.5$ m/s^2, $a_t = 1.4$ m/s^2, total acceleration is: $\hat{\mathbf{a}} = (8.5\,\mathrm{m/s})\hat{\mathbf{r}} + (1.4\,\mathrm{m/s}^2)\hat{\mathbf{t}}$
37. (a) 1.82 rad/s^2 (b) 28.7 rev
39. (a) $a_t = 2.5$ m/s^2; $a_c = 9.7$ m/s^2 (b) at the lowest point; $a_t = 0$
41. $g_M = 1.6$ m/s^2
43. 8.0×10^{-10} N, toward the opposite corner.
45. (a) 100 kg (b) 894 N
47. (a) Applying Newton's second law to the Moon, with m the mass of the Moon, M the mass of Earth, T the orbital period of the Moon, and r the radius of the Moon's orbit: $(GmM/r^2) = mr\omega^2 = mr((2\pi/T)^2)$. Solving for M gives $M = 4\pi^2 r^3/(GT^2)$. (b) 6.0×10^{24} kg
49. 1.5 m/s^2
51. 3.13 km/s
53. 4.4×10^{11} m
55. 1.53×10^9 m

Chapter 8 Multiple Choice

1. (a)
3. (b)
5. (b)
7. (d)
9. (d)
11. (a)
13. (d)
15. (c)
17. (c)

Chapter 8 Conceptual Questions

1. Yes, rolling motion is a good example.
3. According to $v_t = r\omega$. the reading of the speedometer is $v/2$. The point on the top of the tire has twice the radius as the center of the tire (the point at which the tire makes contact with the ground is the axis of rotation), and the speedometer reads the speed of the center of the tire (v_{CM}).
5. Yes, if the net force is zero and it moves at constant velocity.
7. In stable equilibrium, a small displacement results in a restoring force or torque. A mass on the end of a spring is a good example of this. In unstable equilibrium, a small displacement results in a net force or torque tending to move it further away from equilibrium. A good example of this is a wooden beam initially balanced vertically on end that will tend to fall over if displaced from the vertical enough.
9. The moment of inertia depends on how mass is distributed about an axis. Physically, this means that, under a constant torque, the angular acceleration depends on the location of the axis of rotation.
11. The hardboiled egg is essentially a solid object. When it is brought to rest by an external torque, it will have no angular momentum, so when it is released, its angular momentum will remain zero. For the raw egg, the shell loses all its angular momentum, but the liquid center (which is only loosely connected to the solid shell) does not experience enough torque to stop it completely. Thus the liquid part of the raw egg retains some angular momentum. When the egg is then released, the loose connection to the shell transfers some of this angular momentum to the shell, so the egg as a whole will start to rotate, although slower than initially because it has lost some of its original angular momentum.
13. To increase the moment of inertia. If the walker starts to rotate, the angular acceleration will be less, giving more time to recover.
15. Yes. The rotational kinetic energy depends on the moment of inertia, which depends on both the mass and the mass distribution. Translational kinetic energy depends only on the mass.
17. According to the work-energy theorem, rotational work is required to produce a change in rotational kinetic energy. Rotational work is done by a torque acting through an angular displacement.
19. The polar ice caps (with almost zero moment of inertia) will go to the ocean and increase the moment of inertia of the Earth. This results in a slower rotational speed or a longer day.
21. (a) The velocity of the center of mass of the two-skater system is zero before they lock arms because they have equal masses and equal but opposite velocities. Because their collision generates only forces that are internal to their system, the velocity of their center of mass does not change, so it is zero after they lock arms. (b) The initial translational kinetic energy is partially transformed into rotational kinetic energy as the skaters both spin, but some may be lost as they link arms.

Chapter 8 Exercises

1. At the nine-o'clock position, the velocity is straight upward. So it is a "free fall" with an initial upward velocity. It will rise to a maximum height and then start downward.
3. 0.10 m
5. 1.7 rad/s
7. 36 cm
9. 3.3×10^2 N
11. (a) Yes, the seesaw can be balanced if the lever arms are appropriate for the weights of the children (b) 2.3 m
13. (a) No, it is not possible to have the lines perfectly horizontal, because the weight has to be supported by an upward component of the tensions in the lines. If the lines were horizontal, then they cannot support the weight. (b) 1.8×10^3 N > 400 1b.
15. (a) Clockwise (b) 4.66 m·N
17. Bees: 0.20 kg; bees–1st bird combo: 0.50 kg; bees and 1st bird–2nd bird combo: 0.40 kg.
19. (a) 9 books, (b) 22.5 cm.
21. 10.2 J
23. 49 N
25. 16.7 N
27. 0.64 m·N
29. 1.5×10^5 m·N
31. 1.2 m/s²
33. (a) 0.89 m/s² (b) 0.84 m/s²
35. The pennies at the 10-, 20-, 30-, 40-, 50-, and 60-cm marks will be slowed (acceleration-wise) by the meterstick. The pennies at the 70-, 80-, 90-, and 100-cm marks will separate from the meterstick and thus have an acceleration equal to g.
37. 4.5×10^2 J
39. (a) (3) less than $\sqrt{2gh}$ (b) 5.8 m/s

41.   0.47 m·N
43.   2.3 m/s
45.   3.4 m/s
47.   (a) 29% (b) 40% (c) 50%
49.   (a) $\sqrt{gR}$ (b) 2.7R (c) No normal force, which would feel like weightlessness
51.   1.4 rad/s
53.   (a) 2.67×10^{40} kg·m²/s (b) 7.06×10^{33} kg·m²/s
55.   (a) (2) less than (b) 200 rpm
57.   (a) 4.3 rad/s (b) $K = 1.1K_o$ (c) work done by the skater pulling arms inward
59. $d = b(v_o/v)$

Chapter 9 Multiple Choice

1.   (c)
3.   (d)
5.   (d)
7.   (a)
9.   (a)
11.   (c)
13.   (b)
15.   (b)
17.   (c)

Chapter 9 Conceptual Questions

1.   Steel wire has a greater Young's modulus. Young's modulus is a measure of the ratio of stress over strain. For a given stress, a greater Young's modulus will have a smaller strain. Steel will have smaller strain here.

3.   Through capillary action, the wooden peg absorbs water, and it swells and splits the rock.

5.   Bicycle tires have a much smaller contact area with the ground, so they need a higher pressure to balance the weight of the bicycle and the rider.

7.   The absolute pressure is 1 atmosphere but the gauge pressure is zero. The pressure in the flat tire is the same as that of the outside.

9.   Pressure depends only on depth. To get the same pressure, spherical tanks do not need as much water as cylindrical tanks. Also spherical shapes distribute pressure more evenly to reduce the risk of tank damage by water pressure.

11.   (a) A life jacket must have lower density than water, such that the average density of a person and a jacket is less than the density of water. (b) Salt water has a higher density, so it can exert a greater buoyant force.

13.   The pressure increases because the pressure in a liquid increases with depth. The buoyant force does *not* increase because the weight of water displaced does not change (assuming incompressible water) once the rock is submerged.

15.   They will experience the same buoyant force because buoyant force depends only on the volume of the fluids displaced and is independent of the mass of the object.

17.   The airflow between you and the truck is now moving through a narrower opening and therefore flowing faster (compared to the air on the outside surfaces). Thus there is a reduced air pressure in the space between you and the truck, compared to that on outside surfaces. This results in pressure differential forcing both you and the truck toward one another. Since the truck is usually more massive than your car, it is usually you who will feel the effect.

19.   The 5 and the 20 are measures of viscosity at low and operating temperatures, respectively. The W stands for WINTER

Chapter 9 Exercises

1.   0.020
3.   1.9 cm
5.   47 N
7.   (a) (1) a cold day (b) 1.9×10^5 N
9.   (a) bends toward the brass (b) brass: -2.8×10^{-3}; copper: -2.3×10^{-3}
11. 4.2×10^{-7} m
13.   (a) ethyl alcohol (b) water, $\Delta p_w/\Delta p_{ca} = 2.2$
15. -3.2×10^{-6} m
17.   (a) 9.8×10^4 Pa (b) 2.0×10^5 Pa
19. 5.88×10^4 Pa
21.   (a) 1.99×10^5 N/m² (b) 9.8×10^4 Pa
23. 1.07×10^5 Pa
25. 7.8×10^3 Pa
27.   0.51 N toward cabin
29.   37.5 m/s
31.   (a) 1.1×10^8 Pa (b) 1.9×10^6 N
33.   (a) 549 N (b) 1.37×10^6 Pa
35.   (a) (3) stay at any height (b) it will sink
37. 2.0×10^3 kg/m³
39. 1.8×10^3 N
41.   (a) 0.09 m (b) 8.1 kg
43. 1.00×10^3 kg/m³ (probably water)
45.   17.7 m
47.   0.98 m/s
49. -1.69×10^4 N/m² $= -0.167$ atm
51.   0.149 m/s
53.   (a) 0.43 m/s (b) 8.6×10^{-5} m³/s
55. 3.5×10^2 Pa

Chapter 10 Multiple Choice

1.   (d)
3.   (b)
5.   (a)
7.   (c)
9.   (a)
11.   (a)
13.   (b)
15.   (b)

Chapter 10 Conceptual Questions

1. Not necessarily, because thermal energy does not depend solely on temperature. It also depends on mass.
3. Monatomic molecules behave like point-masses so they can have only translational kinetic energy. In addition to translational kinetic energy, diatomic molecules can also have rotational and vibrational kinetic energy because the molecule can rotate and the atoms can vibrate.
5. When the pressure of the gas is held constant, if the temperature increases or decreases, so does the volume. Therefore, a gas's temperature can be measured by monitoring its volume.
7. The balloons collapsed. Due to the decrease in temperature, the volume decreases.
9. Both have the same number of molecules, but oxygen has two atoms per molecule, so the oxygen has more atoms.
11. (a) The ice moves upward. (b) The ice moves downward. (c) The copper should be on top.
13. When the ball alone is heated, it expands and cannot go through the ring. When the ring is heated, it expands and the hole gets larger so the ball can go through again.
15. Most metals have a larger coefficient of thermal expansion than that of glass. The lid expands more than glass, so it becomes easier to loosen the lid.
17. Both have the same average kinetic energy per molecule. Since each atom of helium is less massive than that of neon, helium must have a higher average molecular speed.
19. No. At room temperature a diatomic molecule has five degrees of freedom as opposed to only three for a monatomic molecule. Since each degree of freedom has the same average kinetic energy, the diatomic molecule will possess more average total kinetic energy.
21. $\Delta E_{th} = E_{th,dia} - E_{th,mono} = (5/2)nRT - (3/2)nRT = nRT$

Chapter 10 Exercises

1. 104 °F
3. (a) 248 °F (b) 53.6 °F (c) 23.0 °F
5. −70.0 °C
7. 56.7 °C; −62.0 °C
9. (a) −51.5 °C (b) −47.3 °C
11. (a) Set (3) $T_F = T_C$ (b) −40° on both scales
13. (a) 273 K (b) 373 K (c) 293 K (d) 238 K
15. (a) $(5/9)T_F + 255.37$ (b) 300 K is lower
17. (a) 2.22 mol (b) 5.57 mol (c) 4.93 mol (d) 1.75 mol
19. 1/4
21. 0.0618 m³
23. 574.58 K
25. 2.5 atm
27. 33.4 lb/in²
29. (a) (1) increase (b) $p_2 = 4p_1$
31. (a) (1) increases (b) 10.6%

33. (a) If $T \gg 273$ K, ignore the 273 converting from T_C to T: $T = T_C + 273 \approx T_C$. (b) If $T_F \gg 32$ °F, ignore the 32 converting from T_C to T_F so overall $T \approx T_C = (5/9)(T_F - 32) \approx (5/9)T_F \approx (1/2)T_F$. (c) The Celsius approximation is high by about 2.73×10^{-3} %. Using the Fahrenheit approximation the result is low by about 11%.
35. (a) (1) High (b) 0.060%
37. 0.0027 cm
39. 9.5 °C
41. 5.52×10^{-4} C°⁻¹
43. (a) There will be a gas spill, because the coefficient of volume expansion is greater for gasoline than steel. (b) 0.48 gal.
45. (a) 116 °C. (b) No, because brass has a higher α, the hole will be even larger at a higher temperature.
47. 65 °C
49. (a) (2) Increases by less than a factor of 2. (b) The ratio of the thermal energies is 1.07
51. (a) 6.17×10^{-21} J (b) 1360 m/s
53. 273 °C
55. 7.5×10^{-21} J
57. 1.12 times as fast.
59. (a) (1) $^{235}UF_6$ (b) 1.0043
61. 8.3×10^3 J
63. 224 °C

Chapter 11 Multiple Choice

1. (d)
3. (c)
5. (b)
7. (a)
9. (d)
11. (b)
13. (a)

Chapter 11 Conceptual Questions

1. The Btu is based on the pound which is a unit of weight (not mass) and an object's weight on the Moon is only 1/6 of its weight on the Earth.
3. Heat is not a measure of internal energy of an object but rather the energy that is transferred between objects. So it is the wrong use of the word heat.
5. Heat can move "uphill" temperature-wise, but it requires work to do it, such as by an air conditioner. If this is not the case, then thermal energy always goes from hot to cold.
7. Temperature change is also determined by mass and specific heat. Thus the final temperature of the two objects *can* be different, even if the heat transfers and the initial temperatures are the same.
9. Only that together, the cold water and cup gain 400 J of heat. Since both the cup and water will experience the same temperature change, most of the 400 J will end up in the water because of its high specific heat compared to aluminum.

11. The (heat) energy added is called latent heat. It is the energy required to change the phase of a substance. The energy goes into breaking bonds between molecules rather than into increasing temperature (kinetic energy of the molecules).

13. The water molecules in your breath condense, which looks like steam or fog.

15. The bridge is exposed to the cold air above and below, while the road is exposed only above. Also, the road can receive heat energy from the ground, whereas the lower surface of the bridge is in contact with cold air and transfer of thermal energy (conduction) from gas to solid is less than from solid to solid. Thus the water on the bridge could be frozen, while that on the roadway might still be liquid.

17. The low-pressure gas trapped between the glass walls of the bottle is a poor conductor of heat, so conduction is very low. This gas is a partial vacuum, so it cannot transfer much heat by convection. The interior has a silver film coating that minimizes radiation losses.

Chapter 11 Exercises

1. 5.86×10^3 W
3. 720 cal
5. (a) 60 000 lifts (b) 83 h
7. (a) (1) More heat (b) 2.1×10^4 J
9. 35.6 °C
11. 4.8×10^6 J
13. (a) (1) More heat than the iron (b) 1.8×10^4 J
15. 1.4×10^2 J/(kg·C°)
17. 88.6 °C
19. 1.7 h
21. (a) (1) Higher (b) 3.1×10^2 J/(kg·C°)
23. 34 °C
25. 8.3×10^5 J
27. (a) (1) More heat (b) vaporization by 1.93×10^6 J more
29. 4.9×10^4 J
31. 8.0×10^4 J
33. (a) (2) Only latent heat (b) 4.1×10^3 J
35. 0.437 kg
37. 195 °C
39. (a) (2) water has a high heat of vaporization (b) 1.6×10^7 J
41. (a) 110 °C and 140 °C (b) 1.0×10^2 J/(kg·C°);
    liquid: 2.0×10^2 J/(kg·C°), gas: 1.0×10^2 J/(kg·C°)
    (c) fusion: $L_f = 4.0 \times 10^3$ J/(kg); vaporization:
    $L_v = 60. \times 10^3$ J/(kg)
43. 4.54×10^6 J
45. (a) (3) Slower (b) 0.55
47. 17 J
49. (a) (1) Longer due to higher thermal conductivity (b) 1.63
51. 90 J/(s)
53. 411 °C
55. 171 °C
57. 7.8×10^5 J

Chapter 12 Multiple Choice

1. (d)
3. (d)
5. (a)
7. (b)
9. (d)
11. (b)
13. (d)
15. (b)
17. (d)
19. (a)

Chapter 12 Conceptual Questions

1. An isobaric process is at constant pressure and is represented by a horizontal line on a *p-V* diagram. Since temperature drops, the line should indicate movement from right to left (decrease in volume) by having an arrow in that direction.

3. This is an adiabatic compression. When the plunger is pushed in, the work done goes into increasing the thermal energy of the air. The increase in thermal energy increases its temperature.

5. Since $Q = \Delta E_{th} + W_{gas}$, $\Delta E_{th} = -W_{gas}$ when $Q = 0$.

7. Work: 1, 2, 3. Work is equal to the area under the curve in *p-V* diagram. The area under 1 is the greatest and the area under 3 is the smallest. Final temperature: 1, 2, 3. According to the ideal gas law, the temperature of a gas is proportional to the product of pressure and volume, $pV = nRT$. Since the final volume is the same for all three, the higher the pressure, the higher the final temperature.

9. Upon compression, the molecules on average rebound off the piston with a higher speed. Thus the average kinetic energy of a gas molecule in the sample increases and since temperature is a measure of this energy, it increases also.

11. The conversion of heat to mechanical energy does not violate second law of thermodynamics. The law only requires that the conversion can never be 100% efficient.

13. From the second law, the entropy increases. The cold water gains more entropy than is lost by the hot water.

15. The overall entropy of the universe increases. The Sun's entropy will decrease as it forms but the heat it gives off to the Universe provides a gain in entropy that more than offsets the Sun's decrease.

17. Both pressure and thermal energy return to the original values they had at the start of the cycle, so no overall change.

19. Not necessarily. The amount of work also depends on the efficiency of the engine. Engine A may be less efficient than B and hence waste most of the heat it absorbs, while B could be highly efficient and use most of the heat it absorbs.

21. In practical terms, the COP of a heat pump is the ratio of the heat of interest (that moved into the already hotter reservoir, to heat a room, say) compared to the work required to accomplish this, or
$$COP_{hp} = (Q_h/W_{in}) = (Q_h/(Q_h - Q_c)).$$
However, the work is always less than the heat moved into the hot reservoir (since $Q_h = W_{in} + Q_c > W_{in}$), thus this ratio is always > 1. Fundamentally engine efficiency as defined is always less than one, whereas the COP is defined to be greater than one.

23. Between 100 °C and 300 °C, the efficiency is $\varepsilon_C = 35\%$, and between 50 °C and 250 °C it is $\varepsilon_C = 38\%$. Choose 50 °C and 250 °C for higher efficiency.

Chapter 12 Exercises

1. 5.5×10^5 J
3. (a) (2) The same as (b). There is no change in the thermal energy, so $Q = +400$ J.
5. (a) (3) Decreases (b) -500 J
7. (a) 4.9×10^3 J (b) $\Delta E_{th} = +3.5 \times 10^3$ J
9. (a) (2) Zero, (b) $W_{env} = -p_1V_1$ (on the gas) (c) $Q = -p_1V_1$ (out of the gas)
11. 3.6×10^4 J
13. (a) (1) Positive work (b) 3400 J
15. (a) 17 mol (b) $+5.0 \times 10^4$ J (heat was added to the gas)
17. (a) (3) Negative (b) -1.8×10^5 J (gas is compressed) (c) -480 K (d) -6.0×10^5 J (e) -7.8×10^5 J
19. $\Delta S = -2.1 \times 10^2$ J/K
21. (a) (3) Negative (b) $\Delta S = -6.1 \times 10^3$ J/K
23. 0.20 J/K
25. (a) (1) Increase (b) $+10$ J/K
27. (a) (3) Negative (b) 0.98 J/K
29. 20%
31. (a) 5.6×10^2 J (b) 1.4×10^3 J
33. 1.49×10^5 J
35. (a) 6.6×10^8 J (b) 27%
37. (a) (1) Increase (b) $+0.083$. Thus $R_{25} > R_{20}$, so Q_h/Q_c should increase.
39. (a) 6.1×10^5 J (b) 1.9×10^6 J
41. (a) 2.1×10^3 J (b) 2.0
43. (a) 3.6×10^3 MW (b) 2.7×10^3 MW (c) 4.1×10^5 gal/min
45. 21.4%
47. 158 °C
49. (a) 2000 J (b) 249 °C
51. (a) (3) Higher than 350 °C (b) 402 °C
53. No, because the claimed efficiency is 60% but the Carnot efficiency is only 59.1%.
55. (a) (2) Zero (b) 3750 J

57. (a) 64%, (b) ε_C is the upper limit. In reality, more energy is lost than in the ideal situation.

Chapter 13 Multiple Choice

1. (b)
3. (b)
5. (a)
7. (b)
9. (b)
11. (d)
13. (d)
15. (a)
17. (c)

Chapter 13 Conceptual Questions

1. (a) The energy is four times as large. (b) Maximum speed is twice as large.
3. The mass travels a total distance of $4A$ during T, so it takes $T/4$ to travel A and $T/2$ to travel $2A$.
5. For a spring-mass system, the period is independent of gravitational acceleration. So the answer is no. For a pendulum, the period *does* depend on gravitational acceleration. The period actually increases on the Moon due to the lower value of gravitational acceleration. Thus the answer for a pendulum is yes.
7. Since the period is $T \propto \sqrt{L}$, the period is $\sqrt{2}$ as large, so the ratio is $T_{old}/T_{new} = \sqrt{2}$.
9. For the sine function, the initial position is zero (equilibrium) because $\sin 0 = 0$. Thus the initial speed is a maximum. The acceleration is zero because the restoring force at equilibrium is zero. For the cosine function, things are reversed. The magnitude of the initial displacement and acceleration at a maximum and the initial speed is zero.
11. It is a traveling spring wave. Coils of the spring move in the direction of the wave velocity so it is longitudinal.
13. Their motion is longitudinal, and the direction of the stalk motion is the same as that of the wind.
15. Reflection (*echolocation*) is involved, because the sound is reflected from the prey.
17. In principle, any frequency higher than the fundamental could be generated by pressing the fingers on the bridge to shorten the string. Lower frequencies cannot be generated because we cannot lengthen the string.
19. If the swing is pushed at a frequency of $f_1/2$, it is pushed only once (in one direction near equilibrium) every other oscillation. It is a smooth action since the swing is pushed in phase with its oscillations, and the amplitude of the motion can build up. Similarly, if it is pushed at a frequency of $f_1/3$, it is pushed only once every third oscillation. If it is pushed at a frequency of $2f_1$, it is pushed twice per oscillation, in different directions near equilibrium.

Chapter 13 Exercises

1. $12A$
3. 2.0 Hz
5. Decrease of 2.0 s
7. 0.13 m
9. (a) at (1) $x = 0$ (b) 2.0 m/s
11. 9.26 cm
13. 1.4 m/s
15. (a) 0.140 m (b) 10 cm (original position)
17. (a) 1.7 s (b) 0.57 Hz
19. 444 N/m
21. (a) $y = +A \cos \omega t$ (b) $y = -A \cos \omega t$ (c) $y = +A \sin \omega t$
23. (a) 5.0 cm (b) 10 Hz (c) 0.10 s
25. 0.33 Hz
27. (a) 0.188 m (b) 3.00 m/s^2
29. (a) (2) $\sqrt{2}$ times the old period (b) 3.5 s
31. $f_2 = \dfrac{1}{2\pi}\sqrt{\dfrac{k_2}{m_2}} = \dfrac{1}{2\pi}\sqrt{\dfrac{2k_1}{m_1/2}} = \dfrac{1}{2\pi}\sqrt{\dfrac{4k_1}{m_1}}$

$\quad = 2\left(\dfrac{1}{2\pi}\sqrt{\dfrac{k_1}{m_1}}\right) = 2f_1$
33. (a) (1) Increase (b) 4.9 s
35. (a) $(-0.10\text{ m})\sin[(10\pi/3)t]$ (b) 27 N/m
37. (a) $y = (0.100\text{ m})\cos(6t)$. (b) Starts at $y = +0.100$ m with an initial speed of zero. It will begin moving downward. (c) 1.05 s. (d) 18.0 N.
39. 2.0 Hz
41. 2.59 m
43. 6.00 km
45. 200 Hz
47. 170 Hz
49. 7.5 m
51. 478 N
53. Third harmonic
55. (a) (2) Twice (b) $f_A = 594$ Hz and $f_B = 50$ Hz
57. (a) (3) Shortened (b) 595 Hz

Chapter 14 Multiple Choice

1. (b)
3. (a)
5. (c)
7. (b)
9. (c)
11. (d)
13. (b)
15. (b)

Chapter 14 Conceptual Questions

1. This means that all the different notes (frequencies) from the various instruments will be heard in the proper sequence (as intended by the composer) regardless of the distance from the orchestra.

3. A molecule of helium (just a single atom) is much less massive than a molecule of nitrogen or oxygen (which comprise most of what we call air). Since the speed of sound is inversely proportional to the square root of the (average) molecular mass of the gas, the speed in helium will be faster than in air.

5. The intensity is $I = P/A = P/4\pi R^2$. So if R and P are both doubled, the intensity becomes $2/2^2 = (1/2)$ of its original value. Therefore the intensity is greater at a distance R, by a factor of 2.

7. Yes. For any intensity below the threshold intensity of hearing, β is negative.

9. (a) No, because there is no relative velocity between the observer and the source. (b) An increasing frequency is heard since the source is moving toward the observer, and its speed is increasing.

11. It uses the echo to determine the location of and the distance to the cloud and the Doppler effect to measure the direction and speed of the cloud motion.

13. By pressing on the frets, the player reduces the length of the string that is vibrating. This decreases the wavelength of the standing wave, thereby increasing its frequency and allowing the player to play higher and higher notes. The spacing of successive frets is designed so that the fractional (or percent) change in the frequency is the same from one fret to another in order to preserve the musical intervals of the notes of the scale. As the string gets shorter, the fractional change in frequency remains the same, but the absolute change gets smaller because it is the same fraction of a smaller and smaller length. If the frets were equally spaced, the note *changes* from fret to fret would be different as the string got shorter and shorter.

15. The increased temperature would cause two things to happen: thermal expansion of the pipe and an increase in the speed of sound in the pipe. The pipe would get longer, which would increase the wavelength of the fundamental and other harmonics. A greater wavelength would give *lower* frequencies for the sounds. The increased speed of sound, on the other hand, would result in *higher* frequencies for the sounds. Of these two effects, thermal expansion is usually quite small, so the dominant effect would be due to the increased speed of sound. The net result, then, would be to produce higher frequencies.

Chapter 14 Exercises

1. (a) 337 m/s (b) 343 m/s
3. 1.5 km
5. $\sqrt{\dfrac{N/m^2}{kg/m^3}} = \sqrt{\dfrac{N\cdot m}{kg}} = \sqrt{\dfrac{kg\cdot m^2/s^2}{kg}} = \sqrt{\dfrac{m^2}{s^2}} = m/s$ Since Y has the same units as B, the units of the expression for v in a solid also work out to be m/s.
7. (a) 7.5×10^{-5} m (b) 1.5×10^{-2} m
9. 90 m
11. (a) 0.107 m (b) 1.4×10^{-4} s (c) 4.29×10^{-4} m
13. (a) (1) less than double (b) 100 m (c) 8.7 s
15. -4.5%
17. (a) (4) 1/9 (b) 1.4 times
19. (a) 100 dB (b) 60 dB (c) -30 dB
21. (a) (1) Increases but not double (b) 5.0 W: 96 dB; 10 W: 99 dB
23. 24 W/m^2
25. 2.0×10^{-5} W/m^2
27. 1.0×10^{-3} W/m^2
29. (a) 63 dB (b) 83 dB (c) 113 dB
31. (a) (2) Between 40 and 80 dB (b) 43 dB
33. 379 m
35. No. Removing 25 computers will cut the intensity level to 37 dB.
37. 840 Hz
39. (a) 431 Hz (b) 373 Hz
41. Approaching: 755 Hz; moving away: 652 Hz.
43. (a) (3) Decrease (b) 42°
45. (a) 100 Hz (b) 6 Hz
47. (a) 36.3 kHz (b) 37.6 kHz (c) Yes, if both are moving, then the more general relation applies
49. (a) open-closed pipe (b) 0.675 m
51. (a) 3.4 kHz (b) 3.4 kHz (c) The frequency is lower, as it is inversely proportional to the length
53. For the open-open pipe, 330 Hz; for the open-closed pipe, 165 Hz.
55. 264 m/s
57. (a) (2) an antinode (b) 0.655 m (c) 0.390 m
59. Either 0.249 m or 0.251 m depending on the choice of frequencies from the two possible.

Photo Credits

Figure 1.2	National Institute of Standards and Technology Digital Collections, Gaithersburg, MD
Figure 1.3	National Institute of Standards and Technology Digital Collections, Gaithersburg, MD
Figure 1.7	Jerry Wilson
Figure 1.9	Bo Lou
Figure 1.16	Bo Lou
Figure 1.18	Bo Lou
Figure 2.16b	NASA
Figure 2.17	Jerry Wilson
Figure 4.4	Jerry Wilson
Figure 9.26	Bo Lou

Index

Physical Data*

Quantity	Symbol	Approximate Value
Universal gravitational constant	G	$6.67 \times 10^{-11}\ \text{N}\cdot\text{m}^2/\text{kg}^2$
Acceleration due to gravity (generally accepted value on surface of Earth)	g	$9.80\ \text{m/s}^2 = 980\ \text{cm/s}^2 = 32.2\ \text{ft/s}^2$
Speed of light	c	$3.00 \times 10^8\ \text{m/s} = 3.00 \times 10^{10}\ \text{cm/s} = 1.86 \times 10^5\ \text{mi/s}$
Boltzmann's constant	k_B	$1.38 \times 10^{-23}\ \text{J/K}$
Avogadro's number	N_A	$6.02 \times 10^{23}\ \text{mol}^{-1}$
Gas constant	$R = N_A k_B$	$8.31\ \text{J/(mol}\cdot\text{K)} = 1.99\ \text{cal/(mol}\cdot\text{K)}$
Coulomb's law constant	$k = 1/4\pi\epsilon_0$	$9.00 \times 10^9\ \text{N}\cdot\text{m}^2/\text{C}^2$
Electron charge	e	$1.60 \times 10^{-19}\ \text{C}$
Permittivity of free space	ϵ_0	$8.85 \times 10^{-12}\ \text{C}^2/(\text{N}\cdot\text{m}^2)$
Permeability of free space	μ_0	$4\pi \times 10^{-7}\ \text{T}\cdot\text{m/A} = 1.26 \times 10^{-6}\ \text{T}\cdot\text{m/A}$
Atomic mass unit	u	$1.66 \times 10^{-27}\ \text{kg} \leftrightarrow 931\ \text{MeV}$
Planck's constant	h	$6.63 \times 10^{-34}\ \text{J}\cdot\text{s}$
	$\hbar = h/2\pi$	$1.05 \times 10^{-34}\ \text{J}\cdot\text{s}$
Electron mass	m_e	$9.11 \times 10^{-31}\ \text{kg} = 5.49 \times 10^{-4}\ \text{u} \leftrightarrow 0.511\ \text{MeV}$
Proton mass	m_p	$1.672\ 62 \times 10^{-27}\ \text{kg} = 1.007\ 276\ \text{u} \leftrightarrow 938.27\ \text{MeV}$
Neutron mass	m_n	$1.674\ 93 \times 10^{-27}\ \text{kg} \times 1.008\ 665\ \text{u} \leftrightarrow 939.57\ \text{MeV}$
Bohr radius of hydrogen atom	r_1	$0.053\ \text{nm}$

*Values from NIST Reference on Constants, Units, and Uncertainty.

Solar System Data*

Equatorial radius of the Earth	$6.378 \times 10^3\ \text{km} = 3963\ \text{mi}$
Polar radius of the Earth	$6.357 \times 10^3\ \text{km} = 3950\ \text{mi}$
	Average: $6.4 \times 10^3\ \text{km}$ (for general calculations)
Mass of the Earth	$5.98 \times 10^{24}\ \text{kg}$
Diameter of Moon	$3500\ \text{km} \approx 2160\ \text{mi}$
Mass of Moon	$7.4 \times 10^{22}\ \text{kg} \approx \frac{1}{81}$ mass of Earth
Average distance of Moon from the Earth	$3.8 \times 10^5\ \text{km} = 2.4 \times 10^5\ \text{mi}$
Diameter of Sun	$1.4 \times 10^6\ \text{km} \approx 864\ 000\ \text{mi}$
Mass of Sun	$2.0 \times 10^{30}\ \text{kg}$
Average distance of the Earth from Sun	$1.5 \times 10^8\ \text{km} = 93 \times 10^6\ \text{mi}$

*See Appendix III for additional planetary data.

Mathematical Symbols

Symbol	Meaning
$=$	is equal to
$\neq$	is not equal to
$\approx$	is approximately equal to
$\sim$	about
$\propto$	is proportional to
$>$	is greater than
$\geq$	is greater than or equal to
$\gg$	is much greater than
$<$	is less than
$\leq$	is less than or equal to
$\ll$	is much less than
$\pm$	plus or minus
$\mp$	minus or plus
$\bar{x}$	average value of x
Δx	change in x
$|x|$	absolute value of x
Σ	sum of
∞	infinity

The Greek Alphabet

Name	Cap	Low	Name	Cap	Low
Alpha	A	α	Nu	N	ν
Beta	B	β	Xi	Ξ	ξ
Gamma	Γ	γ	Omicron	O	o
Delta	Δ	δ	Pi	Π	π
Epsilon	E	ε	Rho	P	ρ
Zeta	Z	ζ	Sigma	Σ	σ
Eta	H	η	Tau	T	τ
Theta	Θ	θ	Upsilon	Y	υ
Iota	I	ι	Phi	Φ	ϕ
Kappa	K	κ	Chi	X	χ
Lambda	Λ	λ	Psi	Ψ	ψ
Mu	M	μ	Omega	Ω	ω

Printed and bound by CPI Group (UK) Ltd, Croydon, CR0 4YY

17/10/2024

01775663-0016